The Developing Person Through the Life Span

The Developing Person Through the Life Span

SECOND EDITION

KATHLEEN STASSEN BERGER
BRONX COMMUNITY COLLEGE
CITY UNIVERSITY OF NEW YORK

Worth Publishers, Inc.

The Developing Person Through the Life Span, SECOND EDITION

© 1988, 1983 by Worth Publishers, Inc.

All rights reserved

Printed in the United States of America

Library of Congress Catalog Card Number: 87-51235

ISBN: 0-87901-381-8

3 4 5 – 92 91 90 89

Editors: Peter Deane and Judith Wilson
Picture Editors: Elaine Bernstein and David Hinchman
Illustrator: Demetrios Zangos
Production and Make-up: Sarah Segal and George Touloumes
Design: Malcolm Grear Designers
Composition and Color Separations: York Graphic Services
Printing and Binding: R. R. Donnelley and Sons Company

Cover: Assemblage by Starr Ockenga

Worth Publishers, Inc.
33 Irving Place
New York, New York, 10003

Preface

Students coming to the study of human development for the first time are arriving at the discipline during an exciting period. Until recently, the study of human development was largely confined to the study of children and adolescents. It was generally assumed that once a person was physically grown, he or she was fully developed in most other ways as well. During the past two decades, this assumption has been replaced with the view that people develop throughout the entire life span, and that adulthood and old age offer as much reason to study how and why people do or do not change over time as the so-called formative years do.

This life-span perspective has opened up new vistas for researchers, scholars, teachers, and students alike. For one thing, textbooks such as this one not only "cover" adulthood as well as childhood; they also present childhood in a different light—as the opening act in the drama of development rather than as the main body of the play. Also noteworthy are the many ways in which insights about one stage of development have led to a fuller understanding of other periods of life. For example, the study of childhood has not only revealed ways in which family influences affect the individual in later stages of life, but it has also shown how children shape the family and influence the development of its older generations. As we look at the life span as a whole, the impact of historical and cultural contexts also becomes clearer. How one experiences adolescence, or adulthood, or old age, for example, depends a great deal on when and where one experiences it. Young American adults in the 1980s and 1990s, for instance, face both opportunities and pressures that were rarely experienced by young adults during the 1950s. Likewise, the American elderly of today have options open to them that were largely unknown to their predecessors. At the same time, many also face difficulties in late old age that most of their shorter-lived forebears were spared.

The very newness of the life-span perspective—including its many discoveries and debates—adds excitement to the study of human development. Much of this excitement comes from the topics themselves, which are often controversial and ripe with intriguing implications. Even in the five years since the first edition of this text, many such topics have come to the fore, often raising questions that call for more critical thinking and groundbreaking research. Why, for example, do some infants and parents become securely attached to each other and others not? Why do some children become fluently bilingual and others have trouble in even one language? What are the benefits and shortcomings associated with dual-career parenting? Why do some single-parent families manage to function well and others collapse under the strain? Why is midlife a crisis for some and not for others? What are the myths and realities of old age, and what factors help to make the later years a satisfying period of life? Many other topics currently in the spotlight pose troubling questions that each of us must answer personally—topics that range from the latest innovations in prenatal intervention to the ongoing dispute about the right to die; from child maltreatment to adolescent pregnancy, young adult risk-taking, and abuse and neglect of the aged. Throughout the writing of this text, my purpose has been to explain facts and theories in such a way as to highlight the need for further thought and action—on the part of scholars and citizens alike.

Organization of This Book

A major organizational dilemma confronts every teacher and writer when trying to lay out the essentials of human development. Should the material be organized as life is lived, following development chronologically from conception to death, or should it be organized as most academic researchers study it, topically, with separate sections on language, gender, and so forth? Since both ways have some merit, I have chosen a compromise that, I hope, reflects the best of both.

The four chapters at the beginning, covering methods, theories, genetics, and prenatal influences, and the chapter at the end, death, are topical. Apart from those five chapters, the overall framework is chronological, dividing the life span into seven parts: infancy, early childhood, middle childhood, adolescence, early adulthood, middle adulthood, and late adulthood. Within that framework, development is discussed in a trio of topical chapters covering physical, cognitive, and psychosocial development. I believe this organization makes it easier for instructors to emphasize the whole person at every stage of development, and helps the student to appreciate the ways in which the three domains of development continuously interact.

Changes in This Edition

The first edition of this text was very well received, by colleagues, teachers, and students. For this reason, the overall structure and intention of the book have not changed. However, the text has been extensively revised and updated to reflect rapidly changing research findings. Several chapters have also been reorganized for greater coherence.

One area of life-span study has particularly benefited from increased attention—what Bronfenbrenner once called the "empty middle" of the developmental story, the years from adolescence to old age. In the first edition of this book, these years were covered in three chapters. Because of the proliferation of recent research, they are now covered in six.

Several new features have been added to make the book a better teaching and learning tool. As with the first edition, the entire book reflects the life-span perspective on development, with attention paid to social context, intergenerational relations, and the developmental causes and consequences of various experiences. To reinforce this approach, I have written a number of "Life-Span Perspectives," text discussions in which the connections between development at one period and another are made explicit. Also new to this edition are full-page charts that come at the end of Parts II through VIII of the book and provide an overview of the significant physical, cognitive, and psychosocial events covered in each part. A topical table of contents has been created to make it easier for the student to trace the developmental process throughout life. For students wishing to pursue research in the study of development, a list of Reading Resources for Further Study is provided at the back of the book as a guide to developmental journals, reference works, and book reviews.

The Author

My theoretical roots are diverse. My graduate-school mentors included gifted teachers who studied directly with Erik Erikson, B. F. Skinner, Carl Rogers, and Jean Piaget, and I continue to have great respect for each of these theorists. However, like most developmentalists today, my overall approach is eclectic, influenced by all the theories rather than adhering to any one. The abiding influence of my academic study and training is in my respect for knowledge attained through the scientific method: I believe that the more we know about development, the better we can help all people fulfill their potential.

Equally influential on my work are my personal interests in the study of human growth. As an involved daughter, sister, friend, community member, wife, mother (of four, now aged 19, 18, 11, and 6), as well as a developmental psychologist, an associate professor, and a text author, I find my enthusiasm for the discipline never flagging. I hope this book conveys that feeling.

Supplementary Materials

A *Study Guide,* written by Joan Brown, is available to aid students in their understanding of the key concepts in the text, as well as to help them review the material. Each chapter includes a list of learning objectives, an overview, a detailed summary with accompanying questions, and an extensive battery of true-false and multiple-choice self-tests.

An unusually fine *Instructor's Resource Manual,* reconceived and revised by Richard O. Straub, University of Michigan, Dearborn, is available to every teacher who adopts this text. Redesigned with a new and more useful format, this supplement now provides a set of transparencies, as well as activity sheets and selected readings that can be photocopied for students. An extensive *Test Bank* that includes multiple-choice, fill-in, true/false, and essay questions is also provided. As one who has taught many courses in college and graduate school for twenty years, I know that some instructor's aids are not very helpful, and that many of my colleagues no longer look at them when a new text comes out. If this describes you, I urge you to examine the instructor resources for this book. I think you will be pleasantly surprised.

Finally, I am pleased to say, this edition of *The Developing Person Through the Life Span* has been chosen as the text that will accompany "Seasons of Life," a telecourse produced by The Annenberg/CPB Project. "Seasons of Life" will air on public television beginning in September 1989. Information about the telecourse and its supplements can be obtained by writing to: The Annenberg/CPB Project, 1111 16th Street NW, Washington, DC 20036.

Thanks

This book has benefited from the work of the entire community of scholars involved in human development. I have learned much from conferences, journals, and conversations with fellow developmentalists. Of course, I am particularly indebted to the many academic reviewers who have read various drafts of this book, providing suggestions, criticism, references, and encouragement. Each of them has made the book a better one, and I thank them all:

Margarita Azmítia, Florida International University

Patricia P. Barker, Schenectady County Community College

Eda Bower, Marycrest College

Joseph J. Campos, University of Illinois

Margaret K. Cass, University of Michigan at Ann Arbor

Francine Deutsch, San Diego State University

Vicky Fong, Sacramento City College

Janet Fritz, Colorado State University

Mary Gauvain, Oregon State University

Hill Goldsmith, University of Oregon

Anita L. Greene, West Virginia University

James N. Greene, Ricks College

Sybillyn Jennings, Russell Sage College

Robert Kastenbaum, Arizona State University

Murray Krantz, Florida State University

Gisela Labouvie-Vief, Wayne State University

Dale Lund, University of Utah

John J. Mitchell, University of Alberta at Edmonton

Philip J. Mohan, University of Idaho

Roberta H. Morgan, University of Alaska at Anchorage

Kathryn Quina, University of Rhode Island

Harriette Ritchie, American River College

Bruce Roscoe, Central Michigan University

Gary L. Schilmoeller, University of Maine at Orono

Karen Tee, Vanier College

Ross A. Thompson, University of Nebraska at Lincoln

Joseph Tindale, University of Guelph

Kenneth A. Tokuno, University of Washington

Daniel A. Tomasulo, Brookdale Community College

Cynthia Whitfield, Merritt College

The editorial, production, and sales staffs at Worth Publishers are dedicated to meeting the highest standards of excellence. Their devotion of time, effort, and talent to every aspect of publishing is a model for the industry. When I decided to publish with them, I was told I would have to work twice as hard as I would for other publishers, and that the result would be many times better. It is true, and I am grateful.

I particularly would like to thank Judith Wilson and Peter Deane, my editors, who have worked exclusively on this book for many months. Peter has helped me since the first edition of my first book; he maintains his perseverance, brilliance, creativity, and humor despite sometimes compelling reasons not to. As well as editing, Judith has guided the work on this book from beginning to end. Many of the new features of this edition began as an idea of Judith's. All of the handsome new photographs (more than 80 percent are new to this edition) are the result of Elaine Bernstein's extensive research and selective eye for the "perfect picture." Sarah Segal's careful attention to detail and technical expertise are responsible for the high quality of this book's production.

Finally, I would like to thank my nephew, David Stassen, not only for letting me begin Chapter 1 with his life story, but also for living his life in a way that, for me, illuminates the mission that underlies the study of human development. It is to him, with love and with admiration for his courage, that I dedicate this book.

New York City
January, 1988

Kathleen Stassen Berger

Contents in Brief

Contents

Contents by Topic

This brief index provides a ready reference to the major themes and topics of development through the life span and is intended to be useful for quick reference and review and for researching essays and topical term papers. This index does *not* include all the topics in this text, nor are all minor discussions listed. A complete index appears at the back of the book.

PART I

Beginnings

The study of human development has many beginnings, as you will see in the following four chapters. First, you will learn about research methods and designs, the building blocks, as it were, of the scientific study of development. But building blocks are useless without a master plan for their use. Chapter 2 presents several such plans, in this case the psychoanalytic, learning, humanist, and cognitive theories of development. A good theory is invaluable in helping researchers examine and explain human development, and these are four of the best psychology has to offer.

A different kind of master plan is described in Chapter 3, which traces the workings of heredity. Each human being grows and develops following the instructions carried on the genes and chromosomes. Genes influence everything from the shape of your baby toe to the swiftness of your brain waves, so understanding the basics of genetics is essential to understanding development.

Finally, Chapter 4 details the true beginnings of human life, from the fusing of sperm and ovum to make one new cell to the birth of a new human being, a totally dependent individual who can nevertheless see, hear, and cry, and is ready to engage in social interaction.

CHAPTER 1   **Introduction**

All cases are unique, and very similar to others.

T. S. Eliot
The Cocktail Party

David, my brother's son, is 19 now. Most aspects of his development are typical of many young people his age. For example, he is constantly concerned about his appearance and worries that at 5 feet, 8 inches, he may have reached his full height. In his first year in college, he studies very hard in some courses, not at all in others, and is ready to argue about politics at the drop of a hat. A late-bloomer socially, he is rather self-conscious, and although he has a decided interest in females generally, he has yet to develop a close relationship with any particular one.

In these and many other ways, David's physical, intellectual, and social development is what we might expect of someone leaving adolescence and entering early adulthood. In one very basic way, however, David is not typical. He began life severely handicapped, with little hope even for survival, let alone for a life approaching normality.

Although most of this book is about "normal" development—that is, development that follows the usual patterns of growth and change—we begin with David's story for two reasons. First, the extremity of his handicaps and the extent of his triumphs over them highlight a number of the central issues and concerns of the study of human development. Second, David's story demonstrates what the primary goal of our theories and research is: using an understanding of human growth to help every developing person fulfill his or her potential, from the first days of infancy to the last days of adulthood.

Before looking more closely at David, we need to explain what development means and what the study of it involves.

The Study of Human Development

Briefly, the study of human development is the study of how and why people change over time, as well as how and why they remain the same. Developmentalists not only want to understand why human lives unfold as they do, but want to discover how people might live fuller, happier, more productive lives. In the course of this endeavor, developmentalists study all periods of life, from conception to death, and all kinds of change, from simple growth to radical transformations, in all areas of development.

The Three Domains

To make it easier to study, human development is often separated into three domains: the **physical domain,** including body changes and motor skills; the **cognitive domain,** including intellect, thought processes, and language; and the **psychosocial domain,** including emotions, personality, and relationships with other people (see Figure, 1.1).

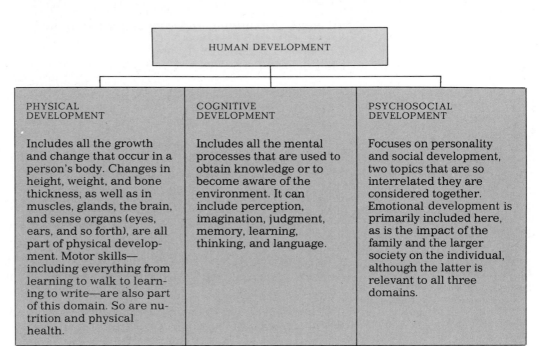

Figure 1.1 *The division of development into three domains makes it easier to study, but we must remember that very few factors belong exclusively to one domain or another. Development is not piecemeal but holistic: each aspect of development is related to all three domains.*

HUMAN DEVELOPMENT

PHYSICAL DEVELOPMENT

Includes all the growth and change that occur in a person's body. Changes in height, weight, and bone thickness, as well as in muscles, glands, the brain, and sense organs (eyes, ears, and so forth), are all part of physical development. Motor skills—including everything from learning to walk to learning to write—are also part of this domain. So are nutrition and physical health.

COGNITIVE DEVELOPMENT

Includes all the mental processes that are used to obtain knowledge or to become aware of the environment. It can include perception, imagination, judgment, memory, learning, thinking, and language.

PSYCHOSOCIAL DEVELOPMENT

Focuses on personality and social development, two topics that are so interrelated they are considered together. Emotional development is primarily included here, as is the impact of the family and the larger society on the individual, although the latter is relevant to all three domains.

All three domains are important at every age. For instance, understanding development in the first year of life involves studying the maturation of infants' motor skills, the range of their curiosity, and the expression of their emotions, as well as dozens of other aspects of physical, cognitive, and psychosocial development. Similarly, to understand adolescence, we consider such things as the physical changes that mark the transition from the body of a child to that of an adult, the intellectual development that leads to thinking about moral issues and future goals, and the changing relationships with friends and family that set the stage for adulthood. Likewise, to understand middle age, it is important to know about a person's health habits, intellectual interests, and social activities. Throughout one's life, development in each of these domains affects development in the other two: one's physical condition, for instance, can affect learning ability and social experiences; one's emotional state can affect cognitive functioning and physical well-being; one's cognitive functioning can affect one's interpretations of events and others' behaviors, and these interpretations, in turn. can affect immediate physical functioning.

Indeed, no moment of life can be fully understood without considering all three domains simultaneously. The constant interaction and overlapping among the domains means that, although different aspects of development are sometimes

Figure 1.2 *This photo reflects the influence of all three domains. While the physical changes of puberty are behind the development of this boy's biceps, the culture's emphasis on masculine strength is behind his urge to display them. At the same time, the increased self-consciousness that is part of adolescent cognitive development was revealed by the question the boy asked when he first saw this photo: "Is my nose really that big?"*

studied piece by piece, human development is conceptualized as **holistic,** that is, as an integrated whole. Similarly, while the study of development is pursued by researchers from a variety of academic disciplines—including biology, education, and psychology—all agree that the field of human development is an interdisciplinary one.

The Context of Development

Until recently, the focus of developmental study was primarily on the individual. The most notable change in developmental studies over the past ten years has been the adoption of a considerably broader focus (Bronfenbrenner et al., 1986). We now are much more aware of how the individual is affected by, and affects, myriads of other individuals and groups of individuals, and also of how the individual is shaped by social forces such as history, culture, politics, and economics. Furthermore, development is increasingly seen as a process that occurs throughout the life span, and people of one age are increasingly seen to affect the development of people of other ages. Consequently, the study of development, much more than before, attempts to understand individuals within the context of their specific multiple environments.

The Ecological, or Systems, Approach This broader approach to understanding development is sometimes called an **ecological approach,** for, just as the naturalist studying a flower or a fish needs to examine the supporting ecosystem as well, so does a developmentalist need to look at the ecosystem in which a human being seeks to thrive. The systems that support human development can be seen as occurring at four levels, each nested within the next (Bronfenbrenner, 1979; Bronfenbrenner and Crouter, 1983):

1. The **microsystem**—the immediate systems, such as the family and the classroom, that affect the individual's daily life.

2. The **mesosystem**—the interlocking systems, such as the parent-teacher communications, or the employment practices impinging on family life, that link one microsystem to another.

3. The **exosystem**—the neighborhood and community structures (including newspapers, television, and government agencies) that affect the functioning of the smaller systems.

4. The **macrosystem**—the overarching patterns of culture, politics, the economy, and so forth.

Another name for this multifaceted way of looking at development is the **systems approach,** a term that highlights the interaction between and among various elements in each system (Sameroff, 1983). Consider this approach in the study of family influences on development, for example. Not too long ago, psychologists investigating this topic focused almost exclusively on how mothers affect their young children (Sigel et al., 1984). Now researchers have broadened their study of family influences to include fathers (Lamb, 1986, 1987), brothers and sisters (Lamb and Sutton-Smith, 1982), and grandparents (Cherlin and Furstenberg, 1986). Each of these family members can influence a child dramatically, sometimes becoming the deciding factor that determines whether development will progress well or badly. Further, developmentalists now recognize that family influences are multidirectional: each child influences the entire family system, and each family relationship (for instance, the relationship between husband and wife, or between siblings, or between grandparent and child) has an impact on all other family members (Bell and Harper, 1977; Lerner and Spanier, 1978; Maccoby and Martin, 1983). Finally, family influences are seen as extending far past childhood. Brothers and sisters, for instance, are often an important source of emotional support in old age.

A systems approach also provides insight into the ways the larger systems of society affect the smaller systems. Thus the interaction within a family is influenced by the gender roles prevalent in the culture, the national and regional economic picture and its impact on employment, the customs and laws governing marriage and divorce, and so on. Take the economic picture, for example. A period of high inflation, such as that in the United States during the 1970s, may negatively affect family stability, which, in turn, may contribute to an increase in the divorce rate and a decrease in the birth rate, with a variety of repercussions for individual family members. Similarly, a high divorce rate may lead to an increase in the number of employed mothers, which, in turn, may affect the economy in ways that alter family patterns.

Figure 1.3 *Because government support for after-school programs has not caught up with changes in the family and the community systems, this boy, who would once have been supervised by his mother or a neighbor, must now either create a private fantasy drama at his parent's workplace—in this case, the Massachusetts legislature—or stay home alone.*

David's Story: Domains and Systems at Work

To better understand the interaction of the three domains, as well as the impact of the various systems that support, direct, or hinder development, let us return to David. David's story begins with an event that seems clearly from the physical domain. In the fourth week of his prenatal life, his mother contracted rubella (German measles), a disease that almost always results in physical damage to a developing embryo, sometimes causing devastating physical handicaps, and sometimes causing more subtle learning problems.

The Early Years: Heartbreaking Handicaps, Slow Progress   In David's case, major problems were obvious at birth. David was born with a life-threatening heart defect and thick cataracts covering both eyes. Other damage caused by the virus became apparent as time went on, including minor malformations of the thumbs, feet, jaw and teeth, as well as brain injury.

From a systems perspective, the larger medical and political systems had already had a major impact. Had David been conceived a decade later, the development and widespread use of the rubella vaccine would probably have prevented his mother's contracting the disease. On the other hand, had he been born a few years earlier, or in a different part of the world, the medical technology that saved his life would not have been available.

As it happened, heart surgery in the first days of life was successful, and it was thought that David would have at least a few years of life. However, surgery to open a channel around one of the cataracts failed, causing damage that completely blinded the eye.

It soon became apparent that David's physical handicaps fostered cognitive and psychosocial liabilities as well. Not only did his blindness make it impossible for him to learn by looking at his world, but his parents overprotected him to the point that he spent almost all his early months in their arms or in his crib. An analysis of the family system would have revealed that David's impact on his family, and their effect on him, were harmful in many ways. Like most parents of seriously impaired infants (Featherstone, 1980), David's parents felt guilt, anger, and despair; they were initially unable to make constructive plans to foster normal development.

Luckily, the larger social system helped rescue David and his family. The first step occurred when a teacher from the Kentucky School for the Blind visited David's home and gave his parents some much-needed encouragement and advice.

Figure 1.4 *When the new baby is physically perfect and developmentally "normal," parents take pleasure and pride in recording landmarks in the child's progress—first steps, first words, month-by-month growth, new friends. For parents with a handicapped child, however, milestones are too frequently marked more by relief than by celebration.*

They were told to stop blaming themselves for David's condition and to stop over-protecting him because of it. If David was going to learn about his world, he was going to have to explore it. To this end, it was suggested that, rather than confining David to a crib or playpen, his parents should provide him with a large rug for a play area. Whenever he crawled off the rug, they were to say "No" and place him back on in the middle of it, thus enabling him to use his sense of touch to learn where he could explore safely without bumping into walls or furniture. David's mother dedicated herself to this and the many other tasks that various specialists suggested, including exercising his twisted feet and cradling him frequently in her arms as she sang lullabies to provide extra tactile and auditory stimulation.

His father helped, too, taking over much of the housework and care of the two older boys, who were 2 and 4 when David was born. When he found an opportunity to work in Boston, he took it, partly because the Perkins School for the Blind had just begun an experimental program for blind toddlers and their mothers. At Perkins, David's mother learned specific methods for developing physical and language skills in multihandicapped children, and she, in turn, taught the techniques to David's father and brothers. Every day the family spent hours rolling balls, doing puzzles, and singing with David.

Thus the family system and the educational system collaborated to help young David develop his physical and cognitive skills. However, progress was slow. It became painfully apparent that rubella had damaged much more than his eyes and heart, for at age 3, David could neither talk, nor chew solid food, nor use the toilet, nor coordinate his fingers well, nor even walk normally. An IQ test showed him to be severely mentally retarded. Fortunately, although 72 percent of 3-year-olds with rubella syndrome have hearing defects (Chess et al., 1971), David's hearing was normal. However, the only intelligible sounds he made mimicked the noises of the buses and trucks that passed by the house.

At age 4, David said his first word, "Dada." Open-heart surgery corrected the last of his heart damage, and an operation brought partial vision to his remaining eye. While sight in that eye was far from perfect, David could now recognize his family by sight as well as by sound, and could look at picture books. By age 5, when the family returned to Kentucky, further progress was obvious: he no longer needed diapers or baby food.

David's fifth birthday occurred in 1972, just when the idea that the education of severely handicapped children could take place in school rather than at home was beginning to be accepted. David's parents found four schools that would accept him and enrolled him in all of them. He attended two schools for victims of cerebral palsy: one had morning classes, and the other—forty miles away—afternoon classes. (David ate lunch in the car with his mother on the daily trip.) On Fridays these schools were closed, so he attended a school for the mentally retarded, and on Sunday he spent two hours in church school, his first experience with "mainstreaming"—the then new idea that children with special needs should be educated with normal children.

Middle Childhood: Heartening Progress By age 7, David's intellectual development had progressed to the point considered adequate for the normal educational system. In some skills, he was advanced; he could multiply and divide in his head. He entered first grade in a public school, one of the first severely handicapped children to be mainstreamed. However, he was far from being a normal first-grader, for rubella continued to have an obvious impact on his physical, cognitive, and social development. His motor skills were poor (among other things, he had difficulty controlling a pencil); his efforts to learn to read were greatly hampered by the fact that he was legally blind even in his "good" eye; and his social skills were seriously deficient (he pinched people he didn't like and cried and laughed at inappropriate times).

During the next several years, David's cognitive development proceeded rapidly. By age 10, he had skipped a year and was a fifth-grader. He could read with a magnifying glass—at the eleventh-grade level—and was labeled "intellectually gifted" according to tests of verbal and math skills. At home he began to learn a second language and to play the violin. In both areas, he proved to have extraordinary auditory acuity and memory.

David's greatest problem was in the psychosocial domain. Schools generally ignore the social skills of mainstreamed children (Gottlieb and Leyser, 1981), and David's experience was no exception. For instance, David was required to sit on the sidelines during most physical-education classes, and to stay inside during most recess periods. Without a chance to experience the normal give-and-take of schoolboy play, David remained more childish than his years. His classmates were not helped to understand his problems, and some teased him because he still looked and acted "different."

David had one advantage that many handicapped children do not. He had two older brothers who treated him normally, playing with him, fighting with him, and protecting him. At one point, David desperately wanted to play basketball, so his brothers took him out to their backyard hoop and showed him the basics. He was not very good, but his enthusiasm was high, and he eagerly joined an after-school basketball program at the Y, which was advertised as "just for fun, not for competition—anyone can participate." On the first day, however, the other kids teased David, calling him a "retard," and the coach asked that David not come back because, he said, he might injure himself.

Adolescence and Beyond: New Problems, New Hopes Because of David's problems with outsiders and classmates, his parents decided to send him to a special school when he was ready for junior high. In the Kentucky School for the Blind, his physical, cognitive, and psychosocial development all advanced: David learned to wrestle and swim, mastered algebra with large-print books, and made friends whose vision was as bad as or even worse than his. For his high-school years, David remained at the Kentucky School, where he mastered not only the regular curriculum but also specialized skills, such as how to travel independently in the city and how to cook and clean for himself. In his senior year he was accepted for admission by a large university in his home state.

Figure 1.5 *Adolescence is a critical time for learning to attract the opposite sex. This young man has apparently mastered this task, while others, self-concious about small flaws, may feel less confident and give themselves fewer chances for success. For handicapped teenagers, whose characteristics may set them apart from others in very noticeable ways, social interaction with the opposite sex can be particularly elusive.*

As this book goes to press, David's severe physical and cognitive problems are a thing of the past. Doctors no longer doubt that he will survive through adulthood, and are helping to improve the quality of his future life: he now has an artificial eye in place of the blind one; he wears a back brace to help correct his posture; and he had surgery to correct his misaligned jaw. In his first year of college, he earned a B⁻ average, partly because the acquisition of a lap-top computer relieved him of having to take class notes in his still-labored handwriting.

His family delights in him, appreciating his many talents and accomplishments. Notably, he is the most musical family member, and he has a gift for speaking foreign languages, an ability that benefited the entire family when they lived in Germany for a year. However, they are concerned about his psychosocial development. Because of the size of his university and the fact that he lives at home, David finds making friends difficult. Further complicating matters, his poor vision inclines him to come very close to people he is talking with—a habit that makes most people draw back. His poor vision also causes him to miss facial cues given by those around him, so he is sometimes unable to "read" how people are reacting to him.

In looking at David's life thus far, we can see how the domains and systems interact to affect development, both positively and negatively. We can also see the importance of research and application of developmental principles. For example, without research that demonstrated the importance of sensory stimulation to an infant's development, David's parents might not have been taught how to keep his young mind actively learning. Nor would David have been educated in schools had not the previous efforts of hundreds of developmental scientists proved that schools could provide effective teaching even for severely handicapped children. David might instead have led an overly sheltered and restricted life, as many children born with David's problems once did. Indeed, many children with David's initial level of disability formerly spent their lives in institutions.

And what of David's future? Development certainly does not stop when adulthood begins. David will experience many more years of adult development that can go well or badly depending on his own determination and effort, as well as on the various systems that help or hamper development. For instance, the macrosystem is now increasingly sensitive to the needs of handicapped people: the laws of the land now safeguard David's right to a normal life, in college, employment, housing, and the like. As you have just learned, the family system often continues to support development throughout the life span, and in this David has been, and will undoubtedly continue to be, fortunate. However, much of the course of David's adulthood will depend on whether he can find satisfying work and begin a family of his own, the two tasks that generally shape much of adult development. (At the moment, David hopes for a career in music or literature and definitely wants to marry.) As you will read later, making good career and marriage decisions are particularly difficult in our society, and David's handicaps may make it even harder for him.

However, in predicting anyone's future or interpreting anyone's past, we must remember that each person is a unique individual who is uniquely affected by, and affects, the constellation of systems that impinges on his or her development. Thus the most important factor in David's past successes may have been David himself, for his determination, stubbornness, and stoic courage helped him weather the physical trauma of repeated surgery and the psychological devastation of social rejection. While thousands of scientists, dozens of teachers, and both parents deserve to be proud of David's accomplishments, the one who should be most proud is David himself. More than anyone else, in the final analysis, David, like each of us, directs his own development.

Two Controversies

As David's case makes abundantly clear, the study of development requires taking into account the interplay of the physical, cognitive, and psychosocial domains, and the effects of familial, cultural, political, and economic forces. Not surprisingly, assessing the relative impact of all these factors is no simple matter. In fact, developmentalists often find themselves on one side or another of two controversies that have been debated since the study of development began.

Nature and Nurture

The central dispute in the study of human development is the nature-nurture controversy. It is the continuing debate over whether the individual's various traits and characteristics are influenced more by inborn factors or by experience.

Nature refers to the range of traits, capacities, and limitations that each person inherits genetically from his or her parents at the moment of conception. Body type, eye color, and inherited diseases are obvious examples. Nature also includes those largely inherited traits, such as athletic ability or memory, that appear after a certain amount of maturation has occurred.

Nurture refers to all the environmental influences that come into play after conception, beginning with the mother's health during pregnancy and running through all one's experience with, and learning in, the outside world—in the family, school, community, and the culture at large.

Figure 1.6 *Most subway commuters spend the waiting time watching others, chatting, or impatiently checking for the train, but this family is obviously different. Why? Is it a genetic tendency that makes them a family of nearsighted readers, or is it nurture—a family culture that encourages intellectual pursuits and discourages "impolite staring," "idle talk," or "time-wasting"?*

Figure 1.7 *It is easy to see the influence of nature in this photo of Kirk Douglas and his four sons. The two offspring of his first wife have her dark hair; the two from his second wife have her light coloring, and all four have their father's distinctive chin. The influence of nurture is undoubtedly also at work, if not quite so obvious. How could we explain the fact that all four brothers are actors and producers without referring to their early experiences, their father's example, and the breaks that come from being at the right place, with the right people, at the right time?*

The controversy about nature and nurture has taken on many names, among them, heredity versus environment, or maturation versus learning. Under whatever name, however, the basic question remains: How much of a person's characteristics, behaviors, and development is determined by his or her genes, and how much is the result of the myriad experiences and influences that occur after conception?

Note that the question asks "How much?" not "Which one?" All developmentalists agree that both nature and nurture are essential to development, and that the interaction between nature and nurture is the crucial influence on any particular individual. They point out, for example, that intelligence is determined by the interplay of heredity and such aspects of the social and physical environment as schooling and nutrition. Despite their acknowledgement of the interaction between nature and nurture, however, developmentalists can get into heated arguments about the relative importance of each.

One of the reasons the controversy over the relative importance of maturation and learning is very much alive is that it is difficult to prove which is more responsible for a particular developmental change. Another is that the practical implications of the controversy are enormous. Consider one example. It is well established that, although boys and girls in elementary school show similar math aptitude, the mathematical achievement of the typical teenage boy is higher than that of the average teenage girl. Furthermore, high-school students who are gifted in math are usually boys, by a four to one ratio (Benbow and Stanley, 1980).

Is maturation responsible for this difference? Perhaps some development in the male brain at puberty permits boys to grasp math concepts more easily (Waber, 1976). Or is learning the key factor? Perhaps girls learn that math ability is not considered feminine, and their parents, teachers, or boyfriends might share this view and discourage them from taking advanced math courses (Sherman, 1982). Whichever the answer, the implications are significant. If the difference between adolescent male and female ability in math is attributable to nature, it is foolish and frustrating to expect males and females to do equally well in math. On the other hand, if the difference is the result of learning, we are wasting half our mathematical potential by not encouraging girls to develop their full math abilities. Furthermore, since careers in science and engineering typically pay more than other careers, one reason adult women on average earn significantly less than adult men may be rooted in the math courses they did not take in high school.

Continuity or Discontinuity

How would you describe human growth? Would you say we develop gradually and continually, the way a seedling becomes a tree? Or do you think we undergo sudden changes, like a caterpillar becoming a butterfly? Furthermore, how much stability is typical of human development, from the beginning of life to the end?

Continuity Many developmental researchers emphasize **continuity** and stability. They believe that, when change occurs, it is usually quite gradual, and often more a matter of form than of substance. Changes that may seem abrupt, such as a baby's first step, or a child's learning to talk or read, or an adult's taking on the role of a parent or grandparent, can actually be seen as the culmination of a long process rather than a sudden shift. Moreover, despite the various modulations in personality that occur over time, many basic personality characteristics endure throughout life. The cheerful, outgoing adolescent is likely to be the senior citizen who frequently invites friends over, helps those in need, and thoroughly enjoys life.

Discontinuity Other developmentalists emphasize **discontinuity.** They see change, including age-related periods of rapid and radical change, as typical of the human life. In the first twenty years of life, many such changes are biologically based, occurring in identifiable stages, each with distinct characteristics and challenges. Erik Erikson, for instance, is one of many theorists who believes that the biological changes of puberty usher in a stage of development in which the young person's chief developmental challenge is to establish a clear sense of identity. This stage is characterized not only by burgeoning sexual development but also by rebellion, confusion, and experimental role-playing that eventually lead to a sense of self-definition and prepare the young person for the next stage and the next challenge, the establishment of intimate bonds. In Erikson's view, as in that of others who see development as occurring in discontinuous stages, the changes in adulthood—whether they concern "intimacy," "career establishment," "the midlife crisis," or the "attainment of wisdom"—are more often triggered by social pressures and personal events than by biologically related ones.

The discontinuity view of development has dominated developmental psychology in the twentieth century. Indeed, this textbook, like most of its kind, reflects a discontinuity view, at least implicitly, by treating development in terms of seven distinct periods, from infancy to late adulthood. There are, of course, some good reasons for using this stagelike organization. To begin with, physical maturation occurs according to a biologically determined timetable, with the result that children of roughly the same age have many physical abilities and limitations in common. Children also share age-related patterns in the way they think about their world and about themselves, and in the way their society treats them. Likewise, in adulthood, people's lives often seem to run according to a developmental clock set by society, as people think and behave in certain ways that society deems "age appropriate."

Despite the apparent evidence for discontinuity, however, a number of developmentalists have cautioned against an overemphasis on distinct stages of development. As Flavell (1982) expresses it, strict stage views "gloss over differences, inconsistencies, irregularities, and other real but complexity-adding features."

A CLOSER LOOK   Stages in History

Whenever we take a stage view of development, we
should bear in mind that the stages themselves are, to
a certain degree, cultural creations. We tend to think of
the stages of childhood and adolescence, for example, as
inevitable consequences of biological maturation, but at
various times, in various cultures, these stages have been
nonexistent, or barely recognizable as such. For example,
the idea that childhood is a tender stage of life, and that
children need protection from hard work, sexual
knowledge, and harsh punishment has not always been
with us. Indeed, Philippe Ariès (1962) contends that
childhood as we think of it was unknown in Europe
through much of history. Children were cared for until
they could take care of themselves—at about age 7—and
then they entered the adult world. This phenomenon was
most apparent during the Middle Ages, when children
dressed as adults, drank with them, worked beside them
at home and in the fields, and could legally be joined as
spouses, crowned as monarchs, or hanged as thieves.
Louis XIV of France, for instance, became king at age 5
and played sexual games with his nursemaids throughout
childhood. Many of the folktales of earlier times were
designed to entertain adults as well as children, and many
children's games were played by people of all ages.

In short, the distinction we now make between activities
that are appropriate for children and those appropriate
for adults did not exist. Even after some educators and
clergy began to see childhood as a special stage several
centuries ago, legal recognition of childhood and of the
special needs of children was a long time in coming. Not
until the twentieth century were laws passed to regulate
child labor, mandate education, and delineate parental
responsibility for minors. (Indeed, some early cases of
child abuse were brought to court under statutes
protecting domestic animals, because laws pertaining
specifically to children did not exist.)

Similarly, adolescence as a separate stage of life—
especially as a stage in which the individual is supposed
to be free of major responsibilities and to gradually
develop a sense of independence—is essentially a
phenomenon of the twentieth-century industrial nation.
In many preindustrial societies, for example, there was
no notion of adolescence at all. Puberty marked the
beginning of adulthood, and the transition from child to
adult was usually accomplished through a single initiation ceremony. In the United States before the twentieth
century, teenagers were expected to labor like adults, but,
if they still lived at home (which most did), to be obedient
like children. Their parents made them pay for their room

(a)

(b)

(c)

(d)

Although there are obvious class differences visible in the historical illustrations shown here (a, b), they also clearly reflect the idea of children as miniaturized adults—either clothed and consorting in the manner of their elders or working full-time as factory work- *ers, frequently under punishing conditions. Today the specialness of childhood is taken for granted, and the child's particular needs are the focus of much parental and scientific attention. Adolescence is a somewhat different story (c). Although teen-* *agers are recognized as a special population, they are often left in limbo with respect to the opportunities they are provided to develop a sense of responsible independence. The final photo (d) shows further changes that have occurred in life's stages in* *recent years. Once preschoolers were expected to be at home and the elderly to be either hard at work or sickly. Now the activities of this foster grandparent and these preschoolers are quite typical of their stages in life.*

and board, told them what to wear and whom to date, insisted on respectful behavior, and accepted no "back talk" (Kett, 1977).

Discipline and dependence, not "self-discovery," characterized the life of most American teenagers until they moved away to begin their own families, an event that often occurred in the mid-teens for girls, and the mid-twenties for boys. Gradually, in the twentieth century, teenagers' obligation to contribute financially to their parents while still obeying their commands gave way to the current notion that teenagers should make more and more of their own decisions, even while their parents pay for them (Green and Boxer, 1986).

In fact, it was not until the rising affluence of mid-twentieth century brought high-school education, the mobility and privacy afforded by automobiles, and the emergence of a distinct "youth market" that adolescence became a time for American teenagers to be "understood" rather than held in check.

In recent years, an even more dramatic shift has occurred in what is expected of late adulthood. In earlier times,

adults continued working, either at home or at a job, until they became too sick or too feeble to do so. Most died soon after that, usually at home, surrounded by their children and grandchildren, and usually before age 60. By contrast, today's elderly, in rapidly increasing numbers, are living longer and much differently from their forebears, achieving the relatively new status and stage of retired senior citizens.

When researchers "discovered" old age as a developmental stage a few decades ago, they described it largely in terms of the problems it presented for the elderly person rather than in terms of the opportunities it offered for further development (Hareven, 1979). Now, in the latest shift, researchers are arriving at a more balanced view of old age, emphasizing the pleasures and rewards of old age as well as identifying, and trying to find remedies for, its problems.

Overall, then, whether or not stages exist and when they might begin or end, as well as how people should act at a particular stage, are more a matter of culture and custom than of universal human experience.

Although it would be convenient to approach human development "as a neat 'ages and stages' developmental story," Flavell cautions that actual development is much more complex, for people grow in varied ways—sometimes in sudden leaps and bounds, sometimes step by step, and sometimes with such continuity that they seem not to change at all. Further, what stages and transitions we do see may not be inevitable. Historically, for instance, there have been many widely different views of the stages of life and their chief characteristics (see A Closer Look, pages 14–15). Childhood, for example, was not always recognized as distinct from adulthood (Ariès, 1962), and the concept of adolescence did not even exist until the twentieth century (Kett, 1977).

The Scientific Method

As the two controversies show, developmentalists have differences of opinion, differences that are partly the result of their own backgrounds and biases. However, as scientists, they are committed to fact-finding and truth-seeking: they therefore are committed to consider insights and evidence from the available research before they draw conclusions, as well as to change their views when new data indicate they should. When doing research, they follow a general procedural model often referred to as the **scientific method,** which helps them overcome whatever biases they have. Procedures and techniques, not theories and assumptions, are what make the study of development a science (Scarr, 1985).

The scientific method involves four basic steps, and sometimes a fifth:

1. *Formulate a research question.* Build on previous research, or on one of the theories of development, or on personal observation and reflection, and pose a question that has relevance for the study of development.

2. *Develop a hypothesis.* Reformulate the question into a hypothesis, which is a specific prediction that can be tested.

3. *Test the hypothesis.* Design and conduct a specific research project that will provide evidence about the truth or falsity of the hypothesis. As the Research Report (see pages 18–19) indicates, the research design often includes many specific elements that help make the test of the hypothesis a valid one.

Figure 1.8 *The scientific method often reveals the unexpected. Popular wisdom usually blames teenage drinking habits on the youth culture and/or state drinking laws. Scientific research has shown that the most influential factor in adolescents' use or nonuse of alcohol or other drugs is the closeness of their relationship with their parents.*

4. *Draw conclusions.* Formulate conclusions directly from the results of the test: avoid general conclusions that are not substantiated by the data.

5. *Make the findings available.* Publishing the results of the test is often the fifth step in the scientific method. In this step, the scientist must describe step three and the resulting data in sufficient detail that other scientists can accurately evaluate the conclusions.

One of the main reasons procedures must be carefully described is so that other scientists can **replicate** the test of the hypothesis—that is, repeat it in an effort to obtain the same results—or extend it, using a different but related set of subjects or procedures. In this way, the conclusions from each test of every hypothesis accumulate, leading to more definitive and extensive conclusions and generalizations.

Ways to Test Hypotheses

There are many ways to test hypotheses. Among the most common are naturalistic observations, laboratory experiments, and interviews. Case studies are also used, not as a definite test of a general hypothesis but more to test the application of a specific hypothesis to one particular individual. Each of these methods has advantages and disadvantages.

Naturalistic Observation Scientists can test hypotheses by using **naturalistic observation,** that is, by observing people in their natural environments—the home, school, or playground, the workplace, the church, or a shopping center —and recording their behavior and interactions in detail.

A Naturalistic Study. The use of naturalistic observation, as well as the steps of the scientific method, become clearer with an example. Two scientists, Harvey Ginsberg and Shirley Miller (1982), were interested in the widely held belief that "males demonstrate a greater willingness to take risks or chances than females." Such a belief is given as an explanation for the higher rate of accidental death for males than for females, a sex difference that is maintained throughout the life span (see Figure 1.9).

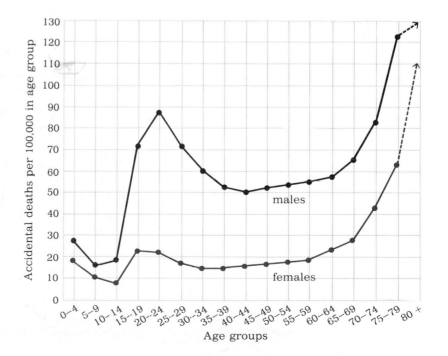

Figure 1.9 *Mortality statistics are hard evidence that males are more likely to have serious accidents than females, but this correlation does not explain causation. Further research is needed to uncover all the reasons for the existence of this sex difference.*

Ways to Make Research More Valid

In scientific investigation, there is always the possibility that the procedures and/or the investigator's biases can compromise the validity of the findings. Consequently, scientists often take a number of steps to ensure that their research is as valid as possible. Five of them are explained here.

1. Sample Size

Since it is usually not possible for scientists to study the entire population, or group, that they are interested in, they begin by studying a portion of that population, called a *sample,* which is representative of the development of the larger group. However, it is important to study a large enough sample so that a few extreme cases will not distort the picture of the group as a whole. Suppose, for instance, that researchers wanted to know the age at which the average American child begins to walk. Using a sample population, they would determine the age of walking for each member of the sample and then calculate the average for the group. The importance of **sample size** can be seen if we assume for the moment that one of the infants in their sample was severely mentally retarded, and did not walk until 24 months. If the sample size were, say, only ten, that one late walker would, relative to the current standard of 12 months, add more than five weeks to the age at which the "average" child is said to walk. However, if the sample were more than 500 children, one abnormally late walker would not change the results by even one day.

2. Representative Sample

The size of the sample is not the only important factor. Since the data collected on one group of individuals might not be valid for other people who are different in

"I'm walking!"

Drawing by Sempé; © 1986 The New Yorker Magazine, Inc.

Do babies walk (or talk) when they are ready, no matter how little attention their parents provide? Only careful research can provide the answer.

significant ways, such as gender, ethnic background, and the like, it is important that the sample population be a **representative sample,** that is, a group of subjects who are typical of the general population the researchers wish to learn about. In the case of a study of when the average American infant begins to walk, the sample population should reflect, in terms of sex ratio, economic and ethnic background, and so forth, the entire population of American children. Ideally, other factors might be taken into consideration. For instance, if it seems that first-born children walk earlier than later- or last-born children, then the sample should include a representative sample of each birth order.

However, this difference might occur because boys are more often in risky situations, rather than that, given an equal opportunity, males take more risks than females. For instance, boys are more likely to play outside and girls inside, and thus boys might be more likely to take a serious fall, be hit by a car, and so forth. Similarly, men are much more likely to be employed in dangerous occupations like heavy construction than women are, and thus might have more accidents because of the nature of their work, rather than because of a greater propensity for risk-taking. (Women, in fact, have traditionally been excluded from "risky" work.) In looking at all the research on sex differences in risk-taking, Ginsberg and Miller found that it cannot yet be determined whether, given equal opportunities, males are actually more daring than females.

The importance of representative sampling is revealed by its absence in two studies of age of walking undertaken in the 1920s (Gesell, 1926; Shirley, 1933), both of which found that the average American infant walks at 15 months. Both studies used a relatively unrepresentative sample (all the children were white and most were middle-class) and both arrived at a norm that was 3 months later than the current one. The present standard of 12 months was derived from a much more representative sample that included black and Hispanic children.

3. "Blind" Experimenters

A substantial body of evidence suggests that when researchers have specific expectations of the outcome of their research, those expectations can affect their perception of the events of the research. As much as possible, then, the people who do the actual testing should be **"blind,"** that is, unaware of the purpose of the research. Suppose the hypothesis being tested is that first-born infants walk sooner than later-borns. Ideally, the examiner who tests the infants' walking ability should not know what the hypothesis is, and would not even know the age or birth order of the toddlers under study. Thus the data the researcher receives for analysis would not be distorted by expectation.

4. Experimental Group and Control Group

When testing a hypothesis experimentally, a researcher can compare two study groups that are similar in every important way except one: one group, called the **experimental group**, receives some special experimental treatment, and the other group, called the **control group**, does not receive the experimental treatment.

Suppose a researcher hypothesized that children who use "walkers" walk earlier than children who do not. In order to find out if this is true, the researcher would select two representative groups of children, and arrange that one group (the experimental group) be placed in a walker for a certain amount of time each day and that the other group (the control group) would not be. If, in fact, the infants in the experimental group walked significantly earlier, on average, than the infants in the control group, then the hypothesis would be confirmed.

5. Determining Significance

Whenever researchers find a difference between two groups, they have to consider the possibility that the difference might have occurred purely by chance. For instance, in any group of infants, some would walk relatively early and some relatively late. When researchers in the foregoing experiment divided the sample population into the experimental and control groups, it would have been possible that, by chance, a preponderance of early walkers ended up in one group or the other. To determine whether or not their results are, in fact, simply the result of chance, researchers use a statistical test, called a **test of significance.** Following a specific mathematical formula, tests of significance take into account many statistical factors and produce a number that indicates exactly how likely it is that the particular difference occurred by chance. Generally, in order to be called significant, the possibility of the results occurring by chance has to be less than one in twenty, which is written in decimals as a significance of .05. Often the likelihood of a particular finding's occurring by chance is even rarer, perhaps one chance in a hundred (the .01 level) or one in a thousand (the .001 level).

They decided to focus their study on children, and started with the hypothesis that boys take more risks than girls. To give the hypothesis a fair test (one in which there would be ample opportunity for finding evidence against the hypothesis as well as for it), they did a naturalistic study in a place where boys and girls would, in roughly equal numbers, have the opportunity to take various risks.

They chose the San Antonio zoo, where, according to an actual count at the admissions gate, boys and girls entered in approximately equal numbers. Observing unobtrusively, the researchers kept careful tabs of the sex and approximate age of each child who participated in one or more of four "risky" activities: riding on the elephants, feeding the animals, petting the burro (a sign said "Careful, he bites"), and climbing up a steep concrete embankment to walk along a narrow ledge.

They found that more than two-thirds of the children who rode the elephant, fed the animals, petted the burro, or climbed the embankment were boys (see Figure 1.10) and that this sex difference was significant at the .001 level (see Research Report, pages 18–19).

Further examination of the results showed that, although older children took more of these risks than younger children did, the sex ratio in risk-taking held no matter what the age: more preschool boys, school-age boys, and preadolescent boys were risk-takers than girls the same age were.

Figure 1.10 *As you can see, the data show that some girls do take risks, but that boys are more likely to take them. Interestingly, the "risks" girls are relatively likely to take are "risks" their parents had to approve, and pay for.*

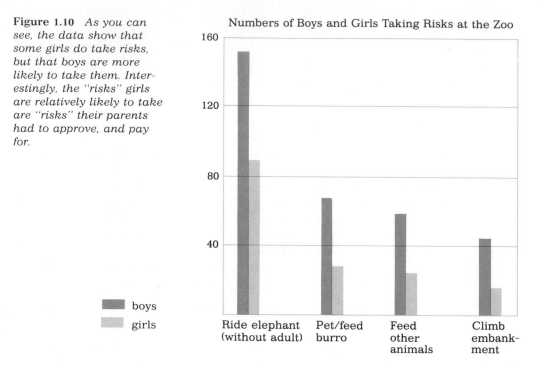

Numbers of Boys and Girls Taking Risks at the Zoo

One way to describe these results is to say that there was a correlation between sex and risk-taking, as well as between age and risk-taking. **Correlation** is a statistical measure used to describe the relationship between two variables. It indicates whether or not one variable (e.g., risk-taking) is likely to occur when another variable (e.g., maleness) occurs, or whether a change in one variable (e.g., increasing willingness to take risks) is related to a change in another (e.g., increasing age). (For a further explanation of correlation, see A Closer Look, next page.)

Limitations of Naturalistic Observation.   Naturalistic observation is an excellent method, but it has one main disadvantage. Because the variables in a natural setting are numerous and uncontrolled, it is difficult to pinpoint precisely which of them could explain a particular event. Perhaps the sex differences in risk-taking are the natural result of biologically governed impulses in males or inhibitions in females. Or the explanations may lie with nurture rather than nature. For instance, even at their young age, many of these children may have been responding to cultural stereotypes of daring heroes and helpless heroines, or perhaps their parents had said such things as "Be a brave boy, pet the burro" or "Be a good girl and hold my hand; I don't want to lose you." It is even possible that the results were influenced by some variable that had little to do directly with the children: for instance, some of the girls may have been dressed inappropriately for attempting daring adventures.

Correlation: What It Does, and Does Not, Mean

Correlation is a statistical term that indicates that two variables are somehow related, that is, that one particular variable is likely, or unlikely, to occur when another particular variable does. For instance, there is a correlation between height and weight, because, usually, the taller a person is, the more he or she weighs. There is also a correlation between education and wealth, and perhaps even between springtime and falling in love.

The fact that two variables are correlated, however, does not necessarily mean that they are related in every instance. Some tall people weigh less than people of average height; some highly educated people are poor; some people fall in love in the depths of winter.

Nor does correlation indicate cause. The correlation between education and wealth does not necessarily imply that more education leads to greater wealth. It may be instead that more wealth leads to greater education, since wealthier people are more likely to be able to afford the expense of college. Or there may be a third variable, perhaps intelligence or family background, that accounts for the level of both income and education.

There are two types of correlation, positive and negative. Whenever one variable changes in the same direction as another variable changes (for example, both increase or both decrease), the correlation is said to be positive. All the examples given so far are examples of positive correlation. Thus, when education increases, income tends to increase as well; when education is low, income tends to be, too.

When two variables are inversely related (one increasing while the other decreases), the correlation is said to be negative. Snow and summertime, maleness and motherhood, and blue eyes and black hair are negatively correlated. When there is no relationship at all between the two variables, the correlation is said to be zero.

Correlations can be expressed numerically. They range from plus one (+1.0), the highest positive correlation (when one event happens, the other always happens, too), to minus one (-1.0), the most negative correlation (when one event happens, the other never does). Halfway between plus one and minus one is zero, which, as you just learned, indicates no correlation at all.

Correlations are one of the most useful tools in psychology and, at the same time, one of the most misused. They are useful because knowing how variables are related helps us to understand and sometimes to predict development. However, as the respected researcher Sandra Scarr (1985) put it, "the psychological world . . . is . . . a cloud of correlated events to which we as human observers give meaning." Unless we are cautious in giving that meaning, we are likely to seize on one or another particular correlation as an explanation, without looking for other possible explanations. For instance, in the 1960s many psychologists noted the correlation between "broken" homes and maladjusted children and concluded that being raised by a single parent necessarily put children at risk. In the 1980s, psychologists looking at the same kinds of homes (now called single-parent families) have noted that many children in them do quite well, and that other factors that correlate with such homes (e.g., low income, bitterness between the ex-spouses, educational disruption) may be the explanation for children's problems when they occur. The lesson is clear: we need to be very cautious in jumping from the discovery of correlations to conclusions about cause (Scarr, 1985).

The people in this group are members of a Polar Bear Club posing before they take an ocean swim in subfreezing weather. As the photo suggests, there apparently is a correlation between being a Polar Bear and being heavier than average. Did some members put on weight to insulate themselves from the cold, or did they pick this sport because they were already well-padded? Since correlation indicates nothing about causation, we cannot say what accounts for a Polar Bear physique.

The Experiment   Another important method scientists use to discover the paths of development is the **experiment.** Unlike naturalistic observation, the experiment tests a hypothesis in a controlled environment, that is, in a setting in which the relevant variables are limited and can be manipulated by the experimenter. Because the variables are controlled, the link between cause and effect is much clearer than it is in naturalistic observation. The importance of this method becomes clear in an example.

Figure 1.11 *Coincidence? Similarities between people with the same genes might be nature, or nurture, or simply chance. A controlled experiment is the best way to explore correlations and their possible meanings.*

"Separated at birth, the Mallifert twins meet accidentally."

Drawing by Chas. Addams; © 1981 The New Yorker Magazine, Inc.

An Experiment with Newborns.   Naturalistic observation suggests that, in addition to the crucial role of nutrition, other factors in the early mother-child interaction also affect children's initial growth and development. This was first made clear by the physical and intellectual deficits found in motherless infants raised in institutions (Dennis, 1973; Spitz, 1945). It is also apparent in some children who are reared in troubled homes and suffer from *psychological dwarfism* (Gardner, 1972; Tanner, 1978), a condition in which growth is retarded even though there is no medical abnormality that would slow development. Some investigators have speculated that, in both cases, stress and neglect might cause the child to eat inadequately even when adequate nutrition is available. Others have suggested that a lack of physical stimulation due to not being held or caressed might lower the production of body chemicals, including growth hormones.

   In an attempt to isolate the key factors that might account for such instances of slow growth, scientists first experimented with rats in the laboratory (Schanberg et al., 1984). One group of mother rats was anesthetized—so that they could nurse their pups but not stimulate them by nuzzling or grooming them as mother rats usually do. Another group of mother rats had their nipples tied off—so that they could groom their infants but not nurse them (the pups were fed intravenously). The researchers found that those pups who had been nuzzled and groomed by their mothers, but not nursed by them, were close to normal in the production of important body chemicals (especially growth hormones), and that those pups who were fed by their mothers, but not nuzzled or groomed by them, had abnormally low levels.

To further isolate the effects of physical stimulation, some rat pups were separated from their mothers and were either left undisturbed or were stroked vigorously on the back and head for 2 hours with a moist, one-inch camel-hair brush in a manner roughly similar to that used by the mother in grooming the pups (Schanberg and Kuhn, 1980). After two hours, the levels of growth hormone in the separated and untouched pups averaged only 25 percent that of a control group left with their mothers, while the stroked group averaged 82 percent of normal level.

This finding was particularly interesting to scientists working with tiny, preterm infants, for whom gaining a few ounces can mean the difference between life and death. Because these infants are deprived of normal mother-infant contact by virtue of being continuously hooked up to intensive-care machinery, it was thought that they might be responsive to a program of physical stimulation. Accordingly, an interdisciplinary team of scientists (Field et al., 1985) selected forty newborns, each born at least a month early and all weighing less than 1,500 grams (about 3 pounds). They divided them evenly into experimental and control groups. The average weight, length, number of birth complications, and so forth were the same for infants in both groups, and infants in both groups were in the same nursery, given the same care by the same nurses.

The only difference between the groups was the experimental treatment, which consisted of gentle back-stroking for 30 minutes per day, and rhythmic moving of the infants' arms and legs for 15 minutes a day. After two weeks of this treatment, the experimental group had gained an average of 25 grams per day, while the control group had gained only 17 grams per day. Thus the experimental group gained 47 percent more weight than the control group, a difference that was significant at the .001 level. The experimental group also tended to spend more time awake and moving around, and to score higher on a test of maturation, given by testers who were "blind" to whether a particular infant was in the control group or the experimental group. Finally, infants in the experimental group were released from the hospital an average of six days earlier than those in the control group, because, in the judgment of their pediatricians, they had reached sufficient maturity and weight to go home.

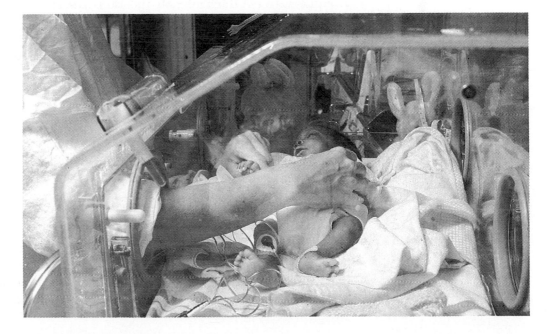

Figure 1.12 *Traditionally, preterm infants were rarely touched, except during the course of medical procedures. However, if the tactile-stimulation experiments described in the text and illustrated here are supported by further studies, programs can be developed to translate research findings into health benefits for these fragile newborns.*

The scientists then tried to determine why the infants had responded so well to the special handling. Had it stimulated their appetites, thus causing greater food consumption and weight gain, or had it triggered more fundamental body changes that affect the growth rate? Tabulating the amount of food all the infants in both groups had consumed, they found no significant differences between the groups. Thus they concluded that the massage and exercises provoked biochemical processes that promoted growth.

Limitations of Experiments.   While the research on the effects of tactile stimulation on infants' growth rate is valuable as an example of the strengths of the experimental approach, it does not reveal two of the limitations that commonly beset experiments with humans.

One of these problems is that, in almost any experiment with people (except, of course, very young infants), the subjects are aware that they are being studied and may act differently than they normally would, thereby invalidating the findings. Second, even if the subjects in an experiment behave normally, the carefully controlled situation of an experiment may not readily apply in the much less controlled circumstances of everyday life. As Urie Bronfenbrenner has pointed out, even the most elegantly designed experiments may be limited in scope, because

> many of these experiments involve situations that are unfamiliar, artificial, and short-lived and that call for unusual behaviors that are difficult to generalize to other settings . . . a variety of approaches are needed if we are to make progress toward the ultimate goal of understanding human development in context. [Bronfenbrenner, 1977]

For this reason, naturalistic observation and laboratory experiments are often used to complement each other. The lack of precision in naturalistic observation is balanced by the control of the laboratory experiment; the artificiality of the laboratory is balanced by the realism of naturalistic observation. Often both methods are combined. For example, a researcher can perform an experiment in a natural setting, or observe, rather than experiment, in a laboratory.

The Interview   A third method, the **interview,** is also often used, either by itself or combined with other methods. In an interview, the scientist gathers information by asking questions of a sample population. This seems to be an easy, quick, and direct research method. For example, in order to find out if men are subject to middle-age crises, one can ask men of various ages questions to find out how happy they are with their lives and how well they are coping with their problems. Then the correlation between age and the answers can be computed to see if middle-aged men are more distressed than other men.

However, perhaps more than any other method, the interview is vulnerable to bias, on the part of both the interviewer and the interviewee. To begin with, the phrasing of the interview questions often affects the answers. A man might give a different response to the question "Have you thought lately that life was not worth living?" than to the question "Have you thought lately about suicide?"—though both questions are intended to ask the same thing.

In addition, many people who are interviewed give answers that they think the interviewer will want to hear, or that they think will make them look good in the interviewer's eyes, or that will perhaps make them feel good about themselves. Even when people wish to give completely accurate information, their responses may be flawed because their opinions and moods vary from day to day, or because their recollection of events is distorted.

Finally, the answers to interviews must be interpreted, and the interpretation depends partly on the interpreter. For example, one study (Farrell and Rosenberg, 1981) concluded that most men had some midlife crisis, even though only 12 percent of the men interviewed expressly indicated they had had one. The basis for the study's conclusion was the fact that another 56 percent reported such symptoms as headaches, poor appetite, and sexual dissatisfaction, which the researchers interpreted as signs of a crisis the interviewees were unable or unwilling to recognize as such. Other scientists, of course, might interpret the same data quite differently. This does not mean that the evidence from interviews is necessarily suspect. It does mean, however, that researchers must be particularly careful in explaining who was interviewed, what questions were used, in what context, and how the results were interpreted. Furthermore, it means that measures to enhance the validity of the results, such as training interviewers, standardizing questions, and recording the results, are often used.

The Case Study   The interview is often the heart of a **case study,** which is an intensive study of one individual. Typically, both the person's background and his or her present thinking and actions are reported, sometimes along with interviews of friends, family members, teachers, and coworkers. Naturalistic observation and standardized tests may furnish additional case-study material.

Case studies can provide a wealth of detail and therefore are rich in possible insights. However, not only are case studies susceptible to the usual problems of the interview, but the interpretation of case-study data depends even more heavily on the expertise as well as the biases of the interpreter. Furthermore, the interpretation may apply only to the particular individual being studied. In general, the case study is not used to do basic research, because no conclusions about people in general can be drawn from a sample size of one.

Figure 1.13 *Reconciling the many possible, sometimes conflicting, views about the subject of a case study requires a talented interpreter, whose views and possible biases must also be taken into account.*

"You are fair, compassionate, and intelligent, but you are <u>perceived</u> as biased, callous, and dumb."

Drawing by Mankoff; © 1985 The New Yorker Magazine, Inc.

**Designing
Developmental
Research**

Thus, there are many ways to test hypotheses. Researchers can observe people in natural settings or experiment with them in a laboratory; they can compare one group with another and find out correlations and significant differences; they can interview hundreds of people or study one or two cases in depth.

However, for research to truly examine development, scientists must discover and describe how people change with time. To learn about the pace and process of change, developmentalists use two basic research designs, cross-sectional and longitudinal.

Cross-Sectional Research   The more convenient, and thus more common, way researchers study development over time is by doing **cross-sectional research** on people of various ages. In this kind of study, groups of people who are different in age but similar in all other important ways (such as their level of education, economic status, ethnic background, and so forth) are compared on the characteristic or tendency that is of interest. Any differences on this dimension that exist among the people of one age and the people of another are, presumably, age-related.

However, in a cross-sectional study, it is always possible that some variable other than age differentiates the group. Consider research on sexual activity, for example. A classic series of cross-sectional studies found that the older adults are, the less frequently they engage in sexual intercourse and the less they think about sex (Kinsey et al., 1948). On the basis of these findings, it was concluded that the aging process gradually takes away sexual drive, resulting in virtual asexuality by old age.

But can we be certain that these cross-sectional differences are solely the result of age? After all, the proportion of people who are married, divorced, and widowed varies with age, and these differences may affect sexual activity more than age itself does. To avoid confounding variables such as these, scientists sometimes use another form of research, called longitudinal research, to study development.

Longitudinal Research   **Longitudinal research** studies the same individuals again and again as they grow older. In this way, longitudinal research allows information about people at one age to be compared with information about those same people at another age, thus enabling researchers to find out how those particular people changed with time.

Figure 1.14 *The long time span of longitudinal research provides a revealing view of how individuals change and how they remain the same. Longitudinal research on this woman would reveal whether her view of woman's role over the years has changed as readily as her sense of hair style, and whether her outgoing personality has remained as stable as her engaging smile would seem to suggest.*

Longitudinal research is particularly useful in studying developmental trends that occur over a long age span. Sexual behavior is a case in point. Longitudinal research reveals that, while it is true that, overall, the average frequency of sexual activity diminishes with time, this average masks important individual differences. Some people remain sexually active throughout life, others become more active with age, others gradually become less active, and still others stop suddenly—usually for health reasons or because a partner is no longer available (Luria and Meade, 1984). Individual attitudes and appetites are at least as important as age. As Alex Comfort (1979) explains, "those whose sexual 'set' is low for physical or attitudinal reasons drop out early—often using age as a justification for laying down what for them has been an anxious business." He adds that the supposed asexuality of the elderly is a social disorder, not a physical one.

Although longitudinal research is "the lifeblood of developmental science," the actual number of large-scale longitudinal studies that have traced development over forty years or more is "woefully small" (Applebaum and McCall, 1983). The primary reason for this is a very practical one: to follow the development of a group of people over a number of years usually requires great effort, considerable foresight, and substantial funding.

Cohort Effects   One fact that researchers conducting both longitudinal and cross-sectional studies must bear in mind as they assess their findings is that each **cohort,** or group of people born in a particular generation, may experience social conditions that are different from those of other cohorts. This is particularly true in a society such as ours, in which rapid social change occurs. For example, sexual attitudes have changed substantially over the past three decades, indeed, even over the past fifteen years. In 1965, 70 percent of women and 33 percent of men felt that premarital intercourse was immoral; in 1980, only 25 percent of the women and 17 percent of the men thought so (Robinson and Jedlicka, 1982). In both cross-sectional and longitudinal research, cohort effects such as these can result in behavioral differences that appear to be rooted in age but that are actually rooted in culture.

Sequential Research   Researchers have a variety of complex ways to take cohort effects into account. One important method, called **sequential research,** uses a

combination of cross-sectional and longitudinal techniques (Schaie and Hertzog, 1985). Essentially, this type of research begins with a cross-sectional study, with several groups of people at various ages. Then, months or years after the first testing, these people are tested again, as would be done in a longitudinal study. At the same time, a new group of people is tested at each age level, to control for any changes that may have occurred in the original group because some had dropped out of the program, or because retesting had affected their performance, or because their particular cohort had experiences that other cohorts did not.

This method has been revealingly applied in research on adult intelligence. Originally, cross-sectional research indicated that adults became less intelligent with every passing decade. Then longitudinal research suggested that people became more intelligent, not less, with age. Finally, sequential research—which allowed researchers to compare given subjects' intellectual ability with their ability years earlier, with the ability of others of their generation, and with the ability of members of older generations when they were the subjects' age—showed that the pattern of intelligence is complex, dependent on which particular abilities are being tested for which particular cohorts, as well as on the education and the health of the individual.

Sequential research is complicated and expensive, but it will probably be increasingly used to investigate many aspects of development. It provides insights that neither cross-sectional nor longitudinal research provides.

One final word of introduction. The study of human development, as objective and scientific as it may try to be, inescapably involves values and goals, those of the scientist, the community, the society as a whole—and your own. Further, each theory, each fact, and each bit of research has implications for all our lives, implications that need to be weighed, evaluated, and accepted or rejected.

This will be apparent, time and time again, as you learn about the mechanisms and patterns of development, and as you are confronted with both practical and ethical issues—from family planning and prenatal care to adolescent sexuality and suicide, from the implications of divorce to care of the frail elderly—which may touch on some aspect of your own life. There will be many times when you will be called upon to set aside your own views for the moment and to think about some problem from an entirely different perspective. Do not resist the challenge. In the final analysis, your own growing involvement in the practical and philosophical questions of development will make your study meaningful, as well as insightful, whether you plan to become a researcher or a practitioner, a teacher or a parent, or simply a better-informed member of the human family.

SUMMARY

The Study of Human Development

1. The study of development is the study of how people change as they grow older, as well as how they remain the same.

2. Development may be divided into three domains, the physical, the cognitive, and the psychosocial. Although research projects often center on one particular domain, development is thought to be holistic.

3. An ecological, or systems, approach stresses the context of development, particularly the influence of family, community, and culture. Each individual is seen as part of many systems, affecting them as well as being affected by them.

4. The interaction of domains is clearly seen in the example of David, a handicapped college student, whose problems originating in the physical domain quickly affected the other two domains. His example also shows how the individual is affected by, and also affects, the surrounding systems of family, society, and culture.

Two Controversies

5. All types of development are guided by innate biological forces as well as by the particular events a person experiences. The relative importance of these factors is a topic of debate, called the nature-nurture controversy.

6. Another controversy exists between those who think that development is smooth and continuous, and those who think it occurs in stages and is discontinuous. While the stage view is common today, in past centuries childhood and adolescence were not recognized as stages.

The Scientific Method

7. The scientific method is used, in some form, by most developmentalists. They observe, pose a question, develop a hypothesis, test the hypothesis, draw conclusions, and make the findings of their research available.

8. One common method of testing hypotheses is naturalistic observation, which provides ecologically valid information but does not pinpoint cause and effect. The laboratory experiment pinpoints causes but is not necessarily applicable to daily life. The interview and the case study are also useful.

9. In developmental research, ways are needed to detect change over time. Cross-sectional research compares people of different ages; longitudinal research (which is preferable but more difficult) studies the same individuals over a long time period. Both are valid for the cohorts under examination, but not necessarily for other cohorts. Sequential research overcomes this problem by incorporating more than one cohort into longitudinal research.

KEY TERMS

physical domain *(4)*

cognitive domain *(4)*

psychosocial domain *(4)*

holistic *(5)*

ecological approach *(5)*

microsystem *(5)*

mesosystem *(5)*

exosystem *(6)*

macrosystem *(6)*

systems approach *(6)*

nature *(11)*

nurture *(11)*

continuity *(13)*

discontinuity *(13)*

scientific method *(16)*

replicate *(17)*

naturalistic observation *(17)*

sample size *(18)*

representative sample *(18)*

blind *(19)*

experimental group *(19)*

control group *(19)*

test of significance *(19)*

correlation *(20)*

experiment *(22)*

interview *(24)*

case study *(25)*

cross-sectional research *(26)*

longitudinal research *(26)*

cohort *(27)*

sequential research *(27)*

KEY QUESTIONS

1. What is developmental psychology?

2. What are the three domains of developmental psychology?

3. Describe the controversy about the roles of nature and nurture in human development. Give examples.

4. Describe the controversy between those who see human development as continuous and those who see it as stagelike.

5. Give examples of the interaction among the various systems that affect the individual's development.

6. What are the steps of the scientific method?

7. What are the advantages of the scientific method?

8. What are the advantages and disadvantages of testing a hypothesis by naturalistic observation?

9. What are the advantages and disadvantages of testing a hypothesis by experiment?

10. Compare the advantages of longitudinal research and cross-sectional research.

CHAPTER 2   **Theories**

"It is a capital mistake to theorize before one has data."
Sherlock Holmes

"There is nothing so practical as a good theory."
Kurt Lewin

When we examine the events of our daily lives and those of people we know, we find that many of the things we do and think require far longer to describe and interpret than we might at first expect. For example, why do you handle challenging situations in the way that you do? Why is it that some of your friends handle these situations quite differently? Why are you attracted to certain types of people and unlikely to begin a conversation with others? Why did you spend the past year of your life in the way that you did—contrary, perhaps, to the advice of at least one of your friends or relatives?

What Theories Do

To begin to answer these questions and many others, we need some way to select significant facts and to organize them in a way that will take us deeper than our first speculations, which are probably biased by being based on our own limited experiences. In short, we need a theory. A theory provides a framework of general ideas that permits a broad and cohesive view of the complexities that may be involved in any given human interaction. Theories can be used to organize our assumptions and guesses into hypotheses that can be tested and proven valid or invalid, an important step in the scientific method.

Many theories, some originating in psychology, others in sociology or biology, are relevant to the study of development. In this chapter, however, we will focus on four of the most comprehensive, influential, and useful theories of developmental psychology—psychoanalytic theory, learning theory, humanistic theory, and cognitive theory. Keep in mind that although each theory provides a somewhat different perspective from which to interpret the life course, each performs the same function: it takes us beyond isolated incidents and toward an understanding of the full context in which human interactions occur.

Figure 2.1 *No matter what interaction developmentalists study, they can make their observations from four theoretical perspectives—psychoanalytic, which emphasizes unconscious drives and motives; learning theory, emphasizing learned responses to particular situations; humanistic, emphasizing innate human potential; and cognitive, emphasizing the individual's understanding of self and others.*

Unfortunately, theory and practice are often discussed as opposites—as though we had to choose between them. In fact, theories arise from experience, and once a theory is formulated, it leads to practical applications (Miller, 1983). The value of a theory can be measured by how "useful" it is—that is, how productive it is in generating hypotheses to test and in inspiring new insights into behavior (Lee, 1976).

Psychoanalytic Theories

The first comprehensive view of human behavior, **psychoanalytic theory** interprets human development in terms of unconscious drives and motives. These unconscious impulses are viewed as influencing every aspect of a person's thinking and behavior, from the crucial choices of a lifetime, including whom and what to love or hate, to the smallest details of daily life, including manner of dress, choice of food, what we say and how we say it, what we daydream about, and how we reason—in fact, the entire gamut of our personal preferences, dislikes, and idiosyncrasies.

Origins

To understand the tenets of this theory, it is helpful to know something about the intellectual climate of the times that preceded it. In the Europe of the mid-1800s, prevailing thought about human behavior included the ideas that men are governed for the most part by reason; that children are "innocent," devoid of all sexual feelings; and that women are particularly susceptible to "hysteria," a nervous disorder that produced such symptoms as paralysis of a limb, or deafness, or loss of feeling, and was believed to originate in the uterus. (The term "hysteric" derives from the Greek *hystera,* meaning "womb." The common treatment for hysteria was a hysterectomy, that is, the removal of the uterus.)

Figure 2.2 *Many of Freud's students and patients spoke about his penetrating gaze, which, they said, helped them uncover their hidden thoughts and fantasies. Indeed, some critics contend that much of Freud's success as a psychoanalyst could be credited to his personality and insight, rather than to his methods or theories.*

By the 1870s, against the grain of these notions, a number of intellectuals were developing an explanation of human behavior that emphasized the controlling power of emotional forces and the subordinate position of reason. In particular, Sigmund Freud (1856–1939), the founder of the psychoanalytic approach, began to evolve a theory that pointed specifically to the irrational basis of human behavior, to the hidden emotional content of our everyday actions, and to the ways in which the individual is driven by powerful sexual and aggressive impulses, and by fear of them.

Freud, who began his career as a medical doctor, formulated his theory while treating some of those women who were considered "hysterics." Freud suspected that the origin of their symptoms was in the mind, not in the uterus. In an effort to uncover the hidden causes of their problems, Freud hypnotized his first patients, suggesting that they talk about events and feelings related to their hysteria. Sometimes these treatments were successful: patients would eventually reveal the events that had precipitated their illness, and when this occurred, their symptoms would be much relieved.

Freud was not satisfied with hypnosis, however. Some patients did not know what troubled them; others did not respond to hypnotic suggestion. Gradually, Freud developed an ingenious new way to uncover the thoughts and feelings of his patients. He would have them recline on his office couch and talk about anything and everything that came into their minds—daily events, dreams, childhood memories, fears, desires—no matter how seemingly trivial or how unpleasant. From these disclosures and such things as the patient's slips of the tongue and unexpected associations between one word and another, Freud discerned clues to the usually unconscious emotional conflicts that paralyzed one person or terrified another. Once the patient, under Freud's guidance, came to understand the nature of these hidden conflicts, the patient's symptoms would frequently diminish or disappear altogether.

The medical establishment ridiculed Freud's "talking cure," especially when he reported that many emotional problems were caused by unconscious sexual desires, some of which originated in infancy. But patients flocked to Freud's door, and as they revealed their problems and fantasies, Freud listened, interpreted, and formulated an influential theory of the human psyche.

Freud's Ideas

One of Freud's basic ideas is that children have sexual pleasures and fantasies long before they reach adolescence. According to this theory of **infantile sexuality,** development in the first five or six years occurs in three **psychosexual stages.** Each stage is characterized by the focusing of sexual interest and pleasure in a particular part of the body, respectively, the mouth (the **oral stage**), the anus (the **anal stage**), and the penis (the **phallic stage**). In each stage, the child strives for satisfaction through activities centered on these organs—feeding and sucking, defecation and toilet training, and masturbation. It was Freud's contention that the parent-child interaction during these stages—especially how the child experiences the conflicts that these stages imply—determines the child's basic personality. In other words, in the Freudian view, personality is essentially fixed by age 5 or so. Then, after a five- or six-year period of sexual **latency,** during which sexual forces are dormant, the individual enters a final psychosexual stage, the **genital stage,** which lasts throughout adulthood.

Figure 2.3 *This girl's interest in the statue's anatomy may reflect simple curiosity, but Freudian theory would maintain that it is a clear manifestation of the phallic stage of psychosexual development, in which girls are said to feel deprived because they lack a penis.*

Id, Ego, and Superego   To help explain the dynamics of the individual's psychological development, Freud proposed three theoretical components of personality: the id, ego, and superego. The **id** is the source of our unconscious impulses toward fulfillment of our needs. It operates according to the *pleasure principle,* that is, the striving for immediate gratification. In other words, the id wants whatever seems satisfying and enjoyable—and wants it *now*. The impatient, greedy infant screaming for food in the middle of the night is all id.

Gradually, as babies learn that other people have demands of their own and that gratification, therefore, must sometimes wait, the **ego** begins to develop. It is the role of the ego to mediate between the unbridled demands of the id and the limits imposed by the real world. The ego is said to operate according to the *reality principle:* it attempts to satisfy the id's demands in ways that recognize life as it is, not as the id wants it to be.

The ego also strives to keep another irrational force at bay. At about age 4 or 5, the **superego** starts to develop, as children begin to identify with their parents' moral standards. The superego is like a *relentless conscience* that distinguishes right from wrong in no uncertain terms. Its prime objective is to keep the id in check. In this regard, it is the function of the ego to mediate between the primal desires of the id and the superego's unbending effort to inhibit those desires.

Figure 2.4 *Freud's theory of personality has proven to be a fertile source of explanations for why we may, at times, not feel at peace with our "selves."*

"Take it up with my ego."

© Sidney Harris 1987.

Defense Mechanisms   When the superego becomes so overbearing, or the id's demands so insistent, that the ego feels in danger of being overwhelmed, people involuntarily defend themselves against the superego's attack and against the frightening impulses of the id by using one of dozens of **defense mechanisms.** Three common defense mechanisms are regression, repression, and displacement.

Regression occurs when someone retreats to a form of behavior typical of a younger person, allowing the individual to avoid dealing with reality in an age-appropriate way. An adult might slip back to dependence on his or her parents; a 10-year-old might start wetting the bed again; a 5-year-old might talk baby talk and want to sleep with a bottle. **Repression** is the pushing of a disturbing idea or memory or impulse into the unconscious, where it will not be actively threatening. Adolescents who are not interested in anything to do with sex, or abused children who forget how they got hurt, are probably repressing powerful feelings that would be deeply distressing if they were to surface. **Displacement** is the shifting of a drive or emotion from a threatening or unavailable object to a substitute object. An infant who is frustrated in the quest for oral gratification during the first psychosexual stage might develop an ongoing need for oral satisfaction. When deprived of a breast or a bottle, the baby might displace the oral drive to thumb-sucking. As the child grows older, the social unacceptability of thumb-sucking might cause the need to be displaced by some activity such as chewing on fingernails or pencil erasers. In adulthood, displacements might take such forms as heavy smoking or drinking, or overeating.

According to psychoanalytic theory, psychologically healthy people gradually develop strong egos, able to cope with the demands of the id and superego. Defense mechanisms help regulate this process throughout life by deflecting or countering the demands of the id or superego when they threaten to become overwhelming. At any point, however, defense mechanisms can be overused, preventing the ego from ever confronting the unconscious drives directly. As a result, the individual is handicapped in dealing with reality.

Psychoanalytic theory holds that each person inherits a legacy of problems from the conflicts of his or her childhood, along with particular ways of coping with them. Depending on our early experiences, some of us are more able to cope with the stresses of daily life than others.

Erikson's Ideas

Figure 2.5 *Erik Erikson has continued to write and lecture on psychosocial development throughout his long life. His most recent work emphasizes psychohistory—the relationship between historical factors and personality development.*

Figure 2.6 *It seems quite clear that the young boy here is in the thick of the psychosocial stage Erikson referred to as autonomy versus shame and doubt. Whether he emerges from this stage feeling independent or inept depends in part on whether his parents encourage his efforts to feed himself or, instead, regularly criticize him for the mess he is making. It seems equally clear that this young woman is trying to negotiate the stage Erikson called identity versus role confusion. In this stage, teenagers try out a number of roles (sometimes mostly through costume) in an effort to discover who they really are.*

Dozens of Freud's students became famous theorists in their own right. Although they acknowledged the importance of the unconscious, of sexual urges, and of early childhood, each in his or her own way expanded and modified Freud's ideas. Many of these neo-Freudians, including Karen Horney, Margaret Mahler, Helene Deutsch, and Anna Freud, are mentioned at various points in this book. One of them, Erik Erikson (1902–   ), formulated a comprehensive theory of development that will be outlined here and discussed in later chapters.

Psychosocial Development   Whereas Freud remained in Vienna for most of his life, Erikson spent his childhood in Germany, his adolescence wandering through Europe, his young adulthood in Vienna under the tutelage of Freud and in analysis with Freud's daughter Anna, and his later life in the United States. In America, he studied students at Harvard, soldiers who suffered emotional breakdowns during World War II, civil rights workers in the South, and Native American tribes. Partly as a result of this diversity of experience, Erikson came to think of Freud's stages as too limited and too few. He proposed, instead, eight developmental stages, each one characterized by a particular conflict, or crisis, that must be resolved.

As you can see from Table 2.1, Erikson's first five stages are closely related to Freud's stages. Freud's last stage occurs at adolescence, however, while Erikson sees adulthood as having three stages. Another significant difference is that all of Erikson's stages are centered, not on a body part, but on each person's relationship to the social environment. To highlight this emphasis on social and cultural influences, Erikson calls his theory the **psychosocial theory** of human development.

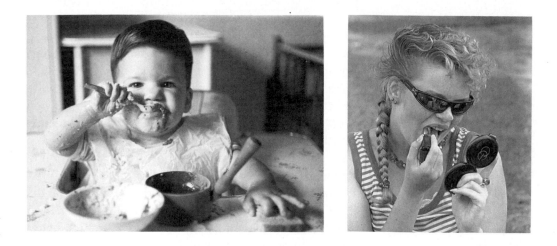

Cultural Differences   Central to Erikson's theory is his conviction that each culture faces particular challenges and, correspondingly, promotes particular paths of development that are likely to meet those challenges. Erikson suggests, for example, that the traditional German stress on early toilet training and cleanliness prepared adults for a society where law and order were paramount, just as the stress on independence and self-assertion in pioneer America prepared adults to explore new territory and ignore traditional laws and conventions—precisely what that society needed at the time (Erikson, 1963). A problem arises when a society's traditional methods of upbringing no longer prepare its children to cope with the demands they face as adults. No culture anticipates the future so well that each child is prepared to live in it without problems. Each society provides better preparation for some crises than for others.

TABLE 2.1 **Comparison of Psychosexual and Psychosocial Stages**

Approximate Age	Freud (Psychosexual)	Erikson* (Psychosocial)
Birth to 1 year	**Oral Stage** The mouth is the focus of pleasurable sensations in the baby's body, and feeding is the most stimulating activity.	**Trust vs. Mistrust** Babies learn either to trust or to mistrust that others will care for their basic needs, including nourishment, sucking, warmth, cleanliness, and physical contact.
1–3 years	**Anal Stage** The anus is the focus of pleasurable sensations in the baby's body, and toilet training is the most important activity.	**Autonomy vs. Shame and Doubt** Children learn either to be self-sufficient in many activities, including toileting, feeding, walking, and talking, or to doubt their own abilities.
3–6 years	**Phallic Stage** The phallus, or penis, is the most important body part. Boys are proud of their penis but ashamed when they masturbate, and girls are envious and wonder why they don't have one. Children of both sexes have sexual fantasies about their parents for which they feel guilty.	**Initiative vs. Guilt** Children want to undertake many adultlike activities, sometimes overstepping the limits set by parents and feeling guilty.
7–11 years	**Latency** Not a stage but an interlude, when sexual needs are quiet and children put psychic energy into learning skills.	**Industry vs. Inferiority** Children busily learn to be competent and productive or feel inferior and unable to do anything well.
Adolescence	**Genital Stage** The genitals are the focus of pleasurable sensations, and the young person seeks sexual stimulation and sexual satisfaction.	**Identity vs. Role Confusion** Adolescents try to figure out "Who am I?" They establish sexual, ethnic, and career identities or are confused about what future roles to play.
Adulthood	Freud believed that the genital stage lasts throughout adulthood. He also said that the goal of a healthy life is "to love and to work."	**Intimacy vs. Isolation** Young adults seek companionship and love with another person or become isolated from others.
		Generativity vs. Stagnation Middle-aged adults are productive, performing meaningful work and raising a family, or become stagnant and inactive.
		Integrity vs. Despair Older adults try to make sense out of their lives, either seeing life as a meaningful whole or despairing at goals never reached and questions never answered.

*Although Erikson describes two extreme resolutions to each crisis, he recognizes that there is a wide range of solutions between these extremes and that most people probably arrive at some middle course.

Evaluations of Psychoanalytic Theories

All developmentalists owe a debt of gratitude to Freud and to the neo-Freudians who extended and refined his concepts. Many of Freud's ideas are so widely accepted today that they are no longer thought of as his—for example, that unconscious motives affect our behavior, that we use defense mechanisms to avoid conflicts, and that sexuality is a powerful drive. While few accept his ideas completely, many have learned from his insights.

Developmentalists have been influenced by three psychoanalytic ideas in particular. Stage theory is one of them. Although poets and playwrights had previously written about the "ages" of man, Freud was the first to use his observations and insights to construct a coherent theory regarding our different needs and problems at various ages, or stages, of life.

Another influential idea concerns the importance of certain parts of the life span. Freud centered our attention on the critical first five years, a time previously neglected in education and psychology. Erikson led us to recognize that adolescents and adults also experience developmental changes.

Finally, psychoanalytic theory has helped us to realize that human thoughts and actions are likely to be far more complicated than might at first be apparent. Indeed, it is difficult to imagine that there was once a world in which it was assumed that human beings act on the basis of reason alone, so clear does it seem to us now that the forces of impulse and fantasy, and the influences and pressures applied by parents and society, shape and direct our behavior throughout our lives.

Although the general concepts and implications of psychoanalytic theory are very much part of our current thinking about development, most contemporary developmentalists find many psychoanalytic ideas to be inadequate or wrong. For instance, Freud's notion that the child's experiences during the first three psychosexual stages form the basis for character structure and personality problems in adulthood has found little support in studies of normal children. Moreover, most researchers agree that, by adulthood, personality characteristics and behavior are affected more by genetic traits, current life events, and the overarching sociocultural context than by the experiences of early childhood (Bengston et al., 1985; Ingelby, 1987; Vandenberg et al., 1986; Whitbourne, 1985).

Some aspects of Freud's theory strike many as anachronistic today. His depiction of the struggle between the id and the superego—that is, between a torrent of impulses seeking immediate release and a ceaselessly judgmental monitor trying to check those impulses—seems more an outgrowth of the Victorian morality of nineteenth-century Vienna than a valid depiction of a universal process.

Erikson's interpretation of development has fared better than Freud's, perhaps because Erikson's ideas, though arising from Freudian theory, are more comprehensive and apply to a wider range of behavior. Even so, most of the sources of Erikson's theory are, like Freud's, subjective, grounded in Erikson's own experiences, the recollections of his patients in therapy, and the insights of classical literature. Psychoanalytic theory does not lend itself to laboratory testing under controlled conditions, which leads to the accusation by some that this theory is more myth than science, with the validity of psychoanalytic ideas "evaluated by dogma, not data" (Cairns, 1983). Certainly, evaluations of psychoanalytic theory are more personal than scientific. And they are also highly varied. Some psychologists find psychoanalytic theory illuminating and insightful; others find it provocative nonsense; most think it a combination of both.

Learning Theories

Early in the twentieth century, John B. Watson (1878–1958) decided that if psychology was to be a true science, psychologists should study only what they could see and measure. In Watson's words: "Why don't we make what we can *observe* the real field of psychology? Let us limit ourselves to things that can be observed, and formulate laws concerned with only those things. . . . We can observe behavior—what the organism does or says" (Watson, 1967; originally published 1930). Many American psychologists agreed with Watson. Thus began a major theory of American psychology, **behaviorism,** now more commonly called **learning theory** because of its emphasis on how we learn specific behaviors.

Laws of Behavior

Unlike the other theorists in this chapter, learning theorists have not developed a stage theory of human development. Instead, they have formulated laws of behavior that can be applied to any individual at any age, from fetus to octogenarian.

Figure 2.7 *Pavlov did not allow even the chaos of the Russian revolution to deter him from his research. Legend has it that he was sometimes forced to fend off hordes of starving peasants who stormed his laboratory to get the food fed to the dogs, and even the dogs themselves.*

Figure 2.8 *B. F. Skinner is best known for his experiments with rats and pigeons, but he has also applied his knowledge to a wide range of human problems. For his daughter, he designed a glass-enclosed crib in which temperature, humidity, and perceptual stimulation could be controlled to make time spent in the crib as enjoyable and educational as possible. He has also speculated and written about an ideal society where, for example, workers at the less desirable jobs earn greater rewards.*

The basic laws of learning theory explore the relationship between **stimulus** and **response,** that is, between any behavior or event (the stimulus) and the behavioral reaction (the response) that it elicits. Some responses are automatic, like reflexes. If someone suddenly waves a hand in your face, you will blink; if a hungry dog smells food, it will salivate. But most responses do not occur spontaneously; they are learned. Learning theorists emphasize that life is a continual learning process: new stimuli evoke new behavior patterns, while old, unproductive responses tend to fade away. This learning process, called **conditioning,** occurs in two basic ways: classical and operant.

Classical Conditioning   More than eighty years ago, a Russian scientist named Ivan Pavlov (1849–1936) began to study the link between stimulus and response. While doing research on salivation in dogs, Pavlov had noted that his experimental dogs began to salivate not only at the sight of food but, eventually, at the sound of the approaching attendants who brought the food. This observation led him to perform his famous experiment in which he taught a dog to salivate at the sound of a bell. Pavlov began by ringing the bell just before feeding the dog. After several repetitions of this routine, the dog began salivating at the sound of the bell even when there was no food. This simple experiment in learning was one of the first scientific demonstrations of **classical conditioning** (also called *respondent conditioning*), in which an animal or person comes to associate a neutral stimulus with a meaningful one, and then *responds* to the former stimulus as if it were the latter. In this case, the dog associated the bell (the neutral stimulus) with food and responded to its sound as though it were the food itself. This part of the conditioning process is called *learning by association.*

There are many everyday examples that suggest classical conditioning you yourself may have experienced: imagining a lemon might make your mouth pucker; reading the final-exam schedule might make your palms sweat; seeing a sexy photograph might make your heart beat faster. In each instance, the stimulus is connected, or associated, with another stimulus that produced the physiological response in the past. For instance, reading the exam schedule might make you sweat, if actually taking tests has made you anxious on earlier occasions.

Operant Conditioning   The most influential contemporary proponent of learning theory is B. F. Skinner (1904–   ). Skinner agrees with Pavlov that classical conditioning explains some types of behavior, especially reflexive behavior. However, Skinner believes that another type of conditioning—**operant conditioning**—plays a much greater role, especially in more complex learning. Whereas in classical conditioning, the animal is merely responding to prior cues, in operant conditioning, the animal learns that a particular behavior produces a particular response and then performs that behavior to achieve that response. (The term "operant" comes from the Latin word for work, emphasizing the work done to get a response.)

In operant conditioning, then, a system of rewards might be used to train a dog to perform a specific behavior that is not in the dog's usual repertoire—fetch newspapers, jump through hoops, capture suspected criminals. Once the behavior has been learned, the dog will continue to do the work—for example, fetching the newspaper—even when a reward is not always forthcoming. Similarly, almost all the adult's daily behavior, from putting on socks to earning a paycheck, can be said to be the result of operant conditioning. (Operant conditioning is also referred to as *instrumental conditioning,* calling attention to the fact that the behavior in question has become an instrument for achieving a particular response.)

Reinforcement   In operant conditioning, the process whereby a particular behavior is strengthened, making it more likely that the behavior will occur more frequently, is referred to as **reinforcement** (Skinner, 1953). A stimulus that *increases* the likelihood that a particular behavior will be repeated is called a **reinforcer.** Reinforcers may be either positive or negative. A **positive reinforcer** is something pleasant—a good feeling, say, or the satisfaction of a need, or something received from another, such as a piece of candy or a word of praise. For a grade-conscious student who has studied hard for an exam, getting an "A" would be a positive reinforcer of scholarly effort. A **negative reinforcer** is the removal of an unpleasant stimulus as the result of a particular behavior. When a student's anxiety about test-taking is reduced by extra preparation or, counterproductively, by "getting high," the reduction of anxiety is a negative reinforcer. Note that a negative reinforcer differs from a **punishment,** in that a punishment is an unpleasant event that makes behavior *less* likely to be repeated.

Figure 2.9 *According to learning theory, the boy on the left is likely to develop good hygiene habits, largely because he is reinforced frequently for his efforts at cleanliness and is aware of the link between behavior and consequence. By contrast, punishment of the boy on the right will result in improved behavior only if he knows precisely why he is being punished and how he can avoid repeating his offense. But an even more important consideration is whether he is often reinforced for good behavior.*

Types of Reinforcement   Learning theorists judge the effectiveness of a reinforcer or a punishment by how well it strengthens or changes behavior. Not all reinforcers or punishments have equal significance for those experiencing them. For instance, some children would work very hard to earn an allowance; others might not be at all motivated by money but would work to earn a special privilege, such as being allowed to stay up an hour later on weekends. Similarly, some adults are motivated by material possessions, others by personal growth, others by service to humanity. At every stage of life, **social reinforcers,** such as praise or affection, can be powerful.

When behaviorists weigh the methods of creating lasting changes, they prefer reinforcers to punishments. Hitting a child or jailing an adult might work for a short while, but research has shown that these are not the best ways to alter behavior permanently, partly because punishment does not teach a desirable alternative behavior to replace the one that is being punished. Punishment can also have destructive side effects: someone who is punished frequently can become an apathetic or aggressive person (Skinner, 1972).

Social Learning Theory

Traditionally, with both classical and operant conditioning, learning theorists have sought to explain behavior in terms of the organism's direct experience, for they believe that each individual's current behavior results from the accumulated bits of learning acquired through past conditioning. Theoretically, this conditioning process could explain complex patterns of human interaction, as well as simpler behaviors. For example, according to this theory, adults smile at other people because in infancy smiling is associated with, and then reinforced by, food and comfort. And then, once the social smile is learned, it is reinforced, at least occasionally, by the pleasant responses of others. Further variations and subtleties are all the result of different patterns of reinforcement (Bijou and Baer, 1978).

However, a more recent group of learning theorists has focused on less direct, though equally potent, learning. They emphasize the many ways in which people learn new behaviors merely by observing the behavior of others, without, themselves, experiencing any personal reinforcement. Smiling, then, particularly the development of variant smiles—"winning" smiles, flirtatious smiles, questioning smiles, sickened smiles—could be the result of seeing other people smile and noting the responses they get. These theorists have developed an extension of learning theory called **social learning theory.**

Modeling   An integral part of social learning is **modeling,** the process whereby people pattern their behavior after that of specific others. We are more likely to model our behavior after people we consider admirable, or powerful, or similar to us, particularly if we have seen them reinforced for what they do (Bandura, 1977). Often the modeling process is patently obvious: when a teacher conducts class calmly and talks in a quiet voice, the children in the class tend to be self-controlled and to respond quietly, while if the teacher seems to be disorganized, frantic, and loud, the children are likely to behave in a similar fashion. Even in the artificiality of a laboratory, a child who sees another child disobey orders, or share a snack, or play with a toy in an unusual way is likely to follow the example. Indeed, as one review concludes, "under the right circumstances, children will imitate almost anything, from physical aggression to moral judgment, from taste in candy to patterns of speech" (Hetherington and McIntyre, 1975). Similarly, adults, too, often seem surprisingly influenced in their choice in everything from leisure activities to hair styles, from personal hygiene to personal philosophy, by political leaders, media newsmakers, or even strangers in a crowd.

Figure 2.10 *Social learning theory tends to validate the old maxim, "Practice what you preach." If the moments pictured here are typical for each child, the girl on the left will acquire the skills and attitudes that contribute to good parenting. On the other hand, the actions of the parents of the girl on the right are likely to speak more loudly than words. Though they may warn her that smoking is an unhealthy habit that should be avoided, it is likely that she will be a cigarette smoker, too.*

RESEARCH REPORT   Children Who Are Out of Control

One of the most serious problems that developmentalists have been called upon to solve is that of disruptive, antisocial children. Such children not only cause havoc at home and trouble at school but also tend to grow up to be juvenile delinquents and even career criminals (Hirschi, 1969).

One social learning researcher, Gerald Patterson, has spent his professional life trying to understand and help "out-of-control" children, that is, children who behave in aggressive and antisocial ways that neither family nor school seems able to control. For the past two decades, Patterson has led a team of scientists at the Oregon Social Learning Center in providing behavioral analysis as well as practical help to families in which one child is disruptively aggressive (Patterson et al., 1967; Patterson, 1980; 1982).

In the tradition of learning theorists, Patterson and his research team have spent thousands of hours observing the moment-by-moment sequences of behavior in hundreds of normal families and in families with an out-of-control child. They have produced a vast amount of data on the frequency of aversive behavior (defined as unpleasant acts such as hitting, yelling, teasing, scolding), as well as on the events leading up to, and the consequences of, such acts.

It was found that out-of-control children behaved aversively at least three times as often as normal children (the record for frequency was set by a 6-year-old boy who behaved aversively, on average, four times a minute). Patterson determined that the problem is not just in the child but also in the social learning provided by the family. For one thing, in problem families the other family members also have higher-than-average rates of aversive behavior, often responding to aggression with aggression in a way that sets up an escalating cycle of retaliation.

Typically, a problem child and a sibling might begin exchanging increasingly nasty names with each other and end up exchanging blows. In time, these patterns of attack and counterattack become so well learned that the parties involved become blind to alternative ways of resolving conflict.

Another factor highlighted by Patterson's research is that mothers of problem children are often unwitting perpetuators of aversive behavior once it occurs. When a child does something aversive, mothers of problem children are about twice as likely to end up responding positively—that is, by giving in to the child—or neutrally, allowing the child to continue the aversive behavior, than they are to respond negatively by punishing the behavior. Thus they buy their children candy to stop them from screaming and crying in the supermarket, or they let a child stay up later if he or she vehemently refuses to go to bed. As Patterson analyzes it, the immediate result of such maternal behavior is reinforcing for both the child and the mother. The child gets candy or the chance to stay up later, and the mother is spared further screaming or other unmanageable behavior that calls attention to her ineffectiveness. In essence, the child becomes operantly conditioned to go out of control in order to get his or her way, and the mother becomes operantly conditioned to avoid confronting the child's behavior in order to avoid intensifying it.

The mother's short-term solution creates a long-term problem, however. Patterson found that mothers are the victims of aversive behavior ten times as often as fathers and three times as often as siblings. As his research clearly shows, the mother's role is typically that of family caretaker and "crisis manager," the one who is almost always at the front lines when problems occur. This is in marked contrast to the role taken by the typical father:

Social learning theory in no way contradicts the findings of earlier classical or operant conditioning. In fact, as the Research Report above shows, when psychologists attempt to apply the insights provided by learning theory, they typically look at all three types of learning.

Evaluations of Learning Theories

The study of human development has benefited from learning theory in at least two ways. First, the emphasis on the causes and consequences of a specific behavior has led researchers to see that many behavior patterns that may seem to be inborn or the result of deeply rooted emotional problems may actually be the result of the immediate environment. And even when the immediate environment cannot be used to explain a problem completely, altering that environment may remedy the problem.

The role most appropriate for fathers might be that of "guest." They expend much effort on activities which they find reinforcing (e.g., reading the newspaper). They may function as reinforcer, spectator, and participant in games, that is, "the resident good guy." They may even enter into some lightweight child management activities. However, given real crisis or high rate of aversives, they tend to drop out. [Patterson, 1980]

Patterson also notes that mothers who do not generally deal effectively with aversive behavior also tend to respond inappropriately to good behavior, either ignoring it or, about 20 percent of the time, actually punishing it. (One explanation for this involves classical conditioning: the mother becomes so conditioned to interpret her child's behavior as negative that she doesn't notice positive behavior.) Since the child is neither reinforced for good behavior nor punished for bad, the child doesn't learn to do anything other than what he or she pleases.

The solution, as Patterson sees it, is for the mothers to become more skilled at management techniques. They must reinforce positive behavior in their children, and, when punishing negative behavior, they must make sure that the punishment is sufficient to stop the outburst, rather than simply escalating and extending it. Here is an observer's account of Patterson's approach to training mothers in appropriate management techniques:

The child went to bed early only when he felt like it, insisted on sleeping with his mother (she had no husband), rarely obeyed even the most reasonable commands, spread his excrement all over the living room walls, was a terror to other children who tried to play with him, and seemed destined to be a terror to his teachers. The first task was to make the mother realize that he was not minding her in important ways because he was not minding her in small ones. Every day for one hour she was to count the number of times the boy failed to obey an order within fifteen seconds of its being issued and report the results to the therapist. This led the mother to become aware of how many times she was issuing orders and how long she was waiting to get results. . . .

At the third session, the mother was taught how to use "time out" as a means of discipline. She was told that whenever her son did something wrong she should immediately tell him why it was wrong and order him to go to time out—five minutes alone in the bathroom. She resisted doing this, because it forced her to confront all of her son's rule-breaking, and to do so immediately. She preferred to avoid the conflicts and the angry protests. She especially resisted using this means to enforce her son's going to bed at a stated, appropriate time; she was a lonely, not particularly attractive woman, and it was clear to the therapist that she wanted her son to sleep with her. In time, the woman was persuaded to try this new form of discipline and to back up a failure to go to time out by the withdrawal of some privilege ("no TV tonight"). As the weeks went by, the woman became excited about the improvement in the boy's behavior and came to value having him sleep alone in his own room. [Wilson, 1983]

However, retraining is not easy. In many families, the parents have developed a marriage relationship that works to encourage aggression rather than to limit it (Morton, 1987). For instance, if a normal family conflict (over whether a child should have a new bike, or where to go for vacation, or if the television can be on during dinner) is solved by one parent outshouting the other, the children never learn how conflicts can be solved in an amicable way. It takes a skilled trainer and several weeks or months to undo the habits learned over many years. Ideally, mother, father, and siblings should be brought into the project to change the social network of the family and to become models of appropriate, rather than inappropriate, behavior. They can also practice specific techniques to condition the problem child. If the entire family works to improve their interaction, a family that has been at war with itself can learn to function in a supportive way for every member.

This realization has encouraged many scientists to approach particular problem behaviors, such as temper tantrums, phobic reactions, or drug addiction, by analyzing and attempting to change the stimulus-response patterns they entail. The success achieved in eliminating such behaviors has astonished many psychologists who, believing these problems to be deep-seated or even inborn, would have regarded them as requiring long-term treatment or perhaps even as being untreatable. Many people trained in the application of learning theory now work in schools, clinics, and hospitals, helping people to change their behavior patterns. One of the more recent applications has been in nursing homes. According to behaviorist analysis, one of the reasons some of the elderly in these institutions seem apathetic and withdrawn is that, in many cases, all decisions about their daily living arrangements are taken over by others, thus reinforcing passive, dependent behavior. However, when reinforcement patterns are changed to encourage the residents to take more control of their lives—for example, by choosing their own room-mates or by participating in decisions about nursing home policy—they become more active, healthy, and happy (Baltes and Reisenzein, 1986).

Second, learning theory has provided a scientific model for developmentalists of all theoretical backgrounds. It encourages them to define terms precisely, test hypotheses, and publish supporting data as well as conclusions.

At the same time, learning theorists are often criticized for ignoring human emotions and ideas because these cannot be readily tested through controlled experiments. They do not accept the existence of the unconscious, and this, their critics believe, limits their understanding of behavior, particularly abnormal behavior. Many think that focusing only on that which can be observed provides too narrow a perspective to allow a full understanding of the complexities of human behavior and development.

Finally, much of the original elaboration of learning theory was based on research with lower animals—generally dogs, rats, and pigeons—on the assumption that all animals, including humans, follow the same laws of behavior. If researchers had begun instead with the study of human learning, their findings might have led them to different conclusions (Stevenson, 1983). Learning theorists ignored, for example, the importance of our assessment of the meaning of others' behaviors and, in turn, the effect that that assessment can have on our own behavior. For instance, verbal praise, such as, "That was a job well done," may be very reinforcing if it is thought to be genuinely given, but not at all reinforcing if it is interpreted as mechanical. The fact that few learning theorists recognized the importance of cognition has meant that their laws of behavior have proved to be much more limited than once believed (Cairns, 1983).

Humanistic Theories

A group of psychologists who take issue with what they see as the emphasis on pathology in Freudian psychoanalytic theory and on the mechanistic orientation of the behaviorists identify their theories as **humanistic.** Essentially, they believe that at the core of each person is a powerful drive to realize his or her full potential, to achieve **self-actualization.** They also take a **holistic** view of human growth; that is, they consider a person to be far more than a collection of drives, instincts, or past conditioning. To them, each person is a whole being, unique and worthy of respect. The two major leaders of the humanistic approach, Abraham Maslow and Carl Rogers, came to their views from deeply religious backgrounds that focused on ethical concerns. They both did most of their writing after World War II, an event that caused many thoughtful people to question the purpose of human life.

Maslow's Ideas

Abraham Maslow (1908–1970) was born into an Orthodox Jewish family in New York City. He earned his Ph.D. from Columbia University in 1934 and began to question the assumptions of the two groups of psychologists who dominated European and American psychology. Instead of studying people who had serious psychological problems, as the clinical psychoanalytic psychologists did, Maslow thought it made more sense to look at mentally healthy people. As he explained in *Toward a Psychology of Being* (1968), "Human nature is not nearly as bad as it has been thought to be . . . It is as if Freud supplied us the sick half of psychology and we

Figure 2.11 *Abraham Maslow had many of the characteristics of a self-actualized person. He was unconventional and intense but not above laughing at himself. He had a deep respect for his fellow human beings and a great love of nature, believing that peak experiences could occur on a solitary walk in the woods as well as through human intimacy.*

must now fill it out with the healthy half." And, unlike the behaviorists who experimented with dogs, rats, and pigeons in order to infer the laws of human behavior, Maslow thought humans were not simply one more type of animal but something greater.

Maslow examined the personalities of people who seemed to live life to its fullest, people such as Abraham Lincoln and Eleanor Roosevelt, as well as some of his personal friends. He found that such people shared several characteristics: they were realistic, creative, spontaneous, spiritual, independent, purposeful, and they had a few intimate friends with whom they felt very close. Although they were not overly concerned about the society's standards of expected behavior, they were deeply concerned for their fellow human beings. They accepted themselves and enjoyed their lives, including the moments of great happiness that Maslow called *peak experiences,* moments of great joy and insight, when they felt in harmony with, as well as in awe of, nature, God, and/or their fellow human beings.

According to Maslow, each of us has his or her own inner nature and a strong motivation to express that nature. First, however, we must assure ourselves that our basic survival needs will be met, needs that all living creatures share. Thus Maslow proposed a hierarchy of needs, beginning with those such as food and water and moving up to higher, more exclusively human needs (Figure 2.12). While Maslow did not propose his hierarchy as developmental, it can be read as such. According to Maslow, if a person's lower needs are not met, that person must spend time and energy trying to meet them, thus stunting the normal drive for love, esteem, and self-fulfillment. But if a person grows up well-fed, safe, loved, and respected, self-actualization becomes more possible. Even children are motivated to fulfill their potential. In Maslow's words (1968), "Healthy children enjoy growing and moving forward, gaining new skills, capacities and powers. . . . In the normal development of the healthy child, . . . if he is given a really free choice, he will choose what is good for his growth."

Figure 2.12 *According to Maslow's hierarchy, unless the basic physiological needs are satisfied, people are unable to fulfill their potential. Some of Maslow's followers have taken this literally and become very concerned about the damage to the human spirit, as well as to the body, brought about by poverty, famine, and war.*

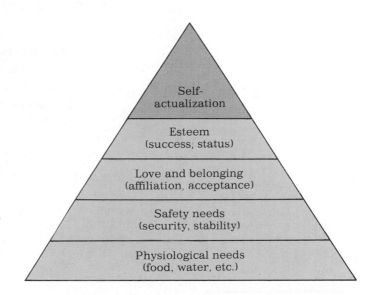

Rogers's Ideas

Carl Rogers (1902-1987) was raised in a very religious Protestant, mid-Western family. In adolescence, he rejected this upbringing, but in early adulthood he decided to go to Union Seminary in New York to become a minister. Gradually he came to feel that psychotherapy was a form of ministry, and he spent a lifetime practicing therapy as well as formulating and expressing many of the themes of humanistic psychology.

Figure 2.13 *Thirty years ago, Carl Rogers became one of the leading advocates of group therapy, a logical outcome of his emphasis on the ability of each of us to help one another. Toward the end of his life, Rogers worked to stop the proliferation of nuclear weapons by suggesting constructive ways for nations to resolve conflicts.*

Rogers agreed with Maslow that all people, even in childhood, try to actualize their potential or, as Rogers would put it, try to become *fully functioning* human beings (Rogers, 1961). Rogers believed that each person has an ideal self—that is, the self that one would like to be—and that healthy people try to come as close to that ideal self as possible. This is done in two ways: first, by improving the real self and, second, by modifying the concept of what the ideal self should be, in order to encompass a wider variety of emotions and behaviors and become more honest and realistic. Generally speaking, guilt, repression, and embarrassment are not part of the fully functioning human's daily life.

Rogers also believed that the process of becoming fully functioning is aided by people who are important in our lives, the "significant others," such as our parents and close friends. If we were fortunate when we were children, our significant others cared for us so much that they gave us *unconditional positive regard*. In other words, they loved and respected us no matter what we did. For example, if a child were to drop a glass of milk, such parents might help the child mop up the spill or point out the necessity for greater carefulness, but they would never call the child stupid or clumsy. Even if the child deliberately dropped the glass, the parents would respect the child's anger enough to try to find out why such behavior occurred, rather than simply reciprocating with anger and punishment.

As adults, Rogers believed, people should try to extend unconditional positive regard to each other. This does not mean condoning everything everyone does; rather, it means respecting the human worth and dignity of all people. As Rogers learned in his seminary days, it is possible to hate the sin but love the sinner.

Figure 2.14 *Unconditional positive regard is most needed (and, often, least likely to be forthcoming) in stressful situations. If this woman's family were to take the ideal humanistic perspective, they would simply express relief that no one was hurt and avoid any urge to parcel out blame for damage to the family car.*

Evaluations of Humanistic Theories

Maslow, Rogers, and the other humanists all agree that it is never too late to become the person one could become. While childhood is important, adulthood is equally so. Indeed, Rogers (1980) wrote convincingly of the significant changes that occurred in his own personality during his 70s. This emphasis on possibilities makes humanistic theory attractive to developmental psychologists, who see that physical, cognitive, and psychosocial growth occur at every period of development, from the first months of life to the last. In addition, the broad vision of the humanists helps researchers look at the whole of development—an important corrective to the narrow views of some behaviorists. Finally, the humanist stress on potential is a philosophical undercurrent of the field of developmental psychology, as the scientific understanding we gain is translated into practical programs to promote optimal development.

However, while humanism is encouraging and optimistic, it is, like all theories, not necessarily true. Indeed, critics argue that humanists are blind to the many ways in which growth is stunted and deflected from the optimal route. Societies, families, and even individuals themselves do not always encourage, or even allow, the full development of human potential. Beyond simply extending unconditional positive regard to all fellow humans, there is the need to provide social reform, constructive discipline, and direct guidance, which may be much more beneficial in the long run than unquestioning acceptance of human behavior. A final criticism of humanistic theories is that they are more likely to formulate what is hoped for in human development than to present predictions based on research.

Cognitive Theories

The prime focus of **cognitive theory** is the structure and development of the individual's thought processes and the ways in which those processes can affect the person's understanding of, and expectations of, his or her world. In turn, cognitive theory considers how these understandings and expectations can affect the individual's behavior.

Piaget's Ideas

Jean Piaget (1896-1980), the most famous of cognitive theorists, first became interested in thought processes while field-testing questions that were being considered for a standard intelligence test for children. Piaget was supposed to find the age at which most children could answer each question correctly, but eventually he became more interested in the children's *wrong* answers. What intrigued him was that children who were the same age made similar types of mistakes, suggesting that there was a developmental sequence to intellectual growth. He began to believe that *how* children think is much more important, and more revealing of their mental ability, than tabulating what they know (Flavell, 1963; Cowan, 1978).

How Cognitive Development Occurs   Piaget held that there are four major stages of cognitive development. Each one is age-related and has structural features that permit certain types of thinking.

According to Piaget, infants think exclusively through their senses and motor abilities: their understanding of the objects in their world is limited to the actions they can perform on them. Preschool children can think about objects independently of their actions on them, and they can begin to think symbolically, as reflected in their ability to use language and to pretend. However, they cannot think logically in a consistent way. School-age children can begin to think logically in a consistent way, but only with regard to real and concrete features of their world. Adolescents and adults, in varying degrees, are able to think hypothetically and abstractly: they can think about thinking and can coordinate ideas (see Table 2.2). Each of these ways of thinking is explained in detail later in this book.

TABLE 2.2 **Piaget's Periods of Cognitive Development**

Approximate Age	Name	Characteristics	Major Acquisitions
Birth–2 years	Sensorimotor	Infant uses senses and motor abilities to understand the world. This period begins with reflexes and ends with complex coordinations of *sensorimotor skills.*	The infant learns that an object still exists when it is out of sight (*object permanence*) and begins to remember and imagine experiences (*mental representation*).
2–6 years	Preoperational	The child uses *symbolic thinking,* including language, to understand the world. Most thinking is *egocentric,* which means that children understand the world from only one perspective, their own.	The imagination flourishes. Children gradually begin to *decenter,* or become less egocentric, and to understand other points of view.
7–11 years	Concrete Operational	The child understands and applies logical operations, or principles, to help interpret specific experiences or perceptions.	By applying logical abilities, children learn to understand the basic ideas of conservation, number, classification, and many other concrete ideas.
From 12 on	Formal Operational	The adolescent or adult is able to think about abstractions and hypothetical concepts and is able to move in thought "from the real to the possible."	Ethics, politics, and all social and moral issues become more interesting and involving as the adolescent becomes able to take a broader and more theoretical approach to experience.

Piaget viewed cognitive development as a process that follows universal patterns. The starting point for this development, according to Piaget, is the need in everyone for **equilibrium,** that is, a state of mental balance (Piaget, 1970). What he meant by this is that each person needs to, and continually attempts to, make sense of conflicting experiences and perceptions.

People achieve this equilibrium through mental concepts, or in Piaget's terms, **schemas,** that strike a harmony between their ideas and their experiences. A schema is a general way of thinking about, or interacting with, ideas and objects in the environment. The infant first comes to know the world through a sucking schema and a grasping schema; by adulthood the schemas through which the individual knows the world are beyond counting, ranging from something as simple as the schema for doing and undoing a button to the abstract moral schema that a human life is more valuable than any material thing. When existing schemas do not seem to fit present experiences, the individual falls into a state of **disequilibrium,** a kind of imbalance that initially produces confusion and then leads to growth as the person modifies old schemas and constructs new ones to fit the new conditions.

These ideas become clearer with an example. In one of Piaget's experiments, a child is shown two identical glasses, each containing the same quantity of liquid (Piaget and Inhelder, 1974). Next, the child pours the liquid from one of these glasses into a third glass, which is taller and narrower than the other two. The experimenter then asks the child which glass contains more. Almost every child younger than 6 says the taller glass contains more, because preschool children use the schema that taller is bigger. They are unshakable in this conviction, even when the experimenter points out that the taller glass is narrower and that the amount of water in each of the original identical glasses was obviously the same.

Most children older than 7, on the other hand, have developed the schema that Piaget called **conservation of liquids:** that is, they realize that pouring the liquid into a taller glass does not change the amount of liquid. They remain steadfast in this conviction even when the experimenter attempts to convince them otherwise.

Figure 2.15 *All his life Jean Piaget was absorbed with studying the way children think.*

Figure 2.16 *My daughter, Sarah, here aged 5¾, demonstrates Piaget's conservation of liquids experiment. First she examines both short glasses to be sure they contain the same amount of milk. Then, after the contents of one is poured into the tall glass and I ask her, "Which has more?" she points to the tall glass, just as Piaget would have expected.*

In both cases, the children's ideas and perceptions are in a state of equilibrium. They have managed to construct a mental structure that enables them to interpret what they see. Children progress from the first state of equilibrium to the second by passing through a transitional period of disequilibrium in which they begin to be able to recognize that some of their ideas conflict with their experiences. In the tall-glass experiment, for instance, they become aware that the idea that the identical glasses originally contained the same amount of liquid is inconsistent with the idea that taller glasses always hold more. They resolve this dilemma in a variety of ways. Some imagine that water was magically added to the tall glass. Some even say that the tall glass contains less because it is narrower, reversing their previous schema in their confusion. Some are puzzled, some are distressed, some say the question is impossible. All these reactions are evidence of disequilibrium.

Thirst for Knowledge Periods of disequilibrium can be disquieting to a child or an adult who suspects that accepted ideas no longer hold true. But they are also exciting periods of mental growth. By seeking out new experiences, children are constantly challenging their current schema. Babies poke, pull, and taste everything they get their hands on; preschool children ask thousands of questions; school-age children become avid readers and information collectors. Adolescents try out a wide variety of roles and experiences, and adults continually increase their knowledge and expertise in areas that interest them—from picking up pointers about car repair to training as a hospice volunteer—all because people at every age want cognitive challenges. Recognition of this active searching for knowledge is the essence of the cognitive theory of human development.

Active searching is also the essence of intelligence, according to Piaget. It is part of human nature to seek to clarify and understand the myriad sensations and ideas that bombard the mind. The result of this quest for comprehension, according to Piaget, is intelligence, which, in his view, comprises two interrelated processes, **organization** and **adaptation.** People organize their thoughts so that they make sense, separating the more important thoughts from the less important ones, as well as establishing links between one idea and another. In the process of learning about various animals, for example, a child may organize them mentally in clusters according to whether they are birds, mammals, or fishes. At the same time, people adapt their thinking to include new ideas as new experiences provide additional information.

This adaptation occurs in two ways, through **assimilation** and **accommodation.** In the process of assimilation, information is simply added to the cognitive organization already there. In the process of accommodation, the intellectual organization has to adjust to the new idea. Thus, in watching a nature film, a child may extend his or her understanding of animals by discovering new animals to add to existing clusters (assimilation); or the child may be led to rearrange old clusters or create new ones as a result of seeing, say, whales, which are mammals that look like, and live like, fishes (accommodation).

These basic cognitive processes are at work even in the first weeks of life. Consider the grasping reflex, for instance. Newborns curl their fingers tightly around anything that crosses their palm. Soon, however, their grasping reflex becomes organized in specific ways as their particular experiences provide them with new knowledge: they grasp Mother's sweater one way, their bottle another way, a rattle another, and the cat's tail not at all. They have thus adapted their inborn grasping schema to their environment, first by assimilation (grasping everything that comes their way) and then by accommodation (adjusting their grasp to the "graspability" of the object.

The process of assimilation and accommodation continues throughout life. Consider the concept of Santa Claus. When 2-year-olds are taken to see a store Santa Claus, many cry or sit mute on his lap, for the typical toddler assimilates Santa into a schema that strange-looking men are frightening. With repeated experiences, however, children quickly come to realize that Santa Claus has something to do with the arrival of presents. They accommodate by forming a more favorable concept of Santa Claus, perhaps using another schema, such as "getting presents is fun." With time, many come to believe that Santa Claus is really the source of Christmas presents. Once formed, this concept often becomes strengthened, as the child assimilates all the evidence that Santa Claus brings presents and resists all claims to the contrary offered by older children. Eventually, however, the notion of a real Santa Claus crumbles, as the child's increasing reasoning ability must confront the obvious impossibility of one man bringing presents to all the children in the world—especially by coming down the chimney. By adulthood, a new Santa schema may develop, perhaps one that centers on the idea of the Christmas spirit or of the commercialization of religious occasions. In any case, intelligence involves the continual adaptation of old organizational structures to assimilate and, if need be, accommodate new ideas and experiences.

Figure 2.17 *Children between 6 months and 2 years are fearful of, or at least wary of, strangers. This fear is most powerful if the stranger is large, male, bearded, oddly dressed, and in a public place.*

Evaluations of Cognitive Theories

Cognitive theory has revolutionized developmental psychology by focusing attention on active mental processes. The attempt to understand the mental structures of thought, and to appreciate the internal need for new mental structures when the old ones become outmoded, has led to a new understanding of certain aspects of human behavior. Thanks to the insights provided by cognitive theory, researchers, parents, and educators now have a greater appreciation of the capacities and limitations of the types of thinking that are possible at various ages—and of the ways these capacities and limitations can affect behavior.

Cognitive theory has also profoundly affected education in many countries, allowing teachers and students to become partners in the educational process once the child's own capacities and needs are recognized. For instance, elementary-school math is now taught with objects the child can manipulate, because we now know that the thinking of school-age children is better suited to working out solutions by concrete activities, such as measuring blocks or counting pennies, than to the more abstract learning tasks involved in reading about, and memorizing, mathematical facts.

Finally, cognitive theory has helped researchers in many fields to reexamine their assumptions, for we have learned that the way we think colors what we see. In order to be more objective as scientists, we need to become aware of our subjectivity—an awareness that cognitive theory has helped to foster.

At the same time, many people think Piaget was so absorbed by the individual's active search for knowledge that he ignored external motivation or teaching. While it is comforting to think that children can develop their own schemas when they are ready, this implies that teachers should not intervene when a child seems uninterested in learning to add or spell. To some extent, the "back to basics" movement in American education is a reaction against Piagetian ideas carried to their logical extreme. And even some of those who most admire Piaget believe that he underestimated the role of society and home in fostering cognitive development. Many psychologists believe that culture and education can be crucial in providing the proper mix of equilibrium and disequilibrium. In Bruner's (1973) words, "Some environments push cognitive growth better, earlier, and longer than others . . . it makes a huge difference to the intellectual life of a child simply that he was in school." In adulthood, variations in personal experience and education may be critical in determining the depth and complexity of an individual's thought processes (Dixon et al., 1985; Rybash et al., 1986).

A number of critics have also found fault with Piaget's depiction of cognitive stages as universal. For example, there are many adults who never develop the capacity for abstract thinking that Piaget described as being typical of adolescence. Further, critics point out, Piaget's description of intelligence focused mainly on scientific and logical thinking and thus has limited applicability in daily adult life. A full understanding of cognitive development, they say, must also take into account the development of abilities such as language, artistic skill, social understanding, and personal judgment (Buck-Morss, 1987; Gardner, 1983; Miller, 1983).

A final point is less a quibble than a qualification. As a number of researchers have pointed out, Piaget's description of cognitive development tends to make it seem as though once a new stage of cognition has been achieved, it will be reflected in all aspects of the individual's thinking. In fact, the cognitive advance may occur in some areas of thinking and not in others. And particularly with children, the advance in a given area may be evident on one occasion and not on another. Most cognitive theorists now generally believe that "unevenness is the rule in development" (Fischer, 1980; Flavell, 1982).

The Theories Compared

Each of the four theories presented here has contributed a great deal to the study of human development. Psychoanalytic theory has made us aware of the importance of early childhood experiences and of the possible impact of the "hidden dramas" that influence our daily lives. Learning theory has shown us the effect that the immediate environment can have on behaviors; humanist theories have given us an encouraging vision of the goals of human development; cognitive theory has brought us to a greater understanding of how our thinking affects our actions.

Each theory has also been criticized: psychoanalytic theory for being too subjective; learning theory for being too mechanistic; humanism for being too idealistic; and cognitive theory for undervaluing the power of direct instruction and overemphasizing rational, logical thought.

Each theory by itself is, in fact, too restricted to grasp the breadth and diversity of human development (Cairns, 1983; Thomas, 1981). As one researcher explains, we now see people as "so complex and multifaceted as to defy easy classification . . . [and] multiply influenced by a host of interacting determinants. . . . It is an image that highlights the shortcomings of all simplistic theories that view behavior as the exclusive result of any narrow set of determinants, whether these are habits, traits, drives, reinforcers, constructs, instincts, or genes, and whether they are exclusively inside or outside the person" (Mischel, 1977).

Because no one theory can encompass all of human behavior, most developmentalists would describe themselves as having an **eclectic perspective,** meaning that rather than adopting any one of these theories exclusively, they make use of all of them. Indeed, many developmentalists who, in the past, had been identified with particular theories have become increasingly aware of the limitations of any one theoretical system and now incorporate some of the ideas from the other perspectives into their thinking (Cairns, 1983; Kuhn, 1978; Sameroff, 1982).

In subsequent chapters, as echoes and elaborations of the psychoanalytic, learning, humanistic, and cognitive theories appear, you can form your own opinion of the validity of each theory. The best challenge you can set for yourself, the same one facing developmental researchers, is the integration of theory, research, and applications into an increasingly comprehensive picture of human development.

SUMMARY

What Theories Do

1. A theory provides a framework of general principles that can be used to interpret our observations. Each theory interprets human development from a somewhat different perspective, but all theories attempt to provide a context in which to understand individual experiences and behaviors.

Psychoanalytic Theories

2. Freud's psychoanalytic theory emphasizes that our actions are largely ruled by the unconscious—the source of powerful impulses and conflicts that usually lie below the level of our conscious awareness.

3. Freud, the founder of psychoanalytic theory, developed the theory of psychosexual stages to explain how unconscious impulses arise and how they affect behavior during the development of the child.

4. Freud interpreted behavior in terms of three components of personality: the id seeks immediate gratification of its desires; the superego acts as a conscience to suppress the id; the ego moderates the demands of the id and superego and copes with the recognition that other people have needs, too.

5. Erikson's theory of psychosocial development describes individuals as being shaped by the interaction of physical characteristics, personal history, and social forces. Culture plays a large part in each person's ability to deal with the most significant tasks, or crises, of psychological development.

Learning Theories

6. Learning theorists believe that psychologists should study only behavior that can be observed and measured. They are especially interested in the relationship between behaviors or events and the reactions they elicit, that is, between the stimulus and the response.

7. Learning theory emphasizes the importance of various forms of conditioning, a process by which particular stimuli become linked with particular responses. In classical conditioning, one stimulus becomes associated with another to produce a particular response. In operant, or instrumental, conditioning, reinforcement makes a behavior either more or less likely to occur.

8. Social learning theory recognizes that much of human behavior is modeled after the behavior of others.

Humanistic Theories

9. Humanists believe that each individual has an inner drive to fulfill the best of his or her unique potential.

10. Maslow theorized that once basic survival needs are met, people can express their growth needs for love, esteem, and finally, self-actualization. Rogers believed that people become fully functioning if they receive unconditional positive regard from significant others in their lives. Humanists stress that potential for growth occurs at all stages of development.

Cognitive Theories

11. Cognitive theorists believe that a person's thought processes—the understanding and expectations of a particular situation—have an important effect on behavior.

12. Piaget, the leading cognitive theorist, proposed that at each age, people develop schemas—general ways of thinking about ideas and objects. When a person becomes aware of perceptions or experiences that do not fit an existing schema, a new schema is created, and cognitive growth occurs. Learning is accomplished by a process of organization and adaptation.

The Theories Compared

13. Psychoanalytic, learning, humanistic, and cognitive theories have all contributed to the understanding of human development, yet no one theory is adequate to describe the complexity and diversity of the human experience. Most developmentalists incorporate ideas from several developmental perspectives into their thinking.

KEY TERMS

psychoanalytic theory *(32)*

infantile sexuality *(33)*

psychosexual stages *(33)*

oral stage *(33)*

anal stage *(33)*

phallic stage *(33)*

latency *(33)*

genital stage *(33)*

id *(34)*

ego *(34)*

superego *(34)*

defense mechanisms *(35)*

regression *(35)*

repression *(35)*

displacement *(35)*

psychosocial theory *(36)*

behaviorism *(38)*

learning theory *(38)*

stimulus *(39)*

response *(39)*

conditioning *(39)*

classical conditioning *(39)*

operant conditioning *(39)*

reinforcement *(40)*

reinforcer *(40)*

positive reinforcer *(40)*

negative reinforcer *(40)*

punishment *(40)*

social reinforcers *(40)*

social learning theory *(41)*

modeling *(41)*

humanistic theory *(44)*

self-actualization *(44)*

holistic *(44)*

cognitive theory *(47)*

equilibrium *(48)*

schema *(48)*

disequilibrium *(48)*

conservation of liquids *(48)*

organization *(49)*

adaptation *(49)*

assimilation *(50)*

accommodation *(50)*

eclectic perspective *(52)*

KEY QUESTIONS

1. What functions does a good theory perform?

2. What is the major premise of psychoanalytic theory?

3. According to Freud's theory, what is the function of the ego?

4. What is the major difference between Erikson's theory and Freud's theory? *man dev even as A stages tied to rel to soc rather than body part*

5. What is the major premise of learning theory?

6. What are the differences between classical and operant conditioning?

7. What are some of the most useful applications of learning theory?

8. How does the humanist approach to development differ from that of other theorists?

9. What is the major premise of cognitive theory?

10. According to Piaget, how do periods of disequilibrium lead to mental growth?

11. What is the difference between assimilation and accommodation?

12. What are the main differences between the psychoanalytic, learning, humanistic, and cognitive theories?

13. Why do most developmentalists describe themselves as having an eclectic perspective?

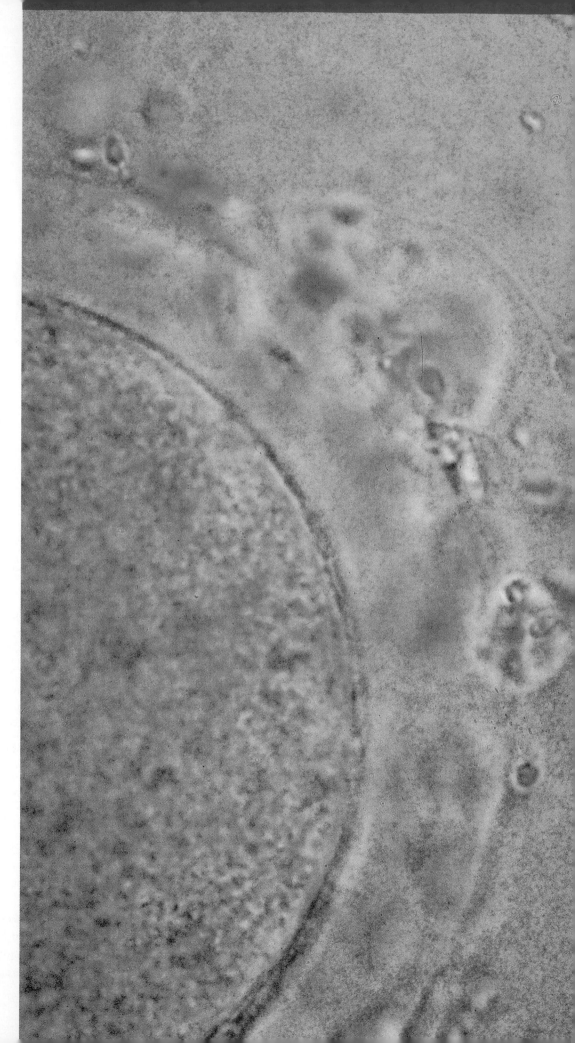

CHAPTER 3

Genetics

A million, million spermatozoa,
  All of them alive
Out of their cataclysm but one poor Noah
  Dare hope to survive
And among that billion minus one
  might have chanced to be
Shakespeare, another Newton, a new Donne
  But the One was Me.

Aldous Huxley
"Fifth Philosopher's Song"

At the moment of conception, a cell called an ovum from a woman and a sperm cell from a man fuse to create one new cell, the **zygote.** That tiny, single-celled organism contains all the biologically inherited information—the genes and chromosomes from each parent—that will affect virtually every characteristic, from hair texture to blood type, from foot size to heart shape, from memory to moodiness. In this chapter, we will explore the mechanism of biological heredity and its influence on development.

Genes and Chromosomes

A **gene** is a segment of a DNA (deoxyribonucleic acid) molecule that provides the specific biochemical instruction a cell needs to perform a particular function in the body. Whether a particular cell "specializes" to become part of the lens of the eye or part of a muscle in the arm depends on which of these many genes are active, or *expressed,* and which are inactive, or *unexpressed.*

Genes are arranged in specific locations like, speaking loosely, beads on a string. The "string" is the long, exquisitely thin DNA molecule that, together with other materials, makes up a **chromosome.** Every normal human body cell has forty-six chromosomes in the form of twenty-three chromosome pairs, one pair member coming from each parent. Since each chromosome carries hundreds of genes, the fusion of sperm and ovum produces an estimated 75,000 genes (Kowles, 1985).

Shortly after the zygote has been formed, it begins a process (*mitosis*) of duplication and division. The single cell first divides into two cells; then the two cells divide into four, the four into eight, and so on. Just before each cell division, the forty-six chromosomes duplicate themselves, forming two complete sets of chromosomes. These two sets separate from each other, move to opposite sides of the cell, and the cell then divides in the middle. Each new cell thus has the same twenty-three pairs of chromosomes, and therefore the same genetic information, that was contained in the zygote. This process continues throughout the development of the individual, creating new cells and replacing old ones. At birth, a baby is made up of about 10 trillion cells. By adulthood, the number has increased to between 300 and 500 trillion. But no matter how many cells a person may have, no matter what their specific function, each cell carries the genetic message inherited by the one-celled zygote at the moment of conception.

The Twenty-Third Pair

Twenty-two of the twenty-three pairs of chromosomes found in human beings are matching pairs. They are called **autosomes,** and they control the development and the functioning of most of the body. The twenty-third pair, which is the one that determines the individual's sex, is a special case. In the male, the twenty-third pair is designated *XY* (so named because of its appearance—see Figure 3.1), and in the female, it is designated *XX*.

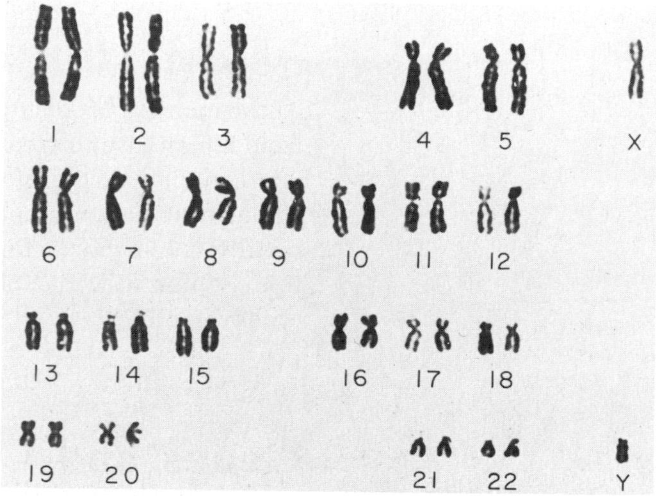

Figure 3.1 *A picture of the forty-six chromosomes from one individual, in this case a normal male. In order to produce a chromosomal portrait such as this one, a cell is removed from the person's body (usually from inside the mouth), processed so that the chromosomes become visible, magnified many times, photographed, and then arranged in pairs according to length of the upper "arms."*

When cells divide to produce reproductive cells (called **gametes**—ova in females, sperm in males), they do so in such a way that each sperm or ovum receives only one member of each chromosome pair. Thus, each sperm or ovum has only twenty-three chromosomes, half as many as the original cell. This process assures that when the chromosomes of the sperm and the ovum combine at conception, the total human chromosome number will remain forty-six. Since in females the twenty-third chromosome pair is *XX*, each ovum carries one *X* chromosome. And since in males the twenty-third pair is *XY*, approximately half of the sperm will have an *X* chromosome and the other half, a *Y*. If an ovum is fertilized by a sperm bearing a *Y* chromosome (*XY* zygote), a male will develop. If it is fertilized by a sperm bearing an *X* chromosome (*XX* zygote), a female will develop.

Genetic Uniqueness

On the basis of what you have been told so far, you might well wonder why all children in a given family do not look exactly alike, inasmuch as they each have twenty-three chromosomes from their father and twenty-three from their mother.

One reason is that each chromosome may carry one of many variations of particular genetic instructions. When the chromosome pairs divide during the formation of gametes, which of the two pair members will wind up in a particular gamete is a matter of chance, and many combinations are possible. In fact, the laws of probability show that there are 2^{23}—that is, about 8 million—possible outcomes. In other words, approximately 8 million different ova or sperm can be produced by a single individual. In addition, corresponding segments of a chromosome pair are sometimes exchanged in the first stages of gamete production, a crossing-over, or "reshuffling," of genes that can create a new genetic interaction because of the new placement (Emery, 1983). And finally, when the genes of the ovum and the sperm combine, they interact to form combinations not present in either parent. All things considered, any given mother and father can form over 64 trillion genetically different offspring, a number "far greater than the number of humans who have ever lived on earth!" (Klug and Cummings, 1983). Thus it is no exaggeration to say that every conception is the beginning of a unique individual or, sometimes, a unique pair.

Twins   About once in 270 pregnancies, a zygote splits apart during its initial division, and the two separate, identical cells begin to develop independently, becoming identical twins. They are called **monozygotic twins** because they come from one (*mono*-) zygote. Monozygotic twins are the same sex, look alike, and share all other inherited characteristics, because each has the genes that were in the original zygote.

Of course, not all twins are identical. In fact, most twins are **dizygotic,** or fraternal. Dizygotic twins begin life as two separate zygotes created by the fertilization of two ova that were ovulated at the same time. Dizygotic twins share no more genes than any other two offspring of the same parents. They may be of different sexes and very different in appearance. Or they may look a great deal alike, just as brothers and sisters sometimes do. Other multiple births, such as triplets and quadruplets, can likewise be monozygotic, dizygotic, trizygotic, quadrazygotic, and so on.

Figure 3.2 *Since nature has provided these monozygotic twins (left) with identical features, only the artifacts of experience (like one boy's chipped tooth) provide ways to tell them apart. While dizygotic twins (right) sometimes look quite different and may not even be of the same sex, they may also look quite alike, as these Cuban-American sisters do. However, a closer look reveals many differences— in noses, hair color, and height, for instance.*

A CLOSER LOOK   Twins and Research

Monozygotic and dizygotic twins have been studied extensively, partly because they provide much-needed data on the interaction between nature and nurture. The basic idea is that if monozygotic twins are much more alike than dizygotic twins for any given characteristic, then genes rather than environment (growing up in the same household) determine that characteristic.

Note, however, that this way of differentiating the influence of genes from the influence of environment makes the assumption that children growing up in the same family will have the same experiences. In fact, children in the same household often have quite different experiences, and the extent of these differences depends partly on how parents and others react to each child (McCall, 1983). It is quite probable that parents and teachers will treat monozygotic twins much more alike than dizygotic twins, if only because it is so hard to tell monozygotic twins apart. Thus, when research data find monozygotic twins to be more alike than dizygotic twins, both genes and environment might be responsible.

One presumed way to separate the effect of genes and experience is to look at monozygotic twins who have been raised in different households. Several studies that have done this (Bouchard, 1981; Juel-Nielson, 1980; Shields, 1962) provide a wealth of suggestive evidence for the influence of genes. In fact, the similarities that they have found between twins reared apart has sometimes been startling.

Take the case of Robert Shafran, who, while walking across the campus of the university in which he had recently enrolled, was suddenly greeted by a young woman who kissed him warmly on the mouth and exclaimed, "Where have you been?" That Robert didn't even know this woman was a fact he admitted somewhat reluctantly, since she was just his type. As it turned out, she was also his brother Eddy Galland's type; in fact, it was Eddy, Robert's long-lost twin, whom the woman had taken Robert to be. And it was soon clear that the resemblance between Robert and Eddy was not limited to physical characteristics. Among other striking similarities, the brothers wore the same kinds of clothes; had similar hairstyles; laughed in the same way at the same jokes; drank the same brand of beer; smoked the same brand of cigarettes (which they held in the same way); engaged in the same sports, including team wrestling (in which they had almost identical records); and listened to the same music, at similar volumes.

When the story of Robert and Eddy hit the press, David Kellman looked at their photo and thought he was seeing

The sources of triple confusion, from top to bottom: Edward Galland, *David Kellman, and Robert Shafran.*

mirror images of himself—and, in fact, he was, for in this case, monozygotic triplets rather than twins had been separated at birth. When the three brothers were reunited, they (and the psychologists who studied them) were amazed at the number of experiences, tastes, and interests they had in common (*New York Times,* 1980).

Similar amazement was recently registered, a bit less publicly, by a group of researchers beginning the Minnesota Study of Twins Reared Apart, an extensive study of monozygotic twins who were separated early in life (Bouchard, 1981; Holden, 1980). One pair of identical twins, Oskar Stohr and Jack Yufe, were born of a Jewish father and German mother in Trinidad in the 1930s. Soon after their birth, Oskar was taken to Nazi Germany by his mother to be raised as a Catholic in a household consisting mostly of women. Jack was raised as a Jew by his father, spending his childhood in the Caribbean and some of his adolescence in Israel.

On the face of it, it would be difficult to imagine more disparate cultural backgrounds. In addition, the twins certainly had their differences. Oskar was married, an employee, and a devoted union member. Jack was divorced and owned a clothing store in Southern California. But, when the brothers met for the first time in Minnesota,

similarities started cropping up as soon as Oskar arrived at the airport. Both were wearing wire-rimmed glasses and mustaches, both sported two-pocket shirts with epaulets. They share idiosyncrasies galore: they like spicy foods and sweet liqueurs, are absentminded, have a habit of falling asleep in front of the television, think it's funny to sneeze in a crowd of strangers, flush the toilet before using it, store rubber bands on their wrists, read magazines from back to front, dip buttered toast in their coffee. Oskar is domineering toward women and yells at his wife, which Jack did before he was separated. [Holden, 1980]

Their scores on several psychological tests were very similar, and they struck the investigator as remarkably similar in temperament and tempo. Other pairs of twins in this study likewise startled the observers by their similarities, not only in appearance and on test scores, but in mannerisms and dress. One pair of female twins, separated since infancy, arrived in Minnesota, each wearing seven rings (on the same fingers) and three bracelets, a coincidence that might be explained by pure chance, but more likely was partly genetic—that is, genes endowed both women with beautiful hands and, possibly, contributed to an interest in self-adornment.

The evidence from monozygotic twins suggests that genes affect a much greater number of characteristics than most psychologists, including the leader of the Minnesota study, Thomas Bouchard, suspected. Says Bouchard,

No matter what trait we look at—psychological interest, personality, temperament, across the whole spectrum— almost everything has a hereditary effect. Some psychologists concluded a while back from other studies that it looked like all traits were about equally genetically influenced. I always thought that was kind of foolish. But

now, with these studies of identical twins reared apart, I'm starting to become a believer. [Quoted in Cassill, 1982]

However, while studies of monozygotic twins astonish researchers with the extent of genetic influence, they also furnish ample evidence for environmental impact. For example, while monozygotic twins often share the same fears, their reaction to the fear may be different; one will feel anxious when riding an escalator or swimming in deep water, while the other will avoid escalators or deep water altogether (Cassill, 1982). Another factor to be considered is that even twins reared apart have generally had quite similar home experiences. Typically, they are raised by close relatives in neighboring communities, and only rarely are they separated by language, culture, and religion, as Oskar and Jack were. Even in their case, says Bouchard, beneath the more dramatic differences in background, their upbringing was basically quite similar. Finally, in all the studies in which one twin had experienced "extreme deprivation or unusual enrichment," the result has been to "lessen the resemblance of identical twins reared apart" (Scarr and McCartney, 1983).

Obviously, the question of the relative importance of nature and nurture cannot be settled by evidence from a few pairs of twins. The Minnesota researchers plan to study dozens of twins over a five-year period before drawing conclusions, and they expect that the data will provide ample evidence for the role of nurture as well as nature. At this point, however, it seems likely that genes have a broader range of general influence than many theorists and researchers had previously thought. At the same time, it also seems clear that the issue of nature and nurture may never be settled precisely, partly because the answer, as you will soon see, appears to vary from individual to individual and from trait to trait.

Since Oskar Stohr (left) and Jack Yufe (right) are monozygotic twins, it is not surprising that they look very much alike. However, since they have been separated almost from birth, it is more difficult to explain their similarities in many of those characteristics that are usually considered to be acquired, for example, their preference for mustaches and their tastes in food and drink.

Dominant and Recessive Genes

In the simplest form of heredity, a pair of genes, one from each parent, determines a particular inherited characteristic. However, the interaction between the two members of a gene pair is not always fifty-fifty. Sometimes one member of the pair has a far greater influence in determining the specific characteristic. Genes that act in this controlling manner are called **dominant genes;** those genes whose influence is obscured when paired with the more powerful genes are called **recessive genes.**

Hundreds of physical characteristics follow the dominant-recessive pattern. Let us consider eye color. To simplify somewhat, let's say that a person inherits two eye-color genes, one from each parent, and that the gene for brown eyes is dominant and that the gene for blue eyes is recessive. (Following traditional practice, we will indicate the dominant gene with an upper-case letter—"B" for dominant brown—and the recessive gene with a lower-case letter—"b" for recessive blue.) If both genes are for brown eyes (BB), the person's eyes will be brown. If one gene is for brown eyes and the other for blue (Bb), the person's eyes will be brown, since the brown-eye gene is dominant. If both genes are recessive genes for blue eyes (bb), the person will have blue eyes.

The total of all the genes a person inherits—that is, his or her genetic makeup—is called the person's **genotype.** The result of the interaction of the genes with each other and with the environment—that is, that part of a person's appearance and behavior that expresses his or her genetic inheritance—is called the **phenotype.** As you can see in the example of brown eyes, although two people have the same phenotype with respect to a particular trait, they may have different genotypes—in this case, one brown-eye gene and one blue-eye gene (Bb) producing the same eye-color as two brown-eye genes (BB).

It is also possible for parents to have offspring whose phenotype for a particular characteristic is quite different from theirs, if the parents both have the necessary recessive genes. For example, if each of two brown-eyed parents has a recessive gene for blue eyes (Bb and Bb), there is one chance in four that a particular child of theirs will inherit the recessive blue-eye gene from both of them and will therefore have blue eyes. (The four possible combinations in their offspring would be BB, Bb, Bb—all yielding brown eyes—and bb, yielding blue eyes.) A person who has a particular gene as a part of his or her genotype is called a **carrier** of that gene. In fact, we are all carriers of dozens of recessive genes that are in our genotypes but not in our phenotypes. Usually we are unaware of which recessive genes we carry until we have a child with a surprising phenotype (see Figure 3.3).

It should be noted that the interaction of dominant and recessive genes just described is a simplified model. The actual interaction of all the pairs of genes that could be inherited is more complicated. In many cases, a dominant gene is not completely dominant and may allow the recessive gene to influence the phenotype (McKusick, 1986). For example, although many textbooks list blue eye color as recessive, this is somewhat misleading. Many a hazel-eyed child has one parent with blue eyes and the other parent with brown. In this case, the child's light-brown eye color bears witness to the recessive gene in his or her genotype. In addition, most people have several pairs of genes that affect eye color, so shades of blue, green, and brown develop, following more complicated laws of inheritance.

X-Linked Recessive Genes   Some recessive genes are called **X-linked** because they are located only on the X chromosome. For example, the genes for most forms of color blindness, certain allergies, several diseases, and perhaps some learning

Figure 3.3 *Phenotype and genotype are not always the same. Since both parents here have curly black hair, a dominant characteristic, we know from looking at their phenotype that they both have the genotype for this characteristic. But we know that they also have the genotype for straight red hair, a recessive characteristic, only because they have a child who has this phenotype. She must have inherited the necessary recessive genes from both parents.*

disabilities and perceptual skills are recessive and are carried by the X chromosome. Since males have only one X chromosome, they are more likely to have these characteristics on their phenotype as well as in their genotype.

For example, if a male inherits a gene for color blindness on the X chromosome he receives from his mother, he will be color-blind, since his Y chromosome carries no corresponding dominant gene for normal color vision to countermand the instructions of the recessive gene for color blindness. On the other hand, if a female (XX) inherits a harmful recessive gene on one of her X chromosomes, but also inherits a corresponding dominant gene for the normal characteristic on her other X chromosome, only the normal characteristic will manifest itself. She will not show the effects of the harmful recessive gene unless she inherits two of them.

Polygenic Inheritance

Most human characteristics are influenced by **polygenic inheritance;** that is, they are affected by many genes rather than by a single pair. One example is human skin color, which is probably the result of the interaction of a dozen or so genes (Kowles, 1985). Thus skin can be any of hundreds of tones. A light-skinned person and a dark-skinned person will usually have children who are some shade between light and dark, each child inheriting half of his or her skin-color genes from the light-skinned parent and half from the dark-skinned parent. However, since each parent's genotype usually contains a range of skin-color genes, the child may inherit mainly light-skin genes, or, alternatively, mainly dark-skin genes, from both parents and therefore have lighter or darker skin than either of them. Similarly, height, weight, and body shape are polygenic, as are almost all complex human characteristics, such as intelligence, behavioral patterns, and special talents.

Figure 3.4 *Skin color is one of the most variable of human genetic characteristics. A child can be lighter or darker than either parent, or have a skin tone that is somewhere in between, even if in many other features the baby seems to take after Mom or Dad.*

Heredity and Environment

Don't be misled by this focus on genes. Heredity is of basic importance, but nurture always affects nature. Both are vital to development. As Scarr and Weinberg (1980) put it, "No genes, no organism; no environment, no organism." Genes influence the direction of development and often set the boundaries for the expression of particular traits, but the impact of environment is crucial.

When social scientists discuss the effects of the **environment** on the individual, they are referring to the impact of everything in the outside world that impinges upon the individual—from food, clothing, shelter, climate, and the like, to social, economic, political, and cultural patterns. Broadly defined in this way, environment affects the expression of almost any genotype.

Figure 3.5 *Is it heredity or environment that explains the fact that several generations of the Flying Wallendas have pursued the perfection of incredible high-wire feats of balance and coordination? Obviously, body type and a hearty attitude toward danger must play a role, together with family encouragement and practice that begins almost in infancy.*

Multifactorial Characteristics

Another way of saying this is that most characteristics are **multifactorial**, that is, they are the result of the interaction of many genetic and environmental factors.

Even physical characteristics are often multifactorial rather than purely genetic. Take height, for example. An individual's maximum possible growth is genetically determined, yet most North Americans are taller than their grandparents, but virtually the same height as their full-grown children. Why? Because to reach the maximum height set by his or her genes, an individual must have adequate nutrition and good health. In the nineteenth century, these two crucial factors were much less common than they are now, and Americans were, on the average, about 6 inches shorter than they are today (Tanner, 1971). Throughout most of the twentieth century, as nutrition and medical care improved, each generation grew slightly taller than the previous one, a phenomenon known as the *secular trend.* Over the past two decades, however, this trend for children to be taller than their parents has stopped. The reasons include the fact that most Americans now receive sufficient nourishment to reach their maximum potential height, and, possibly, that recent developments in medicine have not had a significant effect on growth.

Just as there are environmental effects on physical characteristics thought to be almost completely determined by genetics, the reverse has also been found: psychological traits that have often been considered mainly the result of nurture—shyness and extroversion, phobias, schizophrenia—show definite genetic influence (Goldsmith, 1983; Loehlin et al., 1982; Vandenberg et al., 1986; Walker and Emory, 1983). If one monozygotic twin becomes schizophrenic or manic-depressive, for example, chances are (estimates range from 80 percent to 20 percent, depending on definition) that the other twin will have similar psychological problems—a risk far greater than would be the case for dizygotic twins.

Of course, even though an individual's genetic makeup may play a role in the degree to which that person may be susceptible to a particular disability, environment often provides those conditions that precipitate or prevent the problem. For example, historical studies show a close correlation in the susceptibility of identical twins to tuberculosis, suggesting a genetic link similar to that for schizophrenia (Bracha, 1986). However, modern improvements in sanitation and medical care have made tuberculosis very rare today among all members of the population. Because the environment diminishes or exacerbates the effects of inherited predispositions, we need to be cautious in assuming that genetic susceptibility implies inevitability.

Ages and Stages in Nature-Nurture Interaction

The research question is no longer whether nature and nurture interact for any particular trait, because the answer almost always is that they do. Instead, the question revolves around how that interaction varies from individual to individual and, within each individual, how particular traits are affected by the individual's experiences and stage of development. Consider the importance of one's stage of development, for instance. According to Robert McCall (1981), in the first months of life, nature has a stronger influence on development than at any other period. Perception and cognition develop according to genetic instructions, and only massive environmental deprivation (such as being constantly isolated in a dark room) can prevent the normal development of intelligence. Later, however, variations in nurture can have a substantial impact: quantity and quality of education may profoundly affect how an adult thinks (Willis, 1985).

The relationship between genes and environment also varies from person to person: some individuals, for instance, are genetically more vulnerable than others to certain aspects of their environment—to, say, bee stings, or alcohol, or emotional losses. Some traits—shyness and fearfulness, for instance—have a stronger genetic component than others. And some traits can express themselves in quite different ways, depending on the environment. For example, the inherited characteristics that help to account for abilities with higher mathematics are also involved in musical creativity, but whether an individual possessing these characteristics becomes a mathematician or a composer—or something else altogether—is determined by the interaction of a great number of environmental influences (Gardner, 1983). Whenever we look at polygenic and multifactorial traits, such as emotional health or intelligence, the complexity of nature-nurture interaction becomes apparent.

Abnormal Genes and Chromosomes

Half or more of all zygotes are genetically abnormal. Almost all of these are aborted spontaneously, usually so early that the woman never knew she was pregnant. Most of the others die later in pregnancy. The remainder, some 3 to 5 percent, develop to full-term (National Institute of Child Health, 1979). This means that each year in the United States, between 100,000 and 150,000 infants are born with a chromosomal abnormality, a clearly defined genetic disorder, or a genetically influenced defect, such as a malformation of the spine, the head, or the foot. Chromosomal abnormalities are, in general, the most serious, but they are also the easiest to detect and prevent.

Chromosomal Abnormalities

Sometimes when sperm or ova are formed, the forty-six chromosomes divide unevenly, producing a gamete that has too few or too many chromosomes. About 8 percent of all conceptions involve this kind of gamete (Moore, 1982). In most of these cases, a spontaneous abortion occurs in the first days or weeks after conception. But sometimes, about once in every 200 births, a baby is born with one chromosome too many or one too few (Goad, et al., 1976).

Many chromosomal abnormalities involve the twenty-third pair, the sex chromosomes (see Table 3.1). Males, normally *XY*, are sometimes born with two or three *Y* chromosomes (*XYY* or *XYYY*), or two or three *X* chromosomes (*XXY* or *XXXY*), or two of each (*XXYY*). In addition, males who seem to have a normal pair of sex

TABLE 3.1 **Common Sex-Linked Chromosomal Abnormalities**

Name	Chromosomal Pattern	Physical Appearance*	Psychological Characteristics*	Incidence
Kleinfelter's syndrome	*XXY* (extra sex chromosomes)	Male. At adolescence, secondary sex characteristics do not develop. For example, the penis does not grow, the voice does not change. Breasts may develop.	Retarded in language skills.	1 in 1,000 males
(No name)	*XYY* (extra sex chromosomes)	Male. Prone to acne. Unusually tall.	More aggressive than most males. Mildly retarded, especially in language skills.	1 in 1,000 males
Fragile X	Usually *XY* (but the *X* chromosome does not transmit genetic instructions effectively)	Usually normal; occasionally, enlarged testicles in males.	Variable. Some individuals apparently normal; others definitely retarded.	1 in 1,000 males; 1 in 5,000 females (estimated)
(No name)	*XXX*	Female. Normal appearance.	Retarded in almost all intellectual skills.	1 in 1,000 females
Turner's syndrome	*XO* (only one sex chromosome)	Female. Short, often "webbed" neck. At adolescence, secondary sex characteristics (breasts, menstruation) do not develop.	Mildly retarded, especially in abilities related to math and science and in recognition of facial expressions of emotion.	1 in 10,000 females

*There is some variation in the physical appearance and considerable variation in the psychological characteristics of these individuals. For example, several studies of prison populations have found a higher percentage of *XYY* men than is found in the general population. At the same time, many *XYY* men are normal, law-abiding citizens, and some children in each group seem completely normal. With regard to psychological characteristics, much depends on the family environment of the child.
Sources: Goad et al., 1976; McCauley et al., 1987; Moore, 1982; Vandenberg, 1987).

chromosomes (*XY*) occasionally, in fact, have a "fragile X," an X chromosome that is present in only some of the cells, not all of them. Females, normally *XX*, are sometimes born with only one X chromosome (*XO*) or three, four, or five (*XXX,XXXX,XXXXX*). In every case, these abnormalities result in, among other things, impaired mental abilities.

Chromosomal abnormalities do not always involve the sex chromosomes. Sometimes a sperm or ovum contains twenty-three autosomes instead of twenty-two. In the formation of a zygote, the extra autosome attaches itself as a third chromosome to the eighth, thirteenth, fourteenth, fifteenth, eighteenth, twenty-first, or twenty-second pair of chromosomes, forming syndromes known as trisomy-8, trisomy-13, and so forth.

The most common chromosomal problem that results from abnormalities in the autosomes, rather than the sex chromosomes, is trisomy-21, the main cause of **Down syndrome.** This disorder affects about 1 baby in every 750 (Vandenberg et al., 1986).

People who have Down syndrome usually exhibit distinguishing features of the eyes, nose, and tongue. (Because of the very superficial resemblance between their eyelids and those of people from Mongolia, they are, mistakenly, sometimes called mongoloid.) They often suffer heart defects, and they usually develop more slowly, physically and intellectually, than other children. How well they function as adolescents and adults seems to depend a great deal on early experience. Typically, those who are raised at home are able to care for themselves, to read, and to write, while those who are institutionalized remain much more retarded (Edgerton, 1979). Even if individuals with Down syndrome survive to middle age, they are more likely to develop cataracts, leukemia, and a form of dementia similar to Alzheimer's disease—all because their extra chromosome at site 21 makes them more vulnerable (Patterson, 1987).

Figure 3.6 *This Down syndrome child has the round face, almond-shaped eyes, and thick tongue that characterize those who have an extra chromosome at the twenty-first pair. (His parents are also typical in at least one way, namely, their age: see next page.) This young man is fortunate, however, in that his family's affectionate care and support should help to make him comparatively self-sufficient by young adulthood.*

Causes of Chromosomal Abnormalities   What accounts for chromosome-based defects? Sometimes one parent may have an extra or a missing chromosome in some cells. A parent with this condition—known as **mosaicism** because the person's cells, including reproductive cells, are like a mosaic of different patterns, some normal, some not—has a high probability of contributing an abnormal gamete to the formation of a zygote.

A more common finding is that chromosomal abnormalities, especially Down syndrome and Kleinfelter's syndrome, occur much more frequently when the parents are middle-aged. For example, a 33-year-old woman has one chance in two hundred of having a child with chromosomal abnormalities; for a 39-year-old woman the chances are one in fifty-two; when a woman is 43 years old, the chances rise to one in twenty (Weiss et al., 1984). One possible explanation for this is the aging of the ova. Since the female is born with all the ova she will ever have, a 45-year-old woman has ova that are 45 years old. Perhaps degeneration of the ova leads to chromosomal abnormalities. However, this is not the only reason older parents have more offspring with chromosomal problems, because, no matter what the age of the mother, there is a positive correlation between the age of the father and the birth of a child with an extra chromosome. Perhaps as the male reproductive system ages, it produces a higher percentage of malformed sperm. It may also be that the sexual activity between husband and wife is implicated. Older couples generally have intercourse less frequently than they did when they were younger, increasing the likelihood of a relatively old sperm fertilizing an ovum that has been in the Fallopian tube for a relatively long time and thus the gametes may be defective because they are close to the point of degeneration (Emery, 1983).

The "fragile X syndrome" (see Table 3.1) shows considerable variability in its effects on individuals who inherit it (Brown et al., 1987). Of the females who carry it, most are normal (perhaps because they also carry one normal X chromosome), but a third show some mental deficiency. Of the males who inherit a fragile X chromosome, there is considerable variation in effect, with about twenty percent apparently completely normal, another third somewhat retarded, and the rest severely retarded. The last group is sufficiently large that about half the residents in most homes for the retarded have the fragile X. While the widely variable effects of this disorder are somewhat unusual, some geneticists believe that the more we learn about other abnormal genes and their interactions, the more diversity we will find (McKusick, 1986).

Harmful Genes

While abnormal chromosomes are relatively rare, everyone is a carrier of several genes that could produce serious diseases or handicaps (McKusick, 1986). Over the last twenty years, every three years or so, a researcher named Victor McKusick publishes a new edition of a catalogue of all the known and suspected genetic defects that humans are heir to. The number of known and fully identified genetic problems has increased notably with each edition, from 574 in 1966 to 1,914 in 1986 (McKusick, 1986). This increase does not mean that genetic disease is on the rise; it indicates, instead, that recent research has identified a genetic cause for a large number of disorders, many of them exceedingly rare, whose origins had previously been unknown. Among the more common of these are cystic fibrosis, spinal defects, cleft palate, and club feet (see Table 3.2, page 68, for a detailed listing of genetic disorders). Fortunately, many genetic problems are recessive, so a person will not have a particular condition unless he or she has inherited the genes for it from both parents. In addition, some serious genetic conditions are polygenic, so several specific genes must be present in the genotype before the problem appears in the phenotype. Still others are multifactorial; they do not become apparent unless something in the prenatal or postnatal environment fosters their expression. Thus, most babies have no apparent genetic problems, although all carry some of the destructive genes that their parents have. About one baby in every thirty, however, is not so lucky and is born with a serious genetic problem.

Genetic Counseling

For most of human history, couples at risk for having a child with a genetic problem did not know it. Indeed, if a child of theirs was born with a serious defect or died very young, the couple often had a "replacement" child soon after—unaware of the risk they were taking. More recently, if a genetic problem was suspected in the family, couples could either avoid pregnancy or hope that they would be lucky; there was not much else they could do.

Today, a combination of testing and counseling before and during pregnancy, as well as immediate medical attention at birth, has transformed the dilemmas faced by prospective parents. Through **genetic counseling** couples can learn more about their genes and make informed decisions about their childbearing future.

Two Success Stories

Genetic counseling brings good news to many couples, who learn that their risk of having a child with a genetic problem is no higher than the average. They also learn that, with early diagnosis and treatment, most common serious genetic problems can be avoided or minimized. One particularly encouraging example involves **phenylketonuria (PKU),** a recessive-gene defect that prevents a person from metabolizing protein normally. Left untreated, PKU causes serious mental retardation and emotional disturbance. PKU can now be detected at birth and treated with a special diet that greatly reduces the symptoms.

Even more striking has been the success in treating **erythroblastosis,** or **Rh disease,** a blood disorder that can kill the infant or cause such defects as deafness, cerebral palsy, and mental retardation. Rh disease can occur in conceptions involving a woman whose blood type is Rh negative and a man whose blood type is Rh positive. (This combination exists in about 12 percent of all American marriages.) Since the gene for Rh positive blood is dominant, most children from those marriages inherit the father's blood type. During childbirth, some of the child's positive blood, which had been circulating in the placenta, might enter the mother's bloodstream, especially if the birth is a difficult one. This causes the mother to develop antibodies to the positive blood, in much the same way a vaccination causes a person to develop antibodies against a disease. In any subsequent pregnancy, these antibodies cross the placenta, attacking and destroying some of the fetus's blood. The more antibodies the mother has, the stronger the attack. In the most severe cases, antibodies destroy the fetus months before the baby would have been born. Until 1968, 10,000 Rh positive babies born to Rh negative women died each year in the United States, and 20,000 had serious birth defects.

Medical advances have now made this disease rare. Since the beginning of the 1960s, doctors have been able to give ill newborns a series of blood transfusions, removing all the blood with its destructive antibodies and replacing it with new blood. Most impressive of all, if the fetal blood supply is attacked, even months before birth, the fetus can be given a blood transfusion in the uterus.

Since 1968, there has been a way to avoid the formation of Rh antibodies. In the first days after giving birth, women are given Rhogam (Rh negative blood that already contains antibodies), which stops their bodies from forming additional antibodies. The injected antibodies disappear within a few weeks. When the woman becomes pregnant again, she has no antibodies to destroy the blood cells of the new fetus.

TABLE 3.2 **Common Genetic Diseases and Conditions**

Name	Description	Prognosis	Method of Inheritance	Incidence*	Carrier Detection?[†]	Prenatal Detection?
Alzheimer's disease	Loss of memory and increasing mental impairment.	Eventual death, often after years of dependency.	Some forms definitely genetic; others may not be.	Less than one in 100 middle-aged adults, more than one in 20 adults over age 80.	No.	No.
Cleft palate, cleft lip	The two sides of the upper lip or palate are not joined.	Correctable by surgery.	Multifactorial. Drugs taken during pregnancy or stress may be involved.	One baby in every 700. More common in Japanese and Native Americans; rare in blacks.	No.	Yes, in some cases.
Club foot	The foot and ankle are twisted, making it impossible to walk normally.	Correctable by surgery.	Multifactorial.	One baby in every 300. More common in boys.	No.	Yes.
Cystic fibrosis	Lack of an enzyme. Mucous obstructions in body, especially in lungs and digestive organs.	Few victims survive to adulthood.	Recessive gene.	One baby in every 2,000. One in 25 white Americans is a carrier.	Yes.	Yes, in some cases.
Diabetes	Abnormal metabolism of sugar because body does not produce enough insulin.	Usually fatal if untreated. Controllable by insulin and diet.	Recessive gene, but exact pattern hard to predict because environment is crucial.	About 7 million Americans. Most develop it in late adulthood. One child in 2,500 is diabetic. More common in Native Americans.	No.	No.
Hemophilia	Absence of clotting factor in blood. Called "bleeder's disease."	Crippling and death from internal bleeding. Now transfusions can lessen or even prevent damage.	X-linked. Also spontaneous mutations.	One in 1,000 males. Royal families of England, Russia, and Germany had it.	Yes.	Yes.
Huntington's disease	Deterioration of body and brain in middle age.	Death.	Dominant gene.	Rare.	Yes.	Possible in near future.
Hydrocephalus	Obstruction causes excess water in brain.	Can produce brain damage and death. Surgery can make survival and normal intelligence possible.	Multifactorial.	One baby in every 100.	No.	Yes.

*Incidence statistics vary from country to country; those given here are for the United States. All these diseases can occur in any ethnic group of Americans. When certain groups have a higher incidence, it is noted here.

[†]Studying the family tree can help geneticists spot a possible carrier of many genetic diseases or, in some cases, a definite carrier. However, here "Yes" means that a carrier can be detected even without knowledge of family history.

Name	Description	Prognosis	Method of Inheritance	Incidence*	Carrier Detection?[†]	Prenatal Detection?
Marfan's syndrome	Long bony limbs, heart malformation, hearing loss, eye weakness.	Depends on severity; possibly death.	Dominant gene of varying strength.	Rare.	Yes.	Yes.
Muscular dystrophy (13 separate diseases)	Weakening of muscles. Some forms begin in childhood, others in adulthood.	Inability to walk, move; wasting away and sometimes death.	Duchenne's is X-linked; other forms are autosomal recessive or multifactorial.	One in every 4,000 males will develop Duchenne's; about 100,000 Americans have some form of MD.	Yes, for some forms.	Yes, for some forms.
Neural tube defects (open spine)	Two main forms: anencephaly (parts of the brain and skull are missing) and spina bifida (the lower portion of the spine is not closed).	Often, early death. Surgery may prolong life. Anencephalic children are severely retarded; children with spina bifida have trouble with walking and with bowel and bladder control.	Multifactorial; defect occurs in first weeks of pregnancy.	Anencephaly: 1 in 1,000 births; spina bifida: 3 in 1,000.	No.	Yes.
Phenylketonuria (PKU)	Abnormal digestion of protein.	Mental retardation, hyperactivity. Preventable by diet.	Recessive gene.	One in 15,000 births; one in 80 whites is a carrier.	No.	Yes.
Pyloric stenosis	Overgrowth of muscle in intestine.	Vomiting, loss of weight, eventual death; correctable by surgery.	Multifactorial.	One male in 200; one female in 1,000.	No.	No.
Sickle-cell anemia	Abnormal blood cells.	Possible painful "crisis"; heart and kidney failure.	Recessive gene.	One in 400 black babies is affected. One in 10 black Americans is a carrier, as is one in 20 Latin Americans.	Yes.	Yes.
Tay-Sachs disease	Enzyme disease.	Apparently healthy infant becomes progressively weaker, usually dying by age 3.	Recessive gene.	One in 30 American Jews is a carrier.	Yes.	Yes.
Thalassemia (Cooley's anemia)	Abnormal blood cells.	Paleness and listlessness, low resistance to infection; treatment by blood transfusion.	Recessive gene.	One in 10 Greek- or Italian-Americans is a carrier; one in 400 of their babies is affected.	Yes.	Yes.

Sources: McKusick, 1986; Moore, 1982; Preston, 1986, Vandenberg et al., 1986, Weiss et al., 1984.

Who Should Be Tested?

In spite of successes like these, there are still many genetic disorders that can result in serious problems. In addition, many high-risk couples are unaware of their situation and are therefore unable to avoid potential problems. Even with genetic counseling, many couples must still make difficult choices about conception and pregnancy.

These facts lead to the question of who should receive genetic counseling. Certainly everyone who plans to become a parent should probably know something about his or her genetic inheritance. But genetic counseling is strongly recommended for couples in five situations: those who already have a child with a genetic disease; those who have relatives with genetic problems; those who come from the same genetic stock (as first cousins do) or whose ancestors come from regions where certain genetic problems are common; those who have had previous pregnancies that ended in spontaneous abortion; those in which the woman is 35 or older or the man is 45 or older.

RESEARCH REPORT   ## Methods of Prenatal Diagnosis

Within the last twenty years, literally dozens of methods have been developed to determine whether a particular pregnancy is going well or not. For example, routine blood tests reveal whether the mother has had diseases that might affect the fetus; measurements indicate whether growth and development are occurring on schedule.

Some of these tests, however, are not routine but are used when there is a risk of genetic or chromosomal damage. Depending on the seriousness of the problem as well as on the values of the parents, the results can lead to abortion, or to treatment of a serious defect while the fetus is still in the uterus, or simply to advance notice that the newborn will be ill and in need of special care. The following five tests are helpful when the health of the fetus is in question.

Amniocentesis

In **amniocentesis,** which can be performed after the fifteenth week of pregnancy, about half an ounce of amniotic fluid is withdrawn through the mother's abdominal wall with a syringe. This fluid contains sloughed-off fetal cells that can be analyzed to detect major chromosomal abnormalities and many genetic problems. Amniocentesis also reveals the sex of the fetus (useful knowledge if an X-linked genetic disease is likely) and provides clues about fetal age and health.

Sonogram

The **sonogram** uses high-frequency sound waves to outline the shape of the fetus. Sonograms can reveal problems such as an abnormally small head (anencephaly), fluid on the brain (hydrocephaly), body malformations, and several

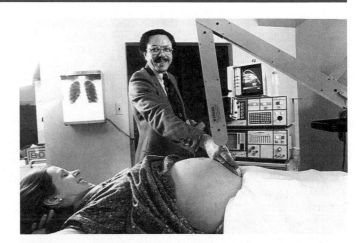

This modern obstetrical examining room provides diagnostic tools that have come a long way from the x-ray shown on the wall at left. This mother and doctor can both see and record the features and movement of the fetus with the sonogram in use here. If necessary, other diagnostic equipment can provide further information about fetal blood composition, growth, nourishment, and health status.

diseases of the kidney. In addition, a sonogram can be used to guide the needle in amniocentesis, diagnose twins, estimate fetal age, locate the position of the placenta, and, if repeated sonograms are performed, reveal the rate of growth. Almost half of all North American pregnancies are now scanned with ultrasound.

Fetoscopy

Fetoscopy is performed using a very narrow tube that is inserted into the woman's abdomen, piercing the uterus.

The process of genetic counseling varies from couple to couple, depending on the genetic background of each prospective parent. However, the basic procedure is the same: testing, predicting, and deciding.

Testing for Genetic Conditions

Detecting some genetic conditions is relatively simple. A blood test can reveal the recessive genes for sickle-cell anemia, Tay-Sachs disease, PKU, hemophilia, and thalassemia (Cooley's anemia). Analyzing a few cells from the prospective parents' bodies (easily obtained by lightly scraping the inside of the mouth) can indicate the possibility of some chromosomal abnormalities. In some cases, minor physical abnormalities, such as an oddly shaped little finger or unusual ear lobes, signal that a person may be a carrier of certain harmful recessive genes (Fuhrmann and Vogel, 1983). Other prenatal diagnostic techniques currently in use are described in the Research Report below.

Then a fetoscope, a viewing instrument, is inserted, allowing the physician to observe the fetus and the inside of the placenta directly. Fetoscopy is most commonly performed when a malformation is suspected. It can also be used for taking a sample of blood from the placenta to test for blood abnormalities.

Alphafetoprotein assay

If a fetus has a neural-tube defect (see page 69), the level of *alphafetoprotein* (AFP) in the mother's blood will be higher than normal. (The level of this protein is determined by a simple blood test.) However, an elevated AFP level does not necessarily indicate a problem, for the level also varies depending on the age of the fetus and the number of fetuses present. High AFP levels do indicate that amniocentesis, a sonogram, or fetoscopy should be performed.

Chorion villi sampling

At this writing, *chorion villi sampling* is an experimental method of analyzing a sample of the membrane that surrounds the embryo, providing information similar to that obtained through amniocentesis. This test, unlike amniocentesis, can be performed in the eighth week of pregnancy, thus allowing an early abortion of a defective fetus. Early abortions are somewhat safer for the woman, as well as being easier psychologically.

Risks

With the exception of chorion villi sampling, none of these five tests is considered particularly risky to either the mother or the fetus. However, none of them is routinely recommended for all pregnancies, for even a low risk should be avoided if possible.

The AFP assay can needlessly alarm a woman, because it may show a high level that is actually due to normal causes. In some cases it has even led to an unnecessary abortion because further tests were not performed (Hooker et al., 1984; Sun, 1983).

The sonogram has no proven risks (most research finds it harmless), but British studies suggest a possible link between repeated sonograms and later childhood leukemia (Boffey, 1983; Stark et al., 1984).

Both amniocentesis and fetoscopy are relatively painless for the mother and usually produce no harm to the fetus. Occasionally, the woman feels cramps for a few hours after the procedure; and though they are rare, complications—including spontaneous abortion—can occur.

For chorion villi sampling, the risk of spontaneous abortion may be higher, so this technique is not yet considered a replacement for amniocentesis.

Most women undergoing these tests feel strongly that they want to know what problems their fetus might have, rather than simply wait and worry. However, if parents learn that their fetus does have serious problems, they need careful and sensitive counseling, ideally, not only from a professional but also from other parents who have experienced the same diagnosis. Fortunately, for most parents, prenatal diagnosis is more likely to bring the welcome news that they can expect a healthy baby.

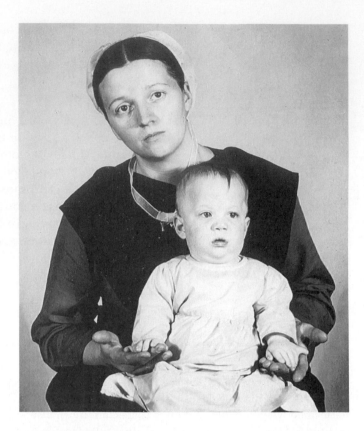

Figure 3.7 *Rare genetic conditions become more common when blood relatives marry, because the chance of a child's inheriting the same recessive genes from both parents increases. This child is a six-fingered dwarf, a condition extremely rare in the general population. However, at least sixty-one cases have occurred among the Old Order Amish, a religious group founded by three couples over 200 years ago. Members of this group are forbidden to marry outsiders, despite the fact that approximately one out of every eight members is a carrier of this gene.*

Family medical history, already-born children, and knowing where one's ancestors came from provide important clues. For example, Tay-Sachs disease is most common among Jews whose ancestors came from Eastern Europe (especially one part of Poland); sickle-cell anemia, among blacks whose roots go back to Central Africa; thalassemia, among descendents of Greeks or Italians (especially from Sicily); PKU, among Scandinavians; and so on.

Predicting Genetic Problems

Once a genetic counselor knows that a problem may be present, the next step is calculating the odds that a child could inherit the condition. Sometimes the prediction is simple. If two carriers of the same recessive gene marry, each of their children has one chance in four of having the disease, because each child has one chance in four of inheriting the recessive gene from both parents. (The same principle applies as in the case (page 60) of the two brown-eyed parents who had recessive genes for blue eyes and a one-in-four probability of having blue-eyed offspring.) It is important to remember that "chance has no memory," which means that *each time* two carriers have a child, the odds of that child's inheriting the disease are one in four. Each child born into the family also stands a one-in-four chance of avoiding the gene altogether and a 50-50 chance of inheriting one recessive gene, making the child a carrier like the parents.

In sex-linked diseases, such as color blindness, Duchenne's muscular dystrophy, and hemophilia, a male who inherits the recessive gene on his X chromosome will always have the disease. A woman who is a carrier will transmit the gene to about half of her ova, so about half of her sons will have the disease, and about half of her daughters will be carriers like their mother. The other half of her children are likely not to inherit the gene at all.

Some genetic diseases are carried by dominant rather than recessive genes. In fact, McKusick's catalogue of genetic disease (1986) lists 1,172 known dominant-gene disorders, as compared with 618 recessive and 124 X-linked problems. Each offspring of a carrier of a dominant-gene disorder has a 50-50 chance of inheriting the disorder. Luckily, deadly dominant diseases are rare, because carriers always have the disease as well as the gene, and therefore usually die before they are old enough to have children. Huntington's disease, a dominant-gene disorder that causes gradual deterioration of the nervous system, leading to physical weakness, emotional disturbance, mental retardation, and eventually, death, is an exception. The symptoms of the disease do not appear until the person is over 30, so a person can have many children before dying. However, carrier detection is now possible, so this tragic disease may become increasingly rare (Pines, 1984).

Many genetic conditions follow neither the recessive nor the dominant gene pattern. Some are caused by mutation, that is, a spontaneous change in a gene's formation. In this case, the problem cannot be predicted from the genotype of either parent. In other cases, dominance is partial, or the problem is polygenic or multifactorial. Cleft palate and cleft lip, club feet, spina bifida (a malformation at the end of the spine), diabetes, emphysema, many forms of cancer, hardening of the arteries, and high blood pressure are probably among this group of genetic diseases. At the moment, they are hard to predict, because the interaction of genes and environment responsible for their appearance is not fully known. New genetic-screening techniques, however, will soon aid markedly in the detection of genetic susceptibility for such problems. Nevertheless, environment and lifestyle will probably continue to exert a strong influence on whether a specific multifactorial disease will develop and how severe it will be (McAuliffe and McAuliffe, 1983).

Figure 3.8 *The first step in genetic counseling is usually the taking of a detailed family history, searching not only for ancestors and descendants with known genetic diseases, but also for relatives with unexplained problems such as infertility, stillborn children, or a seemingly innocuous mental or physical "peculiarity" that might be a marker for a more serious genetic anomaly. The history is typically interpreted as a chart, such as the one here, that helps elucidate inheritance patterns.*

Many Alternatives

The last step of genetic counseling is the most difficult, for, once the couple has been tested for genetic problems and knows the odds of their bearing a child with a problem, they must decide what risks they are willing to take. Most learn that the risks are not great, and they become pregnant and give birth to a normal baby. For others, the risk is substantial. Some avoid pregnancy, perhaps choosing sterilization and adoption. Others simply prepare themselves for the possibility of having a fatally ill or seriously handicapped child.

Others have another option, becoming pregnant and then aborting the fetus if prenatal diagnosis reveals that it is seriously handicapped. Prenatal diagnosis is now possible for chromosomal abnormalities, sickle-cell anemia, Tay-Sachs disease, spina bifida, malformations of the head, and many other major problems. While abortion of a fetus is never an easy choice, many couples would rather face that alternative, knowing they can conceive again, than avoid pregnancy or knowingly give birth to a child who must live a severely limited life.

Still others limit their family to one child, hoping that their first child will not be affected and reconciling themselves, if necessary, to the child's early death or difficult life. Others postpone pregnancy, hoping that medical research will be as successful in other diseases as in PKU or Rh disease.

Genetic engineering, now in an early experimental stage, may someday make it possible to add a dominant, healthy gene to counteract the damage done by recessive genes, thus curing children with genetic defects, or possibly even preventing damage while the fetus is still in the uterus. Progress in this area is sufficiently rapid that some geneticists think that a breakthrough may be near (Baskin, 1984). However, as one expert, looking toward the future, explains, "Direct genetic intervention in animals has already been demonstrated and will be evident in the next several years, but interventions in humans in order to cure genetic disease will be complicated by ethical and legal issues as well as technical ones" (Zimmerman, 1986). While couples of childbearing age in need of genetic counseling have many options, they do not always have easy solutions. Fortunately, as the decision tree below shows, the great majority of couples end up having healthy and normal babies.

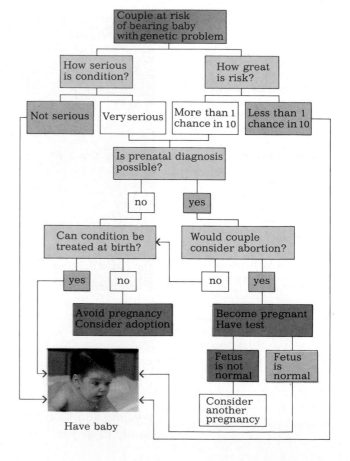

Figure 3.9 *With the help of a genetic counselor, even couples who know they run a risk of having a baby with a genetic defect might decide to have a child. Although the process of making that decision is more complicated for them than it is for a couple with no family genetic illness and no positive tests for harmful recessive genes, the outcome is usually a healthy baby. In each case, the genetic counselor provides facts and alternatives; every couple must make their own decision. In fact, two couples who have the same potential for producing a child with a genetic defect, and are aware of the same facts regarding the situation, sometimes make opposite decisions because they differ in their attitudes about abortion, in their willingness to raise a child with a genetic abnormality, or in their desire to have their own child rather than an adopted one.*

SUMMARY

1. Conception occurs when a sperm penetrates an ovum, creating a single-celled organism called a zygote. The zygote contains all the genetic material that determines the inherited characteristics of the future person.

Genes and Chromosomes

2. Genes, which provide the information cells need to perform specific functions in the body, are arranged on chromosomes. Every normal human body cell contains 23 pairs of chromosomes, one member of each pair contributed by each parent. Every cell contains a duplicate of the genetic information in the first cell, the zygote.

3. Twenty-two pairs of chromosomes control the development of most of the body. The twenty-third pair determines the individual's sex: zygotes with an *XY* combination will become males; those with two *X* chromosomes will become females.

4. Each person has a unique combination of genes, with one important exception. Identical (monozygotic) twins are formed from one zygote that splits in two, creating two zygotes with identical genes. Fraternal (dizygotic) twins are formed from two zygotes, and thus are no more similar, genetically, than other siblings are.

5. Genes can interact in many ways. In the simplest type of heredity, a person who inherits a dominant gene and a recessive gene for a particular characteristic develops the phenotype of the dominant gene. If two recessive genes are present, then the phenotype expresses the recessive form of the characteristic.

6. Males inherit just one X chromosome, and so, through X-linked inheritance, they have a greater chance than females of inheriting certain harmful recessive genes, including the genes for hemophilia and for color blindness.

7. Most inherited characteristics are polygenic, the result of the interaction of many genes, not of a single pair.

Heredity and Environment

8. Most physical and psychological characteristics are multifactorial, the result of the interaction of many genetic and environmental influences. Heredity may have a greater effect early in life, while the environment tends to exert more influence as the individual grows older.

Abnormal Genes and Chromosomes

9. Chromosomal abnormalities occur when the zygote has too few or too many chromosomes. Most of these defects involve the sex chromosomes. The most common autosomal abnormality occurs when an extra chromosome attaches itself to the twenty-first pair, causing Down's syndrome. Middle-aged couples are more likely than younger parents to produce a child with a chromosomal abnormality.

10. Every individual carries some genes for genetic handicaps and diseases. However, since many of these genes are recessive and many of the diseases involved are polygenic or multifactorial, most babies are born without a serious genetic defect.

Genetic Counseling

11. Genetic testing and an evaluation of family background can help predict whether a couple will have a child with a genetic problem. If there is a high probability that they will, they can consider several options, such as adoption, remaining childless, obtaining prenatal diagnosis and, if necessary, abortion. In some cases, appropriate postnatal treatment may remedy or alleviate the problem.

KEY TERMS

zygote *(55)*	environment *(62)*
gene *(55)*	multifactorial characteristics *(62)*
chromosome *(55)*	Down syndrome *(65)*
autosomes *(56)*	mosaicism *(65)*
gametes *(56)*	genetic counseling *(67)*
monozygotic twins *(57)*	phenylketonuria (PKU) *(67)*
dizygotic twins *(57)*	erythroblastosis (Rh disease) *(67)*
dominant genes *(60)*	
recessive genes *(60)*	amniocentesis *(70)*
genotype *(60)*	sonogram *(70)*
phenotype *(60)*	fetoscopy *(70)*
carrier *(60)*	
X-linked genes *(60)*	
polygenic inheritance *(61)*	

KEY QUESTIONS

1. How do chromosomes determine the sex of a zygote?

2. In what ways do identical twins differ from fraternal twins?

3. How does a dominant gene differ from a recessive gene?

4. What is the pattern of X-linked inheritance?

5. What effects can environment have on genetically inherited traits?

6. How does the interaction of nature and nurture change as the individual matures?

7. What are some of the causes of chromosomal defects?

8. How can genetic counseling help those parents who are at risk of bearing a child with genetic problems?

9. Who should receive genetic counseling?

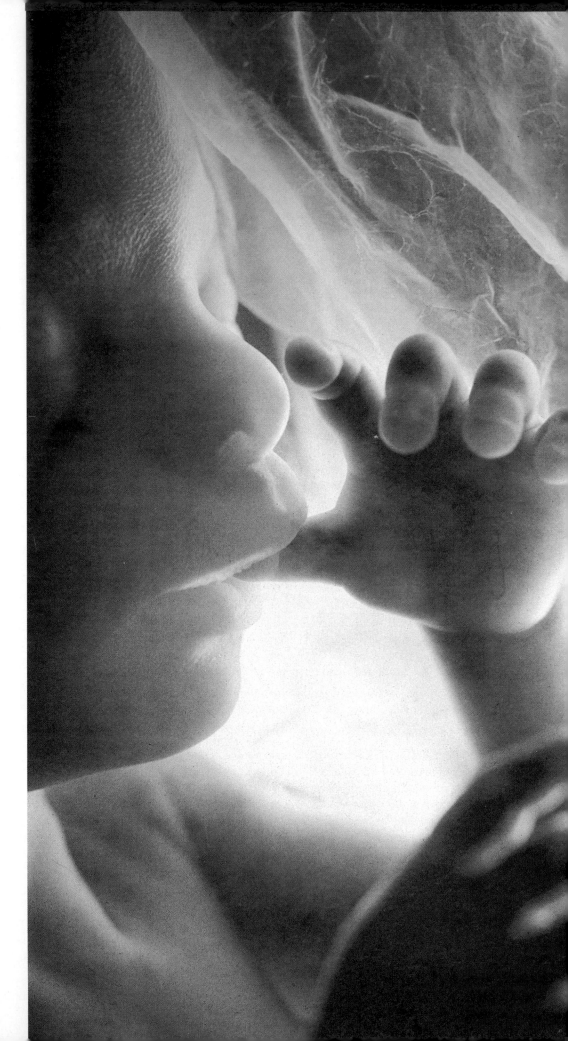

CHAPTER 4

Prenatal Development and Birth

"I don't care whether it's a boy or a girl;
I just want it to be healthy."

Every pregnant woman says that. Or thinks it. Or prays it.
Or incants it like a charm in the secret core of her soul.
So does every prospective father,
every grandparent-to-be.

Virginia Apgar and Joan Beck
Is My Baby All Right?

The first months of life, from conception through birth, can be considered the most critical of the entire life span. Not only is growth more rapid and the developing person more vulnerable than at any other time, but the role of other people is more crucial. The fetus is directly dependent on his or her mother for survival, and indirectly dependent on many others. For instance, the father's role is particularly vital: men who encourage their wives during pregnancy and coach them during birth not only benefit prenatal development and make birth less stressful, they also help to lay the foundation for a strong, supportive family life.

Looking beyond the family for examples of interaction between the developing person and the macrosystem, we find that a society that provides free prenatal care helps to reduce the number of low-birth-weight newborns; health regulations requiring preschool children to be immunized against rubella prevent this virus from being passed on to pregnant women and possibly damaging their fetuses. Let us look, then, at the details and interactions of these critical days.

From Zygote to Newborn

Each newborn begins life as a zygote, a tiny single-celled organism created by the merging of two reproductive cells, one from the father and one from the mother. The growth of this organism into a fully developed baby has been divided into three main periods. The first two weeks of development are called the **germinal period** (also called the period of the ovum); from the third week through the seventh week is the **period of the embryo;** and from the eighth week until birth is the **period of the fetus.***

---

*Technically speaking, the name of the developing human organism changes several times depending on the precise stage of development. While there is no need for the student to know them all, the curious might be interested to note that the organism that begins as a zygote becomes a morula, a blastocyst, a gastrula, a neurula, an embryo, and a fetus before it finally becomes an infant (Moore, 1982).

The Germinal Period: The First Fourteen Days

Within hours after conception, the one-celled zygote divides into two separate cells, which soon become four, then eight, then sixteen, and so on. About five days later, the multiplying cells divide into two distinct masses, a circular mass of outer cells enclosing a mass of inner cells. The outer cells will become membranes that protect and nourish the inner cells, which become the developing person.

Implantation   The organism, now containing more than a hundred cells, travels down the Fallopian tube and arrives at the uterus just at the time in the menstrual cycle when the lining of the uterus is covered with tiny blood vessels. About ten days after conception, it burrows into this lining, rupturing the blood vessels to obtain nourishment and initiating the hormonal changes that will prevent its being expelled during the next menstrual cycle (Moore, 1982).

This process, called **implantation,** is not routine. It has been estimated that 55 percent of all conceptions never achieve implantation, usually because of some abnormality in the organism or in the lining of the uterus (Grobstein et al., 1983). A successful implantation marks the end of the most rapid growth and the most hazardous transition of the entire life span.

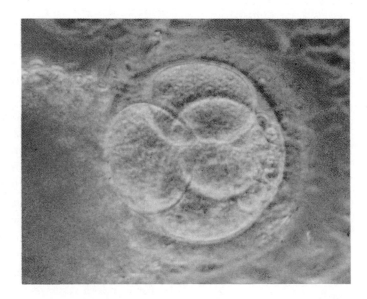

Figure 4.1 *Within 36 hours after fertilization, the one-celled zygote divides into two cells, and then, about one day later, it divides into four cells, as shown here. At the time of implantation, about 6 days after fertilization, the developing organism consists of more than one hundred cells, each one much smaller than the original cell, but each containing exact copies of the genes and chromosomes in the zygote.*

The Period of the Embryo: Two Weeks to Eight Weeks

The early growth of the embryo is rapid and orderly. During the first days of the embryonic period, a fold occurs down the middle of the embryonic disc, which, 21 days after conception, forms the **neural tube**—the beginning of the central nervous system (the spinal cord and brain). This is the point at which the organism is called an embryo. After the neural tube is formed, growth proceeds in two directions: from the head downward—referred to as **cephalo-caudal development** (literally, "from head to tail")—and from the center, that is, the spine, outward—referred to as **proximo-distal development** (literally, "from near to far"). Thus the most vital organs and body parts form first.

Following this pattern, in the third week after conception the head and blood vessels begin to develop. In the fourth week, the heart begins to beat, making the cardiovascular system the first organ system to begin to function (Moore, 1982). At the end of the first month, eyes, ears, nose, and mouth start to form, and buds that will become arms and legs appear. The embryo is now about ⅕ inch long (5 millimeters), about 7,000 times the size of the zygote it was twenty-eight days before.

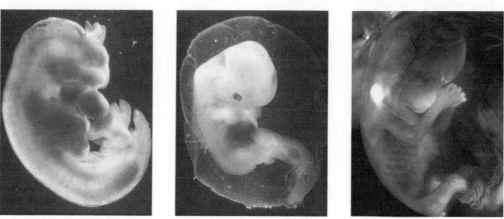

(a)   (b)   (c)   (d)

Figure 4.2 *During the first months of life, growth is rapid and proceeds from the center outward. (a) At 4 weeks past conception, the embryo is only about ⅕ inch long (5 millimeters), but already the head (top right) has taken shape. (b) At 5 weeks past conception, the embryo has grown to twice the size it was at 4 weeks. Its heart, which has been beating for a week now, is visible, as is what appears to be a primitive tail, which will soon be enclosed by skin and protective tissue at the tip of the backbone (the coccyx). (c) By 7 weeks, the organism is about an inch long (2 centimeters). Facial features, the digestive system, and even the first stage of finger formation can be seen. (d) At 8 weeks, the overall proportions of the developing person are close enough to those of a full-term baby that we can recognize this 1½-inch-long (4-centimeter) creature as a well-formed human fetus.*

Even faster growth in these early days occurs in the **placenta,** the organ that makes it possible for the developing organism to have the interaction with the mother's blood system that will allow the embryo (and, later, the fetus) to grow and develop. In the placenta, blood vessels from the mother are interwoven with blood vessels that lead to the developing person. These two sets of blood vessels are separated by membranes that prevent mixture of the two bloodstreams, while allowing some substances to pass from one bloodstream to the other. For example, oxygen and nourishment from the mother pass into the developing organism's bloodstream. Carbon dioxide and other waste products pass from the organism into the mother's bloodstream and are then removed through her lungs and kidneys. The pregnant woman is literally breathing, eating, and urinating for two.

The organism is connected to the placenta by the umbilical cord, which contains three blood vessels—one that carries nourishment and two that remove waste products. The movement of blood through the umbilical cord acts like water at high pressure in a hose, keeping the cord taut and making it almost impossible for the cord to become knotted or tangled or squeezed during prenatal development, no matter how many somersaults the developing baby or its mother does.

The Second Month Following the proximo-distal sequence, the upper arms, then the forearms, the hands, and the fingers appear. Legs, feet, and toes, in that order, follow a few days later, each having the beginning of a skeletal structure. By the end of the second month, the fingers and toes, which originally were webbed together, are separate.

Eight weeks after conception, the embryo weighs about ⅟₃₀ of an ounce (1 gram) and is about 1 inch (2.5 centimeters) long. The head has become more rounded and the features of the face are fully formed. The embryo has all the basic organs (except sex organs) and features of a human being, including elbows and knees, fingers and toes, and even buds for the first baby teeth. It is now ready for another name, the fetus.

The Period of the Fetus: Two Months to Birth

During the third month, muscles develop and cartilage begins to harden into bone. All the major organs complete their formation, including stomach, heart, lungs, and kidneys.

It is also during this period that the sex organs take discernible shape. The first stage of their development actually occurs in the sixth week, with the appearance of the *indifferent gonad,* a cluster of cells that can develop into male or female sex

Figure 4.3 *During the fetal period, cartilage becomes bone, as can be seen in this x-ray of a fetus at 18 weeks. The skull and the spine are most clearly developed and the bones of the fingers and toes are visible. Even buds for the teeth will soon begin to harden, although the first "baby" tooth will not emerge until 6 months after birth.*

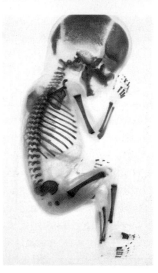

organs (Jirásek, 1976). If the fetus is male *(XY)*, genes on the *Y* chromosome send a biochemical signal late in the embryonic period that triggers the development of male sex organs, first the testes at about seven weeks and then the other male organs during the early fetal period. If the embryo is female and therefore has no *Y* chromosome, no signal is sent, and the fetus begins to develop female sex organs. By the twelfth week after conception, the external male or female genital organs are fully formed (Moore, 1982).

By the end of the third month, the fetus can and does move almost every part of its body, kicking its legs, sucking its thumb, and even squinting and frowning. The 3-month-old fetus swallows amniotic fluid, digests it, and urinates, providing its tiny organs with practice for the day when it will take in nourishment on its own. This active little creature is now fully formed—including its fingerprint pattern— and weighs approximately 3 ounces and is about 3 inches long.*

The Second Trimester Pregnancy is often divided into three-month-long segments, each called a **trimester.** In the second trimester (the fourth, fifth, and sixth months), hair, including eyebrows and eyelashes, begins to grow. Fingernails, toenails, and buds for adult teeth form. The heartbeat can be heard with a stethoscope, and at about the fifth month the woman can feel the flutter, and later the bump, of fetal arms and legs.

During the second trimester, the development of the brain is appreciable. This development, which is essential to the regulation of basic body functions, may be the critical factor in the fetus's attaining the **age of viability** (sometime between the twentieth and twenty-sixth weeks after conception), at which point the fetus has at least some slight chance of survival outside the uterus, if expert care is available.

Figure 4.4 *On Mother's Day, 1983, tiny DeAnna McWhorter was born sixteen weeks early, weighing only 20 ounces, and the odds against her survival were great. According to her mother, DeAnna "was so tiny that I was almost afraid that I could hurt her just by looking at her. Her face was no bigger than a half dollar. Her legs and*

arms were just a few inches long and no thicker than pencils. Her skin seemed thinner than tissue paper." DeAnna was one of the lucky ones. Although she had to spend the first four months of life in an isolette and undergo heart surgery when she was only 3 weeks old, one year later, she was a healthy toddler weighing close to 20 pounds.

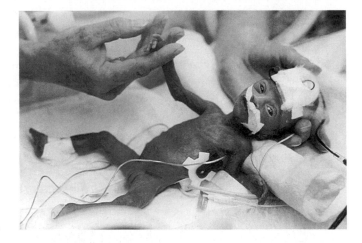

The Third Trimester During the last trimester, the lungs and heart become increasingly capable of sustaining life without the placenta. Brain development is particularly notable during this period, and beginning at about 29 weeks after conception, brain activity gradually takes on patterns of sleeping and waking (Parmelee and Sigman, 1983).

---

*During early prenatal development, growth is very rapid and considerable variation occurs between one fetus and another, so numbers given for length and especially for weight are only rough guidelines. For example, at 12 weeks after conception the average fetus weighs about 1½ ounces (45 grams), while at 14 weeks, the average weight is about 4 ounces (110 grams) (Moore, 1982).

In these final prenatal months, the fetus gains about 5½ pounds (2,500 grams). This gain includes an insulating layer of fat that will protect the newborn in the world outside the uterus, and that will also provide a source of nourishment in the first days after birth until the mother is producing breast milk.

Birth

The birth process usually begins with uterine contractions that many women think may just be movements of the fetus or the irregular contractions they have recently been experiencing. But when the contractions become strong and regular, or when the amniotic sac (the "bag of waters") breaks, it is clear that labor has begun. The drawings and caption in Figure 4.5 describe the major stages of the birth process. Although labor may last from a few minutes to a few days, it usually takes about eight hours in first births and between four and six hours in subsequent births (Danforth, 1977).

The Newborn's First Minutes

People who have never witnessed a birth often picture the newborn being held upside-down and spanked by the attending doctor or midwife to make the baby start breathing. Actually, this is seldom necessary, for newborns, or **neonates,** usually breathe and cry on their own as soon as they are born. In fact, sometimes babies cry as soon as their heads emerge from the birth canal. Nevertheless, there is much for those attending the birth to do. Any mucus that might be in the throat is suctioned out, the umbilical cord is cut, and the baby is wiped clean and wrapped to preserve body heat.

To quickly assess the newborn's physical condition, a measure called the **Apgar scale** is used to assign a score of between 0 and 2 to the baby's heart rate, breathing, muscle tone, color, and reflexes at one minute after birth and again at five minutes (see Table 4.1, next page). If the total score is 7 or better, the newborn is not in danger; if the score is below 7, the infant needs help establishing normal breathing; if the score is below 4, the baby is in critical condition and needs immediate medical attention (Harper and Yoon, 1987).

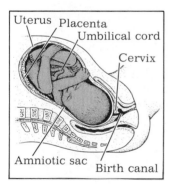

(a)

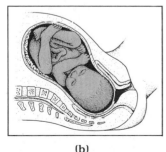

(b)

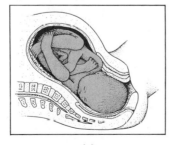

(c)

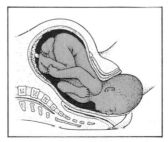

(d)

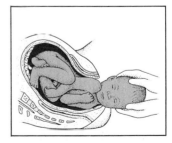

(e)

Figure 4.5 (a) *The baby's position as the birth process begins.* (b) *During the first stage of labor, the cervix dilates about 4 inches (10 centimeters) to allow passage of the baby's head.* (c) *Transition. The baby's head moves out of the uterus and into the "birth canal," the vagina.* (d) *The second stage of labor. The baby's head moves through the opening of the vagina and* (e) *emerges completely. The head is turned and the rest of the body emerges. Minutes after the baby is born, the third stage of labor (not shown here) occurs when contractions expel the placenta.*

TABLE 4.1 **The Apgar Scale**

Characteristic	0	1	2
Heart rate	absent	slow (below 100)	rapid (over 100)
Respiratory effort (breathing)	absent	irregular, slow	good, baby is crying
Muscle tone	flaccid, limp	weak, inactive	strong, active
Color	blue, pale	body pink, extremities blue	entirely pink
Reflex irritability	no response	grimace	coughing, sneezing, crying

Source: Apgar, 1953.

Next the infant is carefully examined for any structural problems, such as a cleft palate, a spinal defect, or a hip dislocation. Silver nitrate or tetracycline drops are put in the newborn's eyes to prevent infection that might result from bacteria picked up in the birth canal. All this can be done as the baby lies next to the mother, allowing her to assure herself that her new child is healthy (Davis, 1983).

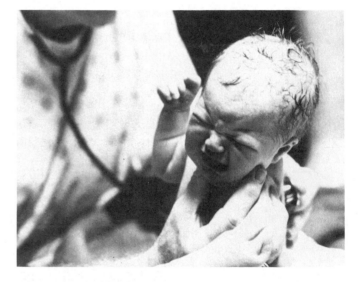

Figure 4.6 *Although he doesn't bear much resemblance to the appealing infants featured in advertisements, this virtually hairless, chinless, splotchy-skinned, squinty-eyed neonate with a somewhat misshapen head is, in fact, quite typical and healthy. The lusty cry and the muscle control in the arms suggest an Apgar score of 7 or higher.*

Problems and Solutions

Because prenatal problems can have possible physical and psychological consequences, and birth that is too stressful or too early can impair later development, we will discuss some of these difficulties and their prevention.

Teratology

The placenta was once thought to be a barrier that protected the fetus against any disease the mother might contract or any harmful substance that might enter her system. However, within the last forty years, we have learned a great deal through **teratology** (from *tera*, the Greek word for monster), the study of birth defects. We know now that hundreds of **teratogens,** that is, substances that cause birth defects, can cross the placenta and harm the embryo (Sever and Brent, 1986).

The Critical Period

The timing of contact with teratogens is particularly important, because as the embryo and fetus develop, there is a particular time span, called the **critical period,** for the formation of each organ and body part (see Figure 4.7). This critical period for each body part is also the time when that part is most susceptible to damage. As can be seen in the chart below, malformations of the heart, central nervous system, and spinal column are most likely to occur in the third through the fifth week; of the eyes, ears, arms, and legs from the fourth through the seventh week; of the teeth and palate in the seventh and eighth weeks. Because most body parts form during the first two months, the period of the embryo is sometimes called the critical period of pregnancy.

Note, however, that the central nervous system develops, and is susceptible to damage, throughout pregnancy. Many learning problems in apparently normal children may be traced to contact with teratogens occurring after the "critical period" (Brackbill et al., 1985).

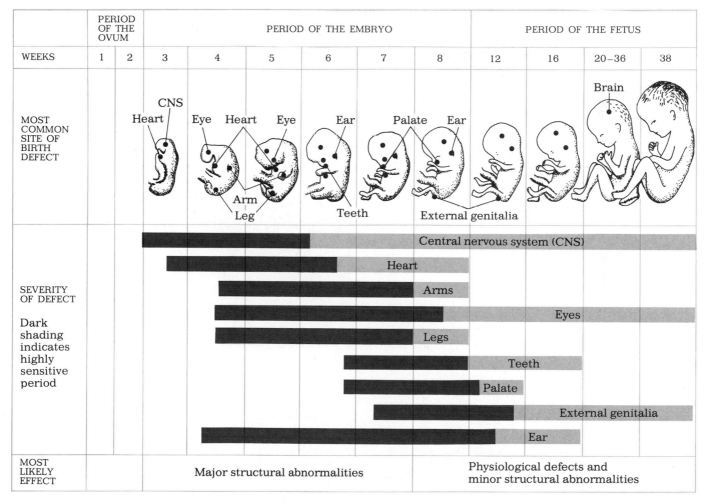

Figure 4.7 *As this chart shows, the most serious damage from teratogens is likely to occur in the first* *eight weeks after conception. However, damage to many vital parts of the body, including the brain,* *eyes, and genitals, can occur during the last months of pregnancy as well.*

Figure 4.8 *Three stages in finger development: (a) notches appear in the hand at day 44; (b) fingers are growing but webbed together at day 50; and (c) fingers have separated and lengthened at day 52. By day 56, fingers are completely formed, and the critical period for hand development is over. Other parts of the body, including the eyes, heart, and central nervous system, take much longer to complete development, so they are vulnerable to teratogens for months rather than days.*

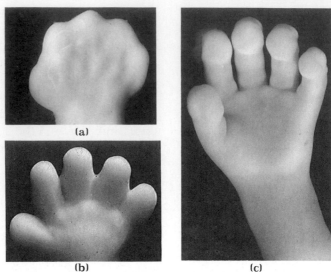

(a)

(b)

(c)

Further, nutrition is particularly crucial in the last trimester of pregnancy when the fetus is growing most rapidly and preparing for birth, storing iron and other nutrients that will be used during the first weeks of postnatal life. In fact, providing extra calories and protein for malnourished women and for teenagers who are still growing themselves dramatically reduces the incidence of prematurity and other problems.

Risk Factors   Indeed, the effects of any particular teratogen are the result of the interaction of many factors. Some babies are more likely to have serious congenital problems than others (see Table 4.2), and it is rare for any teratogen to harm every embryo or fetus that is exposed to it. Among the factors affecting susceptibility are the baby's genetic structure (for instance, some seem especially vulnerable to cleft palate), gender (males are more vulnerable), and the mother's age, health, social class, and nutrition. Birth order is also a factor: later-borns are more vulnerable.

TABLE 4.2 **Risk Factors**

A specific teratogen is more likely to harm a particular embryo or fetus if several of the following conditions prevail.

Family Background	Several children already born to the family. Low socioeconomic status.
Inborn Characteristics	Genetic predisposition to certain problems. The fetus is XY (male).
Mother's Characteristics	Undernourished. Over 40 or under 18.
Teratogen Exposure	Occurs early in pregnancy. High dose or exposure. Occurs over a period of several days or weeks. Other teratogens also present.
Nature of Prenatal Care	Woman is several months pregnant before prenatal care begins. Prenatal visits to doctor are more than four weeks apart.

Diseases

One of the first teratogenic diseases to be recognized was **rubella** (sometimes called German measles), which had long been considered a harmless illness that generally occurred in childhood. It is now well established that rubella, if contracted early in pregnancy, may cause many birth handicaps, among them blindness, deafness, heart abnormalities, and brain damage.

Once the link between rubella and birth defects was established, researchers began looking for other diseases that might affect unborn babies. They found dozens, including mumps, chickenpox, polio, measles, genital herpes (herpes simplex 2), AIDS (acquired immune deficiency syndrome), and some strains of flu.

As is the case with rubella, the woman need not be very ill for the embryo or fetus to be affected. For example, **toxoplasmosis,** a disease caused by a parasite commonly present in uncooked meat and in cat feces, is hardly noticeable in an adult but can cause prenatal blindness and serious brain damage. (If a blood test early in pregnancy indicates that the mother-to-be is among the two-thirds of the population that is not immune [Larsen, 1986], she should not eat rare meat nor change the cat's litter box.) Indeed a woman may not have to be sick at all for prenatal damage to occur: even the disease organisms present in certain vaccines, smallpox and rubella among them, are sometimes teratogenic if the exposure to them occurs just before or during early pregnancy (Korones, 1986; O'Brien and McManus, 1978).

Syphilis is another teratogenic disease for which good prenatal care is critically important. In the first months of pregnancy, the organisms that produce the disease cannot cross the placenta, so syphilis can be cured with antibiotics before the fetus is affected. But later in pregnancy, the fetus may be susceptible to bone, liver, and brain damage, or even death (Grossman, 1986).

Prenatal care is also crucial in the last weeks of pregnancy when there is a possibility of the mother's developing **eclampsia,** a disease that arises when the mother has difficulty ridding her system of fetal wastes. Women who are malnourished or bearing multiple fetuses are particularly likely to develop this problem. The early stage of this disease, called *toxemia* or *preeclampsia*, occurs during the last trimester in about 6 percent of all pregnancies. The symptoms in the mother include sudden weight gain and swollen fingers and ankles due to increased water retention, high blood pressure, and protein in the urine. In its early stages, preeclampsia can usually be controlled by diet and rest. If such measures do not work, the fetus is delivered prematurely, preventing fetal and maternal brain damage and even death.

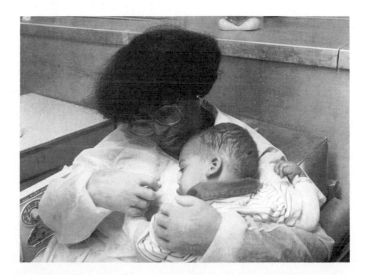

Figure 4.9 *Although prenatal diagnosis can be the first step in preventing or limiting damage from many congenital diseases, this is not true of AIDS. This health worker in a facility for AIDS children born to drug-addicted AIDS mothers can do little for this child except to make him more comfortable. Unless a cure is found, he is likely to die before age 5.*

Drugs

Drugs with proven harmful effects on the human fetus include streptomycin, tetracycline, anticoagulants, bromides, Thorazine, Valium, iodine, most hormones, and phenobarbital (Gupta et al., 1982; O'Brien and McManus, 1978). Several nonprescription drugs, including aspirin, antacids, and megadoses of vitamins C, D, and A have also been implicated in birth defects (Scher and Dix, 1983). Doctors now recommend that pregnant women take no drugs, unless the drug has been proved safe for use during pregnancy and is necessary for the woman's health.

Social Drugs   Many of the drugs that harm the fetus, however, are not purchased at the drug store, but instead are bought at the supermarket, liquor store, or on the street. Unfortunately, if they are part of the woman's lifestyle prior to pregnancy, such drugs are often used during pregnancy as well. In a recent survey of 417 pregnant women, mostly middle-class and fairly well educated (more than half had attended college), three-fourths of the women reported drinking alcohol during pregnancy (five drinks a week, on average), and a third reported smoking cigarettes (an average of sixteen a day) (Streissguth et al., 1983).

Alcohol.   The most prevalent drug in our society, alcohol, can have serious teratogenic consequences. Each year in the United States about 1,500 babies are born physically deformed and mentally retarded because their mothers consumed alcohol during pregnancy (Smith, 1978a). About a third of the infants born to severe alcoholics have the distinctive symptoms of the **fetal alcohol syndrome (FAS),** including small heads, abnormally spaced eyes and malproportioned faces. Many also have abnormalities of the skeleton and joints (Mulvihill, 1986).

In some cases, even moderate drinking may be harmful. One study found that newborns born to women who drink moderately during pregnancy tend to be somewhat more excitable and irritable than the average newborn (Streissguth et al., 1983). Again, animal research suggests an explanation: alcohol in the maternal bloodstream temporarily reduces oxygen to the fetus, a deprivation that can cause minimal brain damage (Mukherjee and Hodgen, 1982).

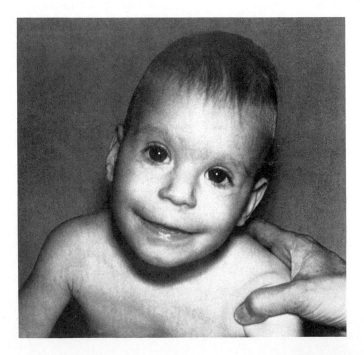

Figure 4.10 *This boy's widely spaced eyes, underdeveloped upper jaw, and flattened nose are three of the typical facial characteristics of children with fetal alcohol syndrome. Many babies born to women who drank alcohol during pregnancy show no signs of FAS; others have more obvious deformities of the eyes and head.*

These studies need to be interpreted cautiously. While no researcher doubts that alcohol abuse sometimes harms an embryo, some researchers believe that serious damage can occur only if the woman drinks heavily and the embryo is genetically predisposed to FAS (Kolata, 1981; Mulvihill, 1986). However, the March of Dimes and the Surgeon General of the United States both advise that pregnant women avoid alcohol completely.

Cigarettes.   There is no controversy about the physical effects of maternal cigarette smoking during pregnancy. Babies born to mothers who smoke are more likely to be underweight, premature, and in need of special care in the hours, days, and months after birth. A massive longitudinal British study finds that many of the common problems of childhood—from temper tantrums to asthma—are also more common in the offspring of smoking mothers (Butler and Golding, 1986).

Nonetheless, about one woman in four smokes during pregnancy. The infants particularly likely to suffer are those born to mothers already at risk, such as women who are malnourished, under age 18, or over age 35 (Werler et al., 1986).

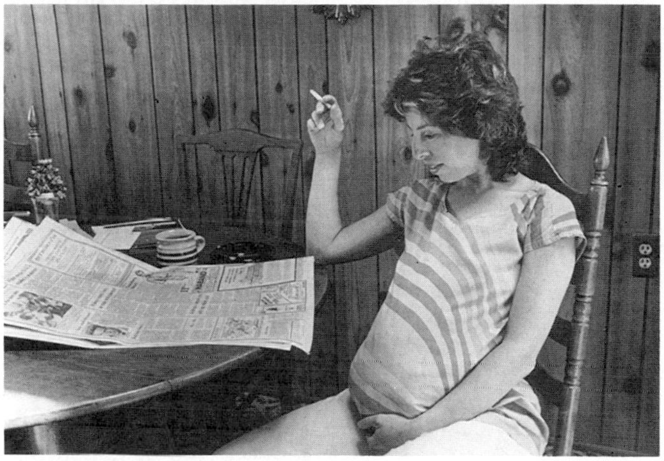

Figure 4.11 *The habit of lighting up a cigarette with the morning coffee and newspaper may be hard to break, but the price of continuing to smoke during pregnancy may be a smaller, more vulnerable newborn.*

Heroin and Methadone.   Babies whose mothers are addicted to either heroin or methadone are born addicted themselves. If they do not receive the drug soon after birth, they may die of severe withdrawal symptoms. Addicted babies also suffer from a variety of problems, among them low birth weight, jaundice, and various malformations. Compared to infants born to other mothers from the same socioeconomic class, addicted newborns are twice as likely to die within days after birth, primarily because they have trouble breathing (Ostrea and Chavez, 1979).

Other Common Drugs.   Definitive research on the teratogenic effects of most common drugs has yet to be done (Brackbill et al., 1985). To the best of our present knowledge, moderate amounts of caffeine (the equivalent to three cups of coffee a day) are not harmful. However, even moderate amounts of marijuana or cocaine have been associated with complications, including low birth weight (Scher and Dix, 1983).

Given our limited present knowledge, and the evidence that drugs once thought safe are not, many expectant mothers are rightly cautious about all the drugs they consume.

Environmental Hazards

Pregnant women who are sufficiently cautious can avoid taking drugs. With a little luck and planning, they may even avoid getting sick. But without adopting a very different lifestyle, it would be virtually impossible for most of them to avoid all cosmetics, food additives, and pollutants in the air and water. Could these be harmful as well? The answer is yes, sometimes. A few are known to cause serious damage. Most of the rest are probably harmless, but no one knows for certain.

Radiation   Massive doses of radiation from the atmosphere can cause many congenital problems. The best evidence of this, unfortunately, comes from the results of the atom-bomb explosions in Hiroshima and Nagasaki in 1945; none of the surviving pregnant women who were within a mile of the center of the explosion gave birth to live babies. Three-fourths of the pregnant women who were between one and four miles of the center had spontaneous abortions, stillborn babies, or severely handicapped infants (Apgar and Beck, 1973).

Another problematic source of radiation is x-rays. Use of x-rays, especially in the first trimester, slightly increases the risk of leukemia during childhood (Stewart and Kneale, 1970). If a medical emergency makes x-rays or radiation therapy advisable, the woman should consult with her doctor about the specific benefits to her and the risks to the fetus. In most cases, doses of radiation are so low that the fetus is unharmed (Brent, 1986).

Pollution   When exposure to them is extensive, several environmental pollutants can be teratogenic, among them, carbon monoxide, lead, and mercury. In fact, even quite low levels of lead—such as the automobile emissions a woman might breathe if she lives near a busy highway or a street with heavy stop-and-go rush hour traffic—correlates with low birth weight and slow neurological development (Raloff, 1986).

For several years, scientists have been aware that the manufacturing chemicals known as PCBs are teratogenic to animals (Allen et al., 1980) and, in high levels, to humans (Rogan, 1986). Recently, evidence of the teratogenic effects of even low levels of PCBs has been found. Pregnant women in Michigan were compared on their levels of consumption of PCB-polluted fish from Lake Michigan, and then their newborns were examined. Women who had eaten more fish had newborns with more problems: their infants tended to be smaller, preterm, and have slowed and depressed reactions to stimuli. As the authors point out, however, the effect of any teratogen depends on a variety of factors, and small amounts of PCB alone may not seriously damage a fetus (Jacobson et al., 1984).

Interaction among Hazards

It is important to remember that the effect of most hazards to development depend on the extent of the exposure and the total number of risk factors (Fein et al., 1983). Most pregnant women experience some of the potential hazards and have healthy babies nonetheless. Limited exposure to most of the hazards—malnutrition, diseases, drugs, pollution, chemicals—probably affects only fetuses that are already vulnerable. The importance of interaction is clearly seen in the most prevalent serious birth problem, low birth weight.

The Low-Birth-Weight Infant

Most newborns weigh about 7½ pounds and are born **full term,** that is about 38 weeks after conception. However, in the United States, about 7 percent of all newborns weigh less than 5½ pounds at birth, and are designated **low-birth-weight infants** (see Table 4.3).

TABLE 4.3 **Percentage of Low-Birth-Weight Infants in Selected Countries**

Bangladesh 50%	Brazil 9%
India 30%	USSR, Poland 8%
Philippines 20%	US, UK, Israel, Egypt 7%
Guatemala, Nigeria 18%	Canada, China 6%
Indonesia 15%	New Zealand, Australia, France, Japan 5%
Iran 14%	
Kenya 13%	Sweden, Finland, Netherlands, Norway 4%
Mexico, Jamaica, South Africa 12%	

Source: Grant, 1986.

Most low-birth-weight infants are born more than three weeks early, and hence are called **preterm,** a more accurate designation than *premature.* Others, born close to the due date but weighing less than most full-term neonates, are called **small-for-dates.**

In deciding whether a particular low-birth-weight infant is to be considered preterm, an evaluation of the neonate's physical maturity is more important than an estimate of the number of weeks of gestation. Preterm infants show many signs of immaturity (Harper and Yoon, 1987). For example, they often have fine, downy hair (lanugo) and a thick coating of vernix on their faces or bodies. If born more than six weeks before term, their nipples are not yet visible, and, if they are boys, their testicles have not descended into the scrotum. Most of these characteristics pose no serious problem. Vernix and lanugo disappear, and nipples and testicles emerge naturally within a few days or weeks.

There is one characteristic, however, that can be critical: the immaturity of such reflexes as breathing and sucking. Even full-term babies sometimes need a few days to coordinate these reflexes to be able to suck and breathe without experiencing such difficulties as spitting up or hiccupping. But preterm babies need more than a few days. They need special equipment and skilled care to sustain life until their reflexes mature. They must be placed in heated isolettes to maintain normal body temperature and to keep them free from infection. If they are more than six weeks early, they must be fed intravenously.

The most critical problem for preterm babies, however, is obtaining oxygen. Because their breathing reflexes are immature, about 60 percent of the infants born three months early and 20 percent of those born one month early suffer from *respiratory distress syndrome,* the cause of about half of all newborn deaths in North America (Behrman and Vaughan, 1983).

Causes of Low Birth Weight Low birth weight can result from many factors. One common cause is multiple pregnancies. Twins usually gain weight normally until eight weeks before the due date, then gain more slowly than the single fetus does. The average twin is born three weeks early, and weighs less than 5½ pounds. Triplets are usually born even earlier and weigh even less.

Often a "small-for-dates" baby is small because the mother was malnourished, or because the placenta and the umbilical cord did not function properly. Prenatal infections, genetic handicaps, and many teratogens can also cause small or immature neonates.

RESEARCH REPORT   The Problems of the Preterm Infant

Although preterm birth is the leading cause of newborn death, medical breakthroughs within the past twenty years have meant that most preterm infants survive. Indeed, increased survival of preterm infants is the main reason the infant mortality rate in the United States in 1985 (about 7 per 1,000 live births) was less than half what it was in 1960.

However, the survival of increasing numbers of very low-birth-weight infants means that there are an increasing number of children with developmental problems. The same medical interventions that have saved lives over the past two decades have sometimes created lifelong problems, among them blindness (from inhaling high concentrations of oxygen administered to aid breathing) and cerebral palsy (from brain damage that occurred during the emergency birth process) (Silverman, 1980). As longitudinal data have revealed the causes of the more extreme developmental problems some preterm infants have experienced, medical procedures have been altered to make obvious damage less common. However, even today preterm infants are likely to develop more subtle problems. For example, during toddlerhood they are slower to put two words together to make a simple sentence and less likely to play in an imaginative way (Ungerer and Sigman, 1983). Indeed, especially in the early years, the average scores of preterm children on many specific measures of cognition are lower than those of full-term children; those with very low birth weights are most likely to show deficits (Field et at., 1981).

What could be the cause of these problems? If maternal malnutrition was the reason for the infant's low birth weight, the infant may continue to be malnourished, and

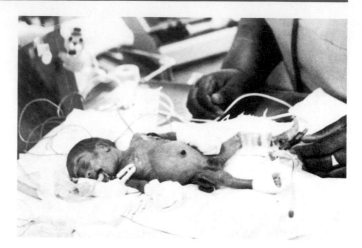

Born weighing only 2 pounds, this infant's heart rate, breathing, temperature, and blood acidity will be monitored continually until he reaches a weight of about 5½ pounds. Although his condition appears to be extremely frag- ile, current medical technologies give him an excellent chance of survival. However, the medical environment required to meet his most critical physical needs may deprive him of subtle, but important, types of stimulation.

thus be susceptible to later learning problems. The birth process itself may cause damage, for low-birth-weight infants are more vulnerable to any stress during birth than full-term infants are.

In addition, the preterm infant's experiences after birth may be an important part of the answer. During the first weeks of life, the daily care of preterm infants is dictated by precautions; they do not get certain kinds of

Figure 4.12 *Even though they weigh much less than average neonates, low-birth-weight babies are typically almost as long as their heavier contemporaries. The result is a scrawny, wizened appearance that may not disappear for many weeks, even when, like this infant, they are almost sufficiently developed to be treated like normal babies.*

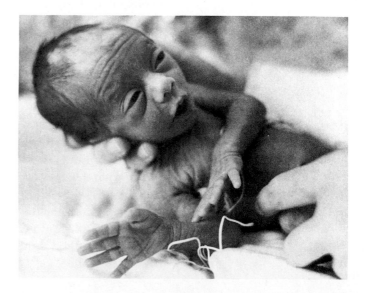

stimulation, such as the regular handling normally involved in feeding and bathing. Until recently, infants in the intensive care nursery were rarely touched and infrequently seen by their parents. At the same time, these infants are subjected to a number of abrasive experiences that normal infants are not, such as breathing with a respirator, being fed intravenously, and sleeping in bright light.

Recognizing these differences, several researchers have attempted to provide preterm infants with substitutes for the soothing experiences and regular stimulation they miss. In one experiment, infants born six or more weeks early were rocked mechanically while being exposed to the sound of a recorded heartbeat for fifteen-minute sessions many times each day (Barnard and Bee, 1983). These infants showed immediate differences in activity level from that of a control group that received the normal hospital treatment; and, in a follow-up at 2 years, they were significantly ahead of the control group in intellectual ability. The authors speculate that the lulling quality of the stimulation, as well as its regularity, "may have aided in the development of crucial, but subtle, aspects of the central nervous system."

Also, in recognition of the special problems of preterm infants and their parents, hospitals have changed some traditional procedures to allow more opportunities for contact, even in the first difficult days. The time spent together may help to establish the early bonds between parent and child. ("Bonding" is discussed in more detail on pages 97–98.)

Studies of parent-infant interaction during the first year of life have found other differences in the typical experiences of preterm and full-term infants. Parents of preterm infants tend to be more active with their babies—rubbing, poking, talking, offering the bottle—than parents of full-term babies are. For their part, preterm babies tend to be more passive than full-term babies (Bakeman and Brown, 1980; Crnic et al., 1983; Field, 1980). It is as if the parents, noticing their babies' relative passivity, try to push them into normal behavior, while the infants, reacting to more stimulation than they can comfortably handle, withdraw. Given the nature of this interaction, it is not surprising to find that parents and preterm infants smile at each other less frequently in the first months of life than do parents and full-term infants (Field, 1980).

Research also points to the importance of the home environment. For instance, most of the learning problems that middle-class preterm children exhibit in the first months of life disappear by school age. Lower-class preterm children, on the other hand, continue to have learning difficulties (Beckwith and Parmelee, 1986; Butler and Golding, 1986; Sameroff and Seifer, 1983). Apparently, families with little education and low income have fewer resources for coping with the special demands of the preterm child. When families of similar class status are compared, one factor that correlates with better cognitive development of preterm children is the mother's ability to adapt to the child's special needs (Sameroff and Seifer, 1983).

Taken together, these studies suggest that relatively simple changes in the hospital environment and a more adaptive parent interaction may lead to better development in preterm infants.

The incidence of low-birth-weight babies follows socioeconomic patterns, with lower-class mothers twice as likely as upper-class mothers to give birth to preterm infants, primarily because poor women are more likely to be malnourished, ill, or have inadequate prenatal care. Age is also a factor: teenage mothers are twice as likely to give birth to babies weighing under 1,500 grams (3 pounds, 5 ounces) than are women between the ages of 25 and 30 (Institute of Medicine, 1985).

Stressful Birth

As we have seen earlier in this chapter, birth is usually a short and natural process that results in a healthy newborn. However, this is not always so. Birth can be long and complicated, or a medical emergency may sometimes result in a lifelong handicap for the newborn. Because the possible physical and psychological consequences of birth complications can be important factors in development, we will discuss one of the major difficulties—the birth that causes too much stress for the fetus.

The major problem that can result from any long and stressful birth is **anoxia,** a lack of oxygen. Moments of anoxia occur even in a normal birth, as strong contractions temporarily squeeze the umbilical cord. This is not harmful to the fetus, any more than momentarily holding one's breath is. However, repeated and prolonged anoxia can cause brain damage and even death, especially if the fetus is undernourished or preterm. In recent years, various medical techniques have helped to reduce the number of long and stressful labors. For example, in virtually all hospital births, the condition of the fetus and the progression of labor is closely monitored minute by minute. Fetal monitoring detects anoxia and indicates when delivery should be speeded up so serious damage does not occur. Fetal monitoring also allows doctors to know when a labor that appears to be difficult is actually creating no unusual stress on the fetus and therefore can continue without intervention (Sher and Dix, 1983).

The most common type of **fetal monitor** consists of a sensing device that is fitted around the woman's midsection to measure and record the fetus's heart rate as well as the strength and frequency of contractions. Another type of fetal monitor is attached to the scalp of the fetus, and can measure the amount of oxygen in the bloodstream.

When monitoring reveals a weak or erratic fetal heart rate, or birth is not progressing as rapidly as it should, or the mother becomes exhausted and shows signs of physical stress (e.g., high blood pressure), several ways of hastening birth can be used. The best known is the **Cesarean section,** performed by making a surgical incision in the mother's abdomen to remove the fetus quickly. Cesareans now account for 23 percent of all United States' births (Plecek, 1986), a dramatic increase from 5 percent in 1968 (Donovan, 1977). Many of these Cesareans are performed even before labor begins, when sonograms and stress tests reveal that vaginal birth will probably be too stressful.

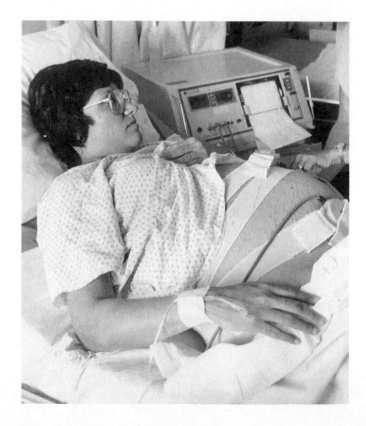

Figure 4.13 *This woman is resting between contractions, as the monitor tracks the regular heartbeat of her fetus. While monitors have been criticized for making women unnecessarily nervous, most women find them reassuring. In addition, the monitor often shows the beginning of a contraction before the woman herself feels it. If she has had training in natural-birth techniques, this signal allows her to begin the special breathing promptly, avoiding some of the pain and tension she would have felt with no forewarning.*

A CLOSER LOOK   The Medicalization of Birth

Modern medical techniques have saved lives and prevented brain damage, probably in millions of cases. Nevertheless, many critics, including some doctors, believe that these procedures tend to be overused in a typical hospital setting, resulting in a more expensive and less humane birth process (Davis, 1983). Moreover, recuperation takes longer if surgery and anesthesia are involved. A Cesarean, for instance, is fairly easy for both mother and child at the moment of birth, but in the days immediately after birth, the mother is much less likely to feel up to cradling, feeding, and bathing her infant than she would have felt if she had delivered vaginally.

The most frequently criticized delivery practice is the use of medication. In North America, more than 90 percent of all deliveries involve the use of drugs to speed up contractions and diminish pain. From a medical perspective, these drugs are usually very helpful, since they reduce some of the stress of the birth process.

As with all medical interventions, however, these drugs carry some risk, especially to the fetus. Virtually all obstetrical medication enters the mother's bloodstream and rapidly crosses the placenta into the baby's blood supply. The dose, the timing, the nature of the drug, and the maturity of the fetus all affect the impact the drug will have. Obviously, smaller doses given late in the birth process are safer for the fetus than larger, earlier doses.

In special circumstances, such as a preterm birth, obstetrical medication can slow down the newborn's breathing reflexes to the point of danger. Physicians are well aware of this risk, and therefore usually give anesthesia sparingly, especially in comparison to a decade ago.

However, even when obstetrical medication causes no apparent harm in the birth process, it may result in negative effects in the hours and days after birth. Even low doses of medication can affect newborn behavior, including making the newborn less alert (Lester et al., 1982), and the greater the dose, the greater the effects tend to be. Mothers are also affected. One study found that mothers who had more anesthesia reported more difficulties and fewer rewards from child care than mothers who had had no medication (Murray et al., 1981). In general, the mothers most likely to experience postnatal depression (the "baby blues") are those who receive the most anesthesia (Davis, 1983).

A growing number of couples in recent years have reacted against the depersonalization and medicalization of hospital birth by deciding to have their babies at home, convinced that home birth is not only safer but also more humane. According to advocates of home birth, the absence of many standard hospital procedures—from

Even though this birthing room is designed to provide some of the comforts and distractions of home (including a television set), it is also equipped for fetal monitoring and intravenous medication, if necessary. In addition, if a serious problem arises, the delivery room is only footsteps away and an emergency delivery can be performed in minutes, thus minimizing risk to mother and child.

dressing the mother-to-be in the regulation hospital "johnnie" to the almost routine use of anesthesia—makes the mother more comfortable, and consequently makes birth more relaxed and natural (Eakins, 1986; Kitzinger, 1983).

Responses to the medicalization of birth vary, not only from person to person but also from place to place. Some countries, notably, England and the Netherlands, and some areas of the United States, such as California, are much more receptive to home births than others. Recently, many American hospitals have responded to the home-birth trend by setting up their own "birthing rooms," pleasant bedroom environments especially equipped for childbirth. Most large cities also have maternity centers that allow for a relaxed and natural birth in an informal setting while at the same time providing equipment and personnel to meet any emergency that might arise.

Nonetheless, many women are suspicious of any medical measure, while some doctors are so understandably wary of potential lawsuits that they intervene at the first hint of a problem. In the United States, malpractice insurance is higher for obstetricians than for other medical specialties, not because these doctors are more negligent, but because, when problems do occur, the tendency is to blame the physician rather than the birth process. As we have seen, however, most complications apparent at birth originate in the months before.

The Parents' Experience

A number of biological and medical factors interact to determine whether a pregnancy and birth is fairly simple or beset with complications. Psychological factors, however, are even more important in determining the mother's and the father's overall experience, and, ultimately, the baby's as well. The effect of a particular pregnancy on the psyche of the parents is subject to enormous variation. Some couples come together with more closeness than they ever experienced before (Kitzinger, 1983). Others fall apart: amazingly, the incidence of wife-beating seems to increase during pregnancy (Gelles, 1975).

Similarly, psychological factors can make a long labor exhilarating, and a short one terrifying. They can make both parents swear "never again" after what physicians would call an easy birth, or can make even an emergency Cesarean so rewarding that the couple are ready to plan a repeat experience. Some factors relevant to the parents' experience predate the pregnancy: for example, the quality of the relationship between the mother and the father and their readiness and willingness to take on the responsibilities of parenthood.

And there is no doubt that pregnancy itself is stressful, even in the best of circumstances. One study of healthy, married women, all pregnant with a desired first child, reports "the overall findings support a view of pregnancy as a turbulent, difficult period rather than one of calm bliss" (Leifer, 1980). In addition, especially for first-time parents-to-be, pregnancy changes the marital relationship and transforms the couple's status in the generational sequence: no longer are they primarily husband and wife and someone's children; they are soon to be someone's parents. First pregnancies may also change the way the parents-to-be think about themselves: many young adults do not consider themselves truly adult until they are cast into the role of parent (Leifer, 1980).

Couples having their second or third child experience stress during pregnancy as well. Indeed, one study found that women experienced second and subsequent pregnancies more negatively than first ones (Westbrook, 1978). One reason for this is that the parents were much more aware of the costs and responsibilities of parenthood than they were as first-time parents-to-be. Another reason is that relatives and friends tend to take a greater interest in, and to be more supportive during, the first pregnancy (Kitzinger, 1983).

We should point out that although fathers-to-be may need as much social support as mothers-to-be, they are less likely to get it. For one thing, their imminent parenthood is not visible, so their friends, family, and co-workers are less likely to express interest in the pregnancy. For another, men at all points of the life span are less likely to reveal their need for help. Especially as fatherhood approaches, many men make a special effort to appear strong and protective, frequently camouflaging their understandable feelings of panic by springing into action. Some build a crib, a room, or even a house; others become intensely involved in physical fitness or sports; others eat too much or develop physical symptoms of their own (Jackson, 1984).

There are many ways friends and relatives can help the expectant couple, and each culture provides many forms of education for pregnancy and birth. We will discuss one that is potentially very useful in allaying the parents' fears.

Childbirth Classes   First-time parents-to-be often approach childbirth with negative feelings picked up from television dramas or novels. Indeed, a recent study (Leifer, 1980) of women who were pregnant for the first time found that almost all had negative attitudes about giving birth. Some attributed their apprehension to television programs in which, as one woman put it, "whenever they have a woman bearing a child, it seems like she's screaming horribly or she's fainting, she can't

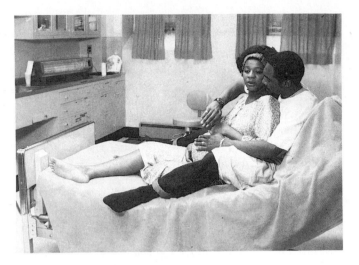

Figure 4.14 *A husband's psychological—and physical—support can help to ensure a safe, speedy, and satisfying birth. In addition to keeping his wife in the position that is most comfortable for her, the husband coaches her in breathing and pushing.*

control herself." Others had picked up their attitudes from their mothers and older women whose view generally seemed to be "It's horrible at the time but . . . you soon forget it."

Fortunately, this situation is changing as more and more parents-to-be prepare for birth, not only by learning about the natural processes and medical techniques involved in birth (knowledge that, in itself, reduces fear, tension, and therefore pain [Dick-Read, 1972]), but also by practicing specific breathing and concentration techniques known as the **Lamaze method** of childbirth. This method has important psychological as well as physical benefits: both parents are often proud of their active involvement in the safe arrival of their infant (Bing, 1983). Indeed, when both partners attend childbirth classes and when the husband as well as the wife is an active participant in the labor and birth process, the result is less pain, less anesthesia, shorter labor, and more positive feelings about birth and about oneself (Felton and Segelman, 1978; Leifer, 1980) (see A Closer Look, next page).

Not all parents-to-be prepare for childbirth, however. Middle-class parents are far more likely to attend classes than lower-class parents, especially when the lower-class parents are also young and from minority groups. Partly for this reason, the less education and income women have, the more likely they are to experience pain, loneliness, and confusion during childbirth (Ball, 1987; Oakley, 1980).

The Father's Participation   Before 1970 fathers were rarely allowed in the delivery room. It was thought that, at best, they would merely be in the way and might even disrupt the birth process by becoming faint or ill.

Fortunately, hospital regulations are changing. Increasingly, fathers are present when their infants are born, even when medical intervention, such as a Cesarean, is needed. This is true, not only in North America, but in many other countries around the world (Lamb, 1987). A review of fatherhood in Britain explains:

the presence of the father at birth is so clearly expected in Britain that it is probably as hard for a man to stay out of the delivery room as it was for him to get in it only a decade ago. The extent to which this has now become conventional was underlined by the widely reported presence of the Prince of Wales at the birth of both his sons. At the time of the Prince's own birth, as several newspapers remarked, the Duke of Edinburgh was playing squash. [Jackson, 1987]

Prepared Childbirth

Forty years ago in North America and Western Europe, it was common obstetrical practice to administer general anesthesia during delivery, so most women who gave birth in hospitals were unconscious when their babies were born. Today, general anesthesia is rarely used in normal births, partly because women are now being taught to use psychological techniques to relieve pain. The specific methods are based on principles of conditioning interpreted and popularized by a French obstetrician, Fernand Lamaze.

Lamaze began his practice using the conventional procedures of his time, delivering babies for women who were too heavily anesthetized to realize they were giving birth. In 1951, Lamaze traveled to the Soviet Union and saw many women giving birth without medication and without the screams of agony he had assumed were inevitable unless drugs were used. He·learned that Soviet doctors had applied Pavlov's theory of classical conditioning (see page 39) to childbirth, teaching women to associate birth with pleasant feelings and mental images, such as a peaceful rural scene, and to lessen discomfort by using specific breathing techniques. These techniques were then practiced so often before birth that they became a conditioned response during birth. In short, women concentrated on the work, rather than the pain, of having a baby.

Lamaze returned to Paris and changed his obstetrical methods, stressing the mother's *active* participation in, and control of, the birth process. By chance, one of his first patients was Marjorie Karmel, an American. Early in pregnancy she "was not interested in natural childbirth— not even if they were giving it away." Later she decided "to string along for a week or two," and joined a class in the Lamaze method. In the end, she felt so "exhilarated and excited" after delivering her first baby through the Lamaze method that she wrote a best-selling book about the experience.

Over the past three decades, millions of North Americans have read Karmel's *Thank You, Dr. Lamaze* (1983, originally published 1959) and classes to prepare for a Lamaze childbirth are now common.

One important part of the Lamaze method is that each woman has a "coach," someone who stays with her throughout labor and birth to time her contractions, help her regulate her breathing, massage her back, and help her focus on the work of giving birth. Although originally the coaches were the women who taught the Lamaze classes, within the past twenty years fathers have increasingly taken over this role and done very well at it, as can be inferred from one father's description of his experience:

. . . Hard contractions had developed and Alice was in back labor . . . I became a one-man band, actually breathing and blowing louder than Alice, my right hand pounding out the rhythm on her arm, while my left hand did a wild effleurage (massage) on her back. Her urge [to push] was almost unbearable, and occasionally she gave way, but I put her back on the track by saying firmly, "Go back! Start over! Stay on top!"

. . .

The most exciting moment came for me at the end of transition, when, with my hand against Alice's back, I felt the baby turn—shoulders, elbows, and all. How fantastic to know what it was!

And then we pushed Alice into the delivery room, where it only took 20 minutes before the baby was born, and Alice claimed it was sheer heaven. [Quoted in Bing, 1983]

As the teacher explains the birth process, these husbands learn how to support their wives when it is time to push the fetus through the birth canal. Many couples have found that classes like this one not only make birth easier but also bring them closer together. In addition, the sharing of exercises, anticipations, and anxieties often begins friendships between couples who otherwise would not have met.

The results of such participation by the father are generally very positive. One father, who had read Maslow, found birth a "peak experience":

> I administered oxygen to her [his wife] between contractions and coached her on pushing, holding her around the shoulders as support during each push. She was magnificent. Slowly I began to feel a kind of holiness about all of us there, performing an ageless human drama, a grand ritual of life. The trigger was probably the emergence of the baby's head—coughing, twisting, covered with blood, as purple as error, so eager for life—that set me into such intensities of joy and excitement that I cannot possibly adequately describe them. It was all so powerful I felt as though my head might come off, that I might simply explode with joy and a sense of participation in a profound mystery. I did explode, was literally reborn myself, saw how my birth, all births, the idea of birth, is profoundly right, good, joyous. [Quoted by Tanzer and Block, 1976]

Not all husbands are so moved by watching their children born, of course, but the overall reaction to birth is almost always good. The same is true for mothers. While actual birth experiences vary, many more find it positive than negative. In fact, in one large British study of women who had, the day before, given birth in hospitals, 29 percent of the new mothers found birth "the best experience of my life," 26 percent found it "a good experience," 37½ percent characterized birth as an important experience; and only 1½ percent deemed it "the worst experience of their lives" (the remaining 6 percent did not reply) (Ball, 1987). In this study, the type of birth—vaginal or Cesarean, spontaneous or induced—did not correlate as strongly with the women's opinion of the birth experience as one other surprising factor: whether or not the woman was allowed to feed her baby in the first hour after delivery. This brings us to a controversial topic—bonding.

Bonding between Baby, Mother, and Father

In recent years, one topic in human development that has captured much attention is bonding between parents and newborn children. The term **parent-infant bond** is used to emphasize the tangible, as well as the metaphorical, fastening of parent to child in the early moments of their relationship together.

Animals and Bonding

Questions about the nature of the parent-infant bond in humans actually arose from animal studies that revealed the formation of a quite specific bond between mother and newborn in virtually every species of mammal. Animal mothers, for instance, nourish and nurture their own young and nearly always ignore, reject, or mistreat the young of others. How, exactly, is the bond formed? Maternal hormones released during and after birth, the smell of the infant, and the timing of the first contact all play a role. In many animals, early contact between mother and infant can be crucial. For example, if a baby goat is removed from its mother immediately after birth, and returned a few hours later, the mother sometimes rejects it, kicking and butting it away. However, if the baby remains with her for the critical first five minutes, and then is separated from her, she welcomes it back. Many other animals react in like fashion (Klaus and Kennell, 1976).

Bonding in Humans

Does a similar critical period exist for humans? While researchers recognize that human behavior is much less biologically determined than animal behavior, nevertheless, they have tried to determine whether the amount of time mothers spend with their newborns in the first few days makes any difference in the mother-child bond. In certain cases, early contact does seem to be important, for example, when mothers are young, poor, or otherwise under special stress (Klaus and Kennell, 1976), or in the case of preterm infants who seem too forbiddingly frail, too dependent on life support devices to hold or play with (Liefer et al., 1972). However, for most experienced mothers with healthy babies, immediate mother-infant contact does not seem to make any lasting difference in the relationship (Macfarlane, 1977). Indeed, several reviews express concern that the importance of early contact between mother and child will be overemphasized, resulting in feelings of guilt and blame and sorrow when a woman, for any reason, does not spend much time with her infant in the early days of life (Lamb and Hwang, 1982; Myers, 1984).

Relatively little research has been done on the question of fathers and bonding. In one study of the effects of early contact, some fathers were allowed to hold their infants for 10 minutes immediately after the birth, while others were granted only the usual glimpse of the infant (Rodhölm, 1981). In a second study, fathers were given 4½ hours alone with their infants during the hospital stay (Keller et al., 1981). Three months later, the fathers in both studies who had had extra contact engaged in more face-to-face play with the babies. However, a review of all the evidence finds mixed results: whether or not a father had extended contact with his newborn may not, in fact, affect his future involvement with his baby (Palkovitz, 1985).

Although early contact between parents and newborns may be beneficial, contact is not essential for bonding to occur. Even Klaus and Kennell, the original researchers in human bonding, now emphasize that the events right after birth are just one episode in a long-term process of bonding between parent and child (Klaus and Kennell, 1982). As another review of the literature concludes: "Separation of the mother and child in the neonatal period may have effects on maternal behavior which last a few months but it is unusual for effects to persist longer than that . . . it is clear that both mothers and fathers can, and commonly do, develop strong attachments to their children in the absence of neonatal contact" (Rutter, 1981). Research increasingly gives evidence that the bond between parent and child begins to grow or atrophy from the first days of pregnancy, throughout infancy, the many years of childhood, and beyond (Ball, 1987; Vaughn, 1987).

SUMMARY

From Zygote to Newborn

1. The first two weeks of prenatal growth are the germinal period. During this period, the zygote grows to an organism more than a hundred cells in size, travels down the Fallopian tube, and implants itself in the uterine lining, where it continues to grow.

2. The period from two to eight weeks after conception is the period of the embryo. The development of the embryo is cephalo-caudal (from the head downward) and proximo-distal (from the inner organs outward). At eight weeks after conception, the future baby is only about an inch long.

3. From the eighth week after conception until birth is the period of the fetus. The fetus grows rapidly; muscles develop and bones harden. The sex organs take shape, and the other organs complete their formation. The fetus attains viability when the brain is sufficiently mature, which happens sometime between the twentieth and the twenty-sixth week after conception.

Birth

4. Birth typically begins with contractions that push the fetus headfirst out from the uterus and then through the vagina.

5. The Apgar scale, which rates the neonate's vital signs at one minute after birth, and again at five minutes after

birth, provides a quick evaluation of the infant's health. Although neonates might sometimes look misshapen, most are healthy, as revealed by a Apgar score of 7 or more.

Problems and Solutions

6. Teratogens (substances that can cause birth defects) can affect fetal development throughout pregnancy. During the critical period, the first two months of development, the embryo is most susceptible to observable harm from teratogens. However, teratogens can be harmful throughout prenatal development. The interaction of risk factors with teratogens further multiplies the chance of birth defects.

7. Several diseases and a large number of drugs are teratogenic. Two of the most commonly used drugs, alcohol and tobacco, have been proven harmful to fetuses. Radiation and certain pollutants in large doses can damage fetal development.

8. Preterm newborns are likely to be low-birth-weight infants, weighing less than 5½ pounds. They often suffer from many medical problems, including difficulty in obtaining oxygen. Other infants are low in birth weight even though they are not born early. Such infants are called small-for-dates. Environmental, prenatal, genetic, and economic factors contribute to the incidence of low-birth-weight infants, and to their future development.

9. Variations in the birth process can result in a birth that is stressful for both mother and child. The use of fetal monitoring provides early warning of these problems. If necessary, delivery can be hastened by Cesarean section, drugs, or other medical procedures. Although none of these procedures is risk-free, they often prevent possible damage to mother and child.

10. While biological factors are the primary determinants of birth complications and the length of labor, psychological factors play a large role in the parents' overall experience of birth. Women who are prepared for birth—who know what to expect and how to make labor easier—and who have the support of their husbands and other birth attendants are most likely to find the birth experience exhilarating.

11. The extent to which fathers participate in the birth process affects how they feel about it. Fathers who are encouraged to attend Lamaze classes and to see their babies being born are more likely to be thrilled by the birth process. Women whose husbands are present throughout delivery feel less pain and use less medication.

Bonding between Baby, Mother, and Father

12. For most experienced mothers with healthy babies, immediate mother-infant contact does not make a lasting difference in the relationship. For first-time mothers or those who are young or poor or under special stress, extended early contact may be linked to more maternal attention and affection later on.

KEY TERMS

germinal period *(77)*	critical period *(83)*
period of the embryo *(77)*	rubella *(85)*
period of the fetus *(77)*	toxoplasmosis *(85)*
implantation *(78)*	eclampsia *(85)*
neural tube *(78)*	fetal alcohol syndrome (FAS) *(86)*
cephalo-caudal development *(78)*	full term *(88)*
proximo-distal development *(78)*	low-birth-weight infant *(88)*
placenta *(79)*	preterm *(89)*
trimester *(80)*	small-for-dates *(89)*
age of viability *(80)*	anoxia *(92)*
neonates *(81)*	fetal monitor *(92)*
Apgar scale *(81)*	Cesarean section *(92)*
teratology *(82)*	Lamaze method *(95)*
teratogens *(82)*	parent-infant bond *(97)*

KEY QUESTIONS

1. What parts of the body develop during the period of the embryo?

2. What parts of the body develop during the period of the fetus?

3 What vital body signs does the Apgar scale measure? What does the Apgar score tell about the health of the newborn?

4. During which trimester of pregnancy is the developing fetus most susceptible to teratogens? Why?

5. What factors make a fetus more likely to be harmed by teratogens?

6. What are the effects of maternal alcohol consumption on the fetus?

7. What are the most serious problems of low-birth-weight infants?

8. What are the advantages and disadvantages of medication administered during childbirth?

9. What are the advantages of Lamaze courses for the parents-to-be?

10. What are the advantages of the father's presence during the delivery?

11. How is the formation of the parent-infant bond different in animals than in humans?

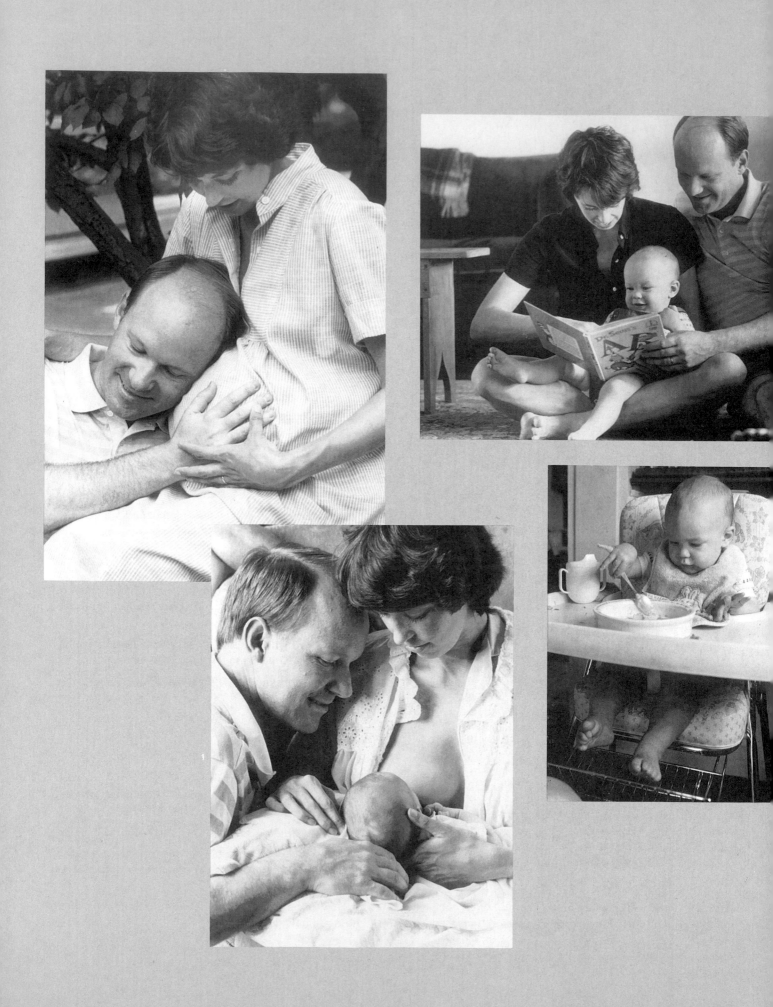

PART II

The First Two Years: Infants and Toddlers

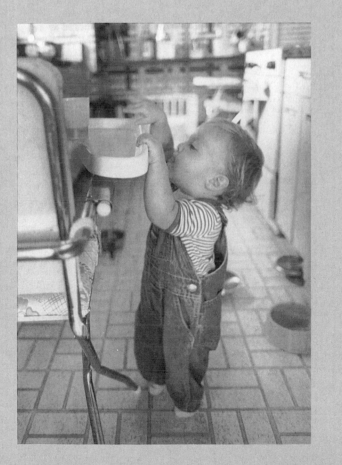

Adults usually don't change much in a year or two. Sometimes their hair gets longer or grows thinner, or they gain or lose a few pounds, or they become a little wiser or more mature. But if you were to be reunited with some friends you hadn't seen for several years, you would recognize them immediately.

If, on the other hand, you were to care for a newborn, twenty-four hours a day for the first month, and then didn't see the baby again until a year or two later, the chances of your recognizing that child are similar to those of recognizing a best friend who had quadrupled in weight, grown 14 inches, and sprouted a new head of hair. Nor would you find the toddler's way of thinking, talking, or playing familiar. A hungry newborn just cries; a hungry toddler says "more food" or climbs up on the kitchen counter to reach the cookies.

While two years seem short compared with the more than seventy years of the average life span, children in their first two years reach half their adult height, complete the first of Piaget's four periods of cognitive growth, and have almost finished the second of both Freud's and Erikson's sequences of stages. Two of the most important human abilities, talking and loving, are already apparent. The next three chapters describe these radical and rapid changes.

CHAPTER 5 | # The First Two Years: Physical Development

He who would learn to fly one day must first learn to stand and walk and run and climb and dance: one cannot fly into flying.

Nietzsche
"On the Spirit of Gravity"

An infant's physical development happens so rapidly that size, shape, and skills seem to change daily. This is no exaggeration. Pediatricians expect normal neonates to gain an ounce a day for the first few months, and parents who keep a detailed baby diary record new achievements every day, such as taking a first step on Monday, taking two steps on Tuesday, and five steps by the weekend. Let us look at some other specifics of this rapid growth.

Size and Shape

The average North American newborn measures 20 inches (51 centimeters) and weighs a little more than 7 pounds (Lowrey, 1978). This means that the average neonate is lighter than a gallon of milk, and about as long as the distance from a man's elbow to the tips of his fingers. In the first days of life, most newborns lose about 10 percent of their body weight before their bodies adjust to sucking, swallowing, and digesting on their own.

Once they have made these adjustments, infants grow rapidly, doubling their birth weight by the fourth month, tripling it by the end of the first year, and growing about an inch longer each month for the first twelve months. By age 1, the typical baby weighs about 22 pounds (10 kilograms) and measures almost 30 inches (75 centimeters) (Lowrey, 1978).

Growth in the second year proceeds at a slower rate. By 24 months, most children weigh almost 30 pounds and measure between 32 and 36 inches (81 to 91 centimeters), with boys being slightly taller and heavier than girls. In other words, typical 2-year-olds are almost a fifth of their adult weight and half their adult height (see Figure 5.1).

Much of the weight gain in the first months of life is fat, providing both insulation and a source of nourishment that can be drawn on should teething or other prob-

Figure 5.1 *These figures show the range of height and weight of American children during the first two years. The lines labeled "50th" (the fiftieth percentile) show the average; the lines labeled "90th" (the ninetieth percentile) show the size of children taller and heavier than 90 percent of their contemporaries; and the lines labeled "10th" (the tenth percentile) show the size of the relatively small children, who are taller or heavier than only 10 percent of their peers. Note that girls (color lines) are slightly shorter and lighter, on the average, than boys (black lines).*

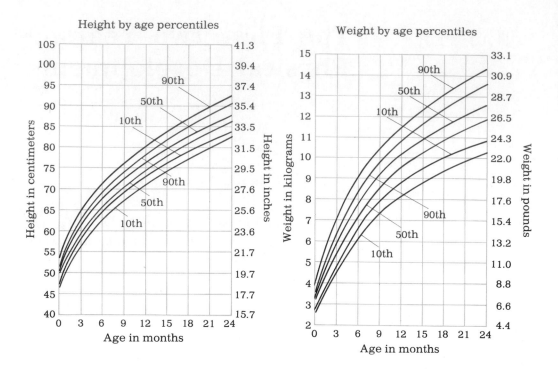

Height by age percentiles

Weight by age percentiles

lems cut down on food intake for a few days. After 8 months or so, weight gain includes more bone and muscle. (Indeed, once they start walking, most children lose fat rather than gain it, and the toddler's pudgy cheeks and potbelly gradually disappear.)

Proportions

The growth that changes the baby's body shape in the first two years follows the head-downward (cephalo-caudal) and center-outward (proximo-distal) direction of development. Most newborns seem top-heavy because their heads are equivalent to about one-fourth of their total length, compared to one-fifth at a year and one-eighth in adulthood. Their legs, in turn, represent only about a quarter of their total

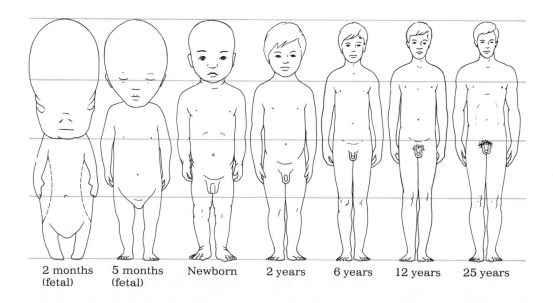

Figure 5.2 *As shown in this figure, the proportions of the human body change dramatically with maturation, especially in the first years of life. For instance, the percentage of total body length below the belly button is 25 percent at two months past conception, about 45 percent at birth, 50 percent by age 2, and 60 percent by adulthood.*

2 months (fetal)   5 months (fetal)   Newborn   2 years   6 years   12 years   25 years

body length, whereas an adult's legs account for about half. Proportionally, the smallest part of a newborn's body is that part farthest from the head and most distant from the center—namely, the feet. By adulthood, the feet will be about five times as long as they were at birth, while the head will have only doubled. Even the head grows more at the bottom than at the top; the newborn's underdeveloped chin will not reach its final proportion relative to the skull until adolescence.

Figure 5.3 (a) *Areas of the brain are specialized for the reception and transmission of different types of information.*

Research has shown that both experience and maturation play important roles in brain development. For example, myelination of the nerve fibers in the visual cortex of the brain will not proceed normally unless the infant has had sufficient visual experience in a lighted environment.

The role of maturation is apparent in the growth and development of the neurons that make up the nerve fibers. These cells increase in size and in the number of connections between them as the infant matures, enabling impressive increases in the control and refinement of actions. The cross-sectional drawings in (b) and (c) show the development of nerve fibers in the visual cortex between birth and 1 year. Drawings (d), (e), and (f) illustrate changes in the neurons themselves.

Brain Growth and Maturation

One reason that the newborn's skull is relatively large is that it must accommodate the brain, which at birth has already attained 25 percent of its adult weight. The neonate's body weight, by comparison, is only about 5 percent of its adult weight. By age 2 the brain is about 75 percent of its adult weight, while the 2-year-old's body weight is only about 20 percent of what it will be in adulthood.

Weight, of course, provides only a crude index of brain development. More significant are the changes in the maturing nervous system, which consists of the brain, the spinal cord, and the nerves. The nervous system is made up of long, thin, nerve cells called **neurons.** At birth, it contains most of, or perhaps all, the neurons it will ever have. Further development consists primarily in the growth and branching of these cells into increasingly dense connective networks that transmit messages— in the form of electrical impulses—between the brain and the rest of the body. Almost every bodily function—from breathing and heartbeat, to seeing and hearing, to sleeping and waking—is regulated by this interchange of information. As the nervous system matures, the neurons become coated with a fatty insulating substance called **myelin,** which helps to transmit neural impulses faster and more efficiently. The myelination process continues until adolescence (Guthrie, 1980; Schwartz, 1978). The development of the nervous system allows children to gain increasing control over their motor functions and to experience refinements in perceptual abilities.

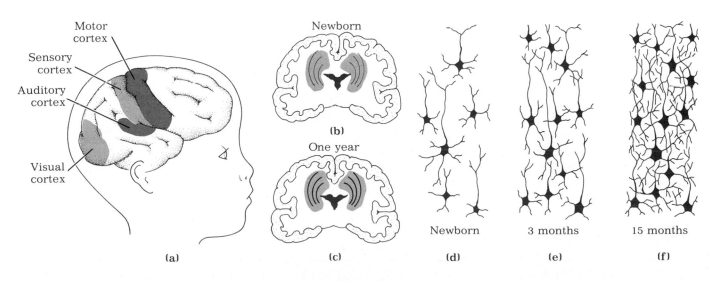

Motor cortex
Sensory cortex
Auditory cortex
Visual cortex

Newborn
(b)
One year
(c)

Newborn
(d)

3 months
(e)

15 months
(f)

(a)

Development of the Sensory and Motor Areas

In the first months, brain development is most rapid in the primary sensory areas and the primary motor areas—the areas that control the senses and simple body movements. While development of these areas of the brain is essential to infants' increasing ability to use their senses and move their bodies, it is also true that sensory stimulation and motor activity are essential to the normal development of these areas. This fact is most clearly shown by research in which animals that were prevented from using their senses or moving their bodies in infancy became permanently handicapped (Parmelee and Sigman, 1983). Chimpanzees who were kept

---

A CLOSER LOOK   **Sudden Infant Death**

Each year in developed countries about one infant in every 350 falls asleep and never wakes up (Bergman, 1986).* These seem to be healthy, normal babies, but then, inexplicably, they die. After examination and autopsy, they are diagnosed as dying of *sudden infant death syndrome* (SIDS) (also called crib death or cot death), but these terms are merely description: the actual cause of death is still unknown.

This uncertainty makes it particularly hard for parents to accept a SIDS death: in the absence of a clearcut explanation, they often blame themselves. Neighbors, relatives, and even the police sometimes cast blame as well, partly because the sudden cessation of breathing often results in bruises and other symptoms that suggest abuse (Bergman, 1986). One woman describes what happened when her baby died of SIDS:

My husband and my father (who were in the house at the time) were both taken to the police station and asked a lot of questions. My husband was brought home and made to re-enact the scene using a teddy bear. The police also interviewed the neighbours whom my husband had gone to for help. It was terrible what my husband went through when the baby died, but it was made much worse by the way he was treated by the police. [quoted in Golding et al., 1985]

Such experiences could be avoided if medical personnel quickly performed an autopsy, if necessary, to confirm that the death was SIDS. Preventing the death itself, however, is not easy at all.

Until the cause or causes of these deaths can be determined, many researchers are attempting to identify potential SIDS infants before the fatal moment. As the accompanying table shows, many diverse factors correlate with a higher incidence of SIDS (Golding et al., 1985; Guilleminault et al., 1982). Furthermore, families that have already lost one baby to SIDS have a much higher risk of losing another, which suggests that part of the cause is genetic.

Factors Correlated with SIDS

	SIDS More Likely	SIDS Less Likely
Characteristics of Mother		
Age	Under 20	Over 25
Blood type	O, B, or AB	A
Personal habits	Smoker	Nonsmoker
Income	Low	High
Characteristics of Infant		
Sex	Male	Female
Birth order	Later-born	First-born
Birth weight	Under 5½ lbs	Over 5½ lbs
Apgar score at 1 minute	7 or lower	8 or higher
	Twin or triplet	Singleborn
Characteristics of Pregnancy		
Mother's health	Urinary or respiratory infection	No complications
Length of pregnancy	Less than eight months	Full-term
Mother's nutrition	Anemic (low iron)	Not anemic
Situation at Time of Death		
Time of year	Winter	Summer
Age of infant	2–4 mos.	Under 1 mo., more than 6 mos.
Infant health	Has a cold, with stuffy nose	Has no cold, no runny nose
Feeding	Bottle-fed	Breast-fed

---

*The rate in poorer countries may be even higher, but there infant deaths are rarely autopsied, so the diagnosis is rarely made.

in castlike restraints to prevent them from moving did not climb when they were released. Dogs who were prevented from experiencing pain actually burned their noses by sniffing a lit candle when they were allowed free rein.

In very simple terms, these abnormalities occur because the deprivation of certain basic sensory experiences prevents the development of the normal neural pathways that transmit sensory information. As researchers explain it metaphorically, the "wiring" of the brain—that is, the basic structures that allow the development of specific capacities—is genetically programmed and present at birth. What

---

Since the characteristics shown in the table provide only a rough profile of infants at risk, researchers are trying to pinpoint the causes of SIDS in order to prevent it completely. The most promising line of research looks at patterns of electrical activity in the brain (*brain waves*), sleep cycles, blood proteins, and breathing patterns in early infancy that correlate with SIDS (Giulian et al., 1987; Golding et al., 1985). Relatively minor neurological abnormalities at birth, as reflected in brain-wave activity and in certain types of behavior (e.g., particular ways of crying [Lester and Zeskind, 1982; Zeskind and Lester, 1978]), may indicate an infant who is at risk for SIDS. Unfortunately, current research finds that few SIDS infants have patterns so clearly abnormal that they can be reliably distinguished from normal infants (Gordon et al., 1986).

If high-risk babies can be identified, measures can be taken to minimize their susceptibility. For example, since the babies who die of SIDS often have a slight cold at the time, parents whose babies are at risk can take such precautions as preventing contact between the baby and anyone who has a cold and using a humidifier in the baby's room during winter to make breathing easier. The parents can also be trained in artificial respiration and in the use of a special monitor that signals when the baby takes an overlong pause between breaths. However, monitors are not always accurate, sometimes ringing when there is no cause and sometimes ringing too late. According to several studies, they provide more interrupted sleep and false assurance than anything else (Golding et al., 1985).

An additional measure comes from an intriguing hypothesis advanced by Lewis Lipsitt. He believes that the learning experiences of the infant may help protect them against SIDS. Lipsitt (1982) points out that "eventual SIDS victims often have generally subdued activity. They suck weakly, move less, respond with higher thresholds to noxious events, and, in general, engage their environment less. Such infants *may* subject themselves to fewer opportunities for learning than normal infants do." This hypothesis would explain several interesting correlations between a baby's experiences after birth and the incidence of SIDS. For instance, breast-fed babies are less vulnerable to SIDS. In part, this may be because they get more practice breathing through their noses than bottle-fed babies do, since breast-feeding requires more constant sucking. This practice may make it easier for them to breathe when they have a cold (Lipsitt, 1982).

Similarly, the fact that first-born babies seem less vulnerable to SIDS may be explained by the fact that parents tend to provide more stimulation for the first child than for later babies. This stimulation may help compensate for the passivity of those infants who are at risk of SIDS because "they engage their environment less" (Lipsitt, 1982).

Lipsitt's hypothesis is as yet unproven. However, by the time an infant is 1 year old, the risk of SIDS decreases markedly, a fact that may be attributed in part to physiological maturation, especially brain development, and in part to the infant's experiences, another example of how inborn vulnerabilities may be affected by events after birth.

There is no need to worry that this alert and active infant may bury her nose too deeply in her pillow to allow proper breathing.

Most infants are endowed with all the reflexes and instincts required to cope with normal activities.

is required is the "fine-tuning" that occurs with the development of the connective networks, and it is this fine-tuning process that can be affected by experience, or the lack of it.

As best we know, the brain development that permits seeing and hearing in humans likewise becomes "fine-tuned" through visual and auditory experiences in the first months (Imbert, 1985; Parmelee and Sigman, 1983). This is not to say that an infant would not be able to see colors or understand speech unless he or she had an opportunity to do so in the first days of life. Nor does it mean that intensive exposure to visual, auditory, and motor stimulation is desirable in the first weeks. (Indeed, many newborns would react to such stimulation by shutting it out with tears or sleep.) However, it does mean that even in the case of a biologically pro-grammed event such as early brain maturation, experience also plays a role.

Regulating Physiological States

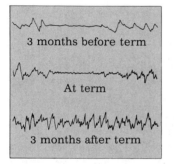

3 months before term

At term

3 months after term

Figure 5.4 *As can be seen from this recording of electrical activity in the brain during quiet sleep, the more mature pattern of brain-wave activity shows many more bursts of electrical activity and greater overall intensity.*

Brain development is also associated with the degree of the infant's responsiveness to the outside world. For example, a sudden loud noise or a moment of pain (as from a pinprick) makes a normal neonate startle and cry. But soon the baby "self-soothes," that is, becomes more relaxed and quiet on its own, as nerve impulses diminish and patterns of electrical activity in the brain, referred to as **brain waves,** display a pattern associated with a calmer state. On the other hand, the less mature nervous system of the preterm infant takes longer, and requires more intense stimuli, to be aroused; and once crying occurs, the preterm infant takes longer to settle down again. Another indication of neurological immaturity is that preterm infants never seem quite awake or quite asleep, and their breathing and muscle activity are irregular. As the brain develops, brain waves and physiological states become more cyclical and distinct. As the weeks go by, for instance, infants are asleep and awake for longer, more regular periods. Brain development may also be the underlying factor in the leading cause of death in babies over 2 months of age, sudden infant death syndrome (see A Closer Look, pages 106–107).

Sensation and Perception

Psychologists draw a distinction between sensation and perception. **Sensation** oc-curs when a sensory system responds to a particular stimulus. **Perception** occurs when the brain recognizes that response so that the individual becomes aware of it. This distinction may be clear to you if you have ever done your homework while playing the stereo and realized that you had worked through an entire album but had actually heard only snatches of it. During the gaps in your "listening," your auditory system was sensing the music—your tympanic membranes, hammers, anvils, stirrups, and the like were vibrating in response to sound waves—but you were not perceiving the music; you were not consciously aware of it.

At birth, both sensation and perception are apparent. Newborns see, hear, smell, and taste, and they respond to pressure, motion, temperature, and pain. Most of these sensory abilities are immature, becoming more acute as the infant develops (Lowrey, 1978).

The perception demonstrated by newborns is very selective. Neonates pay atten-tion to bright lights, loud noises, and objects within a foot of their eyes, and usually screen out almost everything else. Their perceived world is simple—not all the "great, blooming, buzzing confusion" psychologists once believed it to be (James, 1950).

Research on Infant Perception

Over the past twenty years, there has been an explosion of research on infant perception. Technological breakthroughs—from brain scans to computer measurement of the eyes' ability to focus—have enabled researchers to measure the capacities of infants' senses and to gain a greater understanding of the relationship between perception and physiology (Gottleib and Krasnegor, 1985).

The basis of this research is the fact that the perception of an unfamiliar stimulus elicits a physiological response, for example, slowed heart rate, concentrated gazing, and, in the case of infants who have a pacifier in their mouths, intensified sucking. When the new stimulus becomes so familiar that these responses no longer occur, the infant is said to be *habituated* to that stimulus. Employing this phenomenon of **habituation,** researchers have been able to assess infants' ability to perceive by testing their ability to discriminate between very similar stimuli. Typically, they present the infant with a stimulus—say a plain circle—until habituation occurs. Then they present another stimulus similar to the first but different in some detail—say a circle with a dot in the middle. If the infant reacts in some measurable way to the new stimulus (a change of heart rate, a narrowing of the pupils, a refocusing of gaze), the difference in stimulus is taken to have been perceived.

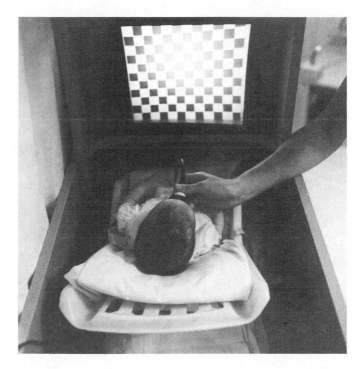

Figure 5.5 *In this experiment in infant perception, the nipple on which the infant is sucking is connected to an element that can focus the image on the screen. Another device records the frequency and strength of the sucking. Typically, infants show their interest by sucking fairly intensely when they see a new image. Then, as they become habituated, their sucking tapers off.*

Vision

At birth, vision is the least developed of the senses. Newborns focus well only on objects that are about 10 inches away. The distance vision of the neonate is about 20/600, which means that the baby can see an object 20 feet (6.1 meters) away no better than an adult with normal 20/20 vision could see the same object 600 feet (183 meters) away. By 4 months, distance vision is 20/150; and by 6 months, visual acuity approaches 20/20 (Salapatek, 1977).

Although we can accurately measure infants' ability to see, it is not easy to tell precisely what they perceive. Do they simply stare at whatever comes into focus? Can they distinguish one object from another? Do they prefer particular things, perhaps a brightly colored rattle or a face?

Figure 5.6 *What do 2-month-olds like to look at? Fantz found that they liked faces best of all. He measured how long 2- and 3-month-olds stared at the six types of disks portrayed here, presented one by one. As this graph shows, the infants stared at the face disk more than twice as long as at the disk of newsprint or the bull's eye, and more than four times as long as at the pattern-less disks.*

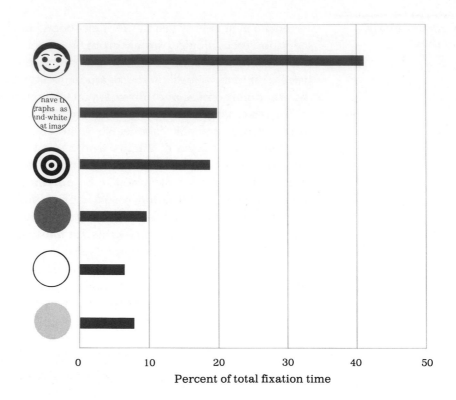

Percent of total fixation time

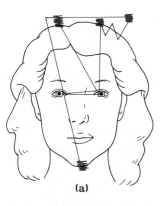

(a)

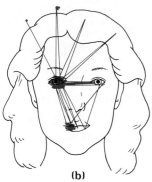

(b)

Figure 5.7 *When researchers recorded patterns of eye movement as infants scanned the human face, differences became apparent between the ages of 1 and 2 months: (a) 1-month-olds concentrated on the edges of the head; (b) 2-month-olds tended to focus on features of the face, especially the eyes.*

Some answers to these questions were provided in a now-classic experiment by Robert Fantz (1961), who showed six types of disks to 2-month-old infants and measured the duration of their gaze at each one. He found that babies looked longer at patterned disks than at plain red, white, or yellow ones, and looked longest of all at the disk of a smiling face (see Figure 5.6). This research led some psychologists to conclude that 2-month-olds can not only see patterns but can also recognize the human face as such, a conclusion that occasioned much celebration of the infant's social skills.

The celebration was premature, however. It is true that very young infants pay more attention to faces than to most other objects they encounter. But further research involving human faces has found that infants are intrigued with faces, not because they recognize them as faces, but because faces have an interesting pattern within an outlined circle (the face framed by the hair) (Caron et al., 1973; Haaf et al., 1983).

There is a developmental shift in precisely what part of the face infants find most interesting: 1-month-old babies look more at the hairline than at the eyes (according to one study, 57 percent compared to 30 percent of total fixation time), but 2-month-old babies look more at eyes than at the hairline (49 percent compared to 33 percent) (Haith et al., 1977). See Figure 5.7.

By 3 months some celebration is in order, for at that age infants respond to their mother's facial expressions and show that they recognize their mother's photograph (Barrera and Maurer, 1981a, 1981b). By 4 months, infants become notably better at using both eyes together (Held, 1985). This makes them much more astute observers, even when looking at photographs. By 6 months, infants who are shown a photo of the same stranger several times are able to distinguish it from a photo of a different stranger. They also can differentiate photos of men from those of women, and they show definite preferences for happy faces over sad or angry ones (Cohen et al., 1979).

Hearing

Relative to his or her vision, the newborn's hearing is quite acute. Sudden noises startle newborns, making them cry; rhythmic sounds, such as a lullabye or a heartbeat, soothe them and put them to sleep. When they are awake, they turn their heads in the direction of a noise (Clarkson et al., 1985), and they are particularly attentive to the sound of conversation. Indeed, by the end of the first month, the infant can distinguish the mother's voice from, and prefers it to, the voices of other women (DeCasper and Fifer, 1980).

By the age of 1 month, infants can also perceive differences between very similar sounds. In one experiment, 1-month-old babies activated a recording of the "bah" sound whenever they sucked on a nipple. At first, they sucked diligently; but as they became habituated to the sound, their sucking decreased. At this point, the experimenters changed the sound from "bah" to "pah." Immediately the babies sucked harder, indicating by this sign of interest that they had perceived the difference (Eimas et al., 1971). It may even be that newborns have some ability to discriminate between vowels (Clarkson and Berg, 1983).

By the age of 4 months, many infants can identify the voices of the most familiar people in their lives. For example, in one experiment (Spelke and Owsley, 1979), infants were placed in a baby seat facing both their parents. While the parents sat expressionless and silent, a loudspeaker alternately played a recording of either the mother's or the father's voice. Infants as young as 3½ months consistently looked at the parent whose voice was being played.

The Development of Motor Abilities

The normal newborn has several important **reflexes,** that is, involuntary physical responses to a given stimulus. Some of these reflexes are essential for life itself, and others disappear in the months after birth.

Three sets of reflexes are critical for survival and become stronger as the baby matures. One set maintains an adequate supply of oxygen. The most obvious reflex in this group is the **breathing reflex.** Normal newborns take their first breath even before the umbilical cord, with its supply of oxygen, is cut. For the first few days, breathing is somewhat irregular, and reflexive *hiccups, sneezes,* and *spit-ups* are common, as the newborn tries to coordinate breathing, sucking, and swallowing.

Another set of reflexes helps to maintain constant body temperature: when infants are cold, they cry, shiver, and tuck their legs close to their bodies, thereby helping to keep themselves warm. A third set of reflexes ensures adequate nourishment. One of these is the **sucking reflex:** newborns suck anything that touches their lips—fingers, toes, blankets, and rattles, as well as nipples of various shapes. Another is the **rooting reflex,** which helps babies find a nipple by causing them to turn their heads and start to suck whenever something brushes against their cheek. *Swallowing* is another important reflex that aids feeding, as is *crying* when the stomach is empty.

The following five reflexes are present in normal, full-term newborns. (1) When their feet are stroked, their toes fan upward **(Babinski reflex).** (2) When their feet touch a flat surface, they move as if to walk **(stepping reflex).** (3) When they are held horizontally on their stomachs, their arms and legs stretch out **(swimming reflex).** (4) When something touches their palms, their hands grip tightly **(grasping reflex).** (5) When someone bangs on the table they are lying on, newborns usually fling their arms outward and then bring them together on their chests, as if to hold on to something, and they may cry and open their eyes wide **(Moro reflex).** All these

Figure 5.8 *At 3 weeks, Joanna displays some of the many reflexes of the newborn—sucking, grasping, running her toes, and raising her arms as she cries.*

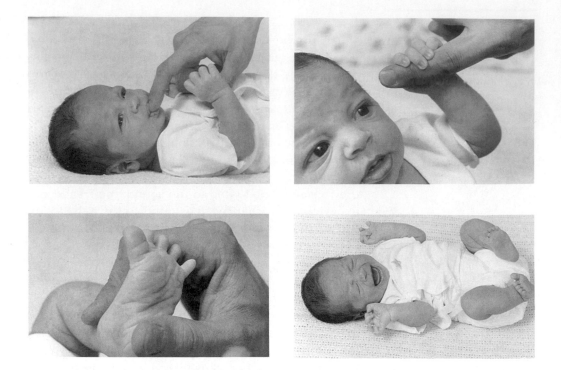

reflexes disappear in the first months of life. (Preterm babies usually develop and lose these reflexes later than full-term babies.) Several of these reflexes are thought to be vestiges of earlier evolutionary development. The grasp reflex, for instance, is crucial for infant monkeys who must hold tight to their mother as she moves from tree to tree. Obviously these reflexes are not necessary in humans today, but they are useful as signs of normal brain and body function.

Indeed, tests of reflexes are frequently used to assess the newborn's physical condition. The most notable of these measures is the **Brazelton Neonatal Behavioral Assessment Scale,** which rates twenty-six items of infant behavior, such as reaction to cuddling, orientation to the examiner's voice and face, trembling, startling, and irritability. The Brazelton scale is a useful diagnostic tool for assessing congenital problems and a good research tool for comparing the physical development of infants born in different cultures or under different circumstances.

Motor Skills

Further evidence of normal brain and body function soon becomes apparent as the infant gradually masters various **motor skills,** that is, control of movements of various parts of the body.

During the months after birth, development in the primary sensory and primary motor areas of the brain transforms infants from dependent creatures who stay in one spot when laid down to active children who wiggle, turn, sit, crawl, stand, and finally walk on their own.

The sequence, if not the timing, of motor skills is the same the world over. To begin with, motor abilities follow the same cephalo-caudal and proximo-distal patterns as physical growth does. Infants lift their heads, and then their heads and shoulders, before they can sit up, and can sit up steadily before they can stand, thus following the head-downward direction of control.

The Sequence of Hand Skills   The specifics of arm, hand, and finger control follow the proximo-distal pattern. As we have seen, at birth, infants have a reflexive grasp, but they seem to have no control of it. Similarly, newborns may wave their arms and hands and feet when they see dangling objects, but rarely do they succeed in hitting them until about their third month. At this point, however, they are still unable to grab the object because they close their hands too soon or too late.

By 6 months, most babies can reach, grab, and hold onto dangling objects (Bower, 1979). But when they successfully grab an object, they have a new problem: they can't let go. A toy seems stuck in their hands until they lose interest in it; then their hands relax and the object, unnoticed, drops out. Generally, letting-go is mastered in the next month or so. By 8 or 9 months, they adjust their reach in an effort to catch objects that are tossed toward them, even when the object is thrown fairly fast and from an unusual angle (von Hofsten, 1983). (Usually the ability to actually catch and hold does not develop until the second year.) Another skill to be mastered is picking up small objects with the fingers. At first, infants use their whole hand, especially the palm and the fourth and fifth fingers to grasp (the *ulnar* grasp). Then they use the middle fingers and center of the palm (*palmar* grasp) or the index finger and the side of the palm (*radial* grasp) (Johnston, 1976). Finally, they use thumb and forefinger together (*pincer* grasp), a skill mastered sometime between 9 and 14 months (Frankenberg et al., 1981). At this point, infants delight in picking up every tiny object within sight, including bits of fuzz from the carpet and bugs from the lawn.

Figure 5.9 *At 1 months, identical twins Adam and Ryan have enough strength and coordination to lift their heads and their shoulders; by 8 months, they can pull their bodies forward in a crawl. With each new motor skill come new opportunities for exploration and pleasure, as Adam and Ryan demonstrate at 9, 10, and 18 months. (Monozygotic twins usually sit up, stand, and walk within a few days of each other. Dizygotic twins often vary by weeks or even months.)*

4 months

8 months

9 months

10 months

18 months

Locomotion   Most babies first learn to move from place to place by lying on their stomachs and pulling themselves ahead with their arms. **Crawling** on hands and knees (sometimes called creeping) involves the coordination of arms and legs and comes later: for some babies, as early as 5 months; for others, as late as 12. Some babies do not crawl at all, achieving mobility instead by either scooting along on their buttocks, rolling over and over, doing the "bear walk" (on all fours, without letting their knees or elbows touch the ground), or even cruising unsteadily on two feet, moving from place to place by holding onto tables, chairs, or bystanders.

On average, a child can stand with support at 5 months, can walk while holding a hand at 9 months, and can walk well unassisted at 12 months (Frankenberg et al., 1981). In recognition of their accomplishment of walking, babies at this stage are given the name **toddler,** although, technically, children are *infants* until they begin to talk. Toddlers are named for the characteristic way they use their legs, toddling from side to side. Since their heads and stomachs are relatively heavy and large, they spread out their short little legs for stability, making them seem bowlegged, flatfooted, and unbalanced.

In fact, 1-year-olds *are* unbalanced, falling frequently. They trip on the edge of a rug, or slip on the grass, or topple over because their head and trunk get too far ahead of their legs when they try to run. But falling doesn't stop them, for they don't have far to fall and they are too excited by their new mobility to be discouraged by a momentary setback. By age 2, most children can walk and run quite well, although they still place their feet wide apart for balance. They can even walk backward and climb stairs, using the two-footsteps-to-each-stair method.

Figure 5.10 *Once the sense of balance develops, there is no end to its possible applications, especially since the excitement of a new accomplishment is not yet accompanied by any thought about possible consequences.*

Variations in Timing

Although all healthy infants develop the same motor skills in the same sequence, the age at which these skills are acquired can vary greatly from infant to infant and still be considered normal. Table 5.1 shows the age at which half of all infants in the United States master each major motor skill, and the age at which 90 percent master each skill. These averages, or **norms,** are based on a representative sample of more than a thousand white, black, Chicano, and Native American infants. Infants known to be brain-damaged or mentally retarded were excluded.

TABLE 5.1 **Age Norms (in Months) for Motor Skills**

Skill	When 50% of All Babies Master the Skill	When 90% of All Babies Master the Skill
Lifts head 90° when lying on stomach	2.2	3.2
Rolls over	2.8	4.7
Sits propped up (head steady)	2.9	4.2
Sits without support	5.5	7.8
Stands holding on	5.8	10.0
Walks holding on	9.2	12.7
Stands momentarily	9.8	13.0
Stands alone well	11.5	13.9
Walks well	12.1	14.3
Walks backward	14.3	21.5
Walks up steps	17.0	22.0
Kicks ball forward	20.0	24.0

Source: The Denver Developmental Screening Test (Frankenburg et al., 1981).

Nutrition

Obviously, nutrition is an important topic in the physical development of the infant, for adequate nutrition is a prerequisite for the changes in size and shape, the brain development, and the skill mastery that we have just described.

The Early Months

At first, infants are unable to eat or digest solid food, but their rooting, sucking, swallowing, and breathing reflexes make them well adapted for consuming the quantities of liquid nourishment that they need.

In these early months, breast milk is the ideal infant food (Eiger, 1987; Jelliffe and Jelliffe, 1977). It is always sterile and at body temperature; it contains more iron, vitamin C, and vitamin A than cow's milk; and it also contains antibodies that provide the infant some protection against any disease that the mother herself has had, or has been inoculated against, from chickenpox to small pox, tetanus to typhoid. In addition, breast milk is more digestible than cow's milk or formula, which means that breast-fed babies have fewer allergies and digestive upsets than bottle-fed babies, even when both groups of babies have similar family backgrounds and excellent medical care (Larsen and Homer, 1978).

Despite the advantages of breast milk, breast-feeding is no longer the most common method of feeding infants. Sterilization, pasturization, the rubber nipple, the plastic bottle, canned milk, powdered milk, and premixed infant formulas available in handy six-packs have meant that millions of modern babies survive and thrive without ever tasting breast milk.

Nutrition after Weaning

In order to thrive during the first two years, infants need about 50 calories per day per pound of body weight (110 per kilogram) (Lowrey, 1978). By 6 months or so, some of these calories can come from "solid" foods that may gradually be added to the diet. Cereals are needed for iron and B vitamins, fruits for vitamins A and C, and, when these first solids are well-tolerated, vegetables, meat, fish, and eggs can be introduced to provide additional nutrition. By the time the infant is a year old, the diet should include all the nutritious foods that the rest of the family consumes.

A CLOSER LOCK   **Breast versus Bottle**

If breast milk is best, why do many women choose to give their infants formula? The reasons have little to do with nutrition directly, but are greatly influenced by the cultural attitudes and social pressures of our modern world. While breast-feeding is usually the better choice in terms of the infant's health, it is often the more difficult choice to make.

Many women find that, even if they want to breast-feed, it may not be easy. Problems may begin in the hospital if procedures make it impossible for the mother to have her newborn near her day and night, so that she can nurse whenever the infant is hungry. Cultural attitudes may make breast-feeding inconvenient, if not impossible, except in the privacy of one's own home.

Another consideration is that the convenience afforded by bottle-feeding is well-suited to many contemporary women, who feel they do not have the time to devote to breast-feeding. In order for breast-feeding to succeed, especially in the early weeks, nursing should occur every two or three hours or even more often if the baby demands it (Riordan, 1983), with each feeding lasting twenty minutes or more. For many women who work outside the home or whose daily activities require them to be out in public much of the time, meeting this kind of schedule is difficult.

In addition, most women today are aware that traces of whatever drug a breast-feeding mother ingests—cigarettes, alcohol, birth-control pills, and so forth—will show up in her milk. Consequently, many women decide that, rather than feeling guilty about compromising the quality of their milk or being as careful after pregnancy as they had to be before, they would rather feed their baby formula.

Fathers are influential, too. Some men are jealous of the close relationship that exists between a nursing mother and her infant, a relationship that seems to exclude them. Some contemporary fathers want to be involved in all aspects of infant care right from the start, and, especially if they are unaware of the advantages of breast-feeding, prefer that their child be bottle-fed so they can sometimes do the feeding.

Nonetheless, while the popularity of bottle-feeding increased dramatically from the beginning of the twentieth century until about 1970, over the past ten years a shift in maternal attitudes, hospital practices, and social approval has resulted in an increase in the number of women who breast-feed.

This trend is applauded and encouraged by virtually all professionals interested in child development, for breast-feeding not only provides good nutrition but also aids in the mother-infant relationship. While breast-feeding does not guarantee a good relationship—any more than bottle-feeding precludes it—the consensus among developmentalists is that breast-feeding generally fosters good maternal care.

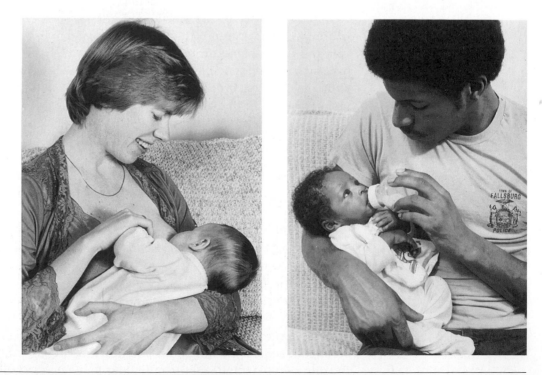

While convenience is one reason for bottle-feeding, some mothers find breast-feeding more convenient. In addition, breast-fed babies, who have less colic, less diarrhea, and fewer illnesses, are easier to care for. However, while breast-feeding obviously means that an infant must rely on his or her mother for most meals, fathers often like to participate in feeding their infants. Interestingly, many babies will accept a supplementary bottle from their fathers much more readily than from their breast-feeding mothers.

In North America, an ample and varied diet is fairly easy to obtain. In fact, a peanut butter sandwich on enriched bread, a pint of fortified milk, one orange, and one egg provide more than the 23 grams of protein, most of the needed vitamins and minerals, and a substantial part of the 1,300 daily calories recommended for 2-year-olds. When problems occur in the nutrition of American toddlers, the cause is not usually inadequate supply but family food practices that allow toddlers to eat when, what, and how much they choose (Eichorn, 1979).

Fortunately, serious forms of malnutrition are uncommon among children in developed countries. Unfortunately, this is not true in many areas of the world; and the consequences for the individual's development are potentially serious (Grant, 1986).

Serious Malnutrition

In the first year of life, severe protein-calorie deficiency can cause **marasmus,** a disease that occurs when infants are severely malnourished. Growth stops, body tissues waste away, and the infant dies. During toddlerhood, protein-calorie deficiency is more likely to cause **kwashiorkor.** The child's face, legs, and abdomen swell with water, sometimes making the child appear well-fed to anyone who doesn't know the real cause of the bloating. Because in this condition the essential organs claim whatever nutrients are available, other parts of the body are degraded, including, characteristically, the child's hair, which usually becomes thin, brittle, and colorless.

The primary cause of malnutrition in developing countries is early cessation of breast-feeding. In many of these countries, breast-feeding was usually continued for at least two years, but now is often stopped much earlier in favor of bottle-feeding, usually powdered formulas. Under normal circumstances, such formulas are an adequate and safe source of nutrition. However,

> For many people in the developing world . . . the hygienic conditions for the proper use of infant formula just do not exist. Their water is unclean, the bottles are dirty, the formula is diluted to make a tin of powdered milk last longer than it should. What happens? The baby is fed a contaminated mixture and soon becomes ill, with diarrhea, which leads to dehydration, malnutrition, and, very often, death. [Relucio-Clavano, quoted in Grant, 1986]

Figure 5.11 *There have been many contributing causes to the recent famine in the countries of Africa, among them, weather conditions, farming practices, and government policies. In each area struck by this disaster, the toll in the health of children has been the same: increases in marasmus and kwashiorkor, and many more deaths from diseases like measles and chicken pox, illnesses that are not often fatal to children who are otherwise in good health.*

A comparison of the survival rates of breast-fed and bottle-fed babies from the same impoverished background reveals that bottle-fed babies have fared much less well: in Chile, for instance, they are four times more likely to die; in Egypt, five times more likely (Grant, 1982). While death and disease rates are obvious evidence of harm, many malnourished infants who survive carry more subtle handicaps that will remain throughout the life span. Specifically, those who are chronically malnourished early in life are physically shorter and intellectually less able than are their peers born in the same community (Bogin and MacVean, 1983). For instance, a longitudinal study of African teenagers who had been chronically malnourished as infants found that their intellectual abilities, especially the ability to reason, were impaired when compared to other teenagers from the same socioeconomic class and tribe who were adequately nourished (Hoorweg, 1976). Similarly, a longitudinal study of children in Guatemala found that those children who were chronically malnourished in early childhood tended to be physically shorter and cognitively slower in middle childhood (Bogin and MacVean, 1983). Other deficits are particularly likely to involve visual and auditory skills, probably because, as we have seen, the early months of life are the ones when these skills are rapidly developing.

Serious Malnutrition in the United States   While severe and extensive malnutrition is not common in the United States, neither is it unknown. As is the case with other types of infant maltreatment, the problem frequently begins in the family and in the community (Sherrod et al., 1984), so the solution is more complex than simply providing food.

Consider one example. An 8-month-old girl was admitted to a Boston hospital. She was emaciated; she could not sit up; she did not respond when spoken to. In the hospital, the baby rapidly gained weight, started to smile and play with the nurses, and was discharged. Two months later she was readmitted, "a small, dirty, smelly infant, with restless movements of her hands, and again, underweight" (Newberger et al., 1976). Clearly, her malnutrition could not be cured without paying attention to the environment that caused the problem.

This time the hospital staff took an ecological approach. Treatment focused on the mother, who had four children, no husband, little money, and few friends or relatives. She was given medical treatment for a chronic infection and dental care for aching teeth. A social worker helped place the two older children in a nursery school and found a homemaker and a public health nurse to give the mother day-to-day support. These measures helped the baby as well as her mother. Four years later the little girl entered kindergarten, normal in her physical and psychological development.

Conclusions   As the authors who described this case have pointed out, treating both mother and child was necessary for a solution of this particular problem. For prevention of such problems, a broader ecological approach, combating social ills such as poverty and isolation, would be necessary.

A similar conclusion could be drawn for all aspects of physical development covered in this chapter. For the most part, babies grow and develop skills as rapidly as their genes allow, as long as their family and culture provide the opportunity. At the same time, when things go wrong, the cause is usually complex. Biological, familial, and cultural conditions all interact to affect the growth of the child.

SUMMARY

Size and Shape

1. In their first two years, most babies gain about 20 pounds (9 kilograms) and grow about 15 inches (38 centimeters). The proportions of the body change. The newborn is top-heavy, for the head takes up one-fourth of the body length, partly because the brain, at birth, has attained a high proportion of its adult size in comparison to other parts of the body. In adulthood, the head is about one-eighth of the body length.

Brain Growth and Maturation

2. Although at birth the nervous system has virtually all the nerve cells it will ever have, these neurons grow and form branching networks during infancy and childhood, resulting in increasing efficiency of communication between the brain and the rest of the body.

3. Fetal brain maturation is responsible for the infant's increasingly regular patterns of sleep and wakefulness, and also for changes in the infant's ability to respond to, and to control responses to, the environment.

4. The development of motor control and perceptual abilities depends on an interaction between brain maturation and the infant's experience.

Sensation and Perception

5. Both sensation and perception are present at birth, and both become more developed with time. Some senses—notably hearing—seem very acute within the first months of life; others—notably vision—develop more slowly throughout the first year.

The Development of Motor Abilities

6. At first, the newborn's motor abilities consist of reflexes. Some reflexes are essential for survival; the purpose of other reflexes is not apparent. However, all reflexes are indicative of brain development.

7. Motor skills follow the cephalo-caudal and proximo-distal sequences; the upper part of the body is controlled before the lower part is, and the arms are controlled before the hands and fingers are. Although the sequence of motor-skill development is the same for all healthy infants, babies vary in the ages at which they master specific skills.

Nutrition

8. Breast milk is the ideal food for most babies; however, commercial formulas, properly prepared, are an accepta-ble substitute. A mother's choice to breast-feed or bottle-feed typically depends on many factors, including education, lifestyle, and cultural pressures.

9. By the time a child is 1 year old, a diet that includes all the major food groups will usually provide adequate calories, iron, and vitamins. In the developed countries, serious malnutrition is uncommon.

10. In developing countries, severe malnutrition can often be attributed to the early cessation of breast-feeding and improper preparation of commercial formulas. Marasmus and kwashiorkor, two major diseases caused by long-term protein-calorie deficiencies, can result in early death. When not fatal, chronic malnourishment may result in intellectual deficits and in shorter physical stature.

KEY TERMS

neurons (105)

myelin (105)

brain waves (108)

sensation (108)

perception (108)

habituation (109)

reflex (111)

breathing reflex (111)

sucking reflex (111)

rooting reflex (111)

Babinski reflex (111)

stepping reflex (111)

swimming reflex (111)

grasping reflex (111)

Moro reflex (112)

Brazelton Neonatal Behavioral Assessment Scale (112)

motor skills (112)

crawling (114)

toddler (114)

norms (114)

marasmus (117)

kwashiorkor (117)

KEY QUESTIONS

1. How do the proportions of the infant's body change during the first two years?

2. How does brain development affect the development of an infant's motor skills and perceptual abilities?

3. How do researchers determine whether an infant perceives a difference between two stimuli?

4. Which reflexes are critical to an infant's survival?

5. What is the general sequence of the development of motor skills?

6. What are the advantages and disadvantages of breast-feeding?

7. What are some of the consequences of serious, long-term malnutrition?

CHAPTER 6

The First Two Years: Cognitive Development

. . . he co-operates
With a universe of large and noisy
feeling states
* Without troubling to place*
Them anywhere special, for, to his eyes,
Funnyface
* Or Elephant as yet*
Mean nothing. His distinction between
Me and Us
Is a matter of taste; his seasons are Dry
and Wet;
* He thinks as his mouth does.*

W. H. Auden
"Mundus et Infans"

Infants begin life knowing nothing about the world around them, but the reflexes, senses, and curiosity that they are born with soon begin to inform them. By age 1 they have already learned about many of the objects in their environment (that some make noise, that others move, that some are fun to play with, that some are not to be touched), about people (that some are familiar and trustworthy, that others are strange and unpredictable), and about experiences (that playing in sandboxes is enjoyable, that visiting the doctor can hurt). Even more impressive, they have learned to communicate quite well. They understand many of the gestures and words of other people, and can make their own needs and emotions known in a variety of ways, including saying a word or two themselves.

All these achievements are part of **cognition,** that is, the interaction of all the perceptual, intellectual, and linguistic abilities that comprise thinking and learning. If cognitive development is "how we acquire and use knowledge in adapting to the vicissitudes of the world" (Caron and Caron, 1982), then the infant's cognitive development during the first two years is impressive indeed.

But what are the precise steps involved in this remarkable development? And how much adult help or environmental stimulation is needed for cognitive development to occur? In other words, is a child's cognitive growth influenced more by internal forces such as genes and maturation or by external factors such as family, teachers, toys, and neighborhood?

Interaction between Maturation and Learning

Although this question has, in the past, been a topic of controversy, over the last twenty years, research has brought a measure of calm to this discussion. Psychologists recognize that *both* maturation and learning are essential. The development of sensorimotor skills—when the baby can focus well visually, or when the baby walks—seems to depend more on physiological maturity than on anything else. Indeed, maturation seems to be a prerequisite for the development of skills of all kinds, even, as we shall see later, for finding objects that have disappeared from sight and for speaking words.

At the same time, experience has also proven to be indispensable. For example, in Chapter 5 we have seen not only that brain maturation is required for the development of motor, sensory, and perceptual abilities, but also that the infant's experiences and new abilities actually promote the development of the brain's capacities.

Although developmentalists now agree that maturation and learning work together, they continue to debate the relative contributions of each during different periods of development. They are also at odds on the questions of when specific cognitive abilities develop, whether cognitive "milestones" develop all at once or gradually, and how individual experiences affect the pace of learning.

To sort out these controversies, we will start with the observations and ideas of Jean Piaget. While many cognitive psychologists now question certain aspects of Piaget's analysis of cognitive development, Piagetian theory and its description of cognitive stages are so basic to our understanding that it is impossible to discuss infant cognition without taking them as the starting point (Brainerd, 1983).

Piaget's Theory

Remember that Piaget emphasized the total organization of intelligence—the process of understanding—through which the individual interprets his or her world. As we saw in Chapter 2, Piaget proposed that the individual's mode of thinking goes through four distinct periods of development, the first occurring between birth and age 2, and the others following at about ages 2, 6, and 12. Within each of these four periods, smaller but discernible shifts in thinking occur.

Sensorimotor Intelligence

Piaget called the first period of cognitive development, the period that begins at birth and lasts until age 2, **sensorimotor intelligence.** He recognized that babies, unlike adults, think exclusively with their senses and motor skills. Present a typical adult with a plastic rattle and that person might refer to it by name, classify it according to function, evaluate it esthetically, while at the same time having thoughts ranging from the fascinations of infancy to the fluctuating costs of petrochemicals. Give a rattle to a baby and he or she will stare at it, shake it, suck it, bang it on the floor. As Flavell (1985) puts it, the infant "exhibits a wholly practical, perceiving-and-doing, action-bound kind of intellectual functioning: he does not exhibit the more contemplative, reflective, symbol-manipulating kind we usually think of in connection with cognition."

Thus, the first two years of cognitive development are called the sensorimotor period because the child both learns about the world and expresses this learning

TABLE 6.1 **The Six Stages of Sensorimotor Development**

To get an overview of the stages of sensorimotor thought, it helps to group the six stages in pairs. The first two involve the infant's own body.

I. BIRTH TO 1 MONTH	*Reflexes*—sucking, grabbing, staring, listening.
II. 1–4 MONTHS	*The first acquired adaptations*—accommodation and coordination of reflexes—sucking a pacifier differently from a nipple; grabbing a bottle to suck it.

The next two involve objects and people.

III. 4–8 MONTHS	*Procedures for making interesting sights last*—responding to people and objects.
IV. 8–12 MONTHS	*New adaptation and anticipation*—becoming more deliberate and purposeful in responding to people and objects.

The last two are the most creative, first with action and then with ideas.

V. 12–18 MONTHS	*New means through active experimentation*—experimentation and creativity in the actions of "the little scientist."
VI. 18–24 MONTHS	*New means through mental combinations*—thinking before doing provides the child with new ways of achieving a goal without resorting to trial-and-error experiments.

chiefly through his or her senses and motor skills. Within this general period of sensorimotor intelligence, there are, according to Piaget, six stages, each characterized by a somewhat different way of understanding the world (see Table 6.1).

Stage One: Reflexes (Birth to 1 Month) Sensorimotor intelligence begins with newborns' reflexes, such as sucking, grasping, looking, and listening. In Piaget's terms, these reflexes represent the only *schemas,* or general ways of thinking about and interacting with the environment, that neonates have. Take sucking as an example: neonates suck everything that touches their lips, using the schema that all objects that can be sucked should be sucked. Similarly, infants grasp at everything that touches the center of their palm, stare at everything that comes within focus, and so forth. Through the repeated exercise of these reflexes, newborns gain information about the world, information that will be used to develop the next stage of learning.

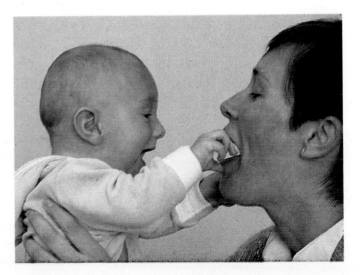

Figure 6.1 *At the sensorimotor stage of development, even Mother's face is a site for active exploration by all the infant's senses and motor skills.*

Stage Two: The First Acquired Adaptations (1 Month to 4 Months)   The second stage of sensorimotor intelligence begins when infants adapt their reflexes to the environment. Again let us take the sucking reflex as an example. By the time infants are 3 months old, they have organized their world into objects to be sucked for nourishment (breasts or bottles), objects to be sucked for pleasure (fingers or pacifiers), and objects not to be sucked at all (fuzzy blankets and large balls). They also learn that efficient breast-sucking requires a squeezing sucking, whereas efficient finger- and pacifier-sucking do not. In addition, once infants learn that some objects satisfy hunger and others do not, they will usually spit out a pacifier if they are hungry, although otherwise they will contentedly suck one.

Even more impressive, between 1 and 4 months infants can begin to coordinate two actions. For instance, while newborns will suck their thumb whenever it happens to come in contact with their mouth, they do not know how to get their thumb into their mouth. When they can do so, thus coordinating the sucking reflex with the motor skill of moving their arms, they are performing a stage-two behavior.

Between 1 and 4 months infants can also begin to coordinate perceptions and actions. They hear a noise and turn to locate it (though not always in the right direction) (McGurk and Lewis, 1974), or they see an object and try to touch (rarely successfully) (Bower, 1977). These behaviors are obviously not very well developed, but they are clearly more advanced than simple reflexes.

Figure 6.2   *This baby's obvious show of delight in having his stomach kissed will probably be interpreted by mother as a request for an encore.*

Stage Three: Procedures for Making Interesting Sights Last (4 Months to 8 Months)   In the third stage, babies become more aware of objects and other people, and they begin to recognize some of the specific characteristics of the things in their environment.

Piaget called stage three "procedures for making interesting sights last," because babies interact diligently with people and objects to produce exciting experiences. Realizing that rattles make noise, for example, babies at this stage shake their arms and laugh when someone puts a rattle in their hands. Vocalization increases a great deal, for now that babies realize that other people can respond, they love to make a noise, listen for a response, and answer back.

Stage Four: New Adaptation and Anticipation (8 Months to 12 Months)   The major intellectual accomplishment that signals the beginning of stage four is the infant's awakening understanding that objects continue to exist even when they cannot be seen. This recognition is called **object permanence.**

For younger babies, "out of sight" is literally "out of mind": if a 5-month-old drops a rattle out of the crib, for example, the baby will not look down to search for it. It is as though the rattle has completely passed from the infant's awareness. However, toward the end of stage three, babies show tentative signs of realizing that when objects disappear, they have not vanished forever. When a toy falls from the crib, the 7-month-old might look for it for a moment rather than immediately lose interest in it. By stage four, the momentary impulse to look for an object that has disappeared becomes an active effort to search for it (see A Closer Look, next page). Piaget interpreted the search for objects as the emergence of the concept of object permanence. It signals the beginning of "goal-directed behavior."

Between 8 and 12 months, as infants become more knowledgeable about the things in their world, they do, indeed, become more deliberate and purposeful in what they do with them. Whereas younger infants are likely to apply the same schema to all objects, either sucking, or dropping, or shaking them, infants in stage four are likely to choose particular schemas to use with particular objects. Here, for example, is Piaget's description of his 9-month-old daughter Jacqueline as she tried out some of her old schemas on a new object:

> Jacqueline looks for a long time at a straw table mat, then delicately touches the edge, grows bold enough to touch [the mat], then grasps it, holds it in the air, . . . shakes it, and ends by tapping it with her other hand. This behavior is accompanied by an expression of expectation and then of satisfaction. [Piaget, 1952a]

Babies at stage four develop definite ideas about what they want. They might see something clear across the room and crawl toward it, ignoring many interesting distractions along the way. Or they might grab a forbidden object—a box of matches, a thumbtack, a cigarette—and cry with rage when it is taken away, even if they are offered a substitute that they normally find fascinating. If they are placed in a playpen when they do not want to be, they may make their feelings known by throwing all the toys out and then demanding to be put on the floor with them.

Figure 6.3 *Once infants become goal-oriented, they tend to become much less tractable, and can make very clear what they do and do not want. Although this boy may not speak his first word for several months, he is obviously quite adept at communicating his preferences.*

A CLOSER LOOK   **Piaget's Test of Object Permanence**

Piaget developed a simple test for object permanence: show a baby an interesting toy and then cover it up with a cloth or blanket. If the baby tries to uncover it, he or she must suspect that the toy still exists. Here is how Piaget tried this test on his daughter Jacqueline toward the end of her seventh month:

I take the duck . . . and place it near her hand three times. All three times she tries to grasp it, but when she is about to touch it I place it very obviously under the sheet. Jacqueline immediately withdraws her hand and gives up.

Any number of researchers have run some form of this object-permanence test on infants of various ages, and nearly all have had the same results as Piaget: children do not search for the hidden object before their eighth month. In one variation of this research, Ann Bigelow (1983) even gave infants a clue to help them find hidden objects: she used noise-making toys that emitted a continuous sound during the test. Nevertheless, as Piaget would have predicted, infants younger than 8 months did not search for the hidden toy, in spite of the steady noise coming from its hiding place. Because object permanence was not developed in these infants, the clue of the noise was, in fact, no clue at all.

I had occasion to run Piaget's test myself. When the art director of this book asked my help in finding a set of photographs illustrating the concept of object permanence, I volunteered my then infant daughter Sarah as the test subject. Sarah was only 7 months old at the time, but because she was precocious-seeming in many

respects, I decided to perform some trial runs of the test to be sure she wasn't ahead of Piaget's predicted schedule. She wasn't. While she watched, I hid various objects under a blanket, and, as soon as they were out of her sight, Sarah completely lost interest in them.

Confident that neither Piaget nor Sarah would let me down, I scheduled the photo session of the test for the following month. Four weeks later, with cameras clicking away, every time I showed Sarah a fascinating object and then covered it with a cloth, she quickly uncovered it— so quickly, in fact, that it was hard to photograph the sequence. Our persistence, however, was rewarded by the photos on the next page.

Object permanence is more than a curious aspect of infant development. Once infants realize that objects exist even when out of sight, and that they can be found by careful searching, infants become more knowledgeable and purposive, demanding specific people and certain toys to which they have become attached. One infant cries for several minutes after his mother leaves; another refuses to go to sleep without a particular blanket; a third finds the teddy bear that fell out of the crib without screaming for parental assistance; a fourth opens the cupboard to get the cookies. In short, as they gain an understanding of object permanence and act on it, infants become less malleable, more self-assertive, and, consequently, much more characteristically human.

Finally, stage-four babies anticipate events. They might cry when they see Mother putting on her coat, and might even hide the hat she usually wears with that coat. If they enjoy splashing in the bathtub, they might squeal with delight when the bath water is turned on. Younger babies often spit out distasteful food after they taste it; 12-month-olds are sufficiently wise to keep their mouths tightly shut when they see spinach on the spoon.

Stage Five: New Means through Active Experimentation (12 Months to 18 Months)
The toddler's ability to anticipate contributes to the development of this stage of sensorimotor development, a time of active exploration and experimentation in which the infant often seems bent on discovering all the possibilities in his or her world.

The Little Scientist.   Because of the explorations that characterize this stage, Piaget referred to the stage-five toddler as a little scientist who "experiments in order to see." Having discovered some action or set of actions that is possible with a given object, stage-five infants seem to ask, "What else can I do with this? What happens if I take the nipple off the bottle, or turn over the trash basket, or pour water on the cat?" Their "scientific method" is trial and error.

Because of their penchant for experimentation, toddlers find new ways to achieve their goals. Through their trial-and-explorations they learn, for instance,

Between 8 and 12 months, finding a toy that has been "hidden" is a cognitive challenge that leads to deliberate action, and to much delight if the challenge is met.

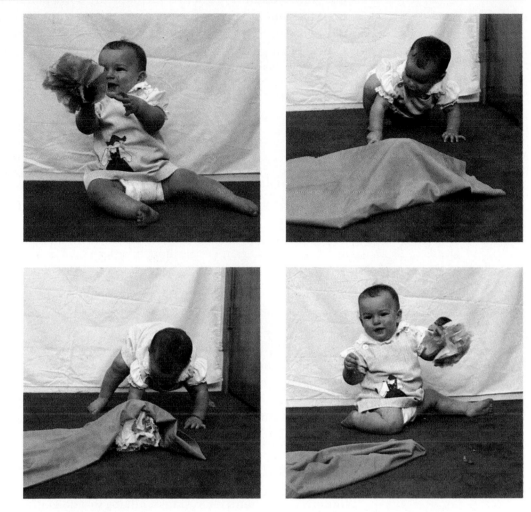

that they can bring a toy closer by pulling on the string tied to the end of it, or that they can "reach" the plate of cookies in the middle of the dinner table by pulling on the tablecloth (Piaget, 1952a). No doubt, parents of toddlers could cite many other examples of the little scientist's experiments, from the delightful to the disastrous.

Figure 6.4 *Do dogs eat flowers? This "little scientist" is out to collect data on this question.*

Stage Six: New Means through Mental Combinations (18 Months to 24 Months)   At this stage, toddlers begin to anticipate and solve simple problems by using **mental combinations** before they act. That is, they are able to try out various actions mentally without having to actually perform them. Thus the child can invent new ways to achieve a goal without resorting to physical trial-and-error experiments. Consider how Jacqueline solved the following problem at 20 months:

> Jacqueline arrives at a closed door with a blade of grass in each hand. She stretches out her right hand toward the knob but sees that she cannot turn it without letting go of the grass. She puts the grass on the floor, opens the door, picks up the grass again and enters. But when she wants to leave the room, things become complicated. She puts the grass on the floor and grasps the doorknob. But then she perceives that in pulling the door toward her she will simultaneously chase away the grass which she placed between the door and the threshold. She therefore picks it up in order to put it outside the door's zone of movement. [Piaget, 1952a]

What makes mental combinations possible at this point is **representation,** the emerging ability to create mental images of things and actions that are not actually in view.

Representation, which is a primitive form of symbolic thought, also allows the child to reproduce behavior seen in the past. For example, one day Jacqueline saw a neighbor's child have a temper tantrum in a playpen: he screamed, shook the playpen, and stamped his feet. She had never seen anything like it. The very next day, Jacqueline had a tantrum of her own, complete with stamping. Piaget referred to this kind of acting-out of a detailed mental image as **deferred imitation.**

Pretending   Perhaps the strongest, and most endearing, sign that children have reached stage six is their ability to pretend. Pretending involves not only deferred imitation (as when a child sings a lullaby to a doll before tucking it into bed) but also mental combinations, as an action from one context (riding in a car) is mentally combined with another (pushing a toy car around a table).

Figure 6.5 *Pretending becomes possible when toddlers reach the stage of mental combinations, at about 18 months. At this point, imagination is expressed primarily through bathing, feeding, or other simple care-giving activities performed with a doll or toy animal; but soon it evolves into the elaborate dramas of the preschool and school-age child. (Judging by the condition of this doll's head, the little girl shown here was an industrious "little scientist" before she became a "little mother.")*

Full Object Permanence   Closely related to the ability to use mental combinations is the full emergence of object permanence. As we have seen, in stage four the infant begins to hunt in earnest for hidden objects. As time passes, the infant's searching for out-of-sight objects becomes longer and more determined. However, the stage-four infant is capable of a fascinating mistake. Suppose a 10-month-old is playing with a ball and it rolls out of sight under a chair. The infant will crawl after it, even if the crawling takes a good deal of effort. But if, after being recovered, the ball rolls out of sight again, this time under a couch, and is not found quickly, the stage-four baby will look for the ball where it was last found—under the chair.

The toddler in stage five will not make that error. As long as the stage-five toddler sees the object disappear, he or she knows where to look. However, the stage-five child cannot imagine "invisible" displacements, that is, hiding places that he or she has not actually *seen* used as such. Whenever Piaget hid a coin in his hand for instance, and then put his hand under a blanket, leaving the coin there, Jacqueline looked in his hand for the coin but didn't think of looking under the blanket for it.

Jacqueline's arrival at stage six was apparent when in a later search for the coin she looked first in her father's hand, and then, without hesitation, under the blanket—still without ever having seen the coin placed there. Her ability to use mental combinations enabled her to imagine where the coin might be, even if she did not actually see it go there. At this point, according to Piaget, object permanence is fully developed.

The Significance of Stage-Six Behavior   Stage-six behaviors all share an important characteristic. They are a step beyond the simple motor responses of sensorimotor thought and a step toward "the more contemplative, reflective, symbol-manipulating activity" (Flavell, 1985) that we usually associate with cognition. As you will see in Chapter 9, mental representation, deferred imitation, and pretending all blossom into the symbolic thought typical of the next period of cognitive development.

Postscripts to Piaget

Piaget's description of sensorimotor thought is generally regarded as a valid and useful analysis of infants' mental development (Kramer et al., 1975; Uzgiris and Hunt, 1975; Wachs, 1975). However, a number of researchers who have followed up on Piaget's work caution against applying his cognitive framework too rigidly (Fischer, 1980; Flavell, 1982; Gelman and Baillargeon, 1983). To begin with, Piaget's sensorimotor stages are descriptions of *developmental functions;* that is, they are general statements about the developmental changes peculiar to the human species as a whole. They are based on "typical" behavior and do not reflect *individual differences* (McCall, 1981). Yet individual differences in native ability and environmental experiences can have a significant impact on cognitive development and its timing. Consequently, the ages Piaget assigned to various stages must be regarded as approximate. Piaget's own children, for example, reached some of the stages of sensorimotor intelligence "ahead of schedule," probably because Piaget, in constantly testing their ability, gave them more practice and experience than most children receive.

In addition, Piaget's delineation of distinct mental stages seems to suggest that the child's cognitive development is discontinuous, moving from one stage to the next all at once. A number of researchers who have enlarged on Piaget's work now believe that the typical child's mental development is much more gradual and continuous than Piaget represented it to be, with the child arriving at each new stage skill by skill, behavior by behavior.

Figure 6.6 *It is interesting to note that an infant develops "person permanence" long before object permanence (recall from Chapter 5 that an infant can recognize the mother's photograph at the age of 3 months). However, the response to a surprisingly different-looking mother depends on whether the child has achieved full object permanence and an understanding of appearance and reality; thus, a 6-month-old may simply be surprised, a 1-year-old, a bit frightened, and a 4-year-old, delighted.*

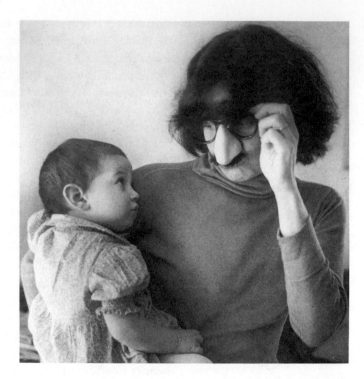

Finally, although most developmentalists accept Piaget's general outline of cognitive growth in infants, several researchers have questioned some of his measures for assessing the infant's development. Object permanence is a case in point (Ruff, 1982; Willatts, 1984). While they agree with Piaget that children gradually develop an understanding of the permanence of objects, and that uncovering a hidden toy is one demonstration of this understanding, they think that Piaget's test of object permanence can be affected by factors that Piaget didn't take into account, such as knowing how to search and the motivation to search. Even the hiding place can make a difference: infants who would not lift a blanket to search for an object might look behind a screen or under a cup for one (Willatts, 1984). A different critique of object permanence is offered by Jerome Kagan, who thinks that 9-month-old infants show an improved ability to search for hidden objects, not because they "possess a new cognitive structure," but because of "an increase in the ability to retrieve schema of prior events, a growth in what contemporary psychologists would call memory capacity" (Kagan, 1979).

Perception and Cognition   Another group of researchers might be said to believe that Piaget's description of sensorimotor intelligence overemphasizes the motor aspects of cognitive development to the neglect of its sensory aspects. Piaget inferred intellectual development from children's actions, but it is the contention of perception researchers that infants, especially in the early weeks, know more than they are able to demonstrate through their very limited motor abilities.

Foremost among these researchers are James and Eleanor Gibson. On the basis of a lifetime of research, the Gibsons (1979, 1982) believe that babies are born with the motivation and competence to begin setting their perceptual world in order, and that they coordinate their various senses to understand the objects and people of their world as soon as they draw breath. On this point the Gibsons are joined by another prominent researcher, T. G. R. Bower, whose investigations indicate that newborns try to look for sounds, grasp at objects, and respond to human faces.

Bower also believes that considerable perceptual learning, particularly aural, occurs even before birth (Bower and Wishart, 1979). Other researchers have also shown that infants have more complex perceptions of the world than their actions would suggest. For example, 2- to 4-month-old infants were shown two lines forming a right angle (Schwartz and Day, 1979). Once they were habituated to this form, half of them were shown the same angle in a different orientation and half were shown a different angle. The infants in the first group stared less at their angle than the infants in the second group stared at their new angle, suggesting that the first group had some understanding that they were looking at the same angle as before, even though it had been rotated.

Similarly, 4-month-old infants demonstrated a recognition of relative size. First they were shown several different instances in which two identical shapes of different size appeared one above the other, the smaller on top. After becoming used to this size ordering, they looked with extra attention when a large shape appeared above a small one (Caron and Caron, 1981). The same general research format has revealed that infants younger than 8 months can distinguish sets that differ in number (two versus three), color, size, density, and shape (Caron and Caron, 1982). Another series of studies demonstrates that infants younger than 4 months are more prone to grasp at objects that are within reach and the right size for holding than at objects that are out of reach or too big. They even know which objects appear squeezable and which do not (Gibson, 1982).

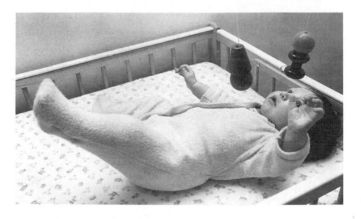

Figure 6.7 *During the first six months of life, infants develop definite ideas about which objects are within reach and the right size for grabbing and pulling. This baby is just about to stop staring and start yanking.*

Although these studies of perception were not designed to prove or disprove Piaget's theory, most researchers now agree that Piaget underestimated early perceptual abilities and hence certain aspects of cognitive development during the first six months of life (Caron and Caron, 1982; Gratch, 1979).

It should be noted, however, that this research on perception substantiates one of Piaget's most important ideas—that infants are active learners. As Eleanor Gibson (1982) writes, perceptual learning

> is not a passive absorption but an active process . . . self-regulating, in the sense that modification occurs without the necessity of external reinforcement . . . Discovery of distinctive features and structure in the world is fundamental in the achievement of this goal.

This indeed sounds very much like Piaget's descriptions of the infant's search for knowledge as "a process of equilibrium," an "internal mechanism" of "self-regulation . . . a series of active compensations."

"The American Question"

The fact that Piaget saw infants as self-motivated, active learners brings us to a disagreement between Piaget and many North American psychologists about how optimal cognitive development occurs. As you can see from Piaget's stages, he viewed infants as actively organizing their experiences themselves. He spoke of newborns adapting their reflexes, toddlers making interesting sights last, finding new ways to achieve their goals—and made little mention of the role of parents or specific learning experiences. He did not think development should be accelerated. Many psychologists in the United States and Canada, on the other hand, believe that parents should feed an infant's curiosity by providing instruction and an enriched environment. They think the infant's powerful desire to learn will benefit from well-organized experiences. In fact, Piaget called "What should we do to foster cognitive development?" the "American question," because it was one of the first questions North Americans asked when he lectured to them. Piaget's implication was that something about American society, rather than something about infant development, makes us ask that question.

Teaching and Learning in Infancy

Whatever the reason, North Americans not only asked the question, they set about to answer it, demonstrating that infants can learn to do many things, if the learning process is correctly programmed. Learning-theory research has shown, for instance, that with proper reinforcements, such as a pleasant taste or an interesting sight, infants can be taught to turn their heads in a particular sequence, kick their legs at a target, vocalize more often, or perform any number of infant behaviors on cue (Fitzgerald and Brackbill, 1976).

It should also be noted that infants learn best when the tasks are related to their most important goals: surviving and thriving. Thus infants can readily be taught skills that help them to get nourishment, attract the attention of their care-givers, and interact with interesting phenomena in their environment (Lipsitt, 1982).

Figure 6.8 *Newborns sometimes reflexively turn their heads toward the side on which their cheek has been stroked. This little girl will exhibit what is called the rooting reflex with much greater frequency than normal because she is being reinforced for doing so with sips of sugar water. In other words, she is being taught that turning her head after a stroke on the cheek produces something pleasurable. The instrument attached to her head records the speed and direction of her head turns, and thus provides an objective measure of her learning rate.*

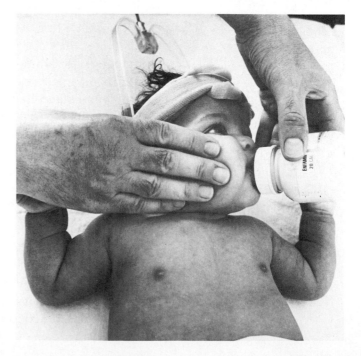

RESEARCH REPORT   **Infant Memory**

Hundreds of experiments have shown that the memory of young infants is quite weak. There is evidence that memory improves at about 1, 4, and 8 months (Lipsitt, 1982), but even in the second half of the first year, retention of information from one day to the next has been hard to demonstrate. A review of infant memory concludes that "most studies have shown retention for only a matter of seconds or minutes" (Werner and Perlmutter, 1979).

Several researchers have wondered if infant memory could be improved. For example, Cornell (1980) let 6-month-old infants look at the photograph of a face for 20 seconds, broken into four 5-second viewing periods separated by 3-second intervals. The infants were then shown the same photograph together with the photograph of a second face. When the pair of faces was presented 5 seconds after they had viewed the first photograph, the infants could distinguish the first face from the second. However, when the pair of faces was presented 1 minute after the first, the infants acted as if both faces were new to them.

In a follow-up test, Cornell decided to add an element that might suit the task to the memory potential of the infant. He applied a well-known rule regarding adult memory, specifically that memory improves if the learning sessions are separated by a period of time, rather than massed together. (This rule of "distributed practice" explains why cramming for a final exam is not as effective as studying on a regular basis throughout the term.) Cornell gave some 6-month-olds breaks that lasted a minute (rather than 3 seconds) between their four learning episodes. Although the total learning time was still 20 seconds, distributed practice helped the infants remember. They could distinguish between the two faces for up to an hour after the training period.

In another experiment, 3-month-old infants were conditioned to make a mobile move by kicking their feet, which were attached to a ribbon. They learned this task quite well, but, two weeks later, when again placed in the crib with the mobile tied to their feet, they showed no evidence of having remembered how to make the mobile move (Sullivan, 1982).

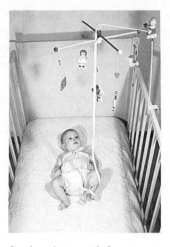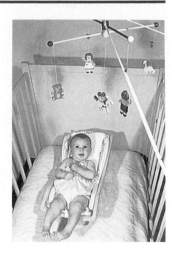

In the photo at left, a 3-month-old infant is learning to kick her foot to make a mobile move. However, after two weeks of being unattached to the mobile and not seeing it move, she will have forgotten what to do when the ribbon is once again tied *to her foot. But if she is given a "reminder session" (photo at right) in which she sees the mobile moving, her "lessons" will likely come back to her, and when the ribbon is attached to her foot, she will start kicking.*

The experiment was then repeated with another group of infants, but this time the infants were allowed a "reminder" session a day before the test. In the reminder session, the infants were able to look at the mobile, but they were not in a position to kick it, nor were they tied to the ribbon that would activate it. The next day, they were positioned correctly and tied to the mobile. At this point, they kicked as they had learned to do two weeks before. Similar experiments have been done with younger infants (8 weeks old) after a long interval (18 days), with the same result (Davis and Rovee-Collier, 1983; Linde et al., 1985).

Evidence such as this clearly suggests that infants can be taught to remember some images and behaviors earlier than had previously been thought.

Some learning theorists assert that in addition to learning through conditioning, infants seem to be primed for social learning. In particular, they point to studies showing that even in the first weeks of life, babies try to imitate facial expressions (Field et al., 1982; Meltzoff and Moore, 1983). However, there is some doubt that this is true social learning, for at least one team of researchers thinks it may be a reflex that disappears, as other reflexes do, when the infants are a month old or more (Abravanel and Sigafoos, 1984). Nevertheless, there is no doubt that from the first moments of life, infants pay particular attention to other people.

Special Learning Programs for Infants

Because of research showing that infants can be taught many skills, some psychologists have tried to improve infants' cognitive development by providing special programs of enrichment and instruction. For the most part, these efforts have been with infants who have been at risk for cognitive problems, either because of birth complications or because they live in an impoverished environment. Programs designed to provide special learning experiences for infants who are at risk for understimulation have shown a measure of success, although the degree of success varies from program to program and infant to infant (Beller, 1979; Halpern, 1984).

The success of some of these interventions has spawned a number of accelerated-infant-development programs available to the public. While some of these have captured the attention of the popular press and the money of middle- and upper-class parents, they make most developmentalists wary. Whatever the actual successes of programs like these, most developmentalists would still question the value of such accelerated learning. A theme of Piagetian theory, backed up by research from many disciplines, is that infants are remarkably able to learn from the experiences that are normally present in a good home, and that they do not need, or benefit from, extra stimulation. Indeed, too much stimulation, like too little stimulation, may be detrimental (McCall, 1981; White, 1975).

Figure 6.9 *A true grasp of the implications of the relationship among these fishes is not likely to develop much before the emerging rationality (and cynicism) of adolescence. However, before age 1, most infants recognize size difference, and, by age 2, with practice and encouragement, they can put a simple series in size order.*

Drawing by Lorenz; © 1987 The New Yorker Magazine, Inc.

"Quick, hon, I think he's got it."

Language Development

Everywhere in the world, in every language, children are talking by age 2, with a grasp of basic grammar and a vocabulary that is frequently surprising in scope. Consider, these sentences uttered consecutively by 24-month-old Sarah:

Uh, oh. Kitty jumping down.
What drawing? Numbers?
[said as I was transcribing her words]
Want it, paper.
Wipe it, pencil.
What time it is? [said upon seeing a watch]

These sentences show that Sarah has a varied vocabulary and a basic understanding of word order (for example, Sarah said "Kitty jumping down," rather than "Down jumping kitty," or "Jumping kitty down," or "Kitty down jumping"). They also show that she has much to learn, for she incorrectly uses the pronoun "it" and its referent together, omits personal pronoun subjects, and puts the predicate nominative before the predicate in the interrogative. Sarah's impressive but imperfect language is quite similar to that uttered by 2-year-olds in many families and cultures. On the basis of detailed studies of thousands of babies, we know quite a bit about the sequence of verbal skills in the first two years of life, and Sarah, both in her early days and here at age 2, is typical.

Theories of Language Development

What accounts for the rapid development of language between birth and age 2? Are infants dependent on the example or instruction of adults to gain an understanding of language? Or are they born with a powerful desire to communicate and a subtle grasp of the underlying principles of language? Each of these positions has been taken by an opposing group of scholars.

Learning Theory: Skinner Some researchers, following the lead of B. F. Skinner (1957), maintain that children learn to talk through conditioning. As Skinner explains it, when a baby babbles "ma-ma-ma" and Mommy comes running with the physiological reinforcers of food or clean diapers, or the social reinforcers of a smile or a hug, the baby is likely to say "ma-ma-ma" again. In addition, since Mother is usually delighted to think that the baby is actually addressing her, she is likely to repeat "mama" whenever the infant babbles it.

Figure 6.10 *Learning the word "nose" is easy when it becomes a mutual game of show-and-tell accompanied by parents' delighted approval.*

As time goes on, the infant becomes conditioned to associate "mama" with the presence of Mother and learns to call "mama" whenever Mother is needed. In a similar fashion, the infant learns to associate various other vocalizations with specific events and things. According to learning theorists, the words and phrases children learn depend on the nature, frequency, and timing of the conditioning process. Some families reinforce almost every utterance: when the child says "juh," for example, the child is given juice; when the child says "ca," someone responds "Yes, cat." Other families withhold rewards until the child gets a word "right" or scold and lecture when the child gets it "wrong"; and still others tend to ignore the toddler's efforts to communicate. Following the general principles of conditioning, it would be likely that a child in the first instance would develop extensive language skills; a child in the second might tend to be reluctant to use much speech; a child in the third might be very limited in language skills. Research bears out this conclusion. Depressed mothers, for instance, tend to ignore their children's attempts to communicate; consequently, their language development occurs more slowly (Breznitz and Sherman, 1987).

At the same time, learning theorists emphasize that learning by association, without specific reinforcement, is also part of the process. Many parents, for instance, habitually name the articles of clothing they put on their infants, the food they put in their mouths, the objects they put in their hands. The infant naturally makes the link between the name and the thing. A more striking example of unreinforced language learning can be seen in the case of toddlers' uttering profanities that their parents never intended them to hear, much less learn and repeat.

A Structural View: Chomsky   Coming at language learning from a very different perspective, Noam Chomsky (1968) and David McNeill (1970) hold that children have an innate predisposition to learn language at a certain age, much like the innate disposition to stand up as soon as their bodies are sufficiently mature.

Chomsky believes that babies have an inborn understanding of the basic struc-

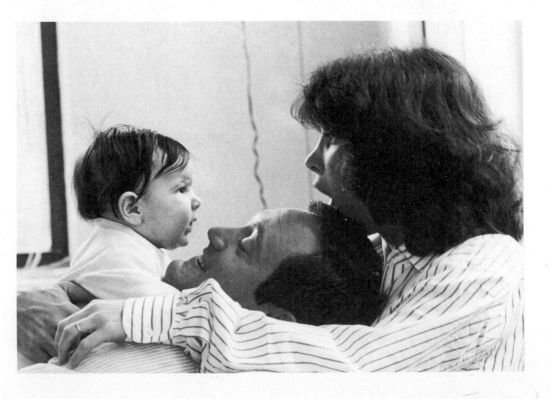

Figure 6.11 *According to Chomsky, this infant was born with the rudiments of language and the will to talk. What remains to be acquired is knowledge of the particular combination of sounds that comprise the surface structure of his own language.*

ture of language, which he calls the **deep structure.** For instance, infants seem to be born knowing that one important way to communicate is by making noises. They also soon demonstrate a recognition that small differences in pronunciation and inflection can change the meaning of a word, and that some sentences are statements and others are questions. These are things that do not have to be *learned.* What does require learning is the particular vocabulary and grammar, or **surface structure,** of a language. Chomsky believes that children learn the surface structure as rapidly as they do because they already understand the deep structure, which helps them grasp the underlying rules of grammar that govern the flawed and incomplete sentences that most adults speak.

To emphasize the inborn, automatic nature of the infant's linguistic ability, Chomsky devised a figure of speech, the **language acquisition device (LAD),** to refer to the human predisposition to learn language. According to Chomsky, infants' innate language-learning ability causes them to listen attentively to speech sounds in the early weeks of life and to imitate speech sounds and patterns throughout infancy. LAD also triggers babbling at about 6 months, the first word at about 1 year, and the first sentences at about 2 years, no matter what the child's language, nationality, race, or socioeconomic background.

Later Research Skinner's and Chomsky's views of language learning represent opposite extremes, which both theorists continue to maintain (Chomsky, 1980; Skinner, 1983). However, substantial research on early language development has shown that neither position tells the whole story. Rather, actual language learning is an interactional process between the maturation that propels cognitive growth generally and the communication that occurs in the parent-child relationship. This will be increasingly clear as we look, first at the steps in language learning, and then at the family context that makes such learning possible.

Steps in Language Development	Children the world over follow the same sequence and approximately the same timetable for early language development (see Table 6.2). The first area in which they become competent is *language function*—that is, the communication of ideas and emotions. Indeed, considering language solely in terms of its function, it can be said that infants are born using language, a language of noises and gestures. As you will see, within the first two years of life, this rudimentary ability to communicate evolves into an impressive command of *language structure,* that is, the particular words and rules of the infant's native tongue.

TABLE 6.2 **Language Development**

newborn	Reflexive communication—cries, movements, facial expressions.
2 months	A range of meaningful noises—cooing, fussing, crying.
6 months	Babbling, including both consonant and vowel sounds.
10 months	Comprehension of simple words; intonation of language; specific vocalizations that have meaning to those who know the infant well. Deaf babies express their first sign.
12 months	First spoken words that are recognizably part of the native language.
12–18 months	Slow growth of vocabulary, up to 50 words.
21 months	First combination of words into two-word sentences.
24 months	Vocabulary of more than 200 words. Grammar apparent in word order, suffixes, prefixes, pronouns (specifics depend partly on the particular native language).

Comprehension   At every stage of development, including the preverbal stage, children understand much more than they express (Kuczaj, 1986). When asked "Where's Mommy?" for instance, many 10-month-olds will look in her direction; or when asked "Do you want Daddy to pick you up?" will reach out their arms. In addition, as the infant learns to anticipate events (stage four of sensorimotor development), words such as "hot!" "no!" or "bye-bye" take on meaning. Of course, context and tone help significantly to supply that meaning. For example, when parents see their crawling infant about to touch the electrical outlet, they say "No" sufficiently sharply to startle and thus halt the infant in his or her tracks. Typically, they then move the child away, pointing to the danger and repeating "No. No." Given the frequency with which the mobile infant's behavior produces similar situations, it is no wonder that infants seem to understand "No" at an early age.

First Spoken Words   At about 1 year, the average baby speaks one or two words, not pronounced very clearly or used very precisely. Usually care-givers hear, and understand, the first word before strangers do, which makes it hard to pinpoint, scientifically, exactly what a 12-month-old can say.

The first words a child speaks are usually the words that are most important to the child. Many are associated with some movement or action the child can do, as might be expected, since the child is at the sensorimotor stage of development. For instance, Katherine Nelson (1973) followed the language development of eighteen children for a year, beginning at about their first birthdays. In their vocabularies, eleven children had "shoe," seven had "clock" (meaning "watch" as well), and six had "key"—all objects that move or that can be held by a 1-year-old—while none had "diaper" or "crib," words referring to objects that are common in infants' lives but that are not normally part of their sensorimotor schema. Other research has also found that early words often express, and, in fact, typically accompany action, for example, "bye-bye" and waving, and "bang" and hitting (Corrigan, 1978).

Vocabulary increases gradually, perhaps a few words a month. By 18 months, the average baby knows somewhere between three and fifty words. Most of these early words are names of specific people and objects in the child's daily world, although some "action" words are included as well (Barrett, 1986; Kuczaj, 1986).

Piaget's research suggests that the child must reach stage six of sensorimotor development, the stage of mental representation, before language learning can become rapid. Piaget's explanation seems plausible, for in order to really grasp the connection between words and referents, a toddler must be able to think about an object or action without using sensorimotor skills. A firm sense of object permanence is also necessary to enable the child to connect the word "ball," for instance, with an actual ball, whether the child is playing with it or is about to get it out of the toy box. Many of the toddler's first sentences seem to show that mental representation ("See dis," "Wha dat?") and object permanence ("Where dis," "Allgone dat") are ideas of such significance that they are often put into words.

At first, infants apply the few words they know to a variety of contexts. This characteristic, known as **overextension,** or overgeneralization, might lead one child to call anything round "ball," and another to call every four-legged creature "doggie." But once vocabulary begins to expand, toddlers seem to "experiment in order to see" with words just as they do with objects. The "little scientist" becomes the "little linguist," exploring hypotheses and reaching conclusions. It is not unusual for 18-month-olds to walk down the street pointing to every animal, asking "doggie?" or "horsey?" or "kitty?"—perhaps to confirm their hypotheses about which words go with which animals.

As children learn their first words, they usually become adept at expressing intention. Even a single word, amplified by intonation and gestures, can express a whole thought. When a toddler pushes at a closed door and says "bye-bye" in a demanding tone, it is clear that the toddler wishes to go out. When a toddler holds on to Mother's legs and plaintively says "bye-bye" as soon as the babysitter arrives, it is equally clear that the child is asking Mommy not to leave. A single word that expresses a complete thought in this manner is called a **holophrase.**

It is important to note that vocabulary size is not the only, nor the best, measure of early language learning. Rather, the crux of early language is communication, not vocabulary. If parents are concerned about their nonverbal 1-year-old son, they should look at his ability and willingness to make his needs known and to understand what others say. If those skills seem to be normal, and if the child hears enough simple language addressed to him every day (through someone's reading to him, singing to him, talking about the food he is eating and the sights he sees), he will probably be speaking in sentences before age 2 (Eisenson, 1986). On the other hand, infants who show signs of language delay (for example, not babbling back when parents babble to them, or not responding to any specific words by age 1) should have their hearing examined as soon as possible. Even a moderate early hearing loss can delay speech acquisition (Butler and Golding, 1986).

Combining Words Within about six months of speaking his or her first words, a child begins to put words together. As a general rule, the first two-word sentence appears between 18 and 21 months (Slobin, 1971). Combining words demands considerable linguistic understanding because, in most languages, word order affects the meaning of the sentence. However, even in their first sentences, toddlers demonstrate that they have figured out the basics of subject-predicate order, declaring "Baby cry" or asking "Rain stop?" rather than the reverse.

Teamwork: Adults and Babies Teach Each Other to Talk

Early research in language development tended to concentrate primarily on either the infant as learner or on the parent as teacher. In recent years, however, researchers have focused on the parent-child interaction, recording and analyzing what each half of the partnership says and does in their communications with each other. This led to the discovery that infants are primed to listen and respond to speech and that care-givers are often quite skilled at facilitating the infant's language learning.

As detailed in Chapter 5, even in the first weeks of life, infants turn their heads and open their eyes wide when they hear voices, show excitement when someone talks to them, and have preferences for certain voices. Even babies under a month old have demonstrated in laboratory experiments that they can hear the difference between very similar speech sounds, suggesting that this ability is inborn (Clarkson and Berg, 1983).

For their part, adults talk to infants even in the first days of life, using a special form of language called **baby talk.** The term "baby talk," as used by researchers, does not refer to the way people think babies talk—the "goo-goo-ga-ga" that few infants actually say. Rather, it refers to the particular way people talk to infants, a distinct form of language that some psychologists have nicknamed *Motherese.* Baby talk differs from adult talk in a number of features that are consistent throughout all language communities (Ferguson, 1977): it is distinct in its pitch (higher), intonation (more low-to-high fluctuations), vocabulary (simpler and more concrete), and sentence length (shorter). It also employs more questions, commands, and repetitions, and fewer past tenses, pronouns, and complex sentences, than adult talk does. People of all ages, parents and nonparents alike, speak baby talk with infants (Jacobson et al., 1983), and preverbal infants prefer listening to Motherese than to normal speech (Fernald, 1985).

The function of baby talk is clearly to facilitate early language learning, for the sounds and words of baby talk are those that infants attend to, and speak, most readily. In addition, difficult sounds are avoided: consonants like "l" and "r" are regularly missing, and hard-to-say words are given simple forms, often with a "-y" ending. Thus, a father becomes "daddy," stomachs become "tummies," and rabbits become "bunnies," because if they didn't, infants and parents wouldn't be able to talk about them. One example of the parents' readiness to adapt language to fit the needs of the infant was shown in an experiment in which mothers and their 13-month-old infants were observed while they played with a variety of toys representing the more ferocious members of the cat family—including a cougar, a leopard, and a lion. Eighty percent of the mothers reduced this variety to "kitty," some even saying "meow" and urging the infant to pet the "nice kitty-cat." Other categories of toys were similarly simplified with inaccurate but familiar labels (Mervis and Mervis, 1982). Another study found that language developed most quickly if the parent noticed what objects or activities had already captured the child's attention and talked about these (Tomasello and Farrar, 1986).

In the earliest stages of baby talk, the conversation is, of course, rather one-sided. However, as the child grows more responsive and communicative, the general interaction between parent and child becomes more like a conversation in its give and take. For instance, the games that parents and babies often play in the last half of the first year, such as peek-a-boo and patty-cake, typically involve a turn-taking ritual. This mutual game-playing may be the first step in language learning (Bruner, 1974–1974).

The "conversational" aspect of the parent-child interaction is strengthened between 5 and 7 months, as parents begin to treat burps, smiles, yawns, gestures, and babbling as part of a dialogue (Snow and Ferguson, 1977). For instance, in the following "exchange," the mother asks and answers several questions, as if the baby's smile and burp were significant communications:

Figure 6.12 *Even a simple parent-infant game like hide-and-seek incorporates some elements of conversation, including turn-taking between players and mutual attention to facial expressions.*

A CLOSER LOOK   **Berger Learns the Rules of Baby Talk**

In the weeks just before the Berlin wall was put up in 1961, I went to work in a day-care center at a refugee camp in West Berlin. The children, aged 4 and younger, spoke several languages, none of which I knew, and as I prepared for my first day, I worried that I would not be able to communicate with them. After a few minutes on the job, however, I realized that communication would be no problem when a 2-year-old tugged my hand and said a two-syllable word I immediately understood, even though I had never heard it before. He wanted me to take him to the bathroom.

I was soon to discover that my young friend's message takes a similar form in any number of languages: a repeated single syllable generally beginning with a hard consonant. Indeed, researchers have discovered a number of commonalities and implicit rules of baby talk that exist across nearly all languages. To begin with, baby talk is made up of a small vocabulary of simple words that are amplified with gestures and vocal expression. Many of these words consist of single repeated syllables (such as "no-no," "bye-bye") and often tend toward onomatopoeia, mimicking a sound or quality associated with the object or action being named ("choo-choo," "bow-wow," "pooh-pooh").

One of the most striking commonalities arises from the fact that the early words of baby talk are made up of the sounds characteristic of babbling—"m," "p," "b," "t," and "d" plus a vowel sound. Consequently, two of the first and most important words in babies' vocabularies—"mother" and "father"—take a nearly universal form (see table).

	Mother	Father
English	mama, mommy	dada, daddy
Spanish	mama	papa
French	maman, mama	papa
Italian	mamma	babbo, papa
Latvian	mama	tēte
Syrian Arabic	mama	baba
Bantu	ba-mama	taata
Swahili	mama	baba
Sanskrit	nana	tata
Hebrew	ema	abba
Korean	oma	apa

Child: (Smiles)
Mother: *Oh, what a nice little smile! Yes, isn't that nice?*
(Pause)
There. There's a nice little smile.
Child: (Burps)
Mother: *What a nice wind as well! Yes, that's better, isn't it?*

Even when the baby seems to do nothing, the parent is likely to carry on the conversation as though it were actively two-way. A mother might say, "Don't you want to take a nap now?", pause a second or two as though allowing the baby to answer, and then say, "Of course you do," in agreement with the imagined response (Stern, 1977). In fact, the response is not all in the mind of the mother. Babies show with facial expressions that they listen to maternal speech, and they indicate when they are ready to hear more. By 9 months, many babies have developed the conversational skill of taking turns, just as adults do, both in vocalization and in gaze (looking intently at the mother when listening, averting the eyes somewhat while "talking"). This skill becomes more efficient and rapid as infants mature, with the number of turns taken per minute increasing from seven at 9 months to sixteen at 24 months (Rutter and Durkin, 1987).

Once the child begins to talk, many conversations between parent and child show the parent interpreting the child's imperfect speech and then responding with short, clear sentences the child can understand, often with special emphasis on important words. Particularly with firstborns, the child's vocabulary expands in the process (Jones and Adamson, 1987). Naturalistic observation is the best way to study this interaction, for facial expression and intonation are as much a part of baby talk as the words spoken. However, recorded dialogues like the following (Halliday, 1979) help give the flavor:

Nigel's mother is putting him to bed.

Mother: And when you get up in the morning, you'll go for a walk.
Nigel: *Tik.*
And you'll see some sticks, yes.
Hoo.
And some holes, yes.
Da.
Yes, now it's getting dark.
I wa [repeated thirteen times].
What?
I wa [seven times]. Peaz.
What do you want in bed? Jamie? [his doll]
No!
You want your eiderdown? [quilt]
(grins) Yeahh!
Why didn't you say so? Your eiderdown.
Ella [three times].

In most episodes of baby talk, the child is an active participant, responding to the speaker and making his or her needs known. In this one, Nigel asked for his quilt a total of twenty times, persisting until his seemingly foolish mother finally got the point.

A Social Interaction   As we have seen repeatedly, infants are motivated to understand the world: the same motivation that makes toddlers resemble little scientists makes infants seek to understand the noises, gestures, words, and grammatical systems that describe the world in which they live. Central to the achievement of understanding language is the verbal interaction of parents and child. As one researcher writes:

> language . . . could not emerge in any species, and would not develop in any individual, without a special kind of fit between adult behavior and infant behavior. That fit is pre-adapted: It comes to each child as a birthright, both as a result of biological propensities and as a result of social processes learned and transmitted by each new generation. [Kaye, 1982]

Thus parent and baby together accomplish what neither could do alone: teach a person to talk. As we will see in the next chapter, the same parent-infant relationship is at the core of the psychosocial development of the child.

SUMMARY

1. Impressive changes in the infant's capacity to understand and communicate with the world are the result of developments in cognition, the interaction of the perceptual, intellectual, and linguistic abilities that result in thinking and learning.

Interaction between Maturation and Learning

2. Although most developmental researchers agree that both maturation and learning contribute to cognitive development, questions remain about their relative contributions during different periods of development.

Piaget's Theory

3. From birth to age 2, the period of sensorimotor intelligence, infants use their senses and motor skills to understand their environment. They begin by adapting their reflexes, coordinating their actions, and interacting with people and things. By the end of the first year, they know what they want and have the knowledge and ability to achieve their simple goals.

4. Infants younger than 8 months believe that when an object disappears from their sight, it no longer exists. But by the age of 1½, when the concept of object permanence is fully developed, toddlers can imagine where to search for an object hidden in their presence, even though the object itself may have been covered from their sight as it was being hidden.

5. In the second year, toddlers find new ways to achieve their goals, first by actively experimenting with physical objects, and then, toward the end of the second year, by manipulating mental images of objects and actions that are not in view.

6. Although recent research confirms Piaget's general description of the stages of human development, some researchers tend to place more emphasis on individual differences and a gradual, rather than abrupt, progression from one stage to another. With regard to the sensori-motor stage of development, Piaget's emphasis on skills requiring motor abilities may not have revealed the full extent of the infant's sensory capacities.

"The American Question"

7. Experiments in learning have proven that infants can be taught tasks that involve abilities they already have. Enriched environments and activities can improve the cognitive skills of deprived infants. For normal infants, the learning experiences available in the home are usually sufficient, and too much stimulation may be detrimental.

Language Development

8. According to Skinner, children learn language through reinforcement and association. Chomsky believes children have an inborn ability to understand and use the basic structure of language and that each culture teaches its children the particular structures, such as vocabulary and grammar, of its own language.

9. Language skills begin to develop at birth, as babies communicate with noises and gestures, and practice babbling. Infants say a few words by the end of the first year, and they understand more words than they speak. By age 2, most toddlers can combine two words to make a simple sentence.

10. Children vary in how rapidly they learn vocabulary, as well as in the way they use words. In the first two years, the child's comprehension of simple words and gestures, and willingness and ability to communicate are more significant than the size of the child's vocabulary.

11. Language learning is the result of the interaction between parent and child. The child is primed to learn lan-guage, and adults all over the world communicate with children using a simplified form of language called baby talk, which suits the child's abilities to understand and repeat.

KEY TERMS

cognition *(121)*
sensorimotor
 intelligence *(122)*
object permanence *(124)*
mental
 combinations *(128)*
representation *(128)*
deferred imitation *(128)*

deep structure *(137)*
surface structure *(137)*
language acquisition
 device (LAD) *(137)*
overextension *(138)*
holophrase *(139)*
baby talk *(139)*

KEY QUESTIONS

1. What is the major controversy about cognitive development in infancy?

2. According to Piaget, what is the basic characteristic of infant thought between birth and age 2?

3. What are the cognitive developments that occur in the first 8 months of life?

4. Between the ages of 6 and 18 months, what changes occur in the infant's understanding of objects that are out of sight?

5. Why did Piaget call 1-year-olds "little scientists"?

6. What cognitive developments make it possible for the toddler to pretend?

7. What are the criticisms of Piaget's description of the first two years?

8. What has recent research contributed to our understanding of the infant's perceptual and cognitive abilities?

9. What kinds of behaviors can most readily be conditioned during infancy?

10. What are the differences between Skinner's and Chomsky's theories of language development?

11. In what way do children's first words reflect their stage of cognitive development?

12. What are some of the ways of determining whether the language development of an infant or toddler is normal?

13. How do the special features of baby talk make it easy for parents and children to communicate with each other?

CHAPTER 7

The First Two Years: Psychosocial Development

*Babies control and bring up their families as much
as they are controlled by them; in fact, we may say
that the family brings up a baby by being brought up by him.*

Erik Erikson
Childhood and Society

*I waited so long to have my baby, and when
she came, she never did anything for me.*

Mother of severely abused 4-year-old girl
The Battered Child

As you remember from Chapter 1, psychosocial development includes not only factors that are usually considered characteristic of the individual's psyche, such as emotional expression, self-awareness, and temperament, but also factors that are clearly social, such as the parents' relationship to their offspring or the culture's impact on the child. Although each of these factors is, of necessity, examined separately, it is in actuality impossible to separate the development and workings of the individual psyche from the social world in which it operates. Thus, as you make your way through this initial chapter on psychosocial development, keep in mind that the development is interactive and holistic, even while we study it topic by topic.

Emotional Development

It was once thought that young infants did not have real emotions. True, they cry when they are hungry and, beginning at about 6 weeks, smile when they see a face peering at them, but these reactions were thought to be simple reflexes or primitive urges not truly related to the emotions felt by older children or adults.

Recently, however, researchers, inspired in part by the surprising discoveries (described in Chapters 5 and 6) concerning infant perception and cognition, have taken a new look at infant emotions. They have discovered that even very young infants may well express, and respond to, many emotions—including joy, surprise, anger, fear, disgust, interest, and sadness (Campos et al., 1983). Indeed, it seems as if there is a developmental "schedule" by which infants acquire the capacity for specific emotions.

The First Half Year

One of the first emotions that can be discerned in infants is fear. When infants a few days or even a few hours old hear a loud noise or feel a sudden loss of support or see an object looming toward them, they often cry and look surprised and afraid

145

Figure 7.1 *This infant's "pleasure smile" is probably an inborn response to the satisfactions of a full stomach and a smoothly functioning digestive system, just as most of the early cries are automatic responses to hunger or indigestion.*

(Izard, 1978; Sroufe, 1979). Slightly older babies have more pronounced fear reactions, and also seem angry at times, as when, for instance, they are forcibly prevented from moving (Stenberg and Campos, 1983).

Sadness, or at least a sensitivity to it, is also apparent early in infancy. In an experiment in which mothers of infants between 1 month and 3 months old were told to look sad and act depressed, their infants responded by looking away and fussing (Cohn and Tronick, 1983; Tronick et al., 1978). More explicit findings were provided by an experiment in which the facial expressions of a 3-month-old girl who had been severely abused were filmed and later shown to "blind" judges for assessment: without knowing anything about her condition, the judges rated her as undeniably sad (Gaensbauer, 1980).

What about the more positive emotions in early infancy? Newborns show the wide-eyed look of interest and surprise, and the accompanying slow-down of heart rate, when something catches their attention (Field, 1982). Smiles also begin early: a half-smile at a pleasant noise or a full stomach appears in the first days of life; a **social smile**—a smile in response to someone else—begins to appear at about 6 weeks (Emde and Harmon, 1972). By 3 or 4 months, smiles become broader, and babies laugh rather than grin if something is particularly pleasing.

8 Months to 2 Years

Figure 7.2 *This mother may, at first, be puzzled, or even embarrassed by her child's response to the kind stranger. However, at this stage, this infant's behavior is simply a sign that she understands the difference between the familiar and the unusual.*

At about 8 months, infants' emotions become much stronger, more varied, and distinct. For example, many new experiences produce fear. In fact, if you were to graph the fears of infancy, the shape of the graph would be an upside-down U (Kagan, 1983). Fears are low in early infancy, begin to rise sharply toward the end of the first year, and then decline as the second birthday approaches.

One fear, **fear of strangers** (also called stranger anxiety), is universal. This emotion is first noticeable at about 6 months and becomes full-blown by 12 months, when every normal infant is at least wary of strangers, and many quite normal infants seem terrified, hiding and crying, when an unfamiliar person comes too close. A related emotion is **separation anxiety,** the fear of being left by mother or other care-giver. Like other fears, separation anxiety emerges at about 8 or 9 months, peaks at about 14 months, and then gradually subsides (Weinraub and Lewis, 1977).

As every parent knows, anger is another emotion that becomes much more common in toddlerhood. When videotapes of infants being inoculated between 2 and 19

months were categorized by "blind" raters who could see only the children's faces, ratings of anger increased dramatically between 8 and 19 months. In addition, the duration of anger increased, from a fleeting expression in early infancy to a lengthy demonstration at 19 months (Izard et al., 1983).

Finally, as infants become older, they smile and laugh more quickly, and more selectively (Lewis and Michalson,1983). For instance, the sight of almost any human face produces a stare and then a smile in the typical 3-month-old, but the typical 9-month-old may grin immediately at the sight of certain faces—and might remain impassive or burst into tears at the sight of certain others.

Emotion and Cognition   What might explain this intensification of emotional development that causes the infant under 6 months to be quite a different creature from the toddler at 12 months or more? Since several emotional shifts occur at about 8 months, which is the same time that a new cognitive stage and new memory abilities appear (see Chapter 6), the emotional changes may be the result of a "cognitive metamorphosis" (Zelazo, 1979). In other words, being able to think and remember in a much more mature way, the infant can recognize more reasons to be happy or afraid, and be quicker to anger or sorrow (Sroufe, 1979).

Self-Awareness   In addition, one of the most important cognitive features of later infancy is the development of a certain measure of **self-awareness.** The emerging sense of "me and mine" makes possible many new emotions, including shame, guilt, jealousy, and pride (Campos et al., 1983). At the same time, a sense of self allows a new awareness of others, and hence allows defiance and true affection (Sroufe, 1979).

The sense of self emerges gradually over the first two years. In the first month, infants have no awareness of their bodies as theirs. To them, for example, their hands are interesting objects that appear and disappear: 2-month-olds, in effect, "discover" their hands each time they catch sight of them, become fascinated with their movements, then "lose" them as they slip out of view. Even 8-month-olds often don't seem to know where their bodies end and someone else's body begins, as can be seen when a child at this age grabs a toy in another child's hand and reacts with surprise when the toy "resists." By age 1, however, most infants would be quite aware that the other child is a distinct person, whom they might well hit if the coveted toy is not immediately forthcoming.

Figure 7.3 *Fingers and toes may already be old hat for this infant; however, increasing self-awareness allows him to make a surprising new discovery— his "belly button."*

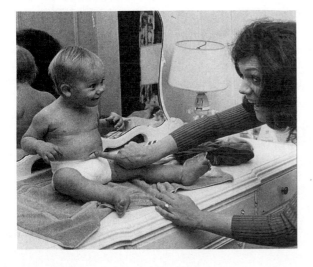
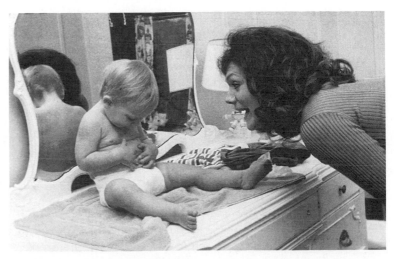

Evidence of the emerging sense of self was shown in an experiment in which babies looked in a mirror after a dot of rouge had been surreptitiously put on their noses (Lewis and Brooks, 1978). If the babies reacted to the mirror image by touching their noses, it was clear that they knew they were seeing their own faces. After trying this experiment with ninety-six babies between the ages of 9 and 24 months, the experimenters found a distinct developmental shift. None of the babies under a year reacted to the mark, whereas most of those between 12 and 24 months did.

Figure 7.4 *Mirror images make young infants smile and try to touch "the baby." It is not until about 18 months that children realize that they are looking at themselves.*

This increased self-awareness is also apparent in the toddler's emotions. The infamous toddler temper develops partly because, as children become more aware of themselves, they become more conscious of being thwarted or hurt, as well as more aware of their ability to thwart others. For instance, one observational study found that conflict between mothers and their toddlers increased over the second year from an average of three per hour at 14 months to six per hour at 24 months. Moreover, many of these conflicts were provoked by the child; and, as the toddlers grew older, their reactions were more self-assertive: in comparison with when they were 14 months old, toddlers at 18 months were four times as likely to express anger, and three times as likely to laugh, when their mothers attempted to remonstrate with them. By 24 months, they were eight times more likely to get angry and four times more likely to laugh (Dunn and Munn, 1985).

Personality Development

Now that we have seen how emotional capacities develop during infancy, the next questions are: How do the infant's emotional responses begin to take on the various patterns that form personality? What happens to evoke or create emotions, personality traits, and social skills during infancy, so that, by age 2 (if not before), the child is a distinct individual?

Traditional Views: The Omnipotent Mother

In the first half of the twentieth century, when psychologists first turned their attention to childhood, they stressed the crucial importance of the mother, who, they believed, created the child's personality through her manner of child-rearing. In their view, the infant was a passive, particularly vulnerable recipient of the mother's ministrations, shaped almost entirely by the mother's powerful influence.

This view was popularized by a host of child-advice books, one of which was, in typical fashion, "Dedicated to all Mothers in Whom lies the power to Create a Healthier and Happier World" (Schick and Rosenson, 1932). Another, written by the director of the Bureau of Child Health of New York, explained that "Father and mother must realize it is largely within their power to determine whether or not a child will be nervous or calm and well poised" (Baker, 1923).

The idea of the "created" personality was endorsed by the leading behaviorist of the times, John Watson, who trumpeted:

> Give me a dozen healthy infants, well-formed, and my own specific world to bring them up in, and I'll guarantee to take any one at random and train him to become any type of specialist I might select—doctor, lawyer, artist, merchant-chief and, yes, even beggar-man and thief . . . . [Watson, 1925]

So certain was Watson of the parents' central influence in shaping the child that he cautioned:

> failure to bring up a happy child, a well-adjusted child—assuming bodily health—falls squarely upon the parents' shoulders. [By the time the child is 3] parents have already determined . . . whether . . . [the child] is to grow into a happy person, wholesome and good-natured, whether he is to be a whining, complaining, neurotic, an anger-driven, vindictive, over-bearing slave driver, or one whose every move in life is definitely controlled by fear. [Watson, 1928]

Like Watson, Sigmund Freud stressed the early years and the influence of the mother (Kagan, 1979). He felt that the experiences of the first four years of life "play a decisive part in determining whether and at what point the individual shall fail to master the real problems of life" (Freud, 1963). He also thought that the child's relationship with the mother was "unique, without parallel, established unalterably for a whole lifetime as the first and strongest love-object and as the prototype of all later love-relations" (Freud, 1947, 1964).

While Watson's pronouncements had substantial influence at the time, Freud's view of the mother's role had a far longer lasting impact, in part because it was a central feature of his theory of personality development. Even today, to understand most theories of infancy, one must be familiar with Freud's basic ideas.

Freud: Oral and Anal Stages

As we noted in Chapter 2, Freud viewed human development in terms of psychosexual stages that occur at specific ages. According to Freud (1935), development begins with the **oral stage,** so named because in the first year of life the mouth is the infant's prime source of gratification. Not only is the mouth the instrument for attaining nourishment, it is also the main source of pleasure: sucking, especially at the mother's breast, is a joyous, sensual activity for babies. Thus Freud viewed the mother's attitudes and actions in feeding her infant, and the timing of weaning, as a crucial part of the infant's psychological development.

In the second year, Freud maintained, the infant's prime focus of gratification becomes the anus, particularly the pleasure taken in stimulation of the bowels. Accordingly, Freud referred to this period as the **anal stage.** This change is more than a simple shift of locus; it is a shift in the mode of interaction, from the passive, dependent mode of orality to the expulsive mode of anality in which the child has some control, and hence some power.

According to Freud, both these stages are fraught with potential conflict for the infant, conflict that can have long-term consequences. If mothers frustrate their infants' urge to suck—by making nursing a hurried, tense event, or by weaning the infant from the nipple too early, or by continually preventing the child from sucking on fingers, toes, and other objects—those infants will become adults who are "fixated," or stuck, at the oral stage, excessively eating, drinking, chewing, biting, smoking, or talking in quest of the oral satisfaction denied them in infancy. Further, in adulthood, oral types may tend to be generous, disorganized, and habitually late; that is, they act "babyish" so someone will "mother" them. Or they may deny these tendencies by adopting an overly tough, independent stance with others, gratifying their oral tendencies by sarcastically "biting people's heads off."

Similarly, if toilet training is overly strict or premature (occurring before the age of 1½ or 2, when children are physiologically ready, as well as psychologically mature enough to participate in the toilet-training process), it will produce adults who have "anal" personalities. They will be either anally retentive—overemphasizing neatness, cleanliness, precision, and punctuality—or they will be anally expulsive, exhibiting messiness and disorganization in nearly all matters. People with anal retentive personality structures are said to have difficulty relaxing and are overcautious about meeting new people or participating in new experiences. Often, when they do abandon their caution, they overreact, becoming excessively emotional, perhaps in an explosive temper. Further, in certain cases Freud traced severe psychological disturbances to the infant's early oral and anal experiences (Freud, 1963).

Although Freud's ideas concerning orality and anality have been extremely influential, research has failed to support the linking of specific conflicts during these stages to later personality traits. Rather, it has shown that the parents' overall pattern of warmth and attention is much more important to the child's emotional development than the particulars of either feeding and weaning or toilet-training (Maccoby and Martin, 1983). This broader perspective is reflected in the theories of Erik Erikson and Margaret Mahler, two contemporary psychoanalytic theorists who have studied infancy.

Erikson: Trust and Autonomy

As you will remember from Chapter 2, Erik Erikson is one of the best known of the neo-Freudians, that is, theorists who have modified many of Freud's insights to create their own version of psychoanalytic theory. Rather than emphasizing the specific events of the oral and anal stages, Erikson focuses on the overall patterns of child-rearing during this period.

According to Erikson, development occurs through a series of basic conflicts throughout the life span (see Chapter 2). The conflict of infancy centers not on experiences related only to oral gratification per se but on the larger pattern of interaction of which these experiences are a part (Erikson, 1963). In Erikson's terms, the crisis of infancy is one of **trust versus mistrust.** Babies learn to trust their world if they are kept well-fed, warm, and dry significantly more often than they are left hungry, cold, and wet.

In taking this broader perspective, Erikson contends that babies begin to develop a secure sense of self when their mothers provide a "consistency, continuity, and sameness of experience," so they learn to know that their needs will be met along predictable and benevolent lines. Like Freud, Erikson believes that the mother should usually be the primary care-giver. However, he notes that mothers from different cultures raise children differently, and therefore he shies away from any specific rules of what the mother should or should not do. He explains that "the amount of trust derived from earliest infantile experience does not seem to depend on absolute quantities of food or demonstrations of love, but rather on the quality of the maternal relationship."

Similarly, Erikson's view of toddlerhood places the idea of anality in the broad context of the conflict of **autonomy versus shame and doubt.** Toddlers want to rule their own actions and bodies. If they fail in their efforts to do so, because their care-givers are too restrictive and forbidding, they come to feel shame and to doubt their abilities.

For many toddlers, the struggle for autonomy does, in fact, center around toilet-training, and Erikson believes that some mothers instill shame and self-doubt in the child who defecates at the wrong time and place. But Erikson notes that as toddlers' increasing mobility allows them to move from safe, boring experiences toward more exciting and potentially dangerous ones, they encounter many other kinds of conflict that lead to the development of self-confidence or doubt. In all the conflicts of toddlerhood, Erikson agrees with Freud that parents should not be too strict, but he also warns against permissiveness:

> Firmness must protect him [the toddler] against the potential anarchy of his as yet untrained sense of discrimination, his inability to hold on and let go with discretion. As his environment encourages him to "stand on his own feet," it must protect him against meaningless and arbitrary experiences of shame and of early doubt. [Erikson, 1963]

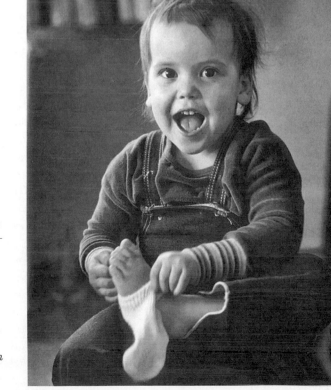

Figure 7.5 *No one is more apt to find all the permutations of dressing and undressing oneself than a toddler. Many parents, however, place a higher priority on the time saved by application of adult expertise than on the creative autonomy involved in finding every way to put on a sock, tie a shoe, or button a button.*

Like Freud, Erikson believed that problems arising in early infancy can last a lifetime. The adult who is suspicious and pessimistic, or who always seems burdened by self-doubt, was an infant who did not develop sufficient trust, or a toddler who did not achieve sufficient autonomy.

Mahler: Symbiosis and Separation-Individuation

The most influential recent reformulation of psychoanalytic theory comes from Margaret Mahler (Mahler, 1968, Mahler et al., 1975). She focuses explicitly on the relationship between the infant and the mother, a relationship that changes as the infant matures.

In Mahler's view, the first weeks of a baby's life are so dominated by primitive needs for food and sleep that most other experiences are filtered out. "The infant spends most of his day in a half-sleeping, half-waking state . . . protected against extremes of stimulation, in a situation approximating the prenatal state . . . " This stage is called "normal autism," because infants are so self-absorbed that they are unaware that anyone else exists in the world, caring for them.

Then, from about 2 months on, the infant enters a period of **symbiosis**, during which dependence on the mother is so strong, and motor skills and intellectual abilities so immature, that the infant feels literally part of the mother. This fusion with the mother is a prerequisite for the later development of the sense of self, according to Mahler. The infant's dependence on the mother forms a secure foundation for later exploration and independence.

Obviously, then, the mother's responses to the infant are crucial to the development of a secure symbiotic stage. Following Freud's lead, Mahler believes that the feeding process is significant during this period, especially the way the mother holds the baby. The infant who is cradled and comforted while sucking on the breast or bottle will have a much more satisfying symbiotic stage than the infant who is held rigidly, or worse, not held at all (Mahler et al., 1975).

Separation-Individuation   At about 5 months, a new period begins that lasts until about age 3. This is **separation-individuation,** when the infant gradually develops a sense of self, apart from the mother. Mahler refers to this as the time of "psychological birth," when babies break out of the "protective membrane" that had enclosed them and "hatch" by crawling or walking away from mother.

During this period, according to Mahler, infants show that they are very much dependent on their mothers even as they show their desire for independence. In the early phases, "all infants like to venture and stay just a bit of a distance away from the enveloping arms of the mother; as soon as they are motorically able to, they like to slide down from mother's lap, but they tend to remain or to crawl back and play as close as possible to mother's feet" (Mahler et al., 1975). Even when they are literally out of touch with their mothers, they repeatedly look back. Thus eye contact replaces the physical contact sought by the younger infant.

Between 15 and 22 months, toddlers attempt greater psychological separation from their mothers, and then become frightened at how independent they have become, perhaps regressing to a period of babyish clinging. Because they are caught between two opposite needs, toddlers can be moody, showing sorrow and dependence, or anger and aggression. Even well-adjusted toddlers show their ambivalence by darting away, hoping to be chased, or following their mother around, hoping to be noticed. Ideally, the mother will recognize the toddler's need for both independence and dependence, allowing a measure of freedom as well as comforting reassurance when it is needed.

Figure 7.6 *From the parent's perspective, the most difficult aspect of separation-individuation is deciding whether or when to encourage or intervene. As this grandfather knows, the dilemma is much the same whether the child is climbing the stairs for the first time or climbing into the driver's seat of the family car for his or her first solo at the wheel.*

Like Freud and Erikson, Mahler (1968) believes that each stage of development is important for later psychological health. Indeed, Mahler thinks severe mental illness results directly from maladaptive mothering in the first six months of life. The resolution of the separation-individuation stage also has lasting implications: adults who avoid intimacy, or fear independence, may still be trying to resolve the tension of separation-individuation and achieve a proper sense of self.

Thus, as you have seen, instead of emphasizing feeding and toilet-training, Erikson and Mahler recognize the need for a much broader view of parental behavior than Freud did. They also note that the older infant has a role in his or her own development, seeking autonomy or separation-individuation.

However, many researchers (e.g., Campos et al., 1983; Horner, 1985; Kagan, 1978; Snyder, 1978) believe that, like Freud, contemporary psychoanalysts still overemphasize the mother's power to shape the infant's development and portray young infants as far more passive and susceptible to the mother's influence than they actually are. This view of infants runs counter to the evidence for early perception and cognition, and, as you soon will see, to evidence that newborns come into the world already possessing strong personality traits that will endure no matter what their parents do.

A Life-Span Perspective

Temperament

As psychologists use the term, **personality** refers to a person's characteristic emotions, moods, actions, and responses—the sum of the ways that an individual usually reacts to events, circumstances, and people encountered in daily life. As we have just seen, the underlying assumption of many early child-care specialists was that personality develops as a result of the parents' attitudes toward, and behavior with, the child. This led, quite logically, to a readiness to blame the mother for any personality problem or character fault a child might have.

An alternative explanation has been sought by researchers who have looked specifically at **temperament,** defined as "relatively consistent, basic dispositions inherent in the person that underlie and modulate the expression of activity, reactivity, emotionality, and sociability" (McCall, in Goldsmith et al., 1987). Many aspects of temperament seem to be largely innate, probably the result of genetic and pre-natal influences. However, although temperament is apparent in the first months of life, as the person develops, the social context and the individual's experiences increasingly influence the nature and expression of temperament.

The most famous and extensive study of temperament is called the New York Longitudinal Study (NYLS), conducted by Alexander Thomas, Stella Chess, and Herbert Birch. According to their initial findings (1963), babies in the first days and months of life differ in nine personality characteristics:

1. *Activity level.* Some babies are active. They kick a lot in the uterus before they are born, they move around in their basinettes, and as toddlers, they are nearly always running. Other babies are much less active.

2. *Rhythmicity.* Some babies have regular cycles of activity. They eat, sleep, and defecate on schedule almost from birth. Other babies are much less predictable.

3. *Approach-withdrawal.* Some babies delight in everything new; others withdraw from every new situation. The first bath makes some babies laugh and others cry; the first spoonful of cereal is gobbled up by one baby and spit out by the next.

4. *Adaptability.* Some babies adjust quickly to change; others are unhappy at every disruption of their normal routine.

5. *Intensity of reaction.* Some babies chortle when they laugh and howl when they cry. Others are much calmer, responding with a smile or a whimper.

6. *Threshold of responsiveness.* Some babies seem to sense every sight, sound, and touch. For instance, they waken at a slight noise, or turn away from a distant light. Others seem unaware even of bright lights, loud street noises, or wet diapers.

7. *Quality of mood.* Some babies seem constantly happy, smiling at almost every-thing. Others seem chronically unhappy: they are ready to complain at any mo-ment.

8. *Distractibility.* All babies fuss when they are hungry, but some will stop fussing if someone gives them a pacifier or sings them a song, while others keep complaining until they are fed. Similarly, some babies can easily be distracted from their inter-est in an attractive but dangerous object and be diverted to a safer plaything, while others are more single-minded.

9. *Attention span.* Some babies play happily with one toy for a long time. Others quickly drop one activity for another.

Thomas and Chess (1977) believe that "temperamental individuality is well es-tablished by the time the infant is two to three months old," before the parents could affect personality much. In terms of various combinations of personality traits, most young infants can be described as one of three types: *easy* (about 40 percent), *slow-to-warm-up* (about 15 percent), and *difficult* (about 10 percent).

In a series of follow-up studies carried into adolescence and adulthood (Carey and McDevitt, 1978; Chess and Thomas, 1986; Thomas et al., 1968), temperamental characteristics showed some stability: the easy baby remains a relatively easy child, while the difficult one is more likely to give his or her parents problems. Similarly, the slow-to-warm-up infant who cried on seeing strangers at 8 months may well hide behind mother's skirt on arriving at nursery school and avoid the crowd in the halls of junior high.

This does not mean that temperament remains exactly the same throughout life. Indeed, some of the NYLS characteristics are not particularly stable. Rhythmicity and quality of mood, for instance, are quite variable, meaning that the infant who has been taking naps on schedule might not do so a few months later, and the baby

Figure 7.7 *At any age, differences in temperament can make one person's treat another person's trauma.*

who has seemed consistently happy might become a malcontent if life circumstances change. The age of the child is also important. In the first few years, stability is more evident from month to month than from year to year (Bronson, 1985; Peters-Martin and Wachs, 1984).

Other Research

Other researchers using a variety of measures, settings, and definitions agree that certain temperamental characteristics are apparent in the first months of life, and remain quite stable as the child gets older (Goldsmith et al., 1987; Hubert et al., 1982; Kagan et al., 1986). Evidence is particularly convincing for three characteristics: activity level (Goldsmith and Gottesman, 1981); sociability or shyness (Daniels and Plomin, 1985; Wilson and Matheny, 1986); and fearfulness (Goldsmith and Campos, 1982). This means that the infant who is active, friendly, and fearless will probably become a busy, brave, and outgoing child, while the relatively quiet, shy, and timid infant will remain so. Taken together, this evidence appears to confirm the idea that some children are indeed "easy," while others are "slow to warm up."

The "difficult" pattern of temperament shows some stability as well, according to studies of children who display behavior problems (Riese, 1987). Many of these children have a poor ability to concentrate and a quick temper, traits that appear early in infancy and tend to continue throughout childhood and adolescence. These cases may be the result of normal heredity, or of prenatal damage, or of brain damage at birth. No matter which of these explanations about origin is correct, however, the result is a child with a distinct, and in this case difficult, pattern of interaction with the world (Ross and Ross, 1982).

Very similar basic dimensions of temperament seem to characterize adults as well. For example, after decades of investigation, two leading researchers have come up with five basic personality characteristics: openness to new experiences, neuroticism (fearfulness and anxiety), extroversion, agreeableness, and conscientiousness (McCrae and Costa, 1987). Like the temperamental characteristics of young children, these adult personality traits seem relatively stable (see Chapter 22), although, of course, specific situations and incidents affect which aspects of temperament will be apparent. For example, agreeableness or disagreeableness depends partly on life circumstances: it is obviously difficult to be pleasant when one is underappreciated, overworked, and unloved.

The research on temperament has powerful implications for parents. It means that parents need not necessarily feel responsible for all the personality traits and characteristics of their developing children. Some children are, because of their inborn characteristics, difficult to direct and hard to live with, while others are, by nature, easy-going and seemingly unaffected by turbulent or unresponsive parenting (Rutter, 1979).

The authors of the NYLS have expressed the belief that the recognition of the impact of temperament

> has a most salutary impact on the "blame the mother" ideology which had previously been so pervasive among mental health professions. Child-care experts from various disciplines could reassure mothers that a child's problems could stem from many causes. Unnecessary and destructive maternal guilt was relieved, with positive effects for both mother and child. [Chess and Thomas, 1982]

Environmental Effects

The fact that temperamental characteristics are partly inborn does not by any means imply that temperament works in cultural isolation. Even a difficult child can become an easy adult. Consider one of the original subjects from the NYLS:

> Carl was one of our most extreme cases of difficult temperament from the first months of life through 5 years of age. However, he did not develop a behavior disorder, primarily due to optimal handling by his parents and stability of his environment. His father, who himself had an easy temperament, took delight in his son's "lusty" characteristics, recognized on his own Carl's tendencies to have intense negative reactions to the new, and had the patience to wait for eventual adaptability to occur. He was clear, without any orientation by us, that these characteristics were in no way due to his or his wife's influences. His wife tended to be anxious and self-accusatory over Carl's tempestuous course. However, her husband was supportive and reassuring and this enabled her to take an appropriately objective and patient approach to her son's development.

> By middle childhood and early adolescent years, few new situations arose which evoked the difficult temperament responses. The family, school, and social environment was stable and Carl flourished and appeared to be temperamentally easy rather than difficult. . . .

> When Carl went off to college, however, he was faced simultaneously with a host of new situations and demands—an unfamiliar locale, a different living arrangement, new

academic subjects and expectations, and a totally new peer group. Within a few weeks his temperamentally difficult traits reappeared in full force. He felt negative about the school, his courses, the other students, couldn't motivate himself to study, and was constantly irritable. Carl knew something was wrong, and discussed the situation with his family and us and developed an appropriate strategy to cope with his problem. He limited the new demands by dropping several extracurricular activities, limited his social contacts, and policed his studying. Gradually he adapted, his distress disappeared, and he was able to expand his activities and social contacts. When seen by us for the early adult follow-up at age 23 his temperament rating was not in the difficult group. [Thomas and Chess, 1986]

Of course, one example does not prove a rule. However, similar findings on the impact of parents comes from a study of 152 infants reared by adoptive parents and 120 infants reared by their biological parents (Daniels and Plomin, 1985). At 12 and 24 months, the infants and their parents were rated on a variety of measures of sociability. As you would expect, a correlation was found between the adopted infants' temperament and that of their biological parents, even when the biological parents had had no contact with their infants since birth. Significantly, however, there was a stronger correlation between the adopted infants' sociability and that of their adoptive parents, suggesting that parental influence in this instance was greater environmentally than genetically. The strongest correlations of all, of course, were found where nature and nurture coincided: when raising their biological children, sociable parents had very outgoing toddlers and shy parents had shy children.

Figure 7.8 *Most grouchy people can give many specific reasons for their mood, from too-tight shoes to too-high prices. However, the problem is often deeper: a "family tradition" of grouchiness can arise when genes and early family experiences conspire to provide an individual with a limited number of interpretations of, and responses to, events in his or her world.*

"Let's face it. The only play you've ever liked is 'Stop the World—I Want to Get Off.'"

Drawing by Price; © 1986 The New Yorker Magazine, Inc.

The outcome of the early temperamental patterns is influenced by the culture as well as by the family. For example, research finds that shy girls are likely to continue to be shy, but that timid boys often outgrow their timidity. Research also finds that the opposite pattern occurs with aggressiveness: girls, but not boys, become less aggressive as they mature. These patterns can be explained by the fact that our culture (and, in turn, parents) tend to value and encourage reservedness in females and assertiveness in males (Kagan, 1983).

Thus, personality depends a great deal on the way people have learned to express their desirable traits and to channel their difficult tendencies. None of us can justifiably blame our parents, our culture, or ourselves for our basic temperament; each of us can, however, find the best way to develop whatever temperamental traits we have.

Parent-Infant Interaction

As we have now seen, neither the parents' method of handling the child nor the infant's innate characteristics are sole determinants of infant psychosocial development. A third factor—the interaction between the parent and the child—is now recognized as crucial. This interaction is affected by the personality of the parent and the temperament of the child, as well as by the child's stage of development. In infancy, these stages, and the nature of the parent-child interaction, can be described by three words: synchrony, attachment, and exploration.

The Beginning of the First Year: Synchrony

State-of-the-art methodology enables researchers to investigate early parent-infant interaction at a level of detail that was previously impossible. Scientists can videotape a parent-infant pair with three cameras; one focused on the baby, one on the parent, and the third on both of them. The resulting tapes are then viewed frame-by-frame and analyzed by carefully trained observers and cleverly programmed computers, producing a wealth of detail about the nature of parent-infant interaction (Lester et al., 1985; Touliatos and Compton, 1983).

Such analysis has revealed that the parent-child interaction is often characterized by a particular type of coordination, called **synchrony,** which leads those who describe it to turn to metaphors. According to various analysts, synchrony is the intricate "meshing" of a finely tuned machine (Snow, 1977); a patterned "dialogue" of exquisite precision (Schaffer, 1984); the interplay of musicians improvising a "duet" or of dance partners executing a skilled "choreography" (Stern, 1977).

To be sure, the specific behaviors of care-givers playing with their babies are not impressive in themselves: mothers and fathers open their eyes and mouths wide in expressions of mock surprise, make rapid clucking noises or repeated one-syllable sounds ("ba-ba-ba-ba-ba," "di-di-di-di," "bo-bo-bo-bo," etc.), raise and lower the pitch of their voice, change the pace of their movements (gradually speeding up or slowing down), tickle, pat, lift, and rock the baby, and do many other simple things. Nor are the infant's behaviors very complex: babies stare at their parent partner or look away, vocalize, widen their eyes, smile and laugh, move forward or back, or turn aside.

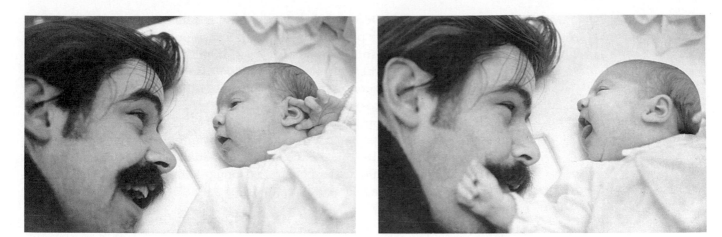

Figure 7.9 *A moment of synchrony!*

But what is fascinating to observers is the coordination of parts in the unwritten script that the parent and infant seem to follow, a coordination that becomes more impressive the more carefully one examines it. Seasoned veterans who have observed and analyzed hundreds of reels of videotaped infants and parents explain: "We have become more and more aware of the . . . balances between attention and nonattention, excitement and recovery, and of reciprocity in both togetherness and separateness" (Brazelton et al., 1979). Although not all observers are as enthusiastic in their descriptions, there is no doubt that care-givers and infants can, and do, respond to each other's emotions, gestures, and noises in the first months of life (Cohn and Tronick, 1987).

While virtually all infants and care-givers synchronize their interaction to some extent (Kaye, 1982), some seem better at it than others. There are "missteps in the dance," which are more likely to occur if the care-givers are preoccupied with their own problems. A depressed mother "may be able to go through all the practical activities of caregiving, but she will not be able to light up her face or voice or movements" (Stern, 1977). Or a mother may be overcontrolling to the point where she ignores the infant's attempt to break off the dialogue, as one mother did:

> Whenever a moment of mutual gaze occurred, the mother went immediately into high-gear stimulating behaviors, producing a profusion of fully displayed, high-intensity, facial and vocal . . . social behavior. Jenny invariably broke gaze rapidly. Her mother never interpreted this temporary face and gaze aversion as a cue to lower her level of behavior, nor would she let Jenny self-control the level by gaining distance. Instead she would swing her head around following Jenny's to reestablish the full-face position. Jenny again turned away, pushing her face further into the pillow to try to break all visual contact. Again, instead of holding back, the mother continued to chase Jenny. . . . She also escalated the level of her stimulation more by adding touching and tickling to the unabated flow of vocal and facial behavior . . . Jenny closed her eyes to avoid any mutual visual contact and only reopened them after [she had moved her head to the other side]. All of these behaviors on Jenny's part were performed with a sober face or at times a grimace. [Stern, 1977]

While this example clearly shows the effects of the mother's personality, it should be noted that the infant's personality and predispositions also affect the ease of synchrony. For example, some infants are constitutionally more sensitive to stimulation than others, and these would have particular problems with an intrusive mother like Jenny's. Further, many early problems with synchrony are overcome as the infant grows older. For example, 5-month-old infants are better able to lead the interaction than 3-month-olds (Lester et al., 1985), so they have a better chance of regulating their parents' behavior. Jenny, the baby in the example above, eventually became more able to adjust to the mother's sudden overstimulation, and the mother, finding her infant more responsive, no longer felt the need to bombard her with stimulation as she had earlier. With time, Jenny and her mother established a mutually rewarding relationship.

The End of the First Year: Attachment

Just as the moment-by-moment harmony between parents and young infants has captured scientific attention, so has the **attachment** between parents and slightly older infants been the subject of extensive research. "Attachment," according to Mary Ainsworth (1973), "may be defined as an affectional tie that one person or animal forms between himself and another specific one—a tie that binds them together in space and endures over time." Not surprisingly, when people are attached to each other, they try to be near one another, and they interact with each other often. Thus infants show attachment by "proximity-seeking" behaviors, such as approaching, following, and clinging, and "contact-seeking" behaviors, such as crying, smiling, and calling (Ainsworth and Bell, 1970). Parents show their attachment by keeping a watchful eye on their infant, even when safety does not require it, and by responding to the infant's vocalizations, expressions, and gestures. (As we shall presently see, attachment is a reliable indicator of the quality of the parent-child relationship and a good predictor of the future competence of the child.)

Measuring Attachment   Attachment can be measured in many ways. In a naturalistic study of mother-infant pairs in Central Africa, Ainsworth (1973) noted when infants cried, smiled, or vocalized differently with their mothers than with other people; whether babies cried or tried to follow their mothers when they left to go somewhere; if babies used their mothers as a base for exploration (going away from mother to explore and then coming back to reestablish contact with her); and how often infants clung to, or scrambled up on, their mothers. She also studied how the mothers fed, comforted, and showed affection to their babies. No single behavior signified attachment. For example, Ainsworth particularly noted up to what age the mothers breast-fed their babies, but found that the duration of breast-feeding did not necessarily predict strength of attachment. Instead, Ainsworth determined that the overall pattern of infant-mother interaction, with responsiveness and alertness in both partners, indicated a secure relationship.

On the basis of her field observations, Ainsworth developed a laboratory procedure in which the infant's reactions to the comings and goings of the mother and a friendly stranger would indicate the security of attachment. In this experiment, infants are observed in a well-equipped playroom in seven successive episodes, with their mother, with the stranger, and by themselves.

In this laboratory test, about two-thirds of all American infants tested demonstrate **secure attachment.** Their mother's presence in the playroom is enough to give them courage to explore the room and investigate the toys; her departure causes some distress (usually expressed through verbal protest and a pause in playing); and her return is a signal to reestablish contact (with a smile or by climbing into the mother's arms) and then resume playing. Securely attached infants also show a clear preference for their mothers over the stranger.

Other infants, however, show one of two types of **insecure attachment.** Some are anxious and resistant: they cling nervously to their mother even before her initial departure and thus are unwilling to explore the playroom; they cry loudly each time she leaves; they refuse to be comforted when she returns, perhaps continuing to sob angrily even when back in her arms. Others are avoidant: they engage in no interaction with their mothers and show no apparent stress when she leaves, and, on her return, they avoid reestablishing contact, sometimes even turning their backs. Insecurely attached infants tend to respond to the stranger no more negatively than they respond to their mothers. In some cases, they even react more positively toward the stranger.

Attachment and Care-Giving   Ainsworth's procedure for measuring attachment has been used in hundreds of studies. From these we have learned that attachment is one indication of the quality of care in early infancy (Ainsworth et al., 1978; Bretherton and Waters, 1985; Sroufe, 1985). Researchers have found that mothers who provide excellent care in the early months are much more likely to have infants who, at 12 months and 18 months, are securely attached than are mothers who are neglectful of their infants (Egeland and Sroufe, 1981) or who have little understanding of their infants' needs (Egeland and Farber, 1984). Among the features of "excellent care" that have been measured are (1) general sensitivity to the infant's need for stimulation as well as for quiet, (2) responsiveness to the infant's specific signals such as fussing or turning away, and (3) talking and playing with the infant in ways that actively encourage the child's growth and development (Egeland and Sroufe, 1981; Sroufe, 1985). Thus synchrony and responsive care-giving in the early months lead naturally to secure attachment in the later months.

Figure 7.10 *Attachment is often evidenced more by eye contact than by physical contact. Parents express their attachment primarily by watching their child, ready to encourage or rescue. Toddlers show attachment by wandering off, with an occasional glance back to mom and dad. For securely attached toddlers, simply knowing their parents are nearby provides the confidence to explore.*

Furthermore, specific deficiencies in the mother's care seem to lead to specific forms of insecure attachment. In general, mothers who overstimulate their infants tend to have fussy, avoidant infants, and mothers who understimulate their infants have anxiously attached ones (Belsky et al., 1984a).

Other research has shown that fathers and other care-givers can, and often do, provide the same secure attachment and base for exploration that mothers, in Ainsworth's classic research, do. Again, sensitivity to the child's needs is key. Fathers who are responsive to their infants tend to have infants who are securely attached to them (Lamb, 1981).

The Importance of Attachment   Why is attachment considered so important? Part of the reason stems from the psychoanalytic emphasis on the developmental importance of the early relationship between mother and child (Bowlby, 1973, 1980). But much more influential is longitudinal research that documents the results of secure and insecure attachment. It clearly shows that secure attachment at age 1 sets the stage for exploration and mature personality development during the preschool years.

For example, observations in nursery school show that 3-year-olds who were rated securely attached at age 1 are significantly more competent in certain social and cognitive skills: they are more curious, outgoing, and self-directed than those who were insecurely attached. The 3-year-olds who are securely attached are also more likely to be sought out as friends and to be chosen as leaders (Sroufe, 1978). Furthermore, securely attached infants become children who interact with teachers in friendly and appropriate ways, seeking their help when needed. By contrast, anxiously attached infants become preschoolers who are overly dependent on teachers, demanding their attention unnecessarily and clinging to them instead of playing with other children or exploring the environment (Sroufe et al., 1983). Even at ages 5 and 6, differences are apparent between children who were securely and insecurely attached as infants (Arend et al., 1979; Main and George, 1985).

Figure 7.11 *Children's behavior in nursery school is often related to their attachment history. If this little girl typically stations herself in the teacher's lap, unwilling to play with the other children, there is reason to suspect that she may not have been securely attached to her mother two years earlier.*

The Second Year: Caring for Toddlers

The synchrony of the early months, and the developing attachment as the first birthday approaches, set the stage for the next challenge to parent-infant interaction, one that even parents who have managed the dependency needs of the infant fairly easily may find difficult (Greenspan and Greenspan, 1985; Mahler et al., 1975; White, 1975). As infants become more secure as well as more mobile, they explore the environment in ever greater scope. Whether this exploration is regarded as a process of separation-individuation, or the crisis of autonomy, or the experimentation carried out by the curious "little scientist," the fact is that this exploration is sometimes disconcerting to the parents and dangerous to the child. Partly because the challenges of toddlerhood are difficult for many parents to meet, a good deal of research has been devoted to discovering which specific parental characteristics and actions encourage or discourage healthy toddler development.

HOME   One answer to that question is summarized by **HOME** (an acronym for Home Observation for the Measurement of the Environment), which is a list of forty-five family and household characteristics that have been shown to correlate with children's development (Caldwell and Bradley, 1984). These characteristics, divided into six subscales, are rated by a trained observer while visiting an infant and care-giver at home.

The Six Subscales of HOME

1. *Emotional and verbal responsiveness of mother.* Example: Mother responds to child's vocalizations with vocal or verbal response.

2. *Avoidance of restriction and punishment.* Example: Mother does not interfere with the child's actions or restrict child's movements more than three times during the visit.

3. *Organization of the physical environment.* Example: Child's play environment appears safe and free of hazards.

4. *Provision of appropriate play materials.* Example: Child has one or more toys or pieces of equipment that promote muscle activity.

5. *Maternal involvement with child.* Example: Mother tends to keep child within visual range and to look at the child often.

6. *Opportunities for variety in daily stimulation.* Example: Mother reads stories to child at least three times weekly.

HOME has been used to evaluate the environment of young children from many racial, cultural, and socioeconomic groups, with better success in predicting children's later cognitive development than conventional intelligence tests (Bradley and Rock, 1985; Elardo et al., 1975, 1977; Mitchell et al., 1985). Contrary to the traditional methods of predicting an infant's competence, HOME suggests that a lower-class toddler of average IQ but with a responsive, involved mother and a safe, stimulating play environment is likely to become a more competent preschooler than a middle-class infant who seems more advanced, but whose HOME scores are low.

Each of the six subscales of HOME are important. However, analysis of which aspects of HOME correlate best with later development reveals that, particularly for boys, the best predictor of future competence is "provision of appropriate play materials" (Bradley and Caldwell, 1980, 1984). Since "appropriate" is meant to indicate toys that encourage the child's motor and cognitive development, play materials do not need to be expensive "educational" toys; large cardboard boxes, pots and pans, and a collection of stones (too big to swallow) and plastic bottles can be

Figure 7.12 *Several studies, including HOME, have found that toddlers who are read to every day are more likely to be verbal, intelligent schoolchildren than are toddlers who are rarely read to.*

Especially for boys, availability of appropriate playthings during toddlerhood predicts later intellectual competence. Many simple objects fit this description—for example, pots and pans stack, make noise, and don't break.

great toys for toddlers. The next most important predictors of future achievement are "variety of stimulation and maternal involvement."

Not surprisingly, then, it seems that parents' greatest impact on personality development as well as cognitive development occurs as they play with their toddlers, or, alternatively, are too busy to play. It is through social play, both the active play typical of fathers (see Research Report, pages 166–167) and the quieter conversational play more characteristic of mothers, that toddlers learn about the social world (Hunter et al., 1987; Slade, 1987).

You will note that we seem to have come back to the idea that the care-giver's behavior is crucial. In fact, there is no denying the importance of the care-giver's role in fostering synchrony, or attachment, or exploration. The difference between the evidence of recent studies and the thrust of traditional theory, however, is the current emphasis on the infant's active participation in the process and ways it can affect the care-giver's behavior. This is clear in two contrasting examples from a study of mother-toddler interaction (Carew, 1980).

The first example is a mother and her 24-month-old daughter Sonja, who has just begun the interaction by saying that she went to a circus:

Mother:	No, you didn't go to the circus—you went to the parade.
Sonja:	*I went to the parade.*
	What did you see?
	I saw . . .
	What?
	Big girls.
Mother smiles.	Big girls and what else?
	Drums.
Mother chuckles.	
Sonja laughs, as if remembering the parade.	
Mother blows up a balloon.	What made all the loud noise at the end?
	Trumpets.
	Yes and fire engines. Do you remember the fire engines?
	You hold my ears a little bit.
Mother smiles:	Yes, I did, just like this. (Puts her hands on Sonja's ears.)
Sonja laughs.	

In this case Sonja begins the interaction and the mother helps her remember and recount her experiences. Indeed, each partner escalates their mutual enjoyment, as when Sonja says, "You hold my ears a little bit" and then the mother playfully holds her ears again. Thus both mother and daughter are creators in this interchange and, judging by their laughter, each enjoys the interaction they are developing together.

The other example comes from 18-month-old Terry, who begins the interaction by his active exploration:

> Terry sits in front of the bookshelf, pulling books out. He pulls out a book and picks up a piece of paper (his sister's school worksheet) and looks at it. Terry pulls out another book. Mother comes over and says "Terry, No" and removes him saying "Don't touch again," and slaps his hand. Terry babbles something back.
>
> Terry goes back and touches the books. Mother "No." Terry throws himself on the floor and whines. He gets up and picks up a doll and throws it on the floor. He throws it again. Terry marches back to the shelf and pulls at the books again. Mother yells "No," and goes to remove him. Terry marches around and then picks up a framed picture from the shelf. Mother: "Terry!" and comes to remove him. "Don't touch." Terry laughs. Mother: "I am not playing with you!" He goes and picks up another picture. Mother tells his sister to get it from him and she does so. Terry tries to get it back. Mother goes and pulls him away from the shelf. Mother: "Don't touch it again, you know it. Don't laugh, fresh kid." Terry laughs and walks to the TV. [Carew, 1980]

One way to analyze this episode is to take the traditional stance and criticize the mother. Terry gave clear signals that he wanted his mother's attention (taking the books down, throwing the doll, and taking the pictures), and his mother's reaction was, at best, ineffectual, and at worst, destructive. For instance, she created a conflict between Terry and his sister, and then intervened on behalf of the sister—an episode that may well lead to greater sibling rivalry. Ideally, Terry's mother should have seen that he needed distraction and attention as well as restriction and prohibition. She should have moved him, not only away from the books, but also toward something he would like—perhaps reading to him from one of his own books, or looking at a family photo album with him.

However, instead of simply blaming the mother, as traditional approaches would do, consider Terry's contribution to this interaction, as well as the overall context. Terry stubbornly persists in doing what his mother doesn't want him to do and laughs at his mother's rebuke. As you remember from the description of infant emotions, Terry's reaction is not unusual for a toddler, but that doesn't make it easier to handle. Further, Terry's temperament may not be particularly easy. At least in this incident, he is difficult to distract, quick to complain, and easy to anger. Finally, the situation is not an easy one: for obvious reasons, many mothers are less patient and creative with their second child than with their first, especially if the first is still relatively young. The recognition that the interaction of care-giver, child, and the overall context are important in understanding a child's behavior and development will help us with the last topic of this chapter, child abuse and neglect.

RESEARCH REPORT   ## Fathers and Infants

While traditional views of infant development typically focused on mothers, the past decade of research on fathers and infants has led to the following conclusion:

With the exception of lactation, there is no evidence that women are biologically predisposed to be better parents than men are. Social conventions, not biological imperatives, underlie the traditional division of parental responsibilities. [Lamb, 1981]

To be specific, fathers are just as capable of sensitively caring for young infants—bottle-feeding, changing, bathing, and soothing them—as mothers are (Parke and Tinsley, 1981). Fathers also show synchrony in play and speak Motherese just as well as mothers. Further, father-infant attachments appear at the same age and are similar to mother-infant attachments (Lamb, 1981).

However, while fathers are capable of acting quite "motherly," there are many interesting differences between typical father-infant and mother-infant interactions. For one thing, mothers usually spend far more time interacting with their babies than fathers do, especially in the early months. This is true partly because, in our culture, fathers are away from the house more than mothers are. But even when both parents are home, such as during the evenings or weekends, the difference holds. While fathers sometimes feed, diaper, and cradle their new babies when both parents are present, mothers do it more frequently (Belsky et al., 1984b).

In addition, the father's involvement seems to be more affected by certain circumstances than the mother's is. Fathers are significantly more involved with first-borns than with later-borns (Belsky et al., 1984b), with sons than with daughters (Lamb, 1986), and when their relationship with the child's mother is strong rather than shaky. This holds true for both adolescent and adult fathers (Belsky et al., 1984b; Lamb and Elster, 1985) and for fathers in many cultures (Lamb, 1987). While mothers are affected by the same factors, their behavior toward their children is more equitable.

Fathers' play is also different from mother's play. Fathers typically are noisier, make bigger gestures, and are more inclined to use the element of surprise in their play. According to slow-motion examination of tapes of parent-infant interaction, mothers' play is much more modulated and contained than fathers', providing "an envelope for verbal interaction," whereas fathers' play is more physical and exciting (Brazelton et al., 1979). Even in the first months of the baby's life, fathers are more likely to play

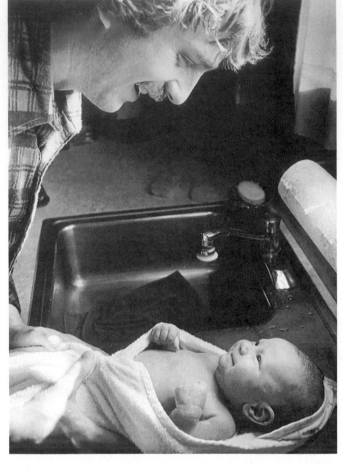

At 8 weeks, this boy already seems to know what to expect from daddy at bath time—lots of fun and games.

by moving the baby's legs and arms in imitation of walking, or kicking, or climbing, or by zooming the baby through the air ("airplane"), or by tapping and tickling the baby's stomach; mothers, on the other hand, are more likely to talk or sing soothingly, or to combine play with care-taking routines such as diapering and bathing (Parke and Tinsley, 1981).

These differences between mothers' and fathers' play are not lost on infants. Even 3-month-olds typically react with more visible excitement when approached by their fathers than when approached by their mothers. In the first months of life infants are more likely to laugh—and more likely to cry—in episodes of play with Daddy.

As infants grow older, fathers generally increase the time they spend with them, and their tendency to engage in physical play becomes more pronounced. Fathers are likely to swing their toddlers around, or "wrestle" with them on the floor, or crawl after them in a "chase."

Mothers, on the other hand, are more likely to read to their toddlers or help them play with toys (Parke and Tinsley, 1981). These differences continue to be reflected in infants' reactions. According to one study (Clarke-Stewart, 1978), 20-month-olds are more responsive during play with their fathers than with their mothers. By 30 months, differences are even more apparent: 2-year-olds are generally more cooperative, involved, and interested in their fathers' games than in their mothers' play, and judging by their smiles and laughter, they have more fun.

The father-infant relationship varies with the sex of the child: fathers and sons play together more, and show greater attachments, than fathers and daughters do. It also varies with the culture. Compared to American toddlers, for instance, Swedish toddlers seem notably more attached to and affectionate with their mothers than with their fathers (Lamb et al., 1983), possibly because, although typically more involved in day-to-day infant care, Swedish fathers are less likely to play with their toddlers than American fathers are.

Of course, it would be a mistake to assess a parent's impact on a child merely by looking at time spent in caregiving, security of attachment, or laughter during play. Indeed, in some cases, a parent's most powerful influence on a child may be an indirect one, as he or she encourages, or sabotages, the mothering or fathering of the spouse (Pederson, 1981). For example, two parents are generally better than one in ensuring good care for the child. However, in some cases, mothers provide better care for their infants when their spouse is gone, because the father was so critical and demanding when he was part of the household. As a leading research team explains:

In our experience, single parents are rather less abusive than couples, which is surprising because one would think that a spouse would provide support in the face of crisis. In fact a spouse who is not supportive is worse than no spouse at all when it comes to childrearing. [Kempe and Kempe, 1978]

Further, the influence of any particular father is affected not only by his individual characteristics, but also by the attitude of his wife and the influence of the culture.

Babies obviously love a good game of catch—but have you ever seen a mother playing one with her infant?

Recent changes in our own society—particularly the increase in the number of mothers in the work force—have meant that today's fathers are much more involved with their infants than they once were. From the perspective of the developmentalist, this is a welcome change: infants thrive under their fathers' care and attention (Lamb, 1986).

Child Abuse and Neglect

Of all the developmental problems discussed in this text, child maltreatment is the most destructive to the child who experiences it, to the care-giver who commits it, and to the society that allows it. Yet child abuse and neglect are serious problems in every nation of the world (Leavitt, 1983).

The actual prevalence is hard to estimate, since maltreatment often goes unreported (Brown, 1983). Even the reported cases, however, show that abuse is all too common. According to information from various sources, in the United States, one out of every twenty-eight American children under age 14 was reported as abused or neglected in 1986. That is more than a million reported cases a year. What, exactly, is **child abuse?** In fact, it is a cluster of different behaviors, each of which results in harm to the child's development.

The most obvious form of abuse is severe physical abuse, the "battered-child syndrome" first described in 1962. Since then, pediatricians and emergency room staff have been trained to examine cases of "accidental" injury and required to report suspected abuse (Solnit, 1980). They look for hidden bleeding from bruises under the skull; burn marks that are round (from cigarettes), or latticelike (from hot radiators), or that stop suddenly part-way up the child's body (from scalding bathwater); partially healed fractures; and many other signs that indicate that a particular injury was not an accident at all. Such severe abuse constitutes only about 4 percent of all reported cases.

However, less extreme physical abuse, producing cuts, welts, bruises, or no marks at all, can sometimes be as destructive as extreme abuse. Vigorously shak-

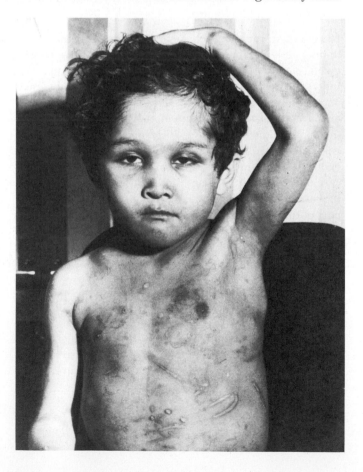

7.13 *While the severe battering of a child may be the most disturbing kind of abuse to witness, long-term and repeated physical and psychological abuse is much more common and more damaging. When asked about his injuries, which included cigarette burns, welts, and bite marks, this 5-year-old explained: "My stepfather sometimes says, 'I'm a lion. You're a piece of meat.' I guess he doesn't like me."*

ing an infant, for example, is a common reason for the brain damage too often found in abused children (Martin, 1980).

Emotional maltreatment and sexual abuse can be even more destructive, in the long term, than physical abuse. Unfortunately, they often go unrecognized and unreported, and therefore do not get treated until years after the damage is done. Particularly in the case of sexual abuse, one reason for underreporting is that people simply had not believed that the sexual abuse of young children could occur. In recent years, increased public awareness of the problem, and the establishment of agencies to deal with sexual abuse, have resulted in a doubling in the sexual-abuse reporting (Brown, 1983). (Emotional and sexual abuse are discussed later in this book, at the stages of development when they become more common, emotional maltreatment in Chapter 10 and sexual abuse in Chapter 16.)

Finally, **neglect** is actually the most common form of maltreatment as well as the most destructive, causing more deaths, injuries, and long-term problems than abuse (Cantwell, 1980; Wolock and Horowitz, 1984). Some instances of neglect are blatant and horrifying: infants who are allowed to starve or freeze to death are examples. Others are less obvious, involving infants who are debilitatingly undernourished, or whose parents rarely cradle, talk, or play with them. Furthermore, many childhood accidents (by far the greatest cause of childhood death and serious injury) can be traced to neglect, although they are rarely reported as such.

Figure 7.14 *Neglect is more common than abuse, and more destructive. It is a factor in most of the accidental deaths that occur to children through age 10 in the United States each year. Neglected children are also more likely to be stunted physically, cognitively, and emotionally, than abused children.*

Causes of Child Abuse and Neglect

Traditionally, the cause of abuse was seen as residing solely in the "disturbed" abuser who lost control. Parents were blamed and sometimes punished, and often the child was removed from the home. However, as the ecological approach has become more widely used to understand child abuse, the focus in searching for causes has moved from the idea of the "pathological parent" to the interaction among parent, child, and society, and the emphasis has shifted from placing blame to early diagnosis and prevention (Cohn, 1983; Thompson, 1983). Let us look first at the overall social and cultural milieu in which maltreatment occurs.

The Social Context   Some cultures have a much lower rate of deliberate parental maltreatment than others. Such cultures have three characteristics:

1. Children are highly valued, as a psychological joy and fulfillment. They also tend to be more of an economic asset than a liability.

2. Child care is shared by many people. If the mother is unwilling or unable to care for her child, other relatives are ready to take over, sometimes for a few moments, other times for years.

3. Young children are not expected to be responsible for their actions. In some cultures, almost any punishment of children younger than age 3, or even age 7, is considered abusive and unnecessary.

The importance of social values and conditions is exemplified by the different rates of child abuse among the Polynesian people who live in their traditional home, the Pacific Islands, and those who have emigrated to New Zealand (Ritchie and Ritchie, 1981). Among the former, abuse is virtually nonexistent. Their society fully meets the three criteria listed above, for children are highly valued, cared for by many adults (in fact, in Polynesian, the same word is used for "adult" and "parent"), and children are considered unteachable until they are at least 2 years old.

However, when Polynesians move to New Zealand, the rate of child abuse skyrockets, surpassing the rate of the Caucasian New Zealanders many times over. The demands of the new lifestyle make it impossible for the parents to continue their old parenting habits of relaxed permissiveness and informal and extensively shared child care. It takes time to develop viable new patterns (such as learning how to enforce new guidelines for children's behavior without resorting to physical punishment, finding ways to replace the freely available care-givers of the past, and limiting family size so that children are not an overwhelming financial burden).

Cultural Factors Affecting Abuse in the United States.   Comparing the three characteristics of nonabusive societies with the patterns common in the United States reveals why abuse is so prevalent in this country. First, children are often considered to be both a financial and personal burden. Not surprisingly, the incidence of abuse and neglect rises as income falls; very poor parents (family income under $5,000 per year) are the most likely to abuse their children (Pelton, 1978), especially if the living conditions are crowded and available friends and helpers are few.

Second, in the United States, social support for mothers and young children is scarce. If she is lucky, a mother has relatives and friends who willingly help her, but many mothers, especially teenage mothers, are not so fortunate. Grandmothers, for instance, once the mainstay of practical help, are now much more likely to live a distance away and to be involved with careers and social worlds of their own.

Social isolation allows child abuse in two ways. First, compared to parents who have an active network of friends and family members, isolated parents are more likely to take out their problems (such as a crisis of divorce or unemployment) on their children. Second, when abuse begins, others are reluctant to intervene until the pattern is well established and considerable harm has been done. It is not surprising, then, that study after study finds a correlation between social isolation and child abuse (Cohn, 1983; Kempe and Kempe, 1984; Thompson, 1983).

Finally, our culture's attitude about young children may add to the problem. The emphasis on the infant's and preschooler's ability to learn may cause some parents to forget that young children are also immature, self-absorbed, and dependent on others. For example, many abusing parents consider irritating but normal infant behavior to be deliberate and therefore amenable to correction: they punish infants for "crying too much" or punish toddlers for being unable to control urination or defecation. Complicating this aspect all the more is the fact that physical force—from an occasional slap to frequent batterings—is a common form of punishment in most American homes (Gelles, 1978).

Problems in the Parents   According to Ruth and Henry Kempe (1978), of every ten parents who severely abuse their children, one is so mentally ill as to be untreatable. Such persons have delusions ("God is telling me to kill this child") or they communicate only by bashing. They are cruel or fanatical, and completely closed to reason. With these parents, the only solution is removing the child from the home.

But most abusive parents are not very different from average parents. They love their children, and want the best for them. In fact, it may be a sense of failure as a parent that triggers their abusiveness. When the infant cries, for instance, they may interpret the crying as a form of accusation. As the Kempes write:

> An average mother will regard a crying or fussy baby as hungry or wet or full of gas. She will proceed to feed, change, burp him, and then put him down in the crib and say, "Baby, you're tired," close the door, then turn on the radio or talk to a friend. The abusive parent is unable to leave the crying child, and tries harder and harder to pacify him until in a moment of utter frustration she is overwhelmed by the thought that the baby, even at two weeks of age, is saying, "If you were a good mother I wouldn't be crying like this." It is precisely because the parent tries to be extra good, to be loved and earn the love of the child, that intractable crying is seen as total rejection and leads to sudden rage. The abuse is clearly not a rational act. It is not premeditated, and it is often followed by deep grief and great guilt. Such parents are seen by doctors and nurses as being very solicitous. Third parties find it hard to believe that so loving a parent could have inflicted such serious injury. [Kempe and Kempe, 1978]

Figure 7.15 *What would be your reaction if she were your baby and had been crying like this for the last 15 minutes (it's now 3:26 A.M.)?*

Factors Associated with Child Abuse

According to several studies (Cohn, 1983; Kempe and Helfer, 1980; Kempe and Kempe, 1984; Thompson, 1983), adults are more likely to be child abusers if:

Background Variables	They were abused or neglected as children.
	They grew up in a culture markedly different from that in which they are now living.
	They have little education.
	They are under age 20.
Social-Setting Variables	They have few friends or nearby relatives.
	They are drug abusers, especially alcoholics.
	They are victims or perpetrators of spouse abuse.
	They have recently experienced family stress (e.g., death of a parent, loss of a job, new pregnancy, divorce).
	They have several children, especially several under age 6.
	Their living space is crowded, with no privacy.
	They are responsible for the care of a child who is not their biological offspring (such as a step-child, or cousin).
Personality Variables	They have difficulty coping with anger.
	They have low self-esteem.
	They have inconsistent and/or unrealistic expectations of children.

Bear in mind that no one of these factors by itself is an inevitable signal of potential abuse. However, the more of these conditions that occur in combination, the greater the likelihood that abuse will occur.

A related problem is that, for some abusive parents, it is not their intentions but their inability to understand the child's communications that is at the root of the problem. For instance, abusive mothers tend to misread the facial expressions and cries of their infants, sometimes mistaking distress for anger (Kropp and Haynes, 1987). Other parents "try too hard," forcing the baby to repel their unwelcome and insensitive interference (Belsky et al., 1984c).

For some abusive parents, the seeds of the problem were sown early in their development, for they themselves were abused or neglected as children. Researchers believe that this type of childhood experience might make abusers out of the abused in two ways: it hinders the development of self-esteem, patience, and social skills and it provides a negative, destructive model of parenting.

An additional factor that increases the likelihood of maltreatment is drug dependency. One study of addicts found virtually all of the parents were neglectful to some degree. In addition, 27 percent of the alcoholics and 19 percent of the heroin addicts physically or sexually abused their children (Black and Mayer, 1980).

Problems in the Child   The systems approach to child development has made researchers look more closely at the other partner in the abusive parent-child relationship. Sometimes something in the infant triggers, or encourages, a destructive pattern.

Babies who are unwanted, who are born too early, who were the product of an unhappy love affair or a difficult pregnancy, who are the "wrong" sex, or who have physical problems, can become victims of their parents' disappointment. Even the baby's appearance, as in the case, say, of the little boy who looks just like the father who left the mother early in pregnancy, or the little girl who reminds her father of his abusive mother, can trigger rejection instead of love. Parents may also be unhappy and frustrated over their baby's temperament, wanting a more quiet one, or a more active one, or a less difficult one (Steele, 1980). All these disappointments may lead to unresponsive and rejecting parenting, which, as we have seen, is likely to make a child much more difficult than he or she otherwise would have been. Note, however, that maltreatment is more likely to create a hard-to-manage child than vice versa. Obviously, abused children are not to be blamed for their fate.

Treatment and Prevention

Most experts agree that early warning is essential to prevention and that the child's first weeks and months should be the focus of particular concern (Schwertzbeck, 1983). Among the warning signs are an absence of synchrony (an infant or parent might repeatedly avoid eye contact), the failure of the infant to gain weight from one pediatric checkup to the next, the failure of a parent to bring the infant in for checkups, and signs of insecure attachment (Lamb et al., 1985). If such signs are noted, a supportive network—either of friends and relatives or of professionals—can be activated before destructive parent-infant interactions become habitual. Even when abuse has already occurred, intervention can halt the process. A four-year British study of severely abused infants whose families had participated in an intensive therapy program found that about two-thirds were doing fairly well—with normal cognitive and psychological development in the children and better family development generally. The other third showed notable retardation and disturbance, but even in these cases the family had typically managed to provide better care for the other children in the family (Lynch and Roberts, 1983). An American team of child-abuse researchers (Kempe and Kempe, 1978) believe that 80 percent of abusing parents can be helped so that they no longer physically punish their children. Both short-term help (a hotline for parents when they find themselves losing control and an around-the-clock crisis nursery where parents in need of a few hours' peace can drop off their children), and long-term help (individual and family therapy) are needed.*

There is encouraging evidence that between 1975 and 1985, the rate of child abuse dropped, probably as the result of better reporting and treatment measures. However, a look at the broader context of abuse suggests that since poverty, youth, and ignorance correlate with poor parenting, measures that raise the lowest incomes, discourage teenage parenthood, and increase the level of education will probably be needed to further reduce the rate of abuse. And since social isolation and unrealistic expectations make it harder to provide good care for young children, any program that fosters friendly contact with others and an accurate understanding of the needs of children should be encouraged. With a little experienced guidance, most parents can become better at appreciating their children and learning how to relate to them with greater respect and with mutual delight.

*An organization called Parents Anonymous, with chapters in every major city, helps abusing parents in much the same way that Alcoholics Anonymous helps alcoholics.

SUMMARY

Emotional Development

1. In the first days and weeks of life, infants are capable of expressing many emotions, including fear, anger, sadness, happiness, and surprise. Toward the end of the first year, the typical infant expresses emotions more readily, more frequently, and most distinctly.

2. In the second year, cognitive advances cause infants to become more conscious of the distinction between themselves and others and thus more self-assertive.

Personality Development

3. In the traditional view of personality development, mothers are the almost omnipotent creators and shapers of infant character. In the first half of the 20th century,

this view was put forth in child-rearing manuals, in the pronouncements of the leading behaviorist, John Watson, and, from a different perspective, in the theoretical assumptions of Sigmund Freud.

4. Freud argued that the child-rearing practices encountered in the oral and the anal psychosexual stages had a lasting impact on the person's personality and mental health.

5. Erikson and Mahler built on Freud's ideas, broadening his concept of the first two stages. According to Erikson, the infant experiences the crises of trust versus mistrust, and then autonomy versus shame and doubt. Mahler envisions three phases of infant development: normal autism, symbiosis, and then separation-individuation. Like Freud, both of these psychoanalytic thinkers stress the lifelong impact of the care-giver's actions during the first two years.

6. Contemporary developmentalists generally do not accept Freud's stages of infant development. Erikson and Mahler have more influence on current thought. However, many critics contend that the psychoanalytic theorists overemphasize the impact of the mother's role in the first two years.

7. Temperament, a group of personality characteristics, some largely influenced by genetics, others more susceptible to environmental influences, is another factor in psychosocial development.

Parent-Infant Interaction

8. In addition to the parents' actions and the infants' temperament, developmentalists now stress the interaction between parent and child.

9. The early parent-child interaction is characterized by synchrony, a harmony of gesture, expression, and timing that can make early nonverbal play a fascinating interchange. Attachment between parent and child becomes apparent toward the end of the first year. Secure attachment tends to predict curiosity, social competence, and self-assurance later in childhood; insecure attachment tends to correlate with less successful adaptation in these areas.

10. During toddlerhood, an important aspect of parent-infant interaction is how the parent encourages, or restricts, the child's exploration of the environment. Developmental researchers agree that the care-giver's responsiveness to the child and the stimulation of the play materials and setting are significant determinants of infant development.

Child Abuse and Neglect

11. Child maltreatment can take many forms—physical abuse, emotional maltreatment, sexual abuse, and neglect. Of these, neglect is the most common, and most devastating, in infancy.

12. The causes of abuse are many, including problems in the society (such as cultural attitudes about children), in the parent (such as drug addiction), and in the child (such as being sickly or difficult). The most effective strategies emphasize prevention and treatment rather than blame. In addition, measures that reduce the stresses and increase the social support for families with young children make child maltreatment less likely.

KEY TERMS

social smile *(146)*	personality *(153)*
fear of strangers *(146)*	temperament *(153)*
separation anxiety *(146)*	synchrony *(158)*
self-awareness *(147)*	attachment *(160)*
oral stage *(149)*	secure attachment *(160)*
anal stage *(150)*	insecure attachment *(161)*
trust versus mistrust *(150)*	HOME *(163)*
autonomy versus shame and doubt *(151)*	child abuse *(168)*
symbiosis *(152)*	neglect *(169)*
separation-individuation *(152)*	

KEY QUESTIONS

1. Which emotions develop in the first year?

2. Which factors influence whether a baby will be afraid of a stranger?

3. What are some consequences of the toddler's growing sense of self?

4. What are the similarities among the theories of Freud, Erikson, and Mahler?

5. What are the three most common tempermental patterns in infancy, and how does nurture affect them?

6. What are the similarities and differences between mother-infant and father-infant interactions?

7. How does attachment affect cognitive development, and vice versa?

8. What are some of the important factors in a child's development of social and cognitive competence?

9. What are some of the reasons parent-infant interaction does not always go well?

10. How common is child abuse and neglect?

11. What are some of the factors in the parent and in the child that may lead to child abuse?

12. What can be done to help abused children and their parents?

Part II

The Developing Person So Far: The First Two Years

Physical Development

Brain and Nervous System

The brain triples in weight. Neurons branch and grow into increasingly dense connective networks between the brain and the rest of the body. As neurons become coated with an insulating layer of myelin, they send messages faster and more efficiently. The infant's experiences help to "fine-tune" the brain's responses to stimulation.

Motor Abilities

Brain maturation allows the development of motor skills from reflexes to coordinated motor abilities, including grasping and walking. At birth, the infant's senses of smell and hearing are quite acute, and although vision at first is sharp only for objects that are about 10 inches away, by 6 months, acuity approaches 20/20.

Cognitive Development

Cognitive Skills

The infant progresses from knowing his or her world only through immediate sensorimotor experiences to being able to "experiment" on that world mentally, through the use of mental combinations and an understanding of object permanence.

Language

Babies' cries are their first communication; they then progress through cooing and babbling. Interaction with adults through "baby talk" teaches them the surface structure of language. By age 1, an infant can usually speak a word or two, and by age 2 is talking in short sentences.

Psychosocial Development

Personality Development

The major psychosocial development during the first two years is the infant's transition from total dependence to increasing independence. This transition is explained by Freud in terms of the oral and anal stages, by Erikson in terms of the crises of trust versus mistrust and autonomy versus shame and doubt, and by Mahler in terms of separation-individuation.

Understanding Self and Others

In the first month, infants have very little understanding of themselves and others as separate persons. Between the ages of 1 and 2, they begin to develop self-awareness and, consequently, become much more attentive to the reactions of others.

Parent-Infant Interaction

Parents and infants respond to each other first by synchronizing their behavior. Toward the end of the first year, secure attachment between child and parent sets the stage for the child's increasingly independent exploration of the world.

The Play Years

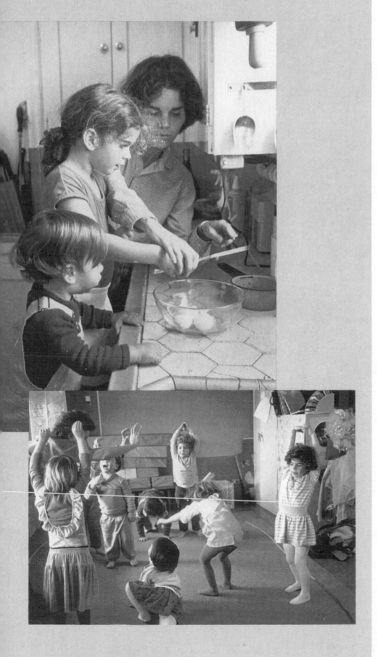

The period from age 2 to 6 is usually called early childhood, or the preschool period. Here, however, these years are called the play years to underscore the importance of play. Play occurs at every age, of course. But the years of early childhood are the most playful of all, for children spend most of their waking hours at play. They chase each other and dare themselves to attempt new tasks, developing their bodies; they play with words and ideas, developing their minds; they play games and dramatize fantasies, learning social skills and moral rules. In the process, they acquire the skills, ideas, and values that are crucial for growing up.

The playfulness of young children can cause them to be delightful or exasperating. To them, growing up is a game, and their enthusiasm for it seems unlimited, whether they are quietly tracking a beetle through the grass or riotously turning their play area into a shambles. Their minds seem playful, too, for the immaturity of their thinking enables them to explain that "a bald man has a barefoot head," or that "the sun shines so children can go outside to play."

177

CHAPTER 8

The Play Years: Physical Development

The lost child cries, but still he catches fireflies.

Ryusui Yoshida
The Lost Child

Physical development during early childhood occurs on many fronts, with the most important—maturation of the nervous system and mastery of skills—the least obvious. Let us begin with what is most obvious: the striking changes in size and shape that cause a 6-year-old to find photos of him- or herself as a chubby, clumsy toddler both amusing and unrecognizable.

Size and Shape

During the preschool years, the child becomes slimmer as the lower body lengthens. The kindergarten child no longer has the protruding stomach, round face, and disproportionately short limbs and large head that are characteristic of the toddler. By age 6, the proportions of the child's body are not very different from those of the adult (Sinclair, 1978).

Steady increases in height and weight accompany the changes in body proportions. From age 2 through 6, children add almost 3 inches (7 centimeters) and gain about 4½ pounds (2 kilograms) per year. By age 5, the average North American child weighs about 40 pounds (18 kilograms) and measures 43 inches (109 centimeters).

Many children are taller or shorter than these averages (see Figure 8.1). Weight is especially variable: about 10 percent of American 5-year-olds weigh less than 35 pounds and another 10 percent weigh almost 50 pounds (National Center for Health Statistics, 1976). Of the many factors that influence height (see Table 8.1) and weight, the two most influential are the child's genetic background and nutrition (Meredith, 1978).

Health care is also an important factor in that children who are repeatedly or chronically sick tend to become malnourished, and this, in turn, affects their growth. Upper-class children, urban children, and first-born children are all somewhat taller than average, largely because their diet and medical care are better (Eveleth and Tanner, 1976).

Figure 8.1 *As these charts show, boys (black line) and girls (color line) grow more slowly and steadily than they did in the first two years of life. Consequently, weight gain is particularly slow, with most children losing body fat during these years. The weight that is gained is usually bone and muscle.*

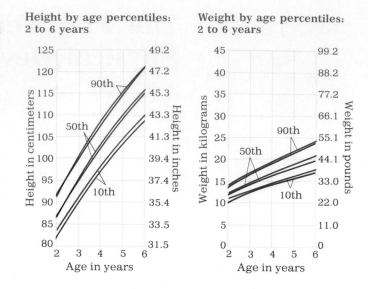

Generally, boys are slightly taller and heavier than girls, although this varies, depending on the culture. Even in the early years, boys in India are markedly taller and heavier than their sisters because boys are much preferred and therefore are given better care (Poffenberger, 1981). In North America, the heaviest children tend to be female because girls are more likely than boys to eat too much and to get too little exercise to burn off the extra calories (Lowrey, 1978). (Obesity will be discussed in more detail in Chapter 11.)

TABLE 8.1   **Factors Affecting the Height of Preschoolers**

Taller than average if	Shorter than average if
well nourished	malnourished
rarely sick	frequently or chronically sick
African or northern European ancestors	Asian ancestors
mother is nonsmoker	mother smoked during pregnancy
upper class	lower class
lives in urban area	lives in rural area
lives at sea level	lives high above sea level
first born in small family	third or later born, large family
male	female

Sources: Eveleth and Tanner, 1976; Lowrey, 1978; Meredith, 1978; Vaughan, 1983.

Eating Habits

Whether a child is short or tall, his or her annual height and weight increase much more slowly from age 2 to 6 than during the first two years of life. (In fact, between ages 2 and 3, an average child adds fewer pounds than during any other twelve-month period until age 17.) Since growth is slower during the preschool years, children need fewer calories per pound during this period than they did from birth through toddlerhood. Consequently, their appetites are smaller, a fact that causes many parents to worry. Indeed, in the view of their parents, 42 percent of the boys and 31 percent of the girls between the ages of 4 and 5 do not "eat well." In most cases, this decline in appetite does not represent a serious problem. Only rarely

Figure 8.2 *Since preschoolers tend to equate size with age, and age with wisdom and power, the question of "who is taller" often leads to some stretching of the facts.*

does it indicate a serious behavioral disorder, and marked improvement almost always occurs by age 8 (Achenbach and Edelbrock, 1981). Serious malnutrition is much more likely in infancy and in adolescence than in early childhood (Abraham et al., 1974).

Of course, as at any age, the diet during the preschool years should be a healthy one. The most common nutritional problem during the preschool years is iron deficiency anemia, the chief symptom of which is chronic fatigue. This problem, which stems from an insufficiency of quality meats and dark-green vegetables, affects mostly the poor (including, in the United States, 34 percent of all black, 13 percent of all white, and 9 percent of all Hispanic preschoolers [Eichorn, 1979]). Although most nutritional problems during the preschool years are related to low family income, it should be noted that too many "sweets" are a contributing factor in every social class. Candy, soda, and sweetened cereals can spoil a small appetite faster than they can a large one, and they therefore should be limited to make sure the child consumes enough of the foods that contain essential vitamins, minerals, and protein.

There is another reason sweets should be limited. Tooth decay is a chronic disease that is directly correlated with sugar consumption. While other factors, such as quality of dental care, frequency of brushing and flossing, and fluoridation of the water are also important, a summary of various kinds of research—cross-sectional and longitudinal, national and international—makes it clear that snacking, especially snacking on sugary foods, is likely to result in tooth decay a few years later (Newbrun, 1982). While this may seem a minor problem in the preschool years, the long-term consequences can include tooth loss and gum disease.

Brain Maturation

The most important physiological development during early childhood is maturation of the central nervous system. As indicated in Chapter 5, during childhood the brain develops faster than any other part of the body. One simple indication of this is weight: by age 5, the brain has attained about 90 percent of its adult weight, even though the average 5-year-old's body weight is less than one-third the average adult's (Tanner, 1978). Part of this increase in brain size is due to the ongoing process of myelination, which provides the nerves with an insulating sheathing that speeds up the transmission of neural impulses. Myelination bears significantly on the child's developing abilities: the areas of the brain associated with hand-eye coordination, for example, do not become fully myelinated until around age 4; those associated with the ability to maintain focused attention, not until the end of childhood; those associated with language and intelligence, not until age 15 or so (Tanner, 1978).

Brain maturation is especially important when one considers the abilities that are required for formal education. Unless the child is neurologically able to concentrate, merely sitting quietly and watching the teacher can be a difficult task. In fact, concentration, especially the ability to screen out distractions, improves markedly between ages 4 and 7 (Higgins and Turnure, 1984). Similarly dependent on a certain level of brain maturation are school skills such as reading, which requires attention and controlled coordination between the eye muscles and several areas of the brain, and writing, which involves mental coordination of sounds, letters, words, and small movements of the fingers.

The Two Halves of the Brain

One of the most important aspects of the development of the brain has to do with its being organized into left and right hemispheres, each controlling one half of the body. Oddly enough, the left hemisphere of the brain controls the right side of the body, while the right hemisphere controls the left half of the body. Thus the right hand, foot, eye, ear, and so on, are "wired" to receive signals from and send them to the left side of the brain; and the left hand, foot, and so on are wired to the right.

As the brain matures, it becomes more specialized; that is, particular parts of the brain tend to be used for certain abilities. In 95 percent of right-handed adults as well as 70 percent of left-handed adults, the left side of the brain is the location of many areas that deal with aspects of logical analysis and language development, including speech, while the right side contains many areas for various visual and artistic skills, including recognizing faces, responding to music, and spatial perception (Kinsbourne and Hiscock, 1983).

Obviously, for a person to be fully functioning, both sides of the brain need to work together. (Despite reports from the popular press to the contrary, no normal person is solely a right-brain or left-brain thinker [Bradshaw, 1983; Gardner, 1982]). Therefore the development of the network of nerves that connect the two sides, in the **corpus callosum,** is particularly important. As the corpus callosum becomes increasingly myelinated between ages 2 and 8 (Yakovlev and Lecours, 1967), the functioning of the two halves of the brain becomes more closely integrated. Interestingly, the corpus callosum is significantly larger in left-handed individuals, which suggests that left-handed children may develop stronger communication networks between the two halves of the brain, and thus develop greater bihemispheric functioning, than right-handed children.

Brain specialization obviously increases the capacity of the person to perform a

Figure 8.3 *Sometimes preschoolers use their left hand to paint or write simply because their right hand is busy doing something else. (One of the author's former undergraduate students had learned to write with his left hand because he usually kept his right thumb in his mouth. As an adult, he is right-handed for everything except writing, the opposite pattern of many left-handed persons.)*

variety of intellectual and motor tasks. At the same time, however, such specialization makes it harder for the brain to compensate for loss of function of particular areas due to injury. This is seen most dramatically in the case of language loss in people who have suffered damage to the language area of the brain. If the victim is a young child, language function may be taken up by another area of the brain, and the child may learn to speak normally again. If the victim is past adolescence, when language specialization is completed, loss of ability to use language is likely to be permanent (Gardner, 1982).

Another indication of brain specialization in early childhood is the clear emergence of hand preference during those years. Some signs of hand preference are evident in early infancy, suggesting that hand preference is genetic (Corballis, 1983). As children begin to become more skilled at manipulating objects with one hand (usually their right), they also become more skilled at sensitively touching objects with the other. A study of fourteen right-handed blind children, for instance, found that they could read Braille more quickly and accurately with the fingers of their left hand than with the fingers of their right (Hermelin and O'Connor, 1971). (This study was prompted by a blind child who had injured his left hand and said he could not do his schoolwork because his "reading hand" was hurt.) Like hand preference for manipulation, hand preference for sensitive touching develops during early childhood, another indication that the brain specialization occurs during these years (Rose, 1984).

Brain Maturation and Vision

Brain maturation probably underlies the improvement of vision that occurs during the preschool years. Children younger than 6 do not usually have sufficiently developed eye muscles to allow them to move their eyes slowly and deliberately across a series of small letters (Vurpillot, 1968). As a result, preschoolers are likely to guess at a word on the basis of the first letter rather than looking at the entire word.

Because preschoolers have not achieved visual maturity does not mean that they cannot see small details: even 3-year-olds can, momentarily, focus on a tiny image (Hillerich, 1983). It is sustained and systematic focusing that is particularly hard. Until age 5 or 6 many children have an additional visual limitation: they are often farsighted; that is, they can see better at a distance than they can up close.

By age 6, most children can focus and scan reasonably well, although they are still much less skilled at scanning than adults are (Mackworth and Bruner, 1970; Van Oeffelen and Vos, 1984). By age 8, most children are able to follow a line of small print. Of course, as with all aspects of physical maturation, individual differences

The discussion of the preschooler's brain maturation raises an important question: When is a child ready for formal education? Until the mid-twentieth century, the American answer to this question was definitive: at age 6. Indeed, according to the laws of the land, 6-year-olds must attend school, but younger children need not.

Before the 1960s, even preschoolers who did attend some sort of school were not taught academic skills. Most kindergartens, for instance, avoided teaching children to read and discouraged parents from doing so (Hillerich, 1983). By first grade, however, in virtually every school district, children were assumed ready for formal education and the three R's were taught to all.

Using age 6 as an indicator of readiness for academic subjects worked well for the average child, but children who matured early became bored and those who matured late became frustrated. In the 1960s educators decided that some children were ready to start learning some basic skills, particularly reading, at age 5 and that others were not ready until age 8. In order to distinguish the two, they developed readiness tests (e.g., de Hirsch et al., 1966; Ilg and Ames, 1965) that asked 5-year-olds to demonstrate gross motor skills, such as hopping on one foot, or fine motor skills, such as drawing a person or copying a rectangle, or intellectual skills such as telling a story or counting to ten. To varying degrees, scores on each of these tests correlated with school achievement later on (Flynn and Flynn, 1978; Lindquist, 1982; Telegdy, 1975). Depending on their scores, some kindergarteners were taught to read and write, while other children were kept at playing with blocks and dolls until age 7 or 8 (Cicourel et al., 1974).

Many parents objected to this solution, for children who were "unready" in kindergarten were likely to be at the bottom of the academic heap throughout their school career (Satz et al., 1978). Indeed, some parents and educators began teaching 4-year-olds to read and add, hoping that an academic headstart would boost their later school achievement (see A Life-Span Perspective: Preschool Education, pages 214-216). On the other hand, many teachers thought it was a mistake to begin formal education with children who seemed unready, for they had seen too many first-graders crying in despair because they could not master the printed page, or could not write without tearing the paper or breaking the pencil point.

The controversy about when children should be taught academic skills is still not settled. Some experts (e.g., Doman, 1980; Emery, 1975) advise parents to begin teaching reading and math during the preschool years, partly because early reading at home will supposedly make a child smarter and less frustrated than children who learn to read in school.

Others take the opposite viewpoint. Dorothy and Raymond Moore (1975) contend that many children, especially boys, are not ready for any kind of formal education until age 8, 9, or even 10. According to them, late starters are better achievers and less frustrated than children who are forced to begin first grade too early. Similarly, the Gesell Institute, the center for the study of child development founded at Yale, maintains that two-thirds of all American children begin formal education before they are really ready and it suggests that repeating one of the early grades is the best solution for many unhappy "overplaced" children (Ilg et al., 1981).

Which point of view is right? Probably most developmentalists would agree with David Elkind (1978), a cognitive psychologist who stresses that "reading English, far from being a simple matter of discriminating letters and

are common. To make sure a particular child can see well enough to do schoolwork, educators recommend a thorough visual examination before the first grade, including tests of near and far vision, eye strain, and binocular vision (the ability of both eyes to work together) (Bond et al., 1979).

Activity Level

Changes in activity patterns from ages 2 to 6 are another indication of the maturation of the nervous system and changes in metabolism. Children during the early preschool years, at about age 2 or 3, have a higher activity level than at any other time in the entire life span (Eaton, 1983). This seems true no matter how "activity level" is measured. In one cross-sectional study, researchers observed children between the ages of 3 and 9 to measure both how frequently they changed activities

Picture books can be very useful for developing a number of skills that are basic to reading, including the importance of looking carefully and of turning pages in sequence. The most important lesson being taught here, however, is that books are fun.

associating sounds, involves complex mental processes from the very start." The child has to be able to regulate the many perceptions, both visual and auditory, involved in word recognition.

Elkind thinks that formal reading instruction probably should be delayed until the child is about 6 or 7. At the same time, he thinks that those 4-year-olds (about 1 in 100) who are eager and able to learn reading should be allowed to, without pressure from parents to master the skill. In fact, one study found that many very gifted children taught themselves to read by age 2 or 3, sometimes despite their parents' efforts to discourage them—including such occasional extremes as removing all children's books from the house (Robinson, 1981). At least until elementary school, then, readiness to read seems to depend largely on the individual child's maturation, interests, and experiences, not on any preset chronological age.

Psychologists Eleanor Gibson and Harry Levin (1975) assert that the essential experiences a child should have before beginning academic instruction include scribbling, looking at books, and playing rhyming games. At the same time, they caution:

Reading, word games, printing, and so on, should not be forced on a child or formalized. They should be fun, not work. . . . "Home learning" kits of various kinds are advertised like patent medicines, and the consumer would do well to be equally wary of them. The only magic in learning to read is the magic that the child supplies himself when a rich and responsive environment gives him the chance.

For reasonably normal children, factors such as chronological age, motor skills, social development, and intelligence are not directly linked to the readiness to do elementary-school work (Hillerich, 1983). Reading to preschool children, listening to them, and helping them learn the vocabulary and concepts that describe their world are far better for developing "readiness" for reading and other academic activities than trying to teach them to recite the alphabet, write their names, or recognize letters and numbers.

and how often they moved from place to place (Routh et al., 1974). For both measures, activity decreased markedly from age 3 to 6, and continued to decrease at a slower rate from 6 to 9.

In the same study, parents were asked about their child's activity at home—whether, for instance, the child wriggled when watching television, fidgeted when eating, or moved a lot during sleep. Again, scores decreased with age, with one exception. Age 6 saw a small increase, perhaps because the children were temporarily more restless at home in response to the new restrictions of school.

Accident Rates

Decreased activity level, as well as increased capacity for judgment, helps to explain why a child's likelihood of being seriously injured in an accident decreases with each passing year after age 3. (Of the childhood accidents that result in death, fatal poisoning occurs most often at age 1; drowning is most frequent at age 2; and death resulting from being struck by a motor vehicle is highest at age 3 [National

Figure 8.4 *Eating even a few flakes of lead-based paint from the walls of an old building can cause brain damage. Such paint is now illegal for toys or interior walls, but many old, poorly maintained buildings have many layers of lead paint. (As soon as the photographer shot this picture, he made sure this little girl played in a safer place.)*

Center for Health Statistics, 1984a].) Older children have more freedom of movement and are exposed to danger more often than are younger children, but they are more able to regulate their behavior. A 3-year-old, for instance, is more likely to run in front of a moving car than an 8-year-old is; indeed, 8-year-olds are sufficiently cautious that they play ball on many less-trafficked streets in relative safety.

This gradual development of caution should not give the wrong impression, however. Throughout childhood, from ages 1 to 11, accidents are by far the leading cause of death, killing more young people than the next five causes combined (National Center for Health Statistics, 1984b). A child has about 1 chance in 1,000 of dying an accidental death before he or she reaches age 16, and most children need stitches or a cast sometime before they are 10 years old.

A child's chances of having an accident depend primarily on three factors: the amount of adult supervision, the safety of the play space, and the child's activity level. Because these three factors differ, some groups of children have many more accidents than others. For instance, boys are more active than girls, and take more risks; consequently, they have more accidents than girls, about one-third more at age 1, and twice as many at age 5. Asian-American children have fewer accidents than any other American ethnic group, probably because their parents traditionally keep them nearby; this makes the children safer, although more dependent (Kurokawa, 1969). Impoverished children have more accidents than wealthier children, probably because their play areas are more hazardous. All these factors become more salient as children grow older, so that for every ten white girls between the ages of 5 and 9 who die in accidents, seventeen black girls, nineteen white boys, and thirty-five black boys die (National Center for Health Statistics, 1984a).

The fact that parents are, obviously, primarily responsible for their children's safety should not blind us to the responsibility of the larger society for protecting all young children (Margolis and Runyan, 1983). Our failure to do this is revealed by comparing accidental deaths and fatalities from disease in childhood: while the latter have steadily declined over the past twenty years, thanks to the efforts of preventive medicine and laws requiring immunization, the accident rate has remained virtually the same (Butler et al., 1984). Such obvious measures as requiring car seats for young children, getting drunk drivers off the road, and providing safe spaces for preschoolers to play in could make a marked difference (Foege, 1985).

Mastering Motor Skills

Between ages 2 and 6, children become more capable of focusing and refining their activity, while their bodies grow slimmer, stronger, and less top-heavy.

Gross Motor Skills

Because of these developments, **gross motor skills** (as large body movements such as running, climbing, jumping, and throwing are called) improve dramatically (Clark and Phillips, 1985; Du Randt, 1985; Kerr, 1985). The improvement is apparent to anyone who watches a group of children at play. Two-year-olds are quite clumsy, falling down frequently and sometimes bumping into stationary objects. But by age 5, many children are both skilled and graceful. Most North American 5-year-olds can ride a tricycle, climb a ladder, pump a swing, and throw, catch, and kick a ball. Some of them can even skate, ski, and ride a bicycle, activities that require balance as well as coordination.

Most young children practice their gross motor skills wherever they are, whether in a well-equipped nursery school with climbing ladders, balance boards, and sandboxes, or on their own, with furniture for climbing, fences for balancing, and gardens or empty lots for digging up (Whiting and Whiting, 1975). On the whole, preschool children learn basic motor skills by teaching themselves and learning from other children, rather than by specific adult instruction. So as long as a child has the opportunity to play with other children in an adequate space and with suitable equipment (none of which is to be taken for granted in today's neighborhoods, especially in large cities [Gump, 1975; Jeavons, 1984]), motor skills will develop as rapidly as maturation and body size allow.

Figure 8.5 *The materials and equipment required for the development of gross motor skills are minimal: someone to play with and something to play on. Unfortunately, not all children are fortunate in having a large, safe, inviting area in which to run and climb. Each year, many children are injured playing in such dangerous areas as debris-filled lots or condemned buildings.*

Fine Motor Skills

Fine motor skills, the skills that involve small body movements, are much harder for preschoolers to master than gross motor skills. Such things as pouring juice from a pitcher into a glass without spilling, cutting food with a knife and fork, and

Figure 8.6 *Activities such as matching shapes and fitting pieces together not only teach fine motor skills, they also help the child understand spatial relationships and discriminate small differences in size and shape. Skills such as these may foster reading readiness.*

achieving anything more artful than a scribble with a pencil are difficult even with great concentration and effort. Preschoolers can spend hours trying to tie a bow with their shoelaces, often producing knot upon knot instead.

The chief reason many children experience these difficulties is simply that they have not developed the muscular control or judgment needed for the exercise of fine motor skills, in part because the myelination of the central nervous system is not complete. For many preschoolers, this liability is compounded by their still having short, fat fingers. Unless these limitations are kept in mind when selecting utensils, toys, and clothes for the preschool child, frustration and destruction can result: preschool children may burst into tears when they cannot button their sweaters, or mash a puzzle piece in their attempt to make it fit into the wrong position.

The Value of Fine Motor Skills   Many educators consider the development of fine motor skills to be an important goal of the preschool curriculum. One of the first and most influential of these was Maria Montessori, who nearly a century ago designed a series of puzzles, peg boards, and small motor tasks that encourage coordination between the eye, the hand, and the brain. Her approach gave rise to Montessori schools, which continue to emphasize development of fine motor skills, and comparison studies suggest that such schools prepare children well for formal learning (Miller and Bizzell, 1983; Miller and Dyer, 1975). Happily, the fine motor skill that seems most directly linked to later development is one that is easy for parents and teachers to encourage—the skill of making marks on paper.

Children's Art

Developmentalists agree that arts and crafts are an important form of play. On the simplest level "the child who first wields a marker is learning in many areas of his young life about tool use" (Gardner, 1980). This is not mere speculation. One classic longitudinal study found a link between fine motor skills, such as drawing a rectangle, and later reading, writing, and spelling abilities (de Hirsch et al., 1966) (see A Closer Look, page 184). One test of children's intelligence asks the child to draw a person, and then measures the complexity of the drawing to obtain an indication of intellectual ability. Scores on the Goodinough-Harris Drawing Test correlate reasonably well with other intelligence scales and are often used to supplement them (Anastasi, 1982).

Finally, a developmental study of children's paintings found a gradual progression of self-correcting behaviors. Three-year-olds often just plunked their brushes into the paint, pulled them out dripping wet, then pushed them across the paper without much forethought or skill. By age 5, however, most children took care to get just enough paint on their brushes, planned just where to put each stroke, and stood back from their work to examine the final result (Allison, 1985). This kind of self-correction, learned in the course of a play activity, will stand them in good stead later in life, whether they must work at copying a letter of the alphabet, check an example of long division, or write draft after draft of a research paper.

Thus, the scribbling of the young child can be compared to the babbling of the infant (Gardner, 1980). Both are a way to master certain raw materials that at some later date will lead to communication. With time, scribbles become the first representational drawings; the written prose, poetry, and sketches of the adult can be seen as a logical outgrowth of these first attempts to express concepts in a visual way. From this perspective, providing a child with pencils, markers, paint, and paper is as important to the development of communications skills as providing things to climb, things to throw, and places to run is to early physical development.

Figure 8.7 *At every age, recognition of discrepancies between one's intention and one's achievement is the first step toward improvement.*

"You moved."

Drawing by Lorenz; © 1987 The New Yorker Magazine, Inc.

Physical Play

As is apparent from the discussion of children's art, developmentalists view children's play as work, a major means through which physical, cognitive, and social skills are strengthened and honed. In Chapter 6, it was apparent that the toddler's newfound ability to pretend is evidence of, and provides practice for, cognitive development (more about this topic in Chapter 9). The varied social interaction that occurs in play is discussed in Chapter 10. Here, let us look at three types of play that are especially well-suited to the development of motor skills.

Sensorimotor Play

Play that captures the pleasures of using the senses and motor abilities is called **sensorimotor play.** We have already seen that infants regularly engage in this kind of play, delighting in such things as watching a turning mobile or kicking the side of the basinette. This pleasure in sensory experiences and motor skills continues throughout childhood. For example, given the chance, preschool children will happily explore the many sensory experiences that can be extracted from their food, feeling various textures as they mix noodles and meat together with their hands, watching peas float after they put them in their milk, listening to the slurping sound they make as they suck in spaghetti, tasting unusual combinations such as cocoa sprinkled on lemonade. Children find similar opportunities for sensorimotor play in almost any context, in the sandbox, the bathtub, or a mud hole.

Figure 8.8 *"Finger painting" frequently seems an understatement for work in this artistic medium, which often requires involvement of the whole arm right up to the elbow, and sometimes even the nose.*

Mastery Play

Much of the physical play of childhood is **mastery play,** a term used to describe the play that helps children to master new skills. Children waste no opportunity to develop and practice their physical skills. A simple walk to the grocery store can become episode after episode of mastery play, as the child walks on top of a wall, then jumps over every crack in the sidewalk (so as not to "step on a crack and break your mother's back"), then skips, or walks backward, or races to the store. Along the way, there may be ice patches to slide across, or wind to run against, or puddles to jump over, or into. Similarly, making a snack, getting dressed, or singing along with music all are occasions for mastery play. Hand skills are also developed in mastery play, as when children tie knots in their shoelaces, put pegs in pegboards, or use a pair of scissors to make snippets of paper out of a single sheet.

Mastery play is most obvious when physical skills are involved, but it includes almost any skill the child feels motivated to learn. For instance, as children grow older, mastery play increasingly includes activities that are clearly intellectual, such as play with words or ideas. While the impulse to engage in mastery play comes naturally to preschool children, their parents' example and encouragement influence which skills a child will master. Parents who enjoy throwing balls will find their children are much better catchers than the children of those who prefer spending their spare time fishing (East and Hensley, 1985).

Rough-and-Tumble Play

The third type of physical play we will describe here is called **rough-and-tumble play.** The aptness of its name is made clear by the following example:

> Jimmy, a preschooler, stands observing three of his male classmates building a sand castle. After a few moments he climbs on a tricycle and, smiling, makes a beeline for the same area, ravaging the structure in a single sweep. The builders immediately take off in hot pursuit of the hit-and-run phantom, yelling menacing threats of "come back here, you." Soon the tricycle halts and they pounce on him. The four of them tumble about in the grass amid shouts of glee, wrestling and punching until a teacher intervenes. The four wander off together toward the swings. [cited in Maccoby, 1980]

One distinguishing characteristic of rough-and-tumble play is its mimicry of aggression, a fact first noted in observations of young monkeys' wrestling, chasing, and pummeling of each other (Jones, 1976). The observers discovered that the key to the true nature of this seemingly hostile behavior was the monkeys' **play face,** that is, a facial expression that seemed to suggest that the monkeys were having fun. The play face was an accurate clue, for only rarely, and apparently accidentally, did the monkeys actually hurt each other. (The same behaviors accompanied by a frown usually meant a serious conflict was taking place.)

In human children, too, rough-and-tumble play is quite different from aggression, even though at first glance it may look the same. This distinction is important, for rough-and-tumble play is a significant part of the daily activities of many children in preschool, especially after they have had to sit quietly for a period of time (Jones, 1976). Adults who wonder when to break up a "fight" may be helped by knowing that facial expression is as telltale in children as it is in monkeys: children almost always smile, and often laugh, in rough-and-tumble play, whereas they frown and scowl in real fighting (Aldis, 1975).

Rough-and-tumble play is a social activity that usually occurs among children who have had considerable social experience, often with each other. Not surprisingly, then, among children in nursery schools, newcomers, younger children, and only children take longer to join in rough-and-tumble play than to participate in any other form of play (Garvey, 1976; Shea, 1981). Rough-and-tumble play is also more likely to occur among boys than among girls—three times more likely according to one carefully controlled study (DiPietro, 1981).

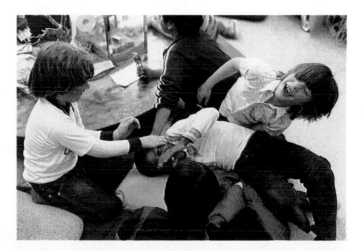

Figure 8.9 *Time to intervene to protect the victim from the attackers? Not as long as the "victim" is smiling. This is rough-and-tumble play.*

Sex Differences and Similarities

The sex differences in types of play that boys and girls tend to engage in raises the larger question of sex differences in physical development. Before looking at these differences, however, it is important to stress the many similarities that exist. Boys and girls follow almost identical paths of physical development during early childhood (Tanner, 1978). They are about the same size, and can do the same things at the same age. Knowing a preschool child's sex provides few clues concerning that child's physical development (Pissanos et al., 1983).

It is also important to stress that the advantage of one sex over the other in most skills is very slight. One researcher studying children's running and jumping skills found that, while the average boy was ahead of the average girl at every age, there was a great deal of overlap (Milne et al., 1976). For instance, while half of the kindergarten boys could jump 35 inches in the standing long jump and could run 400 feet in 50 seconds, so could about 45 percent of the girls. Similarly, the range of ability in running was as great for the girls as for the boys and the overlap was substantial. This means that most boys know several girls their age who can jump farther and run faster than they can. Most of the variations in strength and skill among children are caused by individual differences rather than by any differences that are related to gender.

Figure 8.10 *As with all other areas of development, knowing what is average for a particular group tells us nothing about the abilities of particular individuals within the group or how these individuals will compare with individuals from another group. It is likely that some of the girls pictured here will be better runners than some of the boys. Many boys and girls will simply perform equally well.*

Nevertheless, when averages based on groups of boys are compared with those of groups of girls, several interesting differences appear. Boys are slightly taller and more muscular. Their forearm strength is notably greater than that of girls (Tanner, 1978). They also lose their "baby fat" sooner and have somewhat less body fat throughout life. Girls, on the other hand, mature more rapidly in several ways. For instance, their bone age is usually a year or more ahead of the bone age of boys the same chronological age and they lose their "baby" teeth earlier (Lowrey, 1978).

Some of these physical differences may underlie differences in particular motor skills between girls and boys. By age 4, boys are usually superior in activities that require gross motor skills, especially those like throwing, hitting, and kicking that involve limb strength and control (Ulrich and Ulrich, 1985). At the same time, girls are more coordinated in skipping and galloping (Sinclair, 1973). Some studies have found girls to be better than boys in certain fine motor skills, especially when speed is involved (Ilg and Ames, 1965; Judd et al., 1986). A 6-year-old who cannot print legibly is more likely to be male than female.

Of course, male-female skill differences could be caused by varying amounts of practice rather than by basic body or brain differences. For instance, boys spend more time playing outside, engaging in energetic activities like running, climbing, and rough-and-tumble play, than girls do. Girls spend more time at arts and crafts, and in cooperative and turn-taking games such as jump-rope, jacks, and playing on swings (Crum and Eckert, 1985; Harper and Sanders, 1975).

Further, preschoolers show the same preference as older children for playing with someone of the same sex (see Research Report, page 194). This strengthens whatever sex differences in play patterns there are. If a child wanted to spend a great deal of time in a particular activity favored by the opposite sex, he or she would probably hesitate, since his or her favorite playmates would be doing something else. Meanwhile, the children of the opposite sex might not welcome the lone boy who wants to play hopscotch or the girl who wants to play cops and robbers.

Thus, we do not know to what extent sex differences in play occur because girls are encouraged to do some things while boys are encouraged to do others, and to what extent natural preferences caused by hormonal or brain differences make boys chase each other across the playground while girls sit down and play jacks. But there is no doubt that boys get more experience with gross motor skills, and less with fine motor skills, than girls do, and that boys practice their skills in playful aggression while girls tend to practice their skills in more peaceful contexts.

Figure 8.11 *Why are the girls' faces generally more emotionally expressive than the boys' in response to the story being told? Are girls conditioned to express their emotions more readily, especially in response to a female teacher? Or could this be a reflection of play patterns, with girls more likely to enjoy sitting and listening to a story, while the boys would rather be outside running or playing ball? Questions such as these have important implications, but no easy answers.*

Implications

What are the implications of these sex differences in children's play patterns? Some psychologists (e.g., Gilligan, 1982) have speculated that the play activities of boys prepare them for the largely masculine business world, where self-assertion and competition lead to success. Meanwhile, girls' activities teach them to be cooperative, patient, and relatively passive, as they play in smaller groups and gain skills that might help them in family life but handicap them in business.

Girls and Boys Together

Historical and cross-cultural studies have long noted that, while children of both sexes sometimes play together, generally girls play with girls, and boys with boys. Indeed, even in the 1980s, the typical elementary-school playground during recess reveals sex segregation in play activities—until late childhood, when games that pit boys against girls begin to be played. In recent years, as the more rigid restrictions on sex roles have lifted, developmentalists have wondered how much of the children's continuing preference for the same sex is learned through family, school, and media influences and how much of it is inborn. Two studies shed some light on this topic.

In one, preschool teachers made a deliberate attempt to encourage cross-sex playmates, reinforcing children with praise whenever they played with the opposite sex (Serbin et al., 1977). Over a ten-day period, the incidence of cross-sex play increased markedly. However, when the reinforcements were removed, children quickly reverted to their old habits, the girls playing with the girls, and the boys playing with the boys.

Another study (Jacklin and Maccoby, 1978) explored the extent to which differences in playmate preference would emerge if the children playing together did not know their playmate's sex, and if adults took care not to influence the children's play. Pairs of 2½-year-olds, dressed in T-shirts and jeans with no adornment that would reveal sex, played in a laboratory with a series of toys that would foster either competition or sharing. The children did not know each other, and the adult observers agreed that the sex of the children was not apparent. (When they guessed who was male or female, they often guessed wrong.) Nevertheless, in spite of not knowing if their playmate was a boy or a girl, children played differently with children of the same sex than with those of the opposite sex. The most notable difference was that mixed-sex pairs interacted less. Furthermore, in boy-girl pairs, the girls were likely to stand aside and watch while the boys were active. In contrast, same-sex pairs interacted much more, smiling, talking, and pushing, with boys being particularly likely to engage in a tug-of-war over attractive toys.

Finally, a naturalistic study (La Freniere et al., 1984) of 142 children attending preschools in Quebec found that, as children grew older, more of their friendly approaches to other children were made to children of the same sex,

Even though boys and girls are engaged in very similar activities very close by,   *they tend to gravitate to same-sex groups.*

rising steadily from 51 percent at age 1½ to 72 percent at age 5. In this Canadian study, sex differences appeared even in these same-sex trends. While the 2-year-old boys were equally likely to approach children of either sex, the 2-year-old girls directed their affiliative acts toward other girls 69 percent of the time, a preference that held fairly constant through age 5. As time went on, however, the boys grew more likely to prefer each other, and by age 5, they directed 75 percent of their affiliative acts toward other boys.

While these studies suggest that children choose same-sex playmates because of some innate biological affinity, other explanations are possible. Perhaps parents in early infancy shape their sons and daughters' styles of interaction, so that by age 2, boys have already learned boyish behaviors while girls have learned girlish ones, and each sex may then prefer to play with playmates who have the same patterns. At the moment there is no way of knowing which explanation is the more accurate. Nevertheless, it is quite clear from the research that for one reason or the other, or possibly a combination of both, sex preferences in play partners are formed by early childhood.

This raises the question: What should happen in the future? Given the fact that women are increasingly employed outside the home and that men are taking more active roles in family life, should preschool girls be encouraged to play rougher, more active games, while their male contemporaries are encouraged to spend more time in quiet and cooperative play? Or should we let nature and tradition alone, at least as far as sex differences in early childhood are concerned?

In Chapter 10, where sex-role development is examined in greater detail, these questions will be raised again. Next, however, let us look at cognitive development during the play years, a type of development that virtually no developmentalist is willing to leave to chance or tradition.

SUMMARY

Size and Shape

1. During early childhood, children grow about 3 inches (7 centimeters) a year. Normal variation in growth is caused primarily by genes and nutrition.

Brain Maturation

2. The child's brain and eyes become more mature during these years. This maturation is probably necessary before the child can do typical first-grade work, although the precise relationship between brain, eyes, and learning is not clear. Reading is a complex perceptual and conceptual task that demands much more than left-right distinctions and visual maturity.

Activity Level

3. At about the age of 2 or 3, children have a higher activity level than at any other time in the life span.

4. Because the activity level tends to decline after age 3, children's chances of injuring themselves decrease as age increases. Supervision, play space, sex of the child, and economic status also correlate with the frequency of accidents.

Mastering Motor Skills

5. Gross motor skills improve dramatically during this period, making it possible for the average 5-year-old to do many things with grace and skill.

6. Fine motor skills, such as holding a pencil or tying a shoelace, also improve, but more gradually. Many tasks, including writing, remain difficult and frustrating.

7. Children's attempts at artistic expression not only are a means of practicing fine motor skills, but may also prepare the way for creative self-expression later in life.

Physical Play

8. Play is the work of early childhood. Through sensorimotor play, mastery play, and rough-and-tumble play, children develop their bodies and skills.

Sex Differences and Similarities

9. The average boy plays more active games and is taller and more muscular than the average girl. He is usually better at gross motor skills, such as throwing a ball, than she is, but she is usually better at fine motor skills, such as drawing a person. During these years, however, the similarities between the sexes are much more apparent than the differences. However, the extent to which these differences are influenced by either biological or social factors is not known.

KEY TERMS

corpus callosum *(182)*
gross motor skills *(187)*
fine motor skills *(187)*
sensorimotor play *(189)*

mastery play *(190)*
rough-and-tumble
 play *(191)*
play face *(191)*

KEY QUESTIONS

1. How does the shape of the child's body change during early childhood?

2. What causes variations among children in height and weight during early childhood?

3. What are some of the important brain developments during early childhood?

4. What are the changes in visual development that occur in early childhood?

5. What conclusions can be drawn from statistics on accident rates among children?

6. How do gross motor skills develop?

7. What difficulties do children experience in mastering fine motor skills?

8. Why is play called "the work of childhood"?

9. What are the main similarities and differences in the physical development of boys and girls in early childhood?

CHAPTER 9

**The Play Years:
Cognitive Development**

*"Mommy, I'm so sorry for all the baby horses—
they cannot pick their noses."*

"I'm barefoot all over."

*"Why do they put a pit in every cherry? We just
have to throw the pit away."*

Preschool children quoted in Chukovsky, 1968

The thoughts and verbal expressions of children between the ages of 2 and 6 have always amused, delighted, and surprised adults. A child who wonders where the sun sleeps, or who comforts a parent by offering a teddy bear, or who writes a letter to God about the thunder (see Figure 9.1) is bound to make us smile. Recently, however, researchers studying the intellectual underpinnings of preschool thought have found that they involve much more than charming nonsense. As you will soon learn, early childhood is an impressive period for the development of imagination and language, and even of the basic structures of logic.

Figure 9.1 *The ambitious project of writing a letter to God is a touching and amusing testament to this 4-year-old's faith in the written word and in God's readiness to consider every petition.*

How Preschoolers Think

In explaining the cognitive changes of early childhood, we will begin with Piaget's depiction of them. This is not to say that Piaget's descriptions of the preschooler's cognitive development are universally accepted—in fact, much of the recent research activity in this area centers on modifying and qualifying Piaget's views (Flavell, 1982; Gelman and Baillargeon, 1983). Nevertheless, Piaget remains the most influential theorist in the area of childhood cognition, and although some aspects of his delineation of cognitive development in early childhood appear to need amending, the overall picture Piaget drew provides a firm foundation for understanding the preschooler's thought processes.

Symbolic Thought

The most significant cognitive gain in early childhood is, according to Piaget (1962), the emergence of symbolic thought. As you remember, in the last stage of sensorimotor intelligence, toddlers' thinking is no longer tied to their actions and perceptions; they can begin to "figure things out" mentally. Mental representation, deferred imitation, and pretending all begin at about age 1½ or 2. These budding mental abilities come to full flower between ages 2 and 6, as imagination and language open up new ways of thinking and playing.

To understand how liberating symbolic thought can be, think for a moment about what symbols are. A **symbol** is something that stands for, or signifies, something else. A flag symbolizes a country, a handshake symbolizes friendship, a skull and crossbones symbolizes poison. Each symbol encapsulates a host of emotions and ideas, "packaging" them in a way that allows a person to think quickly and concisely and to move about mentally in the past and the future, as well as in the present. Thus the mind is no longer bound by the limits of immediate sensory and motor experience. Words are the most common symbols, and, whether spoken or written, they are the most liberating. For instance, in English, the sound "dawg" symbolizes a four-legged animal that barks, as do the three written letters "dog." Other languages obviously have quite different sounds, and quite different spellings, but all languages have spoken and written symbols for the general concept of dogness. Knowing the symbols that stand for dog makes it easier to remember dogs, to think about them, and to talk about them. Eventually, the verbal symbol for dog becomes linked with dozens of other symbols associated with dogs and dogness—literal and figurative—and the word itself will call forth a chain of images, ideas, and information. Thus the capacity for symbolic thought opens up much broader and deeper powers of the mind than existed when thought was limited to senses and motor skills (Gardner, 1982).

Pretend Play   One of the most fascinating consequences of symbolic thought is the imaginative play that it makes possible. Indeed, pretending is a favorite activity during the preschool years, when children are old enough to think symbolically but too young to distinguish reality from fantasy very well.

Between ages 1 and 6, pretend play becomes more frequent and more complex with each passing year (Rubin et al., 1983). Whereas the 1-year-old might play with a cup and saucer independently, using them strictly as noisemakers by banging them on the floor, and whereas the 2-year-old is likely to put the cup on the saucer and pretend to drink, and nothing else, the 4-year-old can transform the cup and saucer into almost anything (Lowe, 1975). The child might begin by pretending to pour tea and eat an imaginary cookie, and then, at a moment's notice, turn the cup and saucer into a space ship, a hat, or a domed dwelling. By the end of early childhood, at age 5 or 6, the objects of pretend play can be intangible, existing solely in the child's mind (Rubin et al., 1983). Increasingly, however, in our culture, pretend play is made both more tangible and elaborate by sets of toys, such as those available for Transformers or Barbie dolls, that can be as detailed and extensive as the parents' budget allows (Sutton-Smith, 1986).

Figure 9.2 *In pretend play, preoperational thinking tends to blur the distinction between the real and the imaginary, so for the two girls playing teacher, their "pupils" are as real as the lessons they're teaching from books they can't yet read, and the boys are quite likely totaling up the number of fallen "enemies." Logical inconsistency is another characteristic of preoperational thought— the girl pictured on page 196 has reversed the usual order of events and is bringing her baby to her wedding.*

This growth in the use of symbols occurs as the child gradually develops, and becomes able to mentally coordinate, an increasing number of schemas for the objects in his or her world. This progression is particularly clear when children play with something that could represent a person, such as a doll or stuffed animal. When children first play with such toys, at about age 2, the doll or animal is used to represent one fairly simple action at a time. For example, a doll might be put to bed and left there, ending that particular episode of pretend play. With time, the child will put the doll through a sequence of related behaviors—making her wash her hands, cook the dinner, and eat it. Finally, at age 3 or 4, the child has the doll taking on more complex roles, talking and interacting with other dolls and toys, forming a family, or a school group, or a group of friends and foes. Each of these levels of symbolic play develops in sequence, and is seen as evidence of increasing cognitive development (Case, 1985; Corrigan, 1983; Watson, 1981). By age 4 or 5, much of the symbolic play of children involves other children, as a group acts out "hospital," "store," or "family." (Such social play is discussed in the next chapter.)

Preoperational Thought

Although children's capacity to relate symbols to each other in a meaningful way increases dramatically during the preschool years, it does not include the ability to relate them in a consistently logical way. According to Piaget, preschoolers cannot perform "operations," which are the "schemes of connected relational reasoning" (Isaacs, 1974). In other words, they cannot regularly apply the rule "If this, then that . . . " One example is the idea of **reversibility**—that is, the idea that reversing a process will bring about the original conditions from which the process began. This sounds much more complicated than it actually is, for what it means in practical terms is that the child may know that 3 + 2 = 5 but not realize that the reverse is true, that 5 − 2 = 3. It also means that a preschooler who walks to school every day but gets a ride home would, if asked to walk home one afternoon, probably reply, "But I've never walked home before. I don't know the way." To emphasize the preschooler's inability to understand such logical operations, Piaget refers to the cognitive period from age 2 to 7 as the period of **preoperational thought.**

Centration   The most notable characteristic of preoperational thought, as described by Piaget, is **centration,** the tendency to think about one idea at a time— that is, to "center" on it—to the exclusion of other ideas.

Preoperational children are particularly likely to center on their perceptions, especially on the more obvious aspects of their visual perceptions, rather than consider a broader view of a situation or experience. Thus, in the preschooler's mind, the tallest child or adult is probably the oldest and the best as well (Kuczaj and Lederberg, 1977). Similarly, in the preschooler's view, a cut "hurts" because it bleeds, so covering it with a bandage will make it feel better. Likewise, if the morning sun is streaming through the bedroom window, that means it is time to get up and have breakfast—never mind that it is 5 A.M. on a June morning and Mommy and Daddy are sound asleep.

Because they are so inclined to center on one idea, children have trouble understanding and remembering transitions and transformations: rather, they think in terms of either/or. Their reasoning is static rather than dynamic (Flavell, 1985). For example, they think of themselves as a "good" child or a "bad" one, not as a child who is sometimes good in some ways and sometimes bad in others (Harter, 1983). A related problem is preschoolers' difficulty in understanding cause and effect, partly because they center on one aspect of an event rather than on the relationship between events (Cowan, 1978). This difficulty is particularly apparent when a child has caused some mishap. Children who fall down might blame the sidewalk or another child several feet away. Or a 3-year-old might say, and believe, that a vase fell and broke because it wanted to, or even that it fell *because* it now lies in dozens of pieces on the floor.

The characteristics of preoperational thought are demonstrated most clearly in Piaget's experiments. Let us look closely at experiments in conservation, the most famous of the concepts that Piaget considered impossible for preoperational children to grasp.

Figure 9.3 *Children's tendency to center makes it difficult for them to understand that a person may have several roles simultaneously. For example, they may find it hard to believe that a father is also a son and that a grandfather is a father, too. They may even be uncertain about whether adults also fit into the category of human beings.*

The Problem of Conservation   As you read in Chapter 2, **conservation** (the idea that amount is unaffected by changes in shape or placement) is not at all obvious to young children. Rather, when comparing the amount of liquid in two glasses, they are impressed solely by the relative height of the fluids. If they are shown two identical glasses containing equal amounts of lemonade, and then watch while the lemonade from one glass is poured into a taller, narrower glass, they will insist that the taller, narrower glass has more lemonade than the remaining original.

Preschool children usually have the same problem with regard to conservation of many other sorts (see chart below). Consider **conservation of matter.** Make two balls of clay of equal amount, and then ask a 4-year-old child to roll one of them into a long skinny rope. When this is done, ask the child whether both pieces still have the same amount of clay. Almost always, 4-year-olds will say that the long piece has more.

Tests of Various Types of Conservation

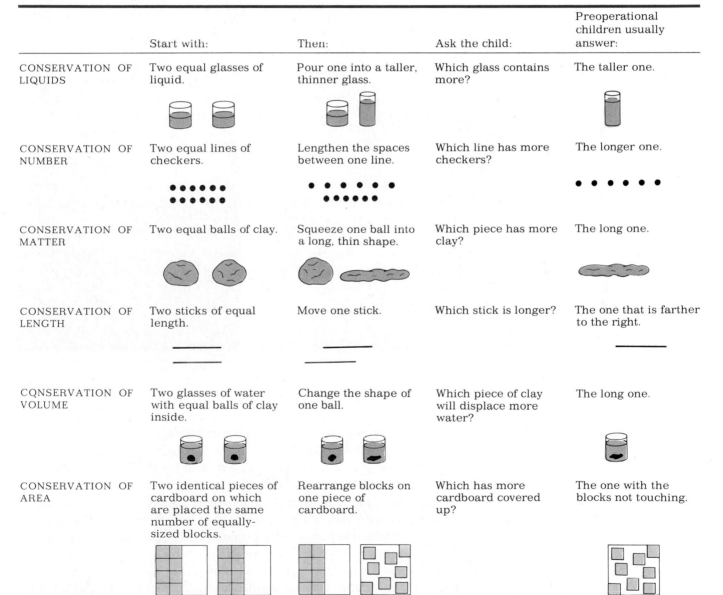

	Start with:	Then:	Ask the child:	Preoperational children usually answer:
CONSERVATION OF LIQUIDS	Two equal glasses of liquid.	Pour one into a taller, thinner glass.	Which glass contains more?	The taller one.
CONSERVATION OF NUMBER	Two equal lines of checkers.	Lengthen the spaces between one line.	Which line has more checkers?	The longer one.
CONSERVATION OF MATTER	Two equal balls of clay.	Squeeze one ball into a long, thin shape.	Which piece has more clay?	The long one.
CONSERVATION OF LENGTH	Two sticks of equal length.	Move one stick.	Which stick is longer?	The one that is farther to the right.
CONSERVATION OF VOLUME	Two glasses of water with equal balls of clay inside.	Change the shape of one ball.	Which piece of clay will displace more water?	The long one.
CONSERVATION OF AREA	Two identical pieces of cardboard on which are placed the same number of equally-sized blocks.	Rearrange blocks on one piece of cardboard.	Which has more cardboard covered up?	The one with the blocks not touching.

Similarly, **conservation of number** is beyond preschoolers. In one Piagetian test (Piaget, 1952b), an experimenter lines up pairs of checkers in two rows and asks the child if both rows have the same number of checkers. The child will almost always say yes. With the child watching, the experimenter next elongates one of the rows by spacing the checkers farther apart, and then asks the child if the rows now have the same number or if one has more. The preoperational child almost always says that the longer row has more.

In such conservation problems, according to Piaget, the problem is that the pre-schooler centers on appearances and thus ignores or discounts the transformation that has occurred. Older children, usually at around age 7 or 8, understand the logical operation of reversibility, and realize that pouring the lemonade back into the shorter and wider glass, or rolling the clay back into a ball, would return things to their original state. Older children would also be able to arrive at these conclu-sions by applying the logical operation of **identity**—the idea that the content of an object remains the same despite changes in its shape. However, because of their tendency toward centration, and their resulting immature reasoning, Piaget con-tends that it is impossible for preoperational children to grasp the concept of con-servation, no matter how carefully it is explained.

Figure 9.4 *When tested on conservation of number, Sarah was not positive that the top row had as many checkers as the bot-tom row, no matter how they were arranged. Her response to this uncer-tainty was to count the rows—a sign that she is beginning to use the logi-cal operations of the school-age child rather than the magical, preoper-ational thinking of the pre-schooler.*

Egocentrism   One of the things preschoolers center on most is themselves, pro-ducing a type of thinking referred to as egocentrism. **Egocentrism** means that thinking centers on the ego, or self. Thus the egocentric child's ideas about the world are limited by the child's own narrow point of view: the child does not take into account that other people may have thoughts and feelings different from the ones he or she is having at the moment.

To say that preoperational children are egocentric means only that because of their cognitive immaturity, they are naturally self-centered, not that they are self-ish (Piaget, 1959). A 3-year-old boy hearing his father crying, for instance, might try to comfort his daddy by bringing him a teddy bear or a lollipop. Obviously, this child is not being selfish: he is willing to give up something of his own. But he is egocen-tric: he assumes that his father will be consoled by the same things he himself finds consoling.

Figure 9.5 (a) *Replications of Piaget's three-mountains experiment are used to measure the ability to imagine a different point of view. The child is first shown a display model of three mountains, and then is shown ten drawings of various views of the mountains. The child is asked to select the drawing that most accurately portrays the point of view of a doll seated at various positions around the table. For instance, if a child were sitting in position 1 looking at the three-mountain display (here shown in an overhead view) and asked how the display would look from sitting-position 4, which picture—(a), (b), or (c)—should the child select? Preoperational children often wrongly select their own view (b) rather than correctly choosing (a).*

Egocentrism is also apparent in **animism,** the idea held by man ___ that everything in the world is alive, just as they are. When thinki ___ children whose play is interrupted by nightfall might get angry at tl ___ to bed'' too early, or they might begin to cry when they drop a st ___ ___ ___ ___ ___ because they think they have hurt it. Animism, as well as egocentrism, is what allows children to accept as literal truth many of the mythical explanations that adults offer them—that Santa Claus, with his one small sled, brings toys to all children, including a large number for the child in question; that Jack Frost paints the windows and makes snow so the child can make snowmen; that Mother Nature brings spring so the child can smell the flowers. Children create their own animistic, egocentric explanations as well: they say that the moon follows them when they walk outside at night or that the thunder comes because they have been bad.

Piaget's Three Mountains   According to Piaget, children are, because of their egocentrism, unable to take another's point of view until at least age 7. The basis for this assertion is Piaget's classic experiment (Piaget and Inhelder, 1963) in which children between the ages of 4 and 11 were shown a large three-dimensional exhibit of three mountains of different shapes, sizes, and colors (see Figure 9.5). First, the children would walk around the exhibit and view it from all sides, and then they were seated on one side of the table that held the exhibit, with a doll seated on another side. Then each child was asked to choose which one of a series of photos showed the scene that the doll was viewing.

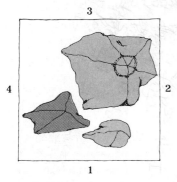

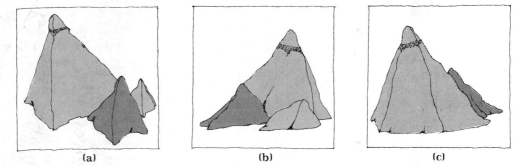

No matter where the doll was seated, children younger than 6 mistakenly chose the photo that showed the mountains as they themselves were viewing them. By age 6 or 7, most children realized that the doll's view would be somewhat different from their own, but they often failed to pick the correct photo. Between 7 and 9, children generally chose correctly, with occasional errors, and finally, at ages 9, 10, and 11, most of the children performed this task perfectly. From these results, Piaget concluded that children under 7 were too egocentric to take the perspective of another, a conclusion many researchers now question.

Revising Piaget

The standard Piagetian experiments have been replicated hundreds of times with children from various cultures, and the results have nearly always been the same: preoperational children make many mistakes with the three-mountains experiment, just as they do with the conservation experiments using glasses of lemonade, or lumps of clay, or rows of checkers (Dasen, 1977; Donaldson et al., 1983).

Perspective-Taking in Preschoolers   However, most developmentalists now believe that the three-mountains task was too complex to be a valid test of the child's

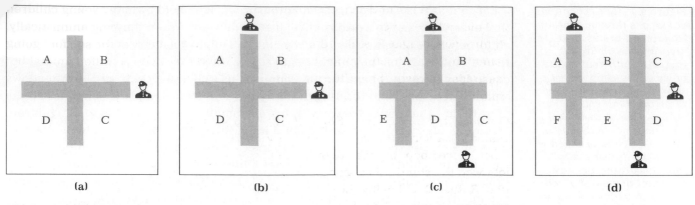

(a)   (b)   (c)   (d)

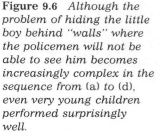

Figure 9.6 *Although the problem of hiding the little boy behind "walls" where the policemen will not be able to see him becomes increasingly complex in the sequence from (a) to (d), even very young children performed surprisingly well.*

ability to take someone else's perspective. Simpler experiments have shown that even preschool children can begin to understand another's viewpoint, at least in an elementary way (Cox, 1986). When perspective-taking becomes part of a game, children between the ages of 3½ and 5 show themselves to be quite capable. This was the general conclusion of one series of experiments in which children were asked to hide a little boy behind a series of "walls," so that policemen who were looking for him could not find him (Hughes and Donaldson, 1979). When the walls made four quadrants (Figure 9.6b) and the policemen were placed so that they could see into three of them, 73 percent of the children correctly picked the fourth quadrant as the hiding place in each of four trials, and 17 percent made only one error in the four trials. In a more complicated set-up (Figure 9.6d), eight out of ten 4-year-olds made no errors, and seven out of ten 3-year-olds made no errors or only one error. (Two of the 3-year-olds who were counted among the failures actually did very well—one by hiding the boy under the table and one by hiding him in her hand—despite the experimenter's best efforts to explain that this was not the way to play the game.)

The researchers ask the key question:

> Why do children find these tasks so easy to grasp, compared with problems like the three mountains task? We believe it is because the policemen tasks make human sense in a way that the mountain task does not. The motives and intentions of the characters (hiding and seeking) are entirely comprehensible, even to a child of three, and he is being asked to identify with—and indeed do something about—the plight of a boy in an entirely comprehensible situation. This ability to understand and identify with another's feelings and intentions is in many ways the exact opposite of egocentrism, and yet it now appears to be well developed in three-year-olds. [Hughes and Donaldson, 1979]

Instruction in Conservation   Researchers attempting to teach the various types of conservation—including conservation of liquids, matter, and number—have achieved greater success by altering the format of the experiment. As in the case of the hide-the-boy-from-the-policemen experiment, simplifying the task, encouraging the child's active participation, and making the experiment seem like a game rather than a test were principles that seemed to help children grasp the concept of conservation. In one study, 4-year-olds learned to conserve after careful instruction, demonstrations, descriptions of what was occurring, and an opportunity for the children themselves to perform the experiments while the experimenter gave them feedback about their performance (Denney et al., 1977). Another study (Field, 1981) found that such training need not be temporary: 79 percent of a group of eighty-five 4-year-olds still understood conservation several months after training, demonstrating this on simple tests of number, length, liquid, and matter. Three-year-olds, however, had difficulty learning conservation, no matter how good the training process (Field, 1981).

Conservation of Number. Similarly, the crucial role of the human context is revealed by a series of experiments in which preschool children seemed to know much more about number than Piaget would have predicted. Several researchers followed the standard conservation-of-number procedures with one telling exception. Rather than having the rearrangement of one row of checkers be a deliberate move on the experimenter's part, they made the change in arrangement seem accidental by having a naughty teddy bear carry it out. Under these circumstances, preschool children were more likely to correctly recognize that both rows still contained the same number (Dockrell et al., 1980; Light et al., 1979; McGarrigle and Donaldson, 1974).

Another series of experiments was designed to discover if children between 3 and 5 years old could recognize when a small group of identical toys was surreptitiously rearranged to make a longer or shorter line, or when one or two toys were taken away from a group of five (these transformations were done "magically," while the display was hidden under a cup). Virtually every child tested showed that he or she noticed the change. Moreover, many of the children gave evidence that they were counting the display—often with idiosyncratic number sequences (instead of 1-2-3, some used 1-2-6 and others A-B-C). Despite their use of these irregular designations, however, the children knew what they were doing: when one toy was missing, they would say something like "Look, it used to be 1-2-6 but now it's only 1-2." Evidence of counting was apparent even among the 2-year-olds who participated: fourteen out of sixteen counted in some fashion or other. Not only did children use their own sequence of numbers, but many of them miscounted, especially when the number of toys was five or more. Nevertheless, it is clear that some understanding of number is present even among very young children (Gelman and Gallistel, 1978).

In looking to these experiments as qualifications of Piaget's findings, it should be remembered that they were *designed* to evoke whatever rudiments of perspective, number, and so forth the young child might possess. Even though preschoolers show some understanding of these concepts in certain conditions, it takes years of maturation and experience before children show mastery of the underlying logical structures and can apply them uniformly in everyday life.

As Rochel Gelman, a leading researcher in cognitive development, points out:

> Despite the many competencies of the young, they nevertheless fail or err on a wide range of tasks that do not seem to be that difficult. . . . The young child, to be sure, has many pieces of competence. However, they are exceedingly fragile. The older child can show that competence across a wide range of tasks. Hence, the idea is that development involves going from the fragile (and probably rigid) application of capacity to a widely based use of these. [Gelman et al., 1982]

A similar pattern of competency and fragility is apparent in the other notable intellectual accomplishment that occurs during early childhood—the development of language.

Language Development

As we saw in Chapter 6, babies normally begin talking at about a year. Language develops slowly at first, as toddlers typically add only a few new words to their vocabulary each month, speak in one-word sentences, and have trouble communicating many simple ideas, sometimes frustrating themselves as well as the most patient care-giver who tries to understand what the child wants. However, during the preschool years, as cognitive powers increase, an "explosion" of language occurs, with vocabulary, grammar, and the practical uses of language showing marked and rapid improvement.

Language and Thought

Piaget believed that cognitive development precedes a child's use of language: a child must first understand a concept before he or she is capable of using the words that describe it (1976). Thus words that refer to the appearance and disappearance of things and people ("bye-bye," "all gone," "where," "no more") emerge after full object permanence has been demonstrated.

Most developmentalists agree with Piaget that infants form concepts first and then learn the words to express them. But most also believe that, at some point during early childhood, language helps form ideas. In Lois Bloom's (1975) words: "There is a developmental shift between learning to talk and talking to learn." Furthermore, most believe that language helps children regulate their social behavior, and, at the same time, that the give and take of everyday social experience is essential to children's language learning (Genishi and Dyson, 1984; Schiefelbusch and Pickar, 1984). Older children and adults are able to learn words and grammatical rules in abstraction, but preschool children depend on practical experience with them. Thus language is at the heart of many aspects of development during early childhood, and for this reason, it is particularly important to understand how language develops.

Vocabulary

The language explosion that occurs during the preschool years is most obvious in the growth of vocabulary—from about 50 words at 18 months, to 200 words at age 2, to between 8,000 and 14,000 words at age 6 (Carey, 1978; Lenneberg, 1967). This means that, from age 2 to age 6, the average child learns between six and ten words a day!

The learning of new words follows a predictable sequence according to parts of speech. Nouns are generally learned more readily than verbs, which are learned more readily than adjectives, adverbs, conjunctions, or interrogatives. Within parts of speech, the order is predictable as well. For instance, the first interrogatives children learn are "where?" and "what?" then "who?" followed by "how?" and "why?" (Bloom et al., 1982). Basic general nouns, such as "dog," are learned before specific nouns, such as "collie," or more general categories, such as "animal" (Anglin, 1977; Blewitt, 1982).

The vocabulary-building process happens so quickly that, by age 5, some children seem to understand and use almost any specific term they hear. In fact, 5-year-olds can learn almost any word or phrase, as long as it is explained to them with specific examples and used in context. One 5-year-old surprised his kindergarten teacher by explaining that he was ambidextrous. When queried, he said, "That means I can use my left or my right hand just the same."

In fact, preschoolers are able to soak up language like a sponge, an ability that causes most researchers to regard early childhood as a crucial period for language learning. This ability is shown in another, less charming kind of "advanced" language usage, when children come home spouting profanity picked up on the street. Any words that seem to shock adults are exciting to a child, so an angelic 3-year-old who has seen the look this new vocabulary produces in Mom may run off to try it on Grandma too.

Grammar

Grammar includes the structures, techniques, and rules that languages use to communicate meaning. Word order and word form, prefixes and suffixes, intonation and pronunciation, all are part of grammar. Grammar is apparent even in children's two-word sentences, since they almost always put the subject before the verb.

By age 3, children typically demonstrate extensive grammatical knowledge. They not only put the subject before the verb, but also put the verb before the object, explaining "I eat apple" rather than using any of the other possible combinations of those three words. They can form the plural of nouns, the past, present, and future tenses of verbs, the subjective, objective, and possessive forms of pronouns. They are well on their way to mastering the negative, progressing past the simple "no" of the 2-year-old ("No sleepy," "I no want it," "I drink juice no"). Preschoolers can use the more complex negatives of "not," "nobody," "nothing," and even "never," and they are progressing toward the impressive negations of the school-age child, such as "I'll come with you, won't I?" or "He hurts no one because he doesn't want to hurt anyone."

Children's understanding of grammar is revealed when they create words that they have never heard, like those in the chart below. Each of the words in the chart shows not only the child's mastery of grammatical rules but the presence of egocentrism as well, in that the children all expected others to understand their linguistic creations.

Children's Knowledge of Grammar in Creating Words

Rule Followed	Word	Context
Add "un" to show reversal.	"unhate"	Child tells mother: "I hate you. And I'll never unhate you."
Use a limiting characteristic as an adjective before a noun to distinguish a particular example.	"plate-egg," "cup-egg" "sliverest seat"	Fried eggs, boiled eggs. A wooden bench.
Add "er" to form comparative.	"salter"	Food needs to be more salty.
Create noun by saying what it does.	"tell-wind"	Child pointing to a weathervane.
Add "er" to mean something or someone who does something	"lessoner"	A teacher who gives lessons.
	"shorthander"	Someone who writes shorthand.
Add "ed" to make a past verb out of a noun (as in punched, dressed).	"nippled"	"Mommy nippled Anna," reporting that Mother nursed the baby.
	"needled"	"Is it all needled yet?" asking if Mother has finished mending the pants.
Add "s" to make a noun out of an adjective.	plumps	buttocks

Sources: Examples come from Bowerman, 1982; Clark, 1982; and the Berger children.

Difficulties with Grammar   Children tend to apply the rules of grammar even when they should not. This tendency, called **overregularization,** can create trouble when their language is one that has many exceptions to the rules, as English does. For example, one of the first rules of grammar that children use is adding "s" to form the plural. Thus many preschoolers, applying this rule, talk about foots, snows, sheeps, and mouses. They may even put the "s" on adjectives, when the adjectives are acting as nouns, as in this dinner-table exchange between a 3-year-old and her father:

> Sarah: I want somes.
> *Father:*   *You want some what?*
> I want some mores.
> *Some more what?*
> I want some more chickens.

Once preschool children learn a rule, they can be surprisingly stubborn in applying it. Jean Berko Gleason reports the following conversation between herself and a 4-year-old:

> She said: "My teacher *holded* the baby rabbits and we *patted* them." I asked: "Did you say your teacher *held* the baby rabbits?" She answered: "Yes." I then asked: "What did you say she did?" She answered again: "She *holded* the baby rabbits and we *patted* them." "Did you say she *held* them tightly?" I asked. "No," she answered, "she *holded* them loosely." [Gleason, 1967]

Although technically wrong, such overregularization is actually a sign of verbal sophistication, since children are, clearly, applying rules of grammar. After children hear the correct form often enough, they spontaneously correct their own speech, so parents can probably best help development of grammar by example rather than explanation or criticism.

Figure 9.7   *This mother has obviously become accustomed to her son's use of overregularization.*

No, Timmy, not "I sawed the chair." It's "I saw the chair" or "I have seen the chair."

Drawing by Glenn Bernhardt

Overregularization sometimes limits a child's understanding of language. In English, the normal word order is subject-verb-object, and children regularly follow this order in their earliest sentences as they say "Mommy give juice" or "I want ball." But when preschool children are confronted with the reversed word order of the passive voice, they generally become confused. For instance, when they hear "The truck was bumped by the car," they often think the truck did the bumping (de Villiers, 1980).

They also think sentence order is a clue to time sequence (de Villiers and de Villiers, 1979). When they are told "You can go outside after you pick up your toys," they think they can play first and pick up later; similarly, if they are told "Before you eat your cookie, you must wash your hands," they are likely to comply by eating the cookie first.

Finally, some grammatical forms demand more logic than most preoperational children are capable of. The conjunction "and" is very simple (though many young children use "and" to string together sentences that have no logical connection). But conjunctions that express conditionality, such as "unless," "although," and "nevertheless," are beyond most 5-year-olds. However, during the preschool years children are able to comprehend more complex grammar, and more difficult vocabulary, than they can produce. In fact, a child's understanding of more sophisticated language forms is facilitated by hearing them in context, so preschoolers might well understand "Tommy was pushed by Billy" if they had just participated in the action. Further, many preschoolers use words such as "before" and "after" correctly in their own speech (French, 1986). However, if a 4-year-old explains "I cut my hand because I am going to the doctor," the most likely diagnosis is preoperational logic rather than a self-destructive reaction to the medical profession.

The preschoolers' impressive attempts to understand and respond to language that they cannot really grasp was shown experimentally when a group of 5-year-olds were asked to answer questions such as "Is milk bigger than water?" "Is red heavier than yellow?" Virtually all of them gave definitive answers, such as "Yes, because it's more thicker" or "Yes, yellow's not bright and red is" (Hughes and Grieve, 1980). Children also gradually master proper listening behavior, such as nodding the head and saying things like "Uh huh" and "Really?" to continue a conversation by indicating that the speaker is heard and understood, even when this is not the case (Garvey, 1984).

Pragmatics

In addition to studying the growth of children's grasp of the meanings and forms of language, developmentalists have recently undertaken the study of **pragmatics,** the practical communication between one person and another in terms of the overall context in which language is used (Rice, 1982). The major emphasis of this study is that a person's communicative competence depends on that person's knowing how to adjust vocabulary and grammar to the social situation.

Children learn these practical aspects of language very early. Evidence of such pragmatic understanding of language can be seen in 2- or 3-year-olds' use of "baby talk" when talking with younger children or with dolls and in their using a deeper voice when giving commands to dogs and cats. Similarly, preschoolers use more formal language when playing the role of doctor or teacher or train conductor, and they use "please" more often when addressing someone of higher status (Rice, 1984).

A CLOSER LOOK   **Difficulties with Communication**

Despite the child's ability to learn new words more rapidly in early childhood than at any other age and rapid progress in formulating ideas and talking about them, preoperational thought and egocentric thinking mean that the preschooler and his or her conversational partner are likely to hit a few snags in communication.

Egocentric Conversation

Piaget (1959) found that about half the statements made by preschool children were instances of egocentric speech, in that the children did not even try to consider the viewpoint of anyone else. Some egocentric speech is a simple **monologue,** in which children talk to themselves or to other people without waiting for a response. Another form of egocentric speech, called the **collective monologue,** occurs when two children have a conversation but neither one listens or responds to what the other says. Here is a collective monologue between two 4-year-olds:

Jenny: They wiggle sideways when they kiss.

Chris: (Vaguely) What?

My bunny slippers. They are brown and red and sort of yellow and white. And they have eyes and ears and these noses that wiggle sideways when they kiss.

I have a piece of sugar in a red piece of paper. I'm gonna eat it but maybe it's for a horse.

We bought them. My mommy did. We couldn't find the old ones. These are like the old ones. They were not in the trunk.

Can't eat the piece of sugar, not unless you take the paper off.

And we found Mother Lamb. Oh, she was in Poughkeepsie in the trunk in the house in the woods where Mrs. Tiddywinkle lives.

Do you like sugar? I do, and so do horses.

I play with my bunnies. They are real. We play in the woods. They have eyes. We all go in the woods. My teddy bear and the bunnies and the duck, to visit Mrs. Tiddywinkle. We play and play.

I guess I'll eat my sugar at lunch time. I can get more for the horses. Besides, I don't have no horses now.

According to the nursery-school teacher who reported this conversation, neither child seemed even aware, let alone upset, that the other child was not responding to the same topic (Stone and Church, 1973).

A hole is to dig

Probably because they are egocentric, preschool children tend to overestimate the clarity of their communications (Beal and Flavell, 1983). A child might say "I want my toy" without specifying which toy. Or a child might begin talking in detail about someone the listener has never met. A further example of egocentrism can be seen in children's thinking they understand more than they do. Young children are well known for jumping to conclusions based on poorly comprehended material and then insisting that they are right. They are particularly likely to do this when statements are open to more than one interpretation. One kindergarten boy resolved never to go back to school because, after being told to sit and wait "for the present," he didn't receive a gift (Donaldson, 1978). Ambiguous communications, either spoken by children or heard by them, are often not recognized as such, with a resulting breakdown in the communication process (Beal and Flavell, 1982; Flavell et al., 1981).

According to one study of 5-year-olds, even children who are trained in recognizing ambiguous statements uttered by someone else are not necessarily very good at recognizing their own ambiguous speech (Sonnenschein, 1984). Specific feedback and suggestions about the clarity of their speech help preschoolers become better at communicating. Rarely, in the normal course of events, do they get such feedback, and for this reason many preschoolers are less adept at communication than they

The world is so you have something to stand on

Little stones are for little children to gather up and put in little piles

might be. In one study (Robinson and Robinson, 1981), those children who were the most precise in describing the clothes on one doll so that a partner could dress another doll in identical garb were also those whose mothers were, in the home, quite specific in helping the children explain exactly what they meant.

Egocentrism and Vocabulary

Because their preoperational thinking is concrete, emphasizing appearances and specifics, preschoolers' vocabulary consists mainly of concrete nouns and adjectives, with very few abstract nouns such as "justice," "economy," "government." For the same reason, when preschool children define words, they usually think about actions a child could do, and come up with egocentric definitions, such as "A hole is to dig."

The preschool child's ability to create and understand metaphors is another example both of increasing cognitive skills and remaining limitations (Vosniadou, 1987). In their early years, many children spontaneously create their own metaphors, for example, comparing a winding stream to a snake or clouds to cotton candy, but they often have trouble understanding figures of speech created by others. When a mother, exasperated by her son's continual inability to find his belongings, told him that someday he would lose his head, he calmly replied:

"I'll never lose my head. I'll find it and pick it up." Another child laughed when his grandmother said winter was coming soon. "Do you mean that winter has legs, and is walking here?" (Chukovsky, 1968). By age 5 or 6, however, they can usually understand these quite well, as long as they comprehend the categories that are being compared. Thus, they could understand the way in which the human body is like a tree, with the arms as branches, but they would be mystified by a phrase such as "the long arm of the law."

Also, because children center on one specific concept and have difficulty with transformations, words expressing a comparison, such as "tall" and "short," "near" and "far," and "deep" and "shallow," are difficult (de Villiers and de Villiers, 1978). Once they know which end of the swimming pool is the deep one, they might obey instructions to stay out of deep puddles by splashing through every puddle they see, insisting that none of them are deep.

Other words expressing relationships are difficult as well. Many children think that all other children are brothers or sisters, and have trouble understanding what makes a person an aunt, or a cousin, or a nephew. Many also confuse "here" and "there," and "yesterday" with "today" and "tomorrow." More than one excited child has awakened on Christmas morning and asked "Is it tomorrow yet?"

Another pragmatic development is shown in children's developing ability to relate an event sequentially, as they become more aware of the usual "scripts" for various everyday events. By age 3 or 4, children can follow the usual script in telling what happened at, say, a restaurant (order food, eat it, and pay for it), or a birthday party (give presents, play games, have cake and ice cream). Here is one 5-year-old's description of how to buy groceries:

> Um, we get a cart, uh, and we look for some onions and plums and cookies and tomato sauce, onions, and all that kind of stuff, and when we're finished we go to the paying booth, and um, then we, um, then the lady puts all our food in a bag, then we put it in the cart, walk out to our car, put the bags in our trunk, then leave. [quoted in Nelson, 1986]

Not only is the sequence correct, enabling listeners to understand exactly what happens, but the child also has the ability to find adequate substitutes for words that are not yet in his vocabulary: the checkout counter becomes "the paying booth."

Children can also relate specific details of those experiences that are different from the usual sequence. Thus, in the middle of a description of birthday parties in general, a child sadly mentions, "It was just at my birthday party. Someone cried there and she ruined the whole party" (Hudson, 1986). From a practical language perspective, the ability to make meaningful conversation about events that happened some time before is even more important than the breadth of one's vocabulary or the correctness of one's grammar.

Figure 9.8 *It is obvious from their body language that these two children have different points of view—perhaps about who might like to ride the bike next. Their ability to communicate their opinions and come to an agreement is an indication of their pragmatic skills.*

Differences in Language Development

By the time children enter kindergarten, differences in language skill are great. While one child seems to know the name of almost every object and action within that child's experience, and can converse in complex sentences, another child has only a basic vocabulary and uses only a few words at a time.

To some extent, one can predict which groups of children are likely to be more advanced in language. On most measures of language production, girls are more proficient than boys; middle-class children, more proficient than lower-class children; first-borns, more proficient than later-borns; single-born children, more proficient than twins, who, in turn, tend to be ahead of triplets (Rebelsky et al., 1967). Note, however, that these group variations are small compared with the differences among individual children from the same group. Thus some lower-class twin boys know far more words than other lower-class twin boys. Similarly, a linguistically advanced lower-class twin boy will have a far better grasp of language than the average middle-class, single-born girl.

It is also important to note that some of the differences between one child and another, as measured by language tests, may be the result of a child's comfort with the testing situation. Edward Zigler and his colleagues (1973) showed that economically disadvantaged children scored much better on a vocabulary test when they were familiar with the examiner and the format of the test, and when they felt relaxed. Courtney Cazden (1976) reports that the language of lower-class children becomes much more elaborate when they are personally involved with the topic. The importance of context is further highlighted by a study (Genishi and Dyson, 1984) that showed children exhibiting much greater language competence when playing with friends than when being questioned by a language researcher.

Nevertheless, even when such factors are considered, there are significant differences among groups of children, and, even more, among individual children (Slaughter, 1983; Snyder, 1984). Researchers who try to explain these differences usually look at the nature and the amount of the language children hear. In general, they have found that mothers talk more to daughters than to sons (Cherry and Lewis, 1976; Goldberg and Lewis, 1969); that middle-class parents provide their children with more elaborate explanations, more responsive comments, and fewer commands than lower-class parents do (Hess and Shipman, 1965; Zegiob and Forehand, 1975); and that parents talk more to first-borns and single-borns than to later-borns or twins (Jacobs and Moss, 1976; Lytton et al., 1977; White, 1979). Particularly important is the parents' attitude about children's language: in cultures where it is thought that children should be seen and not heard, where "idle talk" is frowned on, and where "talking fresh" is punished, children's language production is not encouraged, and, not surprisingly, children do quite poorly on tests of language development (Schieffelin and Eisenberg, 1984; Ward, 1982).

If we look more closely at individual differences, it seems that general parent-child factors, such as strength of attachment and the amount of time spent together, are not particularly good predictors of a child's language competence. However, the amount and quality of conversation between adults and children are relevant. Children become more linguistically competent if the significant adults in their lives encourage them to talk, and reply to their comments with specific and contingent responses (Snow, 1984). (If, for instance, a child says, "I saw a fire engine," a response like "Was it a long red fire engine?" is much more helpful than something like "That's nice.") Adults can also provide experiences that act as a "scaffold" on which to build language skills (Genishi and Dyson, 1984; Schiefelbusch, 1984). Such activities might include looking at picture books together, going on excursions that provide opportunities for new vocabulary and topics of discussion, and pretending together (pragmatic skills are evoked in an imaginary tea party or classroom or trip to the moon). Such measures work as well for learning a second language as for learning a first (McLaughlin, 1984).

Figure 9.9 *These city children visiting a farm for the first time will probably come away with more complex ideas—and feelings—about words such as "cow," "milk," and "barn" than they would have gotten from picture books alone.*

The importance of language encouragement and social opportunities is one reason many early-childhood experts have stressed the quality of the learning environment for the young child. As we will now see, that environment can make a difference not only for the child's language development but for his or her personality and social skills as well.

A Life-Span Perspective

Preschool Education

In recent years, a dramatic change has occurred in the experience of the typical preschooler. Whereas once children almost always stayed at home until about age 6, now most enter some form of school during the time still called the "preschool" years.

There are two factors behind this change. The first has much more to do with parents and grandparents than with children. In today's society, increasing numbers of women are employed outside the home at the same time that they are raising families, and this has created a need to find good child care. In earlier decades, young families could turn to someone further along in the life cycle, often a grandmother who lived at home and was content to care for her grandchildren. These days, however, few grandmothers live with their children, and many are busy with jobs and friends of their own. The second factor is that at the same time that mothers have been returning to the job market in record numbers, developmental research has been providing evidence not only that young children can learn a great deal, but also that they can learn outside the home as well as within it. As a result of both these factors, preschools and day-care centers have been multiplying. In the past decade the number of licensed day-care centers has doubled, to nearly 40,000, with 2.1 million 3- and 4-year olds attending them (Hofferth and Phillips, 1987). All projections are that the increase will continue, since the demand for day care far exceeds the supply.

On the whole, day care has been a boon to working parents, not only providing their children with sound physical care but encouraging their social growth as well

(Clarke-Stewart and Gruber, 1984). But what are the long- and short-term effects of preschool education on children? Some answers to this question are provided by a body of longitudinal research that began with a massive experiment in social change called Headstart (Zigler and Berman, 1983). In the early 1960s in the United States, social scientists and social reformers were particularly concerned that few health or educational services were available to impoverished children during the preschool years, years that were thought to be critical for the child's later achievements. They advocated giving children who might be disadvantaged by their home environment or "culturally deprived" by their community some form of compensatory education, a "headstart" that would remedy the deficiencies of their early upbringing.

Beginning in 1965, hundreds of thousands of "disadvantaged" children, from many racial and ethnic backgrounds, living in many communities, attended a variety of Headstart programs: some full-time, some part-time, some concentrating on classroom activity, some teaching parents how to educate their children in the home. One reason for the diversity of the Headstart program was its rapid, unstructured development, which saw 20,000 children attend programs in about 2,000 different communities in the program's first full year.

The first results from Project Headstart were encouraging. Children learned a variety of intellectual and social skills between September and June of their Headstart year, averaging a gain of five points on intelligence tests. This is a significant gain, and was considered especially important because intelligence tests focus on the language and reasoning skills that are needed for later school achievement. In addition, Headstart children had fewer behavioral problems and greater motivation at the end of their Headstart year than they had had at the beginning of it. Even children who attended only for a summer showed some gains (Horowitz and Paden, 1973). More recent preschool programs designed to improve cognitive development of disadvantaged children replicate these early results: in fact, IQ gains of 10 or more points are not unusual after intensive, well-designed day care (Ramey and Haskins, 1981; Zigler et al., 1982).

Long-Term Benefits

The benefits of preschool programs appear to carry over into later years. As they make their way through elementary school, for example, Headstart graduates score higher on achievement tests and have more positive school report cards than non-Headstart children from the same backgrounds and neighborhoods. By junior high, they are significantly less likely to be placed in special classes or made to repeat a year. In adolescence, Headstart graduates have higher aspirations and a greater sense of achievement. When asked to tell about "something you've done that made you feel proud of yourself," the Headstart graduates were likely to mention an accomplishment at school or on the job, or significant helpfulness to their families. A far higher proportion of non-Headstart than Headstart young people responded that they couldn't think of anything to be proud of.

Since the Headstart programs began, other comparisons of children with and without preschool education have been done. According to several thorough reviews, good preschool education advances the cognitive as well as the social development of children from disadvantaged homes, and probably helps those from more advantaged homes as well (Belsky et al., 1982; Clarke-Stewart, 1984; Rutter, 1982; Scarr, 1984). All reviewers take care to point out, however, that virtually all studies have been done on relatively high-quality schools and day-care centers. Such centers are characterized by a low teacher-child ratio, a well-trained staff, a curriculum geared toward cognitive development rather than behavioral control,

Figure 9.10 *It is likely that intensive, individualized attention to the many skills involved in learning a second language will help this Spanish-speaking preschooler perform well in, and enjoy, his first-grade classes.*

and an organization of space and children groupings that facilitates creative and constructive play. They also tend to be expensive, subsidized by a university, an employer, or the government.

The importance of quality is evident from a number of studies. One was a large comparison done in Bermuda (McCartney, 1984), where 84 percent of the children between 2 and 4 years old spend their days in some sort of preschool. When factors such as parents' socioeconomic status and values were controlled, quality of care in the day-care centers, particularly the amount of adult-child conversation, had a noticeable positive effect on the children's verbal skills and on their overall intellectual development.

The best centers were the ones where teachers spent more time teaching the children (usually in small groups) and less time controlling the children (usually done one child at a time). The best centers also engaged the children in a variety of activities designed to foster creative, motor, social, and language skills. In a comparison study of four different preschool programs for disadvantaged children in Kentucky, the ones that showed the most pronounced long-term gains were a Montessori program, which emphasizes individual work with materials especially designed for self-education, and a Darcee program, which emphasizes language skills. Both programs stress good work habits and attitudes, including persistence, concentration, and the desire to achieve (Miller and Bizzell, 1983).

Preschoolers are not the only ones who benefit from good preschool programs. Other family members may also be positively affected. For example, when parents realize how much the child can learn, they become more active in teaching that child and other siblings. (In fact, one of the original Headstart programs found that siblings benefited as well.) More important, many studies have shown that a working mother's confidence in child-care arrangements is key to her satisfaction in her work and mothering roles, and, in turn, her satisfaction is essential to the happiness of her family (Bradbard and Endsley, 1986). All told, good preschool education can have positive effects that ripple through the preschooler's family for years to come.

SUMMARY

How Preschoolers Think

1. When children become capable of symbolic thought—that is, of using words, objects, and actions as symbols—their ability to understand, imagine, and communicate increases rapidly.

2. According to Piaget, preoperational thought is essentially prelogical. In addition, preoperational children center on one feature of an experience rather than looking at the relationship among several features.

3. The preoperational child believes that other people and even objects think and act the same way he or she does. This general characteristic is called egocentrism, which is quite different from selfishness. One consequence of egocentrism is that preoperational children have difficulty understanding a point of view other than their own.

4. A number of Piaget's critics have shown that in some situations, preschoolers may show some of the cognitive capacities Piaget's standard tests suggest they don't have until the school years. However, it is also true that although young children may demonstrate some rudiments of perspective, number, and so forth, they cannot yet apply them consistently to a wide range of real-life situations.

Language Development

5. Most developmentalists agree that infants form concepts first and then learn the words to express them. At some point in early childhood, language becomes a tool for forming ideas and regulating action.

6. Language accomplishments during early childhood include learning 10,000 words or more, and understanding all grammatical forms. Children at this age, however, are basically prelogical and often misunderstand grammatical rules, metaphors, and abstractions.

7. Children's language learning can be evaluated from the perspective of pragmatics—the use of practical communication in a variety of contexts. The ability to make meaningful conversation often develops before the acquisition of a complete vocabulary and perfect grammar.

Preschool Education

8. Due to the increasing number of working mothers and evidence that young children learn outside the home as well as within it, more and more children today attend preschools and day-care centers.

9. Children who have attended preschool programs designed to improve their intellectual abilities tend to show improvements in social and cognitive skills through adolescence and young adulthood. In addition, preschool education often has a positive effect on other family members as well.

KEY TERMS

symbol *(198)*	identity *(202)*
reversibility *(199)*	egocentrism *(202)*
preoperational thought *(200)*	animism *(203)*
centration *(200)*	overregularization *(208)*
conservation *(201)*	pragmatics *(209)*
conservation of matter *(201)*	monologue *(210)*
conservation of number *(202)*	collective monologue *(210)*

KEY QUESTIONS

1. How does symbolic thinking expand the cognitive potential of the young child?

2. What are the characteristics of preoperational thought?

3. What are several examples of the concept of conservation?

4. Why doesn't the young child understand conservation?

5. How well can young children understand another point of view?

6. How does the increasing capacity to use language affect thought in early childhood?

7. What are the impressive language accomplishments of the young child?

8. What are the limitations of the language ability of the young child?

9. Why do young children have difficulty understanding grammatical rules? How is this difficulty shown in their language?

10. How do preschool programs, such as Headstart, affect young children and their families?

CHAPTER 10

The Play Years: Psychosocial Development

Lady . . . lady, I do not make up things, that is lies. Lies is not true. But the truth could be made up if you know how. And that's the truth.

Lily Tomlin, as Edith Ann

Picture a typical 2-year-old and a typical 6-year-old, and consider the psychosocial differences between them. Chances are the 2-year-old still has many moments of clinging, of tantrums, and of stubbornness, vacillating between dependence and self-determination. Further, many 2-year-olds cannot be trusted alone, even for a few moments, in any place where their relentless curiosity might lead them into destructive or dangerous behavior.

Six-year-olds, by contrast, have the confidence and competence to be relatively independent. They can be trusted to do many things by themselves, perhaps getting their own breakfast before school and even going to the store to buy some more cereal. They also can show affection with parents and friends without the obvious clinging or exaggerated self-assertion of the younger child. Six-year-olds are able to say goodbye to their parents at the door of the first-grade classroom, where they go about their business, befriending and playing cooperatively with certain classmates and ignoring others, and respecting and learning from their teachers.

It is apparent that in terms of self-confidence, social skills, and social roles, much develops during early childhood. This chapter examines that development, looking precisely at what occurs and exploring how it occurs.

The Self and the Social World

Self-concept, self-confidence, and self-understanding, as well as social attitudes, social skills, and social roles, are familiar topics for scientists who study adult personality. Until recently, however, relatively little attention was paid to the development of these aspects of the human character in childhood (Harter, 1983). Now that researchers in child development have begun to investigate these areas, they have found that early childhood is a seminal period for their formation and growth.

The Development of Self-Concept

As we saw in Chapter 7, the idea of self emerges gradually during the latter stages of infancy. By early childhood, children begin to have clearly defined (although not necessarily accurate) concepts of self (Harter, 1983). They assiduously note which possessions are theirs, claiming everything from "my teacher" to "my mudpie"; they repeatedly explain who they are and who they are not ("I am a big girl"; "I am not a baby"); and they relish many forms of mastery play that allow them to show that "I can do it."

Typically, they form quite general, and quite positive, impressions of themselves. Indeed, much research, as well as anecdotal evidence, shows that preschool children regularly overestimate their own abilities. As every parent knows, the typical 3-year-old believes that he or she can win any race, do perfect cartwheels, count accurately, and make up beautiful songs. In a laboratory test, even when preschoolers had just scored rather low on a game, they confidently predicted that they would do very well the next time (Stipek and Hoffman, 1980). Only when it is specifically pointed out to them how poorly they have done will they revise their estimates downward (Stipek et al., 1984).

In addition, most preschoolers think of themselves as able in all areas—competent at physical skills as well as at intellectual ones (Harter and Pike, 1984). This is greatly different from children older than age 8, who make clear distinctions between domains of competence, asserting that they are rather good in intellectual skills but poor in athletic ones, for example (Harter, 1983).

Theories of Self and Others   Each of the three major theories of development recognizes the emerging self-concept of the preschool child. The most encompassing psychoanalytic view is that of Erik Erikson (1963), who notes that the child comes into "free possession of a surplus of energy which permits him to forget failures quickly and to approach what seems desirable . . . the child appears 'more himself,' more loving, relaxed, and brighter in his judgment, more activated and activating." The child initiates new activities with boldness and exuberance.

Indeed, the crisis of this stage, according to Erikson, is **initiative versus guilt.** In this crisis, which is closely tied to the child's developing sense of self and the awareness of the larger society, preschoolers eagerly take on new tasks and play activities and feel guilty when their efforts result in failure or criticism. Their readiness to take the initiative reflects preschoolers' desire to accomplish things, not simply to assert their autonomy as they did when toddlers. Thus, in a nursery-school classroom the older preschoolers take the initiative to build impressive block towers, whereas younger children in the autonomy stage are more likely to be interested in knocking them down.

When initiative fails, according to Erikson—when eager exploration leads to a broken toy, a crying playmate, or a criticizing adult—the result is guilt, an emotion that is beyond the scope of the infant because it depends on an internalized conscience and a sense of self (Campos et al., 1983).

As we saw in Chapter 9, cognitive theory also shows preschool children developing a sense of themselves and then, from that, a sense of others. One result of the child's growing self-concept is a turning away "from an exclusive attachment" to parents and moving toward becoming a member of the larger culture. Erikson sees children at this stage as filled with enthusiasm to learn many things, including the social roles of mother or father, as well as citizen, neighbor, and worker, following the customs of whatever culture the child experiences. As children's new symbolic thinking expands their control over the world, they develop the capacity to imagine all kinds of possibilities and can talk about virtually anything in their experi-

Figure 10.1 *During the initiative stage, children enjoy taking on adult roles, and thus can be quite helpful around the house—setting the table, sweeping the floor, feeding the baby. This big brother has even learned that to encourage an infant to eat, it helps to open your mouth with each bite. Of course, not all the emotions of the preschooler are as advanced as pride in newfound competence; there are likely to be times when this boy's expression matches the stubborn suspicion of his younger brother.*

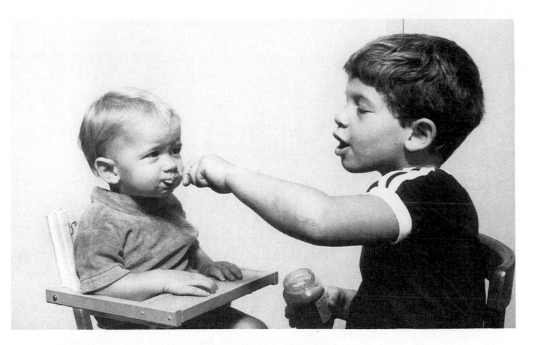

ence. As their cognitive processes mature and their social experiences accumulate, their egocentric sense of self becomes less narrow, and social understanding increases.

Finally, learning theory notes that toward the end of early childhood, praise and blame, as reinforcements and punishments, become powerful as they could not be earlier, because now children are aware of themselves and of how others perceive them (White, 1965).

The differing emphases of the psychoanalytic, cognitive, and learning theories are particularly noteworthy, and intriguing, with regard to the development of sex roles in children, a topic we will take up toward the end of this chapter. For the moment, it should be borne in mind that all three perspectives acknowledge the emerging importance of self-knowledge and self-confidence to interaction with others and to a growing understanding of the wider social world.

Social Skills and Self-Understanding

The close relationship between the child's sense of self and his or her interactions with the social world is easily demonstrated. For example, the first step in social interaction between two preschoolers is often a matter of establishing the individuality of each of them, as they tell each other their names and ages and show off any interesting toy or garment they may have. What may look to adults like bragging and self-preoccupation may actually be a social overture.

One aspect of this relationship between self-awareness and social awareness was shown in a study in which seventy-eight 2-year-old boys were tested for their self-understanding (Levine, 1983). Their comprehension of possessive pronouns was assessed by seeing if they could correctly follow commands such as "Tickle your stomach," "Touch my nose," and "Touch your toes." The accuracy of their use of "I" and "me" was also tested. In addition, their self-recognition was measured by means of the mirror-and-a-dot-of-rouge experiment described in Chapter 7. Then they were paired off, and put in a playroom together. Typically, the first step of interaction was asserting selfhood, usually in terms of ownership, as shown in the following dialogue between two boys who are relatively advanced in development of self-concept. (Each boy is sitting on a toy car, and holding his own nerf ball.)

John: My ball.
Jim: Mine ball.

My ball. [I] have this. No. [The warning came in spite of the fact that Jim has made no move toward the toy.]
My ball.
No.
No ball.
No ball ball. Two ball ball.
Mine.
No.
My ball. Boon ball. [Smiles]
Bump!
Yup!
Car's going bump.

This particular play session began with ten statements of self-assertion before a joke ("boon ball") was used to initiate actual play. Finding this pattern repeated again and again in their study, the authors conclude: "A child's increased interest in claiming toys may not be a negative sign of selfishness but a positive sign of increased self-awareness. . . . With development, a toddler's interactions take on a different character, marked by possessiveness and an attempt to make sense of the other child as a separate social being (Levine, 1983).

In this study, those who had the more firmly established self-understanding as measured by the pretests were also those who engaged in more interactive play. Similar results have been found in a number of contexts: the children who are most social are those who have a better-developed sense of self, as well as a more secure feeling of self-assurance in a given setting (Hartup, 1983). Further, children who are skilled at social interaction tend to be those who are quite confident of their own ability. For example, they are less dependent on teachers in a nursery school or parents at a playground than children who are more awkward at the skills of friendship (Rubin, 1980).

Figure 10.2 *While even infants show their awareness of another's distress by crying along, children do not develop the ability to empathize with others until they develop a better sense of themselves.*

Indeed, self-understanding is closely related to social understanding at many stages of life (Gardner, 1983). This raises an important question: How does personal knowledge—both intrapersonal and interpersonal—develop? In all likelihood, it begins in a social context, through the child's interactions with peers, siblings, parents, and strangers. One of the most important of these interactions in early childhood is play.

Play

As we saw in Chapter 8, play is the work of preschoolers: through it they develop physical skills and expand their cognitive grasp of their world. Play is also important to preschoolers' psychosocial development because of the opportunities it provides for children to develop social skills and roles.

Categories of Social Play

Half a century ago Mildred Parten (1932) studied social play among preschool children and observed that there are five ways children play when other children are around:

1. **Solitary play** A child plays alone, seemingly unaware of any other children playing nearby.

2. **Onlooker play** A child watches other children play.

3. **Parallel play** Children play with similar toys in similar ways, but they don't interact with others.

4. **Associative play** Children interact, including sharing materials, but they don't seem to be playing the same game.

5. **Cooperative play** Children play together, either helping to create and elaborate a game or else taking turns.

Originally, Parten's classification was used as a measure of a child's maturity, since onlooker play is most common at age 2 and associative and cooperative play become common at 5. In fact, these categories devised fifty years ago are still relevant today. Many recent studies have confirmed that the advanced forms of social play are characteristic of the older, more experienced children in a nursery-school setting (Harper and Huie, 1985; Holmberg, 1980).

However, it would be a mistake to apply the developmental progression of social play too rigidly as an indicator of social maturity, erroneously predicting that 2-year-olds are incapable of the more advanced forms of play, or that 5-year-olds who engage in solitary or onlooker play are immature. In fact, quite young children who have had experience with other children are likely to chase each other (rough-and-tumble play), chatter together (language play), and simply enjoy each other's company, all elements of associative play. And children aged 5 and more usually spend some part of their schoolday in solitary play—a sign not of immaturity but of the ability to concentrate on a task (Roper and Hinde, 1978; Smith, 1978). Further, children of any age who are new to a group often begin their attempts at interaction by being onlookers, standing near the play activity and watching, waiting for an appropriate moment to say something like "What are you doing?" or, more bravely, "Can I play?"

Nevertheless, Parten's categories are useful for evaluating both the various kinds of play that children engage in and the types of settings that promote these different kinds of play. Such evaluation can be important, for, as we will see later in the chapter, children who are old enough to regularly engage in associative or cooperative play, but do not, are a cause for concern. Further, some contexts—such as nursery schools in which there are a relatively large group of acquainted children and ample toys and space—foster a higher rate of associative and cooperative play than others and thus are probably better for children's social development (Rubin, 1980).

RESEARCH REPORT   The Importance of Social Play

That social play is enjoyable and interesting proves nothing about its importance. Would there, in fact, be any harm done if a preschool child never played with other children? Say, for instance, that an isolated farm family has only one child, or that a city mother wants to keep her child at home, away from the dangers of the street. Couldn't the child, in either case, learn social skills from adults, or from other children later on?

To some extent the answer is yes. Children are adaptable, and they can learn most essential skills in many ways and at many ages. But there is evidence suggesting that social play in early childhood provides crucial experiences that would be hard for adults to provide or for children to acquire at a later stage of life. For example, learning to play with friends teaches reciprocity, nurturance, and cooperation much more readily than interaction with adults does (Eisenberg et al., 1985; Youniss, 1980). Clinically, the importance of dramatic play is evidenced by the fact that when children have difficulty with social skills, educators and therapists typically use role-playing to help them learn what they apparently did not learn in normal dramatic play. Finally, suggestive evidence on the importance of preschool interaction comes from research on social deprivation in lower primates.

Healthy animals of all species, and both sexes, play when they are young. Researchers have been struck by the similarity between the play of young animals and the skills needed by the full-grown animals of that species. Consider the play of kittens (Egan, 1976). They will sniff and pat any mouse-sized object. If it moves, they crouch and pounce. If it is furry, they will also bite it, carry it around, shake it, and toss it. If the object is alive, and the kitten is hungry, the furry creature is then killed and eaten. But when the kitten is not hungry, the object is played with for a long time, especially if it continues to move. Well-fed kittens play as much as hungry kittens; they just don't eat their prey when they are done. Obviously, kittens who become skilled at kitten's play will become cats who are skilled at catching prey.

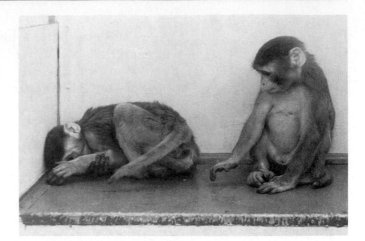

The behavior of the monkey on the left in the first photo is typical of monkeys raised in isolation: they initially cringe from the other monkeys but will attack if approached too closely. Monkeys raised

Similarly, juvenile animals of all species playfully practice the motor skills that they will need as adult animals. They race around (often in circles!); they pretend to fight (with apparent surprise when the game stops if someone is hurt); they build simple beds, or nests, or lairs that will never be used.

In addition, nurturant behaviors are learned through play as well. Primates in early adolescence seem particularly interested in infants, playing with them and caring for them whenever the infants' mothers allow it. Such play is evident in juveniles of both sexes, although male monkeys tend to play more actively, and less gently, with younger monkeys than female juveniles do (Mitchell and Shively, 1984).

While they are engaged in physical play, young animals also playfully practice social skills that may be critically important for survival. For example, most kinds of monkeys live in fairly large groups: in order to live together peacefully, all the adults must know how to

---

In general, developmentalists agree that the higher forms of social play are very important in the life of the preschooler, for through the social interaction that is central to such play, children learn the skills, and gain the self-knowledge they need to become competent in the larger social world (see Research Report above). The intellectual and social sophistication of cooperative play is shown in the most complicated form of social play of all, dramatic play.

Dramatic Play

The beginnings of **dramatic play** coincide with the achievement of symbolic thinking and can be clearly seen, for example, in a child's "feeding," cuddling, and pun-

together without their
mothers often cling to each
other for security, as the
monkeys on the right are

doing. If placed with normal monkeys, they will
gradually develop normal
monkey behavior.

Can socially isolated monkeys be rehabilitated? Harlow and Suomi (1971) tried reinforcing (with food, for example) normal social contact and punishing (with painful shocks) inappropriate behavior, but without success. They then tried social play, putting monkeys raised in isolation for six months together with normal monkeys the same age, but the maladjusted newcomers were attacked and rejected. Finally, the psychologists tried putting an isolated monkey with a younger monkey. Success! The smaller monkey did not intimidate the maladjusted older monkey. The two learned to play together, and by age 1, the disturbed monkey had totally recovered from the early deprivation. Even monkeys who had been isolated for over a year have been rehabilitated by playing with younger "therapist" monkeys (Novak and Harlow, 1975).

assert their rights without antagonizing other adults to the point of a serious fight. So, by pretending to fight with each other, baby monkeys learn complicated behaviors of dominance and submission, using facial expressions, body language, and mock chase and retreat to regulate their interaction.

If rhesus monkeys do not have playmates when they are young, they became socially withdrawn and unusually aggressive during adolescence, sometimes picking a fight with a much larger male monkey or a female monkey with an infant (Harlow and Mears, 1979; Suomi and Harlow, 1976). Such aggression leads to quick, and sometimes fatal, counterattack. If these "unsocialized" monkeys survive adolescence, they sometimes become adults who seem normal, although they remain overly aggressive and relatively unaffectionate. By contrast, orphaned monkeys who play with peers seem much closer to normal in their interactions with other monkeys.

Of course, conclusions from primate research cannot be applied wholesale to people. For instance, humans have intellectual skills, including a measure of self-analysis and direction, that help them adjust to other people if they do not learn social skills from direct early childhood experiences. However, studies of humans who have severe problems with social skills (such as being severely shy or uncontrollably aggressive) reveal that many such people grew up with hostile parents, no siblings, and little interaction with friends and relatives (Gilmartin, 1985). Thus, retrospective data from human adults, as well as experimental research with nonhuman primates, suggest that play in childhood may teach the skills and concepts that help children become mature, fully functioning human beings.

While play and work are often considered opposites, and parents sometimes complain that their children play too much (sometimes even punishing them for "playing around"), developmentalists are convinced that even the most care-free, spontaneous play is clearly part of the productive work of growing up (Garvey, 1977; Rubin et al., 1983; Smith, 1984; Vandenberg, 1978).

ishing a doll or stuffed animal. However, the aspect of dramatic play that is of interest to us here is the social development that occurs when two or more children cooperate in creating their own drama. Numerous studies have found that such mutual fantasy play becomes more common and more complex between the ages of 2 and 6 (Rubin et al., 1983).

For instance, Catherine Garvey (1977) found many examples of dramatic play when she studied forty-eight preschool children ranging in age from 2 years and 10 months to 5 years and 7 months. Each child was paired with a playmate the same age; then both children were placed in a well equipped playroom to do whatever they wanted.

(a)

(b)

(c)

(d)

Figure 10.3 *The toddlers engaged in solitary play (a) will probably remain oblivious of each other unless one grabs a toy the other wants. The observer in onlooker play (b) has sufficient social awareness to hesitate before attempting to join in. The trio playing with sand (c) demonstrate parallel play, for they enjoy being together and playing with the same materials, as each works on a separate project.*

(d) Each child probably has a somewhat different conception of his or her role in the castle-to-be, but by sharing milk crates in this combination of associative and cooperative play, each helps create a fantasy that will satisfy all of them. Finally, a game of tug-of-war demands quite sophisticated cooperative play (e), with everyone not only playing the same game, but also working in close synchrony.

(e)

Many of these pairs, even some of the youngest, chose to engage in dramatic play. The 2- and 3-year-olds often played a simple mother-and-baby game; older children sometimes played a parent-child game or, if they were of the opposite sex, a husband-wife game (which tended to be a somewhat more complicated interaction, since it usually involved making some compromises about who does what).

Garvey found that these simple domestic scenes are one of the standard plots of dramatic play. Others involve such scenarios as one player announcing that a child or pet is sick or dead and the other player automatically becoming the healer, administering food or medicine, or performing surgery, to restore life and health. In a third standard type of drama, one child announces a "sudden threat" (the appearance of a monster, for instance), then both children take the role of victim or defender, attacking or fleeing. The episode can end happily ("I got him!") or unhappily ("He ate me. I'm dead."), unfolding naturally yet without prearrangement.

Dramatic play such as this not only is fun but it also helps children try out social roles, express their fears and fantasies, and learn to cooperate.

Parenting

What kinds of parenting help children become happy about themselves and friendly with others, and what kinds are associated with children who are unhappy, insecure, and hostile? The answers to these questions are elusive, because there is no simple relationship between child-rearing patterns and how a child turns out. As the systems approach makes clear, the outcome of any given child-rearing pattern on any given child depends on many factors that interact with each other, including the child's age, sex, and temperament; the parents' personality characteristics, personal history, economic circumstances, and the like; the needs of all the family members; and the values of the culture.

Hostility and Affection

However, while there is no single path that good parenting follows, researchers have pinpointed several pitfalls to avoid and goals to strive for. They have shown, for example, that early in their children's lives, parents set the stage for later interactions. For example, mothers who are responsive to their children's demands at 10 months tend to have children who are compliant with their mother's wishes at 3 years (Martin, 1981); by contrast, parents who are relatively indifferent to their children at 3 years have children who have poor self-control and difficulty with normal social interaction in adolescence (Block, 1971). Other studies of the effects of parental behavior patterns show that parents who are hostile and rejecting toward their children are likely to have hostile and antisocial children. Thus, if parents are critical, derogatory, and dissatisfied with their children, their children are likely to become dissatisfied with themselves. If parents are insensitive to the child's point of view, the child will be less able to understand the viewpoints of others. The opposite is also true. Warm parents who are understanding and accepting of their children generally raise children who are happy and friendly (Maccoby and Martin, 1983; Martin, 1975).

Patterns of Parenting

That parents should be loving and understanding with their children is, of course, not news. The question is, What is the best way to communicate that love and understanding? What types of child-rearing might most contribute to happy and productive children? Seeking an answer to these questions, Diana Baumrind (1967) set out to determine general patterns of parenting style and the nature of their effects. She began by observing 110 children in their nursery-school activities and, on the basis of their actions, rated their self-control, independence, self-confidence, and the like. She then interviewed both parents of each child, and observed parent-child interaction in two settings, at home and in the laboratory, in order to see if there was any relationship between the parents' behavior with the child and the child's behavior at school.

On the basis of her research findings, Baumrind has delineated three basic patterns of parenting:

1. **Authoritarian** The parents' word is law, not to be questioned, and misconduct is punished. Authoritarian parents seem aloof from their children, not showing affection or giving praise.

2. **Permissive** The parents make few demands on their children, hiding any impatience they feel. Discipline is lax, and anarchy, common.

3. **Authoritative** The parents in this category are similar in some ways to authoritarian parents, in that they set limits and enforce rules, but they are also willing to listen receptively to the child's requests and questions. Family rule is more democratic than dictatorial.

An Example To see the differences among these three patterns more clearly, imagine that it is five minutes before bedtime for a 4-year-old girl. In the authoritarian family, the child would probably already be getting ready, brushing her teeth and putting on her pajamas. If she were to ask "Can I please stay up another half hour?" the parents would probably say, "You know the rule. Get to bed now, right this minute."

In the permissive family, the parents would probably say "Don't you think it's time for bed?" to which the child would likely respond "I want to watch one more TV program." The parents might then suggest that the show is not worth watching, but the girl would probably stay up anyway, finally falling asleep in front of the television.

In the authoritative family, the child might ask to stay up later, explaining that she wants to watch one more program. The parents would probably listen to the request, ask what the program is, and, if there is school the next day, refuse with an explanation. If the program is particularly worthwhile, they might let the child see it—after she has brushed her teeth, put on her pajamas, and promised to get up on time in the morning without complaint.

Figure 10.4 *Since parents tend to raise their children as they themselves were raised, the positive effects of authoritative parenting or, alternatively, the destructive effects of permissive parenting can truly be considered to have long-term consequences.*

"It's perfectly all right. I used to do the same sort of thing when I was a little boy."

Drawing by Gahan Wilson; © 1987 The New Yorker Magazine, Inc.

According to Baumrind (1967, 1971), the following generalizations tend to be true. The sons of authoritarian parents tend to be distrustful, unhappy, and hostile; and neither the sons nor daughters of such parents are high-achievers. The children of permissive parents are the least self-reliant, the least self-controlled, and the most unhappy. (The boys are low-achievers, although the girls do quite well in school.) The children of authoritative parents are the most self-reliant, self-controlled, and content, and are friendly, cooperative, high-achievers.

The basic conclusions of this study have been confirmed by other research. Children who grow up in families that give them both love and limits are most likely to become successful, happy with themselves, and generous with others. Children whose parents are overly strict are likely to be obedient but unhappy; those whose parents are overly lenient are likely to be aggressive and lack self-control.

Generalizations such as these must be cautiously interpreted, however (Martin, 1975). Remember, children affect parents just as much as they are affected by them. Hostile, unfriendly children may produce authoritarian parents rather than the reverse being the case. Likewise, children who have some measure of self-reliance and self-control may produce relaxed, flexible parents (Bell and Harper, 1977). Further, the overall family situation affects the way parents deal with their children. For example, families with many children and little money tend to be more authoritarian, for obvious reasons (Belsky, 1984).

Of course, the temperament of the child as well as the impact of the larger systems of the society can make it easier or harder to raise a cooperative and competent child. This is readily apparent in the Research Report (pages 230–231) on one of the most powerful cultural influences in our society, television.

Punishment

The task of setting and enforcing limits without resorting to the hostile tactics of the authoritarian parent is not easy. One problem is that, at first, physical punishment seems to be effective. It usually stops the child from misbehaving at that moment, and it provides an immediate outlet for parental anger. Unfortunately, parents who use physical force in punishing their children rarely realize the long-term effects that almost every research study of physical punishment has found.

Effects of Physical Punishment When Barclay Martin (1975) reviewed the literature on punishment, he cited twenty-seven different studies on the effect of harsh punishment. The precise definition of "harsh" varied from study to study, but, generally, it referred to punishment that was more severe than that used by other parents in the same community.

Because harshly punished children might be temporarily obedient, their parents sometimes conclude that harsh punishment is good. However, it seems that children store up frustration at this punishment, and when they finally vent it—at school, or against their parents in later years—they are likely to use the mode of expression they have become accustomed to: violence. In twenty-five of the studies reviewed by Martin, harsh punishment at home correlated with aggression against other children, against teachers, and against society. Children who were harshly punished in childhood were more likely to become antisocial delinquents in adolescence. This relationship is especially clear for boys, who, significantly more often than girls, are harshly punished in childhood, are considered "problem children" in elementary school, and are arrested for violent crimes in adolescence. (Apparently the connection between harsh punishment and misbehavior is not immediately apparent during the preschool years, since the only two studies in Martin's review that did not show such a link were based on interviews with mothers of preschoolers.)

RESEARCH REPORT   **Television: A Dilemma for Parents**

Virtually every child in the United States watches television (it is estimated that 99 percent of all households have one or more TV sets). According to the Nielsen survey of preschoolers' TV viewing, in 1984 children between the ages of 2 and 5 watched an average of 27 hours and 9 minutes of television each week (re:act, 1984).

With so much of children's attention being focused on the TV screen, decisions about children's television-viewing can be a frequent source of parent-child conflict. Compounding the difficulties parents may have making these decisions is the controversy over the potential effects of television viewing on the young. Many critics feel that the possible benefits of TV may be purchased at too high a price, and they cite three major problems: the effect of commercials, the content of programs, and the time that could be better spent.

Preschool children usually accept commercial messages uncritically, because, being preoperational and egocentric, they have great difficulty understanding when the truth is bent, or fantasy exploited, in order to sell a product. At times the gullibility of preschool children can be dangerous. One 4-year-old spent two days in intensive care in the hospital after swallowing forty children's vitamins: he had gotten the idea from TV commercials that the vitamins would make him "big and strong real fast" (Liebert et al., 1973). Because young children believe almost everything they see in commercials, they want almost everything they see in them. The parents are therefore placed in the position of resisting the constant demands of their children, or succumbing, buying everything from expensive toys that soon become boring or broken to sugared cereals and drinks that promote tooth decay.

An additional complaint about commercials is that they tend to reinforce certain social stereotypes. For one thing, a disproportionate number of children in commercials are blond males. When girls do appear, they are usually shown in passive roles (Feldstein and Feldstein, 1982). Similarly, minority children are typically among the cast of supporting characters—virtually never the leading characters.

The second major criticism of children's television has to do with the amount of violence it portrays. Many psychologists maintain that TV violence promotes violence in children, primarily through example. Albert Bandura has found that a child who sees an example of violence is not only likely to copy that example immediately but is also likely to refer to that example for future behavior (Bandura, 1977). The effect is interactive and cumulative: children who watch a lot of television are likely to be

The one-eyed monster can be addictive, not only for children who never seem to get their fill of cartoons but also for the parents who find they can spend Saturday morning in bed if the television is on. However, the other "addictions" of television-watching are not so benign: the aggression, the stereotypical heros and villains, the expensive toys that some programs are designed to sell, and the substitution of packaged images for one's own imagination.

more aggressive than children who do not, and children who are aggressive are likely to watch a lot of TV violence (Friedrich-Cofer and Huston, 1986).

Preschool children may be even more likely to be influenced by violence on television than older children are, because they have an especially hard time differentiating reality from fantasy. In television cartoons, which are beamed primarily at young children, physical violence occurs an average of seventeen times per hour. The good guys (Popeye, Underdog, Road Runner) do as much hitting, shooting, and kicking as the bad guys, yet the consequences of their violence are sanitized, never being portrayed as bloody or evil. In cartoonland, demolition, whether of people or things, is just plain fun.

Even the unrealistic tempo of "action" television may have an effect. One study found that television shows with high action and rapid changes of scene (as in cartoons) increase children's aggression, no matter what the content of the program, presumably because the sensory excitement predisposed the children to act without reflection (Greer et al., 1982).

A related concern over the cumulative effects of watching repeated violence on television is that children will become inured to the fact and consequences of violence, a hypothesis that has experimental research to support it (Parke and Slaby, 1983). Children who see a lot of violence on television may thus become more passive when viewing actual violence in real life and may be more inclined to be passive victims themselves when they think they cannot respond in kind.

Many educators and parents have tried to reform children's television, their most successful efforts being through Action for Children's Television (ACT), begun in Boston in 1968. By 1972 they had helped to ban children's vitamin commercials and advertisements with a TV hero directly promoting a product. They have been less successful in reducing violence, which has increased on American children's television during the 1980s (Pearl, 1987).

At the same time, public television has developed special educational programs, such as "Sesame Street" and "Mister Rogers." Although these two programs differ in method and content, research has shown that both succeed in their teaching efforts, especially when parents watch with their children (Cook et al., 1975; Tower et al., 1979). In many other countries, notably Japan, Great Britain, and Sweden, the government requires that at least an eighth of the total viewing time be allotted to children's programs with an educational content (Lesser, 1984).

Nevertheless, a growing group of critics is concerned that even the best television does more harm than good because it robs children of play time, making them less creative, less verbal, less social, and less independent (Singer and Singer, 1977; Winn, 1977). Some support for this idea has come from experiments with certain families who voluntarily gave up television-viewing. The parents of these families reported that their children played and read more, that siblings fought less, that family activities became more common, that mealtimes were longer, and that bedtimes were earlier (Chira, 1984; Winn, 1977). Unfortunately, these experiments lasted only a month, and the families were volunteers. Thus, these results, though provocative, were certainly not the outcome

of a carefully controlled, scientific experiment. The families knew that the experimenters expected to see improvements in family life and no doubt hoped that these expectations would be borne out.

A controlled scientific experiment on the effects of television is almost impossible, because families that do not have television are so rare. It is virtually impossible to find two groups of children who are similar in every way except that some watch television and some do not. However, studies have shown that children who watch a lot of television tend to be lower achievers than children who watch only a little (Gadverry, 1980; Stein and Friedrich, 1975).

One reason for this correlation may be that language skills cannot be mastered in the early years without individualized communication between adult and child. Among the evidence for this hypothesis is the case of Jim, a normal child born to deaf parents. The parents hoped their son would learn to talk by watching television, which he was doing for hours each day. However, Jim did not begin to talk at all until age 2½; and by 3 years, 9 months, his language was immature and abnormal—in fact, his grammatical structures, such as "Be down go" and "Big two crayon," would have been unusual for a 2-year-old. At that point, with intensive, personal language therapy, Jim learned in five months what he hadn't learned in forty-five months of television viewing. At 4 years, 2 months, his language was normal for his age (Bard and Sachs, 1977).

In addition, television tends to cut off social communication, which, as we have seen, is essential for enhancing the social skills that children must develop. A child who enters first grade without the interactive abilities that come from playing with other children may have some difficulties in learning from, and cooperating with, teachers and other students.

What can a parent do? Some professionals recommend no television at all, especially for preschoolers. Others suggest that parents watch with their children, so that they can personalize the learning on educational programs and criticize or censor the content of non-educational television. Many parents have found it easier to impose a simple rule—only an hour a day, or only before dinner, or only on Saturday—than to try to prohibit television completely or censor each program. Although there is no clear consensus on how parents should control television-viewing, one thing is certain: no psychologist who has studied the effects of children's television thinks parents should let their preschoolers watch whatever and whenever they want.

The Effects of Criticism   Parents who do not use physical punishment can be hostile in other ways, especially by being critical and derogatory. Such treatment is more likely to make children become withdrawn and anxious than violent—especially if the child is a boy and the hostile parent is the father (Martin, 1975). Since, again, these consequences are not clear until middle childhood, the man who, for instance, continually tells his shy 4-year-old son to stop acting like a sissy does not realize that he is likely to succeed only in making the boy more frightened and withdrawn.

In much the same way, parents trying to make preschool children less dependent sometimes merely worsen the problem. Many mothers report that after a scolding for being too dependent, their children are even more likely to whine, cling, and make demands (Yarrow et al., 1968). Punishing dependence probably leads to increased dependent behavior because the child, feeling rejected, tries to get reassurance, and one way young children get reassurance is by acting babyish. Usually, the parent of a crying, clinging child gets tired of pushing the child away before the child gets tired of crying.

As with physical punishment, one of the problems with deciding how to control a child verbally (whether to command or cajole, criticize or praise) is that the effects are not immediately apparent. This was shown experimentally when mothers were asked to get their preschoolers to perform a task that would earn money for handicapped or poor children (Chapman, 1979). Some of the mothers chose to command their children to do it; others tried to persuade them to. The children worked under two conditions: with the mother present and with the mother absent. When the mother was present, the children who had been commanded to do the work worked harder than those who had been persuaded. But when the mother was absent, the children who had been persuaded to work worked harder. This experiment, and many other studies, suggest that immediate but transitory obedience from children occurs if the parents are strict and demanding, but that long-term and long-lasting good behavior results from a more authoritative approach.

Suggestions   Obviously, how to reprimand a child effectively without doing physical or emotional harm is a complicated problem. Nevertheless most developmentalists, no matter what their theoretical orientation, almost all agree with the following suggestions (Parke, 1977):

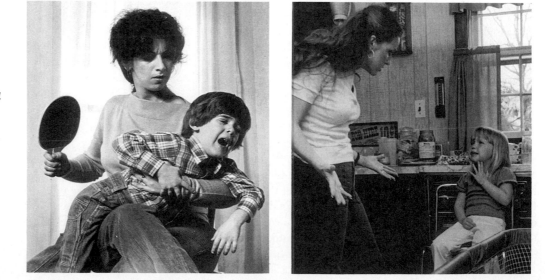

Figure 10.5 *While physical punishment is temporarily more effective in calling a halt to unwanted behavior, in the long run, verbal remonstrances have a greater impact. The boy may remember the pain and be especially careful not to get caught again, but the girl is more likely to feel sorry for what she did and try not to do it again.*

1. Positive reinforcement for good behavior is more effective, in the long run, than punishment for bad behavior. Parents who want their children to be more independent, for instance, should help and encourage them to do things on their own rather than yell at them when they cling.

2. Expected behavior should be within the child's physical, cognitive, and psychological capacities. Since preschool children have trouble understanding situations involving cause and effect or accident and intention, parents should carefully consider whether the child could foresee the consequences of a particular undesirable action before meting out a punishment. The preschooler who is punished for ripping his pants while playing may well be unfairly treated.

3. Rules and expectations should be explained in advance, including the reasons behind them and the consequences for violating them. While preschool children will not always understand the reasons, explaining them emphasizes the necessity of the rule or expectation and gives the child the sense that punishments are not arbitrary.

4. Punishment for breaking an understood rule should be consistent and immediate. Putting a little girl in her room for three minutes the moment she begins to bite her sister is more effective than yelling "Stop fighting—and just wait until your father hears how you've been acting!"

5. Children follow the example set by others. If parents find their children misbehaving, they should make sure they themselves are not providing the wrong behavior model.

6. If the child breaks rules frequently, parents should consider the possibility that the problem lies in some aspect of the home situation rather than in the child. For instance, the rules may be too vague or too difficult, or the child may not get enough attention for good behavior and therefore may misbehave to attract notice, or the child may be tired or hungry or sick.

A number of developmentalists call particular attention to the final item of this list, pointing out that parents should give more attention to the prevention of discipline problems through organization of the environment and through establishment of good parent-child relationships than to the specifics of what to do when the child-rearing system breaks down (Holden, 1983; Patterson, 1980). This is especially important during early childhood, when parents have a great deal of control over the child's environment and when the parent-child relationships are quite intense. A good relationship is like "money in the bank" (Maccoby and Martin, 1983), ready for use in later childhood. As one review explains, "If parents can do what is necessary early in the child's life to bring about a cooperative, trusting attitude in the child, that parent has earned the opportunity to become a nonauthoritarian parent" (Maccoby, 1984).

Possible Problems

No matter how careful and judicious parents are, preschool children sometimes do and say things that cause worry. They seem too ready to strike out at others, or they retreat into their own imaginary world, or they refuse to play. At what point does normal social immaturity become abnormal psychological development? Of course, there is no simple answer to such a question, because there is no one "point or dividing line between normal and abnormal behavior. However, developmentalists have learned a great deal about usual and unusual behavior during early childhood, and about how to help those children who have special problems.

Aggression

When children first start playing with other children in toddlerhood, they are rarely deliberately aggressive. They might pull a toy away from another child or even push someone over, but they do so to get an object, or remove an obstacle, rather than to hurt. As children grow older, the frequency of deliberate physical aggression increases, normally reaching a peak sometime during the preschool years, and then declining (Parke and Slaby, 1983).

Thus a certain amount of aggression in preschoolers is normal. If a pair of 3-year-olds are playing with a toy and a sudden argument about whose toy it is results in a hit on the head or a slap on the face, that is not unusual. Indeed, a certain amount of aggression is not only normal, it is a healthy sign of self-assertion as well as an occasion for social learning, for the reaction of other children and adults may help children learn about the dynamics of social relations (Hay and Ross, 1982).

Figure 10.6 *Young preschoolers show instrumental aggression at its simplest. The goal in the first photo is possession of the toy, and any method of getting it is permissible. If the tears and tugging do not work, the next step is likely to be getting help from a grown-up or hitting or kicking the other child—an effective, but immature, way to win. Hostile aggression, as shown in the second photo, can involve both name-calling and physical force. The goal is not simply to gain an object, but to hurt someone.*

Adult Intervention   However, while a show of force is a normal response to provocation in a 3-year-old, this does not mean that adults should simply ignore preschool aggression entirely. During early childhood, children need to learn how and when to modify their aggressive impulses, partly because, by school age, aggressive children are decidedly unpopular with peers as well as with teachers (Patterson, 1982). In addition, if a pattern of aggression continues into the later years of school, it may be a precursor of delinquency and adult criminality (Rutter and Garmezy, 1983).

Furthermore, if a preschool child is unusually aggressive, this may well indicate that something is wrong in the child's life. It may be simply that a given child has little opportunity for constructive social play and spends too much time watching television (see Research Reports on pages 224–225 and 230–231). Or the child may be reflecting problems at home. The most frightening evidence of this comes from children known to be physically abused. Even at age 2, such children typically are not only more aggressive than other children, but they are also aggressive in contexts in which other children would not be. Simply hearing another child cry—a stimulus that normally makes other preschoolers nurturant—can make an abused child lash out (Main and George, 1985).

Two chilling examples come from a study in which observers in four preschools wrote down everything each child did for half an hour. Some of the children had been abused at home, but the observers were not specifically looking for signs of aggression in them. In one case, when a child started crying for no apparent reason,

> Martin (an abused boy of 32 months) tried to take the hand of the crying other child, and, when she resisted, he slapped her on the arm with his open hand. He then turned away from her to look to the ground and began vocalizing very strongly, "Cut it out. CUT IT OUT!" each time saying it a little faster and louder. He patted her, but when she became disturbed by his patting, he retreated, hissing at her and baring his teeth. He then began patting her on the back again, his patting became beating, and he continued beating her despite her screams.

> Another distress incident . . . began when Kate, 28 months, deliberately swung into the infant Joey, knocking him over with her feet. When Joey lay prone on the ground in front of her, she looked "tenderly down on him" and patted him "gently on the back" a few times. Her patting, however, became very rough and she began hitting very hard. After this, she returned to her swinging while the infant lay prone and still in front of her. Kate stopped swinging to lean forward and hit him hard six or seven times further until he finally crawled away. [Main and George, 1985]

Incidents such as these did not occur among other children who had not been abused themselves. As is clear from these examples, when a preschooler consistently acts with inappropriate and excessive violence, immediate intervention is needed to help the parents as well as the child.

Fantasy and Fear

Another characteristic of preschool children is that they sometimes have vivid nightmares, elaborate daydreams, and imaginary friends and enemies. These flights of fancy can worry their parents, but in fact this active fantasy life is normal and healthy. Children at the preoperational stage of thinking have difficulty distinguishing reality from fantasy; for them, dreams are believable. A child waking up from a nightmare might insist that there is a rhinoceros under the bed and refuse to be comforted by assurances that rhinoceroses are confined to zoos and, in any case, cannot fit under beds. Instead of trying logic, parents should use actions—perhaps turning on the light, looking under the bed, and announcing that there is nothing there except, of course, dirty socks, broken crayons, and cookie crumbs.

In addition, because young children center on physical appearance, many of them are genuinely frightened at the sight of a friend wearing a Halloween costume or of a parent pretending to be a lion. They worry about bodily wholeness and can be quite troubled when they see a person who has an obvious physical abnormality. In addition, everyday objects that make things disappear—the toilet bowl, vacuum cleaner, or bathtub drain—seem ominous to many small children, because they egocentrically wonder if these devices might make them disappear, too.

Because their thinking is so centered, young children can imagine something and then act as though what they have imagined is real. For instance, many children, especially those with few real playmates, create imaginary playmates, who serve as companions for games, reassurance in scary situations, and sometimes as scapegoats for mischief or accidents. Preschool children with such creative imaginations are neither liars nor disturbed; they are simply showing a normal characteristic of preoperational thought. In fact, they may be smarter, more creative, and better adjusted than children who do not have such vivid fantasies (Pines, 1978).

Figure 10.7 *Preschoolers' inability to disentangle the real from the imaginary means that they can become genuinely frightened by movie scenes, sometimes with unexpected results. For example, after seeing the Wicked Stepmother poison Snow White, a 3-year-old might refuse to eat apples or have nightmares about a harmless old lady who lives nearby.*

Sometimes an irrational fear, or **phobia,** becomes so strong that it interferes with the child's normal functioning. In this case, psychologists recommend two techniques (Alexander and Malouf, 1983). The first is **modeling,** through which the phobic child observes another child happily experiencing the feared object (Kornhaber and Schroeder, 1975). For instance, a child who is afraid of the ocean may gain courage by seeing another child playing in the water. The second technique is **gradual desensitization,** in which the child slowly becomes accustomed to the feared object or experience (Wolpe, 1969). In this type of therapy, a child who is afraid of dogs might first be led to sit near a friendly small dog who is docile, quiet, and passive; then to pet the dog; then to become accustomed to larger and more active dogs. These two measures, used together, are much more effective than ridiculing of the child or forcing exposure to what is feared. A child who is terrified of dogs should never be told that only babies are afraid of dogs; nor should a child who is afraid of the water be thrown into it so that he or she will "get over" the fear.

Serious Psychological Disturbances

Although a measure of aggression and of withdrawal are usually nothing to be concerned about in early childhood, severe psychological problems can appear during these years. Generally, they are signaled by abnormal interpersonal relationships (Rutter and Garmezy, 1983). Just as an infant who does not establish an attachment to his or her mother is a source of concern, so the preschool child who is unable or unwilling to play with other children may have a serious problem.

The most severe disturbance of early childhood is called **autism,** from the word element "auto-," meaning self. Autistic children are so self-involved that they hardly recognize that other people exist, and prefer self-stimulation to stimulation from others. They are given to performing the same behavior again and again, and can spend hours on end turning a toy around and around in their hands or even banging their heads against the wall. The least change in their world can produce pathological terror.

The first signs appear in infancy. Autistic babies do not cuddle when held but become rigid or limp, and they cry and avert their eyes, rather than smile, when familiar people greet them. The diagnosis is confirmed if at age 2 or 3 the child still does not play with others or try to converse (Ornitz and Ritvo, 1976).

Autistic children are usually mute, but sometimes they engage in a type of speech called **echolalia,** echoing, word for word, such things as singing advertisements or the questions put to them. If an autistic child is asked "Are you hungry?" the most usual response, if any, is "Are you hungry?"

Other children seem normal in infancy, but show evidence of withdrawal during childhood. Some children reportedly talk normally at home, but are mute when with other children or adults (Kratochwill, 1981). In most cases, this **elective mutism** gradually disappears as the child has experience in a good preschool or some other social setting that allows the child's fear of others to gradually subside (Brown and Lloyd, 1975). However, in some cases elective mutism is a precursor of serious problems with speech and behavior—and a sign of problems in the home.

In children with another serious problem, thought processes and social skills seem to deteriorate. Their speech becomes disorganized, their nightmares grow more terrifying, and their emotions become less predictable. The usual diagnosis in these cases is **childhood schizophrenia,** a disorder that is symptomatically similar to schizophrenia in adults (Achenbach, 1982).

Causes   There is recent evidence to suggest that autism is strongly influenced by congenital factors (Achenbach, 1982). Austistic children often have a history of prenatal rubella, postnatal seizures, or other early signs of possible brain damage. Further, the first symptoms of autism itself appear so early that it is highly unlikely that the parents' behavior could be the cause of the condition. A 4-week-old baby with seemingly normal parents who prefers to look away from faces, for instance, is so abnormal at such a tender age that the cause must be inborn. In addition, it is exceedingly rare for parents to have more than one autistic child, which suggests that something in the child, rather than the nature of the parent's behavior, causes the disorder (Bender, 1973).

The evidence is less clear that parent-child relationships do not contribute to elective mutism and childhood schizophrenia. We know that shyness and schizophrenia in children and in adults are partly genetic, but we also know that identical twins do not always have the same patterns of emotional disturbance or health, so genes are not the sole cause of either of these disorders. However, as with other types of severe disturbance in childhood, it is generally agreed that abnormal parenting is usually not the only cause (Achenbach, 1982; Rutter and Garmezy, 1983).

Treatment   For autism and schizophrenia, the most successful treatment methods usually combine behavior therapy with individual attention: autistic children can sometimes learn to talk, and schizophrenic children, to control their anxiety, if a step-by-step conditioning process is set up and reinforcement is carefully delivered.

Furthermore, while parental behavior is probably not the sole cause of severe psychological problems, it can be influential as either part of the problem or part of the cure. Parental participation in the child's treatment is often the most important factor affecting the future of the child. Ideally, parents should provide the stability, reassurance, and learning environment that disturbed children need.

Sex Roles and Stereotypes

Figure 10.8 *This girl's pleasure at her newly acquired figure is typical of children aged 4 to 6, who usually believe the appearance of femininity or masculinity equals the fact, and that more is better.*

Another important topic that has been the subject of much developmental theory and research is gender identification and sex-role development.

Children learn about gender very early. Most 2-year-olds know whether they are boys or girls, identify strangers as mommies or daddies, and know that Daddy has a penis and that Mommy has breasts. By age 3, children can consistently apply gender labels (Thompson, 1975). They refer to boys, brothers, and policemen as "he" and to girls, sisters, and policewomen as "she." Further, children's behavior acknowledges traditional distinctions between boys and girls at a very early age. By age 3, if not before, children prefer to play with sex-typed toys (dolls versus trucks) and enact sex-typed roles (nurses versus soldiers) (Eisenberg-Berg et al., 1979; Huston, 1983, 1985; O'Brien et al., 1983). Children also have definite ideas of typical male and female behavior and misbehavior, believing girls are more likely than boys to clean the house and to "talk a lot" and boys are more likely than girls to mow the lawn and to hit others (Kuhn et al., 1978).

By age 6, these notions become full-blown prejudices, when most children (even those from liberated homes) express stereotypic ideas of what each sex should do, wear, or feel (Huston, 1983). Indeed, already by age 4, preschool girls and boys react approvingly or disapprovingly toward each other, according to their choice of sex-appropriate toys and play patterns (Roopnarine, 1984). The boy who wants to help the girls dress the dolls or the girl who wants to be one of the space warriors is likely to be soundly criticized by his or her friends.

This does not mean that children understand that gender is biological. Until age 4 or 5, they are likely to think sexual differences depend on clothes, hair, or maturation, rather than on biology, believing that a girl would be a boy if she cut her hair short or that a boy might become a girl if he wore a dress. One preschool girl visited the neighbor's new baby, who was having a bath. Later, her parents asked if the baby was a boy or a girl. "I don't know," she replied, "it's so hard to tell at that age, especially when it's not wearing clothes" (Stone and Church, 1973).

Three Theories of Sex-Role Development

We have already discussed the nature-nurture issue several times, and you are well aware that developmentalists disagree about what proportion of observed sex differences is biological—perhaps a matter of hormones, of brain structures, or of genes carried by the sex chromosomes—and what proportion is environmental. However, even for differences that seem most closely related to nurture, theorists hypothesize various reasons for their existence; specifically, they ask: What is the origin of sex-role preferences and stereotypes that children develop during the preschool years? Each of the three major psychological theories has a somewhat different answer.

Psychoanalytic Theory   Freud (1938) called the period from about age 3 to 7 the **phallic stage,** because he believed its center of focus is the penis. At about age 3 or 4, said Freud, a boy becomes aware of his penis, begins to masturbate, and develops sexual feelings about his mother, who has always been an important love object for him. These feelings make him jealous of his father—so jealous, in fact, that, according to Freud, every son secretly wants to kill his father. Freud called this phenomenon the **Oedipus complex,** after Oedipus, son of a king in ancient Greek mythology. Abandoned as an infant and raised in a distant kingdom, Oedipus later returned to his birthplace, and, not realizing his relationship to them, killed his father and

Figure 10.9 *At this age, boys are given to sticking objects in their pants to exaggerate the size of their penis, and children of both sexes (not yet sure of the internal anatomical differences between male and female) like to stuff the front of their shirts with a "baby."*

married his mother. When he discovered what he had done, he blinded himself in a spasm of guilt.

According to Freud, little boys feel horribly guilty for having the lustful and murderous thoughts that characterize the Oedipus complex and imagine that their fathers will inflict terrible punishments on them, among them blindness and castration, if they ever find out about these thoughts. They cope with this guilt by means of **identification,** a defense mechanism through which people imagine themselves to be like a person more powerful than themselves.

As part of their identification with their fathers, boys copy their fathers' masculine behavior and adopt their moral standards. Through this process, they develop their superego, or conscience, to control the forbidden impulses of the id (see page 34). Freud believed that if a boy does not experience the Oedipus complex—because, say, there is no father in the household—he is likely to identify with his mother too much, thereby increasing his chances of eventually becoming homosexual. (Although there is some evidence that a son raised without a father will have some difficulty with sex-role development, there is no hard evidence for the link between the absence of the father and homosexuality, and few psychologists support this aspect of Freud's theory.)

Freud offered two overlapping descriptions of the phallic stage as it occurs in little girls. One form, the **Electra complex,** follows the same pattern as the Oedipus complex: the little girl wants to get rid of her mother and become intimate with her father. In the other version, the little girl becomes jealous of boys because they have a penis, an emotion called **penis envy.** Somehow the girl decides that her mother is to blame for this state of affairs, so she becomes angry at her and decides the next best thing to having a penis of her own is to become sexually attractive so that someone with a penis, preferably her father, will love her (Freud, 1965; originally published in 1933). (See A Closer Look, next page.)

In both versions, the consequences of this stage are the same for girls as for boys: guilt and fear, and then adoption of sex-appropriate behavior and the father's moral code. By the time this stage is over, children of both sexes have acquired their superego (although, according to Freud, a girl's is not as well developed as a boy's). This strict conscience makes it difficult for people of any age to break the moral codes or transcend the sex roles they learned in childhood.

Learning Theory Learning theorists take another view about sexual attitudes during early childhood. They believe that virtually all sexual patterns are learned, rather than inborn, and that parents, teachers, and society are responsible for whatever sex-role ideas and behaviors the child demonstrates.

Preschool children, according to learning theory, are reinforced for behaving in the ways deemed appropriate for their sex and punished for behaving inappropriately. In some ways, research bears this out. Parents, peers, and teachers are all more likely to reward sex-appropriate behavior than inappropriate behavior (Harter, 1983; Langlois and Downs, 1980).

Interestingly, boys are criticized more often for wanting to play with traditional girls' toys than girls are for wanting to play with boys' toys. Even in toddlerhood, boys are criticized for wanting to play with dolls (Fagot, 1978). Furthermore, fathers are more likely to expect their girls to be "feminine" and their boys to be "masculine" than mothers are. Fathers are more gentle with their daughters and are more likely to engage in rough-and-tumble play with their sons (Maccoby and Jacklin, 1974). Thus, in our society at least, sex-role conformity seems to be especially important for males.

A CLOSER LOOK   Berger and Freud

As a woman, and as a mother of four daughters, I have always regarded Freud's theory of female sexual development as ridiculous, not to mention antifemale. I am not alone in this opinion. Psychologists generally agree that Freud's explanation of female sexual and moral development is one of the weaker parts of his theory, reflecting the values of middle-class Viennese society at the turn of the century more than any universal pattern. Many female psychoanalysts (e.g., Horney, 1967; Klein, 1957; Lerner, 1978) are particularly critical of Freud's idea of penis envy. They believe that girls envy, not the male's sexual organ, but the higher status the male is generally accorded. They also suggest that boys may experience a corresponding emotion in the form of womb and breast envy, wishing that they could have babies and suckle them (see the photo on page 239).

However, my own view of Freud's theory as complete nonsense has been modified somewhat by the following experiences with my three oldest daughters when each was in the age range of Freud's phallic stage. The first "Electra episode" occurred in a conversation with my oldest daughter, Bethany, when she was 4 or so.

Bethany: When I grow up, I'm going to marry Daddy.
Me: But Daddy's married to me.
Bethany: That's all right. When I grow up, you'll probably be dead.
Me: (Determined to stick up for myself) Daddy's older than me, so when I'm dead, he'll probably be dead, too.
Bethany: That's O.K. I'll marry him when he gets born again. (Our family's religious beliefs, incidentally, do not include reincarnation.)

At this point, I couldn't think of a good reply. Bethany must have seen my face fall and taken pity on me.

Bethany: Don't worry Mommy. After you get born again, you can be our baby.

The second episode was also in a conversation, this time with my daughter Rachel.

Rachel: When I get married, I'm going to marry Daddy.
Me: Daddy's already married to me.
Rachel: (With the joy of having discovered a wonderful solution) Then we can have a double wedding!

The third episode was considerably more graphic. It took the form of a "valentine" left on my husband's pillow by my daughter Elissa, who was about 8 at the time. It is reproduced above.

As of this writing, my youngest daughter, Sarah, who just turned 5, has not arrived at a need to dump me. She has,

however, expressed the desire to marry my husband. Her response to my statement that she can't marry him because he is already married to me reveals one of the disadvantages of not being able to ban TV in our household: "Oh yes, a man can have two wives. I saw it on television."

I am not the only feminist developmentalist to find Freud's theories surprisingly perceptive. Nancy Datan (1986) writes about the Oedipal conflict: "I have a son who was once five years old. From that day to this, I have never thought Freud mistaken."

Obviously, these bits of "evidence" do not prove that Freud was correct. But Freud's description of the phallic stage seems not to be as bizarre as it first appears to be.

Theodore Lidz (1976), a respected developmental psychiatrist, offers a plausible explanation of the process evident in my daughters and in many other children. Lidz believes that all children must go through an Oedipal "transition," overcoming "the intense bonds to their mothers that were essential to their satisfactory pre-Oedipal development." As part of this process, children imagine becoming an adult and, quite logically, taking the place of the adult of their own sex whom they know best, their father or mother. This idea must be dispelled before the sexual awakening of early adolescence, otherwise an "incestuous bond" will threaten the nuclear family, prevent the child's extrafamilial socialization, and block his or her emergence as an adult. According to Lidz, the details of the Oedipal transition vary from family to family, but successful desexualization of parent-child love is essential for healthy maturity.

Figure 10.10 *This father is gratified, as most fathers would be, to find that his son wants to be "just like Dad." The father's praise reinforces both the boy's pride in a new-found skill and his acquisition of the masculine sex role of "handyman." The girls on the right, in the swimsuit phase of a beauty contest, are being initiated into the kinds of "female" behavior and appearance that society rewards with trophies and applause.*

Modeling. Social-learning theorists (Bandura, 1977; Mischel, 1970, 1979; Sears et al., 1965) combine the psychoanalytic emphasis on parents and the behavioristic emphasis on learning. They say that children learn much of their sexual and moral behavior by observing other people, especially people whom they perceive as nurturing, powerful, or similar to themselves. For all these reasons, parents are important models during childhood.

Social-learning theorists are not surprised when preschool children seem precociously and dogmatically conscious of sex roles, even when the parents espouse less traditional views. In this case, actions speak louder than words, and most adults are more sex-stereotyped, in their behaviors as well as self-concept, during the years when their children are young than at any other time in the life span (Feldman et al., 1981; Gutmann, 1975). Imagine a typical modern child-free couple who try to avoid traditional sex roles, both of them being employed and both sharing the housework. Then the wife becomes pregnant. The most likely sequel is that she will take a maternity leave, quit her job, or work part-time. This traditional pattern is reinforced by the biological fact that only the woman can breast-feed, the sociological fact that relatives and friends generally expect the woman to provide most infant care, and the economic fact that the man's salary is usually higher than the woman's. If the wife stays home, it is likely that she will do more than half the housework. If she does go back to work full-time, typically she will find another woman to care for her child, either at home or in a day-care center. Further, while most two-career families without children share the housework, once the employed wife becomes an employed mother, and the quantity of housework mounts, she typically does the major portion (Condron and Bode, 1982; Haas, 1981). In addition, influences from the macrosystem, including everything from who runs for president to who does what in television commercials to which characters take the initiative in children's books, teach children those behaviors that are considered sex-appropriate (Barnett, 1986; Huston, 1983).

Cognitive Theory In explaining gender identification and sex-role development, cognitive theorists (Cowan, 1978; Kohlberg, 1966, 1969) focus on the effects of the cognitive limitations and misperceptions that make it difficult for children to understand and apply male-female distinctions. Because of their cognitive immaturity, until they are at least 4 or 5, children do not realize that they are permanently male or female. Once they understand this, however, they try to adopt appropriate sex-role behavior, at first with the rigidity typical of the initial stages of comprehension of an idea, and then with more flexibility.

That an orderly progression of attitudes toward sex roles parallels the child's passages through the stages of cognitive development was demonstrated in a study by Lawrence Kohlberg and Dorothy Ullian (1974), who interviewed boys and girls between the ages of 3 and 18. They found that, at age 3, children know what sex they are, but do not realize that maleness and femaleness are permanent characteristics based on genital differences. Boys think they could become mommies; girls think they could become daddies; and both sexes think boys would be girls if they wore dresses and girls would be boys if they had their hair cut very short. In other words, 3-year-olds tend to be confused about the relationship between appearances and reality in connection with sex roles, the same confusion they experience in connection with conservation experiments.

By contrast, most 6-year-olds realize that sex is a permanent characteristic. Usually, they are pleased to be whatever sex they are, for they believe that their own sex is the "better" one. For example, 6-year-old boys think males are stronger, smarter, and more powerful than females. In addition, children of both sexes believe sex identity is defined by observable physical signs, such as strength, depth of voice, clothes. One boy, asked why he thought men were smarter than women, said they had bigger brains. This emphasis on physical differences is true for 6-year-olds in many cultures and in many types of families, including fatherless ones.

Similar results were found in a study of 5- to 15-year-olds in Australia, England, the United States, Canada, and Sweden. In all five countries, the children's cognitive stage was related to their understanding of sex differences and sex roles, with younger children's ideas being quite stereotyped and inflexible. In both studies, older children displayed a broader, more balanced understanding of masculinity and femininity.

A New Theory: Androgyny

Although almost everyone recognizes that traditional sex roles are too restrictive, many parents hesitate to treat their daughters and sons in the same way, fearing that their girls might become too "masculine" or their boys too "feminine."

Parents, as well as developmentalists, might be helped by a new concept, androgyny. This concept of sex roles counters the misconception that masculinity and femininity are opposites—an assumption that leads a great many people to believe that the more one follows the masculine role, the less feminine one becomes, and vice versa. **Androgyny**—originally meaning the state of having both male and female sexual characteristics—has come to be used in connection with a person's defining himself or herself primarily as a human being, rather than as a male or female.

Several measures of androgyny have been developed in which a person chooses which adjectives are closest to describing his or her own personality (Bem, 1974; Spence and Helmreich, 1978). For instance, someone who scores high in aggression, dominance, competitiveness, and activity, and low in gentleness, kindness, emotionality, and warmth, would be rated as typically masculine; someone with opposite scores would be rated typically feminine. A person who scored high in both sets of characteristics would be considered androgynous.

Thus androgynous men and women share many of the same personality characteristics, instead of following the traditional sex-role patterns. For instance, traditional males rate significantly higher than traditional females on a personality trait

Figure 10.11 *Our culture generally allows greater sex-role latitude to girls than to boys. For example, most parents are delighted to have their daughters learn how to hammer a nail, and would be pleased if, eventually, their girls could safely and deftly maneuver power tools. However, by elementary-school age, it is much more acceptable for boys to create battles between GI Joe and his many comrades than to tenderly cuddle their Cabbage Patch babies and sing them to sleep, as many a girl their age would do.*

labeled "dominant-ambitious," but androgynous males and females score about the same, because the men see themselves as less dominant than the traditional male does while the females see themselves as more dominant than the traditional female does (Wiggins and Holzmuller, 1978). Androgynous people are nurturing as well as independent and try to be neither unemotional nor passive.

They are more flexible in their sex roles, able to display the best qualities of both traditional stereotypes. For this reason androgynous people are generally more competent and have a higher sense of self-esteem than people who follow traditional sex-role behavior (Spence and Helmreich, 1978).

This suggests that it is good to encourage children to develop all their potential characteristics, encouraging them to engage in rough-and-tumble play outside as well as quiet play in the doll corner. Instead of saying such things as "Boys don't cry" and "Girls don't fight," we should apply the same standards to both sexes, teaching children that both crying and fighting are sometimes appropriate behaviors. In fact, it may be particularly important to encourage girls to assert themselves and boys to be nurturant, thus balancing the pull of nature and nurture.

The concept of androgyny is not without its critics, however (Locksley and Colten, 1979). For example, parents who follow relatively traditional male and female roles may do a better job of child-rearing than androgynous parents, according to Diana Baumrind (1982), because they tend to be more child-centered. Certainly in the "chronic state of emergency" that the myriad and never-ending tasks of parenting entail, it may well be more efficient for fathers and mothers to play the traditional complementary parental roles (Gutmann, 1975). Further, the fact that the concept of gender differences comes naturally to children suggests that parents should not be too stringent in trying to deny it.

In fact, Sandra Bem (1985), one of the designers of the androgyny scale, recognizes that the idea of gender differences is useful to help young children organize their perceptions of the adult world. The problem comes, if at all, when children and their parents remain rigid in applying this schema, causing the rigid stereotyping to stifle the full development of the child or the adult. Thus the child's sex-role concepts, like the child's definition of selfhood, mode of play, use of aggression, and all the other themes of this chapter, should change with exposure and maturity. These changes help the preschool child gradually become ready for the next stage of life, the school years, which are presented in the next trio of chapters.

SUMMARY

The Self and the Social World

1. As children develop a more clearly defined idea of self, they become more confident and eager to take on new activities. According to Erikson, the accomplishments of this age help to resolve the crisis of initiative versus guilt. In dramatic play, children can try out social roles, express fantasies and fears, and learn to cooperate.

2. An increasing sense of self-recognition helps children to increase their social understanding and to become more skilled in their relationships with others.

Play

3. Playing with other children prepares preschoolers for the demands of school and the social relationships they will later develop. As children grow older, they spend more time in associative and cooperative play than in the more simple onlooker or parallel play that is characteristic of the younger child.

Parenting

4. Parent-child interaction is complex, with no simple answers about the best way to raise a child. However, in general, authoritative parents, who are warm and loving but willing to set and enforce reasonable limits, have children who are happy, self-confident, and successful. Authoritarian families tend to produce aggressive children. Children from permissive families often lack self-control.

5. Preschool children need clear and consistent rules and to know the consequences for breaking them. The most effective punishments are consistent and immediate.

Possible Problems

6. Normal preschool children sometimes use physical force to get what they want. They may also have vivid fantasies that they think are real. These behaviors are to be expected in a child who is trying to cope with the many ideas and problems of social interaction but whose thinking remains immature.

7. Some children show signs of serious psychological problems, such as autism and schizophrenia. The causes of these disturbances are multifactorial. The best chances for improvement lie in a treatment program in which parents participate.

Sex Roles and Stereotypes

8. While all psychologists agree that children begin to learn sex roles and moral values during early childhood, they disagree about how this occurs.

9. Freud believed that during the phallic stage, the fears and fantasies produced by the Oedipus and Electra complexes lead to adoption of sex-appropriate behavior and development of the superego.

10. Learning theorists think children learn expected sex-role behavior from the reinforcement they receive for acting appropriately, and from the punishment they get for behaving inappropriately. The example set by their parents, as well as cultural role models, is also important.

11. Cognitive theorists remind us that young children are illogical and egocentric. For example, the typical preschool child may think maleness and femaleness are the result of external characteristics, such as clothes and hair style, rather than of biology.

12. According to the theory of androgyny, individuals who are less rigid in their sex roles and seem to have a balance of both male and female characteristics tend to be more confident and have a higher sense of self-esteem than those who follow more traditional sex-role behavior.

KEY TERMS

initiative versus guilt *(220)*	modeling *(236)*
solitary play *(223)*	gradual desensitization *(236)*
onlooker play *(223)*	autism *(236)*
parallel play *(223)*	echolalia *(237)*
associative play *(223)*	elective mutism *(237)*
cooperative play *(223)*	childhood schizophrenia *(237)*
dramatic play *(224)*	phallic stage *(238)*
authoritarian parenting *(228)*	Oedipus complex *(238)*
permissive parenting *(228)*	identification *(239)*
authoritative parenting *(228)*	Electra complex *(239)*
phobia *(236)*	penis envy *(239)*
	androgyny *(242)*

KEY QUESTIONS

1. How does children's increasing self-knowledge affect their relationship with others?

2. Why is social play important?

3. As children grow older, how does their dramatic play change?

4. What are the three basic patterns of parenting?

5. What kinds of punishment are worst for young children and why?

6. What are some ways of preventing or correcting discipline problems?

7. How does the preschooler's stage of cognitive development affect the fantasies and fears that may occur at this stage of development?

8. What are the essential disagreements between Freud and learning theorists about the origin of sex roles during early childhood?

9. How would androgynous individuals describe their personality traits?

Part III

The Developing Person So Far: The Play Years, Ages 2 through 6

Physical Development

Brain and Nervous System

The brain continues to develop faster than any other part of the body, attaining 90 percent of its adult weight by the time the child is 5 years old. Myelination proceeds at different rates in various areas of the brain. This differential neurological development has some bearing on the child's readiness for certain types of activity.

Motor Abilities and Perception

The child becomes stronger, and body proportions become more adultlike. Large body movements, such as running and jumping, improve dramatically. Fine motor skills, such as writing and drawing, develop more slowly. Between the ages of 2 and 3, the activity level is higher than at any point in the life span.

Cognitive Development

Cognitive Skills

The child becomes increasingly able to use mental representation and symbols, such as words, to "figure things out." However, the child's ideas about the world are often illogical and much limited by the inability to understand other points of view.

Language

Language abilities develop rapidly; by the age of 6, the average child knows 14,000 words and demonstrates extensive grammatical knowledge. Children also learn to adjust their communication to their audience.

Psychosocial Development

Personality Development

According to Erikson, increased levels of energy at this stage enable the child to boldly and exuberantly initiate new activities. The outcome of the crisis of this stage of life—initiative versus guilt—will depend on whether the child often succeeds and is praised for his or her endeavors or whether efforts fail and the child is unrewarded, or worse, blamed.

Understanding Self and Others

The child's ability to interact with others depends on a well-developed sense of self. As children's social and cognitive skills develop, they engage in increasingly complex and imaginative types of play, sometimes by themselves and, increasingly, with others.

Parent-Child Interaction

As children become more independent and try to exercise more control over their environment, the parents' role in supervising the child's activities becomes more difficult. Some parenting styles and some forms of discipline are more effective than others in encouraging the child to develop both autonomy and self-control.

The School Years

If someone asked you to pick the best years of the entire life span, you might choose the years from 7 to 11 and defend your choice persuasively. To begin with, physical development is usually almost problem-free, making it easy to master many new skills.

With regard to cognitive development, most children are able to learn quickly and think logically, provided the topic is not too abstract. Moral reasoning has reached that state where right seems clearly distinguishable from wrong, with none of the ambiguities that complicate moral issues for adolescents and adults.

Finally, the social world of middle childhood seems perfect, for most school-age children think their parents are helpful, their teachers, fair, and their friends, loyal. Their future seems filled with promise—at least most of the time it does.

However, school and friendships are so important at this age that two common events can seem crushing: failure in school and rejection by peers. Some lucky children escape these problems; others have sufficient self-confidence or family support to weather them when they arise; and some leave middle childhood with painful memories, feeling incompetent and inferior for the rest of their lives.

The next three chapters celebrate the joys, and commemorate the occasional tragedies, of middle childhood.

CHAPTER 11

The School Years: Physical Development

Growth itself contains the germ of happiness.

Pearl S. Buck
To My Daughters, With Love

In general, physical development in middle childhood seems relatively smooth and uneventful, compared with the rapid growth of early childhood or the transformations of puberty. This is apparent in a number of ways. For one, disease and death are rarer during these years than during any other period. For another, most children master new physical skills (everything from tree-climbing to break-dancing) without much adult instruction, provided their bodies are sufficiently mature and they have an opportunity to practice these skills. In addition, sex differences in physical development and ability are minimal, and sexual urges seem to be submerged. Certainly, when physical development during these years is compared with the rapid and dramatic changes that occur during adolescence, middle childhood seems a period of relative tranquility. Now let us look at some of the specifics.

Size and Shape

Children grow more slowly during middle childhood than they did earlier or than they will in adolescence. Gaining about 5 pounds (2¼ kilograms) and 2½ inches (6 centimeters) per year, the average child by age 10 weighs about 70 pounds (32 kilograms) and measures 54 inches (137 centimeters) (Lowrey, 1978).

During these years children become proportionally thinner as they grow taller. In addition, muscles become stronger, enabling the average 10-year-old, for instance, to throw a ball twice as far as the average 6-year-old. The capacity of the lungs increases and the heart grows stronger, so with each passing year children are able to run faster and exercise longer than before.

These changes can be affected by experience as well as maturation, as shown by studies of girls who train for serious competitive swimming, often beginning at age 8 or earlier to swim three or four hours a day. By adolescence, their lung and heart capacities are significantly greater than those of their peers who were comparable in body type and strength before the training (Eriksson, 1976).

Variations in Physique

In some regions of the world, most of the variation in size is caused by malnutrition, with wealthier children being several inches taller than their contemporaries from the other side of town—whether the town is Hong Kong, Rio de Janeiro, or New Delhi. But most children in North America get enough food during middle childhood to grow as tall as their genes allow. So heredity, rather than diet, causes most of the variation we see (Eveleth and Tanner, 1976).

Not only does size vary, but the rate of maturation also varies from child to child. For instance, a relatively tall 7-year-old might have the muscle maturity and coordination more typical of a 5-year-old. At the other end of middle childhood, some 10- and 11-year-olds begin to undergo the changes of puberty, and may find that they are superior to their peers not only in height but in strength and endurance as well. Thus, various rates of development are quite normal.

These variations follow genetic, and perhaps cultural, patterns. French-speaking Canadian children, for example, tend to be smaller and stronger and to have greater heart and lung capacities than their English-speaking Canadian peers (Shephard, 1976). Black American children tend to mature somewhat more quickly (as measured by bone growth and loss of baby teeth) and to have longer legs than white American children.

While it may be comforting for parents and teachers to know that healthy children come in all shapes and sizes, it is not always comforting to the children themselves. In elementary school, children compare themselves with one another, and those who are "behind" their classmates in areas related to physical maturation may feel deficient. Physical development during this period even affects friendships, for, in part, they become based on physical appearance and competence (Hartup, 1983). Consequently, children who look "different," or who are noticeably lacking in physical skills, often become lonely and unhappy.

Figure 11.1 *Chronological age is the primary reason one school-age child is bigger than another, as shown by the difference in size between the 6-year-old girl (left) and her 11-year-old brother. However, genetic differences also play an important role, as illustrated by the fifth-grade Californians at the right, all close to their tenth birthday. On average, black American children are taller than their contemporaries, while children of Asian descent tend to be shorter.*

Childhood Obesity

The hardest size difference to bear in middle childhood is **obesity.** Although the point at which an overweight child qualifies as "obese" depends partly on the child's body type, partly on the proportion of fat to muscle, and partly on the culture's standards on this question, most experts contend that at least 10 percent of American children are sufficiently overweight to need slimming down (Grinker, 1981; Lamb, 1984). There is no doubt that the North American elementary-school children who are the heaviest 5 percent for their height can be classified as obese. Compared with the average child who weighs about 66 pounds at 53 inches, children in this group weigh 85 pounds or more at the same height (National Center for Health Statistics, 1976). These children are obese by almost everyone's standards.

Obesity is a serious physical and medical problem at any stage of life, for the obese person runs a greater risk of serious illness (Lamb, 1984). It is often a psychological problem as well. Children begin developing negative beliefs about obesity even as preschoolers (Fritz and Wetherbee, 1982), but it is not known if being fat affects their self-concept. In middle childhood, however, fat children are teased, picked on, and rejected. They know they are overweight, and they often hate themselves for it (Grinker, 1981). Obese children have fewer friends than other children (Strauss et al., 1985), and when they are accepted in a peer group, it is often at a high price, such as answering to nicknames like "Tubby," "Blubber," or "Fat Albert," and having to constantly suffer jokes about their shape.

Help for Overweight Children Clearly, an overweight child needs emotional support for a bruised self-concept, as well as help in losing weight. But reducing is difficult, and psychological encouragement is often scarce, partly because obesity is usually fostered by family attitudes and habits that are hard to break (see A Life-Span Perspective, next page). Obese children sometimes try crash diets, which make them irritable, listless, and even sick—adding to their psychological problems without accomplishing much long-term weight loss. To make matters worse, strenuous dieting during childhood can be physically harmful, since cutting down on protein or calcium could hinder important brain and bone growth (Winick, 1975). Unless a child is seriously obese, in which case careful weight reduction is in order, nutritionists generally recommend stabilizing the weight of overweight children to allow them to "grow out" of their fat.

The best way to get children to lose weight is to increase their physical activity. Indeed, inactivity may be as much a cause of childhood obesity as overeating. Overweight children tend to be less active than their peers, burning fewer calories and adding pounds even when they eat less than other children (Mayer, 1968). However, exercise is hard for overweight children, for they are not often chosen to play on the team, and they are likely to be teased and rebuffed when they try to join in group activities.

Parents and teachers can help overweight children to do the kinds of exercise in which their size is not much of a disadvantage—walking to school rather than taking the bus, or doing sit-ups at home—or even an advantage, as it might be in swimming. Parents can also exercise with their children, not only making activity easier and providing a good model, but bolstering the child's self-confidence as well. The importance of changing the child's eating and exercising patterns is apparent when one realizes that if the childhood weight problem reaches the point that the child is obese, and continues at that level throughout the childhood years, it is likely to last a lifetime (Grinker, 1981; Lamb, 1984). Thus, parents of the obese child who do nothing about it are jeopardizing the health as well as the happiness of their child in later years.

Figure 11.2 *Even if they join in physical activities and are accepted by their peers, overweight children often feel painfully self-conscious, especially when the activity requires donning a costume that calls more attention to one's appearance.*

A Life-Span Perspective

Causes of Obesity

Typically, no one explanation suffices for a particular instance of obesity; rather, the problem is generally created through the interaction of a number of influences (Grinker, 1981; Lowrey, 1978; Weil, 1975; Wolff and Lipe, 1978). These influences begin in infancy, continue through childhood, and are significant in adulthood.

1. *Heredity.* Body type, including the amount and distribution of fat, as well as height and bone structure, is inherited. Therefore, not everyone can be "average" in the ratio of height to weight. However, it is unhealthy to be 20 percent or more heavier than average.

2. *Activity level.* Inactive people burn fewer calories and are more likely to be overweight than active people. This is even more true in infancy and childhood than in the rest of life. (For more about the effects of activity level and exercise on weight and health in adulthood, see pages 468–469.)

3. *Quantity of food eaten.* In some families, parents take satisfaction in watching their children eat, always urging them to have another helping. The implication is that a father's love is measured by how much food he can provide, a mother's love, by how well she can cook, and a child's love, by how much he or she can eat. This is especially true when the parents or grandparents grew up in places where starvation was a real possibility. Not surprisingly, in the United States, immigrants from developing countries and their children and grandchildren are more likely to be overweight.

4. *Types of food eaten.* Choice of food is important as well. Besides the obvious culprits, many common foods, from cornflakes to ketchup, have sugar as a major ingredient. Those who eat too much of these foods become overweight. On other

diets—those emphasizing fruits and vegetables—it is hard to become fat. Unfortunately, but understandably, the diet of North American families who are below the poverty line tends to be high in fat, containing more pork than other meats and more fried foods than those that are broiled or steamed (Eichorn, 1979). In addition, research shows that people develop tastes for certain types of foods, depending on what they have become accustomed to eating, as well as on what their body needs (Grinker, 1981). Thus the "treats" and "junk food" that some parents regularly give their children may be creating dietary cravings that will lead directly to lifelong weight problems.

5. *Attitude toward food.* Some people consider food a symbol of love and comfort, and eat whenever they are upset. This pattern may be initiated in infancy, if parents feed their babies whenever they cry, rather than first figuring out if the baby is lonely or uncomfortable rather than hungry (Heald, 1975). The pattern may be reinforced through childhood, if parents, teachers, and even pediatricians use sweets as a reward or consolation, or as compensation for lack of emotional warmth (Lowrey, 1978).

6. *Overfeeding in infancy and late childhood.* For most of life, the number of fat cells in a person's body remains relatively constant, no matter what that person eats. Adults become fatter because each fat cell becomes fuller, or thinner because each cell loses fat. However, in the first two years of life and during early adolescence, when total body fat increases in anticipation of the rapid growth that follows, the number of fat cells is particularly likely to increase. The actual number of cells is related to nourishment, for malnutrition slows down the rate of cell multiplication, and overfeeding speeds it up (Grinker, 1981). This is one more reason why fat babies and adolescents become adults who want more food and gain weight more easily than people who were not overfed as children. Even when these adults diet and lose weight, their bodies still contain those extra cells, just waiting to fill up with fat again, like sponges ready to soak up water.

This would explain why chubby babies are more likely than average-weight babies to become overweight children and obese adults. However, scientists do not believe that there is a direct path from overfeeding to extra fat cells to adult obesity, for fat babies do not always become fat children and adults, and even those who follow this path may be directed more by genes than by childhood diet (Roche, 1981).

7. *Television-watching.* Recent research points the finger at another culprit in obesity, one that exacerbates several of the influences already cited: television-watching. According to a large longitudinal study (Dietz and Gortmaker, 1985), excessive television-watching (more than twenty-five hours a week) by children is directly correlated with being overweight. In addition, those children who watch several hours of TV a day during middle childhood are more likely to become obese adolescents. The researchers suggested three factors that make TV fattening: while watching television, children burn few calories, consume many snacks, and are bombarded with, and swayed by, commercials for junk foods. The same conditions, in all probability, apply to obese adults.

8. *Repeated dieting.* People who fast or severely cut back their calorie intake over a period of days or weeks find that they gain weight even more quickly when they return to their normal eating habits. The reason is that the body reacts to protect itself during periods of severe dieting as it would during periods of famine. The rate of metabolism becomes slower, enabling the body to maintain its weight with fewer

calories. One consequence is that after a certain amount of weight loss, additional pounds become much more difficult to lose. Another problem is that dieting helps the body become more efficient at storing fat. Thus each new round of dieting is harder than the last one and each temporary weight loss is then followed by an even greater and more permanent gain when the dieter returns to normal eating habits (Striegel-Moore et al., 1986). This vicious cycle can begin in childhood; for many women, it becomes unmanageable by adulthood (see pages 404–406).

9. *Physiological problems.*   One more explanation of obesity should be mentioned, even though it is not one of the common reasons. In a few instances, an abnormality in the growth process or metabolism is the cause (Lowrey, 1978). In these cases, obesity is only one sign of a complex physiological problem that usually involves retardation of normal physical and mental growth. It must be stressed, however, that disorders of this type account for less than 1 percent of all cases of childhood obesity. Therefore, parents of the fat school-age child should, in all likelihood, be much more concerned about the effects of diet and exercise than about physiological disturbances.

Motor Skills

The fact that children grow more slowly during middle childhood may be part of the reason they become so much more skilled at controlling their bodies during these years. (Compare this control, for instance, with the clumsiness that typically accompanies sudden changes in body shape and size during puberty.) School-age children can perform almost any motor skill, as long as it doesn't require very much power or judgment of speed and distance. The skills of typical North American 8- and 9-year-olds may include swinging a hammer well, sawing, using garden tools, sewing, knitting, drawing in good proportion, writing or printing accurately and neatly, cutting fingernails (Gesell et al., 1977), riding bicycles, scaling fences, swimming, diving, roller-skating, ice-skating, jumping rope, and playing baseball, football, and jacks. Of course, which particular skills a child masters depends, in part, on opportunity and encouragement.

Differences in Motor Skills

Boys and girls are just about equal in physical abilities during these years, except that boys have greater forearm strength (Tanner, 1970) and girls have greater overall flexibility. Consequently, boys have an advantage in sports like baseball, whereas girls have the edge in sports like gymnastics. But for most physical activities during middle childhood, neither body size nor sex is as important as age and experience. Short children can become fast runners, boys can do cartwheels, and girls can hit home runs.

However, the maxim "Practice makes perfect" does not necessarily hold (Astrand, 1976). Every motor skill is related to several other abilities, some depending on body size, others on brain maturation, others on genetic talent, and others on practice.

For example, brain maturation is a key factor in **reaction time,** which is the length of time it takes a person to respond to a particular stimulus. One study of reaction time in people between the ages of 7 and 75 found that the 7-year-olds took about

Figure 11.3 *Sex differences in motor skills make boys slightly better at pulling themselves up with their arms, and girls slightly better at twisting and leaping.*

twice as long as the typical adult to press a button in response to a flash of light (.75 seconds as opposed to .37 seconds). The 9-year-olds were notably better than the 7-year-olds, and the 11-year-olds were better still, but none of these three age groups did as well as any adult group, even the 75 year olds (Stern et al., 1980). Another study of reaction time in children aged 5 through 14 found that the older children were almost twice as fast as the younger ones (Southard, 1985). Thus in any sport in which reaction time is crucial, the average older child has a decided advantage over a younger one, and the average adult is quicker than the average child.

Other individual and age differences also come into play. Some are obvious, such as the advantage of height in basketball and of upper body strength and size for tackle football. Other differences may not be so obvious to the teacher or parent. For example, the ability to coordinate one's body movements is no more equal in children than in adults, so that some children are not able to aim a kick in soccer, or execute a leap in ballet, nearly as well as others.

Looking closely at the sports that adults value reveals that few are well-suited for children, because they demand precisely those skills that are hardest for them to master. Even softball is much harder than one might think. Throwing with accuracy and catching both involve more distance judgment and eye-hand coordination than many elementary-school children possess. In addition, catching and batting depend on reaction time. Younger children are therefore apt to drop a ball even if it lands in their mitt, because they are slow to enclose it, or to strike out by swinging too late. Thus a large measure of judgment, physical maturity, and experience is required for good ball-playing.

Games Children Play Ideally, the physical activities of elementary-school children should center around skills that most children can perform reasonably well: kick ball and dodge ball more than softball and football; relay races with equally matched teams rather than individual running races. Interestingly, the games that children organize themselves (tag, hide-and-seek, kick-the-can, king of the mountain, ring-a-levio) often include children of varying ages and abilities, and involve relatively simple skills. The pressure on any individual child to perform adequately in these types of games is obviously much less than in games emphasizing one-on-one competition.

Figure 11.4 *Most sports that adults organize for children are scaled-down imitations of professional sports. Unfortunately, many coaches and parents do not scale down their expectations of children's abilities to perform. As a result, many children are excluded from participating, and those who make the team are often subjected to criticism and haranguing that spoil the fun.*

The fact that many sports that adults value are not ideal for children does not mean that physical activity should be deemphasized. In fact, regular exercise, in moderation, is as physically beneficial for children as for adults (Lamb, 1984). In addition, according to a study that compared the school achievement of two similar groups of children, one with a daily program of physical activity and the other without one, spending an hour or more in regular physical activity may well benefit a child's academic work (Bailey, 1977).

Children with Handicaps

Thus far we have been discussing the needs of essentially normal children, who require physical activity as well as social acceptance to function happily and well during middle childhood. But what about atypical children, for whom physical activity or intellectual achievement or social interaction poses special problems, particularly in a school setting?

Traditionally, the education of children with special needs was accomplished in one of two opposite ways. When the differences were not very obvious, children were placed with normal children, to do the best they could. When children could not "pass" as normal, they were educated separately with as little contact as possible with normal children. Let us look more closely at both alternatives.

Separate Education

Because teachers did not feel prepared to handle them, and parents wanted to protect them from loneliness and mockery, many handicapped children were kept at home, educated by their parents or a tutor or not taught at all. Others spent their entire childhoods in schools for special children. From the point of view of the special child, such segregation seemed warranted, since in regular schools, children tend to be intensely compared with the rest of their class, and the children themselves often become openly critical of their classmates.

However, in the 1960s educators and parents began to question whether separate education was always best for children with special problems—either obvious problems like blindness or less apparent difficulties such as moderate mental retardation. For one thing, handicapped children need to learn many cognitive skills that are best learned in a regular classroom. In addition, three other arguments against the separate education of special children seemed compelling:

1. *Social skills.* Most children, handicapped or not, learn social and survival skills by playing and learning with other children. Therefore, many specialists reasoned, normal children and special children should benefit from sharing a classroom.

2. *The effect of labeling.* Children who are labeled as "different," and put in a special environment because of it, learn to see themselves as different and inferior. The child who is deaf will consider deafness his or her most important characteristic, rather than simply one attribute. The retarded child will feel unable to do anything, and, even as an adult, will be plagued with shame and self-doubt. Moreover, once children have become labeled as handicapped and are segregated, their adjustment to special treatment within a self-contained classroom may make it difficult for them to outgrow the stereotype or the special-educational milieu.

3. *Discrimination.* The isolation that was claimed to be for the good of the special child appeared to many observers to be a form of segregation intended to benefit teachers and normal children who would rather not deal with the disabled child. Many also saw another form of discrimination at work, since a relatively high proportion of children in special classes were nonwhite, non-English-speaking, and of low socioeconomic status.

The weight of all the arguments against separate education led to a new movement called mainstreaming.

Mainstreaming

By the end of the 1960s many educators were strongly recommending the integration of handicapped children with normal children as much as possible and as early as possible (Deno, 1973; Dunn, 1973; Gardner, 1977; Nix, 1976). This practice is called **mainstreaming,** because the handicapped children join the main group of normal children. In the United States, mainstreaming occurred gradually, school by school, state by state, until federal law in 1975 mandated that all children, no matter what their handicaps, receive public education in the "least restrictive" environment that is educationally sound. This means that handicapped children are mainstreamed unless it can be proven that mainstreaming is deleterious to them.

To help meet the needs of mainstreamed children, many schools adopted the idea of the **resource room,** a facility in the regular school run by specially trained teachers and equipped with special learning materials. A child with learning problems can be assigned to the resource room for particular help for several hours a day, even in midyear, without disrupting friendship patterns or the child's sense of being part of a regular class. In addition, communication between regular and spe-

Figure 11.5 *(a) Individual therapy sessions provide the special help required for this hearing-impaired boy to derive the most benefit from his classes at a regular public school. (b) Courses at a special school with a curriculum designed for the cognitive and social needs of Down-syndrome children have helped this girl share the sense of achievement celebrated in the graduation ceremony. Intensive instruction (notice the small list of graduates) and encouragement and acceptance by others with the same disability may have made the difference between this child's poise and competence and an unfulfilled potential.*

(a)

(b)

cial teachers, which is essential to the child's optimal learning, is facilitated when both types of teachers have classrooms within the same school.

Despite the many advantages it seems to offer—not the least of which is that teachers and children become more aware of the special needs of handicapped people—mainstreaming for every "special" child has not proven to be the simple solution that some hoped it would be (Carlberg and Kavale, 1980; Meisel, 1986). One problem is that many special children need extensive and expensive supportive services to help them learn in the regular classroom. Especially in recent times, when money for education has been in short supply, many school systems have had to cut back on, or forgo, such services, including resource rooms, and some schools have simply added the special child to the regular teacher's workload, often leaving the special child to "sink or swim." Not surprisingly, few school professionals are currently enthusiastic about mainstreaming (Baker and Gottlieb, 1980).

Another problem is that social skills are not necessarily furthered simply by mixing children together in the same room (Gresham, 1982; Strain and Shores, 1983). Special children tend to be socially isolated, in part because they look and act different; in part because some cannot participate as equals in gym or recess activities while others are obviously not equal in academic work; and in part because the social skills of many have been limited by their handicap.

In addition, although some "special" children, with physical, intellectual, or emotional disabilities, or gifts, could manage in a regular school, they may, at various times, derive greater benefit from some form of special education.

It seems clear, then, that the promise of mainstreaming has to be balanced against the realities of each individual case of special needs. Ideally, decisions about a disabled child's education should be made by special and regular teachers, as well as by the parents, so that each child benefits from the right combination of special and mainstream learning. Of course, educational decisions depend a great deal on the child's cognitive development, the subject of the next chapter.

Figure 11.6 *Mainstreaming works particularly well for children who are physically disabled but who have no social or academic deficits.*

SUMMARY

Size and Shape

1. Children grow more slowly during middle childhood than at any other time until the end of adolescence. There is much variation in the size and rate of maturation of healthy North American children, primarily as a result of genetic, rather than nutritional, differences.

2. Overweight children suffer from peer rejection and poor self-concept. More exercise, rather than severe dieting, is the best solution.

3. Many influences throughout the life span interact to cause obesity. Hereditary factors, overfeeding in infancy and late childhood, repeated dieting, lack of exercise, and other factors, contribute to the incidence of obesity.

Motor Skills

4. School-age children can perform almost any motor skill, as long as adult strength or judgment is not a prerequisite. Physical exercise is beneficial for strength, growth, and coordination. The activities that are best for children are those that demand only those skills that most children of this age can master.

Children with Handicaps

5. Children with obvious physical handicaps often were excluded from regular schools, assigned to special classes, limited to home tutoring, or given no education at all. Now the law requires that these children be provided public education, in the regular classroom if possible.

KEY TERMS

obesity *(251)*	mainstreaming *(257)*
reaction time *(254)*	resource room *(257)*

KEY QUESTIONS

1. What are some of the causes of variation in physical growth in middle childhood?

2. How does obesity affect a child's development?

3. What are the causes of obesity?

4. What are some of the reasons for children's increasing motor skills during the school years?

5. What kinds of considerations come into play when considering the appropriateness of physical activities for children?

6. What are the advantages and disadvantages of mainstreaming?

CHAPTER 12   # The School Years: Cognitive Development

I hear, and I forget.
I see, and I remember.
I do, and I understand.

Ancient Chinese proverb

The thinking and learning of the typical 11-year-old is radically different from that of the typical 6-year-old. For example, most sixth-graders can figure out which brand and size of popcorn is the best buy, can be taught to multiply proper and improper fractions, can memorize a list of fifty new spelling words, and can use irony appropriately—accomplishments beyond virtually every first-grader.

Even their approach to learning is different. Take the first day of school, for example. Six-year-olds enter school filled with excitement and fear, often dressed in their Sunday best and clinging tightly to their mother's or father's hand. By age 11, children arrive at school with new notebooks and sharpened pencils, ready for the serious business of learning. They appear casual and confident, even when they aren't, and they would angrily balk if their parents offered to walk them to the classroom. While first-graders worry about getting lost or wetting their pants, sixth-graders worry about finishing homework or failing a test.

Not surprisingly, there is a vast difference as well in what teachers expect of students over this five-year period. If first-graders can learn to stay quiet when they are supposed to, read simple words, and add simple numbers, that's considered accomplishment enough. Sixth-graders are expected to know their multiplication tables and spelling rules; they are supposed to understand the morals of the stories they read and the general principles underlying the mathematical formulas they use. They are urged to plan ahead and to hand in work that is neat and correct.

These changes in behavior, attitudes, and expectations reflect, in part, the growth of children's cognitive abilities as their thinking becomes less egocentric and more logical. This chapter describes the cognitive processes that produce new skills and reasoning strategies, the language development that expresses them, as well as some of the cognitive difficulties encountered by some children.

Concrete Operational Thought

What underlies the differences we have just described is, according to Piaget, the attainment of **concrete operational thought,** through which children can reason about almost anything specific they perceive. Between ages 7 and 11, children usually come to understand logical principles, as long as the principles can be applied to concrete, or specific, cases. They become increasingly able to **decenter**, that is, to move away from a simple, perceptual focusing on one aspect of a problem. Thus they are able to think more objectively: they are less likely to be misled by mere appearances or to peg their judgments on a single feature of an object or situation. They can watch water being poured from a narrow glass into a wide one and explain why the quantity of liquid remains the same (see Chapters 2 and 9). They are quick to say that 8 is greater than 5, and just as quick to confirm that 11,108 is greater than 11,105. They can look at a scene from one vantage point and imagine it from another point of view (Chapter 9, pages 203–204). Moreover, according to Piaget, their reasoning can be generalized across tasks and situations: thus, if they can apply a logical principle in one context, they can apply it in other analogous contexts. However, they can reason only about the concrete, tangible things in their world: they are not yet able to reason about abstractions, as can the adolescent who has reached formal operational thought.

The 5-to-7 Shift

In Piaget's theory of the sequence of cognitive development, true concrete operational thinking is preceded by a transitional period between the ages of 5 and 7—sometimes called the **5-to-7 shift**—in which the child has not quite outgrown preoperational thought nor firmly reached concrete operational thought. Piaget noted that although children sometimes intuitively grasp the right answer on tests of concrete operational logic, they frequently do not understand the underlying principles that led them to their conclusion (Cowan, 1978).

For instance, many 6-year-olds may answer correctly on the conservation-of-liquids test (see page 201), saying that the taller, narrower glass contains the same amount as the shorter, wider glass, but when asked to explain their answer might act as if reasons are unnecessary because the answer is so obvious and the reasons too difficult to formulate. (The adult's "Why?" is typically answered with "Because.") They also might fail to arrive at the correct answer to another problem that depends on the same underlying principle. In the conservation-of-liquids experiment, for instance, a child might correctly answer that both glasses contain the same amount, but then, if the liquid from one of the glasses is poured into six smaller glasses, insist that the amount of liquid has increased. Piaget (1967) concluded: "Experimentation has shown decisively that until the age of seven, the child remains prelogical."

Logical Ideas

Beginning at about age 7 or 8, children become true concrete operational thinkers. The hallmark of concrete operational thought is the ability to understand certain logical principles—such as identity and reversibility—when these principles are applied to specific, or concrete, cases. (Identity, you will remember, is the idea that an object's content remains the same despite changes in its shape, and reversibility is the idea that a transformation process can be reversed to restore the original form.) A grasp of both identity and reversibility enables the child to realize, among other things, that rolling a ball of clay into a long, thin rope of clay does not alter the amount of clay.

Further, according to Piaget, once these principles are mastered, they can be applied in many contexts. Once children have a firm grasp of identity, for instance, they know that the number 24 is always 24, whether it is arrived at by adding 14 and 10 or 23 and 1; they also know that their mother was the same person when she was a child.

What seems to lie behind the child's understanding of concrete operational principles is the decline in egocentrism and the newly acquired ability to simultaneously hold in mind and relate various characteristics of objects, or persons, or situations (Biggs and Collis, 1982). This is particularly clear with the principle of **reciprocity,** the idea that a change in one dimension of an object effects a change in another dimension. Thus in the conservation-of-liquids test, the concrete operational thinker can mentally relate the differences between the height and width of the two glasses (realizing "It's taller but it's narrower" or "It's shorter but it's wider") and can therefore explain why the different-sized glasses contain the same amount of liquid.

Figure 12.1 *Just because one has reached the stage of "mastering" a concept such as reciprocity doesn't mean that it will never pose a problem again. For example, not all adults respond with the most logical choice when faced with the idiosyncrasies of product packaging: whether to buy the 15½-ounce can for 69 cents or the 18-ounce can for 79 cents?*

Two other concepts become part of the child's way of thinking during middle childhood: *classification* and *seriation*. Mastery of both concepts reveals the child as a logical thinker rather than a thinker heavily influenced by superficial appearances or egocentric perceptions.

Classification and Class Inclusion **Classification** is the concept that objects can be thought of in terms of categories, or classes. For example, a child's parents and siblings belong to the class called *family. Toys, animals, people,* and *food* are other everyday classes. According to Piaget, most preschool children have some understanding of how to apply these labels, but many do not understand the relationships that underlie these categories (Inhelder and Piaget, 1970). Thus a preschool child might tell a sibling "If you hit me, you can't be in my family," and at the same time consider that family to include the pet dog and a litter of gerbils.

A related but more complicated concept is **class inclusion,** the idea that a particular object or person may belong to more than one class. Even when children younger than 8 have learned who is in their family and who is not, they are usually unable to think of a particular family member in terms of more than one category. Thus children who have brothers or sisters might assume that brothers and sisters belong to the category *child,* and that adults, therefore, cannot be brothers and sisters (Piaget, 1962). This confusion was illustrated in an interview between Piaget and a 7-year-old boy:

Child: All boys are brothers.

Piaget: *Is your father a brother?*
Yes, when he was little.
Why was your father a brother?
Because he was a boy.
Do you know your father's brother?
He doesn't have a brother.

Children at this stage have trouble not only with categories and subcategories but with every type of classification. Consider the following experiment, modeled on a series of experiments conducted by Piaget. An examiner shows children seven toy dogs. Four of them are collies, and the others are a poodle, an Irish setter, and a German shepherd. First, the examiner questions the children to make sure that they know that all the toys are dogs and that they can name each breed. Then comes the crucial question: "Are there more collies or more dogs?" Until the concept of classification is firmly established, usually not until age 7 or 8, most children say "More collies." They cannot simultaneously keep in mind the general category *dog* and the subcategory *collie*, mentally shifting from one to the other.

Figure 12.2 *Sorting these eggs may involve complex classification concepts, since the attributes of size, color, and species might all be relevant. It is a challenging task for a school-age child.*

Seriation   **Seriation** refers to the arrangement of items in a series, as in the laying out of sticks from shortest to longest, or of crayons from lightest to darkest. Like other logical operations, seriation begins to be understood toward the end of the preoperational period, but the concept is usually not firmly established until age 7 or 8. Thus, during the 5-to-7 shift, if a typical 6-year-old is asked to arrange a series of ten sticks according to length, the child might first put together three sticks—a short, medium, and long—and then insert the others, rearranging several of the sticks before getting the correct order. A typical 8-year-old, in contrast, would look at the whole jumble of sticks, pick out the shortest, then the next shortest, and so on, systematically and quickly arranging the series.

Mathematics   Many of the concepts of concrete operational thought are said to underlie the basic ideas of elementary-school math. For instance, the concept of

Figure 12.3 *Fractions are one of the most difficult concepts in elementary-school math. Children who have previously learned that each number in the sequence of whole numbers is larger than the one that came before have difficulty understanding that with fractions, this order is reversed: ⅛ is smaller than ¼, which is smaller than ½. As at earlier ages, children can come to understand such concepts more easily by using learning materials they can manipulate.*

seriation may be necessary for a firm understanding of the number system. This would explain why 5-year-olds, who often can recite numbers correctly in sequence, are nevertheless often puzzled by simple questions such as "Which is greater, 6 or 8?" or "What number is one less than 5?" (Piaget and Szeminska, 1965). These concepts take a long time to develop; many first-graders need to do much finger-counting and hard thinking in order to add simple sums. In fact, in most cases, if you add one item (another apple, another penny) to a series of items that a 5-year-old has already counted, the child will have to count the series all over again to find out how many there are.

Other math difficulties may occur because an understanding of classification is needed to comprehend place value. Thus many elementary-school children have trouble understanding the difference between, say, 501, 51, and 15. Similarly, without an understanding of reversibility, the child may not realize that if 6 plus 5 equals 11, 11 minus 5 must equal 6; and without a grasp of reciprocity, the child will have trouble understanding that the area of a rectangle that is 1 inch by 30 inches is the same as the area of a rectangle that is 6 inches by 5 inches.

According to Piaget, until the child becomes a concrete operational thinker, math must be learned problem by problem—mostly by rote—rather than really understood. He wrote, "The indefinite series of numbers and, above all, the operations of addition (plus its inverse, subtraction) and multiplication (plus the inverse, division) are, on the average, accessible only after the age of seven" (Piaget, 1965).

Modifying Piaget

As noted at several points in earlier discussions of Piagetian theory, a number of cognitive researchers have attempted to modify certain of Piaget's ideas by emphasizing the individual differences in, and the unevenness of, children's progress through the cognitive stages described by Piaget. Whereas Piaget tended to describe the child's movement from one stage to the next as occurring fairly quickly once begun, and also as occurring across the board, in all domains, these researchers maintain that the child enters a new stage gradually in certain domains—say, math or science—but not in others, like social understanding.

Particularly in the case of concrete operational thought, these researchers believe that cognitive development is considerably more heterogeneous, or inconsistent, than Piaget's descriptions would suggest. According to Flavell (1982), two of the factors that account for this heterogeneity are the hereditary differences among individuals in terms of abilities and aptitudes and the environmental differences in terms of "cultural, educational, and other task-related experiential background." The sum of these differences, says Flavell, might well produce a great deal of cognitive heterogeneity:

> Imagine, for example, a child or adolescent who is particularly well-endowed with the abilities needed to do computer science, has an all-consuming interest in it, has ample time and opportunity to learn about it, and has an encouraging parent who is a computer scientist (whence much of the aptitude, interest, and opportunity, perhaps). The quality and sophistication of the child's thinking in this area might well be higher than that of most adults in any area. It would also likely be much higher in this area than in most other areas of the child's cognitive life. His level of moral reasoning or skill in making inferences about other people might be considerably less developed, for instance. The heterogeneity could be a matter of time constraints as well as a matter of differential aptitudes and interests; that is, time spent at the computer terminal is time not spent interacting with and learning about people.

As we saw in Chapter 9, some researchers have also suggested that many children demonstrate signs of at least partial entrance into concrete operational thought earlier than Piaget would have predicted. This contention is bolstered by studies (Anglin, 1977; Koslowski, 1980; Winer, 1980) showing that on simplified versions of Piagetian tests for classification and seriation, preschoolers sometimes reveal the same kind of "fragile" grasp of these concepts that we saw them reveal for conservation and perspective-taking in Chapter 9.

The Legacy of Piaget   Despite these revisions and criticisms, Piaget's comprehensive view of children's cognitive development is considered correct in most aspects. With regard to school-age children, three of Piaget's ideas in particular have been supported time and time again as further research has been performed. First, compared with the thinking of the preschool child, the thought of the school-age child is characterized by a more comprehensive logic and markedly broader grasp of the underlying assumptions and ideas of rational thought. While the specific concepts of conservation, classification, and so forth, may not distinguish the thinking of the 8-year-old from that of the 4-year-old as definitively as Piaget believed they do, the typical overall thinking of the 8-year-old on almost any issue is less egocentric and more logical than the 4-year-old's.

Second, children are active learners. They learn best by questioning, exploring, and doing. In fact, Piaget's theories provided the theoretical framework for making most classrooms in the 1970s and 1980s quite different places from those in the 1950s (Ravitch, 1983). The result was interest in what is called *open education,* which encourages individualized learning by discovery, discussion, and deduction.

(a)   (b)

Figure 12.4 *The relative increase in the number of classrooms like the one on the right, as compared to the one on the left, can be attributed to the influence of Piaget. (a) The teacher uses the traditional "talk and chalk" technique, and the children either quietly try to absorb the material or search for distraction. (b) In the open classroom, the teacher does not preside at the front of the room, but, instead, works among the children as they pursue projects that are particularly suited to their needs.*

This approach to learning has tended to overshadow much of the more passive, teacher-centered, lock-step education of a few decades ago.

Finally, how children think is as important as what they know. Piaget's interest in the underlying structures of thought has led to a realization that what distinguishes the thinking of school-age children from that of preschoolers is not new information but new cognitive organizations. What children need to learn, then, is not so much new facts as new ways to assemble facts (Flavell, 1985).

An Information-Processing View

Piaget's emphasis on the structures of cognition, as well as the learning theorists' stress on reinforced learning of specific skills and facts, has guided a new group of cognitive researchers who have synthesized many Piagetian and learning theory ideas (Brown et al., 1983). This group takes an **information-processing** view of human learning, so named because of the observation that some aspects of human thinking bear some similarity to the way computers function. These researchers do not suggest that humans process information exactly like a computer: the human brain is far more complex than the most advanced computer, and no computer can match the mind's capacity for intuition, creativity, and change. However, the information-processing view has led to a more precise understanding of the way the mind functions, and has had important educational implications.

In applying the information-processing model to children's learning, developmentalists emphasize that the output of a computer depends not only on the input, that is, the information fed into it, but also on the program and capacity of the computer. Unless the computer is programmed to understand the input, and has the memory capacity to hold it, the computer cannot process the input correctly. Consequently, say the proponents of this view, instead of simply looking at the general educational milieu of the child, we must also look closely at the capacity and program of the child's mind.

Memory

The topic most studied by those using the information-processing model has been memory. Developmentalists have suggested that there are two aspects of memory

Figure 12.5 *In a few years, the barefoot Pilgrim woman on the left will be able to memorize all her lines. Meanwhile, the technique of linking abstract historical information to concrete experiences improves both memory and understanding. Mnemonic devices are useful, too: When did Columbus sail the ocean blue?*

that should be considered separately. One is **memory capacity**—how much information the brain can hold and how well it can be processed and stored. The other, **metamemory,** involves the understanding and use of various memory techniques that can be used to keep things in mind.

Memory Capacity   Memory capacity involves three levels of storage. The first is called the **sensory register,** which temporarily stores sensory information as it is received (as in the afterimage that occurs like an instant photo when you close your eyes). The sensory register holds material very briefly, for less than a second. As best we can tell, the capacity of the sensory register is about the same in children of all ages and in adults (Hoving et al., 1978).

Once in the sensory register, material fades unless it is processed and stored in **short-term memory,** where it stays for up to a minute. From short-term memory, material may be further processed into **long-term memory,** where it can remain for days, months, or years. A simplified version of this processing and storage sequence can be seen in the following example. Say you are listening to a radio show and hear a call-in phone number announced for a contest. If you have no interest in calling, and therefore pay no attention to the number, the number will get no further than your sensory register: you will not recall it. If you do want to call, and mentally note the number, it will move to short-term memory, probably staying there long enough for you to dial the call. However, unless you have repeated the number to yourself several times, or have noted some memorable peculiarity in the sequence of numbers, the phone number will not have moved into long-term memory. If you get a busy signal and try to dial again, you may find yourself unable to.

Although it is impossible to equate exactly the various memory tests given to children and those given to adults, as well as to measure precisely short-term memory and long-term memory for both age groups, a careful review of memory research concludes that, by school age at least, the basic memory capacity of children is quite similar to that of adults (Stern, 1985). That is, children can store about as many items of information, for about as long, as adults can. Thus differences in capacity are not the primary reason memory improves notably between ages 6 and 11, and continues to improve thereafter.

Metamemory

The main reason adults remember better than children, and older children remember better than younger children, is the difference in their respective metamemories (Brown and De Loache, 1978; Liben, 1982; Ross and Kerst, 1978; Stern, 1985). Let us look at metamemory in more detail.

Selective Attention   One key to being able to remember well is knowing that one should pay attention to certain items and ignore others. A review of the research on **selective attention** (Pick et al., 1975) found that the ability to screen out distractions and to concentrate improves steadily during middle childhood.

One classic experiment showed this development (Maccoby and Hagen, 1965). Children between the ages of 6 and 12 were asked to remember the background colors of a series of pictures—an elephant, a scooter, a water bucket, and so forth. They were not asked to remember the subjects of the pictures. As expected, memory for background colors improved steadily with age: 12-year-olds remembered almost twice as accurately as 6-year-olds. Then the children were asked to name the subjects of the pictures. Ability to remember in this case improved only slightly with age, and then declined. In fact, 12-year-olds apparently had learned to focus their attention and ignore unnecessary information so well that they were actually worse than 6-year-olds at remembering which subjects were on which backgrounds!

Figure 12.6 *What's wrong with this picture? If the boy here is really trying to do his homework, there are several sources of distraction that do not belong in this scene.*

Memory Techniques   Beyond improved ability to concentrate selectively, what else do older children do to help them remember? Apparently, older children are better able, and much more likely, to use **mnemonic,** or memory-aiding, devices (Fabricius and Wellman, 1983). For example, they use **rehearsal,** the repetition of the items to be remembered; or they group items to be remembered into categories, a technique called **chunking.** One way to memorize the fifty states by chunking, for instance, would be to learn them by region, or to learn the eight states that begin with "M," the eight that begin with "N," the four that begin with "A," and so forth. A third technique is memorizing general rules rather than trying to remember every specific case. For instance, if children try to learn irregular spellings simply by trying to memorize them, they will not do very well. They will do much better if, instead, they memorize rules for such irregularities, such as "'i' before 'e' except after 'c,'" "'full' has a double 'l' only when it stands alone, so be careful," and dozens more.

By age 11, most children have learned how to use mnemonic devices, and are ready to apply them when faced with a memory task. While children may teach themselves some metamemory techniques, there is no doubt that training helps (Stern, 1985). Experimenters have found that even a short practice session with instruction in rehearsal or chunking usually results in improved memory performance in school-age children (Hagen et al., 1975). Similar findings come from naturalistic studies comparing children who have radically different educational experiences. For instance, one study found that children from Liberia who had been trained in memorizing the Koran performed significantly better on a similar memorizing task than Liberian children who had not been trained, even though the actual material to be remembered was equally familiar to both groups (Scribner and Cole, 1981).

Learning How to Learn

The investigation into metamemory led information-processing researchers to look more closely at the steps of learning in general. The study of **metacognition,** or techniques of thinking, is now underway in many laboratories and classrooms.

One general principle has been recognized. It is that children and adults learn new material best when they can link it to some schema or cognitive framework, either one they already have in their possession or one they are currently learning. A simple example of this is learning to spell (Henderson, 1985). Children's first attempts at spelling are rife with phonetic fallacies, as they try to link the sounds they hear to the letters they think make those sounds. Typically, they write "likt" for "liked" and "kak" for "cake." Rather than memorizing correct spellings word by word, they need to recognize the general rules that they already sometimes use, such as that "ed," not "t," makes the past tense, or that the "k" sound at the beginning of words is much more often a "c" than a "k." Later on, difficult new words with the same root are better learned together than one by one, as are special cases that are similar, such as silent-"p" spellings like "pneumatic," "pneumonia," and "pneumonic."

The importance of a cognitive framework as an aid to memory was shown by an experiment that compared the ability of second-graders, sixth-graders, and adults to use the sequence of a story to remember the story's details (Buss et al., 1983). All three groups were quite able to remember details of a story that was correctly

Figure 12.7 *Facts are best learned when they are connected to more general concepts—a principle these children seem to have grasped, perhaps a bit too tightly.*

"One plus one equals two. Two plus two equals four. Three plus three equals six, and so on and so forth. Well, that's life."

Drawing by Ross; © 1987 The New Yorker Magazine, Inc.

sequenced, but when faced with the task of remembering details from a story that was out of sequence, only the older children and the adults performed well, because both these groups spontaneously unscrambled the story to aid recall. However, after minimal instruction in how sequence could help, second-graders also were able to create a coherent structure to aid their memories. Research such as this suggests to developmentalists that part of the education process should be helping children create structures, strategies, and links arising from their past experiences to help them integrate the knowledge they are receiving.

In fact, if children are not helped to relate new information to their existing knowledge, they will create their own, often misleading, relationships between facts. As one scientist explains:

> Learners try to link new information to what they already know in order to interpret the new material in terms of established schemata. This is why students interpret science demonstrations in terms of their naive theories and why they hold onto their naive theories for so long. The scientific theories that children are being taught in school often cannot compete as reference points for new learning because they are presented quickly and abstractly and so remain unorganized and unconnected to past experience. [Resnick, 1983]

Information-processing also emphasizes the importance of the sequence in which material to be learned is presented. For example, in one experiment (Hawkins et al., 1984), children were tested on their ability to follow the logic of paired statements. The children were to pretend that each statement was entirely true and then answer the question that followed. Some of the paired statements were about real things that fit into what the children already knew. For example:

> Rabbits never bite. Cuddly is a rabbit. Does Cuddly bite?

Others were pure fantasy. For example:

> Merds laugh when they're happy. Animals that laugh don't like mushrooms. Do merds like mushrooms?

Finally, some statements were against what the children knew. For example:

> Glasses bounce when they fall. Everything that bounces is made of rubber. Are glasses made of rubber?

Not surprisingly, the nature of the statements influenced the children's ability to answer correctly. Statements that were consistent with the children's existing knowledge were generally answered logically; those that contradicted the child's knowledge were generally answered illogically, with the fantasy examples midway between the two. But the most interesting influence was the specific order of presentation of the examples. Those children who heard the fantasy examples first were much more likely to get problems of all kinds correct than were those who heard the incongruent problems first. Apparently, the fantasy problems helped the children learn how to do these logical problems, preparing them for the apparent nonsense to follow.

One lesson from information-processing, as well as from other perspectives, is that children's mental structures may vary, depending on age, learning style, and prior experience, and therefore one step in planning education is to design it for the particular child's "program" (Bennett, 1976; Minuchin and Shapiro, 1983; Rot-

Figure 12.8 *Since it is impossible to learn, once and for all, everything we will ever need to know, it is important to learn how to find out whatever we need to know. Learning to make use of the library's system of organizing information is a key to unlocking needed knowledge.*

berg, 1981). This does not mean, however, that educators need wait until a particular child is ready for a certain bit of knowledge. In fact, the information-processing view suggests that the idea of "readiness" may have been overemphasized; at least by middle childhood, most children are ready to learn almost any concrete skill or concept, including learning how to learn through metamemory and metacognition techniques. Postponing such learning may be a mistake, according to information-processing theory, for trying to teach facts without linking them to concepts, and expecting children to remember without showing them how to do so, will result in children who forget much of what they have been taught and cannot apply what they do remember (Resnick, 1983). It is not unusual, for instance, for children to know how to add a page of sums on a worksheet, but not to know when or what to add when asked simple verbal problems (Bryant, 1985). As you will soon see, these insights about the structures of knowledge apply to language-learning just as they do to learning the concepts of math or science. In the next chapter, we will see how they apply to the knowledge that children gain as their social world expands.

Figure 12.9 *Although computers are no substitute for talented teachers, they make good partners, capable of providing instruction and practice that can be tailored to the child's cognitive development. Another plus is that computers provide immediate, nonthreatening feedback that helps children identify and correct their own mistakes.*

Language

As you saw in Chapter 9, the preschool years are the time of a language explosion, in which children's vocabulary, grammar, and pragmatic language skills develop with marked rapidity. Language development during the years from age 6 to 11 is also remarkable, though much more subtle, as children consciously come to understand more about the many ways language can be used. This understanding gives them greater control in their comprehension and use of language, and, in turn, enlarges the range of their cognitive powers generally. Their understanding of language is a powerful key to new understanding of themselves and their world.

Vocabulary

During middle childhood, children begin to really enjoy words. This is demonstrated in the poems they write (Koch, 1970; Lewis, 1977, 1978), the secret languages they create, and the jokes they tell. This makes middle childhood a good time to explicitly help children expand their vocabularies, thus providing a foundation for more elaborate self-expression.

One of the most important language developments during middle childhood is a shift in the way children think about words. Gradually, during middle childhood they become more analytic and logical in their processing of vocabulary, and less restricted to the actions and perceptual features directly associated with particular words (Holzman, 1983). When a child is asked to say the first word that comes to mind on hearing, say, "apple," the preschooler is likely to be bound to the immediate context of an apple, responding with a word that refers to its appearance ("red," "round") or to an action associated with it ("eat," "cook"). The older child or adult, on the other hand, is likely to respond to "apple" by referring to an appropriate category ("fruit," "snack") or to other objects that logically extend the context ("banana," "pie," "tree").

Similarly, when they define words, preschoolers tend to use examples, especially examples that are action-bound. For instance, while preschoolers understand that "under," "below," and "above" refer to relative position, they define these words with examples such as "Rover sleeps under the bed," or "Below is to go down

Figure 12.10 *With a language as irregular as English, it should come as no surprise that many children (as well as adults) sometimes generate errors by applying logic.*

under something." Older children tend to define words by analyzing their relationship to other words: they would be more likely to say, for instance, that "under" is the same as "below," or the opposite of "above" (Holzman, 1983).

Older children's more analytic understanding of words is particularly useful as children are increasingly exposed to words that may have no direct referent in their own personal experience. This understanding makes it possible for them to add to their conceptual framework abstract terms such as "mammal" (extracting the commonalities of, say, whales and mice) or foreign terms such as *yen* (relating this unit of currency to the dollar), and to differentiate among similar words such as "big," "huge," and "gigantic," or "running," "jogging," and "sprinting." Thus, the cognitive maturation of middle childhood, coupled with the school experiences that children have, encourages children to link words with other words. The combination of maturation and experience leads to more rapid intellectual processing, as well as continued vocabulary development.

Grammar

Similar progress occurs in grammar. Although most grammatical constructions of the child's mother tongue are mastered before age 6, knowledge of syntax continues to develop throughout elementary school (Chomsky, 1969; Romaine, 1984). Children are increasingly able to use grammar to understand the implied connections between words, even if the usual clues, such as word order, are misleading.

For instance, as we saw in Chapter 9, children younger than age 6 often have trouble understanding the passive voice, because they know that the agent of action in a sentence usually precedes the object acted upon. By middle childhood, however, most children realize that "The truck was bumped by the car" does not mean that the truck did the bumping (de Villiers and de Villiers, 1978). The increasing understanding of the passive voice is reflected in children's spontaneous speech as well as in research studies: compared with 6-year-olds, 8-year-olds use the passive voice two-and-a-half times as frequently, and 10-year-olds, three-and-a-half times as often (Romaine, 1984).

Chomsky (1969) also demonstrated that with every passing year of middle childhood, children's grasp of syntactical structure is less likely to get confused by detail that is irrelevant to the meaning of the structure. When Chomsky first presented children with a large doll and asked, " Is this doll easy to see or hard to see?" even preschool children answered correctly. But when Chomsky put a blindfold on the doll, most of the sixteen 5- and 6-year-olds answered the same question by saying the doll was hard to see. Only three of the fifteen 7- and 8-year-olds made this mistake, and none of the nine 9- and 10-year-olds were misled by the blindfold.

TABLE 12.1 **Age-Related Adherence to Syntax**

Age of Child	Wrong (Percentage Answering "Hard to see")	Correct (Percentage Answering "Easy to see")
5	78%	22%
6	57%	43%
7	14%	86%
8	25%	75%
9	0%	100%
10	0%	100%

The school-age child's gradual understanding of logical relations helps in the understanding of other constructions, such as the correct use of comparatives ("longer," "deeper," "wider"), of the subjunctive ("If you were a millionaire . . . "), and of metaphors (that is, of how someone could be a dirty dog or a rotten egg). The ability to use these constructions depends on a certain level of cognitive development that typically occurs during elementary school. This is true even with languages in which the particular construction is relatively simple. For instance, the subjunctive form is much less complicated in Russian than in English, but Russian-speaking children master the subjunctive only slightly earlier than English-speaking children, because the concept if-things-were-other-than-they-are must be understood before it is expressed (de Villiers and de Villiers, 1978).

School-age children have another decided advantage over younger children when it comes to mastering the more difficult forms of grammar. Whereas preschool children are quite stubborn in clinging to their grammatical mistakes (remember the child in Chapter 9 who "holded" the baby bunnies?), school-age children are more teachable. They no longer judge correctness solely on the basis of their egocentric version of the rules, or on their own speech patterns. Assuming that they have had ample opportunity to learn the correct grammar, by the end of middle childhood, children are able to apply the rules of proper grammar when asked to, even if they don't use them in their own everyday speech. Thus, even if they themselves say "Me and Suzy quarreled," they are able to understand that "Suzy and I quarreled" is considered correct.

Pragmatics

You have already seen that preschoolers have a grasp of some of the pragmatic aspects of language: they change the tone of their voice when talking to a doll, for instance, or when pretending to be a doctor. However, preschoolers are not very skilled at modifying vocabulary, sentence length, and nonverbal cues to fit particular situations. The many skills of communicating improve markedly throughout middle childhood, as children become less egocentric.

One of the clearest demonstrations of schoolchildren's improved pragmatic skills is to be found in their joke-telling, which demands several skills not usually apparent in younger children—the ability to listen carefully; the ability to know what someone else will think is funny; and, hardest of all, the ability to remember the right way to tell the joke. Telling a joke is beyond most preschool children. If asked to do so, they usually just say a word (such as "pooh-pooh") or describe an action ("shooting someone with a water gun") that they think is funny. Even if they

Figure 12.11 *Chances are that if the words this girl is whispering were written here, few readers of this book would find them funny. Humor depends not only on context but also on the ability of the joke-teller to use the pragmatic skills of intonation and gesture.*

actually use a joke form, they usually miss the point. As one preschooler said after listening to her older sisters tell jokes on a long car trip, "What happens when a car goes into a tunnel?" "What?" her sisters asked. "It gets dark" came the "punch line." By contrast, almost every 7-year-old can tell a favorite joke upon request (Yalisove, 1978).

The process of asking a riddle shows another pragmatic skill that develops during middle childhood—teasing or tricking someone, especially an adult, verbally. Whereas a 7-year-old is likely to deliver the punch line as soon as the listener says "I don't know," or even before, a 10-year-old is more likely to demand several guesses before giving the correct answer with a self-satisfied grin.

Further evidence of increased pragmatic skill is shown in children's learning the various forms of polite speech. School-age children realize that a teacher's saying "I would like you to put away your books now" is not a simple statement of preference but a command in polite form (Holzman, 1983). Similarly, compared with 5-year-olds, 7- and 9-year-olds are quicker to realize that when making requests of persons of higher status—particularly persons who seem somewhat unwilling to grant the request—they should use more polite phrases ("Could I please . . . ?") and more indirect requests ("It would be nice if . . .") than when they are negotiating with their peers (Axia and Baroni, 1985).

Code-Switching   Changing from one form of speech to another is called **code-switching.** As we will see, children in middle childhood can engage in many forms of code-switching, from the relatively simple process of censoring profanity when they talk to their parents to the complete switch from one language to another.

One of the most obvious examples of code-switching is children's use of one form of speech when they are in the classroom and another when they are with friends after school. In general, the former code, called *elaborated,* is associated with middle-class norms for correct language, while the latter, called *restricted,* is closer to the lower-class norms for pronunciation and vocabulary, as well as for grammar (Bernstein, 1971, 1973). The **elaborated code** is characterized by extensive vocabulary, complex syntax, and lengthy sentences; the **restricted code,** by comparison, has a much more limited use of vocabulary and syntax and relies more on gestures and intonation to convey meaning. The elaborated code is relatively context-free: the meaning of its statements is explicit. The restricted code tends to be context-bound, relying on the shared understandings and experience of speaker

Figure 12.12 *Quite different from the logic expected in the classroom are the nonsense rhymes that are a part of the language code of the playground: "Ring around the rosy," "Miss Lucy had a baby," and "The cheese stands alone" make very little sense but do help to pace the play and synchronize the players.*

and listener to provide some of the meaning. Switching from one code to another, a dispirited student might tell a teacher, "I am depressed today and I don't feel like doing anything," and later confide to a friend, "I'm down, ya know, really down." Research has shown that children of all social strata engage in this type of code-switching, and that pronunciation, grammar, and slang all change in the process (Holzman, 1983; Rogers, 1976; Romaine, 1984).

It seems clear that both elaborated and restricted codes have their place. It is good to be able to explain one's ideas in elaborate and formal terms when necessary. In fact, two of the basic three skills taught during these years, reading and writing, depend on the comprehension of language in a situation devoid of gestures and intonations. At the same time, it is good to be able to express oneself informally with one's peers, using more emotive, colloquial, and inventive modes of communication than those of the standard, accepted code. While many adults rightly stress the importance of children's mastery of the elaborated code ("Say precisely what you mean in complete sentences, and no slang"), the code that is used with peers is also evidence of the child's pragmatic skill.

Nonstandard English

The importance of allowing children more than one code becomes clear with forms of nonstandard English that are a source of group identification and pride. This has been studied with regard to various group dialects in Scotland (Romaine, 1984), the dialect and grammar used by Americans living in Appalachia (Wolfram and Christian, 1976), British Black English (Sutcliffe, 1982), and is apparent in many other English-speaking communities as well (Hughes and Trudgill, 1979; Wells, 1982). Let us look in detail at **Black English,** a form of English spoken by many of this country's 26 million blacks, especially those who live in large urban areas.

Black English used to be considered simply poor English until linguists realized that the so-called errors were actually consistent alternative grammatical forms, some of which originated in African linguistic patterns. For example, the word "be" in standard English is primarily used as part of the infinitive "to be." But in Black English, "be" can also be used to indicate a repeated action or existential state (Labov, 1972). Thus, in Black English, one can say, "I am sick" or "I be sick." The first means "I am sick at this present moment"; the second includes the recent past as well as the present. To express the second concept in standard English, one might say, "I have been sick for a while."

Another difference between Black English and standard English occurs in the expression of negation. Almost all children spontaneously use the double or even triple negative, saying "I don't want to see no doctor" or "Nobody never gives me nothing." But since these forms are wrong in standard English, middle-class children are usually corrected by their parents and older peers, and use only single-negative forms by middle childhood. However, consciously or unconsciously, speakers of Black English tend to resist such correction, perhaps because in Black English (as well as in many languages and dialects other than standard English) the double negative can be correct.

Many black Americans can switch easily from Black English to standard English, depending on the context, just as many other Americans or Britons modify their regional accents and colloquial expressions to fit their intended audience. However, many others, especially children, speak only Black English because that is the form of English spoken almost exclusively by their peers, and often by their parents as well. This can lead to both academic and emotional problems in school. Writing a composition becomes very difficult when one's grammar and pronunciation, constantly reinforced in daily nonschool life, make it hard to formulate a sentence in

Figure 12.13 *In addition to knowing the right answer, this child also realizes that her teacher may pay as much attention to the "correctness" of her phrasing as to the correctness of her facts.*

correct standard English or even to learn correct spelling or punctuation. In addition, if the teacher takes the stance that all nonstandard English is incorrect, the effect on children who speak nonstandard English can be quite destructive.

The opposite approach is not necessarily the solution: if the teacher accepts Black English as a legitimate means of expression in all contexts, the children may be penalized when they enter the larger world where proficiency in standard English is expected. Therefore, the best solution seems to be to help children with the pragmatics of code-switching, so that they will become competent in standard English without being cut off from their original code.

The same basic tactic can be used for children who speak with marked regional accents and colloquialisms, or whose first language is not the one spoken in the school. As we have seen, school-age children are able to analyze and apply language rules, so they can be systematically taught a new code, a new dialect, or even a new language.

Learning a Second Language

No country is without a minority who speak a different language, and a majority of the citizens of the world are bilingual in that they can carry on a conversation in at least two languages. Linguistically and culturally, and probably cognitively as well, it is an advantage for children to learn more than one language (Diaz, 1985). (Although some critics of bilingualism have correctly noted that one language sometimes interferes with another, such interference seems to be either the result of poor teaching or else a temporary condition that ends when both languages are eventually separated in the child's mind [McLaughlin, 1984].)

Thus in the United States the question is how to teach English to the approximately 3.6 million schoolchildren who have another mother tongue, as well as how to help the English-speaking children learn at least one other language before adulthood (Rotberg, 1982).

Unfortunately, although this question has been one of intense concern and emotion, no simple answer is apparent. Almost every approach has been tried—from total immersion, in which the child's instruction occurs entirely in the second language, to "reverse immersion," in which the child is taught in his or her mother tongue until most of childhood is over and the second language can be taught as a "foreign" language. Variations on these approaches present some topics of instruction in one language and other topics in the other language. However, few carefully controlled, longitudinal studies have been done to evaluate the various approaches (McLaughlin, 1985).

Generally, some of the most thorough research comes from Canada, where both English and French are official languages that all children are expected to learn. Immersion programs seem to work best when children are young. In Canada, immersion has proven successful when English-speaking children are taught exclusively in French in their first years of school and then gradually given more instruction in English. These children eventually match their English-taught peers in academic subjects, and surpass them in French (Genesse, 1983).

However, the application of that research to other countries may be limited. In most instances of Canadian immersion, middle-class parents voluntarily placed their children in a special program designed to teach the minority language. Other considerations in learning and teaching a second language are described in the Research Report on pages 280–281.

Figure 12.14 *Bilingual education has a good chance of succeeding in this classroom if the wall chart is an indication of the kinds of teaching techniques that are being used. Explaining words in the three ways—in Spanish, in English, and with pictures—and choosing words that are similar in both languages helps students to bridge the gap from one language to another.*

Children with Special Needs

As we saw in Chapter 11, sometimes special educational measures, like mainstreaming, are needed to help children with physical disabilities develop their cognitive skills, as well as improve their social skills and achieve self-acceptance. But what about the child who looks normal, but who has cognitive needs that become apparent in the classroom, such as the child who has a learning disability or who is

RESEARCH REPORT   Bilingual Education

Most kindergartners, no matter which language they speak, arrive at school knowing such basics of communication as expressive gestures and turn-taking, as well as relevant social and cognitive strategies for "getting along" in a new language (Saville-Troike et al., 1984). In a study of Spanish-speaking children aged 5 to 7 in a largely English-speaking setting, Lily Wong Fillmore (1976) found that children relied on eight strategies for learning English:

Social Strategies

1. Join a group and act as if you know what is going on, even if you don't.
2. Give the impression, with a few well-chosen words, that you speak the language.
3. Count on your friends to help.

Cognitive Strategies

1. Assume that what people are saying is directly relevent to the situation at hand. GUESS.
2. Use some expression you understand and start talking.
3. Look for recurring parts in the formulas you know.
4. Make the most of what you've got.
5. Work on the big things first; save the details for later.

Obviously, the children need to feel fairly self-confident, and their peers need to be relatively receptive, before these strategies can be used effectively. Such is not the case with many minority-language children, especially older children. In addition, the teacher must encourage communication between the children, even if the form is not precisely what he or she would hope. Indeed, in this study many of the first phrases the children used would not be in any textbook: "Lookit," "Whose turn is it?" "All right you guys," "I wanna," "I don't wanna," "How do you do these (little tortillas/flowers/etc.) in English?" and "Shaddup your mouth." In follow-up research, Wong Fillmore (1987) found great variability from child to child and program to program. One critical difference between

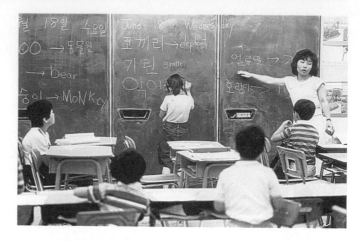

Often learning a new language involves mastering not only vocabulary and grammar, but unfamiliar letters and sounds ("r," "1," and "th" are particularly difficult). Even more important, the learner must feel brave enough to experiment with speaking and writing the new language, the only way to achieve proficiency and overcome the inevitable errors. Thus the bilingual education teacher must not only teach a language but must also encourage self-confidence and social skills.

success and failure was whether or not the children had ample opportunity to converse, either with other children or with the teacher, in the new language. As Piaget would have predicted, large-group instruction and rote learning did not lead to language fluency.

Overall, the context of language learning seems as important as the curriculum. In general, children can successfully take on a second language in the early grades if (Rotberg, 1982; McLaughlin, 1985):

1. The program is specifically designed for language learning, with bilingual teachers who are trained and skilled in language teaching.

intellectually gifted? In these cases, too, extra attention may be needed to provide the best educational and social environment for these "special" children.

Learning Disabilities

Most of the physical problems we discussed in Chapter 9 are evident to some degree before the school years. However, it is not until middle childhood that a more subtle form of handicap becomes apparent, as a child seems to be unable to master one or another skill as readily as other children. If a child of average or better-than-average intelligence as measured by an IQ test (see A Closer Look, page 284)

2. The child already has a mastery of the mother tongue, and wants to learn the new language.

3. The language to be learned, and the native language, are both of relatively high status in the culture.

4. The parents and the community are supportive of the program.

When these conditions do not exist, children have great difficulty learning a new language, especially if that language is simultaneously used for instruction in reading, writing, and math. For such children, bilingual/bicultural education may be best. This type of instruction first reinforces the native language and culture, so children are able to master reading and writing in their first language, as well as to feel proud of themselves and their heritage. At the same time, the children can learn to speak the second language. By age 9 or so, these children are better able to use their logical skills to learn the formal grammar and spelling of a second language.

Bilingual/bicultural education is expensive. So are immersion programs designed to teach language, since classes should be small and teachers specially trained. Moreover, even if money is available, teachers may not be: typically, the minority-language group has relatively few adults who are proficient in both languages and trained to teach. Not surprisingly then, in the United States and in many other countries, most non-English-speaking children are simply placed in standard English classes, with no special language-learning arrangements whatsoever (O'Malley, 1982). How do they do? For the most part, not very well. Not only do they have difficulty mastering the majority language and maintaining their native tongue, they also fall behind in other academic subjects. Far too often, they drop out before high school (Cárdenas, 1977; McLaughlin, 1985).

There are exceptions, however. One such is Richard Rodriguez, a Mexican-American who entered an all-English school in kindergarten, and who eventually mastered the language so well that he studied English literature in graduate school at Berkeley and Columbia.

He and his family chose to learn English, no matter what the cost. His teacher advised his parents to stop speaking Spanish to their children, and even though their understanding of English was minimal, they did so. Richard explains:

as we children learned more and more English, we shared fewer and fewer words with our parents. Sentences needed to be spoken slowly when a child addressed his mother or father. (Often the parent wouldn't understand.) The child would need to repeat himself. (Still the parent misunderstood.) The young voice, frustrated, would end up saying "Never mind"—the subject was closed. Dinners would be noisy with the clinking of knives and forks against dishes. My mother would smile softly between her remarks; my father at the other end of the table would chew and chew his food, while he stared over the heads of his children.

In retrospect, Richard approves of his immersion:

Without question, it would have pleased me to hear my teachers address me in Spanish when I entered the classroom. I would have felt much less afraid. I would have trusted them and responded with ease. But I would have delayed—for how long postponed?—having to learn the language of the public society.

However, Richard never learned to write Spanish, and as he grew older, he lost his ability to speak it as well. He found himself increasingly distant from his parents and their culture, a not uncommon consequence for children who succeed in the majority school.

This raises a basic issue: what is the goal of language education? School-age children need to learn the language of their society, but, as the next chapter explains, they also need to develop their understanding of themselves and of others. Education and family interaction can often play a critical role in determining whether a child develops a lifelong sense of competence or inferiority.

is about two years behind average achievement in a specific area, and there is no obvious explanation (such as a problem with hearing or vision), that child may well have a **specific learning disability.** For example, a child might have **dyslexia,** which is a disability in reading. Dyslexic children may seem bright and happy in the early years of school, volunteering answers to some difficult questions, diligently completing their worksheets, sitting quietly looking at their books. However, as time goes on, it becomes clear that the child is not really reading: the child might guess at simple words (occasionally making surprising mistakes) or try to explain what he or she just "read" by telling about the pictures.

Figure 12.15 *The classroom behavior of children with learning disabilities is usually normal until they are faced with a situation beyond their capacities. (a) This boy listens to his teacher's question, (b) thinks and gets part of the answer, then (c) temporarily despairs of getting the rest. Whether this stressful moment becomes a productive pause, the prelude to an angry outburst, or the beginning of a despondent day depends, in part, on how this child's teacher and classmates respond to his faltering.*

(a)   **(b)**   **(c)**

The specific reading problems of dyslexic children can be many and varied. Sometimes words are read backward ("was" becomes "saw"), although true "mirror reading" is rare. Sometimes the child has difficulty associating the sound of a word with its written form, or cannot readily distinguish between letters that are similar in shape. In fact, the precise problem underlying a particular instance of dyslexia is often hard to specify, which is why the disorder is such a stumbling block for teachers as well as children. Whatever the exact nature of the problem, however, dyslexia usually is apparent—though often undiagnosed—by the second grade, when a child is still trying to master primers, and becomes more and more pronounced as the child grows older.

Another common disability is **dyscalculia,** that is, great difficulty in math. This problem usually becomes apparent somewhat later in childhood, at about age 8, when it is clear that even simple number facts, such as 3 + 3 = 6, are memorized one day and forgotten the next. Soon it becomes clear that the child is guessing at whether two numbers should be added or subtracted, and that everything the child knows about math is a question of rote memory rather than understanding.

A third problem is **dysgraphia,** a difficulty in printing and handwriting that produces the large, uneven letters characteristic of a much younger child. A dysgraphic child might take three times longer than any other classmate to write out an assignment, and still produce a paper that is messy and hard to read.

While the number of children suffering from learning disabilities is hard to estimate, since relatively few are officially diagnosed and counted, a conservative estimate is 10 percent.

What Causes a Learning Disability?   None of the learning problems just described is caused by a lack of effort on the child's part, although, unfortunately, parents and teachers sometimes tell learning-disabled children that they are not trying hard enough. In fact, the precise cause or causes of learning disabilities are hard to pinpoint, although many professionals believe that the origin is often organic. It seems as if some parts of the learning-disabled child's brain do not function as well as they do in most people.

One reason for thinking that some type of brain dysfunction is the cause of learning disabilities is that people with these disabilities often show the same types of cognitive difficulties as people who have suffered brain damage. For instance, a woman with a tumor located in a particular area of the brain was presented with the following problem: "A boy is 8 years old. His father is 30 years older, his mother is 10 years younger than the father. How old is she?" The woman tried three times—first adding 8, 30, and 10 to get 48 and then dividing by 3; then subtracting 10 from 30; then subtracting 10 from 30 again—then finally added 8 and 30, subtracted 10, and got the correct answer, 28. Her difficulty sounds familiar to anyone who has taught people with dyscalcula. The fact is, however, that this woman's disability disappeared when the tumor was removed (Farnham-Diggory, 1978). Victims of other types of brain damage sometimes lose their ability to read, write, spell, or speak normally, even though other aspects of their functioning are normal. This fact led early researchers to think that minimal brain damage (perhaps sustained during prenatal development or in the birth process) was the cause of childhood learning problems, too.

However, a growing number of professionals now believe that, although learning disabilities are sometimes attributable to brain damage, they are more often the result of an inherited difficulty in brain functioning—perhaps involving faulty connections between the left and right hemispheres or between the sensory areas, such as those having to do with vision and hearing (Farnham-Diggory, 1978). The theory that these difficulties are inherited seems plausible because siblings and other relatives of children with learning disabilities often have some learning difficulties themselves.

Nevertheless, it is important to be cautious in citing organic or genetic causes for learning disabilities. First, it is difficult to prove that a particular teratogen or a particular pattern of inheritance causes a particular learning problem. Sometimes, but not always, specific genetic, prenatal, and postnatal factors can be linked to learning disabilities; at the same time, many of the same problems appear even when nothing untoward in the genetic or prenatal history is apparent. In terms of helping a particular child, there is often little to be gained in trying to match up a specific disability with a specific cause (Vandenberg et al., 1986).

More important, sometimes people mistakenly conclude that it is impossible to ameliorate an organic problem. In fact, no matter what the cause of learning disabilities, the way teachers and parents respond to a child who displays difficulties in learning can make an enormous difference to the child's chances of overcoming the problem. If teachers and parents recognize that a child with a learning disability is neither lazy nor stupid, they can help the child become a competent adult. Many such children learn basic skills eventually, especially if they are given patient, individual tutoring. In general, the earlier a disabled child gets special help, and the more that help is tied to the particular problem (not just help in reading, for instance, but help in recognizing shapes for one child, in phonics for another), the better (Achenbach, 1982).

Figure 12.16 *There can be many reasons why a child's attention may wander from schoolwork. If a serious learning problem can be ruled out, and there are no disturbing problems at home, then the child may simply have stayed up too late the night before. In any case, discovering the cause of the problem is more beneficial than simply urging the child back to work.*

A CLOSER LOOK   **The Gifted Child**

While the abilities of children who are unusually talented are in some ways the reverse of the disabilities of other children with special needs, the types of problems experienced by both groups are not altogether different (Zigler and Farber, 1985). As with handicaps, giftedness comes in many forms and benefits from special attention.

Giftedness, or exceptional ability, may be apparent in many different skills or domains, such as music or math, visual or literary expression, dance or athletics. While many gifted children are first spotted by their parents who also are skilled in that area, one traditional sign of intellectual giftedness is an unusually high score on an IQ test. If the child's ability is spotted and encouraged with special teachers, coaches, and performance opportunities, this enhanced education may well lead to professional proficiency by the teen years. From a developmental point of view, the most likely problem is the possible neglect of other basic academic and social skills in the process of fostering a particular talent (Janos and Robinson, 1985).

What IQ Scores Mean

Originally, the intelligence quotient was truly a quotient, arrived at by dividing mental age by chronological age and multiplying by 100. Thus, a 10-year-old who scored at the 12-year-old level would have an IQ of 120; and a 15-year-old who scored at the 12-year-old level would have an IQ of 80. Today the system for determining actual scores is more refined, but the scores are still roughly equivalent to the old system. Specific labels are sometimes used to designate children who score substantially above or below the expected score at their age.

Various types of intelligence tests are widely used to diagnose specific learning problems, as well as to discover indications of mental retardation or intellectual giftedness. Such tests are also frequently and justly criticized, for no one test can accurately measure a child's potential learning ability. Nonetheless, until better measures are found, it can be useful to know what specific IQ scores are supposed to mean. The general categories are given in the table at the top of the next column.

Intellectual Giftedness

About 2 percent of all children are significantly brighter than their peers, perhaps scoring 130 on an IQ test and

Score	Label
Above 130	Gifted
115–130	Superior
85–115	Average*
70–85	Slow learner
Below 70	Mentally retarded
50–69	Educable (can learn to read and write)
25–49	Trainable (can learn to care for self)
0–24	Custodial (will need to be cared for)

* About two-thirds of the population is generally thought to be in the average range, with a score of 100 being the median.

doing second-grade work in kindergarten. These children are usually given extra classroom enrichment, and typically become leaders in school and in later life, with no special problems because of their ability (Sears, 1977; Terman and Oden, 1959). About one child in every thousand is extremely gifted, doing advanced calculus, or reading at the college level while still in elementary school and scoring above 180 on IQ tests. These children tend to have emotional as well as social problems, especially if they are simply given an enriched education with their age-mates. In general, accelerated programs that allow them to learn with older gifted children seem more effective (Robinson, 1981).

Creativity

A final type of giftedness is extreme creativity. For example, when asked "What can you do with a brick?" most elementary-school children think of less than ten uses, almost always incorporating the brick as building material or using it as a weapon. But a few children think of many more uses, and some of their suggestions are highly imaginative: for instance, tying the brick to a raft and using it for an anchor; or grinding it up to make paint; or putting it next to a water fountain so small children can stand on it to drink. Children who give these kinds of responses are considered highly creative. Their thinking processes can be described as divergent, for they lead to many solutions to every problem, in contrast to those of the convergent thinker, whose mind searches for *the* correct answer for any question.

Attention Deficit Disorder

The same note of optimism, and awareness of individual differences, will help us attempt to understand one of the most puzzling problems that sometimes occurs in childhood— **attention deficit disorder,** sometimes called **hyperactivity.** A child with attention deficit disorder is one who is overly active, impulsive, distractible, and

What will teacher do about this artwork created during reading period? Many responses are possible, from proudly pinning it up on the class bulletin board to angrily crumpling it up, throwing it away, and per- haps making the artist stay after school to write, "I will not doodle in class" a hundred times. Unfortunately, creativity, especially if displayed at the "wrong" time, is more likely to be punished than praised.

Of all gifted children, creative children are the ones most likely to get into trouble at school, for their ideas may seem directly opposed to the typical elementary school's emphasis on conforming and knowing the "right" answers. Creative children are often an exasperating experience for teachers, especially when they come up with a new and "better" way to do long division, or when they refuse to eat the school lunch because they conjure up images of what certain ingredients might do to their digestive system, or when they daydream or doodle instead of memorizing their spelling words, or when they ask complicated "What would happen if . . . ?" questions about the history lesson. Creative children often develop a reputation for having "wild and silly ideas," and they are frequently excluded from group activities (Torrence, 1972).

Unfortunately, while it is apparent that some children are much more gifted than others, and clear that creativity

should not be stifled, it is not obvious how creativity should be measured or precisely what should be done once it is spotted (Tannenbaum, 1983; Wallach, 1985). We do know, however, that classrooms and other situations in which everyone is expected to perform in a certain way to produce a certain product thwart and frustrate the creative individual (Amabile, 1983).

Educational Needs

Since many gifted children, once they are grown, have the feeling that they did not fulfill their potential, care must be taken not to make them (or their parents) expect too much (Feldman, 1984). The gifted child, like the disabled child, is at the core a normal person with special attributes.

However, in general, educators and psychologists who study gifted children believe that they need special encouragement, enrichment, and acceleration beginning in the early grades in order to develop their self-esteem and social skills, as well as their talents (Tannenbaum, 1983). Otherwise, they might drop out or fail in their grades. In fact, in their study of famous people from all over the world, most of whom had been gifted children, Victor and Mildred Goertzel (1962, 1978) found that most hated their years in school. Many of them—including Winston Churchill, Thomas Edison, Albert Einstein, Pablo Picasso (who refused to do anything but paint), and Émile Zola (who got a zero in literature and went on to become one of France's most famous writers)—misbehaved, played hooky, failed, or dropped out. Ironically, many of them would have been diagnosed as learning disabled, by today's standards, because of the material they could not, or would not, master. Such responses to standard education were likely to occur in all types of schools (public, private, and parochial) and in every culture.

While it is reassuring to realize that such extraordinary people overcame their early educational experiences, the biographies of such people lead us to wonder how many future inventors, artists, and world leaders are now suffering through boring classes with unsympathetic teachers and peers. Another question is even more disturbing: How many other gifted young people, who may have had a little less persistence, talent, or luck, never learned to appreciate their talents or develop their skills?

excitable, especially at the wrong times and places (Ross and Ross, 1982). The crucial, underlying problem seems to be great difficulty in concentration, the skill known as "paying attention." Attention deficit disorder is considered a form of learning disability, even when it is not accompanied by specific difficulties in learning specific skills (Rutter and Garmezy, 1983; Wender, 1987). About 5 percent of all school-age children are considered to have this disorder by their parents, teachers, or pediatricians, and about 1 percent are considered to have the disorder by all three (Lambert et al., 1978).

Attention deficit disorder appears early in life: many a hyperactive child, as an infant, managed to destroy the hand-me-down crib that older siblings had used without damaging, or seemed much more accident-prone as a toddler than his or her age-mates. However, children are not usually labeled as having attention deficit disorder until elementary school, when it becomes apparent that they cannot sit quietly in one spot and concentrate on schoolwork.

Figure 12.17 *Many children have moments when they seem out of control— too active, too aggressive, and too inattentive for adults to manage. While these behaviors are symptoms of attention deficit disorder, a child is not considered to have this condition unless these symptoms are apparent over a long period of time and in many different settings.*

Causes of Attention Deficit Disorder   When confronted with a child who is considerably more active than other children and cannot concentrate very well, it is not easy to explain that particular child's behavior. However, we do know at least six reasons why some children are much more active than the normal child:

1. *Genetic differences.* As we saw in Chapter 8, activity level is one of the genetic characteristics that clearly vary from person to person. Some children naturally kick a lot in the uterus, run around as soon as they can walk, and want to keep active every minute of the day. About four times as many boys as girls are hyperactive, a sex ratio that may be affected by subtle brain differences that result from the hormones triggered by genes on the Y chromosome.

2. *Prenatal damage.* One of the most common precursors of hyperactivity is prenatal damage of some sort (Hartsough and Lambert, 1985). Thus, a person who was prenatally exposed to a teratogen may have escaped major harm but show minor problems in physical development and learning ability. According to one study, for example, 4-year-olds whose mothers drank moderately during pregnancy (averaging five drinks a week) had more difficulties with concentration than children whose mothers did not drink (Streissguth et al., 1984). On a more general level, children who have a higher-than-average number of minor physical anomalies, such as low-seated ears or widely spaced eyes or a third toe that is longer than the second toe, also tend to be more active and more aggressive than other children

(Bell and Waldrop, 1982; Waldrop and Halverson, 1971). Presumably, prenatal difficulties led to both the physical characteristics and the social difficulties.

3. *Lead poisoning.* Lead poisoning in its early stages leads to impaired concentration and hyperactivity. (If left undetected and untreated, it can lead to severe illness and death.) One of the prime causes of lead-poisoning in children is the ingestion of lead-based paint. Although the interior use of lead-based paint is now illegal, many older buildings, particularly those in run-down areas, are still coated with this potentially harmful paint, which is usually peeling off the walls. If a child were to eat a few chips of this paint each day, the child would suffer lead poisoning within a few months. A second source of lead poisoning is air that is heavily polluted by automobile and truck emissions. For example, the area around Newark, New Jersey, notorious for its heavy-traffic tie-ups, is an "acknowledged lead belt" (Ross and Ross, 1982). Doctors recommend that all young children who have had prolonged exposure to either of these lead sources be tested for the early signs of lead poisoning. If the initial blood test is positive, further testing and treatment can prevent serious damage.

4. *Diet.* Severe vitamin deficiencies, especially of the B vitamins, impair concentration. In addition, certain foods, such as milk, chocolate, sugar, and cola, and some chemical additives, seem to make some children restless (Conners, 1980).

5. *Family influences.* Compared with other children, children with attention deficit disorder come from families who move often, have fewer children, and are less concerned about the child's academic performance than about controlling the child's behavior. Obviously, each of these factors may be the result, rather than the cause, of the child's behavior. However, after elaborate study, some researchers are convinced that these family differences contribute as much to a child's hyperactivity as genetic or temperamental variables (Lambert and Hartsough, 1984).

6. *Environment.* The ecological niche in which some children find themselves may exacerbate hyperactivity. The hyperactive child is especially likely to "misbehave" in an exciting but unstructured situation (such as the typical birthday party) or in a situation with many behavioral demands (such as a long church service, or dinner in a fancy restaurant). Children with no place to play, or who watch television hour after hour, may become restless, irritable, and aggressive. In one study of Puerto Rican children living in overcrowded apartments in New York City (Thomas et al., 1974), the parents' concern for their children's safety led them to keep their children in school or at home virtually all the time. Not surprisingly, 53 percent of these children were considered hyperactive by their parents. In one case, a family with a son described as a "whirling dervish" moved to a new house, which had a small yard. To his parents' delight, the boy quickly "outgrew" his hyperactivity.

Help for Children with Attention Deficit Disorder   Many children with attention deficit disorder continue to have problems in adolescence, not only with hyperactivity but with academic demands and social skills as well. Many become disruptive and angry. In fact, more than half of all children with attention deficit disorder have continuing problems as adults in pacing their work, controlling temper, and developing patience. However, as they grow older, many people learn to cope with these problems, for example, by choosing occupations that suit their skills but that do not place the highest stress on patience and control (Gittelman et al., 1985; Weiss and Hechtman, 1986). In childhood, the most effective forms of help are medication and cognitive and psychological therapy.

Drugs.   For reasons not yet determined, certain drugs that stimulate adults, such as amphetamines and methylphenidate (Ritalin), have a reverse effect on hyperactive children. Approximately 350,000 children take such drugs each day (Ross and

Ross, 1982). For many hyperactive children, the results are remarkable, allowing them to sit still and concentrate for the first time in their lives (Sprague and Ullman, 1981). Indeed, some physicians think that an overactive child who does not respond to such drugs is not really hyperactive.

However, by the time a child has become a candidate for psychoactive drugs, the child's behavior has usually created school, home, and personal problems that no drug can remedy. Psychoactive drugs should never be given as a one-step solution; instead, they should be part of an ongoing treatment program that involves the child's cognitive and psychosocial worlds (Wender, 1987; Werry, 1977).

Psychological Therapy.   Usually, the child with attention deficit disorder needs help overcoming a confused perception of the social world and a bruised ego, while the family needs help with their own management techniques and interaction. As noted in Chapter 2 (see Research Report, pages 42–43), many families with difficult children unwittingly get caught in a vicious cycle of aggression and anger, in which the parents' and siblings' responses to the problem child act to perpetuate that child's problem behavior (Patterson, 1982). Generally, the most effective types of therapy have been those developed from learning theory, such as teaching the parents how to use behavior-modification techniques with their child and helping the child see the effect of his or her own behavior, although other forms of therapy may be better for some children (Dubey et al., 1983; Ross and Ross, 1982).

Teacher Response.   Teachers are often the first professionals to suggest that a particular child might be hyperactive, for they are able to compare these children with their peers in a relatively structured setting. However, teachers, like parents, are often not aware that they may be contributing to the child's difficulties. One study showed that some classroom environments, labeled **provocation ecologies,** made the problem worse, while others, called **rarefaction ecologies,** ameliorated the problem. In the former, structure was either unusually rigid or completely absent, and noise was either completely forbidden or tolerated to a distracting degree. Teachers who managed to diminish hyperactivity were flexible in terms of minor disruptions (for example, allowing children to ask questions of their neighbors so long as they did so quietly), but also provided sufficient structure so that the children knew what they should be doing and when (Whalen et al., 1979).

This is not to say that, with proper teaching, hyperactive children suddenly quiet down and concentrate on their work. On the contrary, as with all physical handicaps and disabilities, no school, family, or neighborhood, no matter how structured or flexible, can make the problem disappear. However, like all "special" children, hyperactive children can be greatly helped or harmed by the particular ecosystem of which they are a part.

Conclusion

It should be clear from our discussion of children with special needs that physiological, educational, and social influences can interact to produce problems, and that all these influences must be understood before the impact of these problems can be reduced. However, our focus on such problems must not blind us to the reality that the same interactional approach should characterize attempts to understand and meet the needs of all school-age children. Further, we must remember that each child has some of the strengths and liabilities typical of children in middle childhood, as well as capabilities and problems that few others share. This is, of course, true whether we are looking at cognitive development, as in this chapter, or at the cultural and social aspects of development, which we shall investigate in the next chapter.

SUMMARY

Concrete Operational Thought

1. According to some developmental psychologists, the years from ages 5 to 7 are a time of transition, when new memory skills, reasoning abilities, and willingness to learn appear. During this period, children sometimes intuit the right answers to logical questions without knowing how they got them.

2. According to Piaget, beginning at about age 7 or 8, children are able to think using the logical structures of concrete operational thought. They can apply their logic to problems involving conservation, classification, and seriation and can distinguish cause and effect. The relationship between time, space, and distance is also better understood, as are others' points of view.

3. While Piaget's ideas about the sequence of cognitive development have been generally acknowledged to be correct, a number of cognitive researchers believe that cognitive changes occur more gradually and more heterogeneously than Piaget's theory suggests.

An Information-Processing View

4. According to information-processing research, cognitive growth during middle childhood is a result of how much information children are given, how efficiently they learn to process it, and how motivated they are to understand and reproduce it. During middle childhood, children learn new strategies for concentrating and remembering, which makes them much more able to master academic skills.

5. Information-processing research has also shown the importance of adjusting the teaching material and sequence of instruction to the child's cognitive structures.

Language

6. Language abilities continue to improve during middle childhood, partly because schools and families encourage this learning, and partly because increased cognitive development makes it easier to grasp difficult grammatical distinctions.

7. The ability to understand that language is a tool for communication makes the school-age child more able to use different forms of language in different contexts, as does, for example, the child who uses Black English on the playground and standard English in the classroom.

8. Teaching children a second language can be accomplished by a number of different methods—from total immersion in the new language to gradually increasing exposure. However, the most important factors seem to be the commitment of home and school.

Children with Special Needs

9. Children with learning disabilities such as dyslexia (severe reading problems) or attention deficit disorder (high activity levels with low concentration ability) need special attention and help to learn to cope with their problems.

10. Some learning disabilities may originate in genetic or physical problems of some sort, but whether the cause is organic or not, many educational and psychological measures can help children with these disabilities. Psychoactive drugs also help some children, but these should be used carefully and cautiously.

KEY TERMS

concrete operational thought *(261)*

decenter *(262)*

5-to-7 shift *(262)*

reciprocity *(263)*

classification *(263)*

class inclusion *(263)*

seriation *(264)*

information-processing *(267)*

memory capacity *(268)*

metamemory *(268)*

sensory register *(268)*

short-term memory *(268)*

long-term memory *(268)*

selective attention *(269)*

mnemonic *(269)*

rehearsal *(269)*

chunking *(269)*

metacognition *(270)*

code-switching *(276)*

elaborated code *(276)*

restricted code *(276)*

Black English *(277)*

specific learning disability *(281)*

dyslexia *(281)*

dyscalcula *(282)*

dysgraphia *(282)*

attention deficit disorder (hyperativity) *(284)*

provocation ecologies *(288)*

rarefaction ecologies *(288)*

KEY QUESTIONS

1. What are some of the cognitive characteristics of the 5-to-7 shift?

2. What are some of the cognitive changes that enable children at 7 or 8 to solve conservation problems?

3. What are some of the concepts that children must be able to apply to perform mathematical operations?

4. What are some of the factors responsible for the improvement in memory during middle childhood?

5. What are some of the language skills that develop in middle childhood?

6. How does code-switching enable children to speak both standard English and Black English?

7. What are some of the factors that encourage learning of a second language?

8. What are the symptoms of learning disability?

9. What are the possible causes of attention deficit disorder?

10. What are the arguments for and against use of psychoactive drugs to control attention deficit disorder?

11. What other types of treatment are helpful in controlling hyperactivity?

CHAPTER 13

The School Years: Psychosocial Development

Ten years from now I will be 19, and
  probably in college, and live away from home . . .
But if I don't go to college, I will romp and roam.
Ten years from now I will be quite pretty,
  and have lots of dough . . .
    and if I don't Oh No!
Ten years from now everything will be fine,
  and I WILL BE MINE.

Rachel, age 9
"Ten Years from Now"

Theories about Middle Childhood

In previous chapters, the three major theories often had quite different perspectives on the nature of children and the impact of the social world on them. Addressing the school years, however, all three theories note the increasing competence of children as their horizons expand from the narrow one of family and home to the wider one of school, peer group, and the community. Let us briefly look at the details of these changes as the three theories portray them.

Psychoanalytic Theory

According to Freud, middle childhood is the period of **latency,** during which the jealousy, passion, and guilt of the phallic stage are submerged, and children's emotional drives are much quieter and steadier. This relative calm frees up children's psychic energy, allowing them to put their effort into understanding their social world and developing their many skills.

Erikson (1963) agrees that middle childhood is a quiet period emotionally and that it is a productive period as well. Once the Oedipal wish to establish a sexual future with the mother or father is over by age 7 or so, "violent drives are normally dormant," giving children new independence from their parents. After the child realizes that "there is no workable future within the womb of his family," says Erikson, the child "becomes ready to apply himself to given skills and tasks." The specific crisis that Erikson sees for this developmental period is **industry versus inferiority.** According to Erikson, as children busily try to master whatever skills are valued in their culture, they develop views of themselves as either competent or incompetent, or, in Erikson's words, as either industrious and productive or inferior and inadequate.

Learning Theory

Like Erikson, learning theorists also emphasize the development of skills during middle childhood. They note that school-age children are particularly easy to teach through the use of the laws of learning theory, if those laws are properly applied.

Figure 13.1 *The need to be industrious can be fulfilled by almost any activity the child can do well and enjoys—from cake-baking to ice-skating, from chess-playing to music-making. Having successfully built a model rocket, this boy turns his attention to a new construction project—an African village.*

To start with, the use of reinforcers in operant conditioning procedures takes on added power when children are better able to understand the relationship between cause and effect. This is true both in the short term ("Be home by 6 o'clock or you can't watch television") and in the long term ("If you earn an A in math, you can have a bike for Christmas"). Of course, some children are better at remembering consequences and waiting for rewards than others, but almost all 9-year-olds have a firmer grasp of cause and effect than they did at age 6.

Furthermore, school-age children become receptive to a wider variety of reinforcements. They still respond to tangible rewards, such as a cookie, a new toy, or a hug. However, school-age children also respond readily to more subtle reinforcements, such as a word of praise or a moment of special attention, as well as to intrinsic rewards, such as pride in work well done (Bandura, 1977).

Indeed, the drive for competence becomes such a potent force in school-age children that it may override the appeal of tangible rewards. As children become bet-

Figure 13.2 *The continuing popularity of the spelling bee, which goes back 200 years in this country, reflects the school-age child's drive for competence (and improved memory capacity). Winners of local spelling bees spend five hours or more each week studying such words as "mansuetude," "eleemosynary," and "syzygy" in preparation for the national contest.*

ter able to evaluate their own performance, at about age 7, they become much more eager to work to master various skills (Bandura, 1981; Harter, 1983; White, 1959). The pride of the child who gets a 100 on a spelling test or hits a home run is almost palpable, making additional "childish" rewards unnecessary. Indeed, whereas the promise of a candy bar might make a 4-year-old work diligently at a task, it might make an 8-year-old lose enthusiasm for a task that he or she might have performed with delight for less overt rewards (Lepper and Greene, 1978).

Social Learning Theory   As contemporary social learning theorists stress, it is crucial to look at the overall context of the learning process rather than solely at the specific details of reinforcement (Zimmerman, 1983). Thus, while praise or tangible rewards may help a child learn to do a specific thing, such as make his or her own bed, the child's overall sense of responsibility is also related to the example set by parents and older siblings, the consistency of the parents' expectations, and the child's own sense of being able to meet those expectations. Factors such as these are much more influential than tangible reinforcements. Indeed, according to a review of effective parenting, if parents overload their children with either rewards or punishments, both promises and threats lose their effectiveness (Maccoby and Martin, 1983). Likewise, if parents' response to their children's behavior is inconsistent or unclear, the children's behavior can get out of control (Patterson, 1982).

Finally, school-age children are very receptive to modeling techniques. They seem to be much more aware of the actions and attitudes of others than younger children are, and can be remarkably adept at imitating other people. In addition, because they notice other people's good and bad qualities, they have more potential "models" available to them: children between 6 and 11 model themselves after not only their parents but also their teachers and friends, choosing people they see as successful (Bandura, 1986). Of course, the specific behaviors that are chosen to be modeled are not always the ones adults might prefer. When my daughter Elissa was 8, she was reprimanded for being mean to her friends; she explained that she was "practicing to be a teenager"—and using her older sister as a model.

Cognitive Theory

Cognitive theory, like psychoanalytic and learning theories, emphasizes the importance of school-age children's enhanced learning abilities. As we saw in Chapter 12, there is no doubt that school-age children are more logical, less egocentric, and more concrete in their thinking than younger children are. These advances affect not only their math and language ability but also their ability to understand themselves and others (Gelman and Spelke, 1981).

Cognitive psychologists were particularly struck by the change in the child's social awareness when they traced children's growing ability to identify and take into account another person's point of view. John Flavell (1975,1985) examined this development in a number of experiments, such as asking children to describe an object to someone who can't see it, to tell a story to someone who has never heard it, or to persuade someone to do something that person would rather not do. Skill at all these tasks develops markedly during the school years. Compare the strategies adopted by third-graders (about age 8) and seventh-graders (about age 12) when they were asked to pretend that they were asking their fathers to buy them a television. Here is one of the third-graders:

> Come on, I want a television for my own room. Come on. Please. Daddy, come on. Buy me a television. I want one for my room. Come on. Come on, Daddy. I want you to. There!

And here is one of the seventh-graders:

> Oh, hello, Dad. How are you today? Do you want your slip-
> pers changed or something? Here, have a cigarette? . . .
> Do you want me to do anything for you? Ah—let's see—
> hey Christmas is coming. . . . Now there's only one thing I
> want and that is a television set. Ah—if you get me that
> you don't have to get me anything else, unless you want
> to, of course. But you know I need one real much, and
> ah—and if I don't feel good I have to come into the living
> room and watch the TV. I can stay right in my room and
> be very rested. . . . And ah—and when my friends come
> over they don't have to pack in front of the television.
> They could stay right in my room and be very quiet. . . .
> And they're real pretty and they make a room look real
> nice, and they're real small ones too, that you can use if
> you want . . . and—well, you could get a pretty cheap one.
> Maybe a used one, a second-hand one.

The most striking difference in the seventh-graders' approach was their ability to take into account the needs and values of the person they were trying to persuade. This enabled them to use reasoning rather than brute insistence or threats; to anticipate and attempt to deflect possible objections to their request; and to try to diminish the apparent self-interest of their efforts by referring to the interests of the beseeched. Examples such as these highlight the development that cognitive theorists find most impressive in looking at the child's psychosocial development during middle childhood: children are able to understand the laws and processes of their social world as well as, or better than, they are able to understand the laws of conservation, classification, and so forth (Hoffman, 1981).

Social Cognition

It is clear that all three major theories lead in the same direction when describing the school-age child. They portray a young person who is much more able to leave the confines of the family (as Freud and Erikson explain); much more open to con-ditioning from the outside world (learning theory); and much better able to under-stand the social milieu in which they find themselves (cognitive theory). Not sur-prisingly, then, developmental researchers from a variety of theoretical per-spectives have come together in a new research area called social cognition.

Social cognition refers to an individual's understanding of the dynamics of human interaction. Since the study of social cognition has emerged from all three major developmental perspectives, it includes emotion, behavior, and thinking processes (Flavell and Ross, 1981). We will survey the findings in this new area of study by looking first at children's understanding of other people, and then at their understanding of themselves, their peers, and the rules of their society.

Understanding Others

Children's social awareness improves in many ways between ages 6 and 11. For one thing, they become more aware of the multiplicity of social roles an individual can have: 8-year-olds, unlike 6-year-olds, understand that a person can be a father, grandfather, and son, all at the same time (Watson, 1984; Watson and Amgott-Kwan, 1983). Eight-year-olds also realize that a person can have a professional role and a set of personal ones as well. (That kindergartners don't realize this is clearly revealed by their bafflement when they run into their teachers or pediatricians

Drawings by Charles Schulz; © 1967, 1968 United Feature Syndicate, Inc.

Figure 13.3 *As children grow older, social cognition improves to the point that not every comment, gesture, or glance is taken personally, because children realize there might be interpretations other than their own. Few, however, are as grateful as Linus to be wrong about having jumped to a first, egocentric conclusion.*

pushing a cart down the aisle of the supermarket or playing with their own children in the park.) Older children are also less likely to cling to sex stereotypes, associating fewer occupations, leisure-time activities, household tasks, or modes of dress exclusively with one sex or the other (Archer, 1984; Goldman and Goldman, 1982; Stoddart and Turiel, 1985). Another advance, for which adults may be grateful, is that older children are better able to recognize and rephrase or avoid potentially offensive statements (Johnson et al., 1984). Thus the 11-year-old is much less likely than the 6-year-old to tell you that your stomach is too fat.

Further, as children grow older, they understand that people have personality traits (Shantz, 1983). Compare the following two descriptions from an extensive study (Livesley and Bromley, 1973) in which children aged 7 and older described other children: the first, from a 7-year-old, focuses exclusively on physical traits, while the second, from a child of 10, emphasizes personality traits and behaviors.

> 7-year-old: Max sits next to me, his eyes are hazel and he is tall. He hasn't got a very big head, he's got a big pointed nose.

> 10-year-old: He smells very much and is very nasty. He has no sense of humor and is very dull. He is always fighting and he is cruel. He does silly things and is very stupid. He has brown hair and cruel eyes. He is sulky and 11 years old and has lots of sisters. I think he is the most horrible boy in the class. He has a croaky voice and always chews his pencil and picks his teeth and I think he is disgusting.

As children become more aware of personality traits, they realize that people who have traits that are different from their own may respond differently than they themselves would in various social situations. For example, in a study comparing 5-, 8-, and 11-year-olds, children at the two older ages predicted that a fearless boy would not run away from a large growling dog, even if they themselves would flee. The 5-year-olds were much more likely to think that any child would do just what they themselves would do (Ross, 1981). In general, at around age 8, there seems to be a "major shift" from a highly concrete evaluation of others to an increasingly abstract understanding that infers motives, beliefs, and personality characteristics from behavior (Shantz, 1983).

The child's increasing ability to recognize others' personality characteristics, and to anticipate how these characteristics might affect their behavior, helps the child in getting along with other people. In one detailed study (Gottman, 1983) in which children between the ages of 2 and 11 were told to play with an unfamiliar peer, the younger children (up to about age 5) tended to just start playing at whatever came to mind. The older children were more likely to introduce themselves and search for some common ground between them to provide a basis for their play. In addition, the older children had a better sense of proper pacing of personal communication: they first discussed the similarities between them before discussing the differences. Unlike the younger children, they also knew whether and when to reveal private information. For example, one 5-year-old told her new playmate that her mother didn't love her anymore—because, she said, her mother wanted to be left alone with Jimmy (her new boyfriend), instead. Older children never shared such personal information on first meeting. Finally, the older children were better at resolving conflict, using humor, for example, rather than confrontation when disagreements occurred. For all these reasons, the older children were more likely than the younger children to be on friendly terms when the play session was over.

Self-Understanding

Closely related to children's understanding of others is their **self-theory,** or understanding of themselves. According to many psychologists, as people grow older, they develop more complexly differentiated theories of who they are—*theories* in the sense that, as it evolves, a person's concept of himself or herself comes to be based on a combination of the evidence of past experiences, the opinion of others, and the person's untested assumptions about himself or herself. Just like other theories, a person's self-theory is subject to change when new evidence or new assumptions emerge, and, again like other theories, it is used to govern future behavior and interpret past experiences (Brim, 1976).

Preschoolers have little, if any, stability in their specific ideas of self, at least in the sense of having a theory about themselves that they can verbalize and use to guide their actions (Harter, 1982). However, children's thoughts about themselves develop rapidly during middle childhood, as their cognitive abilities mature and their social experience widens. In the beginning of the school years, for example, children tend to explain their actions by referring solely to the events of the immediate situation; a few years later they more readily relate their actions to their personality traits and feelings (Higgins, 1981). Thus, whereas the 6-year-old typically says that she hit him because he hit her, the 11-year-old might also explain that she was already upset because she had lost her bookbag and that, besides, he is always hitting people and getting away with it. Further, children's self-theory becomes more differentiated (Harter, 1983), enabling schoolchildren to view themselves in terms of several areas at once. They might see themselves as smart in math but dumb in spelling, with an ability to master geography if they were to put their minds to the task. Similarly, they might feel that they are basically good at making friends, and are considerate of others, but that they have a quick temper that sometimes makes them do things that jeopardize their friendships. Being able to understand various aspects of their personality *sometimes* helps children to modify their behavior. For example, they might apologize for an outburst of anger by referring to their hot temper, or they might even take steps to protect a friend from such an outburst by, say, going for a walk to "cool off."

As their self-theory sharpens, then, children gradually become more self-critical, and their self-esteem dips. Unlike younger children, for example, fourth-, fifth-, and sixth-graders have markedly lower self-esteem if they are aware of their shortcomings in any area (Harter and Ward, 1978). Further, as they mature, children are more likely to feel personally to blame for their shortcomings, and less likely to believe, as younger children often do, that it is bad luck that makes them do poorly (Powers and Wagner, 1984). Girls are especially likely to blame themselves for their difficulties (Stipek, 1984), a tendency apparent throughout childhood. Thus, while children are better able to recognize the diverse areas in which they succeed and fail as they grow older, they are more likely to take failures seriously, and thus their overall self-esteem tends to fall as they understand themselves better. In general, self-esteem, which is usually quite high in early childhood, decreases throughout middle childhood, reaching a low at about age 12 before it gradually rises again (Harter, 1983; Savin-Williams and Demo, 1984; Simmons et al., 1973; Wallace et al., 1984).

Learned Helplessness   These developmental changes affect children's willingness to try to master new skills and learn new material. As one review explains, "developmental change can be a risky business. Greater cognitive capacity for self-reflection can provide the tools for new levels of mastery but can also result in greater inhibition of mastery attempts" (Dweck and Elliott, 1983).

Put another way, compared with younger children, older children are more vulnerable to **learned helplessness:** that is, their past failures in a particular area have taught them to believe that they are unable to do anything to improve their performance or situation. Consequently, whereas a 6-year-old who has failed to master a school lesson several days running is likely to still be eager to try again, an 11-year-old who has failed repeatedly will be likely to quit trying. This has been shown experimentally as well as in naturalistic observation. For example, in an experiment in which children between the ages of 5 and 11 thought they were doing poorly in finding "hidden pictures" in four drawings, the older children were more negatively affected by their alleged failures. They became discouraged and, on the fifth drawing, found fewer pictures and spent less time looking than the younger children (Rholes et al., 1980).

Correspondingly, children who have had several experiences in which their specific performance was judged inadequate—by teachers, peers, parents, or themselves—might well decide "I'm stupid in math" or "I can't play ball" or "Nobody likes me," without giving new math lessons or new ball games or new social interactions a try. As Erikson predicted when he described industry versus inferiority, a child with few successes will develop a sense of inferiority that may lead to anticipation of continued failure and a lower self-esteem for the rest of his or her life. Thus the child who experiences academic problems in middle childhood is likely to become the adolescent who drops out of school and the young adult with low aspirations and bad work habits.

Developing Positive Self-Esteem   The crucial factor in gaining positive self-esteem seems to be feeling that one is competent at varying tasks. Obviously, developing this feeling depends partly on a child's ability: for example, children who are intellectually able at age 7 tend to develop relatively high self-esteem by age 12 (Joreskog, 1973). However, a great deal also depends on the microsystems of family and school, which can make it easier or more difficult for children to develop feelings of competence. Children whose parents are supportive tend to feel more confident and competent than children whose parents tend to criticize and punish

Figure 13.4 *Whether learning to make a campfire or to identify a scarlet tanager, children gain skills and self-esteem (and in this case, an appreciation and understanding of nature) when parents share their enjoyment of their own favorite leisure activities.*

Figure 13.5 *In developing countries the work done by school-age children is essential to their families and thus helps to build the children's sense of competence. These industrious children are at work in* (a) *Mexico,* (b) *Egypt,* (c) *Morocco,* (d) *Sudan and* (e) *Indonesia.*

(Abraham and Christopherson, 1984; Coopersmith, 1967). Further, children feel better about themselves when their school offers a variety of ways for them to succeed (with arts, dramatics, and sports programs, for example, in addition to regular classes) and when their teachers make a point of praising each child for what he or she does well. Children's self-esteem may also be affected by the structure of the classroom and the school, depending on whether the stress is on competition and comparison, or on cooperation and diversity. The school, in effect, is a "social frontier" (Minuchin and Shapiro, 1983), in which children confront unfamiliar adults and children, as well as values and codes of behavior that may be quite different from those of the home. Whether the school attempts to acknowledge and respect these differences, or promotes the idea that there is only one correct way to think and act, can have a profound effect on the child's self-esteem.

The Peer Group

Perhaps the most influential system in which the school-age child can develop his or her self-esteem is the peer group. Acceptance in one's group, and confidence in one's best friend, can go a long way toward building a sense of competence.

Not surprisingly, then, during middle childhood, children become increasingly dependent on their peers. For example, in one study, children between the ages of 6 and 10 were asked whom they would turn to for help in various academic and social situations. As they grew older, they were less likely to turn to their parents, and more likely to turn to their peers, partly because they considered their peers more capable of providing help than their parents were (Nelson-Le Gall and Gumerman, 1984).

The Society of Children   When groups of children play together, they develop particular patterns of interaction that regulate their play, distinguishing it from the activities of adult-organized society. Some social scientists call the peer groups' subculture the **society of children,** highlighting the distinctions between children's groups and the general culture (Knapp and Knapp, 1976; Opie and Opie, 1959). The society of children typically has a special vocabulary, dress codes, and rules of behavior that flourish without the approval, or even the knowledge, of adults. Its slang words and nicknames, for instance, are often ones adults would frown on, and its dress codes become known to adults only when they try to get a child to wear something that violates those codes—as when a perfectly fine pair of hand-me-down jeans is rejected because, by the standards of the dress code, they are an "unfashionable" color, or do or don't have a designer label, or have bottoms that are too loose, or too tight, or too short, or too long. If parents find a certain brand and style of children's shoes on sale, they can bet that they are the very ones that their children would not be caught dead wearing. Sex differences in clothes, behavior, and play patterns and partners become increasingly salient as children move from kindergarten to the sixth grade (Furman, 1987; Hayden-Thompson et al., 1987).

The children's subculture also involves codes of behavior, many of which demand independence from adults. By age 10, if not before, children (especially boys) whose

Figure 13.6 *Friends who share private fantasies and fears, as well as validate each other's preferences in matters ranging from clothing styles to personality types, are particularly important as adolescence approaches. The two kindred spirits on the left found that even their rapidly growing feet were in synchrony, so they bought new shoes, size 8, in friendship pairs.*

Figure 13.7 *A snowball fight is an example of a complex peer-group event: choosing teammates and targets, building forts, and agreeing on rules (no ice-balls, rock centers, or aiming at the face) calls for lots of cooperation and trust.*

parents walk them to school or kiss them in public are pitied; "cry babies" and "teachers' pets" are criticized; children who tattle or "rat" to adults are despised. As they did at younger ages, school-age children (again, especially boys) engage in rough-and-tumble play, but they become increasingly selective in when, where, and with whom they do it, establishing bonds of friendship as well as patterns of dominance (Humphreys and Smith, 1987).

The importance of the peer group to school-age children is perhaps most obvious in their organization of clubs or gangs, in which much attention is given to details concerned with rules, officers, dress, and establishing a clubhouse, often deliberately distant from adult activity. Sometimes the club has no announced purpose, its only apparent function being the exclusion of adults and children of the opposite sex (Minuchin, 1977). From a developmental perspective, however, such clubs serve many functions, including building self-esteem, sharpening social skills, and teaching social cooperation. As one researcher describes his club:

> I was a charter member of a second-grade club called the Penguins, whose two major activities were acquiring extensive information about penguins and standing outside in the freezing weather without a coat for as long as we could. Like most other groups of this sort, the Penguins did not last very long, but in the making and unmaking of such groups, children are conducting what may be informative experiences in social organization. [Rubin, 1980]

Customs and Principles   While they are developing the social codes for their own societies, children also become more aware of the customs and principles of the larger society. This is shown in a series of studies by Elliot Turiel (1983), who found that school-age children not only understand social conventions and ethical principles, but can also distinguish between them.

For example, Turiel asked one typical 8-year-old what rules he knew. The boy cited a rule that the children in his house must clean up the mess that their guests make, and he explained that this rule could be easily changed if his parents decided to do so. He also knew another rule, that children should not hurt each other, and he explained that this rule should not be changed because hurting is wrong whenever and wherever it occurs. Similarly, he felt that the rule against stealing should not be changed because "people would go crazy." He added, with an impressive sense of social justice, "People that don't have anything should be able to have something, but they shouldn't get it by stealing."

Older children also understand that people sometimes obey customs more readily than they follow principles. For example, Turiel (1983) found that 10-year-olds, but not 6-year-olds, were convinced that it is a more serious transgression for a child to steal an eraser than to wear pajamas to school, because stealing is wrong while the question of appropriate clothing is simply a custom. At the same time, however, the 10-year-olds admitted that they personally would be more likely to commit a minor theft than to dress inappropriately. Interestingly, a study of Korean children found similar trends. Despite the greater cultural emphasis on social conforming, Korean children distinguished between behaviors that were universally wrong and those that were a matter of custom (Song et al., 1987).

This increasing awareness of social custom helps explain a surprising finding regarding **prosocial behaviors,** which are defined as actions performed to benefit someone else without expectation of reward for oneself. While one might expect children to become more sharing, comforting, and self-sacrificing as they become cognitively more mature, such age trends are not clear-cut. In fact, some developmental trends go in the reverse direction: nursery-school children are more likely to share generously and indiscriminately than school-age children are (Radke-Yarrow et al., 1983). Similarly, younger children are more likely than older children to respond if a classmate is in distress (Ekholm, 1984; Gottman and Parkhurst, 1980).

To understand this, researchers have looked closely at the many factors that influence prosocial behavior (Eisenberg, 1982; Mussen and Eisenberg-Berg, 1977). One important factor is the example set by adults: children are more likely to cooperate if they have been rewarded for cooperating, or if they have seen someone else cooperating. Within the typical elementary school, however, teachers of older children are more likely to stress individual achievement than cooperation.

Another important factor is who the potential recipient of the prosocial behavior is. As children's social awareness increases, they become more selective in their interactions, and are more likely to cooperate and share only with someone they like or admire, or with someone who they know needs their help, or with younger children (Burelson, 1982; French, 1984). Perhaps most important, prosocial behavior is increasingly seen as a sign and obligation of friendship, and, as we will now see, friendship becomes more selective and exclusive as middle childhood progresses.

Friendship As children grow older, friendships become increasingly important, and children's understanding of friendship becomes increasingly complex. These changes are reflected in a study of hundreds of Canadian and Scottish children, from first grade through the eighth, who were asked what made their best friends different from other acquaintances. Children of all ages tended to say that friends did things together and could be counted on for help, but the older children were more likely to cite mutual help, whereas younger children simply said that their friends helped *them.* Further, the older children considered mutual loyalty, intimacy, and interests, as well as activities, to be part of friendship (Bigelow, 1977; Bigelow and La Gaipa, 1975).

Similarly, in a United States study (Berndt, 1981) children were asked "How do you know someone is your best friend?" A typical kindergartner answered:

> I sleep over at his house sometimes. When he's playing
> ball with his friends he'll let me play. When I sleep over,
> he lets me get in front of him in 4-squares (a playground
> game). He likes me.

The Rejected or Neglected Child

All children sometimes feel left out or unwelcome among their peers. As a children's ditty puts it, "Nobody likes me, everybody hates me, I'm going to eat some worms."

However, an estimated 5 to 10 percent of all school-children are unpopular and friendless most of the time (Asher and Renshaw, 1981). Such children can be grouped into two categories (Younger et al., 1985). Children describe their aggressive classmates according to certain typical behaviors, such as "starts fights," "bothers others," and "is mean and cruel." Other clusters of behavior are used to describe the withdrawn child: "too shy," "chosen last," "feelings easily hurt," a child more often neglected than openly rejected. Both types of children have many problems: not only are they lonely but they have low self-esteem, which affects their learning in school and happiness at home. As they and their classmates grow older, their problems get worse, because children become more critical of themselves and their peers as adolescence approaches.

Several studies have shown that children who are rejected or isolated by their peers tend to be immature in their social cognition. Compared with popular or average children, for instance, they tend to misinterpret social situations, considering a friendly act to be hostile, for example (Dodge et al., 1984), or, especially when they feel anxious, to interpret accidental harm as intentional (Dodge and Somberg, 1987). Typically, they might interpret a compliment as sarcastic, or regard a request for a bite of candy as a demand. They also have difficulty sharing and cooperating (Markell and Asher, 1984) and in understanding what other children's needs might be (Goetz and Dweck, 1980).

Unfortunately, since the way most children develop their social understanding and skill is from normal give-and-take with their peers, rejected children are excluded from the very learning situation they need most (Rubin, 1980; Youniss, 1980).

What can be done to help them? To some extent, a particular ecological milieu encourages or discourages constructive interaction among children. For example, parents who themselves are friendly and have a wide social circle tend to have quite social children; the opposite is true as well. Another influential factor is the school. On the whole, the more informal, open-classroom setting fosters more mutual respect and friendship among the children (Minuchin and Shapiro, 1983). Further, children have more difficulty getting along with children with whom they share few similarities, so it is probably unwise to place a child with poor social skills into a class where few of the children have much in common with the new child.

Unfortunately, many parents change neighborhoods (perhaps moving to a "nicer" house) and many principals reassign students (perhaps to create a class that is easier for a teacher to handle) without considering the friendships that might be disrupted or rejected children who might become even more hostile or withdrawn.

Can anything be done directly to teach children better social skills? Such attempts have been tried, with some success, when the skills taught were quite specific, such as how to make a positive comment on what someone else says (Asher and Renshaw, 1981; Bierman and Furman, 1984).

By contrast, a typical sixth-grader said:

> If you can tell each other things that you don't like about each other. If you get in a fight with someone else, they'd stick up for you. If you can tell them your phone number and they don't give you crank calls. If they don't act mean to you when other kids are around.

Partly because friendships become more intense and more intimate as children grow older, older children demand more of their friends, change friends less often, find it harder to make new friends, and are more upset when a friendship breaks up. They also are more picky: throughout childhood, children increasingly tend to choose best friends who are of the same sex, race, and economic background as they themselves are (Hartup, 1983).

As children become more choosy about their friends, their friendship groups become smaller. Whereas most 4-year-olds say that they have many friends (perhaps everyone in their nursery-school class, with one or two notable exceptions), most 8-year-olds have a small circle of friends, and by age 10, children often have one "best" friend to whom they are quite loyal. Although this trend toward an in-

However, a less direct route may be even better. One surprising study divided socially rejected, low-achieving boys into four groups. One group learned social skills, one academic skills, one both social and academic skills, and one was a control group. Those who concentrated on academic skills improved across the board, in reading and math as well as social acceptance. The boys who learned just social skills improved in only one area, reading comprehension. The authors of this study suggest that children who improve their academic skills are likely to improve in self-esteem as well. This helps them concentrate better in class and feel at ease with their classmates, changes that may be more crucial to improved social status than simply learning how to engage in social interaction (Coie and Krehbiel, 1984).

While academic improvement helped these low achievers, however, academic excellence does not necessarily lead to better social skills or higher status. According to some longitudinal research, high-achieving girls who, at the end of middle childhood, dampened their intellectual accomplishments somewhat felt better about themselves and were more accomplished socially than those girls who remained at the top of the academic heap (Petersen, 1987).

One caveat: too much emphasis may be placed on the need for children to "fit in." As Robert White (1979) cautions:

Historically we have clamored too loud for social adjustment. We have not been sensitive to the dangers of throwing children together regardless of their anxieties and their own social needs. We have been enchanted with peer groups, as if the highest form of social behavior were getting along with age equals . . .

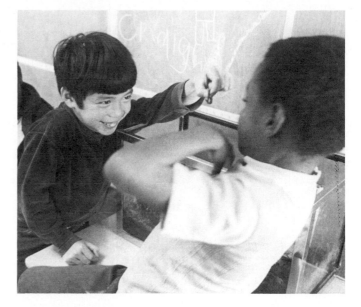

Knowing how to tease without going too far is an important social skill—one noticeably lacking in rejected and isolated children. The glee on the worm-holder's face suggests that he has *found an acceptable way to tease his classmate, who will probably react by finding a worm of his own, rather than insulting, punching, or escaping.*

While it is apparent that unpopular children need to be helped with social skills and social acceptance, all children also need to develop their own interests, talents, and self-confidence apart from the social scene.

Figure 13.8 *While both boys and girls form friendship groups during the school years, girls' friendships tend to be more intimate and the emphasis on conformity more apparent. The sex differences in friendship patterns are especially obvious if one tries to imagine girls posturing like the boys in the first photo, or boys hugging like the girls in the second.*

creasingly smaller friendship network is followed by both sexes, it tends to be more apparent among girls. By the end of middle childhood, many girls have one and only one best friend on whom they depend (Lever, 1976).

Thus, as children grow older, friendship patterns become more rigidly set, so that by age 9 or so everyone knows who hangs out with whom, and few dare to try to break into an established group or pair of friends. With the changes of early puberty (at about age 10), some children come to be more advanced than others, disrupting former social patterns and wrecking many friendships. As one girl named Rachel put it:

> Oh, I feel so horrible about friends. Everybody is deserting their best friend and everybody hates someone else and Paula Davis has been stranded with nobody—except me and Sarah. Christine has run off with Liz and Joan has moved up from being an eleven-year-old . . . and, oh well, I suppose it happens every year.

Problems and Challenges

As we have seen, during middle childhood children become actors on a wider stage, taking on school and community roles as well as family ones. Their expanded social world is full of challenges and opportunities for personal growth. It is also full of potential problems. Here we will look at some of the stresses children may face within their families and within the larger society.

A Life-Span Perspective

Socioeconomic Status

Socioeconomic status, abbreviated **SES** and sometimes called social class, has been revealed to be one of the most powerful influences on a person's life. In research, SES is usually measured by a combination of three variables: the head of household's years of education, income, and type of occupation. In national statistics, SES is usually indexed more crudely according to the per capita income of family members: the total household income is divided by the number of people in the household. If per capita income falls below a certain amount, called the poverty line, the family members are officially designated as poor. In the United States in 1986, when the poverty line was $11,203 for a family of four (U.S. Department of Commerce, 1987), 15 percent of the population was so classified, with more children under age 12 in this category than people of any other age group.

Comparisons among the most likely developmental paths for people at various points along the socioeconomic ladder show many interesting differences. For instance, in general, as socioeconomic status rises, so does the marriage rate, with one exception: after divorce, women of high SES are less likely to remarry than middle-class women. SES is reflected in one's health, too, as can be seen from statistics showing that, for example, increased frequency of dental visits and decreased chances of dying of cancer correlate with high SES.

As you might guess, the most striking and troubling differences in the effects of SES are seen in comparisons between people of low socioeconomic status and those of higher status. At every stage of life, people of low SES are at greater risk of developmental problems. A child growing up in poverty faces risks to health and safety that far surpass, in number and gravity, those that are encountered by people of higher SES. The risks continue into old age—if old age is attained. Statistics show that life is shorter for those of lowest SES. Cancer and heart disease take their toll at younger ages; mental illness is more frequently diagnosed; and the incidence of accidental death and homicide is shockingly high. In the deteriorating neighborhoods where some of our young people are growing up, murder is the leading cause of death among males under the age of 15.

To more fully understand the effects of low SES on a person's life, let us use statistical probabilities to trace a likely path of development. It is important to bear in mind that the following example is a hypothetical case intended to illustrate general conditions; there are many exceptions, and few people will be affected by all the factors described.

The effects of poverty begin in the prenatal period. Consider a fetus, a male fetus since males are more vulnerable throughout the life span. Statistically, his mother is more likely to smoke cigarettes or use drugs, to be malnourished, and to begin prenatal care relatively late in pregnancy, if at all. For all these reasons, he is more likely to experience a difficult birth, to weigh less than average, or to be born with other medical problems.

Once home from the hospital, he probably will get less of his mother's attention than an average boy born to a middle-class family, for not only may his mother lack money and education, but she is more likely to be a teenager or have other children already. Her time may be taken up by completing her education, or by working full-time outside the home, or by attending to the needs of the other young children. One result is that her new son may miss the benefits of breast-feeding.

Later, due to pressures in his mother's life, his socialization in the home will involve more commands and punishment than explanations and rewards. The mother with little money and many unrelieved responsibilities is less likely to be tolerant when her little boy spills a half-gallon of milk or rips his jacket. His early relationship with his father will probably be less close than that of higher SES children with their fathers because poverty and lack of education do not favor active paternal participation in child care. Throughout early childhood, the low SES child is more likely to be neglected, abused, or accidentally hurt.

By middle childhood, the low SES boy enters school and learns about equality and inequality. As he begins to think concretely, he is likely to compare himself with other children in terms of possessions, skills, and achievements. The world that he lives in makes the mastery of the normal skills of middle childhood extremely difficult. In the words of Kenneth Keniston, it is a

> dangerous world—an urban world of broken stair railings, of busy streets serving as playgrounds, of lead paint, rats and rat poisons, or a rural world where families do not enjoy the minimal levels of public health . . . It is a world where even a small child learns to be ashamed of the way he or she lives. And it is frequently a world of intense social dangers, where many adults, driven by poverty and desperation, seem untrustworthy and unpredictable. Children who learn the skills for survival in that world, suppressing curiosity and cultivating a defensive guardedness toward novelty or a constant readiness to attack, may not be able to acquire the basic skills and values that are needed, for better or worse, to thrive in mainstream society. [Keniston, 1977]

Keniston's statement is corroborated by considerable research. Lower-class school-age children are more likely than their better-off peers to be poor achievers, and to perform less well each year.

Partly because their present status seems more dependent on factors over which they have no control, such as their parents' employment or their landlord's decency, lower-class children are more likely to believe that they have little control over their future. To them, luck or other people may seem to have much more power over their destiny than they themselves have. As a result, they try less hard and accomplish less (Bartel, 1971; Maehr, 1974; Maqsud, 1983).

As Erikson would put it, the young man of our story is at risk of settling for inferiority rather than industry, which may hinder him all his life. Or he may find the most promising path to self-esteem is to become the tough guy, a route that gets him into trouble with his teachers, his parents, and many of his peers.

In adolescence, as the child moves into a larger social world, the threats multiply. The low SES teenager is more likely to drop out of school, to use illicit drugs, to get into trouble with the law, to be unemployed. When he encounters these problems, his family is less likely to be able to offer the resources—financial, educational, or personal—that could help him. His parents and siblings are likely to be occupied with overwhelming problems of their own.

In adulthood, a man of low socioeconomic status is less likely to find a stable marriage partner or a satisfying career. In late adulthood, if he lives that long, he is more likely to be lonely, ill, and unable to care for himself.

Again, none of these outcomes is inevitable, and many children from low SES homes grow into healthy adults with the cognitive and social skills that will give them an excellent chance of contributing to society in their work and family lives. Some succeed because they are fortunate in having nurturing families that encourage them and help them with their education, and protect them from the violent streets. (For a discussion of factors that help protect children from stress, see page 313.) Church programs and private efforts also turn some children's lives around. Further, some elements of the macrosystem may help prevent problems that would otherwise occur. The extensive free prenatal and early childhood care provided by visiting nurses in New Zealand is one example; Headstart (see Chapter 9, page 215) in the United States is another; various drop-out-prevention and college-scholarship programs are a third. Nonetheless, given the high probability of there being many strikes against the developing person born in poverty, much more is needed to even the odds. The most beneficial assistance is that which helps a person of low SES achieve the opportunities afforded by middle-class status: a college education, residence in a good neighborhood, and a meaningful job.

Figure 13.9 *The problem of providing the basic elements of a healthful diet on a low income is addressed by the Women, Infants, and Children (WIC) program. Unfortunately, this program reaches just a fraction of those who could benefit. Only pregnant women and mothers who can prove that they or their children have a low income, high risk of malnutrition, and a local residence are eligible for WIC. Ironically, malnourished children whose physical status improves become ineligible—no matter how low the family income— unless a medical exam again reveals malnutrition.*

Problems in the Family

On the basis of two massive longitudinal studies which looked at many possible correlates and predictors of children's functioning in middle childhood, Michael Rutter (1975, 1979, 1982) found that the nature of family interaction was a much more powerful predictor of problems than family characteristics bearing labels such as "broken home" or "father-absent household" or "step-family" or "working mother." As Rutter and many others have pointed out, when looking at the effect of the family on children, we should look at the functioning, not the structure, of the family (Emery et al., 1984; Hetherington and Camara, 1984; Rutter, 1982). This perspective is helpful in evaluating four circumstances that are often blamed for children's problems: divorce, single parenthood, blended families, and maternal employment.

Divorce In the United States in the 1980s, married couples are divorcing at a rate of nearly 50 percent—two and a half times as often as in the 1960s and four times as often as in the first half of the twentieth century. This means that, for today's school-age children, having divorced parents is a common experience: an estimated one third have already experienced their parents' divorce, or will do so before they reach 18 years of age (Glick, 1979).

What effect, then, does divorce have on children? In general, children whose parents have divorced achieve less in school, are less happy at home, and are more disruptive in the community than children from intact families. The negative impact of divorce is particularly apparent in the first year or two after the separation, and then gradually declines (Emery et al., 1984). In the first year, especially, many children become more "aggressive, noncompliant, whining, nagging, dependent, and unaffectionate" (Hetherington and Camara, 1984). However, these generalities gloss over the many differences between one child's experience of divorce and another's, differences that permit some children to come through their parents' divorce unscathed, or even benefited.

Which children will be most negatively affected by a divorce and which will be most likely to make a relatively easy adjustment? While the many variables make prediction difficult, two quite comprehensive, longitudinal studies, one in Virginia (Hetherington et al., 1982) and one in California (Wallerstein, 1984; Wallerstein and Kelly, 1980) have come to similar conclusions, and have been corroborated by a series of studies in the Midwest (Santrock et al., 1982). Thus we are fairly sure that, at least in the United States in the later half of the twentieth century, the following five factors affect the adjustment of the child.

1. *Bitterness and hostility.* Much research has shown that the crucial factor in the adjustment of children to their family, no matter whether the parents are married, separated, or divorced, is the amount of family discord, especially the frequency and severity of the disputes the child witnesses (Hershorn and Rosenbaum, 1985; Rutter, 1982). Typically, the decision to divorce and the stressful legal, emotional, and financial consequences over the next few months exacerbate anger and open hostility between the parents (Emery et al., 1984). In addition, the legal system tends to fan the flames of anger, encouraging parents to argue over custody and child support. When the children are drawn into the parents' disputes, their specific reactions are hard to predict (a child might withdraw from a parent who formerly seemed a favorite, for instance), but almost always they show emotional strain.

2. *Changes in the child's life.* In most cases, children's lives change for the worse during the period of separation and divorce. For one thing, divorced parents usually have much less money to spend on their children, and this makes a difference in lifestyle, sometimes requiring a change of residence and school, and thereby

A CLOSER LOOK   **Single-Parent Households**

Households headed by a single parent are increasingly common, whether as a result of divorce or separation, or of the death of a parent, or of a mother's never marrying. (Unwed mothers represent the fastest-growing group of single parents, the number of children born out of wed-lock having increased nearly threefold between 1973 and 1983 [U.S. Department of the Census, 1985].) In the United States, about one child in five lives in a household headed by a single parent, as does about one child in six in Sweden, and one in nine in Canada. Ninety percent of these heads of households are mothers; the other ten percent are fathers. Of course, some families that are classified statistically as single-parent households are, in fact, characterized by two active parents, each of whom lives in a separate dwelling. For the most part, however, in single-parent families, the absent parent is definitely peripheral, neither of much help to the resident parent nor an important influence in the child's life (Amato, 1987).

Studies of the factors that foster competent or inadequate parenting find that the more stresses parents experience, the harder it is for them to be responsive and supportive of their children (Belsky et al., 1984c). Thus, our efforts to help children in single-parent households should begin with efforts to understand the stresses their parents experience. And certainly single parents have many sources of stress in their lives.

To begin with, most single parents suffer from "role over-load" as they try to provide nurturance, discipline, and financial support all at the same time (Zill, 1983). Having a job and a family is not easy, no matter what one's marital status, but the single parent is particularly likely to have difficulty when a child is sick, or when the job demands overtime, or when school holidays conflict with work obligations. The problems of the single parent increase markedly as family size increases: one child does not put nearly as much strain on the parent as two or more children do (Polit, 1984).

Financial difficulties often seem hardest to overcome. The income of single-parent households is substantially lower than that of two-parent households, even when only one of the parents in those households is employed. About a third of all single parents are forced to rely on public assistance at some point, a solution that does not provide adequate support or self-respect. More than half of all American households headed by women have incomes below the poverty line (Malcolm, 1985).

While single-parent households headed by a man fare somewhat better financially, men as single parents may experience complications that women do not. An exten-sive study found that many men are less prepared than women to handle the simultaneous demands of child-rearing and career (George and Wilding, 1972). Single fathers are also less willing to ask for outside help, financial as well as psychological, when problems become too difficult to handle alone. Further, single fathers are likely to have problems with their daughters (Santrock et al., 1982), perhaps because fathers tend to rely on their daughters to take over some of the housework and caring for younger children that their wives formerly did (Gasser and Taylor, 1976).

One problem that many single-parent households exper-ience is difficulty with sex-role development (Biller, 1981). Boys without fathers at home often become stereotypically "manly"—fighting and generally getting into trouble at quite a young age. It is sometimes thought that this behavior arises because of the absence of a father's discipline, but, in fact, if the father were present, his more moderate example of male behavior might make stern discipline unnecessary (Biller, 1981). Girls without fathers at home also are affected, although the specific effects may depend largely on the reason for the father's absence. According to one study, girls may become prematurely interested in sexual relationships if their mothers are divorced, and unusually shy and withdrawn if their mothers are widows (Hetherington, 1972). Parents

increasing stress. In addition, parents themselves often change in ways that may upset their children. Typically, the mother is initially overwhelmed with the burden of having to run the household and care for the children while worrying about financial problems and trying to repair her damaged self-esteem. Many women become depressed and withdrawn or, alternatively, try to find jobs, develop new skills, and expand their social lives, just when their children need and demand even more attention than before. Correspondingly, mothers frequently become more strict, less playful, and more inconsistent in their disciplining (Hetherington et al., 1982). Fathers also change, especially in the first year. Typically, if they are

themselves may have difficulty with sexual expression. If they have an active sex life, their children may become jealous or may reverse roles, as one boy did when his mother's date picked her up. "Bring her home by midnight," he said (Hetherington, 1987). Some single parents may choose to avoid such problems by not seeking sexual partners. In this case, parents tend to become overly dependent on their children for companionship, finding it difficult to let go when the children become young adults and need to establish independent lives.

Children of single parents are also more likely to be prone to academic difficulties. Both their attendance and their achievements are lower, with a greater number being left back or placed in special classes. Further, growing up in a father-absent house is particularly detrimental to a child's functioning in math. However, some studies find that children in mother-headed families do well verbally (Hetherington et al., 1981; Shinn, 1978). This may be a direct result of having more maternal attention than other children, or it may be indirect, resulting from the greater need to express one's ideas and emotions.

Are single-parent households headed for hard times, then? Not necessarily. If their parents can find ways to cope with the financial and role overloads placed on them, many children do quite well. Children of widows and widowers, overall, achieve as well as their peers from two-parent families, which shows that it is not single parent-hood per se that puts children at risk (Rutter, 1982).

In fact, ecological factors probably make the difference between single-parent households that function well and those that do not (Feiring and Lewis, 1984). The crucial factor is usually a network of social support. Friends and relatives can relieve some of the parent's role overload by helping with child care or financial difficulties as well as simply providing companionship and bolstering self-

Some single-parent families, like this Hispanic one, function very well. In addition to having the advantages of their own personal and family strengths, and the help of critical institutions such as school, church, and neighborhood, *this particular family is fortunate in another way: they are all female. In general, single mothers cope better with their daughters than their sons; the opposite seems true for single fathers.*

esteem. Grandfathers and uncles often become significant role models for children without fathers. Since social support is generally much more readily available to the widow than to the divorced woman or unwed mother, this may well explain many of the differences observed in various types of single-parent households.

the noncustodial parent, they become more indulgent with their children. Many fathers also adopt a more "youthful" lifestyle, including dating a variety of women. Many change their appearance, adopting a new hair style or growing a beard, or taking on a new look in their wardrobe. All such changes, especially in combination, can greatly increase the child's sense of instability in his or her life.

3. *The age of the child.* There is no "good" age for a child to experience divorce, but children do seem more vulnerable at some ages than at others. Divorce during transitional periods, such as the beginning of first grade or the onset of adoles-

cence, may be particularly hard for a child. While younger children are often more disturbed by the immediate changes brought about by divorce, older children tend to feel the effects longer (Wallerstein, 1984; Wallerstein and Kelly, 1980).

4. *The sex of the child.* Boys generally have a more difficult time adjusting to their parents' divorce than girls do, at least as evidenced by their overtly disruptive behavior at home and school. In fact, conflicts between mothers and sons are still common even six years after divorce, while most mothers and daughters have adjusted fairly well by this time (Hertherington et al., 1982). That boys seem to be more negatively affected by divorce may be simply one more example of the greater vulnerability of boys to various stresses throughout childhood, or it may be specific to the fact that most boys typically live with their mothers after the separation, and thus do not have an adult male role model to help them in everyday life (Rutter, 1982). Suggestive evidence for the latter hypothesis comes from studies that suggest that boys tend to adjust better when the father is the custodial parent than when the mother is (Santrock et al., 1982).

5. *Long-term involvement of both parents.* As time goes by, the extent to which continued contact with both parents is maintained is a powerful predictor of the child's well-being. For the 90 percent of children who are in their mother's custody, continued contact with the father is often problematic (see A Closer Look, page 308). While most fathers see their children often in the months immediately after the divorce, only a minority of noncustodial fathers continue to visit frequently and maintain close relationships with their children (Furstenberg and Nord, 1985). This is very unfortunate, particularly because there is now a substantial body of research that shows that a father's attention to his children (whatever his relationship might be to their mother) correlates with their achievement at school, particularly in math, and also with their happiness at school and at home (Biller, 1981; Hetherington et al., 1982; Lamb, 1982; Radin, 1982; Shinn, 1978).

Overall, then, divorce is difficult for children. However, the specific circumstances and the general passage of time can limit the negative effects. For example, in some cases, divorce removes the child from a parent who has lost control of his or her anger, or life, and consequently results in an improvement for the child (Emery et al., 1984). In these circumstances, even academic achievement rises significantly in the year after the divorce compared with the year before (Santrock, 1972; Wallerstein and Kelly, 1980). In most other cases, the problems diminish over the years. Most adults adjust to their new status within a year or two, and, consequently, their children function better as well. Further, the children themselves often find ways to cope with the divorce, developing new friendships and activities. However, even when children are back to functioning normally in school and with their friends, it takes the teachers and the other children time to treat the child normally. In fact, some teachers expect children of divorced parents to be difficult no matter how long ago the divorce occurred (Minuchin and Shapiro, 1983). In many cases, this assumption is unwarranted. If children of divorced parents are still troubled years after the event, it usually is not because of the divorce itself but because of other, related problems, such as financial difficulties, continued feuding between the parents, or the lack of a father's interest and support. Divorce is never easy on a child, but to a great degree the parents can determine whether it is a serious and continuing disturbance or a temporary disruption.

Blended Families   Especially if they are relatively young at the time of the divorce, most divorced adults marry again. How does this affect the children who become part of a new "blended" family with a step-parent and perhaps step-siblings as well? Obviously, there is no pat answer to this question. The change may bring a marked improvement in the child's life or it may create new problems. Much depends on the particular individuals involved, as well as on the kind of family interaction that develops. In addition, the benefits of remarriage for a child depend partly on which parent is remarried, and partly on the age and sex of the child.

Figure 13.10 *On the left, adolescent boys from a first marriage; on the right, their mother and step-father; in the future, a half brother or sister who might cause increased stability or increased bitterness—or moments of both.*

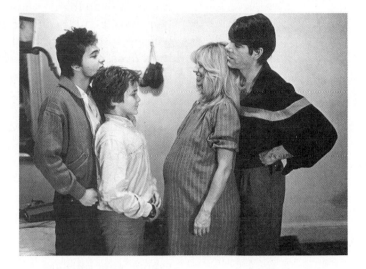

Remarriage of the Custodial Parent. Typically, if the children are relatively young and living with their mother at the time she remarries, their life improves. This is shown by their achievement in school as well as by their subjective reports (Hetherington et al., 1982). Part of the reason for the improvement is that remarriage usually leads to enhanced financial security for the family and higher self-esteem for the mother, both of which benefit the children. The other major factor, of course, is the presence of the step-father: boys, particularly, are helped when their step-fathers develop an active and helpful relationship with them.

This is not to say that the arrival of a step-father is always good for the children. If he has children of his own, their introduction on the scene may create added difficulties for the child who is trying to adjust to the absence of one parent and the presence of a new one (Visher and Visher, 1982). Further, a minority of step-fathers are cruel and abusive to their step-children. While only 10 percent of all children have step-fathers, step-fathers are implicated in 15 percent of cases of physical abuse and 30 percent of cases of sexual abuse (Giles-Sims and Finkelhor, 1984).

When a father who has custody of his children remarries, the results are also usually beneficial. Typically, the step-mother helps considerably with child care and housework, relieving the father and his children of some of the stress of their lives together (Ambert, 1982). The presence of the step-mother seems particularly helpful to girls, who often open up to the new woman in the house. Sons, however, tend to withdraw from the family (Santrock et al., 1982).

Remarriage of the Noncustodial Parent. When the noncustodial parent remarries, difficulties often arise for the custodial parent as well as for the children. Remarriage of a former spouse is often a further blow to self-esteem of the divorced parent and often a threat as well. It is not uncommon for the new spouse to cause alterations in the noncustodial parent's relationship to his or her children—either by reraising custody questions that had been settled, or by insisting that the noncustodial parent see his or her offspring less often than before (Furstenberg et al., 1982; Hetherington et al., 1982).

Overall, the same factors that were relevant to divorce adjustment are relevant to remarriage adjustment. If remarriage means happier parents, more money, and less stress, it benefits the children as well as the adult. Two cautions are in order, however. First, it takes time for children to adjust to a "new" family situation: many react with hostility or withdrawal at first. Second, since the divorce rate of remarriages is even higher than that of first marriages, parents thinking of remarriage "for the sake of the children" should think hard before taking such a step. As you will see at the end of the chapter, most children can adjust to one or two stresses in their lives, especially if the stresses are limited in time, but repeated family disruptions are much more likely to have long-term consequences (Hetherington and Camara, 1984).

Maternal Employment   Most mothers of school-age children are in the job market. To be specific, of the mothers who have school-age children but no children under 6, 63 percent who are married, and 75 percent who are single, are employed. Further, of those who have children under age 6, about half are employed.

While it is often suggested that a mother's working is harmful to her children, in fact, it seems that in many ways it may be beneficial to them (Hoffman, 1984; Lamb, 1982). In most cases, the mother's employment relieves some of the financial pressure on the family, and this obviously benefits the children. Further, employed women generally are more satisfied with their lives than full-time housewives are (Baruch et al., 1983; Newberry et al., 1979), and, as a review of the research on maternal working status concludes, "satisfied mothers—working or not—have the best-adjusted children" (Etaugh, 1974).

It has also been suggested that in two-parent families, children of working mothers benefit because their fathers tend to be more directly involved in the household (Barnett and Baruch, 1987; Carlson, 1984; Hoffman, 1984). Furthermore, children of employed mothers learn responsibility as well as household skills, which enhances their competence and self-esteem. Girls are particularly likely to benefit from their increased responsibilities and independence as well as from a more positive image of the possibilities open to women (Lamb, 1982; Smokler, 1975).

However, some boys may be adversely affected by their mother's employment (Hoffman, 1977; Lamb, 1982). According to one influential Canadian study, boys of lower-class employed mothers are more shy and nervous, dislike school, have lower grades, and are described more negatively by their fathers than are boys whose mothers stay home (Gold and Andres, 1978). The father's attitudes may be crucial to the son's adjustment to maternal employment, for problems are particularly likely to arise when the father is unhappy that his wife is working, and is unwilling to help out at home (Hoffman, 1977).

One final concern is often raised with regard to maternal employment, that of the millions of "latchkey" children who let themselves into an empty house after school because neither parent is home (Turkington, 1983). However, two careful comparison studies of children aged 9 and older who cared for themselves after school found that their self-esteem, social adjustment, and school achievement were just as good as that of children who had some form of after-school adult care (Galambos and Garbarino, 1983; Rodman et al., 1985). The likely reason is that children who care for themselves after school are not simply left to their own devices. They are taught how to manage on their own, what dangers to avoid, and are in telephone contact with their parents.

Obviously, some families would benefit if their employed mothers had more time and energy to devote to them, and measures such as an increase in flextime, greater benefits for part-time workers, and equality of pay for women would facili-

Figure 13.11 *The effect that a mother's employment has on her son may, in many cases, depend largely on her husband's understanding and help. It is likely that this boy will benefit from the father-son comradery of sharing some daily chores.*

tate this. However, the increasing rates of maternal employment under current arrangements should not be seen as a detriment to school-age children. The evidence clearly shows that the opposite is more likely the case.

Coping with Stress

As we have seen, the development of many school-age children may be hindered by problems with their peers and their families. However, particularly in middle childhood, many children cope quite well with a variety of problems. This is apparent statistically: serious problems between children and their parents, and severe emotional disturbance in children, are less common during middle childhood than earlier or later. It is also apparent in longitudinal studies of child development: before age 12, most children encounter some unexpected and potentially handicapping stress, and most develop normally (Garmezy, 1976; Murphy and Moriarty, 1976; Rutter, 1975; Werner and Smith, 1982).

In his study of stresses that lead to severe psychological problems in childhood, Michael Rutter (1979) identified six variables that are strongly and significantly associated with psychiatric disorders:

1. severe marital discord;
2. low social status;
3. overcrowding or large family size;
4. paternal criminality;
5. maternal psychiatric disorder;
6. admission of the child into the care of the local authority.

Rutter found that children with one, and only one, of the six risk factors were no more likely to have psychiatric problems than children with no risk factors. When two of the risk factors occurred together, the incidence of problems more than doubled. When four or more of the risk factors were present, psychiatric problems were four times as likely as when there were two. As Rutter writes, "the stresses potentiated each other so that the combination of chronic stresses provided very much more than a summation of the effects of the separate stresses considered singly" (Rutter, 1979).

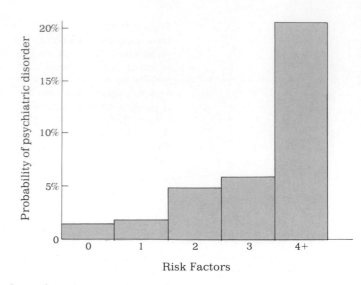

Figure 13.12 *Rutter found that children who had to cope with one serious problem ran virtually as low a risk of suffering a psychiatric disorder as did children who faced no serious problems. However, when the child had two problems, the chances more than doubled. Four or more problems produced about ten times the likelihood of psychiatric disorders as one. About one child in five who experienced four or more serious stresses actually became emotionally disturbed.*

Yet, as Rutter and others have found, some children are able to overcome several serious problems. These children, often referred to as "invulnerable" or "stress-resistant," have been studied extensively to see what it is that helps them cope. Two factors seem especially important. The first is competence in any one area. The second is social support.

Competence   A theme throughout this chapter has been the importance of competence for school-age children. Several studies have shown that children who have well-developed social, academic, or creative skills—or better yet, all three—are much more able to surmount whatever problems they encounter at home or in their community (Block and Block, 1980; Garmezy, 1976; Murphy and Moriarty, 1976; Werner and Smith, 1982). During middle childhood, intellectual competence is particularly important. Children who are intellectually able, or whose schools encourage them to develop their strengths, are often able to overcome handicaps. School achievement makes it possible for a child who lives in run-down housing to aspire to becoming an architect, or for a child who waits for care in an understaffed city clinic to identify with, and become, a physician.

Of course, much depends on the nature of the particular school the child attends. Even more important than the academic quality of the curriculum or the size of the classes is the overall emotional tone of the school. This was found in a study of twelve London schools that served lower-class children, many of whom came from crowded families that were headed by single parents or who had parents with serious psychological problems (Rutter et al., 1979). Some of these schools had markedly more students who passed higher-level exams, fewer students who dropped out, and lower rates of juvenile delinquency than would be expected on the basis of the students' backgrounds. One crucial factor that distinguished the more successful schools was that they cared about the children, as evidenced by such simple things as the student's work being displayed on the walls and the frequency of praise from the teachers. Another was that the administration of these schools had high expectations of their teachers, who, in turn, had high expectations of their students. Apparently, in environments that expect and encourage competence, children tend to meet the challenge, overcoming home and community handicaps to do so.

Social Support   At virtually every point in the life span, the developing person's social network is crucial to overcoming problems and developing his or her potential. The social network includes all the people who could offer assistance of any sort, from a shoulder to cry on to financial aid. Friends and family members are the main sources of social support. And, as is true throughout childhood, children with warm, authoritative parents are likely to become happier, more mature adults (Dubow et al., 1987). Werner and Smith (1982) found that grandparents and siblings often compensated for the neglect or absence of one or even both parents. Neighbors, clergy, and teachers also can be very influential, especially when the child's problems originate in the family, as many children's problems do (Garmezy et al., 1979). During middle childhood and later, a stable and supportive adult sibling can be particularly crucial (Bank and Kahn, 1982).

Figure 13.13 *Parents are usually the most important family members, and schools are usually the most important institutions, in the life of school-age children. However, especially for children who experience extraordinary stress, essential support systems may come from a bit further afield. This boy, leaning on his grandfather while they listen to the pastor of a country church, may be getting the special support he needs.*

The role of social support has been highlighted in ethnic groups in which high-risk conditions are almost the norm (Harrison et al., 1984). For example, 49 percent of all black American children are in single-parent households, and 47 percent of all black American children are living below the poverty line (Malcom, 1985). These problems are never easy to live with, but study after study shows that many black families have extensive networks of family support, with grandparents, aunts and uncles, older siblings, and neighbors frequently providing many forms of help—including taking complete care of the child for days, months, or years when it seems in the child's best interest (Lindblad-Goldberg and Dukes, 1985; Malson, 1983; McAdoo, 1979). Similar patterns are found in Chicano families (Martinez, 1986). While the adults in such families sometimes feel overburdened by advice and requests for assistance from their relatives, children benefit from having several possible sources of comfort and support.

As a child approaches adolescence, then, longitudinal research reveals that the crucial question is not whether the child has any serious problems but whether the child has a store of strengths, such as supportive parents, cognitive competence social skills, or personality assets that will help him or her through to adulthood (Rutter, 1987). As you will see in the next three chapters, while adolescence almost always presents problems and challenges, most young people cope very well.

SUMMARY

Theories about Middle Childhood

1. Freud believed that most of the child's emotions are latent during middle childhood, especially sexual and aggressive urges. This enables the child to develop their skills and their social understanding.

2. Erikson calls middle childhood the industry-versus-inferiority stage, because children like to be busy learning new skills. As they develop the competencies their society values, they, develop their self-esteem, or else come to see themselves as inferior and inadequate.

3. Learning theorists stress the school-age child's increasing ability to learn as well as the growing importance of intrinsic reinforcements, social influences, and ability to control oneself.

4. Cognitive theory notes that children become less egocentric during these years. They become more vulnerable to others' opinions, and more able to understand another point of view.

Social Cognition

5. School-age children develop a greater awareness of the many social roles that individuals can have. They also become increasingly aware that others have preferences and characteristics that are different from their own; at the same time, they become better able to adjust their behavior to interact appropriately with others.

6. Children also develop complex theories about themselves and their behavior. These ideas are based on their past experiences, the opinion of others, and untested assumptions about themselves. As children become more knowledgeable about their own abilities and shortcomings, they become more self-critical and their self-esteem dips.

7. Children develop a healthy sense of self-esteem if they become competent at a variety of tasks and if home and school values encourage cooperation.

8. School-age children develop their own subculture, with language, values, and codes of behavior. The child who is not included in this society often feels deeply hurt, for social dependence on peers is strong at this age, even as independence from adults is valued.

9. Children become increasingly aware of the customs and principles of the larger society as they develop their own social codes. Prosocial behavior develops along with friendships, which become more selective and exclusive as children grow older.

Problems and Challenges

10. School-age children's growing awareness and independence make them more susceptible to the influence of the larger society and the effects of socioeconomic status. SES influences all aspects of children's lives, from educational achievement to self-esteem.

11. Whether or not factors such as low SES, divorce, and single-parent households affect children negatively depends on the functioning of family and school support systems.

KEY TERMS

latency *(291)*

industry versus
 inferiority *(291)*

social cognition *(294)*

self-theory *(296)*

learned
 helplessness *(297)*

society of children *(299)*

prosocial behaviors *(301)*

socioeconomic status
 (SES) *(304)*

KEY QUESTIONS

1. What are the effects of latency on the development of school-age children?

2. How does a child's response to various types of reinforcement change as he or she grows older?

3. How does children's self-theory change from the preschool years through middle childhood?

4. What factors help a child to develop positive self-esteem?

5. What is the role of the peer group during the school years?

6. What evidence is there that the society of children is a powerful force in middle childhood?

7. What are some of the factors that may cause a child to be rejected from the peer group?

8. How does socioeconomic status affect development in middle childhood?

9. What are some of the factors that influence how parental divorce will affect a child?

10. What are some of the problems experienced by single-parent households?

11. What factors tend to make maternal employment beneficial or harmful to a child's development?

12. What are some of the stresses associated with a high risk of serious problems in a child's life?

13. What factors can help a child cope with stress?

Part IV

The Developing Person So Far: The School Years, Ages 7 through 11

Physical Development

Growth

During middle childhood, children grow more slowly than they did during infancy and toddlerhood or than they will during adolescence. Increased strength and heart and lung capacity give children the endurance to improve their performance in skills such as swimming and running.

Motor Skills

Slower growth contributes to children's increasing control over their bodies. However, since brain maturation is not yet completed, the average 7-year-old may take twice as long as an adult to respond to a stimulus.

Cognitive Development

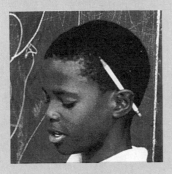

Concrete Operational Thought

Beginning at about age 7 or 8, children develop the ability to understand logical principles. Once acquired, the concepts of reciprocity, classification, class inclusion, seriation, and number help children to develop a more complete understanding of mathematics and of measurement.

An Information-Processing View

Children's expanded ability to understand and learn can be attributed, in part, to enlarging memory capacity and an increasing ability to use metamemory techniques. At the same time, metacognition techniques enable children to organize their knowledge.

Language

Children's increasing ability to understand the structures and possibilities of language enables them to extend the range of their cognitive powers, to become more analytical in their use of vocabulary, and to enjoy the word-play involved in puns, jokes, and riddles.

Psychosocial Development

Personality Development

According to Freud, middle childhood is the period of latency, during which the intense drives of the phallic stage are submerged, freeing children to learn and be productive. Erikson describes the conflict of this stage as the crisis of industry versus inferiority. Learning theorists suggest that children's greater understanding of cause and effect and their awareness of the actions and attitudes of others make them more susceptible to reinforcement and modeling techniques.

The Peer Group

The peer group becomes increasingly important to children as they become less dependent on their parents and more dependent on friends for help, loyalty, and sharing of mutual interests.

Social Systems and the Child

Children are increasingly aware of, and involved in family life, as well as in the world outside the home, and therefore are more likely to feel the effects of family, economic, and political conditions. Whether or not particular situations will be stressful for a child will depend, at least in part, on the child's temperament, competence, and the social support provided by home and school.

PART V # Adolescence

Adolescence is probably the most challenging and complicated period of life to describe, study, or experience. Between the ages of 10 and 20, more physical, cognitive, and psychological changes occur than during any other period. Children's bodies suddenly show the signs of sexual maturing, and childish faces become less round and more angular. These biological changes of puberty are universal, but in their particular expression, timing, and extent, the variety shown is enormous and depends, of course, on sex, genes, and nutrition. There is great diversity in cognitive development as well: many adolescents are as egocentric in some respects as preschool children, while others reach the stage of abstract thinking that Piaget regards as the highest form of cognition. Psychosocial development shows even greater diversity, as adolescents develop their own identity, choosing from a vast number of social, sexual, and moral possibilities.

Yet such differences should not mask the commonality of the adolescent experience, for all adolescents are confronted with the same developmental tasks: they must adjust to their new body size and shape and to their awakening sexuality, think in new ways, and strive for the emotional maturity and independence that characterize adulthood. As we will see in the next three chapters, the adolescent's efforts to come to grips with these tasks are often touched with confusion and poignancy.

CHAPTER 14

Adolescence: Physical Development

*Lots of kids I know, they feel all of a sudden they can't be
kids anymore. They call themselves kids but they don't feel
the same way about it. . . . Their parents may tell them how
they act irresponsibly, like kids, but in their minds they aren't
thinking irresponsible thoughts . . . their bodies aren't kids'
bodies either. It's hard to guess people's age nowadays
because everybody looks older; even if they don't,
they act older. . . . You think, Hey I know I'm not old,
but it's fun to try, because if I want, I can fail miserably at it.
You can pull off this big mask of yours and say "Hi, everybody,
it's really me and I'm really a kid. . . . " You hate to admit it,
but there are lots of times you wish you still were
a little kid, nodding at all the questions people ask.*

Jeannie Melchione, age 16
quoted in Cottle, 1979

Developmentally, between the ages of 10 and 20 young people cross the great divide between childhood and adulthood. This journey occurs in all three domains, with physical changes often preceding and outpacing cognitive and psychosocial ones. No one would call this process of becoming an adult simple or easy. In our culture, at least, adjusting to so many changes in oneself and in the way one is viewed by others can be difficult and stressful.

Before beginning our discussion of adolescence, however, we should acknowledge that no period of life is problem-free, and none—including adolescence—is defined only by its problems. There are moments, it is true, of moodiness, disappointment, and anger in almost every teenager's life, and there are too many adolescents who make serious missteps on the path toward maturity. This chapter, and the two that follow, will examine these problems, focusing on causes and prevention. Keep in mind, however, that the developmental changes that may sometimes be the source of difficulties and usually temporary obstacles are also the source of new excitement, challenge, and growth of many kinds. Let us begin with the growth of the body, the first apparent sign of adolescence.

Puberty

The period of physical growth that ends childhood and brings the young person to adult size, shape, and sexual potential is called **puberty**, which typically begins sometime between ages 9 and 14.

The first signs of puberty are invisible; the next are visible to the individual but not to people in general; and, finally, puberty is apparent to every observer. To be specific, the first indication of puberty is increased concentrations of male and female hormones in the bloodstream—**estrogen** (which increases somewhat in

Figure 14.1 *Increases in height, weight, musculature, and body fat are characteristic of all adolescents, but the range of these changes varies considerably, not only between the sexes but also between individuals of the same sex. While all the teenagers in this photo are developing normally, not all of them may feel that way at the moment. For instance, why is the boy in the center covering his chest, why is the girl on the right turning sideways, and why is the boy on the left covering his armpit? The text on pages 326–328 suggests answers.*

boys and markedly in girls) and **testosterone** (which increases dramatically in boys and slightly in girls). These hormonal changes generally occur at least a year before the appearance of the first signs of puberty that are perceptible as such to the young person—the initial enlargement of the girl's breasts and the boy's testes (Higham, 1980). About a year after that, the first sign of the onset of puberty that is readily observable to others occurs, a period of rapid physical growth called the **growth spurt.**

The Growth Spurt

Typically, parents begin to notice that their children are emptying their plates, cleaning out the refrigerator, and straining the seams of their clothes even before they notice that their children are growing taller, for the growth spurt actually begins with rapid weight gain before rapid height gain. Toward the end of middle childhood, usually between the ages of 10 and 12, both boys and girls become noticeably heavier, primarily through the accumulation of fat, especially on their thighs, arms, buttocks, and abdomen.

Soon after the weight increase begins, a height increase begins, burning up some of the fat and redistributing the rest. (On the whole, a greater percentage of fat is retained by females, who naturally have a higher proportion of body fat in womanhood than in girlhood.) About a year after these weight and height increases take place, a period of muscle increase occurs: consequently, the pudginess and clumsiness exhibited by the typical child in early puberty generally have disappeared by late pubescence, a few years later. Overall, the typical girl gains about 38 pounds (17 kilograms) and 9⅝ inches (24 centimeters) between the ages of 10 and 14, while the typical boy gains the same number of inches and about 42 pounds (19 kilograms) between the ages of 12 and 16 (Lowrey, 1978).

Note, however, that the cross-sectional data, which average out the individual growth spurts, are somewhat deceptive, because the chronological age for the growth spurt varies considerably from child to child. In any given year between ages 10 and 16, some individuals will not grow much at all because their growth

Figure 14.2 *Many of the physical changes of puberty seem to be inevitable sources of embarrassment for the young people experiencing them. The only one at this junior-high dance who appears to be enjoying herself, for instance, is the girl whose growth spurt has not yet started.*

spurt has not begun or is already over, while others will grow very rapidly. Records of individual growth during this period make it obvious why the word "spurt" is used to describe these increases (Tanner, 1978). During the twelve-month period of their greatest growth, many girls gain as much as 20 pounds (9 kilograms) and 3½ inches (9 centimeters), and many boys gain up to 26 pounds (12 kilograms) and 4 inches (10 centimeters).

Sequence of Growth The growth process does not occur in every part of the body simultaneously (Katchadourian, 1977). Just as weight gain often precedes height gain, certain parts of the skeleton develop more rapidly than others. In most cases, adolescents' hands and feet lengthen before their arms and legs do, and their torso is the last part to grow, making many adolescents temporarily big-footed, long-legged, and short-waisted. Thus, growth in puberty, unlike that of earlier periods, is distal-proximal (far to near). In addition, the nose, lips, and ears usually grow before the head itself reaches adult size and shape. The results of this uneven growth rate are often unsettling, as evidenced in the case of a 12-year-old girl who wears a woman's size in shoes while still wearing children's sizes in all her other clothes, or in the case of a 14-year-old boy who fears that his nose is the only part of his body that is getting any bigger.

At least as disturbing to the growing person can be the fact that the two halves of the body do not always grow at the same rate: one foot, breast, or ear can be temporarily larger than the other. None of these anomalies persist very long, however. Once the growth process starts, every part of the body reaches close to adult size, shape, and proportion in three or four years.

Organ Growth. While the torso grows, internal organs also grow. The lungs increase in size and capacity, allowing the adolescent to breathe more deeply and slowly, and the heart doubles in size. In addition, the total volume of blood increases. These organ changes increase endurance in physical exercise, making it possible for many teenagers to run for miles or dance for hours without stopping to rest. However, the fact that the more visible spurts of weight and height precede

Figure 14.3 *Adolescents' quicker reaction time, longer legs, and increased heart and lung capacity enable them to take part in sports requiring endurance, in this case, cross-country running, a sport that is beyond the abilities of younger children.*

the less visible ones of the muscles and organs means that athletic training and weight-lifting should match a young person's size of a year or so earlier. Exhaustion and injury might result if the physical demands on a young person's body do not take this lag into account (Thornburg and Aras, 1986).

One organ system, the lymphoid system, including the tonsils and adenoids, actually decreases in size at adolescence. Consequently, teenagers are less susceptible to respiratory ailments than children. For this reason, about half the victims of childhood asthma improve markedly in adolescence (Katchadourian, 1977).

The eyes also undergo a change, as the eyeballs elongate, making many adolescents sufficiently nearsighted to require glasses.

Finally, the hormones of puberty cause many relatively minor physical changes that, despite their insignificance in the grand scheme of development, can have substantial psychic impact. For instance, oil, sweat, and odor glands become much more active during puberty. One result is acne, a problem for about two-thirds of all boys and half of all girls between the ages of 14 and 19 (Schachter et al., 1971). Another result is oilier hair and smellier bodies, which make adolescents spend more money on shampoo, soap, and deodorants than any other age group.

Nutrition   The rapid height and weight gain of puberty obviously requires additional calories and protein, more than during any other period of life. In addition, the adolescent body needs about 50 percent more calcium, iron, zinc, and vitamin D during the growth spurt than a year or two earlier (Sinclair, 1978). The nutrient most commonly lacking in the adolescent's diet is iron (most commonly found in meat and dark-green vegetables)—a deficiency that poses a special problem for adolescent girls once menstruation begins. Iron-deficiency anemia is more common among adolescent females than among any other part of the population, so if a teenage girl seems apathetic and lazy, she should have her iron level checked before it is assumed that she suffers from a poor attitude or other psychosocial difficulties.

In terms of total protein and calorie requirements, most adolescents meet their needs, often skipping breakfast but nonetheless consuming four or more meals a day (Thornburg and Aras, 1986). Obesity, discussed at length in Chapter 11, is a problem for many adolescents as they try to adjust to their changing nutritional needs and their changing body shape.

Figure 14.4 *The pizza shop has replaced the soda shop as the favorite food hangout of the American adolescent. Fortunately, not only is pizza relatively inexpensive, quick, and easy to share, it is also quite nutritious when compared with the many other sweet, salty, or synthetic snack foods.*

Figure 14.5 *Unlike the retouched photo portraits used for publicity shots and album covers, this candid shot of pop singer Karen Carpenter, taken a year before her death from heart failure brought on by anorexia nervosa, shows the gaunt appearance that hints at the underlying physical problems associated with this disorder.*

However, many adolescents think they are obese when they are not. Teenagers are particularly susceptible to the latest crash diets and food fads, in part because their level of cognitive and psychosocial development leads them to experiment with new foods and ideas. Unbalanced diets are harmful at any age, but they are particularly so during the rapid growth of early adolescence, when the body must have sufficient nourishment for full growth potential to be realized (Mellendick, 1983).

The most baffling and serious nutritional problem is **anorexia nervosa,** a psychogenic eating disorder in which a person severely restricts eating to the point of emaciation, rapidly losing 30 percent or more of body weight and approaching starvation. Typically, the person is female, and the most common ages of onset are at about 11 and 18—the beginning and the end of adolescence (Johnson and Conners, 1987).

Although most victims are above average in intelligence, they insist they are neither hungry nor thin. Typically, they exercise daily "to keep up muscle tone," as one patient put it as she jogged in place beside her hospital bed (Minuchin et al., 1978).

Without psychological help, about a third of all anorexics get better, and about 20 percent die. The remaining have eating problems, either staying too thin, or becoming too fat, or seesawing between the two extremes all their lives. Furthermore, anorexia is often linked to depression, as cause, momentary "cure," and consequence. Thus suicide can be a very real threat (Muñoz and Amado, 1986).

The kind of psychological help anorexics receive depends on the theoretical perspective of the therapist. Psychoanalytic therapists believe that anorexics have a severe disturbance of body image. According to this theory, the anorexic is afraid of becoming a woman, so she maintains a hipless, breastless, childlike form by extreme dieting. One result is that her menstrual periods cease; another is that she becomes the center of family concern and care. In these ways, the anorexic is more like a young child than a young woman. Through psychotherapy, psychiatrists try to help the patient accept her burgeoning femininity, and therefore accept nourishment (Bruch, 1978).

Learning theorists see eating disorders as the result of maladaptive behavior (intended as a means of getting the parents' attention, for example), so they treat it by reinforcing weight gain and punishing weight loss. For example, each day that a hospitalized anorexic gains weight, she is given privileges: freedom to watch television, to make phone calls, to see visitors, to comb her hair. Each day that she loses weight, she must stay in bed, without even the right to go to the bathroom. Her only permitted activity is eating.

From both perspectives, the patient's family is considered part of the problem and thus should be involved in treatment. Often the family is "enmeshed" in their daughter's identity, not allowing normal independence. Consequently, the daughter feels her parents depend on her to be perfect—including perfectly thin.

Also implicated in the problem of anorexia is the current cultural ideal of feminine body shape, which is much thinner than many women naturally are. This cultural ideal is discussed in Chapter 17, as is bulimia, another eating disorder that more typically begins in late adolescence and early adulthood.

Sexual Maturation

While the growth spurt is taking place, another set of changes occurs that transforms boys and girls into young men and women. As we have seen, before puberty, the physical differences between boys and girls are relatively minor. At puberty, however, many significant body differences develop. These include changes in the **primary sex characteristics,** those sex organs that are directly involved in reproduction.

Changes in Primary Sex Characteristics   During puberty, all the sex organs become much larger. In girls, the uterus begins to grow and the vaginal lining thickens, even before there are visible signs of puberty. In boys, the testes begin to grow, and about a year later, the penis lengthens and the scrotal sac enlarges and becomes pendulous.

Toward the end of puberty, the young person's sex organs have become sufficiently mature to make reproduction possible. For girls, the specific event that is taken to indicate fertility is the first menstrual period, called **menarche.** For boys, the comparable indicator of reproductive potential is **ejaculation,** that is, the discharge of seminal fluid containing sperm. Ejaculation can occur during sleep in a nocturnal emission (a wet dream), through masturbation, or through sexual intercourse, with masturbation being the most common cause for the first ejaculation (Kinsey et al., 1948).

Actually, both menarche and ejaculation are simply one more step toward full reproductive maturity, which occurs several years later (Thornburg and Aras, 1986). In fact, a girl's first menstrual cycles are usually *anovulatory;* that is, they occur without ovulation. Even a year after menarche, most young women are still relatively infertile: ovulation is irregular, and if fertilization does occur, the probability of spontaneous abortion is much higher than it will be later, because the uterus is still relatively small. In the case of boys, the concentration of sperm usually necessary to fertilize an ovum is not reached until months or even years after the first ejaculation of seminal fluid (Chilman, 1983). (As many teenagers discover too late, unfortunately, this relative infertility does not mean that pregnancy is impossible at puberty; it is simply less likely than it will be a few years later.)

One problem for more than half of all adolescent girls is that they are unable to carry on their normal activities, at least occasionally, because of menstrual pain, sometimes called dysmenorrhea (Widholm, 1985). In this age group, the incidence of moodiness in the days before a period, and menstrual cramps as a period begins is quite high. Teachers and parents need to know that, while tension may exacerbate the problem, the primary cause of dysmenorrhea is physiological, and medical diagnosis and treatment may help (Dawood, 1985).

Attitudes toward menarche, menstruation, and first ejaculation have changed over the last two decades. Contemporary young people are more worried if they experience these events after, rather than before, most of their friends do. Most feel quite well informed about these first signs of sexual maturity. Furthermore, masturbation, once a source of overwhelming fear and shame, is now commonly accepted by adolescents of both sexes (Chilman, 1983). However, a strong sense of sexual privacy remains, especially in early adolescence. Virtually no young person discusses these private events with friends or parents of the opposite sex. Indeed, few boys tell other boys the details of their own experiences with masturbation or ejaculation, and even girls, who typically promise to tell their close friends when menarche arrives, are usually more reticent than they had anticipated (Brooks-Gunn et al., 1986; Gaddis and Brooks-Gunn, 1985).

Secondary Sex Characteristics   While maturation of the reproductive organs is the most directly sexual development of puberty, changes in many other parts of the young person's body, called **secondary sex characteristics**, also indicate that sexual maturation is occurring. Most obviously, the body shape of males and females, which was almost identical in childhood, becomes quite distinct in adolescence. Males grow taller than females and become wider at the shoulders than at the hips. Females become wider at the hips, an adaptation for child-bearing that is apparent even in early puberty, and becomes increasingly so over the teenage years (Tanner, 1978).

Another obvious difference in the shape of the female body, and the one that receives the most attention in Western cultures, is the development of breasts. For most girls, the first sign that puberty is beginning is the "bud" stage of breast development, when a small accumulation of fat causes a slight rise around the nipples. From then on, breasts develop gradually for several years, with full breast growth not being attained until almost all the other changes of puberty are over (Katchadourian, 1977). Since our culture misguidedly takes breast development to be symbolic of womanhood, girls whose breasts are very small or very large often

Figure 14.6 *Unlike adolescent boys, girls are faced with the question of whether to enhance or diminish the visibility of the most obvious sign of their burgeoning womanhood—their breasts.*

feel worry and embarrassment; small-breasted girls often feel "cheated," even disfigured; large-breasted girls may become extremely self-conscious as they find themselves the frequent object of unwanted stares and remarks.

In boys, the diameter of the areola (the dark area around the nipple) increases during puberty. Much to their consternation, many boys develop some breast tissue as well. However, their worry is needless: about 65 percent of all adolescent boys experience some breast enlargement, usually at about age 14 (Smith, 1983), but this enlargement almost always disappears by age 16.

Hair.   In both sexes, head and body hair usually becomes coarser and darker during puberty. In addition, new hair growth occurs under the arms, on the face, and in the pubic area. For many young people, the appearance of a few light-colored straight strands of pubic hair is the first apparent sign of puberty. As puberty continues, pubic hair becomes darker, thicker, and curlier, and covers a wider area. Girls reach the adult pubic-hair pattern in about three years; for boys, the process takes six years or more.

Facial and body hair are considered signs of manliness in our society, a notion that is mistaken for three reasons. First, the tendency to grow facial and body hair is inherited; how often a man needs to shave is determined by his genes rather than his virility. In addition, facial hair is usually the last secondary sex characteristic to appear, sometimes long after a young male has become sexually active. Finally, girls typically develop some facial hair and more noticeable hair on their arms and legs during puberty—a sign not of masculinity but of sexual maturation.

Voice.   The adolescent's voice becomes lower as the larynx grows, a change most noticeable in boys. (Even more noticeable, much to the chagrin of the young male, is his occasional loss of voice control, throwing his newly acquired baritone into a high squeak.) Girls also develop lower voices, a fact reflected in the recognition of a low, throaty female voice as "sexy."

The Timing of Puberty

As you have just read, the changes of puberty occur in predictable sequence and tempo (see chart, next page). The entire process begins when hormones from the hypothalamus (a part of the brain) trigger hormone production in the pituitary gland (located at the base of the skull), which in turn triggers increased hormone production by the *gonads*, or sex glands (the testes in the male and the ovaries in the female), as well as by the adrenal glands.

For girls, the most important hormones produced by the sex glands are estrogen and progesterone, which cause, usually in sequence, the beginning of breast development, first pubic hair, widening of hips, the growth spurt, menarche, and completion of breast and pubic-hair growth. For boys, the most important hormone is testosterone, which produces, usually in sequence, growth of the testes, growth of the penis, first pubic hair, capacity for ejaculation, growth spurt, voice changes, beard development, and completion of pubic-hair growth. Once the hormonal concentrations start the biological changes of puberty, the process is quite rapid, with most major changes occurring within three years for girls and four years for boys (Rutter, 1980).

However, while the sequence of pubertal events is very similar for all young people, there is great variation in the age of onset. Healthy children begin puberty any time between ages 8 and 16, with ages 10, 11, and 12 being the most typical. The child's sex, genes, body type, nourishment, and health all affect this variation.

Sequence of Puberty

Hormone Signals from the Brain (Hypothalamus → Pituitary) to the Gonads

Girls	Approximate Average Age*		Boys
1. Ovaries increase production of estrogen and progesterone	9	10	1. Testes increase production of testosterone
2. Internal sex organs begin to grow larger	9½	11	2. Testes and scrotum grow larger
3. Breast "bud" stage	10	12	3. Pubic hair begins to appear
4. Pubic hair begins to appear	11	12½	4. Penis growth begins
5. Weight spurt begins	11½	13	5. First ejaculation
6. Peak height spurt	12	13	6. Weight spurt begins
7. Peak muscle and organ growth (also, hips become noticeably wider)	12½	14	7. Peak height spurt
8. Menarche (first menstrual period)	12½	14½	8. Peak muscle and organ growth (also, shoulders become noticeably broader)
9. First ovulation	13½	15	9. Voice lowers
10. Final pubic-hair pattern	15	16	10. Facial hair appears
11. Full breast growth	16	18	11. Final pubic-hair pattern

*Note: Average ages are a rough approximation, with many perfectly normal, healthy adolescents as much as three years ahead or behind these ages. In addition, sequence is somewhat variable. For instance, many girls have some pubic hair before their breasts start to grow, and some boys have visible mustaches before their voices have completely changed.

Factors Affecting When Puberty Occurs Male-female differences are one factor affecting when a particular person will experience puberty, although the average time differential between female and male development depends on which particular event in puberty is being compared. The first signs of reproductive capability appear only a few months later in boys than in girls: the average girl reaches menarche at about age 12½, while the average boy first ejaculates at age 13 (Schoof-Tams et al., 1976; Thornburg and Aras, 1986). However, since the growth spurt appears later in the sequence of pubertal changes in boys, the average boy is two years behind the average girl in this respect. Consequently, between ages 12 and 14, most girls are taller than their male classmates.

Genes are also an important factor in the age at which puberty begins. This is most clearly seen in the case of menarche, the pubertal event that is easiest to date. Although most girls reach this milestone between 11 and 14, the age of onset varies from 9 to 18. However, the difference in the age at which sisters reach menarche is, on the average, only 13 months; and the difference for monozygotic twins averages a mere 2.8 months.

Body weight and the proportion of fat are also important in determining the onset of puberty. Short, stocky children tend to experience puberty earlier than those with taller, thinner builds. Menarche in particular seems related to the accumulation of a certain amount of body fat. Females who have little body fat, such as runners and other athletes, menstruate later than and more irregularly than the

Body Image

In a culture such as ours, which places a premium on physical attractiveness and features beautiful bodies in advertisements for everything from clothes and cosmetics to stereos and auto parts, it is no wonder that most adolescents begin spending a good deal of time every day in front of a mirror, tirelessly checking and rechecking their clothes, their physiques, their complexions, the parts in their hair. As Tanner (1971) explains:

For the majority of young persons, the years from twelve to sixteen are the most eventful ones of their lives so far as their growth and development are concerned. Admittedly, during fetal life and the first year or two after birth, developments occurred still faster, and a sympathetic environment was probably even more crucial, but the subject himself was not the fascinated, charmed or horrified spectator that watches the developments, or lack of developments, of adolescence.

Psychologists generally believe that one of the tasks of adolescence is developing a **body image**, as a person's concept of his or her physical appearance is called.

In the process of acquiring this body image, most adolescents refer to the cultural body ideal—which in our society, currently, is the slim, shapely woman and the tall, muscular man promoted by movies, television, and advertising. Obviously, few people measure up to the standards of physique set by TV commercials for diet sodas or low-fat yogurt, or by the various cheesecake and beefcake stars of television soaps and series. But whereas most adults have learned to accept this discrepancy between the cultural ideal and their own appearance, few adolescents are satisfied with their looks. One nationwide study of young people between 12 and 17 years old found that 49.8 percent of the boys surveyed wanted to be taller, and 48.4 percent of the girls wanted to be thinner. A scant majority of the boys (55 percent) were reasonably satisfied with their build, as were a minority of the girls (40 percent) (Scanlan, 1975).

Other studies show that girls are dissatisfied with more parts of their bodies than boys are. Both sexes, for example, are likely to be concerned about facial features, complexion, and weight, and boys worry about the size of their penis in much the same way girls worry about the size of their breasts. But girls are much more likely to worry about the size and shape of their buttocks, legs, knees, and feet as well (Clifford, 1971).

The intense interest adolescents have in their appearance was shown in a study in which teenagers and adults were blindfolded and asked to identify an object—an upside-down mask of their own face—by touch. The younger adolescents were quicker to recognize the object than the older adolescents and adults were, despite the fact that facial shape changes most rapidly in early adolescence, theoretically making the task harder for the younger subjects. Apparently, the young teenagers' preoccupation with their physiognomy compensated for the unfamiliarity of the stimulus (Collins and LaGanza, 1982).

The adolescent's concern over appearance is more than vanity—it is a recognition of the role physical attractiveness plays in gaining the admiration of the opposite sex. Hass (1979) found that the characteristics that boys sought

---

average girl, while those who are generally inactive menstruate earlier (Frisch, 1983). (This is one explanation for the fact that blind girls, who are usually less active than sighted girls, normally have their first period earlier than sighted girls. It may also explain why menarche is more likely to occur in winter than in spring or summer.)

The consequences of good nutrition, and of better health generally, are reflected in what is called the **secular trend**—the tendency of each new generation over the past hundred years or so to experience puberty earlier than their parents. (A century ago, the average male began puberty at about age 15 or 16 and the average female began it around 15.) In contemporary North America and Europe, nutrition and medical care for most children are now sufficient to allow genes and gender, rather than health and nutrition, to determine the age at which puberty begins. As a result, the average age of puberty is the same today as it was ten years ago. Although most American adults reached puberty at an earlier age than their parents and grandparents, most of today's children will experience their growth spurt and develop their sexual characteristics at about the same age their parents did.

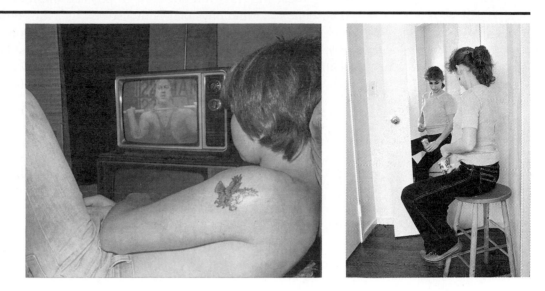

The clash between the cultural ideal and the reality of normal physical development is not easy to cope with for people of either sex, especially in adolescence, when fantasies are so powerful and self-criticism so unforgiving.

in girls were, in order, good looks, a good body, friendliness, and intelligence; girls wanted boys to be intelligent, good looking, have a good body, and be good conversationalists.

The fact that physique is valued by both sexes, and that adolescents can spend hours in front of the mirror, and many dollars on clothes and cosmetics to make themselves appear more attractive, is understandable, for there is a strong relationship between how adolescents feel about their bodies and how they feel about them-

selves. To the adolescent, looking "terrible," or feeling regarded by others as looking "terrible," is often the same as being terrible, a feeling that may be reinforced by the fact that teenagers who are physically unattractive tend to have fewer friends than the average teenager (Rutter, 1980). Consequently, adolescents' concern for their appearance should be an occasion for adults' sympathy rather than their derision. Providing them with practical help, from cosmetic suggestions for minor flaws, to tips on diet and clothes, to medical treatment for physical abnormalities, might have far-reaching benefits for self-concept.

The Stormy Decade?

Now that we have examined the nature of physical growth and related developments during puberty, we are better able to anticipate potential problems during adolescence.

Although most developmental researchers today do not believe that adolescence necessarily conforms to the popular image of an interlude of storm and stress, it is interesting to note that the source of that image is in the views held by most psychologists during most of the twentieth century. This view began to take shape in an influential two-volume treatise titled *Adolescence*, written in 1904 by G. Stanley Hall, the founder of American developmental psychology. (It was this work that introduced the term "adolescence"—coined by Hall from the Latin word for growth—into the English language.) Hall persuasively explained that erratic physical growth coincides with erratic emotional and moral development. According to

Hall, adolescence is a time of rebirth: physical maturation not only changes the adolescent's size and physiology but also changes the young person's way of seeing the world. Consequently, stated Hall, each new generation is capable of surpassing previous generations in moral and intellectual leadership, because the young are idealistic, altruistic, and self-sacrificing. In Hall's view, this "great revolution" is not easily achieved, however, for the young person's development

> often seems insufficient to enable the child to come to complete maturity . . . every step of the upward way is strewn with wreckage of body, mind, and morals. . . . Modern life is hard, and in many respects increasingly so, on youth. . . . Normal children often pass through stages of passionate cruelty, laziness, lying and thievery. [Hall, 1904]

Psychoanalytic Theory

Hall's views were particularly consistent with those of most psychoanalytic theorists, who believed that the stresses of adolescence are inescapable, since the adolescent's rapid sexual maturation and powerful sexual drives would inevitably conflict with the culture's traditional prohibitions against their free expression and their parents' reluctance to accept the maturation of their children. Teenage turbulence seemed only natural to psychoanalysts. Indeed, Freud's daughter Anna, a distinguished psychoanalyst, believed that we should be more worried about the psychological health of the adolescent who does not seem to be emotionally upset than about the one who does. As she explained it:

> We all know individual children who as late as the ages of 14, 15, or 16 show no such outer evidence of inner unrest. They remain, as they have been during the latency period, "good" children, wrapped up in their family relationships, considerate of their mothers, submissive to their fathers, in accord with the atmosphere, ideas, and ideals of their childhood background. Convenient as this may be, it signifies a delay of normal development, and is, as such, a sign to be taken seriously . . . These are children who have built up excessive defenses against their drive activities and are now crippled by the results, which act as barriers against a normal maturational process . . . They are, perhaps, more than any others, in need of therapeutic help to remove the inner restrictions and clear the path for normal development, however upsetting the latter may prove to be. [A. Freud, 1968]

A More Balanced View

The "storm and stress" version of adolescence is not without some support from recent research. Contemporary studies have found that a host of problems, from law-breaking to depression, do, in fact, occur more often in adolescence than earlier (Rutter, 1980). However, deep and sustained problems are the exception rather than the rule. Careful research on large samples of young people over the past two decades (Offer and Offer, 1975; Rutter et al., 1979; Thomas and Chess, 1980) has led to the conclusion that most adolescents, most of the time, are calm, predictable, and purposeful rather than storm-tossed, erratic, and "lost."

How, then, had psychologists arrived at such a firm view of adolescence as a time of trouble? One reason was that they had looked mostly at troubled adolescents. As Joseph Adelson (1979) explains:

Figure 14.7 *At almost any stage of life, the more troubled minority draws the most (often undue) attention. Think of the stereotypes of the "terrible twos," or the senile elderly, or teenage delinquents. Even in this photograph of normal young people, chances are your eyes fastened first, and lingered longest, on the two unusually heavy teenagers.*

Our concentration on atypical factions of the total body of the young—on addicted, delinquent, and disturbed youngsters, or on the ideologically volatile, or on males (far more impulsive and rebellious than females)—has led us to generalize from qualities found among a minority of the young to adolescents as a whole . . . If we examine the studies that have looked fairly closely at ordinary adolescents, we get an entirely different picture. Taken as a whole, adolescents are *not* in turmoil, *not* deeply disturbed, *not* at the mercy of their impulses, *not* resistant to parental values, *not* politically active, and *not* rebellious.

That said, however, we should also say that virtually no developing person in our culture has a carefree adolescence. In terms of physical development, the hardest years are the first ones, when the young person must adjust to having a body that, almost always, takes a form different from the one that the individual expected or wanted. Inevitably, teenagers find something wrong with themselves: they are either too tall or too short, too fat or too thin, big-footed or thin-lipped, or, as one apparently gorgeous 15-year-old complained, "my neck is too short, my eyes are too wide apart, and my littlest finger on both my hands is crooked."

Furthermore, while acceptance of primary and secondary sexual changes is probably easier today than it was a generation or two ago, moments of embarrassment, confusion, and worry are still part of the usual process.

A Life-Span Perspective

Early and Late Maturation

The young people who have the most difficult time of all adjusting to their sexual and physical development are those who must do so later or earlier than their classmates because their bodies happen to be on a somewhat different schedule (Rutter, 1980). Since girls generally mature before boys, the groups at the two chronological extremes of puberty—early-maturing girls and late-maturing boys—are most likely to feel out of step. As you might expect, however, the long-term consequences of being off-time in development are quite different for girls and boys.

Early-Maturing Girls

Girls who are taller and more developed than their classmates discover that they have no peers who share their interests or problems. Prepubescent girls call them "boy crazy," and boys tease them about their big feet or developing breasts. Almost every sixth-grade class has an 11-year-old pubescent girl who slouches so she won't look so tall, wears loose shirts so no one will notice her breasts, and buys her shoes a size too small. There are additional hazards for the early maturer. If she begins dating, it will probably be with boys who are older, and her self-esteem is likely to fall for a number of reasons (Simmons et al., 1983): she may feel constantly scrutinized by her parents, criticized by her girlfriends for not spending time with them, and pressured by her dates to be sexual.

Those who date early (by age 14) are likely to have sexual intercourse before high school is over (Miller et al., 1986). And those who have sexual intercourse early are less likely to use contraception (see Chapter 15), more likely to become pregnant, and, if they keep the baby, more likely to find their educational and career accomplishments deflected downward for at least a decade (Furstenberg, 1976).

Of course, this path is not inevitable. Social pressures and constraints are at least as important as biological ones in determining the age of dating (Dornbush et al., 1981), and many early-maturing girls are prevented from dating by their parents or by their own shyness. Furthermore, dating does not always lead to sexual activity and sexual activity does not necessarily lead to motherhood. Nevertheless, the link is there, and some young women find their adult lives restricted by a chain of events that began when they reached menarche at age 10 or 11 rather than later.

Most early-maturing girls weather their first years of adolescence without premature sexual experiences, however, and most soon find that they benefit more than they suffer from the difference between themselves and their peers. By seventh or eighth grade, early maturation bestows increased status, respect, and popularity (Faust, 1983). Initial problems often lead to more mature thought as well as appearance (Livson and Peskin, 1980). After a few years of awkwardness and embarrassment, the early-maturing girl is able to advise her less mature girlfriends

Figure 14.8 *Which of these sixth-graders has an easier life? Judging just by physical development, chances are the one on the right more frequently feels awkward with her classmates and finds that her parents try to restrict her activities just when she wants more freedom.*

about topics that they find increasingly important, such as bra sizes, dating behavior, menstrual cramps, and variations in kissing. She also becomes more comfortable with her body, and with saying yes or no to the opposite sex.

Further research suggests that, in the long run, girls who are late to mature have more problems than girls who are early. As one review explains:

> The stress ridden early-maturing girl in adulthood has become clearly a more coping, self-possessed and self-directed person than the later-maturing female in the cognitive and social as well as in emotional sectors. . . . By contrast, it is the late-maturing female, carefree and unchallenged in adolescence, who faces adversity maladaptively in adulthood. [Livson and Peskin, 1980]

Late-Maturing Boys

In contrast to temporarily troubled early-maturing girls or to the temporarily carefree late-maturing girls, late-maturing boys may have problems that last into adulthood. They must watch themselves be outdistanced, first by the girls in their class and then by most of the boys, and they are forced to endure the patronizing scorn of those who only recently were themselves immature. Extensive longitudinal data from the Berkeley Growth Study (Jones, 1957, 1965; Jones and Bayley, 1950; Mussen and Jones, 1957) found that in high school, late-maturing boys tend to be less poised, less relaxed, more restless, and more talkative than early-maturing boys, who were more often chosen as leaders. The late-maturing boys were more playful, more creative, and more flexible, qualities that are not usually admired by other adolescents. Another study found more serious problems associated with late maturation: a disproportionate number of late-maturers become juvenile delinquents (Wadsworth, 1979).

Figure 14.9 *Nature often forces different roles on early- and late-maturing boys of the same age. The early-maturing boy generally feels compelled to show his strength, with a cocky self-assurance that is hard to shake. The late-maturer, on the other hand, compensates, perhaps becoming a "brain," a trouble-maker, or the class clown.*

Follow-up studies of the Berkeley late-maturing boys when they were 33 years old, and again when they were 38, found that most of them had reached average or above-average height but that some of the personality patterns of their adolescence had persisted. Compared with early- or average-maturing boys, those who had matured late tended to be less controlled, less responsible, and less dominant, and they were less likely to hold positions of leadership in their jobs or in their social organizations. Some still had feelings of inferiority and rejection. However, in several positive characteristics they scored well. They were likely to have a better sense of humor and to be more egalitarian and more perceptive than their early-maturing peers. Especially in an era of greater liberation from traditional sex roles, later-maturing boys became more adaptive men (Livson and Peskin, 1980).

Of course, not all late-maturing boys feel inferior during adolescence. Some use their talents to become the class scholar, or musician, or comedian. In addition, late maturation today may be less difficult than it was decades ago; for example, because of the greater freedom allowed young girls, the late-maturing boy can now gain experience with the opposite sex by dating girls several years younger than he is. If he attends a school where most students are placed in classes according to interests and abilities, rather than age, he will be less likely to feel inferior and more likely to develop his creative skills.

School, Family, and Cultural Factors

Adolescents are affected not only by how their development meshes with that of others in their class, but also by how it compares with that of others in their school. For example, one study shows that tall, early-maturing sixth-grade girls score higher in measures of social competence when they are enrolled in a middle school where they are among similarly mature seventh- and eighth-graders than when they are the tallest, most mature girls in an elementary school that ends at sixth grade (Nottelmann and Welsh, 1986).

The family context is important as well. No matter what their timing, the physical events of puberty and the budding drives for sexual expression and freedom of action typically increase the distance between parents and their adolescent children, particularly between mothers and daughters. Early-maturing sons increase conflicts as well (Steinberg, 1987). When the family already has a pattern of dysfunction, or when stressful family events such as divorce coincide with puberty, then the adolescent is particularly likely to have difficulty adjusting to the normal changes of adolescence.

In many ways, cultural values can intensify, or lessen, the problems of early or late maturation (Rutter, 1980). The effects of early or late maturation are more apparent, for example, among lower-class adolescents, because physique and physical prowess tend to be more highly valued among lower-class teenagers than among middle- or upper-class teenagers. Correspondingly, alternative sources of status for the early or late maturer, such as academic achievement or vocational aspiration, are less valued among lower-class adolescents.

As always in development, the interaction between the three domains is more potent than changes in any one area. In the next chapter, we will look at the adolescent's ability to understand, and thus cope with, these changes.

SUMMARY

Puberty

1. The growth spurt—first in weight, then height, then strength—is the first obvious evidence of puberty, although some hormonal changes precede it. During the year of fastest growth, an average girl grows about 3½ inches (9 centimeters) and an average boy about 4 inches (10 centimeters).

2. Growth usually begins with the extremities and proceeds toward the torso. The head, lungs, heart, and digestive system also change in size and shape during this period.

3. During adolescence, more calories and vitamins are needed than at any other time of life. Unbalanced diets can prevent normal growth during this period, with anorexia nervosa the most serious, sometimes life-threatening problem.

4. During puberty, all the sex organs grow larger and the young person becomes sexually mature. Menarche in girls and ejaculation in boys are the events usually taken to indicate reproductive potential, although full fertility is reached years after these initial signs of maturation.

5. Most secondary sex characteristics—including changes in the breasts and voice and the development of pubic, facial, and body hair—appear in both sexes, although there are obvious differences in the typical development of males and females.

6. While the sequence of pubertal events is similar for most young people, the timing of the onset of puberty shows considerable variation. Normal young people experience their first bodily changes any time from 8 to 16. The individual's sex, genes, body type, and nutrition all affect the age at which an individual reaches puberty, with boys, thin children, and malnourished children typically reaching puberty later than their opposites.

The Stormy Decade?

7. Until recently, most psychologists believed that adolescence is inevitably a time of storm and stress, or emotional turbulence. Freudians thought that the sexual needs of the adolescent have to conflict with parental authority in order for the young person to develop normally.

8. However, recent research has shown that most adolescents, most of the time, are calm and predictable rather than turbulent and erratic. While only a minority experience a completely smooth adolescence, few are distressed most of the time.

9. While not all teenagers have a difficult adolescence, early-maturing girls and late-maturing boys are more likely to be distressed by their physical development or lack of it. This problem is temporary for most girls, but the lack of confidence of the typical late-maturing boy may continue into manhood.

KEY TERMS

puberty *(321)*	menarche *(326)*
estrogen *(321)*	ejaculation *(326)*
testosterone *(322)*	secondary sex
growth spurt *(322)*	characteristics *(327)*
anorexia nervosa *(325)*	body image *(330)*
primary sex	secular trend *(330)*
characteristics *(326)*	

KEY QUESTIONS

1. What are the main physical changes that occur during puberty?

2. What is the usual sequence of the physical changes that occur in puberty?

3. How can we predict when puberty will occur for a given individual?

4. What was Hall's view of adolescence?

5. What is the psychoanalytic view of adolescence?

6. How "stormy" is adolescence in North America today, according to recent research?

7. How can the age at which puberty occurs affect psychological development?

Adolescence: Cognitive Development

I see no hope for the future of our people if they are dependent on the frivolous youth of today, for certainly all youth are reckless beyond words. . . . When I was a boy, we were taught to be discreet and respectful of elders, but the present youth are exceedingly wise and impatient of restraint.

Hesiod (eighth century B.C.)

The biological changes of puberty are universal and, for the most part, visible, transforming children by giving them adult size, shape, and sexuality. However, another change that typically begins in adolescence is just as important: the intellectual maturation that makes adolescents much more skilled than children in the use of analysis and reason.

Adolescent Thought

Piaget was the first theorist to recognize what many psychologists now consider the distinguishing feature of adolescent thought—the capacity to think in terms of possibility rather than merely concrete reality (Inhelder and Piaget, 1958). As John Flavell (1985) explains, before adolescence, the child has "an earthbound, concrete, practical-minded sort of problem solving approach . . . and speculations about other possibilities . . . occur only with difficulty and as a last resort." The adolescent and the adult, by contrast, are more likely to approach problems "quite the other way around . . . reality is subordinated to possibility."

This means that, on the whole, adolescents are able to speculate, hypothesize, and fantasize much more readily and on a much grander scale than children, who are still tied to concrete operational thinking.

By the end of adolescence, many young people can understand and create general principles or formal rules to explain many aspects of human experience. For this reason, Piaget calls the last stage of cognitive development, attained at about age 15, **formal operational thought.** At this point the adolescent "begins to build systems or theories in the largest sense of the term" about literature, philosophy, morality, love, and the world of work (Inhelder and Piaget, 1958).

Development of Scientific Reasoning

One specific example of formal operational thought is the development of **scientific reasoning.** Inhelder and Piaget (1958) undertook a classic series of experiments that revealed how children between the ages of 5 and 15 reason about certain laws of physics by manipulating specific materials. Children were asked to put objects in a pail of water and explain why some sank and others floated; they were given different weights to hang on a string pendulum and asked to figure out which factors might affect the speed of the pendulum's swing (length of string, size of weight, height of release, force of release); they were asked to roll marbles down an incline onto a flat surface and estimate how far they would go. In all these experiments, Piaget and Inhelder found that reasoning abilities developed gradually in the years before adolescence, culminating at about age 14 with an understanding of the general principles involved.

For instance, the children in one experiment were asked to balance a balance scale with weights that could be hooked onto the scale's arms. This task was completely beyond the ability of most preoperational children (typically, a 4-year-old might put two weights on the same side of the scale and none on the other).

By age 7 (the usual age for the beginning of concrete operational thought), children realized that the scale could be balanced by putting the same amount of weight on both arms, but they didn't realize that the distance of the weights from the center of the scale is also an important factor.

By age 10 (near the end of the concrete operational stage), they were often able, through trial and error, to see that the farther from the fulcrum a given weight is, the more force it exerts (in one child's words, "At the end, it makes more weight"), and to find several correct combinations that would balance the scale. Note, however, that, although they had discovered the importance of the weights' distance from the fulcrum in their trial-and-error experimenting, they still had to try out different weights at different distances to get the scales to balance. They could not hypothesize a general principle to guide their placement of the weights.

Finally, at about age 13 or 14, some children hypothesized the general law that there is an inverse relationship between a weight's proximity to the fulcrum and the force it exerts. Thus, they correctly concluded that if the weight on one arm of the balance is three times as heavy as the weight on the other, it has to be a third as far from the center as the other weight in order for equilibrium to be achieved, and they were able to correctly predict which other combinations of weights and distances would achieve balance.

Logic in Other Domains

Piaget's basic description of the reasoning processes that are characteristic of formal operational thinking has been tested in domains other than that of the natural sciences and, again, confirmed. For example, Peel (1971) examined the development of historical reasoning ("Why were the stones of Stonehenge placed as they were?"); ecological judgment ("What causes and prevents soil erosion?"); and sociopolitical explanations ("Were the people of Italy to blame for the water damage to art masterpieces caused by the floods in Florence?"). In addressing such questions, children younger than 13 could usually be expected to provide simple, one-option answers (for example, "Yes, the people of Florence were to blame because they did not move the masterpieces," or "No, it wasn't their fault because the flood was an act of God"), whereas older adolescents were likely to see events as dynamic and consisting of many interrelated factors, and were constantly thinking of alternative possibilities.

True, False, or Impossible to Judge?   Another way to measure formal operational thinking is to look at children's ability to assess the inherent logic of various statements. In one series of experiments designed to do this, an investigator spread many solid-colored poker chips out on a table and asked adolescents and preadolescents to judge whether various statements about the chips were "true," "false," or "impossible to judge" (Osherson and Markman, 1974-1975).

For instance, the investigator hid a poker chip in his hand without letting the child see what color it was and he asked the child to judge the veracity of the statement "Either the chip is green or it is not green." Almost ever preadolescent replied that the statement was "impossible to judge" rather than saying that it was true. They also thought that the statement "The chip in my hand is green and it is not green" was impossible to judge rather than false.

When the investigator held a red chip so it could be seen and said "Either the chip in my hand is green or it is not yellow," only 15 percent of the preadolescent children correctly answered "true." Beginning at age 11, however, the number of children answering correctly increased. By age 15, about half were able to accurately evaluate the logic of these either/or statements (that is, one of these statements is true if only one of the clauses is true), demonstrating what Flavell (1985) calls **the game of thinking.** That is, they were able to suspend their knowledge of reality (such as knowing that the chip is red) and think playfully about the possibilities suggested by the statement itself. Flavell (1985) gives another example of the adolescent's ability to suspend the real and think about the possible. If an impoverished college student is offered $10 to argue in favor of the position that government should *never* give or lend money to impoverished college students, chances are, he or she can earn the money. By contrast, concrete operational children have great difficulty arguing against their personal beliefs and self-interest. The ability to divorce oneself from what one believes to be the case and argue from other premises makes adolescents much more interesting and adept as participants in intellectual "bull sessions" or as partners in debate. It also opens the possibility that they can modify their own position by perceiving the merits of someone else's.

Figure 15.1 *In an attempt to score a point for his debating team, this young man may be constructing an eloquent and logical argument for a point of view he may truly oppose. He is engaged in the game of thinking, a favorite pursuit of many adolescents but an incomprehensible puzzle for younger children.*

Piaget Reevaluated

Piaget's measures of formal operational thought have been replicated many times, with similar results. Not only Piaget's defenders but also his critics agree that "children with age become increasingly systematic in their exploration of scientific-type problems," with the adolescents notably more logical and systematic than preadolescents or younger children (Braine and Rumain, 1983).

However, many adolescents arrive at formal operational thinking later than Piaget predicted, if at all. In fact, many older adolescents, including college students, do poorly on standard tests of formal operational thought. Many adults likewise have difficulty (Neimark, 1975, 1982).

Piaget (1972) himself acknowledged that society and education are crucial factors in enabling an individual to attain formal operational thought. He believed that the maturation of brain and body that occurs at puberty makes these intellectual achievements possible, but certainly not inevitable. Without experiences such as an education, or social interactions that stress science, math, or logic, adults still think like concrete operational children.

Furthermore, formal operational thinking is more likely to be demonstrated in certain domains than in others, depending upon an individual's intellectual endowments, experiences, talents, and interests. The development of expertise in a particular domain, a characteristic of adulthood even more than of adolescence, is discussed in Chapter 21.

In general, however, as at earlier stages of cognitive development, there are many differences between potential and performance, as well as variations from domain to domain. These discrepancies are much more apparent during adolescence than Piaget's description of formal operational thinking would seem to suggest. While many adolescents and adults have the cognitive competence to think logically, they do not always do so, especially when thinking about themselves.

Figure 15.2 *For parents, one of the most exasperating qualities of adolescents is that they can reason very well in academic subjects but nevertheless may be very illogical about their own lives. The precise, step-by-step thinking that characterizes, say, carrying out a science experiment sometimes seems to be of an entirely different wavelength than that which occurs in establishing personal priorities.*

A CLOSER LOOK   Implications for Education

As adolescents' minds mature, the education that schools, teachers, and parents provide changes accordingly. During elementary school, the science curriculum centers on practical, visible experiences, suited to the thought patterns of the concrete operational child. An important part of elementary-school science is caring for an animal, taking nature walks, or tracing the changes in trees from fall to spring. In junior high, simple experiments such as dissecting a frog, connecting an electric circuit, or turning water into steam become possible.

Because of the young person's growing ability to think abstractly in high school and college, the study of science at these levels can include, among other things, explanation and discussion of the cell theory, calculations about the possible movements of the atom, and the working out of chemical formulas as well as observations of visible transformations. Finally, the very idea that the theories of science are indeed theories, not established facts, is a concept that older adolescents can understand (Lovell and Shayer, 1978).

A similar shift occurs in other areas of study. High-school English no longer need center on the rules of grammar and spelling and the mechanics of reading and writing. Instead, as hypothetical thinking expands, writing becomes more creative and imaginative. Further, as students become less bound to concrete, literal thinking, metaphor, irony, and sarcasm become easier to understand, appreciate, and use. In fact, sarcasm, which is beyond most elementary-school children (Ackerman, 1978), sometimes seems to be the adolescent's favorite mode of communication.

In the social sciences, history becomes the study not only of what was but also of what might have been, and the analysis of current events calls up opinions as well as facts. Anthropology is no longer merely an account of the strange customs of strange people, but a study of the similarities and differences of all human societies, including our own. Psychology also becomes fascinating during adolescence. It is important to remember, however, that adolescents sometimes lose sight of reality in their search for possibilities. Many young people misapply their newly acquired psychological learning to themselves and their families, deciding that they are abnormal and that their parents are to blame.

Further, as Cowan (1978) points out, since a majority of high-school students have not fully reached formal operational thought, it would be a mistake to assume that hypothetical, logical instruction should entirely replace the use of concrete examples and personal experiences.

As scholastic achievement becomes less a matter of rote learning and more a matter of the application of logic and skill, adolescents often take greater pride in their work, thus boosting their self-esteem and motivation.

In fact, a recent survey of the academic achievements of high-school students in the United States found that, while most have learned the basic skills, they have trouble applying what they know. They can read a passage, understanding the language per se, but they cannot draw implications from what they read. Similarly, they know that a square has four equal sides, and that to find the area of a rectangle one must multiply length times width—but less than half of a cross-section of 17-year-olds could find the area of a square given the length of one side (Sizer, 1985). Thus, while they have the potential to use formal operational thought, many of them do not do so, perhaps because English and math are often taught by rule or rote rather than by reasoning.

Clearly, teaching students to think logically and critically about what they know goes hand in hand with helping them to master particular academic skills and to learn specific facts (Kuhn, 1979). During adolescence, as well as later in life, the interplay of the specific and the general, the individual and the universal, and the inductive and the deductive modes of thought is at the heart of most academic, as well as personal, knowledge (Sizer, 1985).

Adolescent Egocentrism

Thinking rationally about oneself is not easy at any age (Nisbett and Ross, 1980). In fact, in some ways, adolescents and adults are more likely to think irrationally than children are, because their increasing sophistication and mental agility allow them to examine all the possibilities, and thus choose versions of reality that suit their needs or confirm their suspicions (Ross, 1981).

Adolescents, in particular, often have difficulty thinking rationally about their immediate experiences. Their thought patterns tend to be flawed by a characteristic called **adolescent egocentrism** (Elkind, 1978). While they are long past the global egocentrism of the preschool child, adolescents tend to see themselves as much more central and significant on the social stage than they actually are. For example, the particular limits of adolescent judgment and logic often lead young people to believe that no one else has ever had the particular emotional experiences they themselves are having—that no one else, for example, has ever felt so angry, or so elated, or so misunderstood.

As David Elkind (1978) explains, this form of egocentrism occurs because adolescents fail

> to differentiate between the unique and the universal. A young woman who falls in love for the first time is enraptured with the experience, which is entirely new and thrilling. But she fails to differentiate between what is new and thrilling to herself and what is new and thrilling to humankind. It is not surprising therefore, that this young lady says to her mother, "But Mother, you don't know how it feels to be in love."

As part of their egocentrism, adolescents often create for themselves an **imaginary audience**, as they fantasize how others will react to their appearance and behavior. For instance, adolescents are so preoccupied with their physical appearance, sometimes spending hours in front of a mirror, that they assume that everyone else judges the final result. Anticipation of a favorable judgment can cause

Figure 15.3 *Nearly all teenagers suffer chronic anxiety about their complexions, even if they don't actually suffer acne itself. At times their thinking is so egocentric that a single blemish is enough to make them want to go into hiding, as though the whole world were waiting to condemn them for a pimple.*

teenagers to enter a crowded room with the air of regarding themselves as the most attractive and admired human beings alive. On the other hand, something as trivial as a slight facial blemish can make them wish that they could enter the room invisibly. Similarly, school phobia, a fear of school that often keeps the student from attending, is particularly likely in adolescence—and often centers more on worries about appearance (particularly on being viewed in the locker room) than about achievement (Rutter, 1980).

Generally, egocentrism peaks at about age 13 (Elkind and Bowen, 1979; Gray and Hudson, 1984; Pesce and Harding, 1986). It often takes several years before this egocentrism declines and the individual can walk into a crowded room without the look of one who thinks he or she owns the world or the lack of confidence of one who imagines disapproval in every gaze. Elkind explains, "Whenever the young adolescent is in public, he or she is—in his or her own mind—on stage playing before an interested critical audience. Some of the boorish behavior of young adolescents in public places has to be understood in [these] terms."

Figure 15.4 *Adolescent egocentrism sometimes makes teenagers seek public attention by almost any means at hand.*

Taking the imaginary audience into account in thinking and behavior can be taken as an indication that the young person is not at ease with the larger social world, thus helping to explain some interesting variations in the intensity of adolescent egocentrism and the power of the imaginary audience. According to a self-report questionnaire designed to measure these concepts, girls are generally more concerned with the hypothetical opinions of onlookers than boys are (Elkind and Bowen, 1979; Gray and Hudson, 1984), and, among middle-class teenagers, younger teens are more self-conscious than older ones. However, delinquent boys think more about the imaginary audience than do nondelinquents of either sex (Anolik, 1981). Further, concern about the imaginary audience, rather than decreasing, *increases* from early to late adolescence in at least two groups: rural, lower-class whites (Adams and Jones, 1981) and Native Americans (Markstrom and Mullis, 1986).

Fantasies and Fables

Egocentrism sometimes leads past the possible into the impossible (Elkind, 1974). One example is what might be called an **invincibility fable.** Many young people feel that they are somehow immune to the laws of mortality and probability: they take all kinds of risks, falsely secure in the notion that they will never get sick, or killed, or caught.

Another example is what Elkind (1974) calls the **personal fable** through which adolescents imagine their own lives as heroic or even mythical. They see themselves destined for great fame and fortune—discovering the cure for cancer or authoring a masterpiece. Piaget recognized the personal fable in the ambitions of one graduating high-school class (a dozen pupils) in a small Swiss town:

> One of them, who has since become a shopkeeper, astonished his friends with his literary doctrines and wrote a novel in secret. Another, who has since become the director of an insurance company, was interested, among other things, in the future of the theater and showed some close friends the first scene of the first act of a tragedy—and then got no further. A third, taken up with philosophy, dedicated himself to no less a task than the reconciliation of science and religion. We do not even have to enumerate the social and political reformers found on both right and left. There were only two members of the class who did not reveal any astounding life plans. Both were more or less crushed under strong "superegos" of parental origin, and we do not know what their secret daydreams might have been. [Inhelder and Piaget, 1958]

Thus, adolescent thought processes are usually a mixture of the abilities to imagine many logical possibilities and to deny reality when it interferes with hopes and fantasies.

Finally, a high degree of egocentrism is usually an indication that an adolescent has not yet mastered formal operational thought (Pesce and Harding, 1986), although, as you have learned, it is possible to think formally in one domain and not in another. By late adolescence, however, as young people become better able to reason logically, they gradually become more secure, more realistic, and more positive in their understanding of themselves (Denny and Thomas, 1986).

Figure 15.5 *Fantasies occur at every age, and are usually more often helpful than harmful. They can get a necessary task done with pleasure or vent negative emotions harmlessly.*

Moral Development

The development of formal operational thought is related to another important aspect of cognitive maturity: the development of moral reasoning. Once people can imagine alternative solutions to various problems in science, or logic, or social studies, they can begin to be able to apply the same types of mental processes to thinking about right and wrong.

In fact, young people from age 10 to 18 are more likely to progress in moral reasoning than people at any other stage in the life span (Colby et al., 1983; Damon, 1984). The likely reasons are that a variety of conditions converge. Cognitive development allows adolescents to think more abstractly; psychological maturation makes them question the moral dicta of their parents; social development exposes them to a variety of ethical values; and personal experiences compel them to make decisions on their own. As a result, adolescents gradually come to see moral questions more broadly, no longer considering only narrow personal interests and gradually looking at the values of their society and beyond. The best way to understand this progress is to look at the work of Lawrence Kohlberg.

A Life-Span Perspective

Kohlberg's Stages of Moral Development

Building on Piaget's theories and research, Lawrence Kohlberg (1963, 1981) studied moral reasoning by telling children, adolescents, and adults a set of hypothetical stories that pose ethical dilemmas. The most famous of these is the story of Heinz:

> A woman was near death from cancer. One drug might save her, a form of radium that a druggist in the same town had recently discovered. The druggist was charging $2,000, ten times what the drug cost him to make. The sick woman's husband, Heinz, went to everyone he knew to borrow the money, but he could only get together about half of what it cost. He told the druggist that his wife was dying and asked him to sell it cheaper or let him pay later. But the druggist said "no." The husband got desperate and broke into the man's store to steal the drug for his wife. Should the husband have done that? Why?

Kohlberg examined the responses to such dilemmas and found three levels of moral reasoning: preconventional, conventional, and postconventional—with two stages at each level.

I. **Preconventional** *Emphasis on avoiding punishments and getting rewards.*
   Stage 1 Might makes right (punishment and obedience orientation). At this stage the most important value is obedience to authority in order to avoid punishment.
   Stage 2 Look out for number one (instrumental and relativist orientation). Each person tries to take care of his or her own needs. The reason to be nice to other people is so they will be nice to you. In other words, you scratch my back and I'll scratch yours.

II. **Conventional** *Emphasis on social rules.*
   Stage 3 "Good girl" and "nice boy." Good behavior is considered behavior that pleases other people and wins their praise. Approval is more important than any specific reward.
   Stage 4 "Law and order." Right behavior means being a dutiful citizen and obeying the laws set down by those in power.

III. **Postconventional** *Emphasis on moral principles.*

Stage 5   Social contract. The rules of society exist for the benefit of all, and are established by mutual agreement. If the rules become destructive, or if one party doesn't live up to the agreement, the contract is no longer binding.

Stage 6   Universal ethical principles. General universal principles determine right and wrong. These values (such as "Do unto others as you would have others do unto you," or "Life is sacred") are established by individual reflection and meditation, and may contradict the egocentric or legal principles of earlier reasoning.

According to Kohlberg's longitudinal research, people advance up this moral hierarchy as they become more mature. For example, 10-year-old Tommy argued that Heinz should not steal the medicine because he could be put in jail, a stage-1 answer. Three years later, Tommy reasoned at stage 2, saying that Heinz should steal the medicine because he needed his wife to help care for him (Kohlberg, 1971).

At every age, *how* people reason morally, rather than what specific moral conclusions they reach, determines their stage of moral development. For example, someone whose moral reasoning is at stage 3 might argue either that the husband should steal the drug (because people will blame him for not saving his wife) or that he should not steal it (because he had already done everything he could legally and people would call him a thief if he stole).

According to Kohlberg, most 10-year-olds reason morally at stage 1 or 2, and many adults never reach stage 5 or 6 (Colby et al., 1983; Kohlberg and Elfenbein, 1975). In fact, in a recent reformulation of Kohlberg's scoring procedures for rating stages of moral judgment, stage 6 was dropped because so few people were in it (Colby et al., 1983). Kohlberg thinks that a person must be at least at the cognitive level of early formal operations (not usually reached until early adolescence) before he or she can reach stage 3, and that a certain amount of life experience and responsibility is a prerequisite for reaching stage 5. For these reasons, Kohlberg generally sees relatively little progress in moral development during middle childhood, and believes few adolescents can go beyond stage 4.

Note, however, that Kohlberg's stages are based on hypothetical situations. Other research, using test cases similar to those children typically experience, has found that children are more advanced in their moral thinking than Kohlberg believed them to be. For example, moral reasoning about how to play games such as jacks or marbles is much more advanced than reasoning about social dilemmas like those in Kohlberg's Heinz story. There are two reasons for this: not only do people reason better about issues in their own experience, they also reason better about issues they have discussed with each other. Many studies have shown that giving children and adolescents a chance to discuss moral issues and make their own moral choices helps them develop more complex ethical thinking (Rest, 1983; Turiel, 1974).

Further, Damon (1984) found that the principle of equality is fairly well understood by age 10, including not only the concept that each person deserves an equal share but also that past deprivations and future opportunities should be taken into account when determining what an equal share is.

Kohlberg and His Critics

Kohlberg's theory of moral development is a good example of the way theories develop in science. As you will remember from Chapter 2, theories help organize and clarify various observations and hypotheses, and they are made to be questioned and tested.

Originally, Kohlberg's ideas were the product of three sets of observations: Piaget's theory of cognitive development; various philosophers' systematic delineation of ethical behavior; and Kohlberg's own research on a group of eighty-four boys, ages 10, 13, and 16, who provided Kohlberg with his original empirical data on the development of moral thinking. From these three elements, Kohlberg created and validated his moral dilemmas, his stages of moral thinking, and his theory of moral development.

His theory attracted a great deal of attention, because many people had apparently been searching for a way to clarify and focus their concern about moral education and growth. However, with this attention came criticism on a number of counts:

1. Kohlberg's "universal" stages seemed, generally, to reflect liberal, Western values (Reid, 1984; Sullivan, 1977). (In some cultures, serving the needs of kin may have a higher moral priority than observing principles that presumably apply to all of humankind.)

2. His original moral-dilemma scheme was validated only on males but was applied to females as well (Gilligan, 1982). (It may be that females and males are socialized to approach moral questions in different ways).

3. His moral stages overemphasized rational thought and underrated religious faith (Lee, 1980; Wallwork, 1980). (Some people believe that divine revelation, rather than intellectual reasoning, provides the best standards for moral judgment.)

4. Kohlberg's scoring guidelines for his moral dilemmas allowed many people to score quite well simply because they were verbally fluent. (Some people are able to persuasively parrot moral positions without really believing or even understanding the reasoning behind them.)

5. Finally, Kohlberg assumed moral development is sequential when, in fact, some people appeared to progress and regress in their moral reasoning (Holstein, 1976; Kuhn, 1976).

Each of these criticisms has some validity. In fact, in response to some of them, Kohlberg's scoring system has been substantially revised to better distinguish moral reasoning from mere mouthing of moral clichés (Colby et al., 1983). (Many of the more verbal adolescents now score at lower stages than they would have under the old system, and no adolescents score at stage 6, the achievement of an under-

Figure 15.6 *Participation in a protest demonstration may be sparked by many reasons, some social, some personal, some superficial. However, for many older adolescents, involvement in political arguments is a response to their greater awareness of moral issues.*

standing of universal principles, a level of reasoning attained only by extraordinary individuals, such as Mahatma Ghandi and Martin Luther King, Jr.) In cross-cultural research, some cultural differences have been found, although not as many as the critics had thought would be found (Snarey et al., 1985). More significantly, longitudinal research using the new scoring system shows moral reasoning developing in a steady, gradual progression over the years, just as Kohlberg had described it (Colby et al., 1983; Snarey et al., 1985).

Sex Differences Now let us consider one criticism in detail—that Kohlberg's stages of development are biased against females. The most compelling and best-known expression of this position has come from Carol Gilligan (1982).

According to Gilligan, girls and women tend to see moral dilemmas differently than boys and men do. In general, the characteristic male approach seems to be "Do not interfere with the rights of others"; the female approach, on the other hand, seems to be "Be concerned with the needs of others." Females give greater consideration to the context of moral choices, focusing on the human relationships involved. Gilligan contends that women are reluctant to judge right and wrong in absolute terms because they are socialized to be nurturant, caring, and nonjudgmental.

As evidence, Gilligan cites the responses of two bright 11-year-olds, Jake and Amy, to the dilemma of Heinz, who must decide whether to steal drugs for his dying wife (page 347). Jake considered the dilemma "sort of like a math problem with humans," and he set up an equation that showed that life is more important than property. Amy, on the other hand, seemed to sidestep the issue, arguing that Heinz "really shouldn't steal the drug—but his wife shouldn't die either." She tried to find an alternative solution (a bank loan, perhaps) and then explained that stealing wouldn't be right because Heinz "might have to go to jail, and then his wife might get sicker again, and he couldn't get more of the drug."

While Amy's response may seem equally ethical, it would be scored lower than Jake's on Kohlberg's system. Gilligan argues that this is unfair, because what appears to be women's moral weakness—their hesitancy to take a definitive position based on abstract moral premises—is, in fact,

> inseparable from women's moral strength, an overriding concern with relationships and responsibilities. The reluctance to judge may itself be indicative of the care and concern that infuse the psychology of women's development. (Gilligan, 1982)

Of course, the difference between male and female moral thinking is not absolute. Gilligan is aware that some women think about moral dilemmas the way men do, and vice versa. Nor does she think that either way of reasoning is better than the other, or even in itself sufficient. If people stress human relationships too much, they may overlook the principles involved and may be unable to arrive at just decisions; if they stress abstract principles too much, they may blind themselves to the feelings and needs of the individuals affected by their decisions. The best moral thinking synthesizes both approaches (Gilligan, 1982; Murphy and Gilligan, 1980).

Gilligan contends that Kohlberg's scoring system tends to devalue the female perspective. However, an exhaustive review of sex differences in moral reasoning (Walker, 1984) finds that, although females may approach moral decisions somewhat differently than males—for example, when asked to describe real moral dilemmas in their lives, females are more likely to discuss interpersonal issues

Figure 15.7 *Arguments between male and female adolescents are sometimes a form of flirtation, a way to make emotional contact while avoiding physical intimacy. However, differences in perspective affect the ways in which men and women approach many questions—from how to spend Saturday night to what to hope for in the future.*

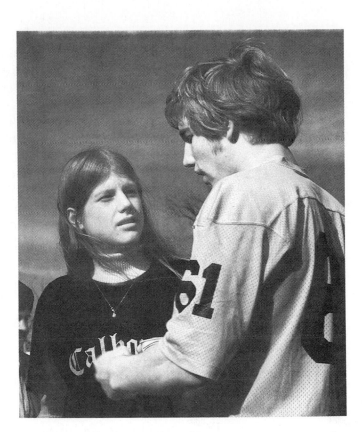

(Walker et al., 1987)—there is no evidence that these differences systematically affect the scores on Kohlberg's dilemmas. Most studies in which males and females are compared find no sex differences at all. Among those few studies that did find adolescent and adult males scoring higher than females, the women generally had less education than the men with whom they were compared, and the old scoring system was used, which tended to favor sophisticated verbal fluency—a product of education rather than gender. Overall, then, Kohlberg's general scheme seems to apply equally to both sexes. Although men and women may analyze dilemmas differently, neither sex is deemed more morally astute than the other.

Conclusion It seems clear that, on the whole, Kohlberg's theory has stood the challenge of criticism, and has proved to be valid and useful in a variety of contexts with people of various ages and backgrounds (Blasi, 1980; Rest, 1983; Walker, 1982).

Moral Behavior

The various findings of the research on moral development raise a crucial and very practical question. What is the relationship between moral thinking and moral behavior? A classic series of studies found that although most children can explain why honesty is right and cheating is wrong, most children cheat under certain circumstances, such as when their friends put pressure on them to do so, and when the chance of being caught is slim (Hartshorne et al., 1929). The same can be said of adolescents and adults, most of whom "bend" the rules when their own self-interest is at stake (Lickona, 1978). Obviously, then, the translation of the intellectual understanding of rules into moral behavior is far from automatic.

On the other hand, most studies have shown that moral thought can have a decided influence on action and vice versa (Rest, 1983). Beginning in middle childhood, children begin to apply their moral standards to their own behavior. They try to figure out what the "right" thing to do is, and feel guilty when they do the "wrong" thing.

This process continues through adolescence and adulthood, with increasing complexity, as Chapter 18 will explain. Now, however, let us look at some of the specific moral dilemmas of adolescence. As you will see, young people are faced with several conflicting sets of values, including their own self-interest, the morality of the peer group, the morality of their parents and teachers, and the principles of the culture (Liebert, 1984; Turiel and Smetana, 1984). It takes substantial maturity, more than many adolescents possess, to coordinate all these values.

Decision-Making: Two Contemporary Issues

Most of the consequences of egocentrism and illogical thought common in the early years of adolescence are benign: the young person who postpones studying in order to go to a party and then bewails a low grade, or who hangs up angrily on a best friend and then must make amends, or who buys a pair of pants that are too tight gradually learns from these experiences and is wiser for it. However, some actions can have serious consequences. One is the decision to leave high school before graduation, a decision made by about 15 percent of today's adolescents in the United States. They will soon discover the personal impact of another statistic: more than half of the American young-adult population who have not graduated from high school are either unemployed (18 percent) or have dropped out of the labor force altogether (34 percent), while most high-school graduates who are not in college are earning regular paychecks (76 percent) (U.S. Bureau of the Census, 1986).

Other disastrous decisions center on delinquency and suicide. More often than is the case with adults, adolescents tend to get caught up in the mood of the moment, affected by a group of friends or a newspaper account of suicide. These topics are discussed in Chapter 16.

Two other topics, however, are best discussed here, for adolescent thinking seems to be a particularly critical factor in adolescent behavior with regard to drug use and sexual activity.

Adolescence is typically a time when a person explores his or her sexual identity and experiments with alcohol and other drugs, often developing patterns and habits that may affect his or her entire life (Coates et al., 1982). To take one blatant and tragic example, alcohol is implicated in most of the fatal or maiming auto accidents involving drivers under the age of 20. To take another, close to 80 percent of the young females who become sexually active by age 14 will become pregnant at least once by the time they are 18.

While sexual behavior and drug use are obviously affected by a host of factors, one of the most important is the way adolescents think about sex and drugs—how they evaluate the various risks and understand the numerous moral questions. We will first look at the facts about adolescent behavior in these two areas, and then at the way adolescents think about them.

Sexual Behavior

Before evaluating the sexual behavior of today's adolescents, let us recognize that their attitudes toward sexuality are probably much healthier than those of previous generations. The decline of the double standard and the increase in sexual understanding mean that relatively few of today's young people will face the negative feelings about sexuality that pervaded our culture only two generations ago.

However, the behavior of today's adolescents leads to two serious problems that make many wonder if today's young people are allowed and even encouraged to have too much sexual experience too soon. Their thinking may not be as ready for sex as their bodies are.

Figure 15.8 *Media promotion of attractive adolescents as double images of innocence and sexuality is one factor affecting how teens think about themselves and their behavior and how they are viewed by adults.*

Sexually Transmitted Disease   The first of these problems is **sexually transmitted disease (STD),** a name that actually covers all the diseases that are spread by sexual contact. With the discovery of antibiotics at midcentury, many physicians thought that the problem of STD (or VD, as it then was called) could be cured once and for all. In fact, the frequency of STD did decline during the 1950s, as infected people were cured with penicillin before they could spread the disease. However, in the past twenty years, the incidence of STD has taken a marked upswing, especially among people under age 25, who now account for three-fourths of all reported cases (Green and Horton, 1982).

The seriousness of this for the future of the young person should not be underestimated: repeated STD is a leading cause of infertility. And, in the case of AIDS (acquired immune deficiency syndrome), even one exposure to this sexually transmitted virus can lead to illness and death.

Adolescent Pregnancy   The second serious consequence is adolescent pregnancy, a particularly common problem in the United States. In this country, about a million teenagers become pregnant each year, which means that almost half of all teenage girls become pregnant at least once before they are 20. Many of these pregnancies are aborted, about 10 percent spontaneously and about 40 percent by induction. Even so, close to half a million teenagers became mothers in 1986.

Figure 15.9 *High-school graduation is traditionally a family celebration, but usually for the graduates and their parents rather than the graduates and their children. This young mother is fortunate, however. Most adolescent mothers quit high school instead of graduating with their class.*

Following a marked increase in the rate of teenage pregnancy in the 1960s, the birth rate among teenagers has been quite stable over the past 15 years (U.S. Department of the Census, 1985). However, the proportion of babies born to unwed mothers has been steadily increasing (see Table 15.1); the current rate of unwed adolescent pregnancy is higher in the United States than in any other major country in the world. A disproportionate number of unwed mothers are black, but the rate of out-of-wedlock births among black teenagers has fallen over the past twenty years. Among white teenagers, however, the rate has more than doubled.

TABLE 15.1 **Teenage Mothers in the United States**

	Early 1960s	Early 1980s
Percent married at conception	*60%*	*28%*
Percent married at birth (but not at conception)	*22%*	*24%*
Percent unmarried at birth	*18%*	*48%*

Possible Explanations   Given the serious risks involved, what can account for the high rate of STD and pregnancy during adolescence? At first, it might seem that the cause is simply increased sexual activity among teenagers. However, increased sexual activity is only part of the explanation, since the overall rate of STD and unwanted pregnancy for sexually active unmarried adults is much lower than for adolescents.

Another explanation commonly offered is that teenagers are poorly informed about sexual matters, or that clinics and doctors are unavailable to them. While this is certainly true in some cases, and while the quality of information and care offered to adolescents can certainly be improved, at this time, most schools provide sex education, and most sexually active teenagers can find clinics that will help them. As a leading researcher explains, "though far from perfect, the system [of education and medical helpl is in place to enable adolescents to control their fertility if that is their intention'' (Dryfoos, 1984).

The crux of the problem seems to be that adolescents are not very logical in their thinking about their sexual activity. As Flavell (1985) explains, many young adolescents have difficulty understanding logical hypothetical arguments. Thus, the questions ''What if your partner has STD?'' or ''What if you become pregnant?'' are not answered in rational, personal ways that would lead the young man to use a condom, or the young woman to take effective precautions, or both of them to avoid sexual intercourse until they are ready to deal with the possible consequences.

In fact, adolescents are less likely to use contraception than adults are (Bachrach, 1984). Boys typically regard pregnancy to be primarily the girl's problem. In a large sample of sexually active boys between the ages of 11 and 19, only a third said that they would be ''very upset'' if they got a girl pregnant, and only 46 percent said they would avoid intercourse if neither they nor their partners were using contraception. This number is especially low when it is considered that a fourth of those who claimed to use contraception were actually using withdrawal as their only form of birth control. The ignorance of this group on certain matters was striking. Half of the boys mistakenly believed that they needed their parents' permission to buy condoms or contraceptive foam at a drugstore; and less than half rated any contraceptive method as ''very good'' at preventing pregnancy (Clark et al., 1984). Girls are somewhat more informed and careful: about one in three sexually active teenage girls uses the most effective contraceptive methods (the pill or an IUD); another one in three uses less effective methods (such as insisting on withdrawal); and the other one-third use no method at all (Dryfoos, 1982). Not suprisingly, 36 percent of all sexually active teenage girls become pregnant within two years of first intercourse. Those under 15 have the highest rate—41 percent (Koenig and Zelnik, 1982).

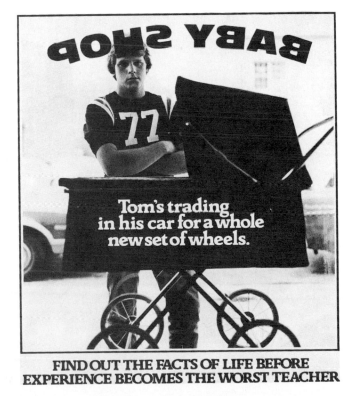

Figure 15.10 *Finding that many boys believe birth control is the exclusive concern of women and that an unintended pregnancy is a girl's responsibility, Planned Parenthood of Central Ohio began a campaign to change young males' attitudes. Prior to the campaign, the Planned Parenthood Hotline was averaging between 200 and 300 calls a month, only about 10 or so of them from boys. After a year of public service promotions that included TV spots like the one shown here, the number of calls had risen to about 800 a month, approximately 100 of them from boys.*

Further evidence of the adolescent's difficulty in thinking logically is shown when a pregnancy occurs. Often, girls delay facing up to the fact that they may be pregnant, which means that if they decide to have an abortion, they tend to have it relatively late, and therefore at greater risk. If they decide to have the baby, this delay means the loss of early prenatal care, and results in a rate of birth complications twice that among adult women. (Teenage girls who do get good early prenatal care have no greater rate of birth complications than adult females do [Makinson, 1985].)

Adolescent egocentrism in the form of the invincibility fable may also play a role. As one pregnancy counselor put it, "The biggest thing is that they just don't think it's going to happen to them. As my mother would say 'They don't believe fat meat is greasy.' They feel they are invincible, and they are risk takers" (Shipp, 1985).

Adolescent Drug Use

Now let us examine the scope of adolescent drug use. Researchers have studied the prevalence of drug use for the past ten years by having a large nationwide cross-section of high-school seniors fill out a detailed questionnaire (Johnston et al., 1986). The most recent results, as you can see in the chart below, show that by the time young people reach their senior year, almost all have used alcohol and a majority have tried tobacco and marijuana. With the exception of heroin, which has become decidedly unpopular among young people (Public Health Services, 1985), between one in five and one in eight have used the other illegal drugs. These rates actually underestimate the drug use of adolescents, for school drop-outs generally have higher use rates, and they are not included in this sample.

Drug use among adolescents varies from cohort to cohort, and even from year to year, with use in the mid-1980s somewhat lower than in the late 1970s. For example, daily marijuana use by high-school seniors almost doubled between 1975 and 1978, then declined each year thereafter, so that the 1983 rate of 5.5 percent was even lower than that of 1975. Use of tobacco and abuse of pills and hallucinogens has also

Figure 15.11 *Prevalence and recency of the use of eleven types of drugs, Class of 1985.*

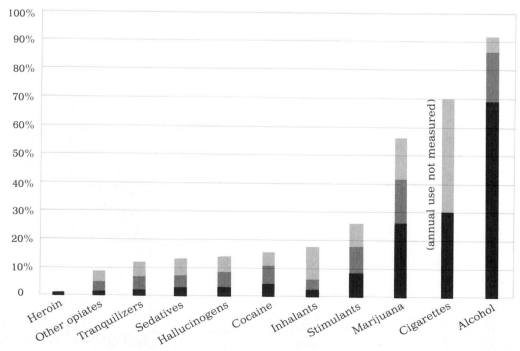

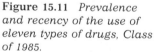 Used drug, but not in past year

Used in past year, but not in past month

Used in past month (30-day prevalence)

declined, most notably with regard to PCP (Angel Dust), the use of which rose and plummeted over a five-year period from 1978 to 1983. (PCP use may be on the increase, however, among school drop-outs [Public Health Services, 1985].)

These encouraging statistics are counterbalanced by two discouraging findings. First, adolescents in the 1980s are generally reporting first use of various drugs at younger ages than in the 1970s (Johnston et al., 1985). Second, alcohol and cocaine use are increasing. More high-school seniors are using alcohol in the 1980s than in the 1970s, and twice as many used cocaine in 1980 as in 1975, with the 1984 rate of cocaine use showing a continued, although more gradual, increase (Johnston et al., 1985).

The specifics on cocaine use are particularly troubling. In general, drug-use rates do not vary a great deal from one part of the country to the other. Cocaine, however, does vary from region to region, with one in every four seniors in the more affluent Far West reporting having used it, compared with one in five in the Northeast and one in eight from the South or North Central states. Further, college-bound seniors are more likely than other young people to have tried cocaine, which is contrary to the general finding that college-bound youths are less than half as likely to use drugs as youths who are not headed for college (Johnston et al., 1985). Finally, while most adolescents consider tobacco, alcohol, and marijuana "easy to get," less than half rate cocaine so. The implications of all this are that, since cocaine is now an expensive, high-status drug that is not as readily available as others, the use of cocaine among adolescents will increase markedly as it becomes more accessible, especially in the less expensive, extremely addictive and dangerous form of "crack."

Further, since wealthier and more intellectual youths are using cocaine, it is clear that the facts about the destructiveness of cocaine are not reaching even those presumably in a position to get the word first. As Lloyd Johnston and his colleagues (1985) point out, part of the problem is that the negative effects of cocaine take longer to be apparent to the user. In the words of one ex-addict:

> At first it's such a wonderful seduction, getting high, doing a few lines; no big deal. Somewhere down the path, maybe a few months or a year, the seduction turns into a nightmare. . . . Paranoia, depression, and thoughts of suicide begin to follow you around like your own shadow. [Miller, 1985]

Figure 15.12 *We do not know what is actually in this pipe. We do know, however, that use of the drugs for which it is designed, including pot, hashish, and crack, puts young people at increased risk for early sex, early school-leaving, and even early and violent death.*

Why Do Adolescents Use Drugs?   Adolescents certainly do not need to be told that drugs are harmful to them. For most of them, this message has come from their homes and schools since they were children. For instance, 97 percent of all adolescents report that their parents would be strongly opposed to their regular use of marijuana or daily use of alcohol, and a majority report parental opposition to even one-time use of most drugs. Given the known and suspected dangers associated with drugs, why do adolescents use them?

Given adolescents' inclination toward the imaginary audience and personal fable, some drug exploration is almost inevitable (Elkind, 1981). Indeed, two tenets of adolescent egocentrism—"I can handle anything" and "Adults don't understand my experiences"—are especially misleading for the adolescent trying to think about drugs. For instance, the cognitive confusion characteristic of regular marijuana use or the slowed reaction time after even one or two alcoholic drinks are particularly difficult to spot if one believes one is invincible. A further explanation, of course, is that adolescents observe adults in their community and conclude, quite logically, that drug use is one of the differentiations between children and adults. Since adult status is the eventual goal of adolescence, markers of that status are particularly appealing, and, by late adolescence, alcohol use is considered more a mark of social acceptance than of deviance (Brown et al., 1986).

The faultiness of adolescent egocentric reasoning about drug use is immediately apparent when children's and adolescents' attitudes about smoking are compared. All have heard the message about the health consequences of smoking many times. Typical school-age children take it quite literally, perhaps breaking their father's cigarettes in half and refusing to kiss their mother if she has been smoking. They nag "You're killing yourself," or cry because they don't want their parents to die. The adolescent who takes up smoking, however, believes that he or she will never have cancer or heart disease, and rather than attacking the parent who smokes, smugly argues that the parent's behavior means that the parent has no right to give advice about any drug-related matter.

15.13 Despite explicit cancer warnings and drunk-driver fatality statistics, tobacco and alcohol maintain their powerful allure for the adolescent. Especially when adolescent egocentrism prevails, the risk of future danger is outweighed by the pleasures of the moment, like feeling "grown up" in the company of friends.

Helping Adolescents Find Answers

What can be done to help adolescents avoid problems with drugs or sex? Given what we know about adolescent cognition, part of the answer may simply be to do whatever might encourage adolescents to postpone drug experimentation and sexual exploration until they are more mature. While measures such as raising the drinking age, closing student smoking lounges, and ensuring the supervision of young teens' parties will obviously not stop drug use and sexual experimentation among adolescents, they may slow the rate of occurrence. Also, role-playing of how to say no is useful for many adolescents who have trouble handling the social pressures to try a drug or to be sexually active.

With regard to drugs, another approach involves the testimony of peers who have given up their drug habits. Research finds that one of the best ways to help adolescents keep from smoking is to have peers who once smoked and quit tell their contemporaries what the immediate problems of smoking are (Evans and Raines, 1982). In typical adolescent fashion, teenagers seem to take the problems of yellow teeth and shortness and badness of breath more seriously than the possibility of eventual cancer or heart disease.

Further, since adolescents model their behavior after that of the adults they know best, it is probably especially important for parents and teachers of young adolescents to model responsible drug behavior. According to a number of studies, there is a significant positive correlation between parents' use of drugs (mainly tobacco, alcohol, tranquilizers, and diet pills) and their adolescents' use (primarily alcohol and illicit drugs) (Feldman and Rosenkrantz, 1977; Kandel, 1974). The same principles can be applied to responsible sexuality. Those young people who are most likely to become involved in early sexual expression are those whose parents are single and relatively young, especially if the parents became parents themselves at a young age.

Interestingly, responsible drug use and responsible sexuality often go hand in hand. The reverse is also true. Many, perhaps most, pregnant teenagers became involved in using drugs before they became sexually active, and drug use (including alcohol) is a frequent accessory to sex without contraception. This link between drugs and sex is even more apparent for white girls than for black girls (Cordes, 1986), partly because white adolescents are much more likely to use alcohol or illicit drugs than are black teenagers (Barnes and Welte, 1986).

Figure 15.14 *The SADD drinking-driver contract is becoming increasingly popular. Such an agreement helps to diminish parents' fears about their child's safety, and it allays adolescents' fear of a bawling out in front of friends—a fear that might otherwise make them decide to try to drive themselves home. The fact that parents make a reciprocal agreement with their children is important, because they are acknowledging their vincibility and indicating their willingness to admit when they themselves have had more than they can handle.*

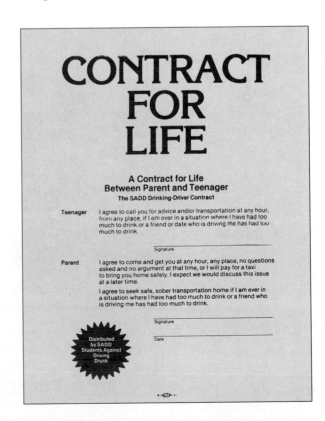

In addition, despite health education in the schools, many young people are not sufficiently knowledgeable about sex (including the risks and benefits of contraception, pregnancy, and parenthood) or about the effects and dangers of drugs (including legal as well as illegal substances). Since many young people start thinking about drugs and sex and begin to experiment even before they are officially teenagers, such facts need to be provided in junior high, and perhaps even grade school, as well as high school.

However, the crucial factor in aiding adolescent cognition does not seem to be providing the facts, but helping the development of the reasoning processes that will allow adolescents to apply those facts to their own lives. Let us look again at the research on moral development, for it provides a clue about helping adolescents develop values.

Almost all developmentalists agree that adolescence, especially late adolescence, is often marked by changing values and standards. Classroom moral debates and late-night bull sessions become lively and compelling, and ethical stands passionately taken one year might be reversed by new reasoning the next (Colby et al., 1983; Damon, 1984). The solutions that young people find to social and moral dilemmas, whether actual or hypothetical, depend not only on their cognitive maturity but also on the dialogues they are able to establish with others. The adolescent who has ample opportunity to state his or her views, and to listen to the views of others, will stand a better chance of developing codes of behavior that are increasingly responsible and comprehensive. Such internal guidelines will continue to be important, especially in young adulthood when risk-taking and violent behavior can take a heavy toll, as we will see in Chapter 17.

This suggests that when adults are confronted with social or moral questions by adolescents, their best long-range strategy is to listen and debate, rather than slamming the door on open discussion or laying down the law. Given adolescent egocentrism, and the climate of our times, the more thinking adolescents can do about social and moral issues before they are confronted with the results of their behavior, the better.

Figure 15.15 *Typically, when teenagers want to explain, parents want to advise, warn, and criticize. This parental response is understandable, but unless parents listen, laugh, and appreciate more often than air their own views, they may find that crucial channels of communication are blocked when help is really needed.*

SUMMARY

Adolescent Thought

1. During adolescence, young people become better able to speculate, hypothesize, and fantasize, emphasizing possibility more than reality. Unlike the younger child, whose thought is tied to concrete operations, adolescents can build formal systems and general theories that transcend, and sometimes ignore, practical experience. Their reasoning is formal and abstract, rather than empirical and concrete.

2. The ability to think logically is the hallmark of formal operational thought. Between the ages of 12 and 15, many young people become able to articulate scientific principles when given Piagetian tests of formal operational thought. They are also much more able to follow logical arguments and reason about social problems.

3. Many adolescents and adults never attain formal operational thinking, as measured by Piaget's tests. However, according to Flavell, unlike children, they are able to understand general laws and logical arguments when presented with them.

Adolescent Egocentrism

4. Another characteristic of adolescent thought is a particular form of egocentrism that leads young people to overestimate their significance to others. This characteristic is sometimes expressed in a personal fable about the grand and glorious deeds they will perform in adulthood.

Moral Development

5. Moral reasoning also becomes more complex during adolescence, for the young person who can grasp general laws of physics or principles of logic is more likely to articulate moral laws and ethical principles.

6. Most adolescents question traditional customs and laws, although none achieve an understanding of universal moral principles, the last of Kohlberg's six stages. However, people exhibit more advanced behavior when tests of moral reasoning are based on familiar experiences.

7. Kohlberg's critics suggest that his theory reflects particular cultural values and sexual socialization rather than universal stages. However, Kohlberg's theory seems to be generally valid, and, although males and females may analyze problems differently, neither sex is deemed to be better at solving moral dilemmas.

Decision-Making: Two Contemporary Issues

8. Cognitive immaturity makes it difficult for adolescents to arrive at rational decisions about sexuality and drug use. Adolescents who believe that they are above many of the normal problems that humans experience may also believe that they will never be confronted with pregnancy or sexually transmitted diseases, even when they don't take precautions against them. They may also believe that their judgment is unimpeded (perhaps even enhanced) by the use of alcohol and drugs.

9. The best way to help adolescents avoid problems with sexual behavior and drug use may be to encourage them to postpone exploration in these areas until they are able to reason more maturely about them. Role models provided by adults and peers are influential, too.

KEY TERMS

formal operational thought *(339)*

scientific reasoning *(340)*

the game of thinking *(341)*

adolescent egocentrism *(344)*

imaginary audience *(344)*

invincibility fable *(346)*

personal fable *(346)*

sexually transmitted disease (STD) *(353)*

KEY QUESTIONS

1. What are some of the tests that determine whether or not a person has attained formal operational thought?

2. What improvements in the ability to think logically occur during adolescence?

3. What are some of the characteristics of adolescent egocentrism?

4. What stages of moral development are reached during adolescence?

5. What are some of the criticisms of Kohlberg's theory?

6. What are some of the explanations for the high rates of sexually transmitted disease and pregnancy in adolescence?

7. What are some of the factors that affect the teenager's decision to experiment with drugs?

8. What are some of the ways of helping adolescents to avoid problems with drugs and sex?

CHAPTER 16

Adolescence: Psychosocial Development

*In youth, we clothe ourselves with rainbows,
and go as brave as the zodiac.*

Emerson
"Fate"

If they give you ruled paper, write the other way.

Ray Bradbury

The physical changes of puberty begin the process of adolescence by transforming the child's body into an adult's, and the cognitive developments described in the preceding chapter enable the young person to begin to think logically. However, it is psychosocial development—such things as relating to parents with new independence, to friends with new intimacy, to society with new commitment, and to oneself with new understanding—that helps the young person eventually attain adult status and maturity. Taken as a whole, these aspects of psychosocial development can best be understood in terms of the adolescent's quest for identity, that is, for answers to a question that never arose in younger years: "Who am I?"

Identity

In the 1940s, Erikson became absorbed with the question of **identity,** the individual's attempt to define himself or herself as a unique person (Coles, 1970). For Erikson, the search for identity represents a basic human need, one which, in modern society, becomes as important as food, security, and sexual satisfaction. In the past four decades, Erikson has written extensively about the search for identity as the primary task and crisis of adolescence, a crisis in which the young person struggles to reconcile a quest for "a conscious sense of individual uniqueness" with "an unconscious striving for a continuity of experience . . . and a solidarity with a group's ideals" (Erikson, 1968). In other words, the young person seeks to establish himself or herself as a separate individual while at the same time maintaining some connection with the meaningful elements of the past and accepting the values of a group. In the process of "finding themselves," adolescents must establish a sexual, moral, political, religious, and vocational identity that is relatively stable, consistent, and mature. This identity ushers in adulthood, as it bridges the gap between the experiences of childhood and the personal goals, values, and decisions that permit each young person to take his or her place in society (Erikson, 1975).

Figure 16.1 *The quest for identity is not always easy for adolescents. Some get themselves into a precarious situation or begin moving in a direction before they really know where it leads. Others can't seem to get going at all. They may think that they should make some forward movement in their lives, but they are without the concrete motivation to do so.*

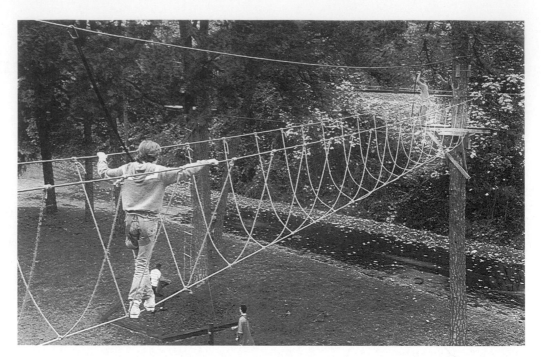

Identity Statuses

The ultimate goal, called **identity achievement,** occurs when adolescents achieve their new identity through "selective repudiation and mutual assimilation of childhood identifications" (Erikson, 1968). Thus, in optimal circumstances, the adolescent abandons some of the values and goals set by parents and society, while accepting others. With identity achievement, adolescents develop their own ideology and vocational goals.

For many young people, however, identity achievement is quite difficult, and even the process of accepting some parental values while rejecting others is problematic. The result often is **foreclosure,** or premature identity formation. In this case, the adolescent accepts earlier roles and parental values wholesale, never exploring alternatives or truly forging a unique personal identity. A typical example might be the young man who from childhood has thought he wanted to, or perhaps was pressured into wanting to, follow in his father's footsteps, as, say, a doctor. He might diligently study chemistry and biology in high school, take premed courses in college, and then perhaps discover in his third year of medical school (or at age 40, when his success as a surgeon seems hollow) that what he really wanted to be was a poet.

Other adolescents may find that the roles their parents and society expect them to fill are unattainable or unappealing, yet be unable to find alternative roles that are truly their own. Adolescents in this position often take on a **negative identity,** that is, an identity that is the opposite of the one they are expected to adopt. The child of a college professor, for instance, might fail high-school English and drop out of school, despite having aptitude scores that show the capacity to do college-level work. The child of devoutly religious parents might begin behaving in blatant opposition to his or her upbringing, stealing, taking drugs, and the like.

Other young people experience **identity diffusion:** they typically have few commitments to goals or values—whether those of parents, peers, or the larger society—and are often apathetic about trying to find an identity. These young people have difficulty meeting the usual demands of adolescence, such as completing school assignments, making friends, and thinking about the future. As one young man said:

> I should be getting out but I don't have the drive. Not motivated I guess. I want to move [slang for being more involved and active] too, you know. It kinda motivates me a little bit to see them [peers] going . . . I want to go too. It might motivate me for a little while. I do care but I just haven't got on it. [Quoted in Gottlieb, 1975]

Finally, in the process of finding a mature identity, many young people seem to declare a **moratorium,** a kind of time-out during which they experiment with alternative identities without trying to settle on any one. In some cases, a society may provide formal moratoriums through various of its institutions. In the United States, the most obvious example of an institutional moratorium is college, which usually requires young people to sample a variety of academic areas before concentrating in any one and forestalls pressure from parents and peers to choose a career and mate. Another institution that performs a similar function is the peacetime military, which makes it possible for many young men and women to travel, acquire valuable skills, and test themselves while delaying lifetime commitments.

Research on Identity Status Following Erikson's lead, many other developmentalists have found the concept of identity a useful one in understanding adolescence. Foremost among these is James Marcia who has defined the four major identity statuses (achievement, foreclosure, diffusion, and moratorium) in sufficiently precise terms that he and other investigators can interview an adolescent and determine his or her overall identity status (Marcia, 1966). Dozens of studies have compared adolescents' identity statuses with various measures of their cognitive or psychological development and have found that each identity status is typified by a number of distinct characteristics (see Table 16.1, next page). For example, each of the four identity statuses correlates with a somewhat different attitude toward parents: the diffused adolescent is withdrawn, perhaps deliberately avoiding parental

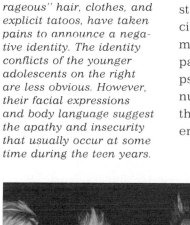

Figure 16.2 *There are a number of ways that young people may signal their involvement in some form of identity crisis. The group at the left, with their "outrageous" hair, clothes, and explicit tatoos, have taken pains to announce a negative identity. The identity conflicts of the younger adolescents on the right are less obvious. However, their facial expressions and body language suggest the apathy and insecurity that usually occur at some time during the teen years.*

TABLE 16.1 **Characteristics of the Various Identity Statuses**

	Foreclosure	Diffusion	Moratorium	Achievement
Anxiety	repression of anxiety	moderate	high	moderate
Attitude toward parents	loving and respectful	withdrawn	trying to distance self	loving and caring
Self-esteem	low (easily affected by others)	low	high	high
Ethnic identity	strong	medium	medium	strong
Prejudice	high	medium	medium	low
Moral stage	preconventional or conventional	preconventional or conventional	postconventional	postconventional
Dependence	very dependent	dependent	self-directed	self-directed
Cognitive style	impulsive	impulsive	reflective	reflective
Cognitive complexity	low	very high (confusion)	medium	medium
College	very satisfied	variable	most dissatisfied (likely to change major)	high grades
Relations with others	stereotyped	stereotyped or isolated	intimate	intimate

Adapted from research reviewed by Marcia, 1980

contact by sleeping or listening to music on headphones when the rest of the family is together; the moratorium adolescent is not withdrawn as much as independent, busy with his or her own interests; both the forecloser and the achiever are loving, but the forecloser evidences more respect and deference, while the achiever treats parents with more concern, behaving toward them as an equal or even as a caregiver rather than a care-receiver.

The table also shows some revealing combinations of statuses and traits. Note, for instance, that both adolescents who have achieved identity and those who have prematurely foreclosed their search for self-definition have a strong sense of ethnic identification, seeing themselves as proud to be Irish, Italian, Hispanic, or whatever. However, those who have foreclosed are relatively high in prejudice, while the identity achievers are relatively low, presumably because they are sufficiently secure in their ethnic background that they do not need to denigrate that of others.

This research, much of which is longitudinal, confirms that many adolescents go through a period of foreclosure or diffusion, and then a moratorium, before they finally achieve identity (Marcia, 1980). The process can take ten years or more (Meilman, 1979; Waterman, 1985). There is no doubt that the ease or difficulty of finding an identity is very much affected by the society, the immediate family, and the friends of the adolescent, topics we will discuss now.

Social Influences on Identity

Erikson was one of the first developmental psychologists to call attention to the role of the wider society, a role now emphasized by ecological and systems psychologists. In adolescence, societies provide an avenue to finding an identity primarily in two ways: by providing values that have stood the test of time and that continue

(a)

(b)

(c)

(d)

Figure 16.3 *In a society
like the United States, a
multiplicity of values and
constantly changing social
conditions make the paths
to identity formation varied and complex. Some
historical examples include
(a) young men enlisting in
the Civilian Conservation
Corps during the depres-
sion of the 1930s; (b) high-
schoolers out on the town
in the prosperous 1950s; (c)
students protesting during
the tumultuous 1960s; (d)
undergraduates on campus
in the conservative 1980s.*

to serve their function, and by providing social structures and customs that ease
the transition from childhood to adulthood. Whether these factors make the search
for identity easy or difficult depends primarily on the degree to which the members
of the society are agreed on basic values, and the degree to which the individual is
exposed to social change.

In a culture where virtually everyone holds the same religious, moral, political,
and sexual values, and social change is slow, identity is easy to achieve. The young
person simply accepts the only social roles and values that he or she has ever
known. In such cultures, the transition to the status of "adult" can be swift, often
designated by a ceremony known as a **rite of passage,** a ritual that fosters, focuses,
and celebrates the attainment of adult identity (see A Closer Look, next page).

A CLOSER LOOK   The Rite of Passage

A rite of passage is a ceremony or event that provides a transition from one social status or life stage to another. Weddings and funerals are two obvious examples. In many cultures, special rites of passage occur at adolescence to initiate the young person into adulthood, easing the transition from dependence on parents to adult independence. Anthropologists and sociologists have looked at adolescent rites of passage and drawn some interesting comparisons between those rites as they occur in traditional, non-Western cultures and in our own contemporary culture.

In many traditional, non-Western cultures, initiation ceremonies were both dramatic and painful, involving tests of bravery and strength, as well as separation from family and from members of the opposite sex. The young people learned religious rituals and social codes known only to adults and emerged from the process with a new status, new responsibilities, and often a new name.

For the most part, male initiation ceremonies were the more painful, often involving circumcision (cutting away the foreskin of the penis) and facial scarring. Typically, a group of boys would be separated from their families and initiated together, emphasizing their bond to their male contemporaries and their separation from their childhood home. The goal of the male initiation rite was to induct the boy as a member of the larger adult community, initially as a warrior.

Female initiation rites were usually intended to prepare the young woman for the more personal experiences of courtship and marriage. Typically, only one girl was initi-

ated during a given ceremony, and her family played a major role in arranging the event. Like the male ceremony, the female initiation rite was an important function, for it announced that a future bride and potential mother was available for match-making (Paige, 1983).

Our own society offers dozens of rites of passage. Some are religious, such as the Catholic confirmation or the Jewish Bar Mitzvah and Bat Mitzvah; some are social, such as debutante or sweet-sixteen parties; some confer legal sanction, such as registering to vote or taking the road test for a driver's license; and some are frowned upon, such as the initiation rites practiced by some street gangs, or, at the other end of the socioeconomic spectrum, the hazing that still goes on in some college fraternities. High-school graduation, preceded by a period of study and anxiety-provoking exams (in preparation for entry into college or the job market), often followed by dancing and drinking until dawn, is probably the most common American rite of passage (Dunham et al., 1986).

Given the complexity of our society, and the difficulty of attaining adult status, it is not surprising that our rites of passage are many and varied, and that some of their practices arise from the adolescent subculture without approval from the adult world, and that some groups of adolescents are more likely to experience them than others. Despite this variety, however, many observers believe that all these rituals further the establishment of identity by helping young people to leave childhood behind (Brown, 1975; Dunham et al., 1986).

Intense and painful non-Western initiation rites such as this circumcision ritual among Australian aborigines propel the individual instantly from the *role of child to full membership in the world of adult religious and social rights and responsibilities. Western rites of passage into adulthood tend to be* *cumulative, with ceremonies in early adolescence marking the child's religious maturity, as in the example of the Jewish Bar Mitzvah shown here. Social* *and legal recognition of adult status, such as the age at which one can marry or assume full responsibility for a bank loan, come later.*

Friends and Family

Given the complexity of forming an identity in our society, whether a particular young person will be able to declare a moratorium on identity formation until he or she is ready to make mature decisions, or whether identity will be premature or confused, depends a great deal on family and friends.

The Role of Peers

The socializing role of peers, which begins to emerge during the latter part of middle childhood, becomes quite prominent during adolescence.

Adolescents help each other in many ways, with identity formation, independence, and social skills. As John Coleman (1980) explains, although "the peer group has a continuous part to play in the socialization process during the whole span of school and college years . . . there are undoubtedly special factors operating during adolescence that elevate the peer group to a position of unusual prominence." Coleman finds three of these factors noteworthy:

1. The physical and social changes typical of adolescence cause the young person to confront new experiences and challenges to self-esteem. At such times, the peer group can function as a self-help group, a sounding board of contemporaries who may be going through the same sorts of struggles.

2. A crucial task in adolescence is questioning the validity of adult standards and authority. As a consequence, "at a time when uncertainty and self-doubt are greatest and support is most needed, many adolescents find themselves in an emotional position where it is difficult, if not impossible, to turn to their parents. Under such circumstances, it is hardly surprising that peers play an unusually important role."

3. Adolescents need to experiment, discovering which of their personality characteristics and possible behaviors will be accepted and admired. "This process of discovery, sometimes rewarding, sometimes painful and embarrassing, is dependent on the involvement of the peer group" (Coleman, 1980).

Figure 16.4 *These photos show an activity critical to healthy adolescent development—spending time with friends. Note the sex differences here. Girls typi-* *cally spend long hours simply talking to their best friends, whereas boys tend to hang out in larger groups. Further, girls tend to be more intimate with* *their friends, both physically and emotionally. Usually, the only physical intimacy acceptable among boys occurs in* *horseplay or in sports, in the form of the congratulatory slap on the buttocks or the bear hug that follows a winning move.*

Some of this peer-group involvement occurs casually, as a group of teenagers "hang out" together, seemingly doing nothing at the local gathering place, be it a park, a parking lot, the street corner, or a shopping mall. Most of it, however, happens through more private, self-revealing interactions with friends.

Friendship   As children become adolescents, their definition of friendship changes to emphasize shared intimacy rather than mutual activities. Further, the nature of the relationship becomes less "tit for tat" and more genuinely reciprocal, involving greater loyalty and helpfulness (Berndt, 1982).

The mutuality of peer friendships helps distinguish them from parent-child relationships (Youniss, 1980). Consider a study (Hunter, 1985) in which groups of adolescents rated how often they discussed various topics with their parents and with their peers, as well as the character of the conversations. As one might expect, most of the conversations about family relationships occurred with their parents, and most of their discussions about their peer relationships were with their friends. However, on academic, vocational, religious, ethical, and political topics, the young adolescents talked more with their parents, whereas older adolescents talked more about these topics with their peers.

More significant, no matter what the adolescent's age or what the issue, parents were reportedly more likely to offer and justify their own thoughts than to try to understand the adolescent's ideas. Peers, on the other hand, listened as much or more than they presented their own views. While having a parent's opinion may often be appropriate, and clarifying, at least about what the parent thinks, having a peer who listens sympathetically while a teenager gives a blow-by-blow account of a social interaction, or thinks out loud about a complex issue, helps the teen gain insight into his or her own ideas, values, and actions. In the process, self-definition, and hence identity, become strengthened. Not surprisingly, then, while school-age children are typically even closer to their parents than to friends of the same sex, as young people leave childhood and enter adolescence, they are much more likely to share personal information and find companionship with peers—of both sexes—than with parents or teachers (Buhrmester and Furman, 1987).

Peer Groups   Peers aid adolescents in their quest for identity in another way as well: they help them define who they are by helping them define who they are not. In our culture, every adolescent is exposed to not one peer group but many, each with distinct preferences in activities, dress, and music, and differing values about school and society (McClelland, 1982). As adolescents associate themselves with this or that subgroup (the jocks, the brains, or the punkers, for instance), they are rejecting others—and the particular self-definitions that would go with them.

Although much has been made of the influence of peer-group pressure in producing unwanted behavior, the pressure to conform to the peer group seems to be strong only for a short period, rising dramatically in early adolescence, until about age 14, and then declining (Coleman, 1980). In addition, the idea of peer-group pressure may be functional in some cases, helping to ease the transition for the young person who is trying to abandon childish modes of behavior, including dependence on parents, but who is not ready for full autonomy (Ausubel et al., 1977).

It is also important to realize that peer pressure can be positive as well as negative. For example, one study of 373 Wisconsin junior- and senior-high-school students found that peer pressure to study hard and get good grades was as apparent as pressure to dress appropriately, and that peers were more likely to discourage

Figure 16.5 *In the formation of peer groups, time tends to erase differences, either by constant conversations and interactions that tend to align opinions and tastes or by weeding out those who don't fit in. This group of elegantly attired Californians arriving at their prom probably no longer includes those who couldn't afford to keep up. Time will make a difference to the grinning American girl with her new-found friends in London (right). At this age, when peer influence is so important, if her visit lasts more than a month, her hair will probably change color, too.*

cigarette-smoking than to encourage it (Brown et al., 1986). In this study and in others, boys were discouraged more from smoking than girls were, which helps explain why today's teenage smoker is more likely to be female than male.

It seems, then, that the teenager who argues that he or she must dress a certain way, or wear a certain hair style, or engage in a particular activity because "everyone else does" feels partially relieved of the responsibility for some "personality" he or she is trying out. In a way, peers act as a buffer between the relatively dependent world of childhood and the relatively independent one of young adulthood. This buffering role may be especially important for minority-group adolescents who become increasingly aware of their position in the society as a whole, once they leave the protection of the family. For them,

> the peer group reduces the total load of frustration and stabilizes the entire transitional period. It can offer compensations not only for the deprivations associated with adolescence per se, but also for the special deprivations that confront certain adolescents by virtue of their class, ethnic, racial, or religious affiliations. [Ausubel et al., 1977]

Figure 16.6 *With hand gestures, body positions, and clothing style, these boys signal their membership in a tight-knit group. Fairly large groups of friends, such as this one, are more typical of adolescent boys than of girls, and seem to be more crucial for minority young people than for others.*

Boys and Girls Together   As you remember, voluntary sex segregation is the common practice among children during most of early and middle childhood. When children approach adolescence, however, this "segregation" begins to disappear, and the peer group plays an important role in the switch.

The specific process was first described over twenty years ago in Australia, when Dunphy (1963) analyzed the social structure of adolescent peer groups. By late childhood, groups are formed of several individuals of the same sex. Then a particular group of girls begins to associate with a particular group of boys, at first merely seeming to appear by chance in the same vicinity at the same time. Gradually, a larger heterosexual group forms from these two groups. The members of this group "hang out" together, with most of the interaction occurring within the large group or smaller subgroups, although occasionally individuals talk briefly with individuals of the opposite sex. Dating, when and if it occurs, is more often done with several couples together. In this context, even physical affection is a semipublic affair.

Figure 16.7 *Adolescents spend a lot of their time waiting—for school to start, for the movie to begin, for independence from their parents. Fortunately, most manage to make waiting itself a social activity, time when informal get-togethers with the opposite sex can occur.*

This gradual transition provides peers with security and role models as well as people to talk to—avoiding the embarrassment of finding oneself alone for any period of time with a member of the other sex without knowing what to do or say. It also provides witnesses and companions of the same sex who will help the young person evaluate whether so-and-so (male) is really nice or a nerd, whether so-and-so (female) is sexy or stuck up, and, equally important, whether a particular attraction is mutual or not. Not surprisingly, then, the typical adolescent friendship circle is quite large and fluid. For example, one study found that the average tenth grader had ten friends of the same sex and seven friends of the opposite sex, and that between the beginning and end of the school year, more than half of the old friends had been replaced by new ones (Fischer et al., 1986).

It is interesting to compare this pattern with that of later adulthood, when friendship circles are typically smaller and much more stable (see page 437). There are

Figure 16.8 *Establishing an intimate relationship is the ultimate goal of much peer-group interaction and identity development among adolescents. Most young people spend a good deal of time talking about, horsing around with, and making eyes at, the opposite sex before reaching the point of kissing one of them.*

obvious psychosocial reasons for this difference: adolescents are experimenting with identities, seeking affirmation, and do not want to be tied down or confined to a particular niche, while older adults presumably know who they are and want friends who share their established interests and clearly defined values.

However, in terms of boy-girl relationships in adolescence, a large friendship circle makes more sense than a small one. Having many friends helps the young person think about, talk about, and associate with members of the opposite sex without the intensity experienced in "going steady" with one person. The effect is positive in at least one sense: premarital sex, pregnancy, and early marriage are less likely to occur (Chilman, 1983).

Finally, in late adolescence or early adulthood, true intimacy with one member of the opposite sex occurs, as people are ready for the close heterosexual friendships and sexual experiences of adulthood. No longer do they need the company of their peers, or their friends' specific reaction to each word and deed of their date, in order to validate their own feelings.

Thus, peers perform a valuable role in developing heterosexual intimacy as well as in developing identity, helping the young person reach maturity. In fact, without friends, adolescents have a much harder time coping with their immediate problems, and are more likely to have a difficult adulthood in store. The most crucial single predictor of an adolescent's future mental health and achievement is his or her ability to get along with peers (Rutter and Garmezy, 1983).

These findings help put the much overemphasized problems of "peer pressure" in perspective. Many parents worry that their children might be transformed during adolescence by the pressure of their friends, becoming, perhaps, sexually promiscuous, or drug-addicted, or delinquent. In some cases, of course, parents are right to worry, and may even have to intervene. However, while certainly some young people do things with their friends that they would not do alone or with a different peer group (the first drag on a cigarette or the first swig of beer is almost always related to the urging of friends), in general, peers are more likely to complement the influence of parents during adolescence than to pull in the opposite direction (Fasick, 1984; Hartup, 1983).

Parental Influence

According to all reports, the "generation gap," as the differences between the younger generation and the older one have been called, is not very wide at all. The younger and older generations have very similar values and aspirations. This is especially true when adolescents are compared, not with the culture as a whole, but with their own parents (McClelland, 1982).

Numerous studies have shown substantial agreement between parents and adolescents on political, religious, educational, and vocational opinions and values (Feather, 1980; Lerner et al., 1972, 1975). In all probability, this means that parental values have a powerful impact on adolescents, although, obviously, adolescents influence their parents as well.

Of course, not all adolescents have opinions similar to their parents', and even those that do tend to differ on some issues (Gallatin, 1980). Indeed, each generation in the parent-adolescent relationship has a psychic need to view that relationship somewhat differently; in effect, each group has its own **generational stake** in the family (Bengston, 1975): because of their different developmental stages, each generation has a natural tendency to see the family in a certain way. Parents are concerned about continuity of their own values, so they tend to minimize the import of whatever conflicts occur, blaming them on hormones or peer influences rather than anything long-lasting. Adolescents, on the other hand, are concerned with shedding many parental restraints and forging their own independent identity, so they are likely to maximize problems. Thus a conflict about a curfew may be seen by the teenager as evidence of the parents' outmoded values, or lack of trust, whereas the parents may see it merely as a problem of management, the latest version of trying to get the child in bed on time.

Furthermore, most adolescents have some conflicts with their parents some of the time (Montemeyer, 1983). Most high-school students report that their parents think they are too lazy, messy, and disobedient (Harris and Howard, 1984), and many parents agree (Costello et al., 1987). Conflicts are most likely to occur between mothers and their early maturing sons, and are especially likely if the boy's father is not living at home (Montemeyer, 1986). There is good reason for this: boys living with single mothers or with mothers and stepfathers are particularly likely to be disobedient (Steinberg, 1987).

Nonetheless, overall agreement is apparent. Most young people, for instance, favor the same candidate for president and attend the same church as their parents do. Interestingly, daughters are even more similar to their parents on numerous value questions—from religion to drug use—than sons are (Feather, 1980).

Other similarities between parents and adolescents are apparent as well. For example, regardless of academic potential, adolescents who do relatively well in high school and college tend to be the offspring of parents who value education and did well in high school and college themselves. By contrast, most high-school dropouts report that their parents do not understand, accept, or care about them or their education (Cervantes, 1965). Similarly, whether or not an adolescent experiments with drugs is highly correlated with his or her parents' attitudes and behavior regarding drugs (Jurich et al., 1985). Indeed, virtually every aspect of adolescent behavior is directly affected for good or ill by the family (Anolik, 1983).

Parenting Styles   Beyond the generality that most parents and adolescents agree quite well, how do differing patterns of parenting affect adolescents? Glen Elder (1962), in a classic study of 7,400 adolescents from intact homes, saw parents' orientation to their children during adolescence as falling into seven categories, ranging from control over every aspect of the adolescent's life to no control at all:

1. *Autocratic.* Adolescents are not allowed to express opinions or make decisions about any aspect of their own lives.

2. *Authoritarian.* Although young people can contribute opinions, parents always make the final decision according to their own judgment.

3. *Democratic.* Adolescents contribute freely to the discussion of issues relevant to their behavior and make some of their own decisions, but final decisions are often formulated by the parents and always are subject to their approval.

4. *Equalitarian.* Parents and adolescents play essentially similar roles, participating equally in making decisions.

5. *Permissive.* The adolescent assumes a more active and influential position in formulating decisions, considering, but not always abiding by, parental opinions.

6. *Laissez-faire.* Parents leave it to their teenagers to decide to consider or ignore parental wishes in making their own decisions.

7. *Ignoring.* The parents take no role, nor evidence any interest, in directing the adolescent's behavior.

Elder found that the young people in his sample rated their parents in the following ways:

	Mother	Father
Autocratic	9%	18%
Authoritarian	13%	17%
Democratic	35%	31.3%
Equalitarian	18%	14.3%
Permissive	24%	17.3%
Laissez-faire	0.6%	1%
Ignoring	0.4%	1%

It is noteworthy that about half the parents were judged to be democratic or equalitarian, the two styles recommended by most psychologists (Maccoby and Martin, 1983). Elder found that parents who were able to share power with their adolescents, neither giving up all control nor insisting on obedience, tended to have adolescents who were high achievers and unlikely to be seriously disruptive or delinquent. It is also noteworthy that when parents of adolescents err, fathers particularly are more likely to be too strict rather than too lenient.

Elder found that autocratic and authoritarian patterns were more common among large families, lower-income families, and familes with younger adolescents. These generalities have been confirmed by other research, which has found reasons for these patterns.

One reason lower-income parents would be likely to cluster at the more authoritarian end of Elder's continuum is that they expect different things of their children than middle-class parents do (Kohn, 1979). While parents of all classes want their children to be honest, happy, and considerate, lower-income parents also particularly value politeness, neatness, and obedience—qualities for which many adolescents do not have equal enthusiasm. Thus lower-class parents may find themselves commanding rather than persuading their children, because this seems the most effective way to make the adolescent behave as respectfully and obediently as the parents want them to (Belsky et al., 1984c). Another, more practical reason is that many lower-class families live in neighborhoods where drugs, violence, and crime are all too prevalent. One critical factor that can keep a particular young person from being caught up in the social destruction of such communities is **parental monitoring,** parents knowing where their child is and what he or she is doing and with whom (Snyder et al., 1986). The worse the social milieu, the more likely monitoring is to result in curfews, restrictions on friends, and other seemingly authoritarian measures. The higher the social class, the less the need for such stringent measures, and the more emphasis there tends to be on self-direction and curiosity—values many adolescents do share with their parents.

Figure 16.9 *Scenes such as this one bring two complementary thoughts to mind. First, common goals and activities and pride in the skills and accomplishments of family members are often shared by teenage children and their parents. Second, although they are temporarily disturbing to all involved, occasional angry outbursts are also typical of healthy family interaction.*

Part of the reason parents of younger adolescents cluster toward greater discipline is that young people begin to become more assertive in family interactions when the physical changes of puberty occur (Steinberg, 1977). In many families, the parents' first reaction is to increase their own assertiveness, trying to insist on the parental authority and respect that their young teenager seems disinclined to give. Gradually, parents tend to yield more often, in part because they recognize that their child is becoming an adult, and in part because the child is beginning to act in more mature ways.

Thus, if there is a "gap" at all, it is more likely to occur in early adolescence, and to tend to center on issues of self-discipline and self-control. During these years, teenagers and their parents are more likely to have disagreements about the adolescent's clothes, domestic neatness, and sleeping habits than about world politics or deep moral concerns. Fortunately, the bickering and alienation that occur in many families in early adolescence diminish with time, although about 20 percent of all families find that conflict appears and reappears throughout adolescence (Montemeyer, 1986; Offer and Offer, 1975). As we will see in the chapters on adulthood, relationships between parents and children tend to improve with time.

Special Problems

Although most adolescents and their parents negotiate the teen years rather well, a sizable minority of adolescents and their parents face serious problems, caused by the confluence of puberty, family difficulty, and inadequate social and cultural conditions. We will discuss three of the potentially most devastating—delinquency, sexual abuse, and suicide.

Delinquency

Minor law-breaking seems to be part of normal adolescence. In both North America and Great Britain, at least, confidential surveys of teenagers reveal that about 80 percent of all adolescents have broken the law—mostly in minor ways, such as smoking marijuana or committing petty vandalism (Rutter and Giller, 1984). Thus, only a minority of teenagers would seem to be completely law-abiding. In fact, a person is more likely to be arrested at age 16 than at any other age.

By no means are all adolescent crimes minor, however. In the United States, 47 percent of the arrests for serious crimes, such as murder, assault, and robbery, involve youths under age 21 (U.S. Department of Justice, 1984).

Crime statistics reveal another interesting fact. Males are much more likely to be arrested than females, and they tend to be arrested for more overt crimes. The relatively rare crime of homicide is one example: males under 18 are arrested for murder eight times as often as females. In addition, boys are arrested for burglary and for robbery fifteen times more often than girls, and for vandalism ten times more often. Thus, troubled boys are more likely to lash out against someone; troubled girls are more likely to avoid confrontation (in fact, running away from home is one of the few things girls are arrested for more often than boys).

It should be pointed out that while most adolescents of both sexes are, at one time or another, delinquent in some minor way, relatively few (about one in twenty) are arrested. These few tend to be arrested repeatedly, and for them, crime is not only a problem for society, but also an indication that they themselves are in trouble. To be specific, young people who become arrest statistics have lower self-esteem, poorer relationships with their family, and more difficulty in school than their peers who have not had trouble with the law. Since delinquency is both a cause and a consequence of such problems, the first-time offender usually needs help in order to prevent these factors from compounding each other. Almost always, when a young person is involved in a serious crime, he or she has a long history of school problems and minor offenses.

Unfortunately, parents and others are all too likely to deny the troubled situation until it is too late. (Typifying this kind of response is the father described by Elkind [1979] as yelling "Why don't you catch some real criminals!" at a policeman who had caught his son stealing a car.) Other parents punish their child's first delinquency, but often make no attempt to strengthen family relationships and build up the young person's self-esteem, steps that would make future delinquent behavior less likely.

School achievement also seems crucial. Academic difficulty in sixth grade is one of the best predictors of delinquent behavior in ninth grade (Magnusson et al., 1975). In fact, those programs that seem to prevent chronic delinquents from becoming criminals are generally focused on developing skills in two specific areas—school and employment. If this is done in a context in which the young person can develop a relationship with a teacher or a counselor, the chances are that delinquency will be an adolescent phase rather than the beginning of a criminal career (Gold and Petronio, 1980).

An additional factor influencing delinquency appears to be a socioeconomic one, since lower-class young people have somewhat higher rates of delinquency and much higher rates of arrest and incarceration than middle-class adolescents do. This difference can be explained in part by the fact that lower-class and racial-minority offenders often get arrested for acts that middle-class whites would merely get lectured for. At the same time, the incentive to commit crime, especially robbery, seems to be linked to chronic joblessness. Statistics show that the unemployment rate among black youth in the inner cities is about 50 percent, and the crime rate for this group is higher than that for any other group in the country.

As one black 17-year-old boy said:

> How they let this happen in a country like this, having all these kids walking around the streets, got their hands jammed down in their pockets, head down, like their necks was bent in half? What do folks think these kids gonna do, when they go month after month, year after year, without nothing that even *smells* like a job? [Quoted in Cottle, 1979]

This leads to a final question. Why are boys so much more likely to become delinquent? One major factor that influences all the other contributors to delinquency is that our culture, in some ways, tends to condone, and even encourage, minor-law-breaking in boys as a way of establishing independence and asserting masculinity. As one review explains:

> If a 15-year-old manages somehow to steal the red flasher (the "bubble gum machine") off a police squad car, that act is a genuine bid for glory . . . [but] stealing the red flasher seems somehow incongruous for girls while eminently suitable for boys . . . Delinquent behavior ordinarily fails to serve adolescent girls as it serves adolescent boys to shore up a failing sense of self-esteem. [Gold and Petronio, 1980]

As with every question, it is impossible to unravel the tangled web of nature and nurture to see precisely what accounts for delinquency. However, virtually all experts agree that poor parental management and a child who did not learn appropriate social skills often combine to create a delinquent (Snyder et al., 1986).

Sexual Abuse

Sexual abuse of the young is a widespread problem that has only recently attracted the public attention it warrants. However, despite the publicity it has received, or perhaps because of it, most people have a distorted view of the problem.

In the public's mind, the typical case of sexual abuse involves a stranger who lures a small child away from a public place such as a playground and then forces the child to participate in some type of sex act. This, however, is the least common form of sexual abuse. In fact, relatives and family friends are the perpetrators in more than 75 percent of all cases of sexual abuse; physical force is used in only about 5 to 10 percent of all instances; and young children are the victims less often than young adolescents are (Holdern, 1980; McCabe, 1985). Incidence statistics show a marked rise of sexual abuse at puberty, and the rates continue to increase until whatever age victims are considered adults.*

As is the case with other forms of child maltreatment, precise data on sexual abuse are hard to come by because abuse is variously defined and its occurrence is, clearly, underreported. For example, some reporting agencies include only cases in which the child has been physically penetrated; most developmentalists, on the other hand, believe that "sexual abuse" is the appropriate label for any act in which an adult uses a child or adolescent for his or her own sexual needs, whether it be through some form of intercourse or a less serious act, such as intentional touching of clothed breasts or genitals. Nevertheless, no matter how it is defined, sexual abuse is far more common than most people believe. As Kempe and Kempe (1984) explain:

---

*The age at which a victim of sex abuse is no longer considered a child ranges from 13 to 18 in the various states of the U.S. and 14 to 16 in Canada.

Figure 16.10 *One of the fortunate outcomes of the recent attention to sexual abuse is that efforts are now being made to teach children the difference between "good touch" and "bad touch" at an early age, as does the book from which these pictures come. Sensitivity is required: children need to remember that caresses and hugs are usually warm and welcome expressions of affection, while learning that unwelcome physical advances are to be rebuffed and reported.*

The lowest estimates based on official reports suggest that . . . the number of women who experienced some form of [child sexual] abuse to be well over 4 percent, or at least 4 million women in the United States. . . . these estimates are far below the actual incidence in both sexes, since most cases of sexual victimization are never reported to anyone.

Much higher estimates have resulted when adults were asked, confidentially, to recall if they themselves had ever been sexually abused before they were adults. In one study of college students in New England, for example, 1 in every 5 (19.2 percent) women and 1 in every 12 (8.6 percent) men acknowledged that they had been abused before age 17. (Abuse in this study included the entire spectrum of sexual exploitation, from fondling and exhibitionism to rape [Finkelhor, 1979a]). A study of adult women in California found even higher rates of adolescent sexual abuse.

Typical Abuse   Typically, abuse is perpetrated by a man the young person trusts, usually the father, step-father, or other close relative or family friend. The abuse often begins in childhood with sexual fondling, and, during early adolescence as the child enters puberty, may include sexual penetration. Overt force is usually not necessary at first, primarily because of the powerlessness the young person feels vis à vis adult male relatives. As one incest victim recalled of her father:

When he would walk into the room it was like a sinister force. That was a world where man is boss, man is king. . . . And all he would do is lower his paper and look at me and I would just freeze. [Quoted in Armstrong, 1978]

Father-Daughter Abuse   Ironically, incest between father and daughter, the most difficult type of sexual abuse for the outsider to understand, is, in fact, the most common type of serious abuse (Kempe and Kempe, 1984). How could a father possibly use his own child as a sexual object? we wonder. In many cases, the father rationalizes his behavior by offering himself the "myth of the seductive child," according to which "children are seductive and willingly participate in or invite sexual activity with adults. This myth is . . . especially applied to adolescent victims" (McCabe, 1985). As Finkelhor points out, this myth is a disastrous fabrication: no matter what their behavior, young people are clearly incapable of the informed consent that is a prerequisite for a healthy sexual relationship:

> They are unaware of the social meanings of sexuality . . . they have little way of knowing how other people are likely to react to the experience they are about to under-take, what likely consequences it will have for them in the future . . . Further, the child does not have the freedom to say yes or no. This is true in a legal sense and also in a psychological sense . . . children have a hard time saying no to adults [because] adults control all kinds of resources that are essential to them—food, money, freedom. [Finkelhor, 1979b]

The Incestuous Family.   Unfortunately, in many incestuous families, incest is an integral part of the family dynamics (Jiles, 1980). The father is often introverted and immature, having little social contact outside his family. Often, he is also alcoholic (Justice and Justice, 1979; Schlesinger, 1982). The mother is usually present in the household but unavailable to her husband as a sexual partner or to her daughter as a protector or confidante, often because she is ill or drug-dependent. Frequently, she is consciously or unconsciously willing to ignore her daughter's well-being to protect herself or her marriage. For example, the husband's attention to the daughter may prevent him from abusing the mother or the other children, or it may keep him from leaving the family entirely. In many cases, a role-reversal oc-curs, with the daughter protecting the mother and younger children, while taking over much of the mother's authority and work.

Fathers manage to continue the relationship in many ways, from buying the daughter extra clothes and otherwise favoring her, to restricting her contact with peers and teachers, to using force or the threat of force. Perhaps worst of all is the use of psychological manipulation—telling the daughter it was all her fault, that she is wrong to complain, and that no one would believe her anyway. Often, there is some truth in this last threat. In about half of all cases in which incest victims tell their mothers of the abuse, their mothers refuse to believe them (McCabe, 1985). In many cases, the daughter simply leaves home as soon as she is able, coming to the attention of officials as a runaway and, far too often, as a young prostitute, rather than as a victim of sexual abuse.

A Life-Span Perspective

Consequences of Sexual Abuse

The psychological effects of sexual abuse depend largely on the extent and dura-tion of the abuse, and on the reaction of other people—family as well as authori-ties—once the abuse is known. If the abuse is a single nonviolent incident, and a trusted care-giver believes and reassures the victim, taking steps to make certain the incident does not happen again, the psychological damage may last only a few days (Schlesinger, 1982). Even with abuse that is more serious, children and adoles-cents can be quite resilient if they are cared for with sensitivity, confidentiality, and respect.

If the abuser is a family member, and the problem is ongoing, much damage may occur before the abuse is uncovered. As Kempe and Kempe (1984) report:

> Longstanding in-house sexual abuse with a loved person and/or relative . . . is particularly damaging for the pre-school child and for the young adolescent; at these two

important times both need to fulfill their sexual development in an orderly and sequential way which this misfortune totally disturbs. As a result these victims have a much higher than normal incidence of poor sexual adjustment and difficulties in sexual identity and preference. As teens they are likely to run away from an intolerable situation, become pregnant, get involved in delinquency such as theft and substance abuse (both alcohol and other drugs), engage in teenage prostitution and, as has been the experience for some of our clients, make a significant number of attempts at suicide. Some have, indeed, killed themselves . . .

Long-Term Consequences   From a developmental point of view, one of the most troubling consequences of incest is that the young person may never learn what a normal adult-child or man-woman relationship should be. Kempe and Kempe (1984) report that

child abuse is more common in mothers who were incest victims. . . . We have noted a teenage mother, who had herself been sexually abused by her father, treat her neglected 3-month-old baby in an inappropriate way during an interview with her. She called him "Lover" several times and kissed him repeatedly on his open mouth. This kind of inappropriate sexualization of a close relationship, in combination with neglectful parenting, may provide a background for future abuse.

Similar findings come from a study of mother-infant attachment patterns that focused on mothers who were under a great deal of stress because they were young, poor, and alone. About 10 percent of these mothers behaved in a seductive manner with their infant sons (Sroufe and Ward, 1980). Follow-up research found that almost half of these mothers had been sexually abused as children by family members, and, as their children grew older and entered nursery school and elementary school, they were more likely to show the kinds of inappropriate interactions with adults and other children that are symptoms of abuse, as well as to act precociously sexual with other children (Sroufe et al., 1985). In addition, adolescent victims of abuse, whether they are male or female, tend to become involved again in violent relationships, either as the abuser or abused (Billingham and Sack, 1986).

Other studies have shown that female victims of sexual abuse may have a distorted view of sexuality, and thus are more likely to marry men who are abusive. If these men begin to abuse their daughters, the mother is less alert to the problem or feels trapped again, unable to help (Kempe and Kempe, 1984). This pattern provides another explanation for the fact that children of women who have been sexually abused are more likely to be abused themselves (Goodwin, 1982). Thus, in several ways, the effects of sexual abuse may be transmitted from generation to generation.

Prevention   Obviously, prevention of sexual abuse requires recognizing factors that foster sexual abuse and putting a stop to, or at least guarding against, them. As the chart on page 382 reveals, these factors begin in the macrosystem—with cultural values and practices that encourage sexual feelings toward children—and continue at each level down to the microsystem of the family.

Preconditions for Sexual Abuse of Children

1. *Adults must have sexual feelings about children.* Such feelings are encouraged by:

childhood sexual experiences;
exposure to child pornography;
exposure to advertising that sexualizes children;
male sex-role socialization that devalues nurturance and encourages sexual aggression;
"successful" adult sexual experiences with children.

2. *Adults must overcome internal inhibitions against abuse.* These inhibitions are weakened by:

cultural values that accept sexual interest in children;
low impulse control;
alcohol;
stress;
low self-esteem;
fear of, or frustration with, sexual relationships with adults;
values that emphasize father's unquestioned authority.

3. *Adults must overcome external inhibitions to committing sexual abuse.* These obstacles to contact with a child are minimized by:

an absent, sick, or powerless mother;
a mother who is neglectful, unaware of her children's need for protection;
crowded living conditions or sleeping together;
opportunities to be alone with the child;
social isolation—family members have few friends;
geographical isolation—family has few nearby neighbors.

4. *Adults must overcome the child's resistance.* Overcoming this barrier is easier if the child is:

emotionally deprived;
socially isolated;
acquainted with the adult;
fond of the adult;
vulnerable to incentives offered by the adult;
ignorant of what is happening;
sexually repressed and sexually curious;
weak and frightened of physical force.

Source: Finkelhor, 1984.

Obviously, not all these factors can be changed or controlled. One that can be changed, however, is the culture's values about sex and about children. Already, rising awareness of the problem of sexual abuse has increased public pressure against the eroticization of children and pubescent young people in pornography and advertising. In certain respects, another, more subtle change has been occurring by virtue of the fact that fathers are more actively involved in the care of their infants than they once were. One researcher finds that incest is rare in families in which the fathers were involved in infant care-giving: presumably, these fathers see their children in protective and nurturant ways, which makes sexual feelings unlikely and inhibition against them high (Herman, 1981). This analysis would help explain why the fathers currently involved in incestuous abuse are more frequently step-fathers, whereas they were more frequently biological fathers in early cohorts (McCabe, 1985; Russell, 1984).

Figure 16.11 *One unfortunate consequence of the intensified publicity about child abuse is that young men are shying away from working with preschoolers. Sexual abuse is, in fact, very rare in day-care centers and does not occur if parents are as actively involved in centers as they should be. And, especially since many children have single-parent mothers, male adults who model the appropriate use of physical touch as well as social interaction are very much needed.*

Certain preventive measures can be established in the social institutions of the community. Since vulnerability is fostered by ignorance, sex education in schools and churches should begin at younger ages, and should include not just the specifics of biology but also discussion of appropriate relationships between adults and children, and between men and women. This may not only prevent young people from being victims; it may also help them become adults who would not permit abuse to occur. A related step would be to make teachers and clergy aware of the preconditions for, and the symptoms of, sex abuse, so they could be alert to help victims early on.*

Adolescent Suicide

One of the most perplexing problems that may occur in adolescence is suicide. From an adult's perspective, the teenager is just at the start of the many wondrous and exciting experiences that life offers. It seems inexplicable that a young person would end his or her life as it is about to really begin. Yet about one adolescent in every ten thousand does that each year, a rate double that of twenty years ago.

As can be seen in Figure 16.12, the suicide rate between ages 15 and 19 is around half that for any subsequent age group. As an index of despair, however, this differential may be misleading, because it probably results from the higher failure rate of teenagers' suicide attempts.† Further, many people who commit suicide as young adults were unsuccessful attempters as adolescents.

Figure 16.12 *The suicide rate for adolescents is half the rate for adults only because adolescents' attempts to kill themselves fail twice as often as those of adults. Unfortunately, the "success" rate for adolescents in the 1970s was double that of the 1960s. Two factors implicated in this increase were adolescent drug use, which increased during the 1970s, and divorces, which doubled between 1970 and 1975. A disproportionate number of suicidal adolescents are from divorced families, and drugs are a factor about half the time.*

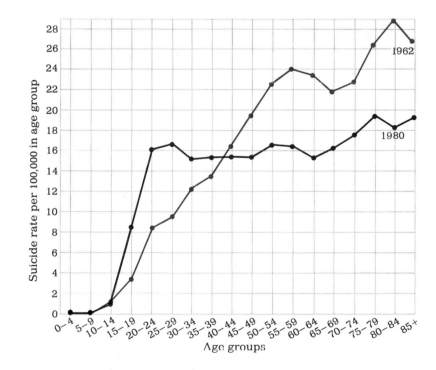

Contributing Factors What factors cause a young person to take his or her own life? Do adolescents who commit suicide differ in personality from normal adolescents or from disturbed nonsuicidal teenagers? And what circumstances drive a young person to the point of self-destruction? Answers to these questions are hard

---

* A national hotline, 1-800-422-4453, or 1-800-"4" A CHILD, is open day and night for questions and problems related to child abuse of any kind.

† Accurate statistics on attempted suicide in adolescence are hard to come by, because many attempts are hidden by embarrassed parents, and many apparent accidents may actually have been suicides. However, it is generally believed that adolescents attempt suicide at least as often as adults do.

to arrive at. For one thing, obviously, information about suicide victims cannot be gotten directly: it must come from those acquainted with the victims, or be inferred from studies of adolescents who have failed in their attempts on their lives. The information from the former source, usually parents, may be tainted for several reasons, including grief, guilt, or a denial of, or blindness to, serious problems that may have existed. The data from the second source may also be faulty, for it simply may not be valid to generalize from studies of failed suicides to successful ones.

Allowing for these limitations in the study of teenage suicides, we can see certain rough patterns emerge from the research. According to a review of the psychosocial and cognitive aspects of adolescent suicide (Petzel and Riddle, 1981), suicidal behavior, and probably suicidal thoughts as well, are not normative in adolescence. Suicidal adolescents tend to be more solitary than normal adolescents and, compared with disturbed nonsuicidal adolescents, they show a greater tendency to be depressed, self-punishing, and emotional. They also seem to exhibit certain patterns according to their sex:

> Suicidal males were tense, jumpy, high strung, perfectionistic, prone to worry and suspiciousness, and had exaggerated needs for affection but few relationships with male peers; distress was channeled into projection, depression, and somatization. Suicidal males were generally less resentful than [control-group subjects] and often were referred for treatment because of impulsive behavior. Suicidal females were tearful, despondent, resentful, weak, unstable, and unpredictable. Weak defenses, poor judgment, deviant ideation, marginal control, flat affect, subjective feelings of depression, few or no friends, and sex difficulties also characterized suicidal adolescent females. [Petzel and Riddle, 1981]

Research has also shown that there is usually no single event that triggers a suicide attempt; rather, it "occurs within the context of long-standing problems," one of the most prominent of which is chronic family conflict, characterized by "anger, ambivalence, rejection, and/or communication difficulties."

Figure 16.13 *No one goes through life without moments of depression. It is, in part, the ability to put those times into the context of past and future that can help to diminish the hurt of painful events in the present.*

Warning Signs of Suicide

A number of warning signs should alert family and friends that a young person may be becoming dangerously overwhelmed with emotional difficulties and may be at increased risk of suicide:

1. *A sudden decline in school attendance and achievement, especially in students of better-than-average ability.* Jacobs (1971) found that while about a third of the young people who attempted suicide had recently dropped out of school, only 11 percent were in serious academic difficulty. Most of them had been doing quite well before their precipitous decline.

2. *A break in a love relationship.* This is the precipitating event for many adolescent suicides. The fact that such events are relatively common in adolescence sometimes blinds parents and teachers to the pain and depression they cause, especially in the egocentric young person who believes that the lost love is the only love he or she could ever have. A sympathetic shoulder to cry on is much more helpful than a statement such as "There are other fish in the sea."

3. *Withdrawal from social relationships, especially if the adolescent seems no longer to care about social interaction.* The adolescent who decides that suicide is the solution sometimes seems less depressed than previously and may cheerfully say something to the effect of "It's been nice knowing you." A joking or serious "goodbye" accompanied by a sudden desire to be alone is a serious sign.

4. *An attempted suicide.* An attempted suicide, however weak it might seem, is an effort to communicate serious distress, and therefore must be taken seriously. If nothing changes in the adolescent's social world, an attempted suicide may turn out to have been a trial run for the real thing. Almost all adolescent suicides follow failed attempts.

5. *Cluster suicides.* Adolescents, given their egocentrism, are particularly likely to be influenced by knowing, or even reading about, another adolescent who committed suicide. Thus, whenever a suicide is publicized, concerned adults and adolescents need to be particularly sensitive to the more vulnerable young people in their midst (Davidson, 1986).

When such warning signs have been detected, they must be acted upon. As Edwin Shneidman (1978) has written:

the act of suicide is an individual's effort to stop unbearable anguish or intolerable pain by doing "something." Therefore, the way to save a person's life is also to do "something," to put your knowledge of the person's plan to commit suicide into a social network—to let others know about it, to break the secret, to talk to the person, to talk to others, to offer help, to put action around the person, to show response and concern, and, if possible, to offer love.

Professional help for suicidal adolescents and their families can often open up channels of communication that had been blocked by the self-absorption of the adolescent, and perhaps by the parents' insensitivity as well. One important goal of therapy is to keep expectations in line with reality. Parents often demand too much. As one pediatrician explains, "A lot of families expect that the minute the youngsters become thirteen or fourteen, they should be capable of making it on their own. In reality, teenagers probably need as much support at that point in their lives as toddlers need, although of a different sort" (Langone, 1981).

Conclusion

In closing our discussion of adolescence, it would seem appropriate to reiterate a point made by Erikson (1964). Essentially, adolescence may be viewed as the dawning of commitment, to others, to ideologies, to their own futures. And it is through these various commitments that young people begin to accomplish the task of achieving adult identity. Each family, each community, each culture, intentionally or unwittingly, works to help or hinder adolescents in their efforts to make adultlike decisions about their lives.

Within the context of the social forces that block some options and encourage others, each individual attempts to chart his or her own course. Just as each infant actively searches for cognitive equilibrium, each adolescent tries to find the identity that expresses his or her own individuality. In the process, adolescents make decisions about how to reach adulthood that may affect their entire lives. This does not mean that a person's entire future is set by the beginning of adulthood, for an individual's destiny is never fully known until the individual's life is over. However, as adolescents chart their course in one direction or another, they foreclose some options and create others. As Robert Frost wrote (1963):

> Two roads diverged in a wood, and I—
> I took the one less traveled by,
> And that has made all the difference.

SUMMARY

Identity

1. According to Erikson, the psychosocial crisis of adolescence is identity versus role confusion. Ideally, adolescents resolve this crisis by developing a sense of both their own uniqueness and their relationship to the larger society, establishing a sexual, political, moral, and vocational identity in the process.

2. Sometimes the pressure to resolve the identity crisis is too great, and instead of exploring alternative roles, young people foreclose their options, establishing a premature identity. Other young people simply choose values and roles opposite to those expected by parents and society, thus forming a negative identity.

3. The process of identity formation depends partly on the society: if its basic values are consistent and widely accepted, and if social change is small, the adolescent's task is fairly easy.

4. Many societies help adolescents achieve identity by providing rites of passage, or initiation ceremonies. In some cultures, these rites are painful and dramatic, helping the young person make the transition from childhood to adolescence in a matter of days or weeks.

Friends and Family

5. The peer group is an important source of information and encouragement for adolescents. The adolescent subculture provides a buffer between the world of children and that of adults, allowing, for example, a social context for the beginning of heterosexual relationships.

6. Many adolescents identify strongly with their generation, believing that some ideas and experiences cannot be properly understood or appreciated by adults.

7. However, parents are the most important influence on adolescents, especially when there is discussion and respect among family members. Children, especially daughters, tend to share their parents' values.

8. Democratic and equalitarian parenting styles are most beneficial for adolescents. Parenting styles vary, however, by social class, size of family, and age of the adolescent.

Special Problems

9. Delinquency, sexual abuse, and suicide are among the most serious problems of adolescence. Many of these problems can be attributed to patterns of family interaction and also to social and cultural beliefs and patterns.

10. The most common types of sexual abuse occur between children and relatives and family friends. Patterns of family interaction may allow such abuse to continue over long periods of time. Abused children tend to develop distorted views of both parent-child relationships and adult sexuality.

11. Researchers have found that most adolescent suicides are preceded by a long sequence of negative events, including family problems and breakdowns in communication among family members. Suicide prevention requires heeding the preliminary warning signs.

KEY TERMS

identity *(363)*	identity diffusion *(364)*
identity achievement *(364)*	moratorium *(365)*
foreclosure *(364)*	rite of passage *(367)*
negative identity *(364)*	generational stake *(374)*
	parental monitoring *(376)*

KEY QUESTIONS

1. What are some of the difficulties adolescents might experience on the way toward identity formation?

2. What is the function of the peer group during adolescence?

3. Which parenting styles seem least helpful to adolescents and why?

4. What social and cultural characteristics tend to be associated with delinquency?

5. What family and social patterns are preconditions for sexual abuse of children?

6. What are some of the psychosocial and cognitive patterns that tend to be associated with adolescent suicide?

Part V

The Developing Person So Far: Adolescence, Ages 10 through 20

Physical Development

Physical Growth

At some time between the ages of 9 and 14, puberty begins with increases in male and female hormone levels. Within a year, the first perceptible physical changes appear—enlargement of the girl's breasts and the boy's testes. About a year later, the growth spurt begins. During adolescence, boys and girls gain in height, weight, and musculature. The growth that occurs during these years usually proceeds from the extremities to the torso and may be uneven.

Changes in Sex Organs and Secondary Sex Characteristics

Toward the end of puberty, the young person's potential reproductive capacity is signaled by menarche in girls and ejaculation in boys. It will take several years before full fertility is achieved.

On the whole, males become taller than females and develop deeper voices and characteristic patterns of facial and body hair. Females become wider at the hips; breast development continues for several years.

Cognitive Development

Formal Operational Thought

By the end of adolescence, many young people can understand and create general principles and use scientific reasoning. For many adolescents, cognitive advancement is also reflected in their ability to reason morally.

Adolescent Egocentrism

Adolescent egocentrism tends to prevent teenagers from thinking rationally about their own experiences. Their feelings of invincibility and uniqueness may prompt them to underestimate risks, for example, with regard to sexual relationships and drug use.

Psychosocial Development

Identity

One of the major goals of adolescence is identity achievement—the development of the young person's own sense of self. Identity formation can be affected by personal factors—including relationships with family and peers—the nature of the society, and the economic and political circumstances of the times.

Peers

During adolescence, the peer group becomes increasingly important in helping adolescents to become more independent, to "try out" new behaviors and explore different facets of their personality, and to interact with members of the opposite sex.

Parent-Child Relationships

Although in early adolescence parents and children may find themselves at odds over issues centering on the child's increased assertiveness or lack of self-discipline and self-control, these difficulties usually diminish as the parents recognize the teenager's increasing maturity and allow him or her more autonomy.

Early Adulthood

As young children, we look forward to the day when we will be "all grown up," imagining that when we attain adult size we will automatically master the roles, privileges, and responsibilities of adulthood. As young teenagers, we likewise impatiently await our high-school graduation or 18th or 21st birthday, anticipating that independence, and the competence to cope with it, will be bestowed when we arrive at these "official" milestones.

But young adults, who must make their own decisions about career goals, intimate relationships, social commitments, and moral conduct, usually find these aspects of independence exciting, but far from easy to deal with. This is especially true today because the array of lifestyle choices seems so vast and varied. And no matter which of the roles of adulthood they choose to take on, or how thoughtfully and eagerly they strive to play them, they are bound to be confronted with stresses, set-backs, and second thoughts. Yet for most young adults, it is problems faced and usually solved, and limitations accepted or overcome that make the decades from twenty to forty an exhilarating period when people often feel they are living to the fullest. The next three chapters describe how many young adults cope with the engrossing, multidimensional realities of early adulthood.

CHAPTER 17

Early Adulthood: Physical Development

Youth is the time to go flashing from one end of the world to the other, both in mind and body.

Robert Louis Stevenson
"Crabbed Age and Youth"

As we have seen again and again, cognitive and psychosocial development from birth through adolescence are closely tied to physical development, that is, to the process of physical maturation. Just as the exploration and autonomy of the toddler must await the development of walking, and the formal education of the schoolchild must await certain levels of brain maturation, so must many of the cognitive and psychosocial aspects of adolescence be preceded by the body growth and hormonal changes of puberty. With the attainment of full maturity, human development is released from the constraints inherent in genetically programmed maturation. At the same time, however, a new aspect of physical development comes into play—decline, decline that in some cases begins even before maturation is complete. Each of the three chapters on adult physical development will, therefore, be centered on the declines that occur with age, but they will also emphasize a crucial fact: how people perceive changes that occur in their bodies over time, and what decisions they make regarding health habits and lifestyle, can have nearly as great an impact on the course of their overall development as the changes themselves.

In terms of physical development, early adulthood can be considered the prime of life. Our bodies are stronger, taller, and healthier than during any other period. The first years of young adulthood (the early 20s) are the best ones for hard physical work, for problem-free reproduction, and for peak athletic performance. As we will see, although the advancing years of early adulthood are accompanied by declines throughout the body, whatever difficulties young adults experience in physical development are usually related to factors other than aging.

Growth, Strength, and Health

For most people, noticeable increases in height have stopped by the beginning of early adulthood, at about age 18 in females and 20 in males, although late-maturers often grow an inch or two during their 20s (Sinclair, 1978). Growth in muscle and increases in fat continue into the 20s, as the body fills out, women attaining their full breast and hip size, and men reaching their full shoulder and upper-arm size. Partially because of these increases, weight typically increases as well, especially during the early 20s. Before middle age, the average man adds 15 pounds, and the average woman, 14 pounds, to their weight at age 20 (U.S. Bureau of the Census, 1986). Women typically have a higher percentage of body fat and a lower metabolism than men do, a sex difference that increases throughout life (Striegel-Moore et al., 1986).

Since more of their body mass is comprised of muscle, men are typically stronger than women. For both sexes, however, physical strength, as evidenced in the ability, say, to run up a flight of stairs, lift a heavy load, or grip an object with maximum force, generally increases during the 20s, reaching a peak at about age 30. In terms of overall health, all the body systems, including the digestive, respiratory, circulatory, and sexual-reproductive systems, function at an optimum level during early adulthood. Visits to the doctor and days in the hospital are significantly lower for this age group than for later ages, and medical attention in early adulthood is more often necessitated by injuries (often sports-related) or by normal pregnancy than by disease. Even the common cold is less frequent in early adulthood than in any other part of the life span (U.S. Bureau of the Census, 1986). Self-reports reflect this healthy state. Seventy-two percent of those in early adulthood rate their health as very good or excellent, and only 6 percent rate it as fair or poor (National Center for Health Statistics, 1985).

Correspondingly, death from disease is rare in early adulthood. Of the fatal diseases, cancer is the leading killer of young adults, yet the annual cancer death rate between ages 20 and 35 is less than 1 person in 10,000, compared with 18 per 10,000 between the ages of 45 and 54, and 82 per 10,000 between the ages of 65 and 74. The data in Table 17.1 make especially clear the relatively low mortality rate that young adults have for disease overall.

All told, then, most adults from age 20 to 40 are strong and healthy. As Lillian Troll (1975) concluded:

> Overall, adulthood is a time of peak physical status. Most body functions reach full growth and development by the middle 20s: height, efficiency, and endurance are at maximum levels. For the next 20 or 30 years — and sometimes even longer — declines from the peak are so gradual as to make them seem a long plateau.

Nevertheless, the gradual "declines from the peak" are already apparent before middle age.

TABLE 17.1 **Death from Disease**

Age	Deaths per 100,000 Americans
15–24	2
25–34	51
35–44	151
45–54	468
55–64	1,215
65–74	2,815
75–85	6,210

Age-Related Changes

By the late 20s, most people notice the first signs of aging in their physical appearance. Slight losses of elasticity in facial skin produce the first wrinkles, usually in those areas most involved in their characteristic facial expressions. As the skin continues to lose elasticity and fat deposits build up, the face sags a bit with age. Indeed, some people have drooping eyelids, sagging cheeks, and the hint of a double chin by age 40 (Whitbourne, 1985). Other parts of the body sag a bit as well, so as the years pass, adults need to exercise regularly if they want to maintain their muscle tone and body shape. Another harbinger of aging, the first gray hairs, is usually noticed in the 20s and can be explained by a reduction in the number of pigment-producing cells. Hair may become a bit less plentiful, too, because of hormonal changes and reduced blood supply to the skin.

Changes that are not so visible or obvious occur in every body system (Brooks and Fahey, 1984). As you can see in Figure 17.1 the efficiency of most body functions begins to decline in the 20s. The decline in efficiency proceeds at a somewhat different rate for each organ system, and is affected, too, by the individual's genetic makeup; however, in general, after adolescence, "there is a nearly linear decline in most integrated body functions at the rate of about 1 percent a year" (Bierman and Hazzard, 1978). Thus the body systems of the typical 40-year-old are already 20 percent less efficient than they were at age 20. Of course, lifestyle, especially exercise, can affect the rate of decline in every individual, making some 40-year-olds more physically fit than some 20-year-olds. (Health habits are discussed in Chapter 20.)

Figure 17.1 *As this chart clearly shows, all body functions steadily decline after age 20. Fortunately, for reasons made clear in the next section, the effects of age-related physical declines are much less dramatic than one might imagine from a glance at these plot lines. In addition, this chart shows averages; individuals actually can do a great deal to change the rate, if not the direction, of physiological change.*

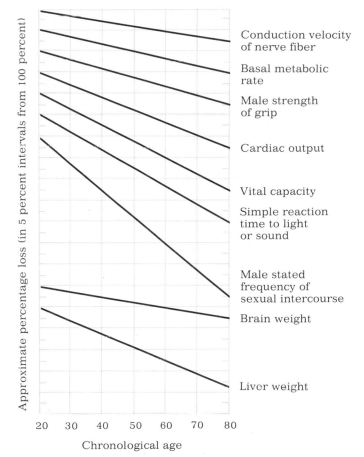

VEGETATIVE FUNCTIONS OF THE BODY

Conduction velocity of nerve fiber

Basal metabolic rate

Male strength of grip

Cardiac output

Vital capacity

Simple reaction time to light or sound

Male stated frequency of sexual intercourse

Brain weight

Liver weight

Approximate percentage loss (in 5 percent intervals from 100 percent)

20 30 40 50 60 70 80

Chronological age

Homeostasis

Many of our body functions serve to maintain **homeostasis;** that is, they adjust automatically to keep our physiological functioning in a state of balance, or equilibrium. For instance, when we are at rest, our breathing and heart rates become slow and steady; when we are very active, both increase to bring us more oxygen. When we are cold, we shiver to activate our muscles and warm up; when we are hot, we sweat to give off body heat.

The older a person is, the longer it takes for these homeostatic adjustments to occur, making it harder for older bodies to adapt to, and recover from, stress (Brooks and Fahey, 1984). For example, even if a younger and an older player are otherwise equally matched in a sport that doesn't demand stamina, the older adult would need a longer warm-up period before the game, and more rest afterward, to allow heart rate, breathing, blood pH (acidity), and blood glucose (sugar) to return to normal. Similarly, older adults might have a harder time adjusting to work that is physiologically stressful. These changes become sufficiently apparent during early adulthood that the average 35-year-old might notice that he or she can no longer skip a night's sleep and still function adequately or can no longer bounce back the next day after a full day of unusually heavy exertion.

Figure 17.2 *As firefighters age, they are increasingly likely to die in the line of duty, but not because they are less able to cope with dangerous situations. Rather, sudden exertion and stress can put an overwhelming strain on an aging body, so the older they are, the more likely professional firefighters are to die of heart attacks than of burns or smoke-poisoning.*

Organ Reserve

In bare outline, the physical declines of adult aging might seem steep. However, the actual experience of aging is usually much less perceptible than it might seem from looking at the graph in Figure 17.1—and much more like the plateau described by Troll. In day-to-day life, most adults of all ages feel that their bodies are quite strong and capable, not much different than they were ten years earlier.

In fact, for most of us, our bodies, if adequately maintained, are capable of functioning quite well until we are at least age 70. The reason is that the declines of aging primarily affect our **organ reserve,** the extra capacity that each organ has for responding to unusually stressful events or conditions that demand intense or prolonged effort (Fries and Crapo, 1981). In the course of normal daily life, however, we seldom have to call upon this extra capacity, and except for when we do, the deficits in organ reserve generally go unnoticed. Thus, while 50-year-olds are somewhat slower than 20-year-olds at, say, running up several flights of stairs, because the reserve capacity of their hearts and lungs is not as great as it once was, they move with ease in normal activity. In the same fashion, a woman in her late 30s might find that pregnancy puts measurable strain on her kidneys, or elevates her blood pressure, more than a pregnancy at an earlier age did, but that when she is not pregnant, these organs function very well.

There is a kind of muscle reserve as well, for few adults develop, or ever need to use, all the muscle capacity that they could develop during their years of peak strength. Whatever one's level of muscle development in early adulthood, maximum strength potential typically begins to decline at about age 30, but so gradually that 50-year-olds can expect to retain 90 percent of the strength they had at age 20, and the 10 percent that is lost is rarely missed (Hodgson and Buskirk, 1981). Consequently, among adults living in developed countries where hard manual work is not a daily necessity, a healthy 50-year-old can perform virtually all the tasks of everyday living as well as a 20-year-old and still have strength to spare.

Figure 17.3 *Construction work in today's urban areas demands planning, judgment, and skill (and, at least in this case, balance), but not much physical strength. Today machines perform most of the tasks that required muscle power in earlier times.*

The most important muscle of all, the heart, shows a similar pattern. The average maximum heart rate—the number of times the heart can beat a minute under extreme stress—declines steadily, as organ reserve is reduced with age. But the resting heart rate remains very stable, as Table 17.2 shows (Brooks and Fahey, 1984). Once again, while peak performance shows declines, this aspect of heart functioning for most of daily life is unaffected by aging until late adulthood.

Thus, most of the age-related biological changes that occur during the first decades of adulthood are of little consequence to the individual in the normal course of

TABLE 17.2 **Heart Functioning**

Average Maximum Heart Rate	Decade of Life				
	20s	30s	40s	50s	60s
Men	195	190	182	175	162
Women	188	185	178	172	152
Average Resting Heart Rate					
Men	75	75	75	75	72
Women	72	72	72	72	70

A CLOSER LOOK   Aging Athletes

If all else is equal, athletes in their late teens and early 20s have a small physiological edge over slightly older athletes, and a distinct advantage over those in their 30s and 40s. Different sports rely on different abilities, so age of peak performance varies from sport to sport (Bromley, 1974). Flexibility and a slight body are great assets in women's gymnastics, so it is no wonder that most world-class champions at this sport are in their teens. Boxing and basketball, which demand quick reaction time, foot and hand speed, stamina and strength, are two sports in which performance tends to peak around the mid-20s. In sports such as bowling and golf—in which aim and accumulated experience are more important than power and speed—peak performance occurs a bit later.

Women athletes tend to reach their physiological prime sooner than men, partly because they begin puberty sooner. This is shown by the greater number of female athletes who turn professional in their teens. It is also reflected in the age of peak performance in speed swimming: female swimmers are their fastest at about age 13; males, at about 18 (DeVries, 1980).

However, for athletes, just as for the rest of us, much depends on the individual's lifestyle and willingness to adapt to changing abilities. Maintaining good health habits and a rigorous training schedule will enable athletes to reach their full potential and extend the years of their star performance (Lamb, 1984). At the same time, experience helps compensate for some degree of physical decline, which is one reason most athletes reach their best performance several years after their physiological peak. Some athletes successfully change their playing style to increase the years they can play. Others attempt to ignore the changes that age brings and fight a losing battle with time.

In general, however, conditioned older athletes can perform so much better than untrained, undisciplined, younger persons that they should serve more as an inspiration than as an example of inevitable physical decline. Practiced marathoners in their 60s have run 26 miles at less than 8 minutes a mile, yet many sedentary 20- and 30-year-olds can't even run around the block. Obviously, many declines that we consider inevitable with age are, instead, the consequences of how we do or don't take care of ourselves.

Mary Lou Retton was only 16 when she won an Olympic gold medal in gymnastics, a victory made possible partly because her body had the wonderful flexibility of youth. As they grow older, many professional athletes find ways to adjust their technique to compensate for purely physical declines. Kareem Abdul-Jabbar, pictured here in the 1987 NBA finals, has lost some of the quickness, speed, and stamina of his youth. However, he has developed the "sky-hook"—a deadly accurate shot that allows him to move away from an opponent rather than having to drive past him. His ability to make adjustments in his game, and his dedication to staying in shape, have allowed Jabbar to continue playing top-level basketball beyond his 40th birthday.

events, especially if the person develops a lifestyle that safeguards health. There are two notable exceptions. The first is athletic performance. Minor differences in strength, reaction time, and lung efficiency can have a notable impact on the ability of professional athletes, as well as on that of weekend players (see A Closer Look, page 396). The second exception is changes in the sexual-reproductive system, changes that we will now discuss.

The Sexual-Reproductive System

The decades from 20 to 40 are the most likely time for adults to begin a long-term sexual relationship and to reproduce. The changes that occur in the sexual-reproductive system during this period therefore have the potential for having a major impact, particularly with regard to reproduction.

Sexual Responsiveness

In both sexes, sexual responsiveness varies for many reasons—innate predispositions, childhood experiences, and cultural preferences and proscriptions among them. Age is also a factor, one that effects men and women somewhat differently.

During the early years of manhood, sexual excitement, which includes a faster heart beat and penile erection, can occur very quickly and frequently, in response to many things—even an idea, a photograph, or a passing remark. Typically, orgasm also occurs fairly quickly. For both young and older men, orgasm usually is followed by a refractory period during which sexual arousal is not possible, but for some young men, a second sexual cycle can regularly follow the first almost immediately.

As men grow older, they often need more direct or explicit stimulation to initiate the excitement phase. In addition, as men age, a longer time elapses between the beginning of excitement and full erection, between erection and ejaculation, and between orgasm and the end of the refractory period (Whitbourne, 1985). Nevertheless, age-related declines in sexual responsiveness are, for the most part, not a concern until middle or late adulthood. Indeed, for most men, frequency of sexual intercourse increases from adolescence until about age 30, because, although their sexual responses are somewhat diminished from when they were teenagers, they are more likely to have a steady sexual partner (Harman, 1978).

Age-related trends in sexual responsiveness are not as clear-cut for women as for men. In general, however, it seems that, as they mature from early adolescence toward middle adulthood, women become more likely to experience orgasm during love-making rather than less likely (Sloane, 1985). Part of the reason for this may be that the slowing of the man's responses makes the sex act likely to last longer, providing the more prolonged stimulation that many women need to reach orgasm. Another possible explanation for women's increased sexual responsiveness during early adulthood may be that, with experience, both partners may be more likely to recognize and focus on those aspects of love-making that intensify the woman's sexual responses.

Fertility

While young adults need not worry about the effects of aging on their sexual responses, the same does not hold true for the effects of aging on their reproductive potential. About 15 percent of all couples have fertility problems, and age is one factor. With every passing year, the number of couples who are **infertile**—usually defined as being unable to conceive after a year of trying—increases. One specific statistic makes the point: about one couple in twenty is infertile when the woman is in her early 20s; about one couple in seven has this problem when the woman is in her early 30s (Menken et al., 1986). It should be noted that fertility statistics are somewhat misleading, since they are usually based on the age of the woman; in early adulthood both sexes contribute about equally to fertility problems (Glass, 1986).

Fertility Problems in Men   The most common fertility problem in men lies in the low number of their sperm or in the sperm's poor **motility,** or ability to swim quickly and far enough to reach an ovum. In order for a single, normal sperm to reach and fertilize an ovum, the mathematical probabilities of conception require that a man produce at least 20 million sperm per milliliter of ejaculate, that at least half of them still be able to move two hours after intercourse, and that at least 60 percent of them be normal in shape (Glass, 1986). For a number of reasons, including genital abnormalities, about 5 percent of young American adult males do not meet these requirements and therefore are infertile (Lipshultz and Howards, 1983). Most young men, however, have more than twice the required number of normal, motile sperm. As these men grow older, the number, shape, and motility of their sperm are adversely affected, but the declines are very gradual. Changes in sperm caused by normal aging usually do not result in sterility, but a man in middle age will probably require more attempts to impregnate a woman than he would have required in early adulthood.

Age is not the only factor that can affect sperm production in normally fertile men, however. Sperm grow in the testes, in tiny long tubes, over a period of seventy-four days. Thus, at any given moment, billions of sperm are in the process of development, and the lengthiness of this process increases the sperm's vulnerability. Anything that impairs normal body functioning, such as an illness with a high fever, or medical therapy involving radiation or a high dosage of prescription drugs, or exposure to environmental toxins, or unusual stress, or an episode of drug abuse, can affect the number, shape, and motility of the sperm for several months (Bardin, 1986; Newton, 1984). Although this type of impairment is not necessarily age-related, age might be a factor: since the incidence of illness rises with age, the chances of a man's fertility being periodically reduced also increase with age.

Fertility Problems in Women   One common fertility problem in women is difficulty with ovulation. A small percentage of women, perhaps 2 percent, do not ovulate naturally, no matter what their age. Most other women ovulate regularly once their menstrual period is well established by late adolescence, but find that ovulation becomes less regular as middle age approaches. There may be cycles with no ovulation, and other cycles when several eggs are released. Thus older women take longer to conceive, and they are more likely to have twins when they do.

The other common fertility problem for women is blocked Fallopian tubes, often caused by pelvic infections—called **pelvic inflammatory disease,** or **PID**—that were not treated promptly. If a woman experiences one or more episodes of PID, she has about one chance in five of becoming sterile (Menken et al., 1986). Although blocked Fallopian tubes are not caused directly by age, if a woman is sexually active with a number of partners over a number of years, she obviously is more likely to get sexually transmitted diseases, such as gonorrhea or chlamydia, which can cause PID.

Finally, if a woman has trouble conceiving, she may have **endometriosis,** a condition in which fragments of the uterine lining become implanted and grow on the surface of the ovaries or the Fallopian tubes, blocking the reproductive tract. Endometriosis is most common between the ages of 25 and 35, and is the likely explanation for about 15 percent of fertility problems (Halme, 1985a).

Age and Fertility   Because successful conception and pregnancy gradually become more difficult with each passing year, most physicians recommend that women begin their childbearing before age 35, and that men realize that fatherhood is less likely to occur once early adulthood is over (Mazor and Simons, 1984). Nevertheless, it is important not to exaggerate the relationship between age and infertility. While age is one factor in many fertility problems, it rarely is the primary cause. Many adults who postpone parenthood until their 30s and then find conception difficult to achieve might have had the same problem if they had tried to have a baby ten years earlier (Menken et al., 1986).

Further, many fertility problems can be solved by modern medical techniques. Minor genital abnormalities that cause infertility in the male are often correctable through surgery. Alternatively, a man with a low sperm count can store his sperm over a period of days, and then his partner can be artificially inseminated with sufficient sperm for conception to occur. In women, an inability to ovulate can usually be treated with drugs to stimulate ovulation. Blocked Fallopian tubes can often be opened. When they cannot be, surgical removal of ova and **in vitro fertilization** outside the uterus may be an option. This technique, experimental in 1978 when the first "test tube" baby was born, is now widely available, with a success rate of about one baby born in ten attempts (Halme, 1985b). Most couples who try in vitro fertilization without success try again, with the success rate of repeated attempts being about one in three.

Overall, of those infertile couples who obtain medical help, half or more eventually have a child of their own. Those who are younger when they obtain medical advice have an even higher chance of conception (Mazor and Simons, 1984). For those who cannot conceive (and for some who can), adoption is a viable route to parenthood.

Figure 17.4   *While pregnancy isn't always as idyllic as it is pictured to be here, many couples, especially those with fertility problems, convince themselves that having a baby is worth any cost, stress, or pain conception might entail. For the one couple in fifteen for whom successful pregnancy is impossible, this conviction makes it particularly difficult to accept their fate.*

Three Troubling Problems

Although the picture of physical development and health in early adulthood has been fairly sanguine so far, it, of course, is not trouble-free for all young adults. A number of diseases that commonly appear in middle or late adulthood—such as cancer, cirrhosis of the liver, coronary heart disease—may have already gotten a toehold in early adulthood, although the symptoms are not yet apparent. The course of these diseases, and of physical development generally, is substantially affected by the individual's lifestyle, and in Chapter 20 we will examine the overall systemic impact of such lifestyle factors as cigarette smoking, years of heavy alcohol consumption, nutrition, exercise, and stress. In this section we will address three problems that are more prevalent in early adulthood than at any other stage—drug abuse, destructive dieting, and violent death.

A Life-Span Perspective

Drug Abuse

Drug abuse—defined as such use of a drug that it impairs one's physical, cognitive, or social well-being—is a topic that touches every point in the life span. To begin with, many of the factors influencing drug abuse begin early in life. Indeed, genetic inheritance is a major factor in determining an individual's propensity for, and response to, any given drug. Some people, for example, are unlikely to become alcoholics because their bodies are so intolerant of alcohol that even a single drink makes them nauseous (Dietrich and Spuhler, 1984; Goodwin, 1984; Zucker and Gomberg, 1986). Others, by their genetic makeup, are quite vulnerable because the way their bodies metabolize alcohol makes drinking more pleasurable, and hence more addictive, than for other people. Genes are also the likely explanation for the fact that children of alcoholics, even if raised by adoptive parents, are more likely to become alcoholics themselves than are children of nonalcoholics (Cadoret and Gath, 1978). That there is a genetic predisposition to alcoholism is strongly suggested by various studies showing that (1) sons of alcoholic fathers are roughly four times as likely to become alcoholics themselves as are sons of nonalcoholics; (2) children of alcoholics, even if raised by adoptive parents, are more likely to become alcoholics themselves than are children of nonalcoholics (Cadoret and Gath, 1978); (3) the concordance rate for alcoholism between identical twins is twice that between fraternal twins (Porjesz and Begleiter, 1985). It has also been suggested that sons of male alcoholics tend to show aberrations in brain-wave patterns prior to any exposure to alcohol (see Figure 17.5). Genetic factors may also account for variations in the ways individuals react to alcohol.

The likelihood of a person's abusing alcohol or drugs in adulthood is also affected by that person's sex, temperament, and experience of family patterns and cultural context during childhood and adolescence. For example, the easy-going, confident child in a warm and stable family who grows up in a subculture that teaches moderation or abstinence with regard to drugs is unlikely to become a drug abuser, especially if that child is a girl. On the other hand, the hostile child with low self-esteem, growing up in a discordant, drug-abusing family, is a prime candidate for becoming a substance abuser, especially if that child is a boy (Leigh, 1985).

Furthermore, the consequences of drug abuse affect people at every stage of life. As we have already seen at various points in the text, the fetus may suffer brain and body damage if the mother is a drug user or an alcoholic; children at every age are at higher risk of abuse and neglect if a parent is a drug user; adolescents may put

Figure 17.5 *Brain responses can be measured by recording the electrical brain-wave activity evoked when various stimuli are presented. In one study, boys aged 7 to 15, whose fathers were alcoholics, and a matched group of sons of nonalcoholic fathers were presented with the stimulus of an unusual tone and their brain waves were measured at the onset of the stimulus. Although the sons of alcoholics were nondrinkers, their brain-wave responses differed from those of the other boys. Similar, but more pronounced, differences are found in the evoked brain responses of many adult alcoholics, whether they were still drinking or long abstinent.*

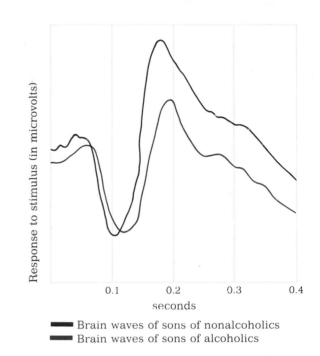

■■■ Brain waves of sons of nonalcoholics
■■■ Brain waves of sons of alcoholics

themselves at risk by using mind-altering substances on a mind that in many ways is not yet matured. Throughout adulthood, too, drug abuse impairs cognition and distorts motivation, resulting in any number of interpersonal problems, from spouse abuse and broken families to job loss and criminal behavior. And, as we will see in Chapter 24, drug misuse can have a particular consequence for many of the aged, who may be dismissed as being senile when, in fact, they are suffering from a drug reaction. Of all the stages of life, however, young adulthood is the time when problem-drinking and drug abuse are not only most likely; they are also most likely to result in damage.

According to a large longitudinal study, use of legal and illegal drugs in the United States increases from late adolescence through early adulthood, peaking at about age 23. This same study finds that more young adults than adolescents are daily drinkers of alcohol, and that 75 to 80 percent of today's young adults have tried an illicit drug, mostly marijuana but increasingly cocaine, with first use of the latter typically occurring at age 18 or later (Johnston et al., 1986). In 1985 about 20 percent of all Americans in their mid-20s used cocaine at least once (Kozel and Adams, 1986), compared with 12 percent for high-school seniors (Johnston et al., 1986). In addition, compared with people of other age groups, young adults are more prone to drinking to excess and to daily use of marijuana (Stephens, 1987).

Why the high rate of drug use and abuse in the first years of adulthood? There seem to be at least four reasons:

1. Many young adults are in transition between families, becoming increasingly independent of their family of origin but not yet established in a family of their own. In fact, for some young adults, drug abuse is a way of striving for independence from parents, even to the point of forcing the parents to push them away (Stanton, 1985). Being single, as most young adults are, correlates with drug and alcohol abuse.

2. A number of life stresses—completing an education, finding a mate, establishing a career—cluster during the 20s. Many people abuse alcohol and other drugs in an effort to escape these stresses, if only for the moment (Yost and Mines, 1985).

3. The need to feel sexually attractive and fulfilled is often very intense during these years, and many young adults believe that alcohol and other drugs enhance sexual responsiveness. In fact, the connection between drugs and heightened sexuality is more myth than reality, more in the distorted perceptions of the drug user than in the functioning of the sexual organs. For the most part, heavy drug use diminishes, rather than increases, sexual responsiveness (Kolodny, 1985).

4. The lifestyle of many young adults encourages alcohol and drug use. For example, young adults, much more than older adults, frequent bars, clubs, large parties, and huge concerts where drinking and drug-taking are often the secondary activity, if not the primary one. Young adults are also more likely to be away from home, at college, or in the military, and more likely to live in larger urban areas than older adults are. Drugs are more readily available in these contexts than in society at large (Harford, 1984).

Figure 17.6 *Partying that includes drinking and smoking is a favorite activity for many single young adults. Ironically, sporting events are a prime occasion for such pastimes. The party-goers pictured here are celebrating the victory of the racing sloop* Intrepid *over the* Kookaburra II, *and appear to be giving added meaning to the title "Americas Cup Race."*

Does drug abuse during early adulthood signal a lifetime of addiction? Not usually. Most young adults who overdrink or abuse drugs gradually realize that such behavior is destructive, and are able to overcome the specific problem. Drug abuse, in fact, becomes less common with every decade of adulthood (see Figure 17.7).

In the meantime, however, many young drug-abusing adults do themselves or others serious harm. Almost always they miss some work or school, and their productivity eventually deteriorates; often they neglect or abuse their sexual partner; sometimes they commit crimes to support their habit; and all too frequently they become involved in serious accidents or commit acts of extreme violence. Drug use is involved in most fatal single-car crashes (Ross, 1984), in most homicides, and in many young-adult suicides. Sometimes the drug itself proves lethal. As more young adults are using cocaine, for example, cocaine-related deaths have risen dramatically (between 1981 and 1985 the rate tripled), largely because cocaine use can put a sudden strain on the heart (Cregler and Mark, 1986; Kozel and Adams, 1986). Furthermore, the ability to master the developmental tasks of young adulthood—getting an education, finding a suitable career, establishing lifelong friendships and love relationships—is impaired by the irrationality and social misjudgment that heavy drug use entails.

Figure 17.7 *This chart reveals a comforting fact. For many reasons, alcohol consumption, like consumption of other drugs, tends to "age-out," that is, become less common with age. However, the data on the far right reveal a distressing truth: far too many young adults drink to the point of drunkenness. This is one of the underlying reasons social problems such as violent fights, unintended pregnancy, and job absenteeism are much more common among adults under age 30.*

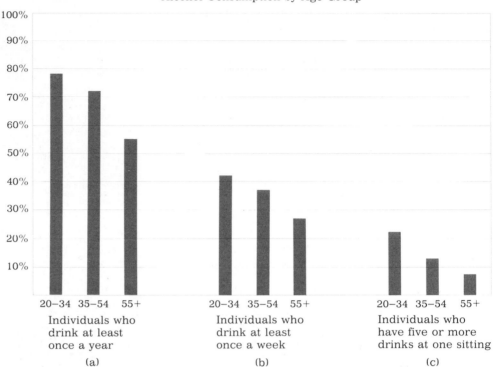

Alcohol Consumption by Age Group

20–34 35–54 55+	20–34 35–54 55+	20–34 35–54 55+
Individuals who drink at least once a year	Individuals who drink at least once a week	Individuals who have five or more drinks at one sitting
(a)	(b)	(c)

Prevention of Drug Abuse

How best can drug abuse be prevented in early adulthood? Certainly education about the consequences of drug abuse is crucial, but it is difficult to evaluate the successfulness of any of the various programs currently being used. One thing does seem clear, however. Providing accurate information is much more effective than resorting to exaggeration and scare tactics. The latter approach was employed in the fight against marijuana use in the early 1970s, and marijuana use only rose; once accurate information about the potential hazards of the drug was widely known, marijuana use fell (Johnston et al., 1986). With respect to illicit drugs, the best source of information is often a former user about the same age as the abuser, rather than an older authority figure. Former users are also best able to help each other in the effort to remain drug-free, although assistance from trained therapists is also needed (Nurco et al., 1983).

In fact, prevention strategies in general seem to work best when they are tailored to the particular person by someone who knows the individual and is familiar with the way a specific drug can be used and abused (Stephens, 1987). At the University of Massachusetts, for example, a general educational campaign about alcohol abuse sponsored by the administration did not reduce problem-drinking, and the passage of a state law raising the drinking age was followed by an increase in drunken behavior and alcohol-related accidents on campus. However, students who participated in a seminar designed to help them develop their own informed opinions about problem-drinking reduced their excessive drinking, and became much more aware of the problem-drinking in their friends (Kraft, 1984). Intensive behavior modification is also a promising strategy with young adult problem-drinkers and drug abusers (Celluci, 1984), especially when it teaches them preventive techniques to apply to themselves. Among the specifics taught in one such program were:

1. self-monitoring to become more aware of one's drinking behavior;

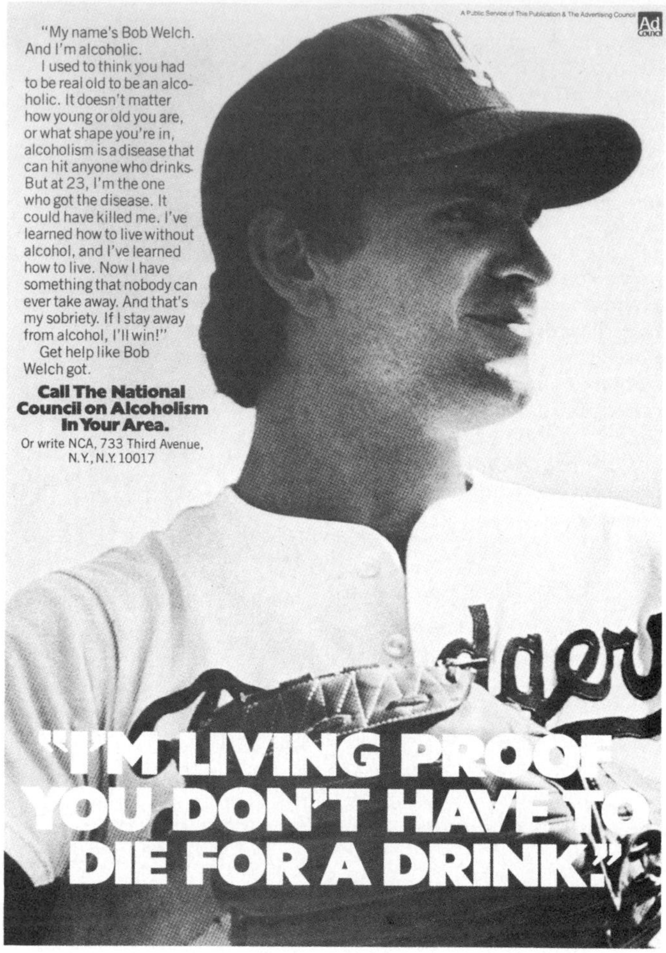

Figure 17.8 *Especially for teenagers and young adults, the image of the alcoholic is the bum on the street or the bleary-eyed, beer-bellied middle-aged man who drinks himself into a stupor at the corner bar. Campaigns such as this one help dispel the stereotype, teaching that addiction to alcohol can begin at any age. Other ads remind young women that they, too, can become alcoholics.*

2. new skills, such as how to say "no" to a drink;

3. estimating alcohol concentration in the blood;

4. specific drinking behavior that lessens problem drinking, such as sipping, rather than gulping, drinks;

5. rewarding oneself for moderation;

6. developing alternative strategies—relaxation, meditation, recreation—to deal with tension (Miller, 1979).

Drug abuse is a problem for both men and women in early adulthood, although men seem more likely to experience serious consequences—like getting into trouble with authorities, and doing themselves and others bodily harm. Now let us turn to a problem that is mostly prevalent among women, one that can be harmful, even destructive, both psychologically and physically.

Compulsive Eating, Destructive Dieting

Feeling fat, dieting, losing weight, putting it back on, counting calories, doing spot exercises, reading diet books, feeling guilty about eating, fasting—all these are experiences that most adult American women have again and again. Indeed, "an overwhelming number of American women currently feel too fat (regardless of their actual weight) and engage in repeated dieting efforts" (Rodin et al., 1985).

All this concern does not result in lower weight or better health. Most American women, dieting or not, gain weight between 20 and 40, and most weigh more than the American woman her age did thirty years ago (Metropolitan Life, 1983). Further, as one physician, writing about women, explains: "There is no evidence that being a little fat is any kind of health hazard, although being considerably overweight . . . or considerably underweight . . . is related to illness and premature death" (Sloane, 1985).

In fact, being "a little fat," perhaps 10 pounds more than the standard height-weight tables, is more natural and healthy for adult women than being very thin. Indeed, if body fat becomes too low, as often occurs in women who overdiet or overexercise, the menstrual cycle becomes irregular or absent altogether.

Unfortunately, cultural standards of female beauty are becoming nearly fat-free. For example, the measurements of Miss America contestants and of *Playboy* centerfolds have decreased over the same thirty years that the average American's woman's weight has increased (Garner et al., 1980). Consequently, many women who do not conform to the current lean, athletic look regard themselves as unsexy. As Kim Chernin (1981) notes, the irony of this situation is apparent when we consider America's most enduring female sex symbol, Marilyn Monroe:

> In 1959, when Marilyn Monroe made the film *Some Like It Hot,* she was voluptuous, as large as a woman in a Renoir painting. For those of us who fell in love with her then . . . that film today is a revelation. She was, by modern standards, fat.

Continual worry about weight and repeated dieting can be destructive, not only for peace of mind and self-confidence, but also for physical health. One problem is crash diets, nearly all of which result in nutritional imbalance and some of which have resulted in death when maintained over long periods. A more general problem is that, ironically, repeated or extensive dieting alters the metabolism in such a way that the body can maintain its weight on fewer calories, with the result that "even normal eating after dieting may promote weight gain" (Striegel-Moore et al., 1986). Frequently, the result is feelings of frustration, guilt, and depression, as one's best efforts are "sabotaged" by the body's natural chemistry (Johnson and Connors, 1987).

Figure 17.9 *This woman's body is very close to meeting the current American ideal for the feminine figure. Whether such leanness is more healthy than having a "little extra" is controversial. What is clear, however, is that in holding themselves to this ideal, and especially in trying to match it, many women subject themselves to physical and psychological harm.*

More destructive is "the yellow brick road to weight reduction"— taking stimulants that work for a few weeks until tolerance builds, requiring the person to quit or get hooked, or using over-the-counter appetite suppressants that can produce such side effects as insomnia, tenseness, anxiety, and, in megadoses, psychosis (Johnson and Connors, 1987). Diet pills are one of the drugs that women are far more likely to abuse than men.

As we saw in Chapter 14, concern with weight can often become pathological, as in the case of anorexia nervosa. Another dangerous eating disorder, which is becoming increasingly common, is **bulimia,** compulsive binge-eating (itself often a consequence of dieting) followed by compulsive vomiting or use of laxatives. Not only is this eating disorder unhealthy; it may be life-threatening, causing severe damage to the gastrointestinal system or creating an electrolyte imbalance sufficient to cause a heart attack (Johnson and Connors, 1987).

In varying degrees, concern with weight seems to be part of the female experience throughout most of the life span. However, the early years of adulthood, a stage of life when appearance and self-control are particularly stressed, are the time of greatest risk overall. The usual bulimic patient is in her 20s, having begun the binge-and-purge cycle when she was 18 (Johnson and Connors, 1987). Young college women are especially likely to diet, take diet pills, binge, and purge (Rodin et al., 1985; Squire, 1983). In fact, one study found 20 percent of college senior women to be bulimic (Pope et al., 1984).

Often, these eating disorders go hand in hand with depression. In fact, suicide is a danger for women with eating problems (Johnson and Connors, 1987).

Violent Death

Just as restrictive stereotypes about "feminine" appearance may promote the eating disorders that afflict mostly women, stereotypes about "manly" behavior may lead to a problem that afflicts young adult males much more than women—violent death.

One American male in every forty dies a violent death sometime between his 15th and 35th birthday, from an accident, a suicide, or a homicide. Such deaths are most likely to occur to young men in their 20s, who are four times more likely to die a violent death than young women the same age are (see Table 17.3). According to one scientist, who titled his analysis "Warning: The Male Sex Role May Be Dangerous to Your Health" (Harrison, 1984), the explanation for this propensity toward an untimely violent end lies in prevailing notions of "masculinity," which lead many young adult males to put themselves at deadly risk. Another scientist drew the same conclusions, listing four esteemed "masculine" traits that place many young men in jeopardy (Brannon, 1976):

TABLE 17.3 **Annual Rate of Violent Death of American Adults Aged 20 to 29**

	Deaths per 100,000 Men	Deaths per 100,000 Women
Suicide	26	6
Homicide	29	7
Accidents	80	19
Total	135	32

Source: National Center for Health Statistics, 1986.

1. "No Sissy Stuff"—the need to be different from women.

2. "The Big Wheel"—the need to be superior to others.

3. "The Sturdy Oak"—the need to be independent and self-reliant.

4. "Give 'Em Hell"—the need to be more powerful than others, through violence if necessary.

Living up to these masculine "ideals" makes it hard for a young man to avoid a conflict, or to back away from a challenge, or to admit that he needs help, especially emotional help, even if doing so would remove him from a life-threatening situation. This aversion to backing down is the likely explanation for an interesting set of facts: young adult women have higher rates of depression, and of suicide attempts, than young men do, but young men kill themselves more than four times as often as young women (see Table 17.3). Instead of using the suicide methods favored by women, such as cutting a wrist or swallowing pills, which allow time for second thoughts or for intervention by someone else, the male's method is likely to be something immediately lethal, such as a gunshot to the head or a leap from a bridge.

The likelihood that a young American adult male will meet a violent death is, of course, affected by other social and cultural factors as well. One very significant factor in the high rate of homicide in the United States—the highest rate among developed nations—is the ready availability of handguns. In most other countries, possession of handguns is severely restricted, but in the United States there are more than 60 million handguns in the population—and nearly two-thirds of all homicides in this country are committed with a handgun or other firearm (Centers for Disease Control, 1986). (Guns are also involved in more than half of all suicides.)

Other social factors can be clearly seen in the fact that, in the United States, blacks are only half as likely to kill themselves as whites are, but they are six times more likely than whites to be murdered—94 percent of the time by another black person (U.S. Department of Health and Human Services, 1985). For young adult black men, homicide is the leading cause of death: a black adolescent has one chance in twenty-one of being murdered before reaching old age. Since the majority of these homicides occur in urban ghettos, it seems likely that the conditions of ghetto life—including disproportionately high rates of crime, drug abuse, broken families, and unemployment—are, at the least, contributing factors to this dismal prospect.

Figure 17.10 *Handguns, rifles, and ammunition are readily available in many places in the United States, among them, mail-order houses in Texas, the neighborhood grocery store in Maine, and, as pictured here, at outdoor flea markets in California.*

Figure 17.11 *When a car leaves the road and smashes directly into a tree, killing the driver, what is the cause of death? Officially, such deaths are usually listed as accidental, but many such "accidents" could be suicides, committed either consciously or unconsciously. Many victims of such accidents are young adult men, rarely wearing seat belts, and often with alcohol or other drugs in their bloodstreams.*

Accident rates are also influenced by social factors. In the United States, for example, the rate of fatal accidents for Native Americans is double the national norm, while the rate for Asian-Americans is only one-third as high. Although there are a number of plausible explanations for these differences, one factor seems clearly implicated: Native Americans have the highest rate of alcoholism among ethnic groups in the United States, and Asian-Americans, one of the lowest. Interestingly, for both groups, genetic factors may play an important role in this difference. Compared with other groups, Asians are more likely to react to alcohol metabolically in ways that make them not want to drink at all, whereas Native Americans are more likely to be genetically predisposed to addiction (Goodwin, 1984).

Comprehensive analyses of the reasons for varying rates of violent death and recommendations for preventive measures are, obviously, beyond the scope of this book, and, in fact, beyond the reach of most current scientific research. Despite a great deal of rhetoric, politicians, physicians, psychologists, and the population at large have yet to work on solving the problem of violent death of young adults with anything approaching the effort or the money spent on controlling the leading killers of middle-aged and older adults. For example, for every productive year of life lost due to cancer, more than $500 is spent in cancer research. By comparison, for every year of productive life lost through violent death, less than $3 is spent on research to learn how to prevent such loss (Foege, 1985).

Given the grimness of the three troubling problems we have been considering, we should end this chapter with a reminder that all the self-destructive behaviors just discussed are problems for only a minority of young adults. Most young adults manage the transition from adolescence to adulthood well and pass through their early adulthood healthy and robust. As we will see in the next chapter, although early adulthood is not easy, most young adults become increasingly capable of understanding and coping with their lives long before they reach the next stage of development.

SUMMARY

Growth, Strength, and Health

1. While young adults do not grow significantly taller in their 20s, they typically grow stronger and fuller as their bodies reach adult size. In terms of overall health, as well as peak physical condition, early adulthood is the prime of life.

Age-Related Changes

2. With each year from 20 to 40, all the body systems gradually become less efficient—declining at the rate of about 1 percent a year — and homeostasis takes longer to reach.

3. However, because of organ reserve, none of these changes is particularly troublesome or even noticeable for most people most of the time. Even athletic performance, while slowed somewhat, can remain at a high level.

The Sexual-Reproductive System

4. As middle age approaches, the speed of sexual responses slows down in men, but not in women. These modest declines usually have no negative effect on the man's sexual experiences, and, in some cases, they may enhance the woman's experience.

5. When they reach their early 30s, about 15 percent of all couples have fertility problems. One common reason is that the man's sperm are insufficient in quantity or motility. Another common problem is that the woman's ova do not reach the uterus because the Fallopian tubes are blocked, or because ovulation itself does not occur.

6. The normal aging process is rarely the primary cause of infertility in the two decades of early adulthood, but age can be a contributing factor. While most couples can conceive a child even in their early 40s, those who have fertility difficulties should get medical assistance early so that the age-related declines in the reproductive system do not make existing problems worse.

Three Troubling Problems

7. Young adults are more likely to use alcohol and illicit drugs than are people of any other age, often doing themselves serious harm. Prevention of drug and alcohol abuse in early adulthood works best when it is personalized and based on accurate information.

8. Eating disorders are also more common during young adulthood than at other ages, as some young women feel a compulsion to be thinner than their bodies naturally tend to be.

9. Suicide, homicide, and fatal accidents are a serious problem for young adults, especially for young men in our society. The reasons are at least as much cultural as biological, as revealed by ethnic differences in the rates of these three causes of violent death.

KEY TERMS

homeostasis *(393)*	endometriosis *(399)*
organ reserve *(394)*	in vitro fertilization *(399)*
infertile *(397)*	drug abuse *(400)*
motility *(398)*	bulimia *(406)*
pelvic inflammatory disease (PID) *(398)*	

KEY QUESTIONS

1. In what specific ways is early adulthood the prime of life?

2. How is the physical performance of a 20-year-old athlete likely to be different from that of a 40-year-old?

3. As a person ages, what are the changes that occur in organ reserve? How do these changes affect a person's activities?

4. What are some of the factors that tend to diminish fertility toward the end of early adulthood?

5. What can be done to prevent and remedy the main causes of infertility?

6. Why are young adults particularly susceptible to drug use and abuse?

7. How can concern about being fat become a health hazard?

8. What are the sex differences in the rate of violent deaths, and how do you explain them?

CHAPTER 18

Early Adulthood: Cognitive Development

*The knowledge of the world is only to be acquired
in the world, and not in a closet.*

Lord Chesterfield
Letters to His Son

*Life is a succession of lessons which must be
lived to be understood.*

Ralph Waldo Emerson
"The Illusion"

Over the course of adulthood, there are a number of changes that occur in our thinking processes. Age-related changes occur in how fast we think, in how efficiently we process new information, in how much old information we have to draw upon, and in what we think about. We may also think more deeply, broadly, and wisely as we grow older, but evaluating such differences involves value judgments, so an objective assessment of this aspect of cognition is difficult to arrive at.

Although it is clear that there are cognitive changes over the entire span of adulthood, it is not clear which of the major approaches to studying cognition yields the most satisfying picture of those changes. The developmental approach, which emphasizes cognitive stages and processes that shift predictably with age, has recently been applied to adulthood in an effort to see if there is cognitive development beyond the formal operational thinking described by Piaget. The psychometric approach, which focuses particularly on changes in the components of intelligence as measured by IQ tests, addresses the highly controversial question of whether specific cognitive abilities improve or decline over the course of adulthood. The information-processing approach, which studies the moment-by-moment mental processes of input, encoding, memory, and output, is primarily concerned with how the efficiency of these processes may be affected as the individual grows older.

Each of these three perspectives is valid and useful throughout adulthood, and each provides somewhat different insights into the nature of cognition throughout the life span. However, to use all three in each of the three chapters on adult cognition might be confusing, and would certainly seem repetitive. Therefore, we will concentrate on each separately. Our primary focus in this chapter will be on the developmental approach. The psychometric approach will be emphasized in Chapter 21, and the information-processing approach, in Chapter 24.

Adult Thinking

To nearly every adult, adult thinking seems different from adolescent thinking in many ways. By common definition, adult thinking is less self-centered, broader, more reasonable, more practical. But what are the actual components of adult thinking, and what encourages their development? Although the answer to the first part of the question is complex and controversial, many investigators agree that the cognitive patterns that emerge in adulthood are largely initiated and directed by the individual's taking on the responsibilities and commitments of adult life. As Gisela Labouvie-Vief (1985) explains:

> Much as the theme of youth is flexibility . . . that of adulthood is commitment. Careers must be started, intimacy bonds formed, children raised. Here, amidst a world of multiple possible alternatives, there is a need to adopt one course of action. This conscious commitment to one pathway . . . may indeed mark the onset of adult cognitive maturity.

As we will see in this chapter and Chapter 21, one way that the taking on of commitments affects cognitive development is to channel and focus the individual's cognitive abilities and interests in specific directions. It also deepens the individual's ties to others and to the surrounding social world, requiring the person to more seriously consider differing points of view and to find and negotiate solutions to the many conflicts and obstacles that come in the way of his or her commitments.

Schaie's Stages

The idea that personal commitment is one hallmark of adult thought is stressed particularly by K. Warner Schaie (1982), who has proposed four stages of adult cognition that correspond to the patterns of commitment and social emphases in adult life (see Table 18.1). Schaie begins with a basic distinction: cognitive development before early adulthood is, for the most part, unrelated to the specifics of the individual's life; thereafter, it is rooted in them. Schaie believes that, cognitively, childhood and adolescence constitute a **period of acquisition**, during which information is absorbed and problem-solving techniques are learned, with little regard for their actual importance to the young person's life. Thus a bright high-school senior could, with equal enthusiasm, devote time and energy to learning nuclear physics, analyzing Elizabethan poetry, and taking quilt-making lessons, all in the same day. The acquisitive thinker learns a subject because it is taught, in much the same way that the famous mountaineer Lord Mallory climbed Mount Everest "because it is there."

TABLE 18.1 **Schaie's Stages of Cognitive Development**

Childhood and Adolescence	Early Adulthood	Middle Adulthood	Late Adulthood
Acquisition (Piaget's four stages)	Achieving (goal-directed learning)	Responsible (concern for others)	Reintegrative (wisdom)
		Executive (concern for social systems)	

Beginning in the late teens or early 20s, a shift occurs as young people move away from an indiscriminate acquisition of knowledge and enter the **achieving stage**, in which they *use* knowledge to establish themselves in the world. According to Schaie, the thinking of young adults is "much more goal-directed," displaying "more efficient and effective cognitive function with respect to tasks which have role-related achievement potential." In contrast to the high-school dilettante, a young adult might well strive to excel at nuclear physics or Elizabethan poetry or quilt-making, but, unless a very unusual career goal is in sight, he or she is unlikely to specialize in all three.

As middle adulthood approaches, many people enter a third stage, called the **responsible stage,** in which "the goal-directed entrepreneurial style of the achieving stage will be replaced by a pattern which facilitates integrating long-range goals as well as consequences for one's family unit in the solution of real-life problems" (Schaie, 1982). In other words, while the young businessperson might have a clear goal of personal achievement—perhaps to be a corporation president by age 35—in ten years or so he or she tries to adjust personal goals to fit with family goals. Being rich and powerful may no longer seem as important as having well-nurtured, happy children, and the personal risk-taking or 80-hour work weeks that seemed an acceptable part of striving for personal achievement may be tempered by a greater concern for the needs of the people one loves.

For some adults, according to Schaie, the middle years also bring an unusually broad and deep sense of responsibility, which leads to a particular new stage of cognition called the **executive stage.** People at this stage are concerned about larger social systems: they may well be in charge of a company, a school, a town government, and so forth. Their concerns and obligations are more complex and stretch further than those of the usual person in the responsible stage, for they must coordinate the needs of various social groups, some of whom may have conflicting interests.

Finally, in late adulthood, a **reintegrative stage** appears, when thoughts turn to making sense of life as a whole. At this point, people may turn inward to focus on their own lives or turn outward to the cosmos, seeking the purpose of life in general.

Figure 18.1 *According to Schaie, thinking during early adulthood is largely achievement-oriented. Being focused on the attainment of specific goals, it tends to be narrower and more intense in its concerns than the "responsible" mode of thinking that may emerge as the individual approaches middle adulthood.*

Schaie's description of adult thinking is, of course, very general, and obviously does not apply to everyone in every culture. Taken as a whole, however, the direction of adult thinking described by Schaie points to a type of thought that can be distinguished from the concrete or formal operational thinking of the younger person. Several other researchers have described adult thought in ways that fit in with Schaie's overview. The resulting depiction has been called postformal thought, to call attention to the fact that it follows formal operational thinking.

Postformal Thought

As we saw in Chapter 15, Piaget's hierarchy of cognitive development culminates in formal operational thought, the stage in which the individual can engage in hypothetical deductive reasoning. Usually, this stage is not attained until a combination of adolescent maturation and educational experience makes it possible. (Indeed, even many adults do not pass Piaget's tests.)

Once the capacity for formal thinking appears, a person can begin to follow the reasoning of logical arguments and apply logical processes to solve various problems, particularly problems involving mathematical and scientific principles. When fully developed, the capacity for formal thinking enables a person to coordinate such logical relationships as combination, negation, proportion, reciprocity, correlation, incompatibility, and reciprocal exclusion, and therefore to be able to consider all the hypothetical possibilities of a given logical system. Whether one considers something as majestic as Einstein's special theory of relativity or as commonplace as the design of an automobile engine, it is clear that formal thinking is a powerful problem-solving tool.

But is it the kind of thinking that adults generally use in their daily lives? Is it even the most suitable for solving real-world problems? A number of researchers say no to both questions. As they point out, formal thinking is best suited to solving problems in a **closed system,** that is, one in which a finite number of known variables result in a specific outcome. In such a system, the relationships among all the variables can be analyzed separately, and then in combination until a single, correct answer is arrived at. Most of the problems of daily adult life, however, occur in a multiplicity of overlapping **open systems**—family, work, friends, community—that are characterized by ambiguity, partial truths, and an infinite number of variables, many of them unknown. Some researchers have found that formal thought can be too formal and rigid to be effective in dealing with open systems. Their work has led them to describe a type of thinking beyond formal thought, a dynamic, in-the-world cognitive style that is generally referred to as **postformal thought.**

Postformal thought is less abstract, and less absolute, than formal thought: it can adapt to life's inconsistencies and, at its best, is dialectical, that is, able to combine contradictory elements of a thought or situation into a more comprehensive whole. We will explain both of these characteristics of postformal thought in detail.

Adaptive Logic   Labouvie-Vief (1985, 1986) points out that the traditional models of mature thought stress objective, logical thinking and devalue the importance of subjective feelings and personal experience. This kind of thinking, she maintains, is adaptive for the schoolchild, the adolescent, and "novice adult," because it permits them to "categorize experience in a stable and reliable way." However, it may be maladaptive in trying to understand, and deal with, the complexities and commitments of the adult world. For the adult, subjective feelings and personal experiences must be taken into account, or the result will be reasoning that is "limited, closed, rigidified." In this view, truly mature, adaptive thought involves the interaction between abstract, objective forms of processing and expressive, subjective forms that arise from sensitivity to context.

To demonstrate the development of this form of thought, Labouvie-Vief and her colleagues presented subjects between the ages of 10 and 40 with brief narratives that tested problem-solving logic. Because the researchers were more interested in their subjects' problem-solving approach than in their problem-solving competence, the tests were designed to be superficially simple and logical but to allow for deeper interpretations outside the straightforward propositions of the text. One such story went as follows:

> John is known to be a heavy drinker, especially when he goes to parties. Mary, John's wife, warns him that if he comes home drunk one more time, she will leave him and take the children. Tonight John is out late at an office party. John comes home drunk. —Does Mary leave John?

In arriving at their answers, all the young adolescents and many of the older ones reasoned strictly according to the basic premise of the story: in the case of the drunken husband, it was evident to them that Mary would leave John because that is what she said she would do. Older respondents, of course, recognized the explicit logic of the story, but they resisted the limitedness of the narrative's logical premise and explored the real-life possibilities and extenuating circumstances that might apply—whether, for example, Mary's warning was a plea rather than a final ultimatum, whether John was apologetic or abusive upon his return home, whether Mary has somewhere to go, what the history of the marriage relationship might be, and so forth. At the most advanced level, adults tried to "engage in an active dialogue" with the text, forming multiple perspectives as a result (Adams and Labouvie-Vief, 1986). This appreciation and reconciliation of both objective and subjective approaches to real-life problems is the hallmark of adult adaptive thought.

Dialectical Thought   A related approach in describing adult thought comes from theorists who consider **dialectical thought** the most advanced form of cognition (Basseches, 1984; Leadbeater, 1986; Riegel, 1975). The word "dialectical" refers to the philosophical concept that every idea, every truth, bears within itself the suggestion of the opposite idea or truth. In the terms used by philosophers, each new idea, or *thesis*, implies an opposing idea, or *antithesis*. Dialectical thinking involves considering both poles of an idea simultaneously and then forging them into a *synthesis*, that is, a new idea that integrates both the original idea and its opposite. The idea of the dialectical process also emphasizes that, because in-the-world systems are open rather than closed, constant change is inevitable, and the dialectical process, continual.

For our purposes here, we may say that in daily life, dialectical thinking involves the constant integration of one's beliefs and experiences with all the contradictions and inconsistencies they encounter. The result is a continuously evolving view of oneself and the world, a view that recognizes that few, if any, of life's most important questions have single, unchangeable, correct answers.

This does not mean that dialectical thinkers adopt the idea that "everything is relative" and stop there, unable to commit themselves to values. On the contrary, a dialectic view explicitly recognizes the limitations of relativistic positions such as, "If you think it is true, then it's true for you" or even "If people in one particular culture believe it, then it's true for them" (Leadbeater, 1986). Truly dialectical thinkers, in fact, acknowledge both the open, inconsistent, subjective nature of reality *and* the need to make firm commitments to values that they realize are likely to change over time.

Let us see how the dialectical process might work in a simple example. Take the aphorism "Honesty is the best policy," which many people accept uncritically. A dialectical thinker, too, might begin by agreeing with this thesis, but would then consider the opposite idea: honesty can cause hurt feelings, or foolish behavior, or destructive emotions. Whereas a relativistic thinker might go on from this point to conclude that honesty should be employed, not universally as a set policy, but, rather, selectively, as a means to specific ends, the dialectical thinker arrives at a synthesis of the two ideas, deciding, say, that honesty is a valuable goal in itself, even when it seems hurtful at the moment. Of course, as the dialectical thinker realizes, the dividing line between short term and long term is not easy. In the short term, and in some emergency situations, honesty may be the worst policy, but, in the long term, honesty may indeed be the best policy because people are better able to relate to each other if they know they can count on each other for the truth, and the more of the truth one knows, the better one can make decisions.

The dialectical process does not stop here, however, for this synthesis is constantly refined in an endless succession of real-life situations. In each new case, the dialectical thinker attempts to ascertain how, why, and in what form of expression honesty is best, recognizing all the while that whatever choices he or she makes may have to be reconsidered in light of new information or changing circumstances. You can see that dialectical thought is a complicated process, seeking answers that are adaptive rather than simple and fixed.

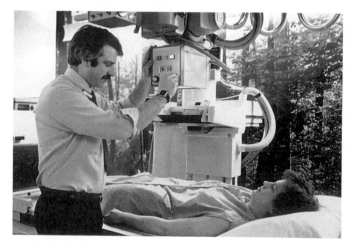

Figure 18.2 *Dialectical thinking is often encouraged by confrontation with ideas and experiences that challenge our everyday assumptions. For example, exposure to a foreign culture sometimes causes people to question certain of their own culture's values, and then to appreciate them again in a new light. Having to face a medical crisis can also lead to a dialectical process in which the meaning of one's life is questioned, reexamined, and then reaffirmed.*

Now let us look at an example of dialectical thought as it might apply to an intimate experience that is probably familiar to everyone, the fading of a love affair (Basseches, 1984). A nondialectical thinker is likely to see the partners in any relationship as having fixed traits, and to see the relationship itself as arising from the static interaction of those traits. Faced with a troubled romance, then, the nondialectical thinker would most likely arrive at a single, static explanation—perhaps that one partner or the other is at fault, or that the partners are basically incompatible and the relationship was a mistake from the very beginning. In this view, either the partner "at fault" must change or the relationship must be terminated. By contrast, the dialectical thinker sees personal relationships as dynamic and interactive, the relationship creating the partners as much as they create the relationship. Thus the dialectical thinker realizes that the personalities, needs, and circumstances in any relationship change, making alterations in that relationship inevitable. Taking this view opens a new option to the partners—that they attempt to accommodate their relationship to the changes that have occurred in it and in themselves, in effect creating a new relationship between them.

A recognition of the continually evolving nature of human relationships gives the dialectical thinker a broader and more flexible perspective on all aspects of personal and social interaction, making that person better able to adapt to the flux of life.

Adult Thinking in the Moral Realm

According to many researchers, the responsibilities and concerns of adulthood can also affect moral reasoning, propelling a person from a lower moral stage to a higher one (Kohlberg, 1973; Rest, 1983; Rest and Thoma, 1985). In fact, Kohlberg states that in order to be capable of "truly ethical" reasoning, a person must have "the experience of sustained responsibility for the welfare of others and the experiences of irreversible moral choice which are the marks of adult personal experience." The development of faith follows a similar path, for the same reasons (see A Closer Look on pages 418–419).

Carol Gilligan has looked particularly at the relationship between adult life experiences and a broader understanding of moral issues. As we saw in Chapter 15, Gilligan believes that in matters of moral reasoning, males tend to be more concerned with the question of rights and justice, whereas females are more concerned with the human relationships involved, tending to put personal needs above abstract principles. According to Gilligan, as people's experience of life expands, and especially as they become involved in the care and nurture of others, they often shift from ideological or personal reasoning to ethically responsible reasoning. They realize that both strictly formal reasoning about ethical dilemmas and reasoning based solely on personal context are limited (Gilligan, 1981; 1982).

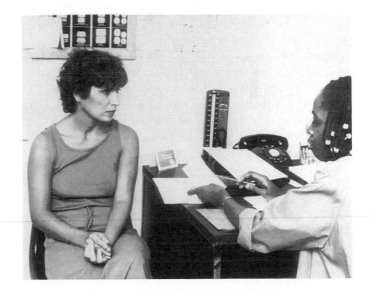

Figure 18.3 *For many women, decisions about reproduction can involve difficult moral choices. Family-planning counselors, who are trained to listen and inform rather than to direct, can assist in clarifying the issues and heighten women's sense of responsibility for their own decisions.*

One young man whom Gilligan studied illustrates this shift very well. In late adolescence he was able to reason abstractly, and at an advanced level, about the moral dilemma involving Heinz and his dying wife (see page 347), citing the principle that life is more important than money. But seven years later, in early adulthood, he had become much more aware of the personal implications of his answer:

The Development of Faith

Thinking about religious matters is another aspect of adult cognitive development that has interested some researchers. Like morality, faith obviously is not only a cognitive process: it involves practice as well as preaching; it arises from religious insight as well as religious education. Nonetheless, one view of faith is as a developmental process: as a person has more experience trying to reconcile religion with daily life, his or her faith may reach higher levels.

The most detailed description of the development of faith comes from James Fowler (1981, 1986), who delineates six stages of faith. It should be noted that when Fowler describes "faith," he does not necessarily mean religious faith. He agrees with Paul Tillich (1958) that all humans need to have faith in something, whether that something is God and church, philosophical principles, country, or simply oneself. Faith gives humans a reason for living their daily lives, a way of understanding the past, a hope for the future. It is whatever each person really cares about, his or her "ultimate concern" in Tillich's words.

Stage One: Intuitive-Projective Faith

Stage-one faith is magical, illogical, imaginative, and filled with fantasy, especially about the power of God and the mysteries of birth and death. It is typical of children, ages 3 to 7.

Stage Two: Mythic-Literal Faith

At this stage, the individual takes the myths and stories of religion literally and believes simplistically in the power of symbols. In a religious context, this stage usually involves reciprocity: God sees to it that those who follow his laws are rewarded, and that those who do not are punished. Stage two is typical of middle childhood, but it also occurs in adulthood. For example, Fowler cites the case of a woman who says extra prayers at every chance, in order to put them "in the bank." Whenever she needs divine help, she thinks she can withdraw some of her accumulated credit.

Stage Three: Synthetic-Conventional Faith

Interpersonal relationships and a tacit, nonintellectual acceptance of cultural or religious values are typical of stage three. Unlike stage-two faith, stage-three faith serves to coordinate the individual's involvements in a complex social world, providing a sense of identity and adding significance to the rituals and symbols of daily life. For example, Fowler describes one man who puts his faith in his relationship with his family, a man whose personal rules include "being truthful with my family. Not trying to cheat them out of anything . . . I'm not saying that God or

anybody else set my rules. I really don't know. It's what I feel is right." Because of his commitment to his family, he has learned to accept the "rat race" of his daily work. These responses are typical of the conformist stage of faith, which is conventional, concerned about other people, and values "what feels right" more than what makes intellectual sense.

Stage Four: Individual-Reflective Faith

By contrast, stage-four faith is characterized by intellectual detachment from the values of the culture and from the approval of significant other people. The experience of college can be a springboard to stage four, as the young person learns to question the authority of parents, teachers, and other powerful figures and to rely, instead, on his or her own understanding of the world. An unexpected experience in adulthood, such as a divorce, the loss of a job, the death of a child, can also lead to stage four. The adult's understanding of faith ceases to be a matter of acceptance of the usual order of things and becomes, instead, an active commitment to a life goal and lifestyle that differs from that of many other people.

Fowler's example of someone at the fourth stage of faith is Jack, whose time in the army provided him with a chance to think and to talk with people from other backgrounds, and gradually to develop a philosophy. Jack explains:

I began to see that the prejudice against blacks that I had been taught, and that everybody in the projects where I grew up believed in, was wrong. I began to see that us poor whites being pitted against poor blacks worked only to the advantage of the wealthy and powerful. For the first time I began to think politically, I began to have a kind of philosophy.

Jack's ability to articulate his own values, distinct from those of family, friends, and culture, makes his faith an individual-reflective faith.

Stage Five: Conjunctive Faith

This type of faith incorporates both powerful unconscious ideas (such as the power of prayer and the love of God) and rational, conscious values (such as the worth of life compared with that of property), and is characterized by a willingness to accept contradictions. It involves a synthesis of the magical understanding of symbols and myths that characterized stage two and the conceptual clarity of stage four. Fowler cites one woman at this stage who believes strongly in God, but adds, "I don't think it matters a bit what you call it. I think some people are so fed up with the word God that you can't talk to them about God." Her recognition that the word "God" may be distracting and misleading is typical of the ability of the stage-five thinker to articulate paradoxes and con-

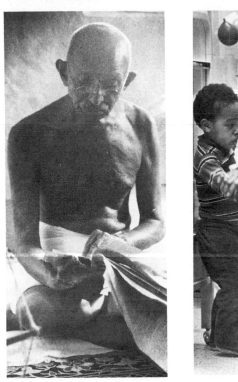

Neither Gandhi (who helped lead India to independence from Great Britain through his campaign of nonviolent resistance) nor Mother Hale (who has cared for hundreds of infants born addicted to drugs) planned to take on the heroic roles they eventually did. As Fowler and Kohlberg explain, people often develop a deeper and more universal faith or moral vision, and sometimes transform their lives, as events and experiences lead them to feel a greater connectedness to their fellow beings.

traditions in faith. Also typical of this stage is an openness to new truths; this woman explains her beliefs by referring to Jesus, George Fox, Krishna-Murti, and Carl Jung. Fowler says this cosmic perspective rarely comes before middle age.

Stage Six: Universalizing Faith

People at stage six have a powerful vision of universal compassion, justice, and love that compels them to live their lives in a way that, to most people, seems either saintly or foolish. They put their own personal welfare aside, and sometimes even sacrifice their lives, in an effort to enunciate universal values. Often, a transforming experience converts an adult to stage six, as happened to Moses when he saw the burning bush, and to Paul as he traveled on the road to Damascus. Fowler mentions some twentieth-century people who have reached this level, among them Gandhi, Martin Luther King, Jr., and Mother Theresa, each of whom radically redefined their lives after a particular experience produced a new understanding of human brotherhood. Clearly, a person reaching stage six of faith is an exceedingly rare individual.

Indeed, the scarcity of people at the upper stages of Fowler's hierarchy might make one wonder how useful

it is. Moreover, it may be galling to read that there are "higher" stages of faith than most adults, probably including the readers and the writer of this book, are likely to reach—especially when it is considered that some of the "lower" levels of thinking can be seen as no less valid than the "higher" levels. It is easy to imagine, for instance, Fowler's students coming to class, as Kohlberg's once did, wearing buttons proclaiming "Stage Two and proud of it" and insisting that what Fowler characterizes as "childish" stage-two faith is actually the purest kind.

In Fowler's defense it should be noted that he never explicitly says that the higher stages are better. In fact, Fowler explains:

Each stage has its proper time of ascendancy. For persons in a given stage at the right time *for their lives*, the task is the full realization and integration of the strengths and graces of that stage rather than rushing on to the next stage. Each stage has the potential for wholeness, grace and integrity, and for strengths sufficient for either life's blows or blessings.

If Fowler is correct, faith, like other aspects of cognition, may progress from a quite simple, self-centered, one-sided perspective to a more complex, altruistic, and multisided view.

This is a very crisp little dilemma and you can latch onto that principle pretty fast and say that life is more important than money. But then, when you reflect back on how you really act in your own life, you don't use that principle, or I haven't yet used that principle to operate on. And none of the people who answer that dilemma that way use that principle to operate on because they were blowing $7,000 a year for their education at Harvard instead of giving it to the Children's Fund to give porridge to the kids in Botswana and to that extent answering the dilemma with that principle is not hypocritical, it's just that you don't recognize it. I hadn't recognized it at the time, and I am sure they didn't recognize it either. [Gilligan and Murphy, 1980]

Similarly, a woman stated that she once thought there were no absolute principles of right and wrong: "I went through a time when I thought things were pretty relative, that I can't tell you what to do and you can't tell me what to do, because you've got your conscience and I've got mine." But at age 25, she held these views:

Just seeing more of life [led me to recognize] that there are an awful lot of things that are common among people. There are certain things that . . . promote a better life and better relationships and more personal fulfillment than other things, and . . . you would call [those things] morally right . . . I have a very strong sense of being responsible to the world, that I can't just live for my enjoyment, but just the fact of being in the world gives me an obligation to do what I can to make the world a better place to live in, no matter how small a scale that may be on.

In general, Gilligan says, women must come to see the importance of establishing principles to help guide them between the conflicting needs of their relationships to others and their obligations to themselves, while men need to give greater recognition to their own personal responsibility for their world. As adults are confronted with important moral decisions in their own lives, some of them, at least, are able to achieve this synthesis.

A Fifth Stage?

It has been suggested by some theorists that the type of thinking described as postformal thought represents a new stage of thinking. Other theorists maintain that postformal thinking cannot be considered a stage in the Piagetian sense of the term, because it is not age-based, it is not universal, and it is not totally different in structure from earlier thinking. They believe that it is more accurate to regard postformal thought as a style of thinking, a style that seems to be related more to the complexity of one's life experiences, and perhaps to the degree of one's education, than to factors connected to age itself.

Basseches (1984) has done substantial research on dialectical thought, in part by asking people thought-provoking questions (such as "What is the nature of education?") and then scoring their answers on the basis of twenty-four characteristics of dialectical thought. He found that both life experiences and education tend to promote dialectical thinking, although neither guarantees its development. Indeed, the picture of adaptive, dialectical adult thinking presented by postformal theory is one that should perhaps be looked on as much more ideal than normative. Not only do some adults never use a postformal style of thinking, but many who do, do so irregularly, or only in specific areas.

Cognitive Growth and Higher Education

Of particular interest to many academics doing research, and to readers of this text, is the relationship between college education and adult thinking processes. Although most people today attend college to secure a better job (cited as a primary reason by 76 percent of the incoming students in 1984) and to learn specific information, such as how to program a computer or how to prepare and interpret balance sheets (Higher Education Research Institute, 1985), the avowed goal of most colleges and universities is the intellectual development of the students who attend them. How successful are they in this objective? Do people think deeper and better because they have been to college? To be more specific, is there any sign that they are more likely to reach postformal thinking, combining the practical and the theoretical in a flexible, dialectical way?

Figure 18.4 *Many parents today wonder whether a college education is worth the cost. Purely in terms of enhancing an offspring's earning potential, it usually is. But a full answer to the question involves intangibles that are impossible to measure. The value of the cognitive growth that can result from the college experience cannot be expressed in dollars and cents.*

There is no doubt that, in general, education powerfully influences cognitive development. Years of education are strongly correlated with virtually every measure of adult cognition, even more so than other powerful variables such as age and socioeconomic status (Labouvie-Vief, 1985; Reese and Rodeheaver, 1985).

In addition, past longitudinal research on students at Vassar and elsewhere has shown that college education in particular leads people to become more tolerant of political, social, and religious views that differ from their own and to be more flexible and realistic in their attitudes (Chickering, 1981; Webster et al., 1979). Indeed, some research even finds a year-by-year progression in this process. It begins with freshmen year, when students believe that there are clear and perfect truths to be found, and are disturbed if they do not discover them or if their professors do not provide them. This phase is followed by a wholesale questioning of personal and social values and of the idea of Truth itself. Finally, after carefully considering many opposing ideas, students become committed to certain values, at the same time realizing the need to remain open-minded and prepared for change (Perry, 1981; Sanford, 1970).

This progression has been described in detail by William Perry, primarily on the basis of an intensive study of Harvard students. As you can see from the chart on page 423, Perry found that, over the course of their college careers, the thinking of his subjects progressed through nine levels of complexity, going from a simplistic either/or dualism (one is either right or wrong, a success or a failure) to a relativism that acknowledged a multiplicity of perspectives. According to Perry, this progression was a product of the college environment: when students reached a new level, their peers, professors, reading, or classwork stimulated new questions that opened the way to the next level.

These findings do not mean that students, en masse, switch from conservative to liberal values during their college years. In fact, the difference is not so much a change in attitudes as a change in the way one's attitudes are held—with greater confidence and tolerance. College experience seems to make people more accepting of other people's attitudes and ideas because it makes people less threatened by them (Katz and Sanford, 1979).

Research that focuses specifically on dialectical reasoning finds less clear-cut steps of development but nonetheless suggests that the more years of higher education a person has, the deeper and more dialectical that person's reasoning is likely to become. This was shown in a detailed study of students at Swarthmore (Basseches, 1984), as well as in more general research elsewhere (King et al., 1983; Rest and Thoma, 1985).

Can we say, then, that the years of early adulthood usually show a marked broadening and deepening of thinking, a movement toward dialectical thought, especially if the individual attends college? Not necessarily.

The College Student of Today

Much of the research that showed clear development in dialectical thinking among college students was done some twenty years ago, when fewer young adults went to college and when the majority of these students were pursuing a liberal arts education. In addition, the sample populations for most of the research were full-time residential students at quite selective institutions. As Figure 18.5 shows, in recent years the number of students in higher education has multiplied significantly in the United States and in virtually every country worldwide (Geiger, 1986). Correspondingly, the demographic characteristics of the student body have changed both in

Figure 18.5 *Some observers charge that as the percentage of high-school graduates attending college has risen, the standards of a "higher education" have fallen. In some respects, they are correct, since there has been a corresponding trend away from the classic liberal arts education. At the same time, many of these critics fail to recognize the positive cognitive impact that a "nontraditional" education can have.*

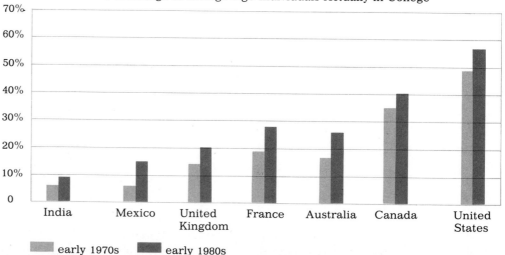

Percentage of College-Age Individuals Actually in College

early 1970s   early 1980s

Scheme of Cognitive and Ethical Development

	Position 1	Authorities know, and if we work hard, read every word, and learn Right Answers, all will be well.
DUALISM MODIFIED	Transition	But what about those Others I hear about? And different opinions? And Uncertainties? Some of our own Authorities disagree with each other or don't seem to know, and some give us problems instead of Answers.
	Position 2	True Authorities must be Right, the others are frauds. We remain Right. Others must be different and Wrong. Good Authorities give us problems so we can learn to find the Right Answer by our own independent thought.
	Transition	But even Good Authorities admit they don't know all the answers *yet*!
	Position 3	Then some uncertainties and different opinions are real and legitimate *temporarily*, even for Authorities. They're working on them to get to the Truth.
	Transition	But there are *so many* things they don't know the Answers to! And they won't for a long time.
	Position 4a	Where Authorities don't know the Right Answers, everyone has a right to his own opinion; no one is wrong!
	Transition *(and/or)*	But some of my friends ask me to support my opinions with facts and reasons.
RELATIVISM DISCOVERED	Transition	Then what right have They to grade us? About what?
	Position 4b	In certain courses Authorities are not asking for the Right Answer. They want us to *think* about things in a certain way, *supporting* opinion with data. That's what they grade us on.
	Transition	But this "way" seems to *work* in most courses, and even outside them.
	Position 5	Then *all* thinking must be like this, even for Them. Everything is relative but not equally valid. You have to understand how each context works. Theories are not Truth but metaphors to interpret data with. You have to think about your thinking.
	Transition	But if everything is relative, am I relative too? How can I know I'm making the Right Choice?
COMMITMENTS IN RELATIVISM DEVELOPED	Position 6	I see I'm going to have to make my own decisions in an uncertain world with no one to tell me I'm Right.
	Transition	I'm lost if I don't. When I decide on my career (or marriage or values), everything will straighten out.
	Position 7	Well, I've made my first Commitment!
	Transition	Why didn't that settle everything?
	Position 8	I've made several commitments. I've got to balance them—how many, how deep? How certain, how tentative?
	Transition	Things are getting contradictory. I can't make logical sense out of life's dilemmas.
	Position 9	This is how life will be. I must be wholehearted while tentative, fight for my values yet respect others, believe my deepest values right yet be ready to learn. I see that I shall be retracing this whole journey over and over—but, I hope, more wisely.

Source: Perry, 1981.

the United States and elsewhere. There are more women students, more low-income students, more ethnic-minority students, more older students, more students who choose quite specific career-based curricula (computer programming, health services, engineering, accounting, business) rather than a broad liberal arts education, and more part-time students, who now make up almost half of the American college population (American Council on Education, 1984).

As both a cause and a consequence of this shift, attendance at four-year residential colleges has held steady, while nonresidential, public, and/or two-year colleges have expanded their enrollment dramatically (Frances, 1980). At such colleges, faculty tend to have fewer advanced degrees, to teach more classes, and to spend more time developing basic skills than discussing great philosophical ideas.

It would seem likely, then, that the earlier picture of the impact of college on cognitive growth, drawn twenty years ago at highly selective, residential, four-year schools, may not be as applicable today. It is certainly possible that part-time attendance at a local community college might have less cognitive impact than full-time residential attendance, partly because remaining within the familiar home and neighborhood setting might blunt the potential influence of the college culture, and partly because the academic milieu may be less intellectually challenging (Gaff and Gaff, 1981; Weis, 1985).

But before drawing that conclusion, consider several lines of research. First, it is impossible to use type of college as the sole basis for making assumptions about the impact any given school might have. The specific characteristics and background of a student and the specific nature of the institution are more important in determining intellectual growth than the general category the institution fits into (Alwin, 1976; Fleming, 1984; Riesman, 1980). Second, a great deal of the impact that any college might have on its students depends on the availability of professors as mentors and role models. This is particularly crucial in whether the student will become open to other viewpoints and to intellectual conflict (Haan, 1985b). Not only must the faculty, and the college milieu in general, challenge students to think more deeply, but they must also make students feel sufficiently supported so that they don't become intimidated or defensive when confronted with new ideas (Gaff and Gaff, 1981; Whiteley, 1982). This means that the proper match of students with teachers is more important than the students' aptitude scores or the professors' credentials.

Figure 18.6 *The instructor in this photo happens to be discussing one of the factors that can be highly influential in the cognitive growth of college students—self-esteem. Students who are made to feel that their opinions are respected, and that their ignorance is forgiven in advance, are much more likely to be open to new ideas, constructive criticism, and disconcerting questions.*

One specific example of this has been studied with great care—predominantly black colleges. These institutions typically have fewer resources and a higher student-to-faculty ratio than predominantly white colleges. However, many black students experience greater cognitive development (as measured by tests of intellectual depth and by plans for future education and career) at such colleges than black students at predominantly white schools, especially those that have only recently enrolled a significant number of blacks (Fleming, 1984). One likely explanation is that the faculties of predominantly black schools, as teachers and as role models, are more sensitive to the concerns and needs that black students in particular may have. Given a similiar faculty sensitivity to students' needs, it seems reasonable to assume that community colleges, colleges designed for older students, and other schools not considered among the academic "elite" may all have the potential to exert a greater impact on the intellectual development of their students than simply looking at the students' admissions folders or professors' academic achievements would predict (Fleming, 1984; Riesman, 1980; Weathersby, 1981).

Figure 18.7 *Much of the learning that occurs in college takes place outside the formal setting of the classroom. This conversation among American students and their classmate from Africa is likely to lead to personal insights that no lecture or text could provide.*

At the moment, we do not know how commonly colleges of today promote dialectical thought. Obviously, more research needs to be done on students who are representative of the contemporary college population. However, it is certainly possible that they, like the students of earlier generations, find their thinking processes broadening between the day of orientation and the day of graduation.

Cognitive Growth and Life Events

Research on one final topic is spotty, but the tentative conclusions are intriguing, especially from a developmental perspective. It has been suggested that many **life events,** or specific notable occurrences, can trigger new patterns of thinking and thus cognitive development.

Parenthood is a prime example. From the birth of a first child, which tends to make both parents feel more "adult"—thinking about themselves and their responsibilities differently—through the unexpected issues raised by adolescent children, parenthood is undoubtedly an impetus for cognitive growth (Feldman et al., 1981; Flavell, 1970; Galinsky, 1981). One aspect of that growth is that parenthood

Figure 18.8 *Some events that lead to new patterns of thinking are commonplace. For many people, entering into marriage begins a series of cognitive shifts that intensify the individual's sense of commitment to others. Brief, dramatic events can also result in long-term cognitive changes. The photo on the right is of Jessica Mc-Clure and her mother after the Texas toddler was pulled from the well shaft in which she was trapped for fifty-six hours while the nation breathlessly awaited her rescue. Jessica's parents may never think of their lives, or of Jessica's, in quite the same way they did before the accident.*

offers "insights into the interdependence of the larger social system" (Veroff and Veroff, 1980), which is, as we have learned, an important aspect of mature thought.

Similarly, other life events also might make people think more deeply about the nature and meaning of their lives and their relationship with others. A new intimate relationship or the end of an old one, a job promotion or dismissal, being the victim of a violent attack or, unexpectedly, someone's savior, exposure to a radically different lifestyle, an intense religious experience or in-depth psychotherapy, experiencing the death of a loved one—all these can occasion cognitive disequilibrium and reflection, which, in turn, can result in a new view of oneself and the meaning of one's life.

Evidence for this abounds in biographical and autobiographical literature and in personal experience. Probably every reader of this book knows someone who seemed to have a narrow and shallow outlook on the world in early adulthood, but who later developed a deeper and broader perspective as experience and insight accumulated.

A Case of Cognitive Maturity

One particularly poignant example, reported in a study of adult psychosocial development (Kotre, 1984), is the life of a woman called Dorothy Woodson. She was the last of seven children, born to a poor rural family who, she thought, never loved her. She grew up feeling "horribly ugly," stupid, and neglected, and her perspective on life as she approached adulthood was constricted and bleak. She had no confidence in herself nor any plans for the future. As she later said of herself, "Inside I was very fearful and uncertain. I always wanted somebody just to hold my hand through something. I didn't want to do anything by myself." Dorothy's early adult life reinforced her narrow life view. At 18 she had married a man on whom she felt desperately dependent but who rarely seemed to care about her. Within a year her first baby was born, weighing only 3 pounds. Neither her family nor her husband seemed much interested in him and gave Dorothy no help or encouragement in the difficult task of caring for a premature child. The infant developed slowly and died before his first birthday. Her second child, born soon after that, was "an extremely beautiful and precious" girl named Diana. Shortly after Diana's birth, Dorothy's husband moved to a city 200 miles away, leaving Dorothy alone with their daughter, afraid to protest his absence or to follow him. At one point when Diana was 2, Dorothy, desperate to see her husband, went to visit him, leaving Diana with a 13-year-old babysitter. The sitter's inexperience led to a tragic accident that killed Diana.

Not suprisingly, Dorothy's childhood experiences, her empty marriage, and the deaths of her first two children numbed Dorothy in many ways. Her view of life was shallow and pessimistic. She just barely managed to get through each day and was not at all ready to make commitments or to try to build any kind of future for herself. When she had her third child, James, she thought, "You're just going to die anyway, I'm not going to love you."

But then, gradually, Dorothy began to develop a new perspective on herself and her life. With the help of a friend, she became involved with a religious community, where she found not only emotional support but also encouragement to read and study. The real turning point for Dorothy came when she found a mentor who spurred and nurtured her intellectual and spiritual growth:

> I began to discuss my thoughts with him. He made me aware of studying critically. He started me wondering about a lot of things. He planted a lot of seeds in my mind, things I never even considered or aspired to. He would question me about everything and make me think about how I really felt, my gut reaction to things rather than parroting what I thought somebody wanted to hear. That opened up a whole new world for me.

As a result of this cognitive awakening, Dorothy began to see herself as worthwhile and her life as meaningful. More particularly, she began to be able to clarify her values and to see the issues in her life in a fuller dimension.

This is not to suggest that her whole life suddenly changed for the better. To the contrary, she was faced with continual struggle—with the self-doubt and pain that she carried from the past; with the difficulties of raising a child alone; with the tedium of the jobs she was restricted to because of her lack of formal education; with disillusionment in personal relationships; with various medical crises. What did change, increasingly over time, were the depth and complexity of the thinking she brought to these struggles. Listening to her speak of her life, and especially of her son, one is struck by the sense of balance she eventually achieved:

> I feel life is a process and that your experiences bad or good, specifically bad experiences, can either build character or destroy character. And if individuals are seeking a higher good or a higher reason for life, then the experiences will add to their character and add to their life. . . . One of my thoughts is that I've got to make every day count with my son. I mean every day. Not a day should pass that he's not loved and that his ideas and he as a person are not acknowledged and guided in some way.

. . . I see family traits that I have, that I've seen in my sisters and my mother and my son. It's like ancestral influences I'm trying to overcome and replace with stronger, more transcendental values. I guess that's what I'm trying to do with him, to let him see that I've passed things on to him—and his dad has too. He's got some pretty good stuff from both of us, but he's got some weaknesses too. If he doesn't focus on them and just concentrates on those strong areas and replaces the weak areas with the creative things he likes to do, then he's going to be that much more valuable to society. . . . He's turning out to be such a fine, fine individual that I want to be a part of that and his posterity. . . . I guess what I'm saying is that there is a reason to fight for life.

Thus certain key life experiences enabled Dorothy to break through her self-defeating, helpless view of herself, and to experience considerable cognitive growth. Her view of life, of parenthood, of her role in the larger society had become more complex, and at the same time more adaptive and more responsive to the contradictions inherent in personal experience.

Case studies, of course, do not prove general trends: they simply indicate what can sometimes happen. However, longitudinal studies that include many cases point in the same general direction. While there are always exceptions, the general movement of thinking about one's own life in adulthood is toward a more responsible and less self-centered view of the world (Haan, 1985a; Vaillant, 1977).

While much more longitudinal research needs to be done before firm conclusions can be drawn, the general theme of this chapter seems plausible. As people move from late adolescence toward middle adulthood, the interplay between thought and experience, between the logic of formal operational intelligence and the sometimes erratic, sometimes confusing challenges of daily life, may propel adults to a new, postformal style of thought. At that point, they may be able to recognize and adjust to the contradictions and conflicts of adulthood, conflicts that become more apparent in the next chapter.

SUMMARY

Adult Thinking

1. Adult cognition can be studied in several ways, from a developmental, stage perspective; from a psychometric perspective, focusing on measurement of specific cognitive abilities; or from an information-processing perspective. This chapter focuses on the developmental perspective.

2. A number of researchers believe that adult thinking is rooted in, and shaped by, the specific commitments the individual takes on. According to Schaie, as a person makes the commitments of adulthood and sets goals, he or she may pass through several cognitive stages: achieving, responsible, executive, and reintegrative.

3. Piaget's last stage of cognitive development, formal operational thought, is characterized by the ability to

solve problems in math and science. Many researchers believe that, in adulthood, the complex and often ambiguous or conflicting demands of daily life produce a new type of thinking, called postformal thought, which is better suited than formal thinking to coping with problems that have no correct solutions. Postformal thought is adaptive, integrating thinking processes and experience.

4. Postformal thinking may also be characterized as dialectical, capable of recognizing and synthesizing complexities and contradictions. Instead of seeking absolute, immutable truth, dialectical thought leads to a flexible, ever-changing approach.

5. Thinking about questions of faith and ethics may also progress in adulthood, along the lines of postformal thought. For example, Gilligan suggests that gender differences in moral thinking may be muted in middle age, as

men and women come to recognize the limitations of basing moral reasoning solely on abstract principles or personal concerns and try to integrate the two.

6. Although postformal thinking is sometimes described as a stage, it is not the same kind of universal, age-related stage that Piaget described. Its appearance is more gradual, dependent on particular experiences and education rather than on a universal, chronologically determined restructuring of mental processes.

Cognitive Growth and Higher Education

7. College education tends to make people more tolerant because they are less threatened by conflicting views. This has been shown with some certainty for full-time students at selective residential colleges in the 1970s; it may also be true for the broad range of college students today.

Cognitive Growth and Life Events

8. Life events probably also promote cognitive growth. Parenthood is one example, unexpected life events another. However, while life events certainly affect the cognition of some people, they have not been shown to have this effect universally.

KEY TERMS

period of acquisition *(412)*	closed system *(414)*
achieving stage *(413)*	open system *(414)*
responsible stage *(413)*	postformal thought *(414)*
executive stage *(413)*	dialectical thought *(415)*
reintegrative stage *(413)*	life events *(425)*

KEY QUESTIONS

1. What are three approaches to the study of adult cognition?

2. What are the limitations of formal operational thought?

3. What are the main characteristics of postformal thinking?

4. Can you describe an instance of dialectical reasoning in addition to the two cited in the text?

5. How might the moral thinking of adults be different from that of children and adolescents?

6. In what ways is postformal thinking not a stage in the Piagetian sense of the term?

7. According to research, how does college education affect the way people think?

8. According to your own observation, how does college education affect the way people think?

9. How might life events affect cognitive development?

CHAPTER 19

Early Adulthood: Psychosocial Development

*Just as the cautious businessman avoids tying up
all his capital in one concern, so, perhaps,
worldly wisdom will advise us not to look for the
whole of our satisfaction from a single aspiration.*

Sigmund Freud
Civilization and Its Discontents

In terms of psychosocial development, the hallmark of adulthood is diversity. No longer limited by the process of biological maturation or bound by parental restrictions, adults are much freer to choose their own developmental path from the many options open to them. In the final decades of the twentieth century, the possible paths have become incredibly varied. Both men and women have numerous career choices, marriage and parenthood possibilities, lifestyle and friendship options. Let us examine several themes that underlie and help to organize the complexity and diversity of adulthood.

The Tasks of Adulthood

No matter what specific pattern a particular adult life may form, two basic psychosocial needs drive its development. Various theorists describe these needs in somewhat different terms. For example, Maslow (1968) wrote about the need for *love and belonging,* which, if met, is followed by the need for *success and esteem.* Other psychologists have described these needs in terms of *affiliation* and *achievement,* or social *acceptance* and *competence.* Freud (1935) put it more simply, explaining that a healthy adult was one who could *love* and *work.*

Building on Freudian theory, Erik Erikson suggested that there are two basic crises, or tasks, in adulthood. One is **intimacy versus isolation,** which involves the need to share one's personal life with someone else or risk profound aloneness. As Erikson (1963) explains:

> . . . the young adult, emerging from the search for and the insistence on identity, is eager and willing to fuse his identity with others. He is ready for intimacy, that is, the capacity to commit himself to concrete affiliations and partnerships and to develop the ethical strength to abide by such commitments, even though they call for significant sacrifices and compromises.

In Eriksonian theory, the crisis of intimacy is followed by the crisis of **generativity versus stagnation,** which involves the need to be productive in some meaningful way, usually through work or parenthood. Without a sense of generativity, says Erikson, life seems empty and purposeless; adults have "a pervading sense of stagnation and personal impoverishment" (Erikson, 1963).

These two themes of adult life are recognized by almost all developmentalists, although researchers find considerable diversity in the particular method, as well as in the sequence, of meeting these basic needs.

Ages and Stages

Several authors have suggested that, within the broad themes of love and work, adults typically shift back and forth between periods of openness and change and periods of commitment and constancy. They explore alternatives in their lives, particularly those involving love and work, make choices among them, and follow through on those choices; later they reexamine their earlier choices and either recommit themselves to them or make changes in their lives. The most clear-cut delineation of this pattern has been put forth by Daniel Levinson (1978, 1986). Levinson believes that the years of adulthood can be seen as a chronological sequence in which life structures are built, reexamined, and rebuilt (See Table 19.1). According

TABLE 19.1   **Levinson's Stages of Adulthood**

Ages 17 to 22	*Early Adult Transition* Leave adolescence, make preliminary choices for adult life.
Ages 22 to 28	*Entering the Adult World* Initial choices in love, occupation, friendship, values, lifestyle.
Ages 28 to 33	*Age 30 Transition* Change in life structure. Either a moderate change, or, more often, a severe and stressful crisis.
Ages 33 to 40	*Settling Down* Establish a niche in society, progress on a timetable, in both family and career accomplishments.
Ages 40 to 45	*Midlife Transition* Life structure comes into question, usually a time of crisis in the meaning, direction, and value of each person's life. Neglected parts of the self (talents, desires, aspirations) seek expression.
Ages 45 to 50	*Entering Middle Adulthood* Choices must be made, a new life structure formed. Person must commit to new tasks.
Ages 50 to 55	*Age 50 Transition* Further questioning and modification of the life structure. Men who did not have a crisis at age 40 are likely to have one now.
Ages 55 to 60	*Culmination of Middle Adulthood* Build a new life structure. Can be time of great fulfillment.
Ages 60 to 65	*Late Adult Transition* Reappraisal of life. Moments of pride in achievement are interspersed with periods of despair.
Ages 65 to 80	*Late Adulthood* Make peace with oneself and others. Fewer illusions, broader perspective on life.
Age 80 plus	*Late Late Adulthood* Final transition. Prepare for death.

to Levinson, transitions, or periods of questioning and change, occur every decade, and then

> . . . as a transition comes to an end, one starts making crucial choices, giving them meaning and commitment, and building a life structure around them . . . When all the efforts of the transitions are done—the struggles to improve work or marriage, to explore alternative possibilities of living, to come more to terms with the self—choices must be made and bets must be placed. [Levinson, 1986]

As we will now see, Levinson has quite explicit ideas about the characteristics of each chronological period, as do several other authors who describe similar trends (Gould, 1978; Haan, 1981; Vaillant, 1977). Briefly, this is how they depict the most common patterns of early adulthood.

The 20s   The early 20s are considered the time to finish breaking away from one's parents (a task begun in adolescence, but now given new form through economic and residential independence) and to begin making choices concerning affiliation and achievement—choices regarding marriage, parenthood, further education, employment specialization, political, community, and church membership, and so forth. These choices are not necessarily considered lifetime ones; indeed, they are often explicitly provisional in nature, particularly those involving employment and social-group membership.

By their mid-20s, most adults have made some important commitments and decisions, and have a sense of who they are (Gould, 1978; Levinson, 1978). They have completed the move from the period of late adolescent exploration to adult stabilization, refining the patterns of their lives by accepting and accommodating to social norms. In choices regarding both the superficial aspects of life—such as, say, clothing style and entertainment—and the more abiding aspects—such as holding down a steady job and getting married—people in their 20s become recognizably adult. Often this is in marked contrast to the experimental and sometimes rebellious adolescents they once were (Haan, 1981). Both affiliation and achievement needs now begin to be met.

Figure 19.1 *What a difference a decade can make! Pictured here is Jerry Rubin, founder of the Yippie (Youth International Party) movement, who advised young people in the late 1960s to smoke pot, protest the Vietnam war, and "Never trust anyone over 30." On the left, he wears vintage hippie garb; on the right, in 1980, he wears the traditional costume of the establishment. The reason for the switch is that, at about age 40, he decided to take on a new job—as an investment analyst on Wall Street.*

The 30s As age 30 approaches, many young adults reexamine and question the commitments they have taken on. In Levinson's (1986) words:

> At about age 28 the provisional quality of the 20s is ending and life is becoming more serious, more "for real." A voice within the self says: If I am to change my life—if there are things in it I want to modify or exclude, or things missing I want to add—I must now make a start, for soon it will be too late.

Young adults seem to perceive the early 30s as being much older than the late 20s, and they therefore become more conscious of the limited time available for accomplishing their goals.

Age 30 has particular salience for many women (Reinke et al., 1985). Those who have no children must confront the fact that their childbearing years will soon be running out. For many, this necessitates a reexamination of career goals, of self-image, and of their relationship to their husbands or potential husbands and to their parents. This rethinking might lead to new directions in their lives (Notman, 1980; Wilk, 1986).

Similarly, at about age 30, many women who are mothers realize first-hand how quickly their children become more independent. Once these young mothers can again envision having time to themselves, they may renew or begin educational or career plans. Another major decision that many women in their 30s make is to avoid future childbearing forever: for American women, age 30 is, in fact, the peak age for having a last child (Koo et al., 1987), and by age 35, more than a third of married American women are surgically sterilized, with the early 30s being the most common age for this operation (Bachrach, 1984).

Men also undergo a period of questioning at age 30, although often they focus more on vocation than on questions of parenthood. Typically, they seek greater independence from the sponsors and advisors who helped them in their 20s (Bray and Howard, 1983), entering a stage Levinson calls BOOM (Becoming One's Own Man). They also reassess their position at work, asking for more responsibility, money, and power.

After resolving whatever questions age 30 brings, adults typically spend the rest of their 30s following through on work begun: men and women continue up the career ladder, bear their last child, and in many ways come as close to the stereotype of the dedicated, hard-working, career and family men and women as they will ever be (Haan, 1981; Levinson, 1978; Vaillant, 1977). According to a number of stage theories, the relatively conservative stage of career and family consolidation in the 30s might evolve into a new questioning phase as age 40 approaches, a topic discussed in Chapter 22.

The Social Clock

Although these rough chronological stages seem broadly applicable, they need to be put into perspective. Because biology no longer propels development, whatever stages of adulthood there might be are determined in large measure by a kind of **social clock,** a culturally set timetable that establishes when various events and behaviors in life are appropriate and called for. One of the reasons that development in adulthood is so diverse is that the social clock is not set the same for everyone. Each culture, each subculture, and every historical period has a somewhat different social clock, with variations in the "best" age to become independent of one's parents, to choose a vocation, to marry, to have children, and so on.

For example, some research has found that people of lower socioeconomic status generally see middle age as beginning around age 35, whereas people of higher status think of it as starting between 45 and 55 (Kuhlen, 1968). Not surprisingly, then, adults of lower socioeconomic status generally leave school, begin work, marry, and become parents at younger ages than middle- or upper-class adults do (Boocock, 1978).

The influence of socioeconomic background is particularly apparent with regard to the age at which women become wives and mothers. Women from the lower classes may feel pressure to marry at 18 and may decide to stop childbearing by age 25, while wealthy women may not feel pressure to marry until age 30 or to stop childbearing until age 40. Internationally, the social clock varies even more: in Venezuela, for instance, marriage is legal at age 12 for females and 14 for males, and more than half of all women are married before they are 19. By contrast, men and women in Sweden cannot legally marry until they are at least 18. Most Swedes wait much longer, with the median age of marriage being 26 for women and 30 for men (UNESCO, 1985).

The social clock's regulation of marriage varies historically as well. To pick one specific example, in the United States, the median age for becoming a new bride was 22 in 1910, 20 in 1956, and then 23.3 in 1985, the oldest since Americans have been recording this statistic (*New York Times*, 1986; U.S. Bureau of the Census, 1976).

Within each subculture, individuals are made aware of their personal social clock by pressuring from friends and family who, in various ways, admonish "It is time for you to . . ." or "You are getting too old to . . ." or, simply, "Act your age." As Neugarten explains:

> Age norms and age expectancies operate as prods and brakes upon behavior, in some instances hastening an event, in others delaying it. Men and women are aware, not only of the social clocks that operate in various areas of their lives, but they are aware also of their own timing, and readily describe themselves as "early," "late," or "on time" with regard to family and occupational events. [Neugarten et al., 1968]

Figure 19.2 *According to her culture's social clock, this young Moroccan adolescent has become a mother "right on time."*

Figure 19.3 *The social clock ticks less insistently for many of today's middle-aged adults. At left, mother and daughter congratulate each other on their college graduation— one "on time" and the other about thirty years "late." At right, a father delights in the daughter born years after the "usual" time. Over the past twenty years, the numbers of those who attend college or become parents "late" have risen dramatically.*

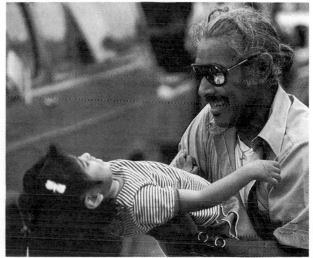

In contemporary American culture, the social clock allows more diversity than it once did. However, prescribed ages for many adult roles still tend to cluster in the early and mid-20s. Young adults experience considerable social and personal pressure to take on new roles and get on with adult life. By age 30, many people think that they should be educated, employed, married, and raising children or be prepared to explain why not. This can make early adulthood a highly stressful period for psychosocial development (Pearlin, 1982).

Let us look now at some of the ways young adults attempt to satisfy their needs for intimacy and generativity.

Intimacy

To meet the need for intimacy, or affiliation, an adult may take on many roles—friend, lover, and spouse among them. Each role demands some personal sacrifice, a giving of oneself to others. In the process of becoming more open and vulnerable to others, the individual gains deeper self-understanding and avoids the isolation caused by too much self-protection. As Erikson explains, the young adult must

> face the fear of ego loss in situations which call for self-abandon: in the solidarity of close affiliations, . . . sexual unions, in close friendship and in physical combat, in experiences of inspiration by teachers and of intuition from the recesses of the self. The avoidance of such experiences . . . may lead to a deep sense of isolation and consequent self-absorption. [Erikson, 1963]

For most adults, marriage is the favored path to intimacy, so it is marriage that we will focus on primarily in this section. Given the contemporary American social clock, which sanctions a period of living independently from one's parents before marriage (Goldscheider and Waite, 1987), and given the current divorce rate, which is about 50 percent in the United States and is rising in most countries worldwide, our discussion of marriage will begin with the antecedents of marriage, then look at marriage itself, and conclude with an examination of divorce and remarriage.

The Single Life

In the first years of young adulthood, the majority of American adults are single (including 75 percent of the men and 57 percent of the women between the ages of 20 and 25). Most enjoy this period as a time when they do not have to be responsible to or for another person and are therefore free to pursue their educations and careers. For women, particularly, being single makes it more likely that they will attend college and find a satisfying job, even compared with married women the same age who have no children (Haggstrom et al., 1986).

Especially in early adulthood, most single people are quite happy and busy, pursuing personal goals and interests and socializing with a wide circle of friends and acquaintances (see A Closer Look, next page). Those who live alone are particularly likely to develop an extensive network of friends to lend them support (Alwin et al., 1985). The typical single person spends almost as much leisure time with friends as alone, and when alone, is usually engaged in fulfilling activities. In short, satisfying independence is the norm for the single adult; loneliness is not (Cargan, 1981).

Developmental Trends in Friendship

Friends are very important at every stage of development. Even more than family members, they are a buffer against stress and a source of positive feelings about oneself (Gottlieb, 1983; Norris and Rubin, 1984; Vega et al., 1986). One probable reason for this is that friends *choose* each other, often for the very qualities (understanding, tolerance, loyalty) that make them good helpers in time of trouble.

On the whole, people tend to have more friends and acquaintances during the first years of adulthood than during any subsequent period (Antonucci, 1987). One reason may be that, beyond these early years, adults become too busy filling the demands of their careers and children to maintain extensive friendship networks. A second reason is that friendships with the opposite sex tend to be much less common after early adulthood, probably because, once an adult is married, friends of the opposite sex are perceived as potential threats to the marriage, so cross-sex friendships are kept at a distance (Huyck, 1982).

Sex Differences in Friendship

Males and females tend to have different friendship patterns throughout the life span, differences that are particularly apparent in young adulthood (Fox et al., 1985). Compared with the friendships of men, which are often based on shared activities and interests, friendships between women tend to be more intimate and emotional, based on shared confidence and assistance. Typically, a man might go bowling with his friends and talk about sports or work, while a woman might ask a friend over for coffee and discuss personal matters (Huyck, 1982; Tognoli, 1980). One interesting study (Hacker, 1981) found that, compared with men, women had twice as many "very close" relationships with friends of the same sex. This study also found that women were much more likely to reveal their weaknesses to friends than men were. In fact, about 25 percent of the women said they revealed only their weaknesses, and about 20 percent of the men (but none of the women) said they revealed only their strengths. This suggests that, for men, friendship may be more a way of maintaining a favorable self-concept, and that for women, it may be more a way of coping with problems.

Several researchers have wondered why men's friendships are less intimate than women's. One reason may be that the competitive pressure on men makes them fear showing weakness to a potential rival (Tognoli, 1980). Another may be that men fear expressing their affection for other men because they are afraid of being considered homosexual (Huyck, 1982). Ironically, the closest bonds between men in North America and Western Europe develop in an aggressive situation, such as between teammates on a football squad or comrades in war, perhaps because in such situations, few people would question a man's masculinity.

Friendship patterns vary from person to person, of course, but, in general, they reflect notable differences according to age and sex. In the beginning of adulthood, large, active social groups are typical, especially for men. Intimate friendships are more characteristic as adulthood progresses, and, as at every age, are more common among females than among males.

Figure 19.4 *For many single young adults, dancing at a crowded club is a source of fun and new friends. Others enjoy spending time more quietly, alone or with one or two carefully chosen companions. Either way, in early adulthood, singles are unlikely to spend much time lonely, bored, or unhappy.*

Many unmarried young adults also spend several months or years living with a nonrelative of the opposite sex. **Cohabitation,** as this living pattern is called, has become increasingly common in virtually all industrialized countries over the past twenty-five years (Davis, 1985). In the United States, the number of cohabiting adults in 1985 was over 2 million, five times the number in 1960; and by the mid-1980s, almost a third of all American single women between the ages of 20 and 29 had lived with an unrelated man (Tanfer, 1987). Virtually all cohabiting young adults consider cohabitation a prelude to marriage rather than a replacement for it. About a third of cohabiting couples do eventually marry (Clayton and Voss, 1977), while most of the rest break up within three years.

For some couples, cohabitation may be an attempt to test their compatibility. If so, it may be an inconclusive one. After investigating the satisfaction of newly married couples and the quality of their communication, a recent study determined that "while living together before marriage is increasingly becoming a common phase of courtship, cohabitation had no particular advantage over more traditional practices in assuring couple compatibility in marriage" (DeMaris and Leslie, 1984).

Choosing a Partner

Several researchers have studied the mate-selection process, and according to a synthesis of their findings, choosing a mate occurs in three stages: the stimulus stage, the values stage, the roles stage (Murstein, 1982).

In the first stage, one or more initiating stimuli—appearance, intellect, popularity, social status—make two parties notice each other. Physical attractiveness, of course, is a powerful stimulus, especially for men. Women look for attractiveness, too, but they are also drawn by status, preferring a leader over a follower, or a slightly older man over a slightly younger one, or a man with more education, or a better job. For both sexes, of course, preferring a particular person does not necessarily mean choosing that person. People compare their perceptions of themselves with their perceptions of someone they prefer, and are likely to approach the other person only when these perceptions seem about equal.

In the second stage, the couple compares their values. They discuss their attitudes toward work, marriage, religion, culture, and society. The more similar their views are, the more likely it is that the attraction will deepen, since such concurrence enhances each party's self-image. Similar values and interests also lead to shared activities through which the couple may further confirm their suitability.

Finally, as the couple interacts more frequently and becomes more intimate, they begin to develop their roles with each other. Role development is far deeper than the comparison of general values in the previous stage: it is a testing of the relationship in which each partner discovers how the other copes with the situations of daily life—finding out whether the other person shoulders or shirks responsibility, is honest or deceitful, is basically happy and stable or moody and unpredictable, and so on. The partners also discover how they cope together in moments of anger, jealousy, or depression, and in periods of family stress or work overload, as well as in the more conventional contexts of dating.

On the whole, men tend to think they are compatible with their partners before women do. One reason may be that men and women tend to have different attitudes about love (Peplau and Gordon, 1985). Men are more likely to be "romantics": for example, they are inclined to believe in love at first sight, to feel that each person has one true love they are destined to find, and to regard true love as magical, impossible to explain or understand. Women are more likely to be "pragmatists," believing that financial security is as important as passion in nourishing a close relationship, that there are many possible individuals that a person could learn to love, and that love does not necessarily conquer all—especially barriers such as economic or ethnic differences. Women tend to be more cautious than men before deciding to take the final step. As we will soon see, men and women respond somewhat differently to marriage as well.

Figure 19.5 *Social custom is not the only reason men are usually the ones to propose marriage. They are usually quicker to decide that "popping the question" would be an appropriate step in the relationship.*

"I'd like you to be the co-star in the melodrama that is my life."

Drawing by Weber; © 1986 The New Yorker Magazine, Inc.

The Developmental Course of Love and Marriage

Marriages, just like people, develop with time. Obviously, each marital relationship is different, and much depends on the personality and history of the individuals involved. However, some general trends can be identified.

One of the clearest is that marriages between people who have a lot in common are more likely to be satisfying and long-lasting than marriages between people who are quite different. Of course, it is not always the case, but all other things being equal, **homogamy,** that is, marriage between individuals who are similar in age, socioeconomic background, religion, ethnic group, and the like, has a better chance of survival than **heterogamy,** that is, marriage between individuals who are dissimilar on these variables (see A Closer Look, page 440).

A CLOSER LOOK   **Birds of a Feather**

As a guide to successful mate selection, the adage "Birds of a feather flock together" is much more valid than "Opposites attract." Homogamy, or marriage between people of similar backgrounds, is significantly more common and long-lasting than heterogamy, or marriage between people who have social differences.

Surprisingly, people even tend to prefer someone who looks like them. As one scientist explained:

If you measure enough things about enough couples . . . on the average, spouses resemble each other slightly but significantly in almost every physical feature. That goes not only for the traits you'd think of first, like skin and eye and hair color, but also for an astonishing variety of other traits, such as thickness of lips, breadth of nose, length of ear lobe and middle finger, circumference of wrist, distance between the eyes, and lung volume. Experimenters have made this finding for people as diverse as Poles in Poland, Americans in Michigan, and Africans in Chad. . . . So don't be discouraged by those old generalizations about sex appeal—gentlemen prefer blondes, men seldom make passes at girls who wear glasses, etc. Each such "rule" applies only to some of us . . . beauty is in the eye of the beholder. [Diamond, 1986]

What accounts for the preponderance of homogamy? The three-stage theory of mate selection helps explain

it. People from the same background are likely to have similar values, and to more easily develop complementary roles with each other. In fact, within successful, heterogamous relationships, one or both of the partners often have rejected some of the ideas or values held by most other people of their own race, religion, and social class.

The differences that separate husband and wife in heterogamous couples often involve values that may not be immediately apparent. For example, a courting couple might have reconciled religious differences between them, but might not realize that celebrating birthdays and holidays is very important in one tradition and insignificant in another. Another couple from different socioeconomic classes might discover that wrapping up and preserving leftover food is a waste of time to one of them but nearly sacred to the other. While both may value education, they may be at loggerheads about the specific approach to take when their child fails math or skips school.

When a couple who were raised differently disagree, their problems may be exacerbated because each has learned his or her family's way to resolve arguments. One may expect an active shouting match, with insults and criticisms that "clear the air" and lead to a passionate resolution,

---

Another factor that strongly affects the likelihood of a marriage's enduring is the maturity of the individuals who enter the relationship. At least until young adults are in their mid-20s, the younger marriage partners are when they wed, the less their chances of having a successful marriage. One reason for this is that, as Erikson explains, intimacy is hard to establish until identity is reached. Many older adolescents and young adults are still figuring out their values and roles, so a young couple might initially see themselves as compatible only to find their values and roles diverging as they become more mature. Further, the compromise and interdependence that are part of establishing intimacy are hard to achieve until one has a clear notion of self and has experienced independence.

More practical reasons are also apparent. Both partners are generally happier in marriage when the husband, in particular, has some financial security, and most young men have little until their mid-20s. This factor can be particularly detrimental when the reason for the marriage is a pregnancy. A disproportionate number of marriages between younger people do, in fact, occur because the bride is pregnant, and almost half of such marriages break up within five years (Kelly, 1982).

Whatever the age of the newly wed couple, in the early months and years, marriage is at its most intense, which can be a good or bad thing. During this "honeymoon period," newly married people typically spend more time together, talking, going out, establishing their marital roles and routines, arguing, making up, and making love. When the marriage is going well, the intensity of the honeymoon period results in a higher rate of reported bliss. In fact, marriages characterized by

while the other expects terse, intellectual coldness that leads to a gradual thaw in a day or two. In addition, to aid a failing marriage, one might consult a priest, the other, a psychiatrist; one might hope for a baby, the other, an extramarital affair.

None of these problems is insolvable, especially if the partners learn to break down the barriers that divide them. But statistics on marital success suggest that people from dissimilar backgrounds should take even more time to develop shared interests, comfortable patterns of interaction, and mutual respect before marriage than people who want to marry the boy or girl next door.

This middle-aged couple, proudly displaying photographs of each other when young, seem to prove a point. Not only do young adults tend to marry someone quite similar in age, attractiveness, and background, but couples who are happily married also tend to become more similar to each other over the years, often drawing closer together in views and tastes, and sometimes even taking on each other's habits and facial expressions.

extreme closeness, in which the partners share many activities and are open to new experiences together, tend to be most satisfactory to both husband and wife in the early years (Olson and McCubbin, 1983). When the marriage is troubled, however, this high intensity can lead to greater unhappiness (Swenson et al., 1981).

One of the major tasks for the couple during the honeymoon period is adjusting to the different perceptions and expectations that each partner may have of their life together. For example, women often are much more concerned about maintaining close relationships with their friends and relatives than men think is necessary or wise. Women are also more likely to find problems in the marriage relationship and to want to talk about them. Men are not as likely to recognize problems, and when they see them, tend to avoid the very confrontation that women seem to seek (Peplau and Gordon, 1985). The couple may also have to adjust to differences in the way they handle confrontation. Women typically try to evoke sympathy, while men use logic and get angry, to win an argument. Thus a wife might start to talk about an issue and begin to cry, provoking the husband to leave the room in a huff, telling her to stop sobbing and start thinking as he slams the door behind him. This pattern is obviously counterproductive, and successful couples usually find more constructive ways to resolve disputes.

After the honeymoon period, intensity diminishes and satisfaction with marriage generally dips, especially for wives. The most commonly cited reason for this change is the arrival of children. For most couples, the time and effort spent on parenting usually comes out of the marital relationship. As Norval Glenn explains:

Figure 19.6 *The negative effect children may have on marital happiness is, obviously, moderated by other factors. For example, if both spouses want a large family and can afford it, their happiness tends to increase when they achieve their goal. Here, both parents believe that children make life worthwhile, a belief that may acquire some of its depth from a special circumstance: the mother has multiple sclerosis, an incurable, sometimes degenerative, disease.*

Negative effects of children on marital happiness grow to a large extent out of interference with the companionship and intimate interaction of the spouses. Taking care of children requires time and energy which the husband and wife could use to sustain their own relationship. Also conflict may ensue from disagreements concerning child-rearing, and of course, adding persons to a dyad creates a more complex social system in which the potential for jealousy and competition is greater. [Glenn and Weaver, 1978]

Does this mean that marriages without children are more satisfying? This, of course, is a complicated question. In general, when both partners agree to postpone or avoid child-raising, they tend to be much more pleased with each other than couples who have several young children. On the other hand, those couples who are involuntarily childless experience considerable stress until they decide what to do—adopt, discover the reason for their infertility and attempt to remedy it, develop an alternative plan for their life together, or divorce (Matthews and Matthews, 1986).

As time goes on, the interests and needs of the partners may move in different directions or they may converge more strongly, and the marriage relationship may evolve along new lines (see A Closer Look, next page). An intensive study of more than 400 "successful" marriages (ones that had lasted ten years or more) found five quite different patterns (Cuber and Harroff, 1965):

1. *Conflict-habituated.* Husband and wife constantly argue, nag, fight, and belittle each other. Such battling is not considered a reason for divorce, however. On the contrary, for some couples, conflict may be a way of expressing their attachment to each other. When one woman was asked if she ever considered divorce, she answered, "Divorce never. Murder, every day."

2. *Devitalized.* The husband and wife, once loving and close, have drifted apart. They continue to get along with each other, but share few activities and interests.

3. *Passive-congenial.* For these couples, marriage is, and was from the start, comfortable and convenient. They seem more like compatible roommates than lovers.

A CLOSER LOOK · The Development of Love

According to a theory developed by Robert Sternberg (1986), love has three distinct components: (1) passion, an intense physiological desire for another person; (2) intimacy, the feeling that one can share all one's thoughts and actions with another; (3) commitment, the willingness to stay with a person through thick and thin. Sternberg believes that the relative presence or absence of these three components typifies seven different forms of love (see chart). Ideally, marriage is characterized by all three components in healthy amounts, although the balance among the three shifts as time goes on.

Early in a relationship, passion is usually high, which may be one reason new marriages are most intense. Intimacy, however, is not as high, because the partners have not spent enough time together or shared enough experiences and emotions to be able to understand each other completely. Passionate love without intimacy carries with it a high risk of misunderstanding and jealousy about any person or activity that seems to interfere in the relationship. Commitment also tends to be on the low side early in a relationship. Interestingly, these trends seem to hold true for all kinds of couples, married and unmarried, heterosexual and homosexual (Kurdek and Schmitt, 1986).

As the relationship continues, companionate love, characterized by greater intimacy and commitment, becomes stronger, but passion fades. As two researchers put it, "passionate love is a fragile flower—it wilts in time. Companionate love is a sturdy evergreen; it thrives with contact" (Hatfield and Walster, 1978). According to Sternberg, passion is like an addiction: in the beginning a touch of the hand, a smile, even a mere glance will produce excitement. Gradually, however, one needs a greater dose of stimulation to get the same feeling. Finally, just as an addict becomes habituated to a drug, lovers get used to the pleasures of passion with the same person, and passion weakens.

Seven Forms of Love

Liking	Intimacy, but no commitment or passion.
Infatuation	Passion, but no commitment or intimacy.
Empty Love	Commitment, but no intimacy or passion.
Romantic Love	Intimacy and passion, but no commitment.
Fatuous Love	Commitment and passion, but no intimacy.
Companionate Love	Commitment and intimacy, but no passion.
Consummate Love	Commitment, intimacy, and passion.

It should be noted that the depiction of these changing types of love has been derived from cross-sectional research; longitudinal research finds exceptions to these trends. Some people are strongly committed to their marriage partner from the moment they take their vows; others waver in commitment even twenty or thirty years into the marriage. Likewise, for some people, passion remains strong throughout a relationship; for others, passionate love never occurs (Traupmann and Hatfield, 1981).

According to Sternberg, knowing about the three components of love and their developmental trends can help couples in building their relationships. For example, they might be more careful to arrange time for intimate sharing, especially when they are busy raising children and developing careers. Or, being aware of the addictive pull and narrow focus of passion, they might be less likely to leap into, or out of, marriage because of it. They also might be less likely to think that something is wrong with their relationship if, after a while, the fireworks fade a bit.

"How do I love thee? Let me count the ways." If these couples are typical, the older one makes up for fading passion by enjoying a strong and steady commitment to each other.

4. *Vital*. Both spouses are very involved in all the family activities, economic and emotional, recreational and social.

5. *Total*. These couples are very involved, not only with the family activities but with each other's lives, sharing work interests, personal confidences, fantasies.

According to this study, about 80 percent of all marriages fall into one of the first three categories.

Much of the pattern of a particular marriage over the years is determined by the nature of the dependence of each spouse on the other (Kelley, 1981). When dependence is high, and each spouse is equally dependent on the other, the marriage is usually close and strong. However, when the dependence of one partner is much higher than that of the other, the marriage is likely to be characterized by conflict, stress, and anger. Changes in individual lives over the life span often make one partner more dependent, and the other more independent, than originally (Kelley, 1981). If one partner becomes much better educated or gains higher social status, or conversely, if one becomes significantly less healthy, less attractive, or loses status, the balance between them may suffer. In some couples, such imbalance is endured with resignation; in others, a way is found to restore the mutuality; in many others, the inequality leads to divorce.

Divorce

Divorce statistics worldwide, and from different decades, reveal that, in general, marriages founder rather quickly. Internationally, the peak time for divorce is three or four years after the wedding; in addition, divorce is most likely to occur when the couple are in their late 20s. The United States shows similar trends, with half of all divorces occurring in the first seven years (Fisher, 1987).

Although the stage of marriage at which divorce may occur seems to follow universal patterns, how frequently divorce actually does occur seems to be affected by specific historical and cultural conditions. Married adults are now divorcing two and a half times as often as adults did twenty years ago, and four times as often as fifty years ago. In fact, for every two American marriages that occurred in 1985, there was approximately one divorce.

Does this increase in failed marriages mean that the present cohort of young adults are less able than earlier cohorts to maintain a good marriage relationship? In fact, the case is probably the opposite. A study comparing families in the 1920s with their counterparts in the 1970s found substantial improvement in the communication and affection between spouses (Caplow, 1982).

What accounts for the rise in divorce, then? One major factor, clearly, is that spouses today expect a great deal more from each other than spouses in the past did. In earlier decades, earning the money was considered the man's responsibility, and housework and child care, the woman's. As long as both partners did their jobs, the marriage usually survived. As one woman, married in 1909, advised newlyweds on her seventy-first wedding anniversary:

> Don't stop on the little things. Be satisfied whatever happens. Ben didn't commit adultery, he's not a gambler, not a liar . . . So what's there to complain about? [Elevenstar, 1980]

In addition, husbands and wives in the past usually did not expect to really understand each other: they generally assumed that masculinity and femininity are opposites, and that the sexes therefore are naturally a mystery to each other.

Today, marriage partners have a much more flexible view of marriage roles and responsibilities and are likely to expect each other to be a friend, lover, and confidant as well as a wage-earner and care-giver.

Evidence for this shift in what is expected from marriage partners is seen in the changing reasons given for divorce. In 1948, recently divorced women were asked what had caused the break-up of their marriage. Cruelty, excessive drinking, and nonsupport were among the most common reasons cited (Goode, 1956). A comparable survey in 1975 found lack of communication and poor understanding to be the most common reasons (Kitson and Sussman, 1982).

Research suggests that now both sexes have quite similar complaints. For example, in one county in Wisconsin, everyone who files for divorce is asked the reasons the marriage failed. Table 19.2 lists the ten most common reasons cited by each sex and the percentages of respondents of each sex citing those reasons (Cleek and Pearson, 1985).

TABLE 19.2 **Reasons Given for Divorcing**

Women		Men	
1. Communication problems	70%	1. Communication problems	59%
2. Basic unhappiness	60%	2. Basic unhappiness	47%
3. Incompatibility	56%	3. Incompatibility	45%
4. Emotional abuse	56%	4. Sexual problems	30%
5. Financial problems	33%	5. Financial problems	29%
6. Sexual problems	32%	6. Emotional abuse	25%
7. Alcohol abuse—spouse	30%	7. Women's liberation	15%
8. Infidelity—spouse	25%	8. In-laws	12%
9. Physical abuse	22%	9. Infidelity—spouse	11%
10. In-laws	11%	10. Alcohol abuse—self	9%

It should be noted that, while women generally cited more reasons than men, both men and women tended to see their marriage problems in similar terms. Two notable exceptions to this pattern may be accounted for by the persistence of some men's stereotypic views regarding appropriate sex roles: whereas 22 percent of the women considered physical abuse a significant cause of their divorce, only 3 percent of the men did; and whereas 15 percent of the men considered women's liberation a significant cause of their divorce, only 3 percent of the women did.

Contributing Social Factors Two changes in the social setting have also contributed to the recent increases in the divorce rate. The first was the easing of divorce laws. Beginning about 1970 in the United States, traditional divorce laws— which had previously "preserved" a vast number of marriages by permitting divorce only on proof of serious misconduct—gave way to "no fault" laws that permitted divorce simply on the assertion of incompatibility. The impact of these changes is clearly seen in the case of California, which inaugurated no-fault divorce in 1970 and saw its divorce rate increase 34 percent over that of the preceding year. As other states followed suit, their divorce rates jumped in a similarly dramatic fashion.

The other change that has increased the divorce rate over the past several decades involves the alternatives to staying married. For men, divorce once meant the possibility of paying alimony for a lifetime, and if there were children, of supporting them but rarely seeing them. For women, divorce often meant a lonely, stigmatized existence, cut off from many of the friends she had in common with her husband. Now, alimony is less common, custody laws are more flexible, and the higher proportion of unmarried young adults means a better social life for the divorced person, and a good chance for a new marriage that is happier than the first. The current alternatives to marriage may, in fact, be the key to understanding why divorce occurs so frequently. One longitudinal study of married people found that the best predictor of eventual divorce was not the person's satisfaction with marriage but the person's estimate of the advantages and difficulties of being single again (Udry, 1981). Thus people who are relatively happy in marriage, compared with other married people, might nevertheless get a divorce if they think that they would be even happier out of the marriage. Correspondingly, unhappily married people might stay married if they see no better options.

The Impact of Divorce   Although the general picture of life after divorce has improved over the past two decades, the impact of separation and divorce on the psyche of the divorcing partners is often substantial, and can continue to be so for a long time. A nationwide survey revealed that divorce can impair an individual's well-being for at least five years after the event, producing a greater variety of long-lasting negative feelings than even the death of a spouse (Nock, 1981). Economic and social reasons are part of the problem: divorce generally reduces a person's income and social circle. However, even taking those factors into account, a longitudinal study found that divorced people tended to be more depressed than when they were married (Menaghan and Lieberman, 1986).

This does not mean that most divorced adults wish they had stayed married. An intensive longitudinal study found that five years after the divorce, only 20 percent of the former partners felt the divorce was a mistake. Most approved of the divorce, including many who initially had been opposed to it. However, most also acknowledged that they had underestimated the pain the divorce would cause (Wallerstein and Kelly, 1980).

Short- and Long-Term Effects of Divorce.   The first emotional impact of divorce is often that the former spouses become even angrier and more bitter with each other than they were in the marriage. In many cases, this is because the legal system fosters contention over property and/or child custody. In addition, many divorcing spouses underestimate their attachment to each other, and therefore may be more vulnerable to an insult from the other and more ready to enter into a petty quarrel than before. This increased hostility is often followed by, or interspersed with, periods of depression and disequilibrium, as patterns of eating and sleeping, drug and alcohol use, and work and residence change (Kelly, 1982; Simenauer and Carroll, 1982).

In the short term, divorce is usually particularly devastating to men (Chiriboga, 1982). For one thing, it is more likely to be the wife who first finds fault with the marriage and files for divorce. In fact, many men are surprised and shocked by the break (Kelly, 1982). In addition, it is the husband who is likely to be blamed for the problems with the marriage, to accept that blame (Kitson and Sussman, 1982), to move out, and thus to find his personal and social life disrupted. Consequently, although women may be more distressed prior to the separation, men have more symptoms of psychic and physical stress immediately following it (Bloom and Caldwell, 1981). Over the long term, however, women are more affected, primarily because they are likely to have less money and fewer marriage prospects than divorced men. If they are mothers with custody, the impact of divorce is particularly strong.

The Role of Children   Indeed, whether or not a divorcing couple has children is often critical in the adjustment to divorce. Childless divorced people adjust much more readily than those with children, partly because they can make a cleaner break from the relationship.

For divorced people with children, many of the difficulties they may experience arise from custody arrangements. Despite much public discussion of joint custody, in most cases one parent has custody, leaving that parent to play two parental roles, while the other one is thrust into the role of occasional visitor. More than 90 percent of the time the custodial parent is the mother, and the burdens for her can be considerable. At the same time that her share of child care is increasing, her standard of living is decreasing. On average, divorced women with dependent children experience a 73 percent decline in their standard of living in the year after a divorce, while their ex-husbands experience a 42 percent rise (Weitzman, 1985). The reason is that child care is much more expensive than most people realize, and divorced fathers typically contribute far less to this expense than they did before the break.

While fathers may fare better financially, they suffer psychologically, for they gradually become distant and alienated from their children. Although many of today's fathers want to be active in their children's lives, a nationwide study finds that, over the long term, few divorced fathers are (Furstenberg and Nord, 1985). Two or more years after the divorce, only one child in five ever stays overnight with the father, and half of all children of divorce have not even seen their father over the past year.

The hostilities that may be attached to issues of support, visitation, and the like can be harmful to all concerned. In a Wisconsin study, a fourth of all divorced couples were bitter enemies, unable to agree about anything, to the detriment of themselves as well as their children (Ahrons and Wallisch, 1986).

Of course, it is not inevitable that former spouses with children remain angry and alienated from each other. In the same Wisconsin study, an eighth of former couples were very good friends, telephoning each other frequently, having dinner together occasionally, and cooperating fully in the upbringing of their children; and

Figure 19.7 *As a single mother contemplates her future, she is likely to envision considerable stress, insufficient money, and children who wish their father were around. Hopefully, she will develop a network of supportive friends and relatives who will help her solve whatever practical and psychological problems she must face.*

in a study of middle-class Californians, 30 percent of divorced couples had developed a comfortable pattern of involvement with each other and their children (Wallerstein and Kelly, 1980).

In these studies and in other research, divorced parents who are able to overcome their anger and cooperate not only help their children but also are much happier and better adjusted themselves. What factors predict this good adjustment? Having had a relatively good marriage and a relatively smooth divorce are two obvious ones. Another factor is remarriage. Generally the adjustment is easier if both ex-spouses, or neither, remarries than if one does. The new spouse's attitude about the previous marriage is important as well. Interestingly, it seems easier for a new husband to accept his wife's friendly relationship with her former husband than it is for a new wife to accept her husband's good relationship with his former wife.

Remarriage

Almost 80 percent of divorced people remarry—on average, within three years of being divorced (Glick and Lin, 1986b). Remarriage is more likely to occur if the divorced person is relatively young, in part because there are more potential partners still available. For the same reason, especially at older ages, men are more likely to remarry than women, because men tend to marry younger women, whereas women tend to marry older men, which means that there is a substantial sex difference in the number of available partners. For every divorced man between the ages of 30 and 34, there are three women in his age group, and twelve women between ages 20 and 29, who are unattached; for every divorced woman between the ages of 30 and 34, there are only two men in her age group, and 2.5 between ages 35 and 44, who are unattached. Older women who are relatively well educated have the lowest rate of remarriage (Glick and Lin, 1986b), presumably because they have more difficulty finding a comparably educated man, and also because they have less need of the financial security that remarriage brings.

Popular wisdom to the contrary, there is no guarantee that marriage will be better the second time around: the divorce rate for remarriages is higher than that for first marriages. One reason is that some lonely divorced people marry too quickly, "on the rebound" (sometimes to the first person who seems interested). In many instances, however, an important factor is the disruptive effects of stepchildren. Within three years after marriage, the divorce rate of remarriages involving stepchildren is 17 percent, compared with 10 percent for childless remarriages and 6 percent for first-time marriages (White and Booth, 1985).

Stepfamilies It is not easy to create a harmonious stepfamily. One problem is that children sometimes blame the stepparent for their parents' divorce, and refuse to accept the stepparent's role in their lives. Especially if both spouses have children from previous marriages, coordinating visiting schedules, dealing with new forms of sibling rivalry, and simply managing a group of often critical children while trying to build a new marriage are far from easy.

For their part, many stepfathers become more disciplinarian than they had expected to be, or they retreat from the parental role altogether, in reaction to the children's refusal to accept their authority (Hetherington et al., 1982). Many stepmothers resent the fact that they must help feed, clothe, and care for another woman's children. Interestingly, however, stepmothers generally are more satisfied with their role when the father has custody, for they can then develop more of a relationship with their stepchildren, getting some of the rewards as well as the burdens of mothering (Ambert, 1986).

Figure 19.8 *The bride and groom are obviously absorbed in each other, but the future of their marriage may depend on three somewhat inattentive attendants, their children from former marriages. Stepchildren do not necessarily welcome, respect, or even listen to their stepparents, a problem particularly likely to disrupt the early months of marriage.*

Teenage stepchildren are, apparently, particularly hard to deal with, and remarriages involving them are especially likely to founder. When such marriages do succeed, it may sometimes be for reasons other than good family relations: in one large study, half of all teenage children left home within the first three years of their custodial parent's remarriage—going to boarding school or to college, living with another relative, or simply setting forth on their own (White and Booth, 1985).

While the findings on the potential pitfalls of remarriage signal the need for caution, the advantages of remarriage usually outweigh the disadvantages, not only for childless adults but also for divorced parents, especially if the children are under 10 or are grown. Most remarried people have learned from their earlier marriage experiences, talking problems out more than they formerly did (Benson-von der Ohe, 1984), and find that their second marriage is better than their first (Cherlin, 1981; Wallerstein and Kelly, 1980).

Generativity

The motivation to achieve is one of the strongest, and most frequently studied, of human motives. The expression of achievement motivation varies a great deal, with some people being much more competitive than others, for example. But in one way or another, for our self-esteem as adults, we all need to feel successful at something that makes our lives seem productive and meaningful. Adults meet their need for achievement, confronting the crisis of generativity versus stagnation, primarily through their work and through child-rearing. Let us look first at career development, then at family development, to see what the possibilities and problems with both are.

The Importance of Work

For many people, their job is central to their life for reasons other than economic ones. This fact is highlighted by studies in which workers have been asked what they would do if they were to become sudden millionaires. In study after study, in many countries, more than 80 percent of the workers replied that they would keep on working (Harpaz, 1985). The centrality of work is also revealed in research on unemployment, which finds that many laid-off workers feel lost, depressed, and empty without a job (Kelvin and Jarrett, 1985). In addition to putting bread on the table, work clearly serves other functions. As Marie Jahoda (1981) explains, employment

> imposes a time structure on the waking day . . . implies regularly shared experiences and contacts with people outside the nuclear family . . . links individuals to goals and purposes that transcend their own . . . defines aspects of personal status and identity, and . . . enforces activity.

Other activities could perform a similar function, of course. Attending school or volunteering to help others also meets similar needs. But, in our culture, paid employment is the most common source of generativity.

Young Adults and Work   Finding satisfying work, however, is not easy. Typically, a person's first job is taken without much forethought. Many young adults tend to model their career goals on those of their parents (Auster and Huster, 1981; Mortimer and Kumka, 1982), sometimes irrespective of their own abilities and inclinations, and of the changing demands of the job market. Most often the first job is chosen because it is readily available and because friends or family members suggested it. Entering a new job, workers usually have fairly high expectations of what the job will be like, and experience "reality shock" when the job turns out to be less exciting and purposeful than anticipated (Reilly et al., 1981). Since almost every job involves more dull moments, and more bureaucratic hassles, than the naive worker imagines, it is not surprising that young workers tend to be less satisfied with their jobs than older workers are, and are more likely to switch jobs and careers. In fact, the average American working adult has had three distinct occupations by age 30 (Goldstein, 1984). Naturally, some of these shifts occur in response to changes in the job market, but, in early adulthood, most shifts occur as the young worker gains a better understanding of his or her own abilities and aspirations and therefore becomes better able to adapt them to the realities of the work world.

Figure 19.9 *Which would you rather be, a nurse working in the trauma room of a hospital or an engineer figuring out a design problem? Your answer is likely to depend to some degree on your sex, because in addition to talent, personality, and working conditions, gender expectations affect the probability of a particular young adult's choosing a particular career. In America in 1980, more than 95 percent of all nurses were women and more than 96 percent of all engineers were men.*

According to Levinson, this adaptation is a part of the modification of the person's original "dream," or vision, of the kind of life he or she wants to live. One of the men Levinson (1978) studied, for example, began college wanting to be a psychiatrist, for mixed reasons typical of his age:

I'd done a lot of reading in psychiatric literature, and had done very well in the sciences in high school. I was in love with a girl whose father was a doctor. So it seemed to me medicine would be a good career.

Then he decided he wanted to be a famous novelist, and switched his major from premed to literature. After college, however, he found he could not earn a living as a fiction writer, so he entered the business world, eventually earning hundreds of thousands of dollars. Business, however, gave him no time to write, so he quit and turned to teaching—only to discover he missed the respect and excitement that high finance had given him. For the next several years he tried to juggle his business and writing careers, finally choosing writing, even though he realized he could probably never become the renowned novelist he had dreamed of becoming. By middle adulthood, according to Levinson, he was "entering a genuinely productive season of his life"—writing biographies rather than novels.

While this man made more switches than most, Levinson found that the majority of the people he studied changed jobs, accomplished less than they had aspired to, and, to the extent that they were psychologically healthy, adapted their "dream" to fit reality.

Stages of Careers

This research lends support to a theory that a person's entire work history, called the work "career," goes through stages of development (Super, 1957; Van Maanen and Schein, 1977). The first period, as we have just seen, is one of **career exploration,** in which the person attempts to match his or her skills, talents, and needs to a specific occupation and job.

Once the period of exploration is over, usually at about age 30, a worker enters a new phase called **career establishment.** During this phase, as workers demonstrate increasing commitment and competence, they are given tasks or titles that require more responsibility and skill and entail more money and respect. During the phase of career establishment, workers tend to work harder, and to be more satisfied with their work, than earlier (Slocum and Cron, 1985).

Interestingly, satisfaction with work is less related to variations in wages and benefits than to workers' feelings that their work is meaningful and that their supervisors and coworkers recognize their efforts (Dawis, 1984). Also important are the chance to learn new skills and the opportunity to realize one's potentials. For many workers the degree to which their job allows them to mesh work with the rest of their life is a major determinant of job satisfaction. For example, the opportunity to choose one's own work hours, called flextime, generally increases worker satisfaction and thereby reduces absenteeism (Ralston and Flanagan, 1985). This trend is particularly apparent for women, who often must balance family and work needs.

A key factor for many young workers is a relationship with an older, more experienced worker who becomes a **mentor.** Ideally, the mentor fills many roles— teacher, sponsor, coach, protector, role model, confidant, counselor, and friend (Krim, 1985). In these various capacities, the mentor passes on not only "formal" know-how about the technical aspects of the job but also "informal" tips, such as how to deal with a difficult superior, what office rivalries to beware of, where to find shortcuts through the bureaucratic maze, and the like. Given the closeness of the mentor-mentee relationship, most workers feel more comfortable with mentors of the same sex, because cross-sex mentoring raises the possibility or suspicion of a sexual relationship. Difficulties in finding a suitable mentor may be one important

Figure 19.10 *This woman is a trailblazer: until recently, female stockbrokers did not exist. After seven years as a licensed stockbroker, she is now sufficiently well-established in her career that she can become a role model as well as a mentor for the next generation of women. However, since she still experiences sexual discrimination in her work, she is willing to help women who want to follow her example, but not willing to encourage the undecided.*

reason many women have a hard time advancing in male-dominated occupations (Keele, 1986). Men and women may encounter other problems in the establishment phase of career development, as well as in the next phase, career consolidation. Several of these difficulties, including burn-out and alienation, will be discussed in Chapter 22.

Parenthood

When one stranger asks another, "What do you do?" the answer is rarely "Raise children." Yet adults are as likely to be parents as to be employed, and many of them consider the successful rearing of their children to be their most important achievement. As Erikson points out, while generativity can take many forms, its chief form is "establishing and guiding the next generation," usually through parenthood. Caring for children fulfills an important adult need, according to Erikson (1963), for "the fashionable insistence of dramatizing the dependence of children on adults often blinds us to the dependence of the older generation on the younger one. Mature man [and woman] needs to be needed." The interdependence of parents and children is a lifelong process, which begins at conception and continues throughout late adulthood.

A Life-Span Perspective

Stages of Parenthood

Like the course of the career path, parenthood can be seen as developing through several fairly distinct stages, which are based primarily on the age of the oldest child in the family (Aldous, 1978; Duvall, 1971; Galinsky, 1981) (see Table 19.3, next page).

Each stage of the family life cycle has its own characteristics and requirements. How parents respond to them varies greatly because of the diversity of both children's and parents' personality structures and psychological needs. Some parents, for instance, like the dependence of infants, enjoying the feeling of being totally needed; others feel much more comfortable in the later stages when children are more independent. Some parents feel lost, sometimes even purposeless, when their last child leaves home; others feel as though they have been given a new lease on life.

TABLE 19.3 **Stages in the Family Life Cycle**

Honeymoon Period	Lasts from the wedding until the birth of the first child.
Nurturing Period	Lasts until the first child is 2.
Authority Period	Spans the years between 2 and 5.
Interpretive Period	Spans the years between 5 and 12.
Interdependent Period	Occurs when children are adolescents.
Launching Period	Lasts from the home-leaving of the first child to the home-leaving of the last.
Empty-nest Period	Begins when the last child leaves home.

It is also true that many families do not follow the stages straightforwardly. The many parents who are separated, divorced, single, or remarried experience any number of variations in the family life cycle, from going through all or parts of the cycle without a partner to going through different stages of the cycle simultaneously with children from different marriages. Despite all the diversity in family life cycles, however, some general trends are apparent.

The Honeymoon Period As we saw earlier, the first stage of this family cycle, the honeymoon period, occurs as the newly married couple establish their relationship to each other. During this stage, the couple supposedly decide whether or not to become parents. In actual fact, most couples simply assume they will have children, and the only remaining decision is when. For many couples today, the answer to "When?" is simply "Not now"—not until education is completed, financial security is assured, or vocational goals are within sight. The approximately 5 percent of all couples who eventually decide not to have children usually "slip into childlessness" when, after years of postponement, they eventually realize that they have little time or interest for parenthood in their lives (Veevers, 1980).

Because most couples take it for granted that they will have children, they often enter into parenthood without much forethought, or for reasons that have little to do with wanting to rear a child. Some have a child to stabilize a shaky relationship, to remedy an ailing self-concept, to affirm their masculinity or femininity, to please their parents, or to quiet relatives and friends who admonish that not having children is selfish and advise that children are "the ultimate life experience" (Bombardiere, 1981; Whelan, 1975). One woman "figured that as long as I was bored, why not have a baby now instead of waiting a couple of years? . . . I've heard that it always brings the couple closer together and sometimes it seems like we, we just need something" (Liefer, 1980). Those who drift into child-bearing for reasons that have little to do with child-rearing will be surprised. Parenthood, particularly in the early years, is more likely to cause frustrations that pull a couple apart than to bring marital harmony and personal satisfaction.

For whatever reasons a couple decides to become parents, the realities of parenthood usually come as a shock. As one review notes, "research showing the stress and problems of parenting and its negative consequences in relation to other adult roles . . . does not seem to have reached most adults who decide to have a child" (Alpert and Richardson, 1980).

The Nurturing Period: Birth to Age 2 The joy of this period is getting to know a new human being whose creation one has shared in. The drawback is that, from virtually the moment of birth, most parents are overwhelmed by the necessity of meeting the infant's need for twenty-four-hour care (Galinsky, 1981; Leifer, 1980). In a longitudinal study that followed women from early pregnancy through the first year of motherhood, Myra Leifer (1980) reports that many of the mothers-to-be looked forward to quitting their jobs to be full-time mothers and wives, imagining that they would then have time to prepare gourmet candlelit dinners for their husbands, fashion clever decorations for their homes, and sit in rocking chairs, crooning lullabies to their sleeping infants. The reality was quite different. Their babies needed to be fed, changed, burped, bathed, carried, and cuddled so often that there was little time for anything else in the mother's life, day or night. As one woman said, "For the first time in my life, I'm doing everything in a mediocre way. I don't have enough time, energy, or ability to do things I want to do."

Many new fathers find that their wives are making more demands on them and giving them less love and care than they did before the baby's arrival. In addition, couples often quarrel over who should do what. One father described a shouting match in which both he and his wife insisted that they were contributing more than their fair share to maintaining the family:

> The answer was that we were both giving a lot, and it was a drain . . . We never did the simplest things to give ourselves time together. We never got a baby-sitter, never went out together. Our lives together just ceased to exist.
> [Galinsky, 1981]

Figure 19.11 *At every stage, parenthood demands a great deal of sacrifice, both great and small. To a degree, most parents anticipate some of the larger sacrifices of raising a family, like those involving money, time, and emotional energy. What often comes as a complete surprise is the endless number of "little" sacrifices, from eating "kid food" more often than one might like to vacationing at the beach and having to keep a constant eye on the children when one might rather run, swim, or take a nap.*

The Authority Period: Ages 2 to 5 When children become preschoolers, the issue of authority, over the child and within the marriage, becomes crucial. This is often the period of greatest direct confrontation between husband and wife, due to mounting pressure from increased financial burdens, multiplying household tasks, and shifting roles, as well as to the child's growing need to assert his or her independence, sometimes in strident, destructive ways. In addition, couples often have a second child while the first one is small, in effect doubling the stresses related to child care during these years. Other couples make shifts in their employment patterns, which can often lead to new stresses (see A Closer Look, pages 456–457).

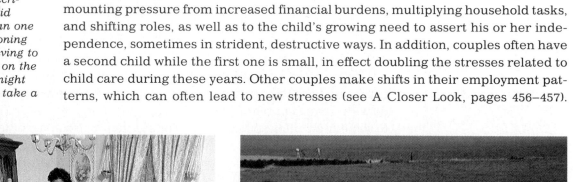

The Interpretive Period: Ages 5 to 12 In many ways, parenting grows easier as children become increasingly self-sufficient and as parents' experience at child-rearing accrues. In general, from the time the first child enters school until adolescence, parents need only set realistic goals for their children and provide encouragement and guidance, allowing their children to develop all their latent competencies. During this period, parents' main task is to interpret, adjudicate, and modify the broadening experiences that come from the outside world, as well as from the child's own expanding mind. This period is the best time for parents to establish open communication with their child to prepare for the potentially difficult period of adolescence.

The Interdependent Period: Adolescence   Adolescence brings new challenges, as the young person who is developing the body of an adult demands the privileges of adulthood before knowing how to handle the responsibilities that go with them. New alliances are formed in the family: parents feel the need to stand together, even though a teenager's strivings for independence might provoke quite different responses from mother and father. Meanwhile siblings, who once took opposite sides on every issue, may unite to defend their own common interest. Each generation has a different stake in the outcome of the various conflicts between adult and teenager, as parents seek to ensure continuity from generation to generation, and adolescents seek their own independent identity. While harmony is more prevalent than hostility in this stage, parents tend to underestimate, and adolescents to overestimate, the gap between their values (Bengston and Troll, 1978).

Figure 19.12 *As detailed in Chapters 14, 15, and 16, adolescents sometimes give their parents considerable worry and distress. They can also give valuable advice, understanding, and help. This daughter is providing encouragement as she searches the want ads, while her mother, preparing to reenter the job market after many years, types up her resumé.*

The Launching Period and the Empty-Nest Period   The remaining stages of the family life cycle are the launching stage, in which one's grown children begin to set out on their own, and the empty-nest stage, in which the last of the "fledglings" have flown. For many middle-aged parents, the successful launching of an adult child, and eventually all of the children, is a time of rejoicing. According to several surveys, the "empty-nest" period of marriage is generally more satisfying than any of the previous stages since the honeymoon, with love between the partners remaining steady, and the problems with the marriage declining.

A CLOSER LOOK   **Parenthood and Career**

Until the Industrial Revolution, most married couples worked together all their lives. Wives would help husbands on the farm or in the shop, and husbands helped their wives oversee the children, especially once the children had begun to participate in the family work (at about age 6). However, with industrialization, workplaces became distant from living places, and women more often stayed home tending to house and children while men earned their living outside the family setting.

In recent decades, smaller families, greater financial pressures, more employment opportunities, and higher education have put many women in the labor market. Virtually all contemporary women are employed at some point in their lives, and slightly more than half continue with their careers, either full- or part-time, throughout motherhood. The other half stop temporarily while their children are young. Neither choice is ideal, for the women themselves or for their families.

Being able to have lunch with a parent during the week can provide children with important emotional, as well as dietary, nourishment.

When the Mother Is "Only a Housewife"

If a new mother quits her job, she gives up an external source of self-esteem, social support, and status. While most mothers who have left the work force believe that the sacrifices are worth it (O'Donnell, 1985), in general, employed mothers enjoy their work more than mothers who are not employed (Newberry et al., 1979).

Part of the problem, of course, is that homemaking is "not recognized as real work": it lacks standards of accomplishment, so many women feel the job is never completed; it fosters social isolation; it makes wives financially and psychologically dependent on their husbands (Ferree, 1976). Even the lowest-status job has standards, coworkers,

and a paycheck, each a potential source of self-esteem. A job well done in the workplace generally gets acknowledged; a job well done in the home generally gets ignored. For example, most young children are far more demanding than appreciative (Patterson, 1980), and tears or anger at "No more cookies" is more typical than smiles of gratitude for a healthful dinner. Worse yet, most household tasks, such as providing fresh laundry or a neat house, soon need to be done all over again.

Fathers also experience stress when they become sole breadwinners for a growing family. They sometimes feel trapped and overburdened and, if they work overtime to help pay the bills, they see less of their family, usually to

---

Parenthood in Perspective

Clearly, raising children is no easy task, no matter how benign the circumstances. Even the smoothest stages of the family life cycle are rife with possible pitfalls. However, we must guard against making too much of the difficulties of parenthood. Most parents are glad they had children; they just wish parenthood were a little easier. As one mother said, "The baby has been a delight; it's motherhood that's so damn difficult" (Leifer, 1980).

Further, the dip that occurs in marital satisfaction when children arrive is generally a slight one, not a drastic drop. Indeed, some research has found that marital happiness does not dip at all if the children are planned, and if the parents are ready for the responsibilities of child care (Feldman, 1981). Generally, parents become dissatisfied with their parental roles only when pressures build and exhaustion sets in, as when small children, financial demands, and full-time employment for both spouses all occur together (Schumm and Bugaighis, 1986).

Certainly parenthood brings rewards along with its responsibilities, rewards that may be easier to see in retrospect. Separating the daily work of parenthood from the anticipated fun or the final sense of achievement, E. E. LeMasters (1977) says,

the distress of all concerned (Piotrkowski, 1979). Further, hard-working fathers are likely to think that their wives are less appreciative than they ought to be, while the most common complaint of unemployed wives and mothers is that their husbands do not help enough around the house (Rhyne, 1981).

When the Mother Is an Employee

Because of significant increases in the cost of living in recent decades, most young families find they cannot live on one income and still give their children everything that they need. It is not surprising, then, that approximately 53 percent of all married women with children under age 6 are employed (U.S. Bureau of the Census, 1986). This helps with the financial squeeze, of course, but it creates other sources of friction.

One common problem of two-paycheck families is the division of housework and child care, which usually has employed wives doing much more of the housework and child care than their employed husbands do. Many other factors correlate with the division of labor. As you might expect, men who are better educated or younger tend to be more helpful, and the wives who work fewer hours at the job find themselves doing more work at home. Surprisingly, however, the more children a couple has, the less likely they are to share equally in the household labor, even if both are working an equal number of hours outside the home (Haas, 1981). Thus many employed mothers feel overworked and underappreciated.

Another problem may involve the husband's wounded ego. While most men welcome the income their wives add to the family, many were raised believing firmly in the idea

that the man's role is to be the major breadwinner. Especially if his wife's income equals or surpasses his, the man may lose some of his sense of self-worth (Staines et al., 1986).

Furthermore, stresses of employment and parenthood impinge on each other (Hoffman, 1986; Moos, 1986). The worst situation occurs when both spouses are under pressure at work (a condition that is typical in the first stages of career development), and, at the same time, worried about their children's welfare because good child care for infants and preschoolers is hard to find. Unfortunately, the mesosystem and macrosystem have not yet done much to help relieve these dual pressures. To pick one example, although most toddlers and preschoolers are sick several times a year, virtually no day-care center allows sick children to attend, and few employers let parents take time off to care for a sick child. Consequently, parents must either lie to their employers or find an available caretaker (not easy) and then worry how their child is doing. Nor surprisingly, in a national survey, 38 percent of employed fathers and 53 percent of employed mothers would prefer to work fewer hours in order to "spend more time with their spouse and children, even if it meant having less money" (Moen and Dempster-McClain, 1987).

Fortunately, most young dual-career families find ways of meeting the demands of family and work. Overall, while there are times (especially when children are small) when parents are overwhelmed by the conflicting needs of career and family, adults who combine all three roles—spouse, parent, and employee—are healthier and happier than those who do not (Baruch et al., 1983; Moos, 1986; Vailliant, 1977; Verbrugge, 1983).

the truth is—as every parent knows—that rearing children is probably the hardest, and most thankless, job in the world. No intelligent father or mother would deny that it is exciting, as well as interesting, but to call it "fun" is a serious error.

For LeMasters, parenthood was like the three years he spent in the armed forces during World War II. Looking back, he was proud and glad that he had had the experience. But during the actual process, there were many times he would have been grateful to have received an honorable discharge. A similar comment could be made concerning many of the experiences of young adulthood. The ups and downs of the single life, of career choice and mate selection, of working and marrying, of childbearing and child-rearing, of career consolidation and marriage building, and perhaps of changing jobs and divorcing a spouse are not easy. There are moments, and even long stretches, when the pressures associated with these experiences can be overwhelming, debilitating, heartbreaking. Nonetheless, the sum total of all these experiences, good and bad, may be a richer, more rewarding life.

SUMMARY

The Tasks of Adulthood

1. Adult development is remarkably diverse, yet it appears to be characterized by two basic needs. First is the need for affiliation, also called the need for love, belonging, and/or intimacy. Second is the need for achievement, as the need for success, esteem, generativity, and work can be called.

2. Adulthood can be described as occurring in stages. Levinson, for example, believes that periods of notable reevaluation and change occur every decade, with more stable periods occurring between these transitions.

3. Once young adults break away from dependence on their parents, usually by the early 20s, they begin to make commitments to career, marriage, and parenthood.

4. Age 30 can be a turning point, as many men and women reevaluate their earlier decisions and begin what tasks of adulthood, in work, family life, or education, they have not yet started.

5. Adult stages are propelled more by social standards than by biological processes. The social clock varies from time to time and place to place, and this is one reason for the many variations in adult developmental paths.

Intimacy

6. Most young adults plan to marry eventually, but spend several years unmarried. Single adults are far from lonely, having friends and, sometimes, lovers, with whom they share their lives.

7. Mate selection is a three-stage process, beginning with the stimulus of relatively superficial attributes and moving through a comparison of values and role compatibility.

8. Marriage itself tends to follow a developmental path, with passion and anger highest in the early years, and commitment higher later on. If divorce occurs, it typically happens quite early in marriage—usually within the first 5 years.

9. The divorce rate has risen dramatically over the past thirty years, partly because people expect more of marriage and partly because divorce laws have eased.

10. Divorce is emotionally draining on both partners and is particularly difficult for those who have children. While some divorced parents manage to maintain good relationships with each other and with their children, more typically the mothers have more work and less money than before, and fathers become estranged from their children.

11. Most divorced people remarry, usually attaining more happiness than in their former marriages or than in their divorced state. Especially if there are children from former marriages, however, a new marriage also creates additional stresses.

Generativity

12. For most adults, work is an important source of satisfaction and esteem, as it helps meet the need to be generative.

13. Finding appropriate work is not easy. Most young adults go through a period of career exploration, during which their jobs and vocational goals change. Ideally, the young worker finds not only a fulfilling job but also a mentor, who will guide him or her through the first steps of the career ladder.

14. Parenthood is the other common expression of generativity. Families go through various stages, depending on the ages of the children, with the hardest adjustment to parenthood usually occurring when the children are very young.

KEY TERMS

intimacy versus
  isolation *(431)*

generativity versus
  stagnation *(432)*

social clock *(434)*

cohabitation *(438)*

homogamy *(439)*

heterogamy *(439)*

career exploration *(451)*

career
  establishment *(451)*

mentor *(451)*

KEY QUESTIONS

1. How does the social clock affect life choices?

2. What are some age and sex differences in friendship patterns?

3. What are the stages in the process of mate selection? How does this description fit with your observations?

4. Why has the divorce rate risen dramatically over the past few decades?

5. What factors make some divorced people more unhappy than others?

6. Why is work important beyond supplying income?

7. What similarities frequently exist between the young adult's decision making in taking a first job and in beginning a family?

8. What is the typical path of career development in early adulthood?

9. How does the arrival of children affect the honeymoon period?

10. What are the rewards and costs of parenthood?

Part VI

The Developing Person So Far: Early Adulthood, Ages 20 through 40

Physical Development

Growth, Strength, and Health

Noticeable increases in height have stopped by about age 18 in females and age 20 in males. Physical strength increases through the 20s and peaks at about age 30. Although all body systems function at optimum levels as the individual enters early adulthood, declines in body functions begin to diminish the efficiency of most organ systems at the rate of about 1 percent a year.

The Sexual-Reproductive System

Sexual responsiveness remains high in early adulthood; the only notable changes are that men tend to experience some slowing of their responses with age, and women tend to become more likely to experience orgasm. Problems with fertility, however, become increasingly frequent after the early 20s.

Cognitive Development

Adult Thinking

As an individual takes on the responsibilities and commitments of adult life, thinking may become more reasonable, practical, and dialectical to take into account the inconsistencies and complexities encountered in daily experiences.

Years of education is the variable that has been shown to have one of the most powerful effects on the depth and complexity of adult thinking. Significant life events can also occasion cognitive development.

Psychosocial Development

Intimacy

The need for affiliation is fulfilled by friends and, usually, at some point, by a marriage partner. The developmental course of marriage depends on several factors, including the presence and age of children and whether the interests and needs of the partners converge or diverge over time. Divorce, if it is to occur, is most likely three or four years after the wedding.

Generativity

The need for achievement can be met both by finding satisfying work and establishing a career and by parenthood.

Middle Adulthood

Popular conceptions of middle adulthood are riddled with clichés like "midlife crisis," "middle-aged spread," and "autumn years" that conjure up a sense of dullness, resignation, and perhaps a touch of despair. Yet the tone of these clichés is far from reflecting the truth of the development that can and often does occur between the ages of 40 and 60. Many adults feel healthier, smarter, more pleased with themselves and their lives during these two decades than they ever did.

Of course, such a rosy picture does not apply to everyone. Some middle-aged adults are burdened by health problems, or a decline in intellectual powers, or unexpected responsibilities for aged parents or adult children. Some feel trapped by choices made in early adulthood. But the underlying theme of the next three chapters is that in middle age, much of the quality of one's life is directly related to how one views it and to decisions, sometimes new ones, about how to live it. There are still many turning points ahead where new directions can be set, new doors opened, and a healthier and happier life story written.

CHAPTER 20

Middle Adulthood: Physical Development

As we grow old, . . . the beauty steals inward.

Emerson
Journals

In Chapter 17 we saw that the first two decades of adulthood can be considered the prime of life as far as physical development is concerned, with the effects of aging being, for many people, barely perceptible. Between the ages of 40 and 60, aging continues at the same steady rate, but the levels of change that are now reached are more difficult to ignore. As we will see, however, adults in middle age can do a great deal to safeguard their vitality and remedy, or compensate for, many of the physiological declines they experience, discovering in the process that though aging is inevitable, it is not inevitably bad.

Normal Changes in Middle Adulthood

As people advance in age past 40, their hair usually turns noticeably gray and thins appreciably; their skin becomes drier and more wrinkled; and their body shape changes, as "middle-age spread" develops and pockets of fat settle on various other parts of the body—the upper arms, the buttocks, even the eyelids.

For some people, the size as well as the shape of their body changes. As back muscles, connecting tissues, and the bones lose strength, the vertebrae collapse somewhat, causing some individuals to lose nearly an inch in height by age 60 (Whitbourne, 1985). In addition, overeating and underexercising result in more people being noticeably overweight during middle age than during any earlier period.

For the most part these changes in appearance have no significant health consequences. The consequences for self-image, however, can be substantial, particularly for women, because the cultural link between youthful beauty, sexual attractiveness, and social status is much stronger for them than for men (Katchadourian, 1987). Simone de Beauvior probably spoke for many middle-aged women when she confessed in her 50s:

> I loathe my appearance now: the eyebrows slipped down toward the eyes, the bags underneath, and the air of sadness around the mouth that wrinkles always bring. Perhaps the people I pass in the street see merely a woman . . . who simply looks her age, no more, no less. But when I look I see my face as it was, attacked by the pox of time for which there is no cure. [de Beauvoir, 1964]

Most developmentalists would find these feelings both understandable and unfortunate. As we will see throughout the rest of this book, the overall impact that aging has on the individual in middle age, and in late adulthood, depends in large measure on the individual's attitude toward growing old. For those who develop a constructive, adaptive attitude, the difficulties of the aging process can be greatly diminished, and the pleasures, significantly enhanced. Regarding the "pox of time" and the superficial physical changes of middle age, many developmentalists might wish for both men and women a view closer to the one expressed by another writer in her middle years, Germaine Greer (1986):

> Now, at last, we can escape from the consciousness of glamour; we can really listen to what people are saying, without worrying whether we look pretty doing it . . . We ought to be turning ourselves loose, freeing ourselves from inauthentic ideas of beauty, from discomfort borne in order to be beautiful.

Figure 20.1 *Advertising images like this one daily reinforce our culture's emphasis on youthful appearance. As the baby-boomers approach middle age, the market for skin-care products to mask the signs of aging is booming. Some people go even further: in 1986, approximately 720,000 Americans had surgery solely to improve their appearance, one of the more common procedures being a face-lift.*

This is not to suggest that developmentalists are in favor of people just "letting themselves go" in middle age. On the contrary, they emphasize that throughout adulthood, the benefits of staying in shape are far more than cosmetic. What they are in favor of is people's developing a more balanced view of the changes of aging, one that is, literally and figuratively, more than skin deep.

Now let us take a look at some of the less obvious but more crucial physical changes that occur during middle adulthood.

The Sense Organs

Sometime before age 60, virtually all adults notice that their sense organs are not as acute as they once were. Although all the sensory systems decline at about the same rate, age-related deficits are most obvious in the two most crucial systems, hearing and vision.

Hearing Most of the hearing losses that adults experience depend on their sex (men are more likely than women to develop difficulties), genetic tendencies, and age. However, some decline is environmentally caused. Performing very noisy work for long periods without protective headphones, for example, or habitually listening to music at ear-splitting levels, increases hearing loss, both in the short and the long term.

Although some hearing loss is inevitable in middle age, most middle-aged adults can still hear quite well, as the statistics on the ability to hear speech at age 50 suggest (see Table 20.1). It should be noted, however, that these data may underestimate overall hearing losses, because ability to distinguish pure tones declines faster than ability to understand conversation: often the first sign of a hearing loss is difficulty hearing a doorbell or telephone ring in the distance, or a tendency to turn up the stereo. Hearing losses are also more evident in the case of high-frequency noises, especially for men.

TABLE 20.1 **Hearing Loss at Age 50**

	Men	Women
Can understand a whisper	65%	75%
Cannot understand a whisper, but can understand soft conversation	28%	22%
Cannot understand soft conversation, but can understand loud conversation	5%	2%
Cannot understand even loud conversation	2%	1%

Fortunately, most hearing losses in middle adulthood are easy to remedy. Usually only a minor accommodation is needed, such as asking others to speak up, adjusting the ring of the telephone, and the like. In the case of more serious loss, today's tiny wireless hearing aids can usually correct the problem much more efficiently than the hearing aids of even a few years ago could.

Vision The standard measure of visual acuity, which is the ability to focus on objects at various distances, shows great variation from person to person across adulthood, largely because, after puberty, heredity affects focusing much more than age does.

Nevertheless, older adults are more likely to need corrective lenses, and age-related changes in the shape of the cornea affect the kind of lenses they need (Whitbourne, 1985). Young people tend simply to be nearsighted, whereas older adults tend also to be farsighted and astigmatic, necessitating bifocals or two pairs of glasses.

Several other aspects of vision, among them depth perception, eye-muscle resilience, and adaptation to darkness, decline steadily with age. Each of these changes can affect daily life. Decreasing depth perception makes people more likely to trip going down stairs; muscle weakness makes it harder for them to focus on small print for several hours; slower adaptation to darkness increases the time it takes them to begin to find their way in a dark room after coming in from the bright light, or, more ominously, to see the road at night after experiencing the momentary blindness caused by oncoming headlights. These changes are particularly likely to become apparent after age 50 (Whitbourne, 1985). It is noteworthy, however, that middle-aged adults seem to adjust to these changes without major difficulty. Serious accidents, either in a fall or while driving a car, are much more common in early or late adulthood than in middle adulthood.

Although most age-related vision problems are minor, and relatively easy to correct or compensate for, one can be very serious. **Glaucoma,** an eye disease characterized by an increase of fluid within the eyeball, becomes increasingly common after age 40, and is the leading cause of blindness by age 70. About 60,000 people in the United States are blind because of it (Johnson and Goldfinger, 1981). Luckily, the serious consequences of glaucoma can usually be prevented by early treatment; unfortunately, the disease has no obvious early warning signs. There is, however, a simple optometric test (a puff of air to detect increasing pressure within the eyeball) that spots glaucoma in the early stages, and it should be part of routine health care for every middle-aged person.

Changes in Vital Body Systems

Systemic declines, as outlined in our discussion of early adulthood, continue in the efficiency and the underlying organ reserve of the lungs, the heart, the digestive system, and so forth, making people more vulnerable to disease (see Table 20.2). However, for most middle-aged people, none of these changes is critical. Indeed, in a nationwide survey of people between the ages of 45 and 64, more than half rated their health as excellent or very good, and only one-fifth rated their health as merely fair or poor (National Center for Health Statistics, 1985).

Reflecting this self-reporting of good health is the fact that, thanks to better health habits and disease prevention, death rates among the middle-aged have declined dramatically over the past thirty-five years, especially those relating to the two leading killers of this age group, heart disease and cancer. Overall, the death rate of people between ages 40 and 60 decreased by a third between 1950 and 1984. Current estimates are that only 3 out of every 100 40-year-olds will die before age 50, and only 8 out of every 100 50-year-olds will die before age 60 (U.S. Bureau of the Census, 1986).

Figure 20.2 *In addition to maintaining better health habits, such as cutting down on fatty foods and not smoking, many middle-aged people are coming to realize the benefits of active conditioning. A moderate diet and a program of regular exercise can produce wonders of restoration, even for a facade like this.*

"Structurally, you're sound. It's your facade that's crumbling."

Drawing by Ross; © 1986 The New Yorker Magazine, Inc.

Variations in Health

Of course, overall statistics about the aging and health of the middle-aged are generalities that cover over many specific variations. For example, those with more education and income report better health, and have lower mortality rates, than those less well-off. Race, and the socioeconomic factors associated with it, are also relevant: the death rate for black middle-aged Americans is twice that of whites, and black middle-aged Americans are more likely to say that their health is fair or poor. Sex can be an important factor as well: men, for example, are much more likely to suffer fatal illnesses such as heart diseases, and women are more likely to suffer chronic disabilities such as arthritis. As a result, while middle-aged women are more likely than men to rate their health as fair or poor, the overall death rate for middle-aged men is three times that of women.

Indeed, throughout the life span, differences between the health of one individual and another are affected by genes and by social factors that correlate with education, income, and race. In middle age, another variable becomes particularly salient—health habits that can be critical in preventing or slowing the progress of virtually every major disease and disability (Michels, 1985). (See Table 20.2 and the Research Report on the next page).

TABLE 20.2 **The Increments of Chronic Disease**

Age	Stage	Atherosclerosis (Hardening of Arteries)	Cancer	Arthritis	Diabetes	Emphysema	Cirrhosis
20	Start	Elevated cholesterol	Carcinogen exposure	Abnormal cartilage staining	Obesity, genetic susceptibility	Smoker	Drinker
30	Discernible	Small lesions on arteriogram	Cellular metaplasia*	Slight joint space narrowing	Abnormal glucose tolerance	Mild airway obstruction	Fatty liver on biopsy
40	Subclinical	Larger lesions on arteriogram	Increasing metaplasia	Bone spurs	Elevated blood glucose	Decrease in surface area and elasticity of lung tissue	Enlarged liver
50	Threshold	Leg pain on exercise	Carcinoma *in situ*	Mild articular pain	Sugar in urine	Shortness of breath	Upper GI hemorrhage
60	Severe	Angina pectoris	Clinical cancer	Moderate articular pain	Drugs required to lower blood glucose	Recurrent hospitalization	Fluid in the abdomen
70	End	Stroke, heart attack	Cancer spreads from site of origin	Disabled	Blindness; nerve and kidney damage	Intractable oxygen debt	Jaundice; hepatic coma
Prevention or Postponement		No cigarettes; no obesity; exercise	No cigarettes; limit pollution; diet; early detection	No obesity; exercise; minimize stress on any one joint	No obesity; exercise; diet	No cigarettes; exercise; limit pollution	No heavy drinking; diet

*Abnormal replacement of one type of cell by another.
Source: Adapted from Fries and Crapo, 1980.

RESEARCH REPORT   Five Lifestyle Factors and Their Impact on Disease

Over the past twenty years, several large longitudinal studies have followed the lives of thousands of healthy adults, noting the relationship between their lifestyle—particularly health habits—and the later incidence of disease. As you will see, the links between certain lifestyle factors and health are many more than previously thought.

Cigarette-Smoking

About 37 percent of all middle-aged men, and 32 percent of all middle-aged women, smoke cigarettes, at significant peril to their health. Smoking is a known risk factor for most serious diseases that beset adults, including cancer of the lung, bladder, kidney, mouth, and stomach (Engstrom, 1986), as well as heart disease, stroke, and emphysema. Marijuana and low-nicotine cigarettes increase the risk of the same diseases, although researchers are uncertain whether they are equally, more, or less harmful (*Surgeon General's Report,* 1982). Second-hand smoke is also deleterious. For example, nonsmokers have a 30 percent higher risk of lung cancer if they are married to smokers than if they are married to nonsmokers (National Research Council, 1986).

Alcohol

In some respects, alcohol in moderation may be good for overall health; at least wine drinkers are less likely to have heart attacks than teetotalers are (St. Leger et al., 1979). One possible reason is that alcohol increases the blood's supply of HDL (high-density lipoprotein), a protein that aids in ridding the body of excess fat and cholesterol, thereby helping to prevent clogged arteries and reducing the chance of blood clots. Mitigating against the possible beneficial effects of alcohol is the clear fact that many people do not drink in moderation. Indeed, alcohol abuse is considered the leading health problem in the United States, one that in 1986 cost the economy an estimated $117 billion, primarily in the loss of productive labor due to workers' being too ill to work (Holden, 1987). Heavy drinking is the main cause of cirrhosis of the liver (the seventh most common cause of death in the United States), and it also puts a stress on the heart and stomach and destroys brain cells. Further, it hastens the calcium loss that causes the bone-weakening known as osteoporosis, and it is a risk factor in many forms of cancer, including breast cancer, the second most common form of cancer for women (Edwards, 1987). Given all these risks, "the difficulty many people have in controlling their alcohol consumption makes it unwise to encourage drinking for the sake of a small and uncertain benefit" (Johnson and Goldfinger, 1981).

Nutrition

As we have seen, nutrition plays a central role in development throughout the life span. During middle adulthood, it may play an important part with regard to the onset and progress of the two major killers of the middle-aged, heart disease and cancer. Adults in industrialized countries typically consume 40 percent of their calories as fat. Much of it is animal fat (in whole milk, cheese, butter, beef, pork, and eggs), and therefore is high in cholesterol, a contributor to coronary heart disease, particularly in middle age. High-fat, low-fiber diets also correlate with several types of cancer. Consequently, the National Cancer Institute recommends that adults reduce the fat content of their daily diet to no more than 30 percent and increase their consumption of fiber, from the typical current average of 20 grams per day to 30 grams (Engstrom, 1986).

Weight

A recent survey of Americans found that a third of all middle-age people are obese, that is, 20 percent or more above the average weight for their particular height (National Center for Health Statistics, 1985). Obesity is a definite risk factor for heart disease, diabetes, and stroke, and is a contributing factor for arthritis, the most common disability for older adults. For obese persons in middle age, then, weight reduction should be a clear priority.

As long as a person is not obese, however, physicians disagree about how thin he or she should be. Some believe that it is better to be slightly overweight than to be underweight (Andres, 1981), and the National Institute of Aging, on the basis of data from twenty-five insurance companies, recently raised the weight standards that are considered safe for middle-aged and older people (see table opposite).

However, other experts agree with Dr. William Castelli, the director of the renowned Framingham heart study, who maintains that every excess pound increases the risk of disease and premature death (Shell, 1982).

Exercise

There is no dispute among physicians about the importance of exercise. Active people have much lower rates of serious illness and death than inactive people (Bortz, 1980; Whitbourne, 1985). Just being in an active job seems beneficial. For example, British bus conductors, who are constantly moving through the bus and up and down the stairs of double-deckers to collect tickets, are generally in better health than their coworkers who drive the buses. Similar findings come from a comparison of San Francisco

Desirable Weight Ranges (Men and Women Combined) at Various Ages

Height	Ages 20–29	Ages 30–39	Ages 40–49	Ages 50–59	Ages 60–69
4'10"	84–111	92–119	99–127	107–135	115–142
4'11"	87–115	95–123	103–131	111–139	119–147
5'0"	90–119	98–127	106–135	114–143	123–152
5'1"	93–123	101–131	110–140	118–148	127–157
5'2"	96–127	105–136	113–144	122–153	131–163
5'3"	99–131	108–140	117–149	126–158	135–168
5'4"	102–135	112–145	121–154	130–163	140–173
5'5"	106–140	115–149	125–159	134–168	144–179
5'6"	109–144	119–154	129–164	138–174	148–184
5'7"	112–148	122–159	133 169	143–179	153–190
5'8"	116–153	126–163	137–174	147–184	158–196
5'9"	119–157	130–168	141–179	151–190	162–201
5'10"	122–162	134–173	145–184	156–195	167–207
5'11"	126–167	137–178	149–190	160–201	172–213
6'0"	129–171	141–183	153–195	165–207	177–219
6'1"	133–176	145–188	157–200	169–213	182–225
6'2"	137–181	149–194	162–206	174–219	187–232
6'3"	141–186	153–199	166–212	179–225	192–238
6'4"	144–191	157–205	171–218	184–231	197–244

Source: Katchadourian, 1987.

longshoremen who had active jobs with those who did not (Oberman, 1980).

Even better than having an active job is engaging in exercise that is sufficiently strenuous to raise the pulse to about 75 percent of its maximum capacity—three or more times a week for at least 30 minutes per workout. (The maximum heart rate for a normal healthy adult is about 220 minus the age of the adult.)

Such exercise increases heart and lung capacity, lowers blood pressure, increases HDL in the blood, and, even if weight remains the same, reduces the ratio of body fat to body weight. Each of these results helps prolong life (Williams, 1980). Regular exercise even enhances cognitive functioning, especially in middle-aged and older people, probably because it improves blood circulation in the brain (Elsayed et al., 1980). Exercise is also the best

method of weight reduction: it not only burns calories but decreases the appetite and increases metabolism, so the person continues to benefit for several hours after a workout is over (Sloane, 1985). Finally, for many people, exercise is the pathway to establishing control over other poor health habits, including smoking and drinking (Brownell et al., 1986). As one physician put it:

Probably the least understood but most important and universal benefits of a regular exercise program are the associated behavioral changes. Innumerable reports confirm that persons in physical activity programs develop a sense of well-being, increased tolerance to anxiety and various psychological stresses, and an improved self-image. These acquired characteristics may lessen the propensity toward CHD [coronary heart disease] or strengthen other health habits, such as cessation of smoking, modification of diet, and general health behavior. [Oberman, 1980]

Establishing a regular exercise routine in early adulthood is a good way to help ensure a healthy middle age. However, people who choose weight-lifting as that routine may be making a fatal error. The most important muscles to build as one gets older are not biceps but those of the heart and lungs, and the best way to strengthen them is through steady, rhythmic motion lasting twenty minutes or more. Pumping a bicycle, for instance, is much better exercise than pumping iron.

Developmental Changes in Health Habits

As signs from their bodies, advice from their doctors, and the celebration of a 40th or 50th birthday drive home the reality of aging, middle-aged adults become more likely to improve their health habits than younger adults are (Katchadourian, 1987; Tough, 1982). For example, although the percentage of cigarette smokers is quite similar across adult age groups (about one in three), today's younger nonsmokers are likely to have never smoked, while middle-aged nonsmokers include the highest percentage of former smokers, 40 percent of the men and 19 percent of the women (National Center for Health Statistics, 1985). Alcohol and drug abuse also become less common as adulthood progresses (Robins et al., 1984).

A similar pattern may be emerging with regard to exercise. Although most middle-aged people exercise less than they did when they were younger, cross-sectional research finds that they are exercising more than their counterparts did in decades past, when the thought of a 50-year-old on a bicycle or a 60-year-old in a jogging suit would raise eyebrows and provoke chuckles (Bortz, 1980). One survey of regular outdoor exercise found that 53 percent of middle-aged people walk for pleasure, 41 percent swim, 22 percent bicycle, 13 percent jog, 11 percent engage in outdoor team sports, and 10 percent play tennis (U.S. Bureau of the Census, 1986). Some middle-aged adults even exercise more than they did when they were younger. Middle-aged Jane Fonda, movie star and aerobic guru, says, "I can run farther, stretch deeper, climb steeper, lift heavier, stand taller, and dance longer than when I was twenty" (Fonda, 1984). Jane Fonda, of course, is representative of a very small minority, since women especially are likely to exercise less in middle age than earlier. However, the current cohort of young women may change this developmental pattern, because, compared with earlier cohorts, they know much more about the benefits of exercise, and have been able to participate in many physical activities that, for the most part, were once restricted to males.

Figure 20.3 *Increasingly for middle-aged people, not engaging in regular exercise means being out of step with one's peers.*

"Anything wrong?"

It should be noted that many more middle-aged adults attempt to change poor health habits than actually succeed. Some of this group try once, fail, become discouraged, and never try again. Their early resignation is unfortunate. Research has shown that in gaining control of one's health habits, as in many things in life, to the persistent goes the prize (see A Closer Look, next page).

Figure 20.4 *To quit smoking, people have resorted to dozens of different techniques, from hypnosis to acupuncture. Shown here is aversive conditioning. Would-be quitters sit together in a small room and smoke continually, puffing hard, until they literally get sick of it. Later, if they are tempted to light up, feelings of revulsion and even nausea usually stop them. When this method works, the new nonsmokers often become the most dedicated opponents of second-hand smoke.*

Stress

Stress is a critical but extremely complex factor in the overall determination of health (Eisdorfer, 1985). Even the definition is somewhat complicated. **Stress** is the adverse physical and emotional reaction to demands put on the individual by unsettling conditions or experiences, which are called **stressors** (Selye, 1982). Important in this definition is the recognition that what is a stressor for one individual on any given occasion may not be one for someone else, or even for the same person on some other occasion. In other words, what makes a potential stressor in fact stressful is the individual's reaction to it.

For example, attending a noisy, crowded party is, obviously, fun for many people, a chance, perhaps, to make new acquaintances, to show off, to flirt, or to just relax and let go of the day's tensions. For other people, the same type of party may, to varying degrees, represent a stressor, because they hate crowding and noise, or worry that they are dressed inappropriately, or feel awkward among strangers, or have difficulty making "party conversation," or regard such gatherings as a waste of time, or fear that the arrival of a former love is going to stir old feelings—the list of possible reasons is virtually endless. And so is the list of possible stress reactions that one person or another might experience: a vague sense of unease; restlessness and an inability to concentrate; headache; indigestion; a feeling of deep fatigue; streams of self-conscious, troubling thoughts; elevated blood pressure; excessive talkativeness; an outbreak of hives; sudden fits of stammering or slips of the tongue; a case of hiccups; a chronic urge to urinate; an onset of nervous laughing; blushing or the appearance of a rash; a pounding heart and sweaty palms; fainting. Any of these reactions, and a host of others, are typical stress responses.

Also endless are the ways various people might try to cope with their feelings of stress. One person might develop a migraine and never get to the party; another might make up an excuse for an early departure; another might stay glued to a close friend for the entire evening; another might sulk in a corner all night; another might retreat to the kitchen to "help out"; another might get drunk; another

Try, Try Again

Ask a psychiatrist, psychologist, or clinical social worker what the chances are of changing a destructive health habit and you are not likely to be cheered by the answer. Generally, the percentage of people who go into treatment for addiction to cigarettes or alcohol, and who remain drug-free for at least a year, is only about 20 or 25 percent. This low cure rate is just about the same for heroin addicts, a surprising revelation for anyone who believes that people who are addicted to alcohol or nicotine have greater self-control with regard to their drug habit than heroin addicts (Brownell et al., 1986; Maisto and Caddy, 1981; Parloff et al., 1986).

Attempts to treat obesity are similarly discouraging. One classic review of treatment programs puts it bluntly:

As is well-known, the usual results of outpatient treatment of obesity are extremely bad. Most obese people won't even come in for treatment. Those who do often drop out. The ones who don't drop out don't lose much weight. Finally, those who do lose weight usually regain it. [Skunkard, 1972]

Over the past two decades, an increasing number of residental treatment programs have been developed for obesity and other addictions; but while inpatient care gets better immediate results than outpatient care, the long-term results are no better than those of outpatient care (Miller and Hester, 1986; Orford, 1985).

Recently, researchers have turned their attention from people trying to "kick the habit" in treatment programs to people trying to kick it on their own, and they have found quite a contrast (Ockene, 1984).

One particularly encouraging study was conducted by Stanley Schachter (1982). He wanted to find out how many people in a given population had managed to stop smoking or lose weight permanently. To do this, he interviewed all the members of the psychology department of Columbia University (including faculty, graduate students, and secre-

taries), and all but one of the people who worked in sixteen stores in a small town where he lived and shopped. He chose these two groups (admittedly a nonrepresentative sample) because he wanted to limit the number of people who would refuse to be interviewed. He reasoned that many people might be reluctant to tell a stranger about their smoking or overeating, especially if they were ashamed of their habits.

In the total sample of 161 people, 94 had smoked cigarettes, most for ten years or more, and 40 people had a history of obesity. As you can see from the charts (next page), most of the people in both groups had tried to break their habit, and most of those who tried had succeeded.

These impressive success rates were apparent in both men and women, and in both the urban academics and rural shop people. Those who were heavy smokers or very obese had just as good rates of success as those who were light smokers or slightly overweight.

Interestingly, only 2 of the 77 smokers who tried to quit, and only 10 of the 40 overweight people who tried to slim down, sought some professional help—either from individuals such as psychotherapists or from groups such as Weight Watchers or Smoke Enders. Of these 12 help-seekers, 6 were among those who eventually succeeded.

Other studies confirm that most people who, on their own or with informal help from family or friends, try to change their habits for the better eventually succeed, with middle-aged people being particularly likely to change a health habit (Tough, 1982). The most surprising research on self-help comes from studies of Vietnam veterans who became heroin addicts during the war. Most of these men, who were psychologically and physiologically dependent on the drug, permanently kicked the habit when they reached the States, without any formal treatment (Robins, 1974). A follow-up study found that only 12 percent of

might decide to look at the party as a challenge and figure out ways of making the best of it. As you can infer from this lengthy but rather simple example, the variables of stress are usually hard to pin down. What, in fact, causes stress in any given individual, and how that person responds to a particular stressor physiologically and behaviorally, depend on the individual's temperament, past experiences, physical vulnerabilities, resources and strategies for coping, the overall context, and, perhaps most important, how all these factors influence the way the individual interprets a potential stressor.

Despite this variation, however, study after study finds that psychological stressors ranging from the devastation of war to the hassles of everyday life correlate with a wide range of health problems. Indeed, virtually every physical disease and psychological disability, from arthritis to alcoholism, heart attacks to headaches,

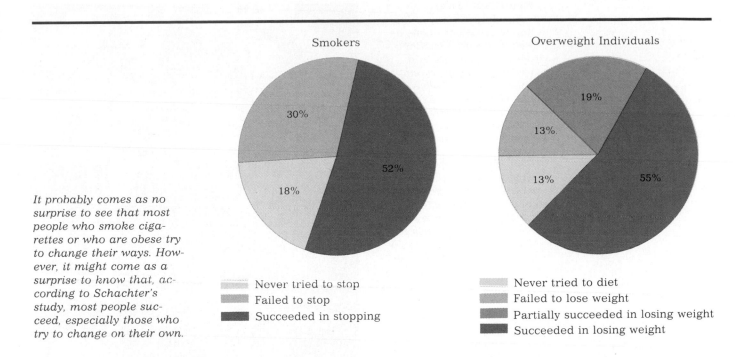

Smokers

- Never tried to stop
- Failed to stop
- Succeeded in stopping

Overweight Individuals

- Never tried to diet
- Failed to lose weight
- Partially succeeded in losing weight
- Succeeded in losing weight

It probably comes as no surprise to see that most people who smoke ciga- rettes or who are obese try to change their ways. How- ever, it might come as a surprise to know that, ac- cording to Schachter's study, most people suc- ceed, especially those who try to change on their own.

them had become readdicted at any time during the three years after they left the service (Robins et al., 1980).

Why are these self-help results so much more promising than those reported in follow-up studies of various treat- ment programs? Schachter has two explanations:

First, people who cure themselves do not go to therapists. Our view of the intractability of the addictive states has been molded largely by that self-selected hard-core group of people, who, unable or unwilling to help themselves, go to therapists for help, thereby becoming the only easily available subjects for studies of recidivism and addiction.

Second, the inferences drawn from studies of therapeutic effectiveness are based on single attempts to cure some addictive state or other. In fact, people do try to quit repeatedly. It may be that, with or without professional help, success rates with multiple attempts to quit are greater than with single attempts to do so.

There is an important implication in this for developmen- tal psychology. In looking at an adult's behavior, we should take a long-term developmental perspective. The fact that someone has tried to improve and failed does not mean that he or she will never succeed. Similarly, the fact that someone has successfully broken a habit for a time, and then gone back to it, does not mean total failure. A lapse is not necessarily a relapse; a relapse is not neces- sarily permanent (Brownell et al., 1986). For many people, each attempt to break a habit, each partial success or partial failure, is instructive: it teaches them which strate- gies work and which don't work, making the next attempt more likely to succeed.

strokes to suicides, is more likely to occur in people who experience significant stress.

How exactly might disease and stress be linked? In addition to the numerous physiological stress reactions like the increased heart rate and elevated blood pressure mentioned in our party example, the experience of stress produces changes in the concentration of many of the hormones and other chemicals in the body (Zales, 1985). Any of these physiological changes could certainly take a toll on the various organs over an extended period of time, and many could harm the body in the short term as well. In fact, the immediate biochemical responses to stressors might explain a curious finding: heart attacks and strokes are far more likely to occur at 9 A.M. on Monday morning—the beginning of the work week—than at any other time (Kolata, 1986).

Figure 20.5 *Stress can occur during many of the happy occasions of life, as well as during troubling ones. These guests at a wedding reception seem to be enjoying themselves, but, chances are, some of them had "butterflies" in their stomachs or sweat on their palms as they were getting ready for the wedding, and others may be experiencing similar signs of stress as they get ready for the celebrating. The people most likely to have physiological or psychological stress reactions are, of course, the "happy couple."*

Stress may also be less directly involved in causing disease, in that stress over time reduces the effectiveness of the immune system (Stein and Schleifer, 1985). Thus people who are under constant or severe stress are more vulnerable to a variety of illnesses, ranging from the common cold to cancer.

While some of life's stressful experiences are unpredictable and uncontrollable, in many ways, people often help to create the very conditions they experience as stressful. One highly pronounced example is found in people who have a Type A personality pattern. These are people who characteristically strive for high achievement and quick action, frequently feeling hostile toward others, especially others who compete with them in some way. They seem to have stress-seeking personality characteristics that may help them succeed in competitive and demanding careers, but that may become destructive in the long run, especially if, for reasons of genetic inheritance or lifestyle, they are at risk for coronary heart disease (see A Closer Look, next page).

Developmental Changes in Coping with Stress   As we have seen, there are many ways to cope with stress, from altering the stressor, to denying it, to escaping from it, to reevaluating it. Each coping strategy is well suited to some stresses and poorly suited to others. For example, some people, especially young adults, have a very active coping strategy, attacking problems head on. This may serve them well when the conditions creating stress can readily be changed, but not when they are beyond individual control. Conversely, some people have more passive coping strategies, down-playing or even ignoring problems that are stressful. While this strategy may be successful when the conditions creating stress cannot be changed, it can also leave the person continually vulnerable to stressors that could be adjusted or eliminated.

A CLOSER LOOK Type A Behavior

Do you feel that you are in "an incessant struggle to achieve more and more in less and less time," with a sense of "time urgency and hurry sickness and . . . a free-floating but well-rationalized hostility"? More specifically, do you

1. Explosively accent various key words when there is no need to do so?

2. Have nervous habits, such as clenching your jaw or biting your knuckles?

3. Always move, walk, and eat rapidly?

4. Feel impatience at the rate at which most events occur?

5. Frequently try to accomplish two or more things at once, like doing your homework while eating breakfast?

6. Always find it difficult to talk about things of little interest to you?

7. Almost always feel guilty if you relax for a few hours or do nothing useful while on vacation?

8. No longer notice the interesting and beautiful objects in your world?

9. Believe that whatever success you have attained is due, in part, to your ability to get things done faster than others can?

10. On meeting someone like yourself, feel the need to challenge that person?

If your honest answer to most of these is yes, you have a Type A temperament, which puts you at risk for a heart attack (Friedman and Rosenman, 1974), especially if you live in a challenging, competitive environment, as most Type A people do (Rosenman and Chesney, 1983). Although physicians agree on this point, whether Type A is a major cause of coronary heart disease or simply a contributing factor is still under debate.

Detailed analysis of Type A behavior finds two components, an achievement-striving component and an impatient-aggressive component. The impatient-aggressive component, which includes masked hostility and self-centeredness, is the more destructive (Fischman, 1987). Not only does anger stimulate potentially harmful physiological responses, such as increased heart and breathing rates, higher blood pressure, and increased adrenalin in the bloodstream, but hostile people have less of two factors that reduce stress—friends and a sense of satisfaction with life. Interestingly, these personality traits may predispose people to other diseases as well. Several longitudinal studies have found, for example, that people who tend to be tense and withdrawn socially are significantly more

"It always takes Howard a couple of days to relax."

Type A people tend to pack their hard-driving traits along with their vacation gear.

Drawing by Stevenson; © 1986 The New Yorker Magazine, Inc.

likely to develop cancer than those who are more easy-going and sociable (Bower, 1987).

Although it is not known how people become Type A, it does seem that early influences may be important, since even children can exhibit Type A traits. To what degree these influences are genetic or environmental is not at all clear, however, especially given the fact that while children with Type A traits do not necessarily become Type A adults, adolescents with Type A traits generally do (Steinberg, 1986).

In adulthood, Type A personality patterns are difficult to change. Even Type A individuals who have already had a heart attack find it difficult to recognize and change their behavior. They are likely to deny that they have a problem, or to defend their behavior as necessary for their work, or to become angry at those who suggest that they need to change. One of the most effective remedies, it seems, is group therapy, in which several Type A personalities help each other recognize and change their maladaptive attitudes and habits. As one therapist put it, "when you put ten Type A's in a room, all competing and interrupting each other, they quickly begin to see how obnoxious this behavior is" (quoted in Fischman, 1987).

Figure 20.6 *Stress can take many forms. A brief, usually harmless episode of stress is likely to be generated by the anxiety and confusion that usually accompany moving to a new home, as the couple on the left are doing. Much more debilitating is the prolonged stress of having no home, as is the case for the mother and daughter on the right. The physical and psychological damage of living in run-down, overcrowded, often dangerous welfare hotels has been documented in studies and can be readily observed: many school-age children in temporary shelters still suck their thumbs, wet their beds, and cry themselves to sleep.*

Ideally we all should have several strategies available so that we can approach some problems directly and avoid or ignore others, and ideally we should assess each stressor in its full context so that the strategy we choose is the one most likely to succeed. Unfortunately, many people tend to be locked into one coping pattern or another, without considering all the alternatives (Roth and Cohen, 1986).

Are middle-aged people any better than younger people at reducing or coping with stressors? Overall, research finds such great individual variations that it would be a mistake to say that aging per se results in better adaptation to stress. Indeed, because the body's organ reserve gradually declines with age, physiological adjustments to stress take longer as people grow older.

Psychologically, however, many people do seem to experience stress less, or to be better at handling stressors, as they grow older (Lazarus and DeLongis, 1983). A longitudinal study that followed men from college days through middle age found that they gradually became more likely to use adaptive defense mechanisms, such as humor, rather than maladaptive ones, such as anger, when they were in potentially stressful situations with other people (Vaillant, 1977). Another longitudinal study that followed college women found that in middle adulthood they too used better coping methods and thus experienced less stress (Helson and Moane, 1987). Notably, in early adulthood, many of the women tended to interpret demands made on them as exploitive, and to feel resentful and depressed. In middle adulthood, on the other hand, they tended to see the demands made on them as legitimate tasks, and to feel competent and productive in meeting them. In both studies, one of the most effective coping strategies was to try to fully understand others' points of view and to depersonalize conflict, concentrating on the issues involved rather than on the emotions surrounding them.

Since stress is a major contributor to most forms of mental illness, additional evidence for improved stress responses over time comes from a large, carefully constructed study of the prevalence of fifteen mental disorders (as defined by DMS-III, a detailed diagnostic manual) among a cross-section of adults living in three major U.S. cities (Robins et al., 1984). As you can see from Table 20.3, "the total rates of disorders drop sharply after age 45" (Robins et al., 1984). Virtually every specific disorder became less common with age, not only ones that you might expect, such

as drug abuse (ten times less likely in middle than in early
might not seem age-related, such as schizophrenia (less †
age 45 as under). Similarly, depression among the mic
that of adults between the ages of 25 and 44 (4.6 percent

TABLE 20.3 **Incidence of Psychologi**

Age	Men	Women
18–24	23%	23%
25–44	21%	22%
45–64	13%	14%
65+	12%	13%

There are several possible explanations for the apparent developmental changes
that may occur in the way stress is experienced and dealt with:

1. As you read in Chapter 19, compared with young adults, middle-aged people on
the whole experience fewer psychosocial demands, since they generally already
have chosen a career, a spouse, and so forth.

2. Because they tend to be more settled with family, friends, and work, middle-
aged people may be better buffered by their social networks against whatever
stresses they encounter.

3. Middle-aged people may be more likely to seek, and more able to afford, professional help for their problems, and therefore may have fewer long-standing difficulties.

4. As Chapters 18 and 21 make clear, over time, many people evidence a more
flexible cognitive approach, and this flexibility no doubt enhances one's ability to
both appraise and deal with potential stressors.

Certainly it would seem that the combination of these psychosocial and cognitive
factors could result in a general pattern of reduced stress during middle adulthood. In addition, recent, startling research suggests that there may be a physiological factor involved. Autopsy studies show that during middle age there is a
major loss of cells in the *locus coeruleus,* an area of the brain stem that is associated with the experiencing of panic and anxiety.

The Sexual-Reproductive System

As the sexual-reproductive system continues to age during middle adulthood, sexual responses gradually become slower and less distinct. We will soon discuss the
consequences of these ongoing changes, but first let us look at one change that is
definitive: the cessation of reproductive potential in women.

Menopause

After about age 35, many women find that the time between their menstrual periods becomes shorter, from about every twenty-eight days at age 35, to twenty-five
days at age 40, to twenty-three days by the mid-40s (Whitbourne, 1985). Then toward
the end of the 40s, periods become erratic, with some coming fairly close together,
and others being missed entirely.

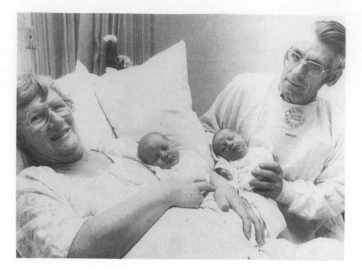

Figure 20.7 *Although fertility begins to decline in the 30s, motherhood is still possible until a year after the last menstrual period. This 50-year-old woman and her husband had eight children and five grandchildren by the time twins James and Justin arrived to complete the family.*

Sometime between her late 40s and early 50s (the average age is 51), a woman reaches **menopause,** as ovulation and menstruation stop and the production of the hormone estrogen drops considerably. Strictly speaking, menopause is dated one year after a woman's last menstrual cycle.

The Climacteric   The term "menopause," and the misnomer "change of life," are also sometimes used to refer to the time, lasting about three years, during which the woman's body adjusts to much lower levels of estrogen. A more inclusive term for these years is the **climacteric,** which refers to all the various biological and psychological changes that accompany menopause.

The most obvious symptoms of the climacteric are hot flashes (suddenly feeling hot), hot flushes (suddenly looking hot), and cold sweats (feeling cold and clammy). These symptoms are all caused by **vasomotor instability,** that is, a temporary disruption in the body mechanisms that constrict or dilate the blood vessels to maintain body temperature. Lower estrogen levels produce many other changes in the female body, including drier skin, less vaginal lubrication during sexual arousal, loss of some breast tissue, more brittle bones, and an increased risk of heart attack. Many women also find that, during the climacteric, their moods change inexplicably from day to day. Despite these changes, however, the prevalent conception of menopause as a time of difficulty and depression is largely myth. For most women, the anticipation of menopause is usually worse than the actual experiencing of it.

Some women who do have difficulty with the symptoms of the climacteric take synthetic estrogen to compensate for the diminished production of this hormone. However, **estrogen-replacement therapy,** abbreviated **ERT,** is controversial, because it is associated with a greater risk of uterine cancer, and it is now not generally recommended for minor problems or long-time use (Katchadourian, 1987). In recent years, ERT was commonly given to women who had a hysterectomy (surgical removal of the uterus) that included the removal of the ovaries, which are the producers of estrogen (Notman, 1980). (Between 1972 and 1982 more than 5 million American women had hysterectomies, mostly of this type.) A hysterectomy is still the most common surgical procedure for women in their 30s and 40s; however, since the early 1980s, routine removal of healthy ovaries is no longer standard practice, thus ERT is no longer always necessary after a hysterectomy.

Low doses of estrogen are also recommended for women who are at risk for **osteoporosis,** the condition of thin and brittle bones that leads to increased fractures and frailty in old age. High risk is suggested by a woman's being thin, Caucasian, and postmenopausal, and is confirmed by X-rays of bone density. For most women, a childhood diet high in calcium, regular weight-bearing exercise (such as walking or running), and avoiding cigarettes and alcohol significantly lessen the likelihood of their developing osteoporosis, particularly if their daily diet in adulthood includes 1,000 milligrams of calcium. It should be noted that the best sources of calcium are food products such as milk, cheese, and yogurt—and that calcium supplements are probably of little help, despite advertising to the contrary (Culliton, 1987).

Male Menopause?

For men in middle age, there is no sudden downward shift in reproductive ability or hormonal levels, as there is with women. Thus, physiologically, men experience nothing like the female climacteric. Most men continue to produce sperm indefinitely, and although there are important age-related declines in the number and motility of sperm (see Chapter 17), men are theoretically (and in some cases, actually) able to father a child in late adulthood. Similarly, the average levels of testosterone decline gradually, if at all, with age (Whitbourne, 1985).

Although strictly speaking there is no "male menopause," this phrase may have been coined to refer to another phenomenon: testosterone can dip markedly if a man suddenly becomes sexually inactive or unusually worried, as might happen if he were faced with unemployment, marital problems, serious illness, or unwanted retirement. Levels of testosterone correlate with levels of sexual desire and speed of sexual responses, so a man with low testosterone might find himself unable to have an erection when he wanted to. Thus, the effects of this dip, especially when added to whatever age-related declines have already occurred, may make a man highly anxious about his sexual virility, which, in turn, may reduce his testosterone level even more.

Reproductive Changes in Context

How strong is the psychological impact of the sudden loss of reproductive potential in women, and the related, more gradual decline in men? In this day and age, not very.

Until about a hundred years ago, the more children a couple had, the more fortunate they were considered to be. Children cost little to raise, helped their parents at home and at work, and supported their parents in old age. Traditionally, childbearing was particularly important for women, since most women attained social status directly from their role as mother. Men wanted children too, especially sons, to carry on the family name and tradition. In these circumstances, the psychological impact of declining fertility and menopause may have been substantial. Especially if a couple had only one or two children, menopause may have been greeted with sorrow as the final "closing of the gates" of reproduction, as psychoanalyst Helene Deutsch (1945) described it.

However, historical changes have meant that the end of childbearing is now determined less by age than by personal factors, such as the number of children a couple already has or the couple's financial situation. In fact, for the most part, the end of childbearing occurs through a conscious decision that is usually made when a woman is in her 30s, long before reproduction becomes biologically impossible. Menopause, then, as the time when sexual activity is no longer accompanied by fear of pregnancy, is more often welcomed than regretted (Luria and Meade, 1984).

A CLOSER LOOK   **Attitudes about Menopause**

Bernice Neugarten has been a pioneer in the study of adults. In the 1960s, when good research on adult development was scarce, most of what did exist was coauthored or inspired by Neugarten; in the 1970s, almost every major study of adult development referred to Neugarten. Neugarten's work reflects a number of gifts, but one in particular stands out: her ability to ask questions no one else thought to ask.

Until Neugarten, virtually no one (except psychiatrists working with troubled patients) had asked women about their psychological adjustment to menopause. Neugarten's initial interviews with women about this topic formed the basis of a questionnaire that she gave to 267 women between the ages of 21 and 65. The results (Neugarten et al., 1968) showed many significant differences between the younger women and the older women in their opinions about menopause. Younger women tended to have very negative attitudes about menopause and its consequences. Older women, by contrast, found menopause a liberating experience, and a third of the women between the ages of 45 and 55 (presumably right at the age of menopause) thought women are more interested in sex after menopause than before.

This is not to say that menopause is a joy. Most of the women found it somewhat unpleasant and troublesome. Significantly, many initially worried about their husband's reactions and about whether they themselves might lose their sanity. As one woman said, "I wondered if I would come through it whole or in pieces." But the oldest women, who had completed menopause, agreed that neither of these potential problems was as great as they had anticipated.

Indeed, on a multiple-choice question concerning the

"worst thing in general about the menopause," the answer that got the most response (29 percent) from the women between 45 and 55 was "Not knowing what to expect." Discomfort and pain bothered 19 percent most; "It's a sign you are getting old" was chosen by 18 percent. "Not being able to have more children" and "Loss of enjoyment in sex relations" were checked as worst by only 4 percent each.

There is still a dearth of research on the psychological effects of menopause. However, it seems more and more clear that culture, rather than nature, determines a woman's reaction to it (Parlee, 1984). Compare the women of the Rajput caste in India with the women of Western countries. Rajput women must live a sheltered life, avoiding any contact with men except their husbands, until menopause. Then they are free to socialize with men and to appear in public. Not surprisingly, they welcome menopause, and experience no unpleasant symptoms (Flint, 1982).

In our culture, while women who have experienced menopause realize that it is more benign than its reputation, this is often not understood even by their doctors. In a survey of midlife women and their physicians (almost all male), 21 percent of the doctors thought that menopause was a major health problem, and 83 percent thought that it was a major concern for midlife women. By contrast, none of the women thought it was a major health problem, and most of them rated menopause as no concern at all (DeLorey, 1984).

Neugarten suggests that only middle-aged and older women, who have experienced menopause, are able, in the words of one of her interviewees, to "separate the old wives' tales from that which is true of old wives." Cer-

---

Age-Related Changes in Sexual Expression

One usual way to measure sexual activity is in terms of the frequency of intercourse and orgasm. By this measure, sexual activity usually declines during middle age (Martin, 1977), though with wide individual differences, including some people who stop having intercourse altogether and others who continue to have intercourse on a regular basis (Comfort, 1979). One large longitudinal study found that by age 50, about two-thirds of the men and three-fourths of the women had intercourse once a week or less often, a substantial drop from the several times a week typically reported by younger people. By age 60, about 10 percent of the men and 50 percent of the women had stopped having intercourse entirely (Pfeiffer et al., 1972). Three-fourths of the couples who had stopped having intercourse cited impairment of the man's health or a decline in his sexual interest as the reason. Since women tend to

tainly it is time for the facts about the psychological impact of menopause to be better understood. As a psychic marker, menopause quite often signals new zest rather than the end of vitality and energy. As one woman remarked, "Since I have had my menopause, I have felt like a teenager again. I can remember my mother saying that after her menopause she really got her vigor, and I can say the same thing myself. I'm just never tired now."

Women's Attitudes Toward Menopause

Questions*	Percent Who Agreed with the Item (by Age)			
	21–30	31–44	45–55	56–65
1. Women generally feel better after menopause than they have for years.	32[†]	20[†]	68	67
2. A woman gets more confidence in herself after the change of life.	12[†]	21[†]	52	42
3. After the change of life, a woman feels freer to do things for herself.	16[†]	24[†]	74	65
4. Many women think menopause is the best thing that ever happened to them.	14[†]	31	46	40
5. If the truth were really known, most women would like to have themselves a fling at this time in their lives.	8[†]	33	32	24
6. After menopause, a woman is more interested in sex than she was before.	14[†]	27	35	21
7. Going through menopause really does *not* change a woman in any important way.	58[†]	55[†]	74	83
8. A woman is concerned about how her husband will feel toward her after menopause.	58[†]	44	41	21[†]
9. Women worry about losing their minds during menopause.	28[†]	35	51	24[†]
10. A woman in menopause is apt to do crazy things she herself does not understand.	40	56	53	40
11. In truth, just about every woman is depressed about the change of life.	48	29	40	28

*Not all questions are reprinted here. These are illustrative items that show the trend of responses.

[†]The difference between this group and the 45–55 group is statistically significant, that is, it is highly unlikely (one chance in twenty) that it would have occurred by chance.

be younger than their mates, this means that women find their sex life changing at a younger age than men do.

Even for the sexually active, however, the specifics of their activity change with age, especially for men. Sexual stimulation takes longer, and needs to be more direct than earlier. Further, as Herant Katchadourian (1987), a physician who studies sexuality, writes about men, "orgasmic reactions become less intense with age . . . contractions are fewer, ejaculation is less vigorous, and the volume of the ejaculate is smaller." Changes in the woman's orgasmic ability are harder to measure, but many researchers think a woman's eroticism is at least as strong in middle age as in early adulthood (Masters and Johnson, 1966; Van Keep and Gregory, 1977). After menopause, signs of arousal, including sexual lubrication, may be less appar-

Figure 20.8 *Throughout adulthood, continued, pleasurable sexual relations depend much less on the partners' age than their attitudes toward each other and toward sex itself. As one expert put it, the most important human sexual organ is the brain.*

ent, but none of these changes need impair a sexual relationship. As Katchadourian explains it, "while the intensity of the physiological responses clearly diminish over time, the subjective experience of orgasm continues to be highly satisfying, though not as explosive as in previous years." Summarizing the changing nature of sexual activity during adulthood, Katchadourian concludes:

> Sex for the young is fast and furious. It is ignited easily and fizzles out like fireworks. It is turned on by appearance and tends to stay skin deep. Some of us grow old but never outgrow this style. Hence, as we age we find ourselves and our partners increasingly less satisfactory. Others develop a deeper eroticism with the passage of years. The sensual quality of the person, rather than the body as such, becomes the main course.

One final fact seems to be suggested by the research: couples do not move from active happy sex lives to passive troubled ones unless their relationship is plagued by other problems that are reflected, but not caused, in the bedroom. Quite simply, people who have active sex lives in young adulthood are most likely to have active sex lives in middle and late adulthood, and couples who were never comfortable with their sexual relations are likely to end them in later years. For middle-aged and older adults, present interest and enjoyment in sex are much more strongly correlated with past interest and enjoyment than with variables such as overall health, or satisfaction with life, or education, or income (Pfeiffer and Davis, 1972).

Throughout life, it seems that sexual activity itself helps promote sexual interest and excitement; correspondingly, absence of sexual activity results in lower levels of sex hormones and a loss of sexual interest (Katchadourian, 1987; Masters and Johnson, 1970). Putting it into a broader context, Pfeiffer (1977) concludes:

> Examination of all the available data makes it clear that successfully aging persons are those who have made a decision to stay in training in the major areas of their lives. In particular, they have decided to stay in training physically, socially, emotionally, and intellectually. We have every reason to believe that staying in training sexually will also help improve the quality of life in later years.

Again, it seems that, as adults grow older and biological development is much less tied to chronological age, personal choices become increasingly important. In the next chapter, we will see that choice can influence our intellectual skills as well. At least in some abilities, to some extent, we can choose to be smarter, wiser, or more expert.

SUMMARY

Normal Changes in Middle Adulthood

1. A person's appearance gradually changes as middle age progresses, with more wrinkles, less hair, and new fat, particularly on the abdomen.

2. Hearing gradually becomes less acute, with noticeable losses being more likely for high-frequency sounds, particularly in men. Vision also becomes less sharp with age. Two particular difficulties for many middle-aged people are reading small print and adjusting to glare at night.

3. Overall, health is generally quite good during middle age, with a death rate significantly lower today than for earlier cohorts.

Variations in Health

4. Genetic, educational, and socioeconomic factors, along with gender, are partially responsible for variations in the health of middle-aged persons.

5. The most important reason for individual variations in health during middle age, however, is personal lifestyle. Cigarette-smoking, heavy alcohol consumption, high-fat diets, obesity, and lack of exercise are all risk factors for heart disease and cancer, as well as for other ailments.

6. Stress also correlates with virtually every physical and psychological impairment that middle-aged people can have. Great individual variation exists, however, in what constitutes a stressor, in how the person copes with it, and in the particular impact that a stressor will have on the body or the psyche. In general, middle-aged people may experience less stress, or cope with it better, than younger adults.

The Sexual-Reproductive System

7. At menopause, as a woman's menstrual cycle stops, ovulation ceases and levels of estrogen are markedly reduced. This hormonal change produces various symptoms and possible problems, although most women find the experience of menopause much less troubling than they were led to expect.

8. Men do not have sudden age-related drops in hormone levels or in fertility. In this sense, there is no "male menopause."

9. In modern society, most couples have decided to stop childbearing by middle age if not before, so the infertility caused by menopause is more often welcomed than regretted.

10. As a man's sexual responses slow down with age, many couples find that they engage in intercourse less often. However, active sexual relationships can, and often do, continue throughout adulthood, to the satisfaction of both sexes.

KEY TERMS

glaucoma *(466)*

stress *(471)*

stressor *(471)*

menopause *(478)*

climacteric *(478)*

vasomotor instability *(478)*

estrogen replacement therapy (ERT) *(478)*

osteoporosis *(479)*

KEY QUESTIONS

1. What changes in appearance typically occur during middle age, and what is their impact?

2. What are the reasons one person might have a greater hearing loss than another in middle adulthood?

3. What are the likely changes in a person's vision during middle adulthood?

4. What characteristics and health habits would you expect a middle-aged person in excellent health to have?

5. If an overweight, underexercising, cigarette-smoking, alcohol-drinking adult could change just one bad habit, which one would you recommend changing and why?

6. Do you think you will be less stressed ten years from now? Why or why not?

7. What are the differences between the changes men and women experience in their sexual-reproductive systems in middle adulthood?

CHAPTER 21

Middle Adulthood: Cognitive Development

*The latter part of life is taken up
in curing the follies, prejudices, and false
opinions contracted in the former.*

Jonathan Swift
Thoughts on Various Subjects

Overall, would you say that adults become more intelligent, or less, as they grow older? No matter which way you may answer, you are in good scientific company. Investigators who have spent their professional lives gathering and reviewing research on this question have reached opposite conclusions.

On one side, researchers such as John Horn and Gary Donaldson assert that intellectual decline during adulthood is inevitable (Horn, 1985; Horn and Donaldson, 1976, 1977). These investigators maintain that although increases in knowledge may mask it, there is an undeniable decrease in learning potential with age. On the other side, researchers such as K. Warner Schaie and Paul Baltes contend that, throughout life, intelligence is quite plastic, molded by health, education, life experiences, and many other factors, and that, depending on the impact of these factors, intelligence can either increase or decrease. The idea that intelligence necessarily declines with age is, this team argues, a dangerous and debilitating myth (Baltes and Schaie, 1974, 1976; Schaie, 1983; Schaie and Baltes, 1977).

How could researchers looking at the same issue, and at much of the same evidence, come to such different conclusions? Which view is more accurate? Is there any synthesis that combines these opposing points of view? This chapter attempts to answer these questions.

Decline in Adult IQ?

For most of the twentieth century, psychologists were convinced that intelligence reaches a peak in adolescence, and then gradually declines during adulthood. This belief was based on what seemed to be solid evidence. For instance, all literate American draftees in World War I were given an intelligence test, called Alpha, that tested a variety of cognitive skills. When the scores of men of various ages were compared, one conclusion seemed obvious: the average American male reached an intellectual peak at about age 18, stayed at that level until his mid-20s, and then began to show a decline (Yerkes, 1923).

Similar results came from a classic study of 1,191 subjects between the ages of 10 and 60, chosen from nineteen carefully selected, insular, New England villages. (The point of the sampling procedure was to achieve an ethnically homogeneous group of adults who had had fairly similar life experiences. Thus age would be the only salient difference among the test-takers.) IQ tests from this group showed intellectual ability peaking between ages 18 and 21, and then slowly and steadily declining, to the point that the average 55-year-old scored the same as the average 14-year-old (Jones and Conrad, 1933). The case for age-related decline in intelligence was considered proven beyond a reasonable doubt.

Contrary Evidence

The first evidence to contradict the assumption that intelligence declines with age was uncovered by Nancy Bayley and Melita Oden (1955). They were analyzing the adult development of the children originally selected by Lewis Terman in 1921 for his study of child geniuses, a group that has been studied by a succession of researchers over the past sixty years. As Bayley later explained, she fully expected to find a decline in these subjects' cognitive development, because in previous cross-sectional studies "the invariable findings had indicated that most intellectual functions decrease after about 21 years of age" (Bayley, 1966). But on several tests of concept mastery, including questions that involved use of synonyms, antonyms, and analogies, the scores of these gifted individuals increased between ages 20 and 50.

Bayley decided to follow this clue by retesting a more representative group of adults who had also been tested as children. (These subjects, as members of the Berkeley Growth Study, had been selected in infancy to be representative of the infant population of Berkeley, California.) Bayley's results again showed a general increase in intellectual functioning from childhood through young adulthood. Instead of reaching a plateau at age 21, the typical person at age 36 was still improving on the most important subtests of the Wechsler Adult Intelligence Scale, specifically, Vocabulary, Comprehension, and Information. On only two of the ten

Figure 21.1 *Intelligence testing of immigrants to the United States early in this century led to the assumption that people from such countries as Poland, Hungary, and Italy were less intelligent than those from, say, Great Britain, France, and Germany. Only much later was it recognized that the tensions inherent in the testing situation were much greater for people from remote and "alien" cultures than for those from cultures more like our own. The notion that intelligence varies by nationality was only the first of a number of false conclusions to be drawn from intelligence tests.*

subtests—Arithmetic, which measures speed and accuracy of mathematical ability, and Picture Completion, which requires the person to spot an element that is missing from a picture of a common object (such as an ankle from a foot, or a pair of legs from a bee)—did scores decline. Bayley (1966) concluded that "intellectual potential for continued learning is unimpaired through 36 years."

K. Warner Schaie had a similar experience. In 1956 he expected to find a progressive decline in adult intellectual ability when he tested a cross sectional sample of adults aged 25, 32, 39, 46, 53, 60, and 67, a total of 490 people, with about 70 in each group. He examined five "primary mental abilities," essentially verbal comprehension, spatial visualization, reasoning, mathematical ability, and word fluency. His initial results did, in fact, show some gradual decline in ability with age, a pattern that was apparent even when the two youngest groups, the 25-year-olds and the 32-year-olds, were compared. However, the data did not show an inexorable decline in all abilities among all age groups. On some abilities, for example, vocabulary, an older group did better than the next younger age group, a clue that certain intellectual abilities might improve with age.

Cohort Differences

At this point in the history of developmental psychology, the shortcomings of cross-sectional research, especially when applied to adults, were becoming very clear. First and foremost, it is impossible to select adults who are similar to each other in every aspect except age: the very age differences entail experiential and learning differences. Each generation, or cohort, has its own characteristics. In Schaie's study, adults who grew up, say, during World War II, or during the Great Depression, or, more crucially, before every 16-year-old was expected to be in high school rather than in the work force, may well have developed cognitive skills different from those of other cohorts.

Schaie, in fact, was one of the earliest researchers to recognize the distorting effects of cohort differences. Wondering if such differences accounted for the cross-sectional picture of declines in intelligence, he retested his original subjects in 1963, 1970, and 1977 (Schaie, 1983). He found just what Bayley had found: most people improve in primary mental abilities during most of adulthood. Cross-sectional research had, apparently, given misleading testimony regarding adult IQ.

Figure 21.2 *Most 12-year-olds know more about computers and video games than their parents do, but this is true because of the varying experiences of the different cohorts, not the changing ability of the aging brain.*

"Roger, it won't kill you to help your own father with his computer."

But longitudinal research of this kind has drawbacks too. One of them is that being retested several times on similar tests might improve a person's performance. To correct for this, Schaie went one step further, developing a new research design called sequential research (described in detail in Chapter 1). Each time he retested his original subjects he also tested a new group of adults at each age interval, and then followed each new group longitudinally as well. The cross-comparisons that Schaie was able to make with this accumulation of longitudinal and cross-sectional data on all seven age groups allowed him to analyze the possible effects of retesting, as well as to track the impact of cohort differences.

Schaie's findings are quite encouraging. On all five of the primary mental abilities, adults in their 30s and 40s scored higher than adults in their 20s. By age 50, math ability had started to decline, but the other four remained strong. By age 60, average scores on all five abilities showed some decline, but not until age 74 were the averages lower than the average scores of 25-year-olds.

Schaie also found notable cohort differences. In general, more recently born cohorts outperformed earlier cohorts when they were the same age. There were also cohort differences in specific abilities. For instance, recent cohorts of young and middle-aged adults were much better at reasoning ability, but worse at math, than those who were young and middle-aged in previous decades, a trend that was especially apparent when adults born at mid-century—when progressive education was on the rise—were compared with adults born at the beginning of the century—when rote learning was stressed.

Figure 21.3 *Standardized intelligence tests, such as the Scholastic Aptitude Test (SAT) pictured here, are intended to measure nature, not nurture; potential and ability, not learning and achievement. They should therefore be impervious to cohort differences. Nonetheless, average SAT scores in the United States have been falling since about 1960, and only recently have held steady, about 60 points lower than they once were.*

Overall, the results of this and other research have lead to two general conclusions about the assessment of cognitive development during adulthood (Bray and Howard, 1983; Cunningham and Owens, 1983; Schaie, 1983; Shock et al., 1984):

1. While cross-sectional research comparing adults of various ages shows a gradual decline in intellectual ability, longitudinal research shows an increase in most abilities throughout early adulthood, and usually throughout middle age.

2. Cohort differences affect test scores more powerfully than age differences until about age 60.

Many researchers also point out that the most commonly used IQ tests were primarily designed for, and standardized on, children and young adults, and therefore might not be a valid measure of the abilities of middle-aged and older people.

Certainly adults who for years have not thought about, say, mathematical principles and formulas—and may not even have had to do much adding and subtracting—are at a disadvantage on tests of math ability that were designed for young people who are being drilled in the fundamentals of math nearly every day. In addition, adults' cognitive abilities may grow in ways that are relevant to their personal and work lives, but that are not measurable through standard IQ tests. Thus even longitudinal and sequential research may underestimate intellectual increases over adulthood. We might all become more intelligent with age, or at least maintain our intellectual abilities throughout adulthood (Rybash et al., 1986; Willis, 1985). However, before being tempted to draw that happy conclusion, let us see why John Horn, an expert psychometrician, calls this view "wishful thinking" and insists that crucial aspects of intelligence decline steadily as adulthood progresses.

Fluid and Crystallized Intelligence

Horn began his work on intellectual development with an idea, originally suggested by Raymond Cattell (1963), that the various types of intellectual ability can be divided into two categories, fluid and crystallized. As its name implies, **fluid intelligence** is free-flowing and can move in any direction. It is made up of those basic mental abilities—short-term memory, abstract thinking, speed of processing, and the like—required for learning and understanding any subject matter, particularly material that is unfamiliar. Someone high in fluid intelligence is quick and creative with words, numbers, intellectual puzzles, and the like. Among the questions that might be used to test fluid intelligence are the following (Horn, 1985):

> What comes next in these series?[*]
> 4 5 6 3 4 5 6 2 3 4 5 6
> B D A C Z B Y A

Another standard type of item that is used to measure fluid intelligence is the timed assembly of puzzles, with credit given for completion within 2 minutes, bonus points for completion within 1 minute, and no credit after 2 minutes.

   Crystallized intelligence is more "solid." It is the accumulation of facts, information, and learning strategies that comes with education and experience within a particular culture. The size of vocabulary, the knowledge of chemical formulas, and long-term memory for dates in history are all indications of crystallized intelligence (Horn, 1982). Test items designed to measure crystallized intelligence might include questions such as these:

> What is the meaning of the word "temerity"?
> What do you do with a mango?
> What word is associated with bathtubs, prizefighting, and weddings?
> In what year was the Magna Carta created?

Originally, Cattell and Horn thought fluid intelligence was primarily genetic, and crystallized intelligence, primarily learned. Now Horn (1985) thinks this nature-nurture distinction is invalid, in part because the acquisition of crystallized intelligence is affected by the quality of fluid intelligence. For instance, the strength of

---

[*]The correct answers to these two test items are "1" and "x."

your present vocabulary is partly the result of your reading speed and of your ability to make logical associations among words—both of which are related to fluid intelligence.

In adulthood, fluid intelligence declines markedly according to Horn's research. This decline is temporarily disguised by an increase in crystallized intelligence, which continues to build on the base set in childhood. Take verbal ability as an example. Once you have acquired, via fluid intelligence, a working knowledge of your native language, you are likely to remember it all your life, as long as you continue to use it. This crystallized ability to speak, read, and write your native language makes it easier to expand your understanding of it, because you can relate new words to, or define them with, words already familiar to you. For instance, the term "fluid intelligence" is probably new to you, but you can infer its meaning and remember it because of your previous knowledge of the words "fluid" and "intelligence." However, if, as an adult, you had to learn to write and speak a new phrase in a language quite different from any you knew, such as, perhaps, Hindi or Korean, the older you were, the longer it would take you. The reason is that your crystallized knowledge of language would be of little use to you, so you could not compensate for the decline in your fluid ability to learn new material.

Figure 21.4 *Throughout adulthood, people can learn new and complex intellectual skills. Among the many who prove this to be true are the workers on the left, who are being retrained as accountants, and the Asian immigrants on the right, who are learning English. However, since fluid intelligence declines in adulthood, the first group will probably have an easier time of it, since in their tasks, they can make greater use of their crystallized intelligence, relating what they are learning to what they already know.*

Is Faster Thinking Better Thinking?

It is significant that speed of thinking is a critical component of fluid intelligence. In fact, many standard intelligence tests include speed of response as one component, limiting how much time can be spent on any one problem, and giving bonus points for quick answers. This is one aspect of intelligence testing that many specialists in adult development consider unfair.

The reason they think it unfair is that older adults are slower than younger adults at almost everything. Between ages 20 and 60, the average adult slows down about 20 percent in basic reaction time (Birren et al., 1980), and that is with tasks that involve quite simple behaviors, such as pushing a button in response to the sound of a buzzer. With more complicated activities, reaction time is even slower. Handwriting, for instance, takes twice as long at age 60 than at age 30 (Salthouse, 1985). Critics argue that speed of thinking should not be confused with quality of thought. In fact, slower thinking may be deeper and better thinking. As Botwinick (1973) has noted, "if the young are to be compared to the hare, the old may be compared to the tortoise—slow and steady, sometimes to win the race."

However, speed of thinking may, in fact, be related to quality of thought. Timothy Salthouse (1986) argues that slower thinking is also less efficient thinking: if one cannot process information or ideas quickly, then one cannot remember or think about many facts or ideas at once. The result, he contends, is not only slower thinking but simpler, shallower, more superficial thought.

Problems with Longitudinal Research

Whether slower thinking is diminished thinking remains controversial. However, Horn maintains that even if speed of thought were to be subtracted from the definition of intelligence, longitudinal research does not necessarily reach more accurate conclusions than cross-sectional research. True, it avoids most cohort difficulties, but it has another serious flaw. It is impossible, over a testing period lasting several decades, to find all the same individuals for retesting. Subjects move away without leaving a forwarding address, or they get sick or die, or they refuse to come back for another round of testing. In Schaie's research, for example, only a third of the original sample were retested twenty-one years later. Moreover, it is quite possible that many of those who could not be retested were the very people whose intellectual performance declined the most: people who feel their intellect is fading would be particularly likely to avoid retesting. Schaie tried to overcome some of this problem with his sequential design, but whether he succeeded or not is still disputed.

A related problem is that most longitudinal research tries to mitigate the follow-up problem by using a select group of subjects—perhaps all college graduates or people who are unusually intelligent or who come from very stable families—who are likely to be willing to be tested again and again. Yet such subjects may well show a pattern of intellectual development quite different from that of the average person. In addition, in Schaie's study, and in most longitudinal research, the participants were volunteers. It is likely that people who volunteer to take intelligence tests do so partly because they think they will do well on them.

Horn has one more criticism that we should ponder:

> There are powerful reasons for wanting to believe that intellectual decrement does not occur. . . . Humans have a well-developed ability for wishful thinking, and most humans who derive their livelihood and status from exercise of their intellectual abilities have a strong wish that these abilities will not wane . . . The audience for abilities research is thus set to hear what it wants to hear, and what it wants to hear is that intelligence does not decline with age. [Horn, 1976]

Despite these objections, most developmental psychologists think that Schaie is more on target than Horn. Moderate increases in adult intelligence, at least through middle age, seem to be the norm (Dixon et al., 1985; Eichorn et al., 1981; Willis, 1985). It is safe to say, beyond a reasonable doubt, that we do not all reach an intellectual peak in adolescence and then decline. Nonetheless, the jury has still not reached a final verdict as to whether intellectual decline begins in late, late adulthood or long before, and on how widespread that decline is whenever it does begin. As a recent review states, "whether the course of adult intelligence is typified by progression, decrement, or both, after more than half a century of systematic research is still an unresolved issue" (Dixon et al., 1985). However, as we will now see, many researchers, including both Schaie and Horn, have recently taken a different and more fruitful approach to this topic.

Intelligence Reconsidered

In the social sciences, breakthroughs are likely to occur, not because someone discovers an astonishing new answer to an old problem, but because someone asks an important new question that transforms the problem. This may be happening in the study of adult intelligence. The question, according to a number of researchers on both sides of the controversy, should not be "What happens to intelligence in adulthood" but "What happens to intelligences in adulthood" (Denney and Thisson, 1983–1984; Dixon et al., 1985; Horn, 1985; Willis and Baltes, 1980).

Historically, psychologists, as well as laymen, have thought of intelligence as a single entity, a single ability that people have more or less of. A leading theoretician, Charles Spearman (1927), argued that there is such a thing as general intelligence, which he called "g." Although it cannot be measured directly, he contended, it can be inferred from various factors that can be tested, such as vocabulary, memory, and reasoning. Just the fact that psychologists give IQ tests with various subtests, and then calculate an overall IQ for the person tested, implies that intelligence can be thought of as a whole. To be sure, a number of psychologists have seriously questioned whether anything like "g" exists. Nevertheless, the idea that there is such a thing as "intelligence," and that "intelligence is that which intelligence tests measure" (Boring, 1950), continues to influence research.

Recently, however, several researchers have proposed that there are many quite distinct cognitive abilities. They also maintain that these abilities, or intelligences, follow different developmental paths depending not only on the age of the person but also on the individual's life experiences, interests, and education. In one summary of this view, Willis and Baltes (1980) describe adult intellectual competence as being multidimensional and multidirectional, characterized by interindividual variation and plasticity. Let us look at each of these four characteristics in some detail.

Multidimensionality

The closer researchers look at adult cognitive abilities, the clearer it becomes that they are **multidimensional,** that is, that they involve several distinct dimensions. As we saw, Cattell and Horn proposed that intelligence is made up of two independent sets of abilities, fluid intelligence, which declines during adulthood, and crystallized intelligence, which increases. Recently Horn added to this conception, suggesting that intelligence may consist of as many as ten distinct elements, including visual processing, auditory processing, short-term memory, and long-term memory (Horn, 1985). Schaie, too, asserts that intelligence involves distinct abilities, each of which should be conceptualized and measured differently (Schaie and Herzog, 1983).

Other theorists have proposed different lists of distinct factors. For example, Howard Gardner (1983) believes that there are seven autonomous intelligences: linguistic, musical, logical-mathematical, spatial, bodily-kinesthetic, self-understanding, and social understanding, each with its own neurological network in the brain. One reason he sees it this way is that brain-damaged people sometimes lose one or more of these abilities while the others remain intact. Similar evidence comes from *idiot savants,* who are gifted in one of these abilities and severely retarded in the others.

The value placed on each of these seven intelligences depends on the personal

Figure 21.5 *Each adult gradually discovers and develops his or her individual capacities and benefits from the specialized abilities of others, whether it be their skill with clay or with circuitry.*

gifts and cultural values of the evaluator. For instance, in contemporary Western civilization, linguistic and logical-mathematical skills are highly valued. Therefore reading and math are the core of our school curriculum, and any hint of lessened reasoning ability with age is viewed with alarm. By contrast, some other cultures stress social relationships much more than we do. Therefore they train their children to be good listeners and careful observers of human behavior, and they take it for granted that their elders have greater social wisdom.

Multidirectionality

No matter which particular list of intelligences one considers, it is clear that the abilities involved are **multidirectional** in that they can follow different trajectories with age. Thus short-term memory generally falls quite steadily, while vocabulary generally rises. Other abilities—such as mathematical reasoning—might rise, fall, and rise again, depending on how much they are used in daily life. Still others might hold steady until a sudden drop occurs because of such factors as illness and depression. Since virtually every pattern is possible, it is misleading to ask whether intelligence, in general, either increases or decreases. An either/or answer is not possible.

Interindividual Variation

One reason many patterns exist, of course, is that each individual is genetically unique and has unique experiences, and both of these factors affect the pattern of a particular person's intellectual development. The result of this **interindividual variation,** as shown by longitudinal research, is that some individuals decline in some or all mental abilities by age 40, while others are just as capable in some or all at age 70 as they were at earlier ages. Often such variations are related to changes in family and career responsibilities, as well as to changes in health (Eichorn et al., 1981; Schaie, 1983). The housewife who finds that an "empty nest" brings empty days, and who suffers from, say, hypertension, might begin to show noticeable declines in certain kinds of intellectual functioning. On the other hand, someone who replaces a stimulating career, as a parent or a worker, with some equally stimulating form of retirement is likely, given continued good health, to maintain fairly steady levels of cognitive functioning well into late adulthood. Adding to these individual differences are cohort differences, which may be quite strong, as we have seen, when generations differ not only in the average number of years of education they received but also in the emphasis of that education. Probably for this reason, current cohorts of young adults score higher on conventional IQ tests, even while their arithmetic skills are lower (Bower, 1987).

Plasticity

Instead of being innate and immutable, the actual pattern of the various intelligences is shaped by experience, not only the individual's general educational experience but also the person's specific training and motivation. This characteristic is called **plasticity,** to suggest that intelligences can be molded in many ways. Abilities can become enhanced or diminished, depending on how, when, and why a person uses them. Numerous studies of middle-aged and older adults find that training in a specific area—such as techniques of memory improvement or mathematical problem-solving—can result in proficiency levels comparable to those of young adults (Baltes and Willis, 1982). This is especially true when the area of training is relevant to the older person's actual life. (Some of the specifics of this research are reviewed in Chapter 24.)

It should be clear at this point that the recognition of adult cognition as multidimensional, multidirectional, variable, and plastic greatly defuses the debate about the course of intelligence throughout adulthood. The dispute now can take a much milder form: Who develops which intelligences for what reasons? It also moderates the common argument about who is more intelligent, the "absent-minded" professor or the "street-smart" illiterate. They are both intelligent in different ways, because each has fostered a different type of intelligence.

Expertise

Given the variability in adult intelligence, researchers have now begun to ask why a particular person develops cognitively as he or she does. Some developmentalists, recognizing the plasticity of intelligence, believe that, as we age, we each develop **expertise** at whatever is important to us, in our work, in our leisure activities, in our relationships with others. That is, we tend to become selective "experts," learning more and more about less and less, becoming very good, for instance, at fixing one kind of machine, or curing one type of disease, or researching one particular aspect of human development (Dixon et al., 1985). Thus our intelligences increase in adulthood in quite specific areas, through a process of **encapsulation,** whereby our intellectual strengths become concentrated and contained in one small area (Rybash et al., 1986). Of course, as one area of expertise "captures" our intelligences, other areas become neglected, and fail to grow or even decline. Flavell (1982) cites the example of an engineer whose

> use of measurement concepts and other well-practiced components of his engineering knowledge may be highly consistent across a wide variety of problems and situations in that field. However, this level of thinking is liable to appear both less mature and less consistently same-level in other areas—areas where he possesses less expertise, where his learning experiences have been less systematic, coherent, and rational, where affective and cognitive biases are likelier to intrude, etc.

The idea of encapsulation helps to partially explain why many of the adults we know seem to be brilliant experts in some respects and outright novices in others.

What Makes an Expert?

When developmentalists use the term "expert," they do not mean someone who is extraordinarily gifted at a particular task. They simply mean someone who is significantly better than people who have not put time and effort into doing that task.

The difference between experts and novices cannot be reduced to merely quantitative differences in experience, however. Research suggests that there are at least five qualitative distinctions to be made between those who are experts and those who are less skilled (Charness, 1986; Rybash et al., 1986; Salthouse, 1985).

Differences in Quality   First of all, novices tend to rely on formal procedures and rules to guide them. Experts, on the other hand, rely more on their accumulated experience and on the circumstances of the immediate context, and they are therefore more intuitive and less stereotyped in their performance. For example, experienced physicians, compared with newly certified doctors, interview patients with more varied questions, following up on cues, both verbal and nonverbal, that seem to pass unrecognized by the less experienced physicians (Elstein et al., 1978; Leaper et al., 1973). Similarly, when they look at x-rays, expert physicians interpret them more accurately, though often they cannot verbalize exactly how they arrived at their diagnosis. As one pair of researchers explain:

> The expert physician, with many years of experience, has so "compiled" his knowledge that a long chain of inference is likely to be reduced to a single association. This feature can make it difficult for an expert to verbalize information that he actually uses in solving a problem. Faced with a difficult problem, the apprentice fails to solve it at all, the journeyman solves it after long effort, and the master sees the answer immediately. [Rybash et al., 1986]

In much the same way, the expert artist, or musician, or scientist is not simply a practiced technician but is an intuitive creator as well (Charness, 1986; John-Steiner, 1986).

Second, many elements of expert performance are automatic, almost unconscious. A very obvious example is driving an automobile. Whereas a student driver must concentrate fully on steering, regulating speed, negotiating traffic, and so on, the experienced driver does these things automatically. Thus the experienced driver can enter into a lively conversation, while the novice driver not only cannot participate in talk, but even insists that everyone else be quiet. Automaticity of the basics allows the expert driver to react more quickly to changing situations and to notice small but significant details (the dog off the leash 100 yards ahead, or the police car half a mile back), as well as the big picture (the overall flow of traffic and the changing road conditions), that the novice is not cognitively "free" to notice.

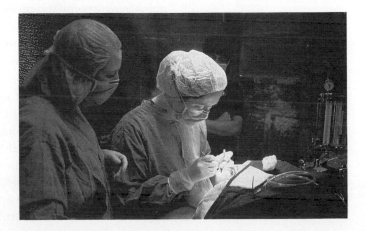

Figure 21.6 *The importance of experience is obvious for a surgeon. Almost any one would agree that if a child needs to have a cleft palate fixed (as shown here), the more practice the surgeon has with this particular operation, the better. Less obvious, but equally valuable, is expertise in other areas. For example, for the auto mechanic, the airplane pilot, the cook, and the parent, years of practice produce a combination of intuition, creativity, and wisdom that make the job easier and the results better.*

Similar automaticity seems to occur in many kinds of expertise—from that of, say, the expert athlete to that of the expert craftsperson or jet pilot—making it appear that most aspects of the work in question are performed "instinctively." Professional athletes, in fact, are noted for saying that they play their best when they are able to "stop thinking and just react." In some skills, the shift from a conscious, deliberate processing of material to an automatic one seems to involve a shift in the neural pathways in the brain (Salthouse, 1985). Novice typists, for instance, must keep looking at the paper they are typing on to see if they have made a mistake. Expert typists, however, can tell when they have made an error without looking—their fingers sense that the wrong key has been pushed.

Third, certain aspects of the expert's cognitive processes appear to become specialized as expertise is acquired, making the expert better at processing new information in his or her area of expertise than the novice is. This is revealed most clearly in memory studies comparing chess experts and novices (Chi, 1985). If both are shown a chess game in progress, allowed to look at it for a few seconds, and then are asked to reproduce the positions of the pieces, the experts can correctly reproduce far more positions than the novices can. However, on general tests of memory not related to chess, and even on tests involving chess pieces placed in a random manner on a chess board, chess experts score no better than novices. For other kinds of experts too, specialization of cognitive processes allows more efficient processing of specialized information (Chi et al., 1982).

Fourth, the expert has better strategies, and more of them, for accomplishing a particular task. In fact, this may be the crucial difference between a skilled and an unskilled person (Welford, 1980). For example, when Salthouse (1984) gave typists of various ages a test in which they had to press a button as soon as they saw a light, he was not surprised when older typists were slower, since reaction time slows with age. However, when he tested the same typists on typing speed, he found no correlation between speed and age. Obviously, the older typists had found successful strategies to compensate for their physiological slow-down.

Finally, perhaps because of all these qualitative differences, experts are more flexible. The expert artist, musician, or scientist, for example, is more creative and curious in his or her work, deliberately experimenting and enjoying the surprise when things do not go according to plan (Arlin, 1984; John-Steiner, 1986). Another example of the flexibility arising from expertise comes from surgeons (Salthouse, 1985). Obviously, since no two patients are exactly alike, every type of operation has the potential for sudden, unexpected complications. Not only will the expert surgeon be more likely than the novice to notice little telltale signs (an unexpected lesion, an oddly shaped organ, a rise or drop in a vital sign) that may signal a possible major problem, but the expert will be better able to deviate from standard textbook procedure to devise strategies to overcome the problem. Similarly, experts in all walks of life seem better able to adapt to individual cases and exceptions to the rule—somewhat like an expert chef who adjusts ingredients, temperature, technique, and timing as things develop, and virtually never follows a recipe exactly.

How, now, can we best answer the question with which we began this chapter? Perhaps with cautious optimism. The longitudinal evidence reviewed in the first half of the chapter is encouraging: certainly on some basic abilities, many adults improve over most of adulthood, and show no decline by age 60. Indeed, as you will see in Chapter 24, some people show no evidence of decline at all, and continue to master new areas of knowledge even in late adulthood.

The suggestion that we are most likely to improve and perhaps develop expertise in those areas that capture our attention and intelligences is also heartening. And to the degree that some of the declines that may occur can be attributed to encapsulation, we can take some comfort. To have cognitive powers fade in areas that are of no interest to us, or that we seldom use in daily life, does not seem great cause for alarm.

Still, the evidence for cognitive growth is not definitive, and there is strong evidence for cognitive declines in some people. However, on the whole, it does seem clear that in middle adulthood individual differences are much more critical in determining the course of cognitive development than chronological age alone is. As we will see in Chapter 24, even in old age, many of the cognitive declines that are apparent seem more closely related to particular personal circumstances, such as health and social context, than to age per se.

SUMMARY

Decline in Adult IQ?

1. On the basis of many large cross-sectional studies, psychologists once believed that intelligence inevitably declined in adulthood. Within the past twenty-five years, longitudinal research has led to the opposite conclusion, that in many ways intelligence may improve during adulthood.

2. Sequential research, which attempts to distinguish the general aging process from the specific experiences of each generation, finds that cohort effects have at least as much impact on average IQ during middle adulthood as chronological age does.

Fluid and Crystallized Intelligence

3. Despite evidence for cognitive growth during middle adulthood, the impact of age on intelligence is still controversial. Some psychologists believe that while crystallized intelligence, which is based on accumulated knowledge, increases with time, the basic, fluid abilities of the mind inevitably decline with age.

4. All researchers agree that speed of thinking, as well as speed of behavior, slows down with age. Tests that measure rapidness of thought therefore show an age-related drop. Theorists disagree, however, about whether a slowing of mental processes necessarily affects quality of thought.

Intelligence Reconsidered

5. Most researchers now think that, rather than there being one entity called intelligence, there are several distinct intelligences.

6. Each intellectual ability may increase, decrease, or remain stable with age, depending on such factors as education and experience. Adult intellectual competence is multidimensional and multidirectional, characterized by interindividual variation and plasticity.

Expertise

7. As people grow older, they may become more expert in whatever types of intelligence or skills they choose to develop. Meanwhile, abilities that are not exercised fall into decline.

8. In addition to being more experienced, experts are better than novices in many ways. Experts are more intuitive and flexible, and use better strategies to perform whatever task is required. Their cognitive processes are more specialized, as well as automatic, often seeming to require little conscious thought.

KEY TERMS

fluid intelligence *(489)*
crystallized
  intelligence *(489)*
multidimensional *(492)*
multidirectional *(493)*
interindividual
  variation *(493)*

plasticity *(494)*
expertise *(494)*
encapsulation *(494)*

KEY QUESTIONS

1. What evidence suggests that intelligence declines during adulthood?

2. What evidence suggests that intelligence increases during adulthood?

3. What are the advantages of longitudinal research on IQ?

4. What are the advantages of sequential research on IQ?

5. What differences would you expect to find in your own intelligence ten years from now?

6. What are the four recently recognized dimensions of intelligence and how do they broaden the picture of "intelligence" and of cognitive development in adulthood?

7. What are the differences between an expert and a novice?

8. What do you think you are an expert at and why?

CHAPTER 22

Middle Adulthood: Psychosocial Development

Whoever, in middle age, attempts to realize the wishes and hopes of his early youth, invariably deceives himself. Each ten years of a life has its own fortunes, its own hopes, its own desires.

Goethe
Elective Affinities

The social clock—that cultural timetable that suggests when various stages and landmark events of adulthood should occur—continues ticking in middle adulthood as it did in early adulthood. Although no dramatic biological shifts occur to signal it, our society recognizes a point called **midlife** at about age 40, when the average adult has about as many years of life ahead as have already past. Midlife ushers in **middle age**, which lasts until about age 60.

As you will soon see, however, while it may be easy to designate an age for midlife, it is not simple to describe what occurs during these years. Much depends on the cultural context and the particular ecological niche in which individuals find themselves at age 40.

Midlife: Crisis, Shift, or Transition?

Midlife is often thought of as a period of crisis and dramatic change (Brandes, 1985). Indeed, one of the first popular books to broach this idea, *Passages*, describes the "age 40 crucible," when various psychological pressures supposedly compel people to take apart their lives and then put them together into "a whole new puzzle" (Sheehy, 1976).

That middle age might be a time of crisis is certainly a plausible idea, for a number of potentially troubling personal changes often do cluster in the 40s. The most obvious of these is simply the awareness that one is beginning to grow old. Birthdays tend to be seen in a new perspective—a measure less of time lived than of time remaining. This perceptual shift is often highlighted by the death or serious illness either of a close relative from the next older generation—perhaps a parent or a favorite aunt or uncle—or of a friend or colleague, perhaps someone only a few years older than oneself. Such events bring not only feelings of personal loss but also thoughts about one's own mortality.

For many middle-aged parents, a potential source of upheaval is the need to make important, and not always easy, adjustments in their parental roles. Typically, at just about the time parents enter midlife, children become adolescents, demanding greater independence and, to varying degrees, putting their parents' values and authority to the test. No sooner have parents adjusted to these changes than their children set out on their own as adults, perhaps distancing themselves from their parents initially and then calling forth a different form of nurturance and closeness when they make their parents become parents-in-law and grandparents.

Achievement-related shifts may also hold the possibility for difficulty. At about age 40, many individuals reach the limits of their vocational potential and enter a career plateau. Whatever choices adults have made about generativity, midlife is an occasion to question those decisions. Those who have been single-mindedly climbing the career ladder often become concerned about the things in life they may have missed out on; those who have concentrated on child-rearing often worry about what they will do when their children are gone, and wonder if it is too late to begin a new career.

All three of these midlife shifts are recognized by most developmentalists who study adulthood. However, there is substantial debate about whether such changes produce a **midlife crisis**, provoking radical reexamination and sudden change in people's public or personal lives, or simply provide an occasion to reaffirm that one's life is heading in the right direction.

Levinson's Findings

The researcher offering the strongest evidence for midlife crisis is Daniel Levinson (1978, 1986), who originally studied forty men intensely over several years and then did follow-up research on additional adults of both sexes. According to Levinson's findings, sometime between ages 38 and 43, virtually everyone is faced with a basic question:

> "What have I done with my life? . . . What is it I truly want for myself and others? What are my central values and how are they reflected in my life? What are my greatest talents and how am I using (or wasting) them? . . . Can I live in a way that combines my current desires, values, and talents?" [Levinson, 1978]

In addressing this question, the majority of the men in Levinson's study concluded that much of their life had been based on illusions, that they had mistakenly placed too much importance on money, or public achievement, or a particular personal relationship. Although some men maintained psychic equilibrium during this period, making only minor adjustments in the structure of their lives, 80 percent of Levinson's original group experienced

> tumultuous struggles within the self and with the external world. . . . Every aspect of their lives comes into question, and they are horrified by much that is revealed. They are full of recriminations against themselves and others.

As a result, many of these men made major changes in their lives, ranging from external changes such as divorce, or remarriage, or a shift in occupation, or an alteration in lifestyle to internal changes such as a modification in social outlook or a reordering of personal values. Levinson maintains that while the specifics vary from individual to individual, few can avoid this midlife reappraisal.

Figure 22.1 *The hand-shaker to the right in both pictures is the same man—Ronald Reagan. At age 39 (left), he stars in "Law and Order"; at age 77 (right), he presides at an arms-agreement summit with Mikhail Gorbachev. Was it a midlife crisis, increasing maturity, or a response to changing times that propelled the changes he made at about midlife—from New Deal Democrat to conservative Republican, from marriage to Jane Wyman to marriage to Nancy Davis, from actor to politician—and led to his greater personal success?*

Other Research

Other studies also have found that many people, especially men, experience some form of midlife distress (Bergler, 1985; Gould, 1978; Vaillant, 1977). For instance, a study of a representative group of 300 middle-aged men (Farrell and Rosenberg, 1981) found that a higher proportion of them had psychological problems of one sort or another than a comparison group of younger men. As you can see in Figure 22.2, only 12 percent had an obvious, classic midlife crisis, openly wrestling with feelings that their life was goalless and empty, analyzing what they really wanted in life and wishing they could start over afresh. However, another 30 percent expressed even greater dissatisfaction, but rather than questioning their own life choices and trying to make changes, "they attributed their unhappiness to external circumstances and other people," blaming their jobs, their children, their marriages, and people from other races and ethnic groups.

A third group, 26 percent of the total, tended to deny that anything was generally wrong with their lives, but they had many specific complaints that could well have been caused by their inability to cope with psychological difficulties. For example, they were likely to feel that their intellect was fading, that their children needed

Figure 22.2 *About two-thirds of all the middle-aged men in Farrell and Rosenberg's study experienced significant levels of distress, anger, or depression regarding their lives, but most of them did not acknowledge themselves as being in a crisis: in one way or another, they managed to mask their lack of self-satisfaction. The same study also investigated younger men and found that some, but not as many, had similar feelings.*

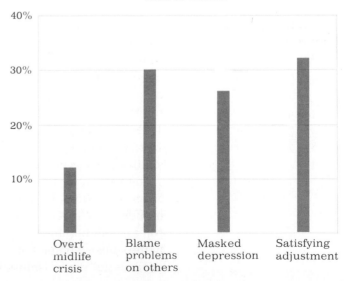

Men at Midlife

authoritarian control, that their enjoyment of food had diminished, and that their sex lives were unsatisfying. The researchers concluded that for this group, "the effort to avoid experiencing their own feelings and changing selves" took its toll "in immobilization, loss of appetites, and depression."

Finally, only 32 percent seemed completely crisis-free, satisfied with their work, their marriages, their children, and their health. This group, rather than experiencing midlife as a disorienting crisis, were able to deal actively and constructively with whatever problems they encountered.

As you see, this study found that men's responses to midlife depended largely on their personal style of coping with problems. Interestingly, education and socioeconomic status seemed to affect coping style. Those who were relatively affluent and well educated were more likely either to have a crisis or to cope effectively; those with less education and less status were likely either to blame others or to "punish" themselves psychosomatically.

Gender Differences in Middle Age

Whereas the idea of a male midlife crisis has been around only twenty years or so, it has long been assumed that, for biological reasons, women inevitably experience a crisis at midlife. If a woman in her 40s or 50s seemed sad or angry, or behaved in unexpected ways, it would automatically be ascribed to the "change of life." (In fact, the assumption that women had a crisis at midlife was so strong that it led to the misnaming of a midlife crisis in men as the "male menopause.")

Surprisingly, when researchers began to study the lives of adult women, they found relatively little age-related stress in middle age, with tension more likely to occur at age 30 than at 40 or 50 (Baruch et al., 1983; Livson, 1976; Reinke et al., 1985; Rossi, 1980). Instead of being a time of depression brought on by the fear of growing old, an empty nest, or menopause (see Chapter 20), midlife for women may be, as one review concludes, "a time of great growth, expansion, and satisfaction" (Brooks-Gunn and Kirsh, 1984). During the same years that Levinson found men's dreams shattered and their illusions exposed, another study (Baruch, 1984) found many women (especially those who focused their early adulthood years on motherhood) discovering new opportunities. They went back to school, began careers, became more involved with their community. Even Levinson, while insisting that women as well as men restructure and reappraise their lives every decade, agrees that the full-blown midlife crisis seems more often to be a masculine affair.

Cohort and the Midlife Crisis

One explanation for this unexpected finding is that women may once have had more midlife and middle-aged difficulties than they have now. Surveys confirm this hypothesis. A nationwide American survey in the 1950s found that, as women grew older, their self-esteem fell (Gurin et al., 1960). Twenty years later, women responding to the same questions no longer felt lower self-esteem with age (Veroff et al., 1981). Similarly, the rate of psychological problems among middle-aged women was substantially higher during the 1950s than more recently, according to two similar surveys done twenty years apart (Srole et al., 1962; Srole and Fischer, 1980).

The reason for these differences clearly seems to lie in changes in social conditions. Most women who reached middle age several decades ago had been homemakers exclusively, and for most, their chief sources of self-esteem were their roles as mother and wife and their physical attractiveness. When their children left home, these women had no jobs and no marketable skills. Nor were they encouraged to get them. Unless their husbands, for some reason, could not provide for them, they were supposed to stay at home, keeping house and waiting to become needed again as grandmothers. No wonder middle-aged and older women, their children grown, their beauty fading, experienced a fall in self-esteem, and a rise in depression, with age.

However, beginning about 1970, women started to enter or reenter colleges and the labor market in far greater numbers than before. Today, women in middle age are likely to have trained for a career, to have been employed for much of their adult lives, to have work opportunities available to them at midlife, and therefore to have a source of self-esteem that women in earlier cohorts did not have (Brooks-Gunn and Kirsh, 1984; Masnick and Bane, 1980). Of course, this does not mean that motherhood and wifehood have necessarily diminished in importance. Most middle-aged women who have both career and family believe that their family roles are the more important ones (Lopata and Barnewoolt, 1984). However, when women have both work and family roles, each role provides an important buffer to protect against whatever midlife strains may arise in the other (Baruch, 1984). If midlife difficulties occur at all in this cohort, unmarried women without children (Mellinger and Erdwins, 1986) or married women without jobs (Baruch, 1984) are more likely to experience them than women who have combined both roles.

Men reaching midlife in recent years may also have been affected by the particular experiences of their cohort, but, in their case, historical change may have worked to make their lives more difficult. As you know, psychologists' interest in adult development is fairly recent. Their collection of data on midlife in men began in earnest primarily during the 1960s and 1970s, so the generation of men that provided most of the data for the midlife crisis were born in the 1920s. For the men in this cohort, early adulthood went particularly well, especially if they were veterans of World War II (receiving such benefits as free college education and low-cost mortgages), or if they were starting up the career ladder in the prosperous 1950s. Most of them were happy and successful both at work and at home: they married full-time housewives who took care of them, their homes, and their children.

However, when these men reached midlife during the 1960s, many of the nation's youth were seriously questioning authority—not only the authority of parents and teachers but also the authority of church (on sex) and country (on civil rights, civil liberties, and Vietnam). Thus many fathers found their fundamental values and assumptions being challenged by their own children.

Figure 22.3 *When he expects to be welcomed home with a kiss and a home-cooked dinner in the oven, and she hopes for the same, both "honeys" need to do some adjusting.*

"Honey, I'm home!"   *"Honey, I'm home!"*

Further, society as a whole was shifting in a way these men had never antici-pated: women of all ages began questioning traditional gender roles, often fiercely, and middle-aged men found their wives, daughters, and female employees chang-ing in disconcerting ways. Wives insisted on taking jobs, and on having equal mari-tal rights; daughters wanted to live on their own, or worse, with a boyfriend; secre-taries refused to make coffee, and began to protest sexism in the office. Many women who were dissatisfied with the progress of their lives blamed the "male chauvinist pigs," a term applied most freely to the very type of middle-aged, middle-income men that researchers were studying.

Obviously, these were not comfortable times for middle-aged men, and because things had previously been going so well for them, this cohort of men was particu-larly unprepared for the issues that were being raised. It should hardly seem sur-prising, then, that researchers suddenly found that many midlife males were under stress.

So, then, is there a midlife crisis after all? The answer is yes—for some people, some of the time, under certain conditions. For others, because of the particular events in their lives, midlife and middle age are a very rewarding period. Now let us examine the two significant areas of family and work, and identify patterns and events that are likely to be salutary or stressful.

Family Dynamics in Middle Adulthood

Family ties across the generations are particularly important for the middle-aged generation, precisely because they are the generation in the middle between aging parents and adult children. Often they have grandchildren as well. While middle-aged adults are now less likely to share a roof with their elderly parents or grown children than in earlier decades (see A Closer Look, next page), their family ties are typically strengthened during these years.

Middle-Aged Adults and Their Aging Parents

The relationship between most middle-aged adults and their parents improves with time. One reason is that, as adult children mature, they develop a more bal-anced view of the relationship as a whole, especially with regard to their years of growing up. This change is particularly apparent in men who had a tendency as adolescents and young adults to blame their fathers for not having been sufficiently helpful and understanding. In middle age they are likely to reevaluate the "old man" and become more appreciative of his good qualities and more accepting of his limitations (Farrell and Rosenberg, 1981). Women, too, become more apprecia-tive of the older generation (Helson and Moane, 1987).

The improvement in relationships between middle-age children and their par-ents is particularly likely to occur between current generations because most of today's elderly are healthy, active, and independent. They typically prefer not to live with their adult children, and, thanks to a changed economic picture in the past several decades, most of them can afford to live on their own, giving both themselves and their grown children a measure of freedom and privacy that en-hances the relationship between them. Indeed, the current financial picture is such that many of the older generation continue to provide financial assistance for their children and grandchildren, just as they did in middle age.

Figure 22.4 *Like the middle-aged woman on the right, many adults find their understanding and love of their parents increasing as they themselves become older. Father-son and mother-daughter relationships are particularly likely to become closer than they were when the child was an adolescent.*

A CLOSER LOOK   Variations in Family Structure

When anthropologists and sociologists study family structure throughout the world, they distinguish two basic family types. One is the **extended family,** which includes several generations and family branches in close and frequent contact. Usually many members of an extended family live together, sharing income, housework, meals, and child care. In the extended family, children grow up not only with siblings and parents but also with cousins, aunts, uncles, grandparents, and perhaps great-grandparents.

By contrast, the **nuclear family** includes only married couples and their dependent children. Once those children are old enough, they are expected to leave the household to start families of their own, no longer dependent on their parents for shelter, income, or emotional support.

Especially when the crucial indicator of family type is taken to be the number of generations, and the variety of relatives, living under the same roof, the extended family is more typical of traditional societies (as in China, India, and most regions of Africa and Latin America), while the nuclear family is more common in modern societies, such as twentieth-century North America (Schneider, 1980).

By the mid-1980s, in many developed countries, even the nuclear family seems to be in the minority. For example, whereas in the 1950s the nuclear family made up the ma-

jority of households in the United States, less than a third of all households today include husband, wife, and dependent children. As such households have been declining, there has been a steady increase in the number of married couples without dependent children, and in single-parent households, most of them composed of a divorced or unwed adult with young children. The category that has increased most, however, is "households" consisting of one person, often a single young adult, who a few decades ago would already have been married, or an elderly widow, who a few decades ago would have been living with the middle generation.

Does this shift away from the extended family, and now away from the nuclear family, mean that families are less close than they were? Not necessarily. Researchers have found that family members who do not live together typically have a great deal of contact with each other nevertheless, and exchange a great deal of support. They stay in touch by visiting, telephoning, or corresponding regularly, and provide each other with help ranging from gifts and loans to babysitting, advice, encouragement, comfort in times of distress, and assistance in times of emergency. As one social scientist put it, what we have now is a "modified extended family" (Troll, 1980), including three and four generations more often now than ever before. The family is certainly different from what it was, but it still is an important part of the social network

Middle-Aged Adults and Their Adult Children

Likewise, the relationship between middle-aged parents and their children improves throughout middle age, especially if the children have emerged from adolescence successfully. Of course, it seems as though there is almost always some aspect of a young adult's life that parents find fault with, particularly in the case of fathers and their young adult sons (Nydegger, 1986). About a third of all fathers, for example, complain about their sons' lack of achievement. Interestingly, when parents complain about their daughters, it more often relates to the daughter's marriage than to the daughter herself. A third of all fathers feel that their sons-in-law are not worthy husbands. When the relationship between young adults and their parents becomes uncomfortably tense, the usual strategy is for the younger generation to limit contact with the parents, a step that may make the relationship less close but often, in fact, makes it more affectionate (Green and Boxer, 1986).

Despite the possible conflicts, being the parent of a young adult child who is beginning college or a career, and perhaps marriage and parenthood, is usually a source of pride. In addition, the younger generation serves as a "cohort bridge," a source of information and advice about new developments in the culture (Green and Boxer, 1986). Many a middle-aged woman has gone to college because her adult children urged her to; many a middle-aged man has begun to take better care of himself because his children provided specific health information. For their part, most young adults benefit not only from the material aid and advice their parents provide, but also from the self-confidence that results from realizing their parents treat them as adults.

Finally, for many middle-aged adults, a new intergenerational tie is formed when their adult children have children themselves. Most grandparents are pleased with their new status, happy to give presents and provide occasional babysitting, and hesitant to give too much advice. In today's society, grandparenthood brings pride and companionship to many middle-aged adults, without the day-to-day responsibility and worries that they had with their own children (Cherlin and Furstenberg, 1986). (Grandparenthood and great-grandparenthood are discussed in Chapter 25.)

In terms of midlife fulfillment and midlife distress, it should be noted that women, as they do throughout adulthood, tend to focus more on family than men do. Women tend to be **kinkeepers,** the people who celebrate family achievements, gather the family together, and keep in touch with family members who no longer live at home. For this reason, family accomplishments are particularly likely to be rewarding to the middle-aged woman who sees her parents happy in retirement, her grown children become responsible adults, and her grandchildren reaffirm the significance of human growth and family continuity.

The Sandwich Generation

Thus far we have outlined the many ways family ties provide stability and gradual improvement in the lives of middle-aged adults. However, under some circumstances, family ties can become quite burdensome, a possible source of stress and crisis. Because of their position in the generational hierarchy, the middle-aged feel obligated to help both the older and younger generations. Especially if a family member becomes divorced, as many of the younger generation do, or widowed, as many of the older generation do, the middle generation is often called on to provide crucial emotional support, and often financial support as well (Cherlin and Furstenberg, 1986; Troll, 1986). In recent years, such demands have so increased on the middle-aged that they are commonly referred to as the **sandwich generation,** a generation that is often caught between, and squeezed by, the needs of two adjacent generations.

Figure 22.5 *Few of today's middle-aged parents expected to support their young adult children as much, and for as long, as they do.*

Figure 22.6 *The male bond can be a strong one across the generations. The elderly man at the left enjoys both the company and the respect of his son and grandson, who are visiting him in a nursing home. However, when an elderly man needs daily nursing care, the relative most likely to provide it is a female. For example, when the man at the right lost the use of his legs, the family decided that the best place for him was in his son's home, at least in part because the best care-giver was his son's wife, shown here.*

For example, the number of adults in their 20s living with their parents has risen sharply in recent years, reaching 30 percent in 1984 in the United States (Glick and Lin, 1986a). About a third of these young adults are not employed, and many of those who are employed are not yet earning enough to launch out on their own. Of divorced adults in their early 20s, 40 percent live with their parents, often finding the emotional and practical support at home that they could not find elsewhere. Usually this state of affairs is temporary. By their early 30s, only 17 percent of divorced adults live with their parents (Glick and Lin, 1986a), presumably because adults in their 30s are likelier to have the social and economic security necessary for remaining independent.

The middle-aged generation often is called on to furnish another kind of support to the younger generation. Especially when a young family's income is low, many grandparents provide major care for their grandchildren. For example, almost a fourth of all the employed American women who have infant children use grandparents as their chief alternative care-givers (Klein, 1985).

Another, usually more stressful form of care-giving is often required of middle-aged adults. Almost everyone, sometime before age 60, is called on to provide care for a frail, elderly relative. Such care-giving is required much more often of the current generation of middle-aged adults than of the previous generation, primarily because the numbers of frail elderly, and the number of years that they live, have increased (Troll, 1986). Indeed, the latest statistics indicate that nearly a third of the elderly spend some portion of their last years unable to care for themselves, and that they are far more likely to be cared for by family members at home than in nursing homes or hospitals. If the frail elderly person has a spouse who is capable of providing normal daily care, the middle-aged generation may be called on only for occasional assistance. Often, however, a spouse is frail, too, or the elderly person is widowed, divorced, or has never married. In such cases, it is most often the middle-aged generation that assumes the burden of daily care.

In earlier generations, such care was typically shared by several adults in the middle generation. Many of today's middle-aged adults, however, were born during the Great Depression, when the birth and marriage rates plummeted. Consequently they have relatively few siblings to share the responsibility of caring for elderly parents or unmarried aunts and uncles.

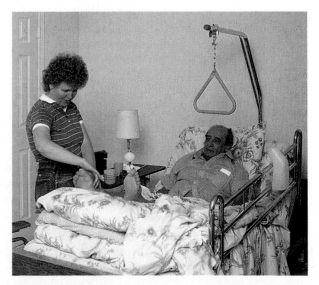

Partly because of their role as kinkeeper, women are much more likely to be cast in the care-giver role than men, and the unexpected divorce of a child or illness of a parent can well become a source of crisis and depression in a middle-aged woman. In some cases, just when a woman has began to appreciate the freedom of an "empty nest" or the satisfaction of her own career, she finds that she is called on to babysit for grandchildren and care not only for her own parents but for her husband's as well (Stueve and O'Donnell, 1984). While middle-aged men are less often care-givers, they do, however, feel financially obligated to help the older and younger generations if necessary. This is one of the factors that contributes to the "financial squeeze" of middle age (Estes and Wilensky, 1978).

We should not minimize these burdens: they have a high potential for precipitating a crisis of some sort, especially if they disrupt established family patterns and relationships. (Causes of, and solutions to, stressful care-giving relationships with the elderly are discussed in Chapter 25.) However, it is important to keep in mind that most men and women welcome their continuing family responsibilities during middle age; even when they provide major care for aging parents, most feel more rewarded than disturbed by their family obligations (Stueve and O'Donnell, 1984; Troll, 1986).

Marriage in Middle Adulthood

Throughout adulthood, marriage is the family relationship that seems most closely tied to happiness and companionship. As we saw in Chapter 19, the "launching" of adult children is cause for some celebration, and is often accompanied by an increase in marital satisfaction. Women are particularly likely to report an improvement in the marriage (Rhyne, 1981; Swenson et al., 1981).

There are three possible reasons for this improvement. The first is that, particularly for women, financial security is an important correlate of marital satisfaction, and disagreement over finances is one of the chief sources of marital tension (Berry and Williams, 1987). Overall, families at the empty-nest stage typically have more income, because employed wives can devote more time and energy to their work, thus increasing their pay, and fewer expenses, because the children are, at least partly, self-supporting.

The second reason is that, for many married couples, one of the major goals of the relationship was to raise a family, again a goal that women, particularly, make sacrifices to attain. Once the children are successfully raised and launched, the parents are relieved of many anxieties concerning their children's development and are able to share with each other important feelings of achievement.

Finally, in typical marriages, doing things together—everything from fixing up the house to taking a vacation—contributes to marital satisfaction. Once children leave home, there is more time for such activities. As a result, married couples often recapture some of the close companionship and marital intimacy that were not possible in the hectic child-rearing years.

This does not mean that all marriages improve in middle age. Obviously illness and unemployment of either spouse, or of parents or adult children, increase strain while reducing the time and money for doing things together. Further, some relationships survived to middle age not because the couple did things together but because they learned to do them apart, in effect, becoming "emotionally divorced" (Fitzpatrick, 1984). If middle age means that such a couple must now spend more time together, their marriage may be sorely tested. Also, at every age, one's partner's physical appearance correlates with marital satisfaction and sexual interest, particularly for men (Margolin and White, 1987). In most cultures, in-

FAMILY DYNAMICS IN MIDDLE ADULTHOOD

Figure 22.7 *These lovers going off on a picnic happen to be a long-married couple whose children have grown. For many couples, the "empty nest" period is a time of rediscovering and renewing the marital relationship.*

cluding our own, age-related changes in appearance are noted more readily, and judged more harshly, in women than in men (Katchadourian, 1987). Partly for this reason, many middle-aged couples experience a drop in the intensity of their relationship.

The Gray Divorcée

Most couples overcome these strains: the longer a couple has been married, the less likely they are to divorce (see Figure 22.8). However, when divorce does occur, the older the partners are, the more disruptive—psychologically, socially, and financially—it is (Chiriboga, 1982), because both partners have typically invested years of their emotional and practical lives in each other. The divorce may be made even worse by the fact that long-standing mutual friends often turn away, rather than decide which of the two ex-spouses to side with.

Figure 22.8 *The chart of divorces in the United States shows that most divorces occur within the first five years of marriage, with the peak at three years. Interestingly, the same pattern is found in most other societies, ranging from contemporary Sweden to the hunting and gathering groups of southwest Africa.*

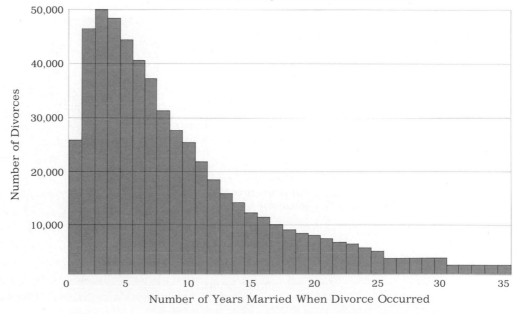

Duration of Marriages That Ended in Divorce

Number of Divorces

Number of Years Married When Divorce Occurred

Men and women tend to find different solutions to the problems of divorce in middle age. Men often remarry, typically to a woman several years younger. Women find that the odds of remarriage are stacked against them and consider divorce a personal failure as well. Adjustment to divorce is particularly difficult for those older middle-aged women who centered their entire lives on their role as wives. One woman, who was stunned when her husband of many years left her for a younger woman, explained:

> The only thing I had been sure of was how to run a house and keep a good marriage. Well, it was difficult to believe in *that* any more . . . I was a total failure . . . I thought "Lillian, either you kill yourself on the spot or you find some way out of this." [quoted in Goodman, 1979]

Fortunately, after a period of numbness and bitterness, this woman began to replace the center of her life: she went to work for a friend in a small store and then began a mail-order business on her own. Whereas earlier, when she was married, she had eagerly awaited the birth of grandchildren, now, when the first grandchild actually arrived, she was pleased that she no longer *needed* a grandchild to keep her busy.

Many women are not so fortunate as to find a new source of income, and this can necessitate drastic adjustments. Although middle-aged wives typically have spent years cooking, cleaning, and otherwise helping their husbands, ex-wives are not legally entitled to any of their former husband's Social Security unless the divorce occurs after he has stopped working, nor do ex-wives share in the ex-husband's health or pension benefits. Many older divorcées who have no income of their own find themselves dependent on welfare, something they never imagined would happen to them (Cain, 1982). However, grown children are often very supportive of their divorced mothers (and often angry at their fathers). As the years pass, divorced women find that many of their friends become single too, usually because of the death of their husbands rather than divorce. Finally, time once consumed by housework and husband can now be devoted to work and social activities. With careers, children, grandchildren, and friends, women who are divorced in midlife are usually able to rebuild their social networks and self-esteem.

Figure 22.9 *A typical American domestic scene: a mother munching on pizza with her teenage daughters. What makes the scene less typical is that this woman is getting divorced after nineteen years of marriage. Even more atypical is the fact that this housewife of nearly two decades is able to go back to college partly because her ex-husband backs her decision and is willing to provide full financial support for his daughters.*

Career Dynamics in Middle Adulthood

The need to achieve continues in middle adulthood as earlier. Especially if status, seniority, and salary continue to rise, a career provides a buffer that makes it easier to adjust to changes in the family. However, there are several shifts that may occur in work lives as people grow older, shifts that can be a transition in some cases and a crisis in others.

The Career Plateau

Some adjustment is natural in the work lives of many middle-aged people who have been doing the same work for twenty years or more. They no longer change jobs or climb the career ladder as they once did because now they have reached the limits of their abilities and/or career opportunities. They have entered the "maintenance phase" of career development. When they reach this phase, they frequently feel pressure to redefine their work goals and their relationship to work (Farrell and Rosenberg, 1981). Often this redefinition makes them see their career in a more balanced perspective, no longer wanting to achieve for achievement's sake. When this happens, workers generally invest less of their ego and energy in the work world; they can afford to be more relaxed in their approach to their job, more satisfied with the work itself, more helpful to others (Karp, 1985–1986). This is a likely time for them to become mentors to their younger fellow workers.

According to some theorists, this transition may be related to more than merely having reached a career plateau. David Bakan (1966), for example, proposes that young adults are motivated by **agency**, the need to be an active agent, pushing for one's own success and personal accomplishments. As a result, coworkers are seen as facilitators of, or obstacles to, one's own advancement; work assignments are judged by whether they will develop and show off one's talents; a job is taken because it increases one's income, or status. Bakan suggests that, with maturity, another need, the need for **communion**, grows stronger. The worker is more concerned about the welfare of others, about the company as a whole, about the role of that company in helping the larger human community. K. Warner Schaie, you will recall (Chapter 18), proposed a similar idea in his delineation of adult cognitive development, suggesting that in midlife adults move from concern about their own accomplishments (the achieving stage) to concern for the welfare of others (the responsible stage) and sometimes to concerns for the wider community (the executive stage).

Figure 22.10 *One way that people express a need for communion is to volunteer to help others in some way. By middle age, most people know their talents and interests well enough to find a volunteer job that suits them, including such unusual tasks as working for a volunteer fire-and-rescue squad, battling a Texas flood, or for the Audubon Society, checking the effects of pollution on the bird population.*

Of course, these patterns are not inevitable. Some workers in their 40s are still rising in status and power; others are dissatisfied with their progress at work but see no hope for improving their position. Workers in each of these categories would, understandably, have attitudes toward their job quite different from those of workers who were comfortably settled on a career plateau.

These differences were demonstrated by a twenty-year longitudinal study of managers at AT&T (Bray and Howard, 1980). As you can see from Figure 22.11, work became more important as a source of satisfaction to those who were rapidly gaining authority, remained somewhat important to those who were promoted at a more gradual pace, and became less important to those who did not advance very much. In middle age, this last group, particularly, reported that they valued family, recreation, and community service more than they did when they were younger. These men also became more nurturant and helpful, in a word, more communal, while those at the highest levels were more likely to be dominating and less likely to be sympathetic and helpful to their fellow workers (Howard, 1984).

Figure 22.11 *At American Telephone and Telegraph (AT&T), young managers at the beginning of their careers ranked the importance of work about equally. By midlife, however, those who were at the top of the heap, or were still rising to it, thought work was very important, while those who had less success valued other aspects of their life more than their work. One likely explanation is that once people no longer feel the pressure to be intensely competitive and to continually prove themselves in the workplace, they are able to develop other dimensions of their personality more fully.*

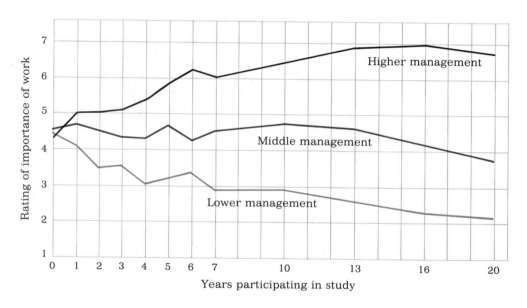

Note that, for both the successful and less successful workers, midlife brought a shift, but not necessarily a crisis. Under what conditions, then, is work likely to create a crisis? In general, a crisis occurs when a worker no longer feels that his or her contribution to the work world is useful or productive. Two specific conditions in which such feelings occur are burn-out and alienation.

Burn-out

Burn-out is a state in which the worker feels depleted of the energy and enthusiasm for doing the job. Although burn-out can occur in almost any occupation, it seems particularly prevalent in the helping professions. Typically, dedicated workers such as teachers, nurses, doctors, and social workers enter their field determined to change the lives of those they serve, solving such social problems as illiteracy, poor health care, and poverty. However, these problems are complex, not amenable to easy solutions, and often the helped do not readily appreciate the helper. After a time, frustration turns high ideals, great expectations, and dedicated action into bitter disillusionment and emotional exhaustion.

Two remedies for burn-out are to reduce expectations and to share responsibilities (Cherniss, 1980). For instance, the teacher who had once hoped to reform the school system or ignite an intellectual fire in each and every student might settle for helping some of his or her students become a little more skilled, and might attempt, with the help of other teachers and parents, to change one aspect of the system—perhaps the process for choosing a new principal or new textbooks. A more encompassing, but less accessible, remedy for burn-out is to change the overall structure of the institutions in which service workers function (Pines and Aronson, 1981). Consider the hospital system, for instance. In many cases, its high specialization means that few nurses are able to regularly tend and befriend the same patients from one day to the next. Without this repeated contact, nursing may become mechanical, and nurses may begin to feel more like robots than people. If the system were changed so that a nurse were responsible for the same patients every day, including most of their care from admission to discharge, burn-out would be less likely to occur.

Figure 22.12 *Burn-out is less likely to occur if the worker experiences deep personal satisfaction. The woman on the right is a social worker in a state institution for mentally retarded adults, a job she has held happily for many years. One reason for her long stay on the job is moments such as this one: after months of working with this man, she is about to see him leave the institution to enter a halfway home, where mentally retarded adults are freer to make friends, enjoy some independence, and still get the care they need.*

Alienation

Another problem occurs when workers feel that their work is uninteresting and unimportant. **Alienation,** as this feeling is called, is theoretically likely to occur when workers do not see the relationship between their work and the final product. Such a feeling is particularly likely to occur in large factories or corporations in which the individual worker is not recognized as a person but is simply regarded as a means of performing some task—usually the same task for hours and years on end.

The feeling, and consequences, of alienation, are revealed in the following comments from disaffected blue-collar workers. A steelworker reports:

> You can't take pride anymore. You remember when a guy could point to a house he built, how many logs he stacked. He built it and he was proud of it . . . It's hard to take pride in a bridge you're never gonna cross, in a door you'll never open. You're mass-producing things and you never see the end result of it . . . Sometimes out of pure meanness, when I make something, I put a little dent in it. I like to do something to make it really unique. Hit it with a hammer.'' [Terkel, 1974]

A GM auto assembly-line worker tells an interviewer:

> "There's a lot of variety in the paint shop. You clip on the color hose, bleed out the old color, and squirt. Clip, bleed, squirt, think; clip, bleed, squirt, yawn; clip, bleed, squirt, scratch your nose."
>
> I asked about diversions: "What do you do to keep from going crazy?"
>
> "There's always water fights, paint fights, or laugh, talk, tell jokes. Anything so you don't feel like a machine."
>
> But everyone had the same hope: "You're always waiting for the line to break down. . . ." [Garson, 1979]

Alienation is particularly likely when the relationship between management and worker is structured as a hostile one. Workers demand pay increases and threaten to strike; bosses watch suspiciously for worker laziness, cheating, and sabotage. Eventually, many younger workers respond to alienation by quitting their job or by getting themselves fired. Middle-aged workers, however, are generally more concerned with job security and are reluctant to risk long-term unemployment, which, in fact, can be particularly devastating to the 40-to-60 age group in terms of health, self-esteem, and income.

By middle age, alienation makes many men vulnerable to a variety of probelms, among them, drinking too much, sleeping too little, and becoming seriously depressed. Their problems at work affect their relationships with their family as well. They are likely to be dissatisfied with their marriages, especially sexually, and to feel that their children are bound to be a disappointment (Farrell and Rosenberg, 1981).

On the individual level, it is hard to fight alienation productively. The worker above who joins in paint fights and hopes that the production line will break down is coping in his own way, but it is a counterproductive one. On a larger level, alienation can be mitigated by techniques that make the worker feel more involved in the total production process. Borrowing and refining from the Japanese, many companies have instituted programs in which employee groups decide how a job is to be done, and how improvement can occur (Ouchi, 1981). Cooperation between man-

Figure 22.13 *These Tennessee employees are working on two Japanese imports: one is the Nissan cars and trucks they are building; the other is the management techniques adapted from the Japanese. Workers at all levels, from low-level labor to management, who are engaged in the same project meet together daily to offer suggestions for improving work procedures and to exchange news, jokes, and announcements. One result is less alienation; another, apparently, is better quality.*

agement and worker, or, better yet, restructuring work so that the worker is the manager, has improved productivity as well as worker satisfaction (Hackman, 1986).

Even if neither alienation nor burn-out occurs, many adults in the maintenance phase of their career find that once further advancement is unlikely, the excitement fades from daily work activities. One solution, of course, is to change jobs or careers. However, few middle-aged workers choose this path, unless changes in the economy or the job market require it. The "one-life, one-career" imperative is very strong in our culture (Sarason, 1977), as is the typical employer's preference for hiring a young trainee rather than a middle-aged one. Thus many workers remain stuck in a less than satisfying job.

Personality throughout Adulthood

So far we have discussed some of the many circumstances and patterns that might lead to continuity or crisis, stability or change, in middle adulthood. However, we have yet to consider the most crucial variable of all, individual personality. Personality traits lead us to interpret and react to life events in ways that are distinctly our own. If personality traits remain stable throughout life, the impact of the age-related events and changes we encounter will be powerfully affected by the continuity of who we are.

Although the degree of stability of personality traits is controversial (Mischel and Peake, 1983; Porvin, 1985), longitudinal research over the decades of adulthood has, for the most part, found notable continuity in many personality characteristics. As two leading researchers explain:

> Personality forms part of the enduring core of the individual, a basis on which adaptation is made to an ever-changing life . . . Lives surely change, perhaps in stages; personality, we maintain, does not. [McCrae and Costa, 1984]

This may come as a surprise to many middle-aged adults, partly because people tend to overestimate whatever personality changes they think they have undergone. A study that clearly showed this (Woodruff, 1983) began with a group of college students, who rated themselves on various personality traits—such as whether they saw themselves as aggressive, tender, cheerful, and so on. Twenty-five years later, the same people rated themselves again twice, first as they currently saw themselves, and then as they thought they had been twenty-five years earlier. There was a clear similarity between the original college rating and the current one, but little similarity between either of those ratings and the way people remembered themselves as having been.

This is not to deny that people behave differently as their experiences and responsibilities change. Underlying the many behavioral changes, however, are certain core personality characteristics. One extensive longitudinal study of personality structure (McCrae and Costa, 1984) found that many of the various personality traits could be grouped into three basic personality dimensions. The first they call **neuroticism**—characterized by feelings of anxiety, worry, hostility, and depression. The second is **extroversion**—characterized by tendencies to be outgoing, active, and assertive. The third is **openness**—characterized by receptiveness to new experiences, new ideas, and changing one's life as well as one's mind.

Figure 22.14 *Laid-back, outgoing, and willing to go with the flow—all characteristics of the nonneurotic, open, extrovert. Someone who exhibits these characteristics as an older adult most likely possessed them as a young person, too.*

These three clusters of attributes remain quite stable as people age. Thus anxious, neurotic people are likely to be so throughout life. If things go badly, they are likely to despair; if things go well, they are likely to worry that their luck might run out. Such people tend to have periodic crises, not only at midlife because they have reached age 40 and are worried anew about the future, but also earlier and later in their lives.

Similarly, the very outgoing college student is likely to be, in middle age, the kind of person who speaks to everyone at work, who spends a lot of time interacting with friends, family members, and neighbors, and who is likely to be involved in community activities ranging from local politics to volunteer work and amateur-sport leagues. Especially for this kind of person, an important source of stability in middle age is an extensive social network. When a social network is large and of long-standing duration, as it usually is for extroverts, it tends to minimize radical shifts in life patterns and to protect against marked depression in times of stress.

Finally, people who as young adults are open to new experiences continue to be so. Such people are even open to experiences that may be stressful: they are more likely to change jobs, spouses, and neighborhoods than people who are more closed. While such shifts may give the appearance of discontinuity, in terms of personality structure, they, in fact, reflect continuity (McCrae and Costa, 1984). Fortunately, because openness and flexibility are part of their makeup, such people are less stressed by change than a more conventional person would be (Eidison and Forsythe, 1983). What others might consider a crisis, they might consider a challenge, an adventure, or a breath of fresh air.

This research also finds two more characteristics that are quite stable over time: agreeableness and conscientiousness (McCrae and Costa, 1987). Stability of all five personality characteristics is found in other longitudinal research as well, although the terms that are used to describe these traits vary from study to study. These studies have found a consistency in the appearance, from childhood on, of traits such as fearfulness, anger, sociability, cheerfulness, and creativity (Block, 1981; Caspi et al., 1987; Maas and Kuypers, 1974; Mussen et al., 1980; Ryff, 1984; Vaillant, 1977).

Figure 22.15 *The similar emotions in mother and babe may well be both genetic and environmental—sunny dispositions tend to enhance—and produce—pleasant experiences and satisfying relationships.*

Reasons for Stability

What are the reasons for this apparent consistency? In part, it may be inborn. As you will remember from the Life-Span Perspective on pages 153–158, some temperamental characteristics are probably inherited. Also important are the experiences of childhood. In study after study, having had a strong, supportive, and affectionate relationship with one's mother and/or father enhances one's innate tendency to be outgoing and open to new experiences. The reverse is also true. Growing up in a critical, unstable, and unloving family makes it more likely that, in adulthood, one will be fearful and friendless, and have neurotic difficulty coping with the stresses of life (Antonovsky, 1986; Honzik, 1986; Vaillant, 1977).

Furthermore, choices made in early adulthood tend to foster one set of traits more than another. For instance, the 20-year-old extrovert might decide to follow a career working directly with people, and marry a spouse and choose friends who appreciate and foster sociability themselves. Moreover, the demands of the particular job the extrovert is likely to choose will probably foster an extension of the social contacts and interests that encouraged the career choice in the first place (McCrae and Costa, 1984).

Changes in Personality

Do the findings of research on inborn temperament and family influences mean that we are restricted to whatever personality characteristics we have as we enter adulthood? Not necessarily. First, the choice of a marriage partner can have a notable impact in reinforcing or lessening tendencies to be open and outgoing or closed and withdrawn (Elder, 1986; Vaillant, 1977). Second, as depicted in three quite different longitudinal studies, there is a general trend toward improvement in personality (Haan, 1985b; McCrae and Costa, 1984; Vaillant, 1977). Individuals gradually recognize and attempt to remedy the weak parts of their personality: some learn to control their tempers, while others begin to express their anger constructively; some learn to become better organized, while others learn to loosen up a bit. In general, traits that are undesirable within a particular culture tend to become less prominent with age, while desirable traits show more consistency (Moss and Susman, 1980). In our culture, as people mature they generally become less neurotic and more open to new ideas.

Figure 22.16 *Research on lawbreakers reveals the interplay of three factors: inborn temperament, the tendency for personality to improve with age, and the role of life experiences. Most criminals commit their first crime in adolescence but tend to become relatively law-abiding by age 30. A minority persist in crime as long as they are free to do so. A prison term does not greatly affect reform unless something unusual happens. For example, the opportunity to earn a college degree can make a return to crime less likely.*

The Unisex of Later Life   Another way in which people often begin to change in middle age concerns gender roles: as they age, both men and women tend to become more androgynous. According to one aspect of Carl Jung's personality theory, we all, to some extent, are bisexual, with both a masculine and a feminine side to our character. In early adulthood, the side that conforms to social expectations is dominant; consequently, men suppress their feminine aspects while women suppress their masculine ones. During middle age, however, many individuals are sufficiently well established in their sexual identity that they feel free to explore their opposite side, becoming more flexible and less superficial in the process. Jung's theory finds some support in personality research. Especially in the presence of certain life experiences, there is a tendency in middle age for men to become less masculine, as traditionally defined (McCrae and Costa, 1984), and women, less feminine (Allen, 1987). In many ways, each sex moves closer to a middle ground between the traditional gender roles (Troll and Perron, 1981). As David Gutmann (1977) describes it: "The sharp sex distinctions of earlier adulthood break down, each sex becomes to some degree what the other used to be, and there is ushered in the normal unisex of later life."

Gutmann (1985) believes that the changing demands of daily life as people grow older are the main reason for whatever blurring of the gender roles and traits may occur. In childhood, of course, boys and girls are quite similar until puberty, when adolescents of both sexes exaggerate their masculine and feminine traits in order to establish their own sexual identity and to attract the opposite sex. Then, during the early years of family and career development, men and women are pressured by the demands of daily life into quite traditional male and female roles. Particularly critical is the "parental imperative," the pressing demands of raising young children, demands that make many mothers more nurturant and self-effacing, and many fathers more competitive and ambitious, than they might otherwise have been.

However, with time, the urgency of child-rearing lessens, and more latitude is possible, allowing some middle-aged adults to develop broader gender roles, especially if events push them in this direction. For example, as their children become more independent and they are able to devote more time to their careers, women

Figure 22.17 *The older man on the right is a volunteer in a nursing home, a task he does with a sympathetic gaze and touch more often associated with women. However, we should not be surprised: whatever sex differences in personality there may be are more apparent at age 30 than 60. During later adulthood, many men are much more nurturant, tender, and emotional than those who knew them in early adulthood would have ever predicted.*

tend to become more assertive and self-confident. Men, on the other hand, may become more considerate and tender as career goals become less pressing and their need to be competitive diminishes. Later on, the birth of a grandchild, especially if it coincides with retirement, may make a man become more nurturant. Similarly, the illness or death of her husband may cause a woman to become more assertive, especially if she must earn a living for the first time.

The idea that life experiences are critical in affecting gender roles is supported by research that shows that, in later adulthood, while most people do indeed become more androgynous, and happily so, others remain quite traditionally feminine or masculine, and function happily as well (Sinnott, 1986). Conflicts, dissatisfaction, and potential crises occur only when older adults feel forced to fit into gender roles that they are not happy with.

Our discussion of adult psychosocial development began, in Chapter 19, with the recognition that adults follow much more diverse developmental paths than children and adolescents do, partly because biological constraints and precipitants are much weaker after physiological maturation is complete. We have seen this diversity in middle adulthood as well, as the typical changes and events of middle age provoke a crisis and depression in some people and nothing of the sort in others. We now turn to late adulthood, a time of continued diversity, but, as Chapter 23 explains, also a time when biological limitations may again play a dominant role.

SUMMARY

Midlife: Crisis, Shift, or Transition?

1. Midlife is often thought of as a period of crisis, a time when self-doubt and unhappiness lead to radical change. Reasons for such a crisis include shifts in family and work responsibilities, as well as a growing awareness of the limited time left to live.

2. Evidence that a midlife crisis is normative comes from Levinson's longitudinal research. He finds that, at about age 40, most men begin a serious reexamination of their lives, and many make radical changes in connection with their career or family. Other research also suggests that

distress is not unusual in midlife men, although in most cases it does not constitute a full-blown crisis that is directly attributable to factors relating to middle age.

3. While women may once have been likely to become depressed during middle age, contemporary women seem more likely to experience middle age as liberating. One explanation for this is that women may now be buffered against a crisis by having careers of their own. Men in recent cohorts, on the other hand, may have suffered from changed expectations of the male role, and thereby may have been made more susceptible to whatever stresses occur in midlife.

Family Dynamics in Middle Adulthood

4. Middle-aged adults generally find their relationships with their own parents, and with their young adult children, improving, particularly if each generation lives under separate roofs. Middle adulthood is also the time people are likely to become grandparents for the first time, usually a welcome event.

5. However, under some circumstances, family ties can be particularly burdensome in middle age, making adults feel like the sandwich generation, squeezed by the financial and emotional needs of both their parents and their adult children.

6. Young adults today often cannot afford to leave the family home, and, if they have children, often need their parents' help with child care. This is particularly likely if they become divorced, as about a third of all married young adults do.

7. Even more draining on the middle generation can be the care of an elderly relative, a task many adults—usually women—eventually perform.

8. Marriage usually improves with the "launching" of adult children and the availability of more time and money for the couple to spend together once children become self-supporting.

9. Divorce, if it does occur, is likely to be psychologically, socially, and financially disruptive, especially for women.

Career Dynamics in Middle Adulthood

10. For most adults in midlife, work continues to be a source of satisfaction. At some point, adults typically reach a career plateau, a point at which further advancement no longer seems desirable and/or possible. At this juncture, many adults experience a greater need for communion than for agency, and are more willing to help younger adults find their way.

11. However, for some adults, work itself becomes a reason for dissatisfaction. Burn-out or alienation makes them feel that their efforts are unproductive and meaningless.

Personality throughout Adulthood

12. Throughout life, several personality traits tend to remain quite stable, and these influence the course of development. For example, midlife crises are most likely to occur in those who have a neurotic personality pattern, and probably least likely to occur in extroverts whose wide social circle protects them against strain and minimizes the likelihood of sudden change. People who are open to new experiences seem least likely to experience such change as disruptive and may readily adapt to stressful circumstances.

13. This stability of personality is partly genetic, partly the result of early life experiences, and partly the result of choices made in early adulthood. However, personality changes occur as well. Two changes are notable. People generally improve in personality traits, and both sexes typically become more androgynous, ushering in the unisex of later adulthood.

KEY TERMS

midlife *(499)*	agency *(511)*
middle age *(499)*	communion *(511)*
midlife crisis *(500)*	burn-out *(512)*
extended family *(505)*	alienation *(513)*
nuclear family *(505)*	neuroticism *(515)*
kinkeepers *(506)*	extroversion *(515)*
sandwich generation *(506)*	openness *(515)*

KEY QUESTIONS

1. What factors can be seen as contributing to a crisis at midlife?

2. How has the study of different middle-aged cohorts affected views about midlife crisis?

3. What are some of the stresses involved in being a member of the "sandwich generation"?

4. What are some of the career developments that are more likely to occur in middle age?

5. What are some of the personality characteristics that tend to be stable through the life span?

6. What factors contribute to stability of personality?

7. How do male and female gender roles tend to change in middle age?

Part VII

The Developing Person So Far: Middle Adulthood, Ages 40 through 60

Physical Development

Normal Changes in Appearance and the Senses

Changes in the appearance of the skin, hair, and body shape are benign, but can be disconcerting. Losses of acuity in hearing and vision are usually gradual, and individuals usually learn to compensate quite easily.

Health

Overall health is influenced by variables such as sex, race, and socioeconomic status and by long-term habits, such as exercise, diet, and smoking. Susceptibility to stress may also play a role.

The Sexual-Reproductive System

In their late 40s or early 50s, women experience the climacteric during which the body must adjust to declining estrogen levels. Menopause, the cessation of menstruation, signals the end of a woman's reproductive capacity. Men experience no comparably dramatic decline in reproductive ability.

Cognitive Development

Types of Intelligence

Fluid intelligence tends to decline over time, while crystallized intelligence tends to remain steady or increase. Reaction time and speed of thinking slow. Some abilities improve with age, while others decline.

Adult intelligence tends to become encapsulated in areas of interest, leading to the development of expertise, characterized by cognitive processes and responses that are intuitive, automatic, specialized, and flexible.

Psychosocial Development

Crisis, Shift, or Transition

Whether or not one experiences a midlife crisis depends, in part, on one's sex and cohort. Men are more likely to experience some form of midlife distress than women are.

Family Dynamics

The relationship between middle-aged adults and their parents generally improves with time as does their relationship with adult children. However, middle-aged adults may feel "sandwiched" between the needs of both the older and the younger generations.

With the extra time and money available during the "empty-nest" period, satisfaction with marriage usually rises.

Work

Adults reach the "maintenance phase" of their career. Individuals may sometimes experience negative feelings of "burn-out" and alienation.

PART **VIII**

Late Adulthood

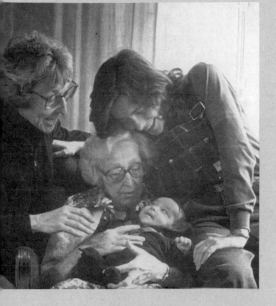

What emotions do you anticipate experiencing as you read about development in late adulthood? Given the myths that abound regarding old age, you may well expect to feel discomfort, depression, resignation, and sorrow. Certainly there are instances in the next eighty pages when such emotions would be appropriate. However, your most frequent emotion in learning about late adulthood is likely to be surprise. For example, you will learn in Chapter 23 that most centenarians are active, alert, and happy; in Chapter 24, that marked intellectual decline is the fate of only a minority of the elderly who are victims of conditions that can often be prevented; in Chapter 25, that relationships between the older and younger generations are neither as close as some sentimentalists idealize them to be nor as distant as some critics claim.

If surprise, indeed, dominates your reactions in learning about late adulthood, you are not alone. Few people, young or old, know much about the actual experiences of most of the elderly, and most people, of every age, fill the gaps in their knowledge with dated clichés and misleading prejudices.

Why does this period of life, more than any other, seem to be a magnet for misinformation and mistaken assumptions? Think about this question when the facts, theories, and research of the next three chapters are not what you expected them to be.

523

CHAPTER 23 **Late Adulthood: Physical Development**

Youth is cause, effect is age; so with the thickening of the neck we get data.

Djuna Barnes
Nightwood

Which photo do you think more accurately reflects the situation of elderly Americans—the one on the opposite page or the one below? If you say the one below because you believe that most of the elderly are idle, lonely, and in poor health, you are in the majority—and wrong.

According to a nationwide poll of adults between the ages of 18 and 65, most Americans think that the elderly spend "a lot of time" doing very little—for example, 67 percent said that the elderly spend a lot of time watching television, 62 percent said sitting and thinking, 39 percent said sleeping, 35 percent said "just doing nothing." However, when the elderly themselves were asked how they spent their time, the figures dropped by at least half. (In point of fact, more people *under* 65 said they themselves spent a lot of time sitting and thinking than did those over age 65.) Similarly, many younger adults thought that loneliness is a serious problem for the elderly, yet only 12 percent of those over age 65 cited it as such (National Council on Aging, 1975; Schick, 1986).

Figure 23.1 *A typical "senior citizen" or one of the unlucky few?*

The most telling misperception of all concerns the proportion of the elderly the public believes are in nursing homes. The usual estimate is one out of every three. The actual number is one in twenty, most of whom are women over age 75 who have no living spouse or children who could care for them (United States Bureau of the Census, 1987). Two-thirds of those over age 65 in the United States rate their health as good or excellent. Most were *not* hospitalized over the past year, and most find that their daily activities are *not* limited in any way by their health (United States Department of Health and Human Services, 1985).

Ageism

Why are people's perceptions of the elderly so much worse than the reality? The reason is that most of us tend to be prejudiced about older people. **Ageism,** as this prejudice is called, is similar in many respects to racism and sexism, and is equally harmful. It fosters a stereotype of the elderly that makes it difficult to see them as they actually are, and, by permitting laws and attitudes that discourage the elderly from participation in work and leisure activities, it prevents many older people from living their lives as happily as they might. Ageism also isolates the older generation socially, reducing their ability to contribute to the larger community and perpetuating the fear of aging in all of us.

Why Is Ageism So Strong?

Perhaps the main reason for the strength of ageism is our culture's emphasis on growth, strength, and progress. Our veneration of youthfulness is so great that, as Butler and Lewis (1982) point out, any sign of a person's "beginning to fail" is feared and exaggerated. Furthermore, for many people, interaction with the very old is a reminder of their own mortality. They develop self-protective prejudices against the old as a way of "avoiding and denying the thought of one's own decline and death."

Another factor that reinforces ageism is the increasing age segregation of society, which limits contact between the generations. Fewer and fewer Americans live in multigeneration families; and more and more neighborhoods, voluntary organizations, and workplaces tend to involve one age group of adults more than others. This makes contact between the older generation and the younger ones increasingly limited, so young people seldom have a chance to compare the myths of old age against its realities.

Also contributing to ageism is the human tendency to generalize about any group on the basis of its most noticeable members, often those who draw attention because they belong to a problem-prone minority. Here we can make a useful distinction between the **young-old,** and the **old-old,** a distinction based not on age but on characteristics related to health and social well-being. The young-old, who make up the large majority of the elderly, are, for the most part, "healthy and vigorous, relatively well-off financially, well integrated into the lives of their families and communities, and politically active" (Neugarten and Neugarten, 1986). The old-old are those who suffer "major physical, mental, or social losses" and who are likely to require supportive services or to be spending their days in nursing homes or hospitals. It is the latter group on whom the ageist stereotype is based. Although the young-old are much more numerous than the old-old, we tend not to notice them, precisely because they do not fit the stereotype.

Figure 23.2 *A cross-section of the population, like these people waiting for a parade, always includes some old, some young, and some in between. In the future, however, when the baby-boomers mature, a representative sample of parade-watchers will include a great many more of the elderly than are present here. It seems safe to predict that as the elderly come to represent an ever-increasing proportion of the general population, the culture's extreme emphasis on youth, and its attendant ageist biases, will fade considerably.*

Professionals and Ageism

Ironically, many professionals who work with the elderly have inadvertently strengthened ageist prejudices (Quinn, 1987). The majority of these professionals are in the field of medicine and social work. They spend most of their time working with the minority of the elderly who are sick and infirm, and seldom see those who are well and active. This quite naturally leads them to develop, and to report, a rather dismal view of old age. Even gerontologists—professionals specializing in **gerontology,** the study of old age—have contributed to the stereotype. Until recently, they focused, almost exclusively, on the difficulties and "declines" of old age, ignoring strengths and stability. In addition, they usually studied the aged who were in nursing homes or retirement communities, ignoring the majority who lived in their own homes in the general community and who often maintained the activities, friends, and interests they had at younger ages. As Butler and Lewis (1982) summarize the problem:

> Medicine and the behavioral sciences have mirrored societal attitudes by presenting old age as a grim litany of physical and emotional ills. Decline of the individual has been the key concept and neglect the major treatment technique. Until 1960 most of the medical, psychological, psychiatric, and social work literature on the aged was based on experience with the sick and the institutionalized, even though only 5 percent of the older people were confined to institutions.

In recent years, this picture has been changing. Many contemporary gerontologists are among the leading foes of ageism. The cultural bias that associates youth with health and vigor, and age with disease and fragility, is weakening. As A Closer Look (next page) reveals, the number of older adults in the population has been increasing dramatically, and this fact has been instrumental in getting professionals, politicians, and the general public to reconsider their ageist assumptions (Naisbitt, 1982). Now let us look at what the aging process actually entails.

A CLOSER LOOK   Squaring the Pyramid

One reason that ageism used to go relatively unchallenged was that the number of elderly was small. Consequently, there was little incentive for professionals, politicians, and others to try to look beyond the stereotypes of old age. Their lack of numbers left the old politically powerless and easy to ignore. In recent years, this "numerical weakness" has been fading fast, worldwide, as **demography**—the study of population—reveals.

In the past, when populations were sorted according to age, the resulting picture was a **demographic pyramid,** with the youngest and largest group at the bottom and the oldest and smallest at the top. There were two reasons for this picture. First, each generation of young adults gave birth to more than enough children to replace themselves, thus ensuring larger cohorts at the bottom of the pyramid. Second, a sizable number of each cohort died before advancing to the next higher section of the pyramid.

Today, however, because of falling birth rates and increased longevity, the shape of the population is becoming closer to a square (Pifer and Bronte, 1986a). In the United States in 1900, there were 5 million people over age 60, a total of 6 percent of the population. In 1984 there were more than 39 million over age 60, about 16.5 percent of the population. The age group growing the fastest in recent years is those over 75, who were 3 percent of the population in 1960, and 5 percent in 1985. During the same period, the percentage of those under age 14 decreased, from 30 percent to 20 percent (U.S. Bureau of the Census, 1986).

The increase in the numbers, as well as in the proportion, of aged citizens will continue well into the future, although it is impossible to predict exactly at what rate. However, if (1) the birth rate remains level, and (2) improvements in health habits and medical care continue to add to life expectancy at the rate they have in recent decades, and (3) the ages of new immigrants are not appreciably lower than the current population, then by the year 2025, when the baby-boom generation are elderly, the American population will be divided roughly into thirds— one-third below age 30, one-third aged 30 to 59, and one-third aged 60 and older.

What will be the result of this "squaring of the pyramid"? Some experts warn about new social problems, such as increased expense for medical care and decreased concern for the quality of education for children. Others envision social benefits, such as lower crime rates and more civic involvement (Pifer and Bronte, 1986b). This much is clear: in order to successfully meet the changes and challenges brought about by a growing older population, we must all understand what the effects of aging actually are, so we can separate the ageist myths from reality.

As the demographic pyramid becomes more square, people of every age are affected. For example, as more people reach late adulthood, the political power of this age group increases, ensuring that their needs will be included in the national public agenda. However, as the number of people who are parents of dependent children declines, the needs of the youngest, non-voting age group may be neglected, unless those not directly responsible for their care become concerned.

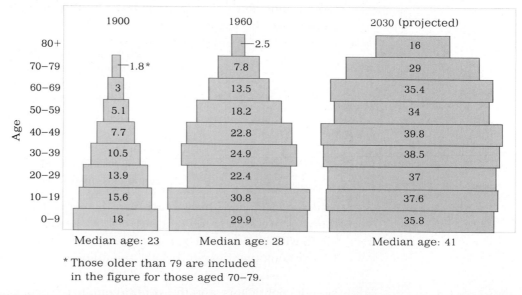

The Changing Population of the United States (in millions)

* Those older than 79 are included in the figure for those aged 70–79.

The Aging Process

In old age, as in earlier periods of adulthood, **senescence**, or the weakening and decline of the body, continues at a gradual but steady pace. Appearance continues to change, senses continue to lose acuity, major body systems continue to slow down as organ reserve continues to shrink. As in earlier periods, too, diet, exercise, and other aspects of lifestyle can do much to soften the impact of senescence. Barring the effects of serious illness, no sudden downturn in health or activity occurs simply because a person is now 60 or 70 or 80.

Nonetheless, for many people, sometime after age 60, the cumulative effects of the aging process reach a point at which self-concept may be affected or adaptive measures in daily activities may be required. Let us now reexamine the major changes of aging in terms of their physical and psychological consequences.

Appearance

Looking "old" is something most adults dread and try to avoid, by many means, no matter what age they are. However, there comes a time for everyone when the superficial changes of aging are beyond concealment. One of the most obvious of these changes occurs in the skin, which becomes dryer, thinner, and less elastic, producing marked wrinkling and making blood vessels and pockets of fat much more visible (Klingman et al., 1985). In addition, the dark patches of skin known as "age spots" are visible in about 25 percent of adults by age 60, about 70 percent by age 80, and in almost everyone by age 100. Hair also undergoes very apparent changes, continuing to become thinner and grayer, and, in many people, eventually becoming white.

Changes in overall body height, shape, and weight also occur in late adulthood (Whitbourne, 1985). Most older people are more than an inch shorter than they were in early adulthood, because their vertebrae (the small bones that make up the spine) have settled closer together. Further, the muscles that hold the vertebrae have become less flexible, making it harder to stand as straight as in earlier years. Body shape is affected by redistribution of fat, which collects less in the arms, legs, and upper face, and more in the torso (especially the abdomen) and the lower face (especially the jowls and chin).

Body weight is often lower in late adulthood, partly because muscle tissue becomes reduced, a difference particularly notable in men, who have relatively more muscle and less body fat than women. For both sexes, the reduction in muscle strength is especially apparent in the legs, necessitating slower walking and sometimes use of a cane or walker. Another reason for lower body weight is osteoporosis, the loss of bone calcium that causes bones to become more porous and fragile. By late adulthood, "relatively massive bone loss" has often occurred, particularly in women (Hall, 1976). Osteoporosis is the main reason some older people walk with a marked stoop, and also the reason the elderly are much more likely to break a bone, particularly the hip bone, when they fall.

All these changes in appearance have serious social and psychological implications: in an ageist society, those who look old are treated as old, in a stereotypic way. This fact is poignantly underscored by the reactions of the elderly themselves. Most older people consider their personality, values, and attitudes quite stable, and except for acknowledging that they may have slowed down a bit, do not feel that they have changed all that much from their younger days (Kaufman, 1986). Therefore when older people see a recent photograph of themselves, or catch an unguarded glimpse of themselves in the mirror, or merely notice how others treat them, they are often taken with surprise and regret, even in late-late adulthood.

As one 92-year-old woman described this experience:

> There's this feeling of being out of one's skin. The feeling that you are not in your own body . . . Whenever I'm walking downtown, and I see my reflection in a store window, I'm shocked at how old it is. I never think of myself that way. [quoted in Kaufman, 1986]

Similar feelings were expressed by a man, also 92, who needed a cane to get around:

> I look like a cripple. I'm not a cripple mentally, I don't feel that way. But I am physically. I hate it . . . You know, when I hear people, particularly gals and ladies, their heels hitting the pavement . . . I feel so lacking in assurance—why can't I walk that way? . . . I have the same attitude now, toward life and living, as I did 30 years ago. That's why this idea of not being able to walk along with other people—it hurts my ego. Because inside, that's not really me. [quoted in Kaufman, 1986]

When elderly people associate appearance and identity, or depend on the reactions of others to validate their self-concept (as we all do sometimes), the realization that they look like, or are treated like, an old person may make them act and think like the stereotype of the elderly—with harmful consequences. As you will learn in Chapter 25, activity and feeling young promote psychological as well as physical health in the aged.

Sense Organs

The incidence of impaired vision and hearing increases with each decade. A nationwide survey of Americans over age 65 found that 90 percent had some visual impairment, and that only 10 percent had good vision without glasses (U.S. Bureau of the Census, 1986). Virtually no one hears as well in late adulthood as they did earlier (Olsho et al., 1985), and about a third of all the elderly find that inadequate hearing hampers them in their daily lives (Whitbourne, 1985).

Most of the visual and auditory losses of the aged can be corrected. Approximately 80 percent of the elderly have glasses and can see well when wearing them; only 10 percent still have serious vision problems when using corrective lenses. Even most of the very old can see fairly well. In a study of more than a thousand Americans who had reached age 100, 9 percent could see well without glasses, 62 percent could see well with glasses, and 29 percent (including 4 percent who were blind) had vision deficits that glasses did not correct (Segerberg, 1982).

Unfortunately, far fewer people who need them use hearing aids than use glasses. Although most people using the new types of hearing aids are well satisfied with them, and although nearly a third of the elderly could benefit from them, less than 10 percent use them (Olsho et al., 1985). One important reason is that the hearing aid is regarded as a symbol of agedness, and many people would rather miss some of the sounds of daily life than risk being considered old. Ironically, this is usually what their poor hearing causes them to be thought of as anyway, often with unanticipated consequences. The person who frequently mishears and misunderstands conversation or who often asks "What did you say?" is likely to be thought of as a bit dottering and to be excluded from much social give-and-take, thereby being deprived of important cognitive stimulation (Whitbourne, 1985). Furthermore, the hard-of-hearing suffer more from their impairment than do those with poor vision: they tend to withdraw socially when others become annoyed at

Figure 23.3 *The man on the left seems eager to hear what these children have to say. However, he is making a mistake common among many of the hard-of-hearing elderly—trying to get by with guesswork and cupped hands rather than using a state-of-the-art hearing aid.*

not being heard properly and to become suspicious of inaudible conversation that seems to be about them (Busse, 1985).

Body Systems

At some point in old age, the depletion of organ reserve and the ongoing slowdown of body functioning reaches a level where daily routines need to be adjusted. For the healthy young-old, the changes may be minor: eating smaller, more frequent meals, devoting more time to stretching and warm-ups before heavy exercise, or giving more attention to forms of sexual expression other than intercourse. For some of the old-old, energy and effort may need to be conserved, requiring that, beyond the basic routines of life, each day be limited to only one or two activities—having lunch with a friend *or* working in the garden *or* visiting a grandchild.

Exercise Even for the very old, however, some physical activity is beneficial, not only for the cardiovascular system but also for the respiratory, digestive, and virtually every other body system (Buskirk, 1985). In late adulthood, of course, the pace of exercise must be carefully adjusted to match the declines that have occurred in heart and lung functioning. For some, this means that jogging replaces running; for

Figure 23.4 *Regular exercise in old age not only maintains and prolongs one's physical abilities (and even one's cognitive performance, as we will see in the next chapter), but also helps to sustain a vital and active self-image.*

others, that brisk walking replaces jogging; for others, that strolling replaces brisk walking. Nonetheless, regular exercise—three or more times a week for half an hour—is even more important in late adulthood than earlier to help maintain the strength of the heart muscle. If an older person does not exercise regularly (and, unfortunately, the elderly are much less likely to exercise than the young), a sudden, unusual exertion—which would have been absorbed by the organ reserve and homeostatic reactions of a younger person—can result in a heart attack. Not only is regular, measured exercise more likely to strengthen the heart and protect against potentially fatal effects of exertion; it is likely to improve the overall quality of life as well (Harris, 1986).

Figure 23.5 *Many of the elderly are embarrassed to exercise with younger people. One advantage of retirement communities is that there are sufficient numbers of the elderly to form their own exercise class. Water exercise, shown here, is one of the best, since the total body is in motion without putting too much strain on any one part. This particular community also provides three exercise classes on dry land: a vigorous aerobic one for the healthy; a slower, stretching one for those with cardiovascular impairment; and an upper-body one for those in wheelchairs.*

The Aging Brain   Physiologically, the brain in late adulthood is notably smaller and slower in its functioning than in early adulthood, a fact that once led many people to jump to the conclusion that aged minds are necessarily feeble. The evidence for the physical decrements was fairly clear: the brain loses at least 5 percent of its weight and shrinks in size by about 15 percent by late adulthood. One reason for this change is that neurons, which do not reproduce themselves, die throughout life, and die at an increasing rate after about age 60. The amount of cell death varies across different areas of the brain: in the primary visual area of the cerebral cortex, it is about 50 percent; in the motor areas, between 20 and 50 percent; in the areas related to memory and reasoning, 20 percent or less (Whitbourne, 1985).

Changes in the brain are not solely in the size and the number of the cells. There are many indications that the neural processes in the brain become markedly slower with age, noticeably so beginning in the late 50s. The most obvious measure of this slowdown is reaction time: in laboratory tests the elderly are much slower to push a button in response to a light flashing on than younger adults are. Slowed neural processes are also reflected in changes in brain-wave patterns, as measured by the EEG (Obrist, 1980). For example, alpha waves, the waves that appear when a person is awake and relaxed, occur at the rate of eleven cycles per second in young adulthood, nine cycles per second by age 65, and eight cycles over age 80.

This slowing of various neural processes may be related to reduced production of **neurotransmitters**—chemicals in the brain that allow a nerve impulse to be communicated from one cell to another. Several neurotransmitters are present in smaller concentrations in the aged brain than in the young one. The slowing may also be related to reduced blood flow to the brain. It is interesting to note that exercise, which increases this blood flow, has been shown to significantly improve reaction time (Botwinick, 1979).

However, the practical implications of all this evidence are not obvious. Except in cases of extreme damage and loss, there is no proof that brain activity is directly related to the brain's size or weight or number of cells. When brain cells die, existing cells routinely take over their function. Further, the dendrites reaching out from the remaining neurons continue to develop throughout adulthood (Cotman and Holets, 1985). It is quite possible that, while discrete brain cells are dying, the connections among the remaining cells increase, a compensation that might allow older adults to think as well as they did, although somewhat more slowly.

Overall, on the basis of research on brain development in aging animals and on the learning potential of older humans (see Chapter 24), developmentalists now believe that the brain may "retain a considerable degree of cerebral plasticity in extreme old age" and that "the adaptive potential of the geriatric brain may have been considerably underestimated" (Walsh, 1981).

Sleep Brain-wave patterns reveal another interesting difference between young and old adults. Although older adults get about as much light sleep and REM (dreaming) sleep as younger adults, they get significantly less deep sleep, which is the most restorative kind. Either as a cause or a consequence, older adults tend to wake up numerous times each night, an average of twenty-one times according to one study (Hayashi and Endo, 1982). Among the immediate causes of such wakefulness are frequent leg movements, extended pauses in breathing, and an increased sensitivity to noise. As a result of such interruptions to their sleep, many of the elderly have sleep patterns—including being fully awake several times in the middle of the night—that would be considered pathological in a younger person (Dement et al., 1985). If an older person doesn't accept these changes as normal, and adapt by taking naps in the daytime, or by listening to the radio or reading when having trouble sleeping at night, these sleep difficulties can create psychological problems as well as chronic fatigue.

Aging and Disease

The relationship between aging and disease is a complex one. On the one hand, aging and disease are not synonymous: it is inevitable that a person's body will gradually weaken overall with age; it is not inevitable that the person will develop any particular disease. Most aged people, most of the time, consider their health good or excellent and, on physical examination, are found to be quite well. Whether a particular elderly person is seriously ill, somewhat ailing, or in fine health depends, not on age, but on genetic factors, past and current lifestyle (including eating and exercise habits), and psychological factors such as social support and a sense of control over one's daily life (Rowe and Kahn, 1987).

On the other hand, it is undeniable that the incidence of chronic diseases—long-standing diseases that are generally irreversible—increases significantly with age. One reason is that the older a person is, the more likely he or she is to have accumulated several risk factors for such diseases. For example, smoking, drinking, and inactivity eventually lead to osteoporosis in many elderly men as well as women.

In addition, many of the biological changes that occur with aging reduce the efficiency of the body's systems, making the older person more susceptible to disease. Because of these changes, aged people are not only more likely to develop a disease; they also take longer to recover from illnesses, and are more likely to die of them (see Figure 23.6). As one gerontologist explains, "the most important feature of aging is . . . a steadily increasing vulnerability to fatal damage. Old men are killed by mishaps that a fit young man would hardly notice" (Burnet, 1974). For example, if a younger person contracts pneumonia, he or she almost always is fine again in a few weeks, but if pneumonia comes to a person already seriously weakened by very old age, it is often the immediate cause of death.

Figure 23.6 *The death rate from the eight leading causes of death is significantly higher for elderly people than for younger people. This table shows approximate ratios of the death rate for Americans aged 65 and older and those under 65. A finer analysis reveals some interesting age differences. For example, elderly pedestrians are much more likely to be killed in auto accidents than other adult pedestrians are, whereas younger adults are more likely to be killed in auto accidents when they are the drivers or passengers.*

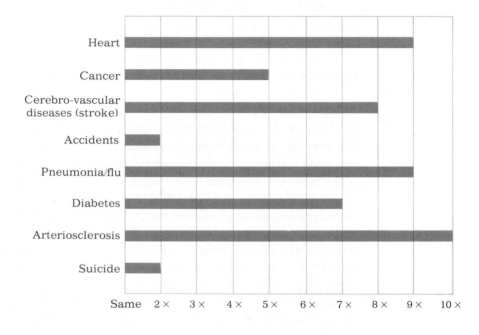

Not surprisingly, then, the elderly see doctors more often, and spend more time in hospital beds, than younger adults do. On this point, it should be noted that, contrary to the ageist stereotype, the elderly are not more likely to be hypochondriacs, imagining or exaggerating health complaints (Shock, 1985). In fact, compared with younger adults, the elderly are less likely to seek medical attention for problems that can be treated (Rowe and Minaker, 1985; Siegler and Costa, 1985), largely because they interpret the signs of disease as the signs of normal aging, and try to overlook them.

Let us look more closely at the complex relationship between aging and disease with two examples, the leading killers of the aged, heart disease and cancer.

Heart Disease and Cancer

Normal aging reduces the functioning of the heart—especially in times of exercise and stress—partly because it reduces the strength of the heart muscle and lengthens the time the heart needs to relax between contractions. Normal aging also reduces the elasticity of the cardiovascular system. But aging in itself does not cause heart disease. Most aged people have quite healthy, although aging, hearts, capable of sustaining life for many more years (Harris, 1986). However, many older people also show a number of risk factors related to heart disease, including elevated blood pressure, a high cholesterol level, obesity, lack of exercise, a history of smoking. Over time, the interaction of these accumulating risk factors with the general weakening of the heart and relevant genetic weaknesses makes the elderly increasingly vulnerable. Thus it is not surprising that heart disease causes about 40 percent of all deaths over age 65.

A similar relationship exists between aging and cancer, the cause of about 20 percent of all deaths in the aged. The predisposing factors that may lead to cancer—genetic vulnerability and environmental insults (such as exposure to asbestos or tobacco smoke) among them—typically predate old age by half a century or more, and the cancer process probably begins years before it is evident. As people age, however, the latent potential for cancer is more likely to become manifest, and cure is more difficult: a person older than 85 is a hundred times as likely to die of cancer as a person aged 30.

Thus far we have discussed two of the reasons that the aged are more vulnerable to disease: their bodies are weaker overall and they have been exposed to environmental stresses for a longer period of time. There is a third factor, however, which may be particularly important in affecting the person's ability to recover from whatever illnesses may strike: diminished immunity.

Immunity

The immune system determines the body's ability to defend itself against invaders from without (such as viruses) and from within (such as cancer cells). The immune system works by recognizing foreign or abnormal substances in the circulatory system, and then isolating and destroying them, mainly with two types of "attack" cells. The first are called **B-cells** because they are manufactured in the bone marrow. B-cells create antibodies that attack specific invading bacteria and viruses. Since these antibodies remain in the system, we do not get measles, mumps, or specific strains of flu more than once. The second type are called **T-cells** because they are manufactured by the thymus gland. T-cells produce specific substances that attack infected cells of the body. T-cells also help the B-cells produce more efficient antibodies and strengthen other aspects of the immune system as well.

The first notable change in the immune system involves the thymus gland, which begins to shrink during adolescence and by age 50 weighs only about 10 percent of what it did at age 15 (Hausman and Weksler, 1985). The production of T-cells declines as well, although not as dramatically. With age there also is a reduction in the power of both T- and B-cells, as well as in the efficiency of the mechanisms that regulate them.

These changes help explain why most forms of cancer become much more common with age, and why various other illnesses—from chicken pox to the latest strain of influenza—are much more serious in an adult than in a child. By late adulthood, the "flu" can be fatal. In fact, according to one theory, the gradual breakdown of the immune system accounts for the aging process itself, allowing genetic vulnerabilities (such as diabetes) or disease processes (such as the growth of cancer cells) that had been kept in check earlier in life to take over (Hausman and Weksler, 1985; Segre and Smith, 1981).

Changes in the immune system are just one of several explanations for the aging process. Let us look now at the other major theories of aging.

Causes of the Aging Process

The attention of many researchers and physicians is turning toward making all the years of the life span good ones. The hope is that most people will be active, healthy, alert, and enthusiastic until age 85 or so, when they will die, quickly and easily, a natural death, caused by a relatively sudden accumulation of aging effects in all the parts of the body (Fries and Crapo, 1981). Increasingly, this is the case, as medical science attacks diseases once thought to be inevitable with age, and as people develop healthier lifestyles throughout adulthood. This raises another question: What actually causes the aging process, and might we ever be able to control it sufficiently that an average person might live 95 or 100 healthy years instead of simply 75 or 80? There are many answers to this question, some implicating the environment, others pointing to our genetic makeup (Conn, 1987; Rosenfeld, 1985; Schneider and Reed, 1985).

Wear and Tear

The oldest, most general theory of aging, the **wear-and-tear theory,** compares the human body to a machine. It maintains that just as, say, the parts of an automobile begin giving out as mileage adds up, so the parts of the human body deteriorate with each year of use, as well as with accumulated exposure to pollution, radiation, inadequate nutrition, disease, and various other stresses. According to this theory, we wear out our bodies just by living our lives.

Although appealing in its simplicity, the wear-and-tear theory seems to be of limited usefulness. Certainly it is true that an athlete who puts repeated stress on elbows or knees is likely to have damaged joints by middle adulthood; or that an outdoor worker whose skin is continuously exposed to sunlight is likely to have damaged skin; or that an industrial worker who inhales asbestos and smokes cigarettes over many years will eventually have damaged lungs. But, overall, the analogy to a machine's wearing out simply doesn't hold up. In many respects, the human body is its own repair shop, replacing or mending many of its damaged parts. In addition, unlike most machines, many parts of the human body benefit from use. As we have seen, the heart functions better if the person regularly makes

Figure 23.7 *Stretching and bending regularly, in any context, at any age, are much more likely to maintain and extend one's physical capabilities than to diminish them.*

it work faster than normal; the respiratory system benefits from routine exertion; the sexual arousal system is more likely to function in old age if the person has been sexually active throughout adulthood; the digestive system benefits from raw fruits and vegetables that require vigorous digestive activity. It seems clear, then, that the notion of wear and tear applies to some diseases and problems in some organs and body parts, but it is not very helpful in explaining the aging process overall.

Cellular Theories

More promising theories of aging begin at the cellular level, suggesting that some occurrence in the cells themselves causes aging.

Cellular Accidents One cellular theory of aging proposes that senescence is the result of the accumulation of accidents that occur during cell reproduction. With the exception of certain types of cells, notably those of the nerves, the cells of the human body continue to reproduce throughout life. An obvious example is the outer cells of the skin. Under normal conditions, these cells are entirely replaced every few years; the process occurs much more rapidly when a cut or scrape is healing. Thanks to precise functioning of DNA and RNA, each replacement cell is the exact copy of an old cell, or ought to be. However, radiation from the sun and other sources bombards our cells throughout the life span, gradually causing mutations in the structure of the DNA in more and more of our cells. Mutations may also occur in the process of DNA repair. As a consequence, the instructions for creating new cells become imperfect, so the new cells are not quite exact copies of the old. Over time, such changes may result in aging of the skin, or benign skin changes, or possibly cancer. Throughout the body, cellular imperfections and the body's declining ability to detect and correct them can result in harmless changes, small declines in function, or sometimes, potentially fatal damage.

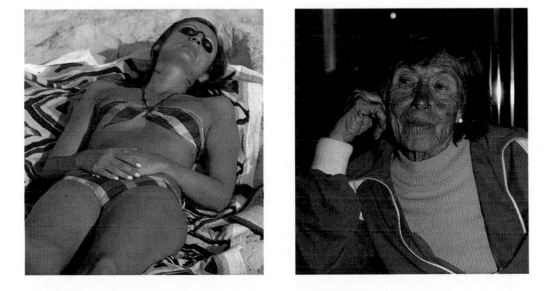

Figure 23.8 *The "beauty benefits" of devoted sunbathing are transient; the damage is cumulative. Age spots, wrinkles, and leathery skin texture associated with aging become exaggerated from lengthy exposure to the sun.*

Free Radicals Another aspect of this theory of aging began with the observation that some of the body's metabolic processes can cause electrons to separate from their atoms, resulting in atoms with unpaired electrons. These atoms, called **free radicals,** are highly unstable and capable of reacting violently with other molecules in the cell, sometimes splitting them or tearing them apart. The most critical damage caused by free radicals occurs in DNA molecules, producing errors in cell maintenance and repair that, over time, may eventually contribute to such diseases as cancer, diabetes, and arteriosclerosis (Harman, 1984). Free radicals formed from oxygen molecules (the most common kind) have been observed to aggravate these diseases, so much so that doctors hesitate to give their patients pure oxygen, which increases the number of free radicals (Marx, 1987). It seems, then, that since free radicals damage cells, affect organs, and accelerate diseases, the gradual accumulation of damage as the individual ages may be one of the causes of the aging process (Harman, 1984).

Error Catastrophe   When the systems of the body, especially the immune system, are in shape, the effects of cellular damage are minor, held in check by other cells that destroy seriously damaged cells and take over the work that imperfect cells no longer perform (Schneider and Reed, 1985). However, according to some theories, as the immune system declines and the processes of repair and healing become less efficient, the constellation of errors can become so extensive or affect such critically important cells that the body can no longer control or isolate the errors, leading to what has been called an **error catastrophe** (Cohn, 1987). At this point the normal, healthy aging process gives way, disease overtakes the person, and death occurs. This theory of accumulating errors in cell reproduction is one possible explanation for the fact that cancer, a disease in which normal cell reproduction somehow goes awry and cells reproduce so rapidly that malignant tumors are formed, is much more common in humans, and indeed all mammals, as they reach old age.

Cross-Linkage   Another theory of cellular aging focuses on **cross-linkage,** the biochemical process through which certain kinds of molecules become linked with other kinds. This process is most commonly seen in the manufacture of plastic and varnish, and is essential to the hardening of those products. With time, certain proteins of the body, notably collagen and elastin, which form the connective tissue that bind the cells of the body together, become cross-linked to other protein molecules. As this cross-linkage occurs, the proteins become less elastic and more brittle (Kohn, 1979).

Different parts of the body have different types and rates of cross-linkage, yet all of them show this phenomenon to some degree. It is most apparent in the skin, which becomes less elastic, more wrinkled, and feels dry and leathery with old age, precisely because more of its collagen is cross-linked. Cross-linkage may also be the underlying cause of major changes in the cardiovascular system, such as hardening of the arteries (Kohn, 1977). In fact, increasing cross-linkage may be synonymous with aging, making each cell of the body less efficient. As one biologist wryly put it, cross-linkage is "extremely useful in the paint and varnish industry but . . . may be highly detrimental to biological systems over the long run . . . The gradual accumulation of a layer of varnish over various intracellular structures is an unpleasant prospect" (Strehler, 1977).

Programmed Senescence

Certainly it seems plausible that the changes in cellular processes that occur over time eventually lead to aging and death. However, many scientists suspect that the aging process, while reflecting these changes, is not caused by them. They contend that, just as we are genetically programmed to reach various levels of biological maturation at fixed times, we are genetically programmed to die after a fixed number of years. Thus, even if no particular illness or accident occurs, **programmed senescence** will inevitably bring about death at a certain age.

That "certain age," the maximum number of years that a particular species is genetically programmed to live, is called **maximum life span,** a number that is based on the oldest age to which members of each species have been known to live (Kirkwood, 1985). For the house mouse, the maximum life span is 3 years; for the house cat, 28; for both the horse and the golden eagle, 46; and for the elephant, 70. For humans, the maximum life span is 115 years.

Maximum life span is quite different from **average life expectancy,** which is the number of years the average newborn of a particular population of a given species is likely to live. In humans, life expectancy varies according to historical, cultural, and socioeconomic factors that affect frequency of death in childhood, adolescence, or middle age. In the United States in 1985, average life expectancy at birth was about 71 years for men and 78 years for women. Americans who already are 60 years old, and thus no longer at any risk of an early death, are expected to live to 78 if male and to 83 if female. Current life expectancy is about four times what it was for people of ancient times, and twenty-eight years more than it was at the turn of the century. The reason for this improvement is not that the maximum life span has increased but that infants and children are less likely to die.

A Genetic Clock? According to one theory, the DNA that directs the activity of every cell in the body also regulates the aging process. Our genetic makeup acts, in effect, as a **genetic clock,** triggering hormonal changes in the brain (similar to the hormonal changes that produce puberty, for instance) and regulating the cellular reproduction and repair process. As the genetic clock gradually "switches off" the genes that promote growth, there is speculation that genes that promote aging are switched on. The damage associated with aging continues to accumulate until one or more body systems can no longer function, and a natural death occurs.

Genetic regulation of aging is suggested by several genetic diseases that include premature signs of aging and early death as part of their symptoms. Down syndrome is the most common: people with Down syndrome who survive childhood almost always die by middle adulthood, with symptoms of heart disease and Alzheimer's disease, a type of dementia that occurs most frequently in old age. Children born with a rare genetic disease called **progeria** have a normal infancy but by age 5 stop growing and begin to look like old people, with wrinkled skin and balding heads. They develop many other signs of premature aging during middle childhood, and die by their teens, seemingly of heart diseases typically found in the elderly.

The Hayflick Limit Evidence for a genetic clock that limits the life span also comes from laboratory research, particularly from the work of Leonard Hayflick (Hayflick, 1979; Hayflick and Moorhead, 1961). In this research, cells cultured from human embryos were allowed to "age under glass" by providing them with all the necessary nutrients for cell growth and replication and protecting them from external stress or contamination. In such ideal conditions, it was believed, the cells would multiply forever. Instead, the cells stopped replicating after about fifty divisions. Cells similarly cultured from children showed a smaller number of doublings before they stopped dividing, and cells from adults divided even fewer times. The total number of cell divisions was shown to be roughly related to the age of the donor.

A CLOSER LOOK   Nutrition and Aging

There is a clear consensus regarding three key aspects of nutrition during late adulthood (Chen, 1986; Kart and Metress, 1984; Maruyama, 1986):

1. Vitamin and mineral needs do not decrease with age. If anything, they increase, as the body's ability to break down food and use the nutrients in it becomes less efficient.

2. Calorie requirements decrease by about 20 percent from those of early and middle adulthood. Although variations in height, weight, and activity are obviously relevant, the average older man should take in about 2,200 calories a day, and the average woman, about 1,700. During the years of late adulthood, calorie consumption should continue to decrease with every decade (National Research Council, 1980).

3. Because more nutrients need to be packed into fewer calories, a varied and healthy diet, emphasizing fresh fruits and vegetables, lean meats and fish, and complex carbohydrates (cereals and grains), is even more important in late adulthood than earlier.

For many reasons, getting enough nutrients is more problematic for the aged than for younger adults (Chen, 1986; Kart and Metress, 1984). The chief age-related factor is the reduced efficiency of the digestive system, which makes absorption of nutrients more difficult. The senses of smell and taste also diminish with age, making food less appealing. A number of external factors may also affect certain segments of the elderly: (1) poverty (high-quality nutrients are more expensive); (2) living alone (those who eat alone tend to eat quick, irregular meals); (3) dental problems (missing teeth and gum disease make people eat softer food and less of it).

Furthermore, many of the elderly take drugs that affect nutritional requirements. For example, aspirin (taken daily by many who have arthritis) increases the need for vitamin C; antibiotics reduce the absorption of iron, calcium, and vitamin K; antacids can reduce absorption of protein; oil-based laxatives deplete vitamins A and D; and so on (Kart and Metress, 1984; Rikaus, 1986). Alcohol, especially in large amounts, is very detrimental to good nutrition, depleting B vitamins, calcium, magnesium, and vitamin C in particular (Yunice and Hsu, 1986).

In an effort to maintain health and vigor, about two-thirds of the elderly take vitamin and mineral supplements (Read and Graney, 1982), a practice encouraged by many of the advocates of life extension (Demonpoulus, 1982; Rosenfeld, 1985; Walford, 1983), some of whom recommend massive doses of certain vitamins as well as a low-calorie diet.

Some believe that antioxidants, which reduce the number of free radicals of oxygen (page 537) by forming a bond with their unattached oxygen electron, might slow down

A hot, balanced meal, a chance to socialize, and a good reason to take a mid-day walk—all provided by this senior-citizen lunch program, which serves the inner-city elderly who otherwise might eat less nourishing food, alone, at home.

the disease and aging process. Antioxidants include vitamins C and E, and beta-carotene (most commonly found in carrots), as well as some enzymes. However, a direct link between these substances in the human bloodstream and a reduction in free radicals is not proven (Cutler, 1984; Weber and Miquel, 1986), and most nutritionists warn that taking megadoses of antioxidants may do more harm than good.

Many nutritionists find that most of the elderly are relatively well nourished, and that they are more likely to overspend and overdose on vitamins than to be deficient in them (Baily, 1986; Chen, 1986; Kart and Metress, 1984). At the moment, there is no clear and compelling evidence one way or the other for the usefulness of vitamin supplements in old age. However, enough is known to advise caution. Large doses of vitamins may well harm an individual by upsetting the natural nutritional balance in the system, creating vitamin needs and dependencies where there were none before (Cutler, 1984; Hsu and Smith, 1984).

Given this state of affairs, it seems fair to conclude that the elderly should consume a healthy diet, and beyond that, proceed cautiously. Diagnosis of particular individual vitamin deficiencies should be the only basis for undertaking any expensive or unusual vitamin regimen, and acceptance of any life-extending recommendations should await controlled longitudinal research that uses a sufficient sample size, control groups, and blind experimenters.

This research has been replicated by hundreds of scientists, using many techniques and various types of cells from people and animals of different ages. The result is always the same. A **Hayflick limit** is found: even in ideal conditions, the cells of living creatures reproduce a finite number of times, with each species having a particular number of cell divisions. Cells from people with progeria, Down syndrome, and other genetic conditions characterized by accelerated aging show fewer numbers of doublings than would be expected given the age of the donors (Tice and Setlow, 1985).

In all cases, at the point where cell division stops, analysis of the cells shows that they are different from young cells in many ways. This provides new support for the idea that DNA and RNA are responsible for cell death, not only because of random errors but, more important, because of programmed senescence.

The fact that aging and death seem to be genetically programmed does not mean that we can predict when an individual will die. Three factors make such a prediction impossible:

1. Not only is the maximum life span predetermined for each species, but how close each particular individual comes to reaching that maximum is also genetically influenced. Longevity, for example, runs in families; and compared with dyzygotic twins, monozygotic twins are much closer in rate of aging and age at death (McClearn and Foch, 1985; Tice and Setlow, 1985). Most people, if they escape specific diseases in their younger years, die in their 80s, but some individuals live quite naturally without debilitating chronic diseases until their 100s.

2. Environment is also involved in the timing of aging and death. Consequently, recent environmental changes (mostly in lifestyle and medical advances) have resulted not only in more people surviving to age 80, but also in more 80-year-olds surviving to 90 and 100 (Kirkwood, 1985).

3. The genetic clock itself may be resettable. Genetic engineering is on the brink of allowing children to escape the consequences of genetic diseases they inherit (see page 74); the same engineering techniques may also be able to reprogram the aging genes to extend the "fixed" limit of human life.

Life Extension

The idea that the human life span may be lengthened in the near future is supported by a number of gerontologists (Rosenfeld, 1985; Walford, 1983). One scientist writes about "the almost certain extension of the healthy human life span by 10 to 20 years" and suggests that there is a 40 percent chance of extension by 50 or 100 years (Strehler, 1979).

Whether or not extending the life span might be beneficial to the individual or to society is debatable: many people now shudder at the idea of late adulthood lasting fifty years or more, or at the concept of a society in which the elderly outnumber the young by two to one. However, before drawing your own conclusions, remember that ageism clouds our perceptions of late adulthood. A look at some people who have lived to be very old might be instructive.

May You Live So Long

Three remote areas of the world—one in the Soviet Union, one in Pakistan, and one in Peru—have become famous for having large numbers of people who enjoy unusual longevity. In these places, late adulthood is not only long but is also usually quite vigorous.

Figure 23.9 *Three remote regions of the world are renown for the longevity of their people. In Vilcabamba, Ecuador, (a) 87-year-old Jose Maria Roa stands on the mud from which he will make adobe for a new hose, and (b) 102-year-old Micaela Quezada spins wool. In Abkhasia of the Georgian Soviet Socialist Republic, companionship is an important part of late life, as shown by (c) Selakh Butka, 113, posing with his wife Marusya, 101, and (d) Ougula Lodara talking with two "younger" friends. In this remote area of the U.S.S.R., the elderly continue to work long past the usual American retirement age, so (e) Gumba Tikhed, age 98, still picks tea on a collective farm. (f) Temir Tarba rides his horse daily, as he has for almost all of his 100 years. Finally, Shah Bibi (g), at 98, and Galum Mohammad Shad (h), at 100, from the Hunza area of Pakistan, spin wool and build houses. Alexander Leaf, the physician who studied these people, believes that the high social status and continued sense of usefulness of the very old in these cultures may be just as important in their longevity as the diet and exercise imposed by the geographical conditions in each region.*

(a)

(b)

(c)

(d)

(e)

(f)

(g)

(h)

One researcher describes the Abkhasia people in the Soviet Union:

> Most of the aged [those about age ninety] work regularly. Almost all perform light tasks around the homestead, and quite a few work in the orchards and gardens, and care for domestic animals. Some even continue to chop wood and haul water. Close to 40 percent of the aged men and 30 percent of the aged women report good vision; that is, they do not need glasses for any sort of work, including reading or threading a needle. Between 40 and 50 percent have reasonably good hearing. Most have their own teeth. Their posture is unusually erect, even into advanced age. Many take walks of more than two miles a day and swim in mountain streams. [Benet, 1974]

Among the people described in this report are a woman said to be over 130 who drinks a little vodka before breakfast and smokes a pack of cigarettes a day; a man who sired a child when he was 100; and another man who was a village storyteller with an excellent memory at a reported age of 148.

A more comprehensive study (Pitskhelauri, 1982) finds that all the regions famous for long-lived people share four characteristics:

1. Diet is moderate, consisting mostly of fresh vegetables and herbs with little consumption of meat and fat. A prevailing belief is that it is better to leave the dining table a little bit hungry than too full.

2. Work continues throughout life. In these rural areas, even a very elderly adult is likely to help with farm work and household tasks, including child care.

3. Family and community are important. All the long-lived are well integrated into families of several generations, and interact frequently with friends and neighbors.

4. Exercise and relaxation are part of the daily routine. Most of the long-lived take a stroll in the morning and another in the evening (often up and down mountains), most take a midday nap, and most spend several hours socializing in the evening, telling stories and discussing the day's events.

Beyond these four aspects of the individual's life, geography and tradition may be influential as well. All three places famous for long-lived people are in rural, mountainous regions that are at least 3,000 feet above sea level. This minimizes pollution and maximizes lung and heart fitness, for even walking in these regions can be considered aerobic exercise. Furthermore, in all three, the aged are respected, and strong traditions ensure the elderly an important social role.

Some researchers suggest that another factor may account for some of these cases of unusual longevity—lying. None of the communities of the long-lived have birth or marriage records from the nineteenth century that are verifiable, at least to the satisfaction of critics. In fact, beginning at about age 70, many people in these areas systematically exaggerate their age, although how greatly is debatable (Leaf, 1982; Mazess and Forman, 1979; Medvedev, 1974; Pitskhelauri, 1982). It may be that persons who claim to be 100 are only in their 80s, and that those who are supposedly long past 100 are only a little bit past it.

This does not render the earlier reports useless, for no one doubts that an unusual number of very old and healthy people thrive in these isolated areas of the world. While their genetic clocks almost certainly do not allow them to live to age 148, their habits and culture do allow many of them to reach 100.

Research on those who reach late-late adulthood in countries where records are accurate show lifestyle patterns similar in many ways to those of the long-lived of

Peru, Pakistan, and the Soviet Union: a lifetime of moderate diet, hard work, and social involvement is typical. For example, between 1955 and 1968 more than a thousand Americans on Social Security reached their 100th birthday and were interviewed by a Social Security representative (Segerberg, 1982). Time after time the interviewer arrived dreading the encounter (and cursing the administrator who had ordered it), and left amazed and heartened that the centenarian still chopped wood, or did the housework, or, in the case of one physician, still saw patients. Very few were in wheelchairs, or had serious illnesses such as cancer or heart disease; only one suffered from senile dementia.

A general impression of these centenarians is conveyed by the following description of one of their members, Francisco Guerra:

> His close cropped hair is black except on the sides, where it is white. His lean body looks hard and sinewy, not an excess ounce of flesh on it. His skin, broiled for years under the Southwest sun, is like leather—but malleable enough for smiling eyes and mouth.
>
> "I have never been sick a day in my life," he says. "I feel good."
>
> Why? How has he been able to do it?
>
> "I have lived so long because there's a Chief up there. Maybe it's punishment, maybe it's glory. I am happy."
> [quoted in Segerberg, 1982]

Taken together, these and many other studies of the old-old lead to a ready conclusion: if people reach late adulthood in good health, their attitudes and activities may be even more important in determining the length and quality of their remaining years than purely genetic or physiological factors.

SUMMARY

Ageism

1. Prejudices about the elderly are common and destructive, for they result in the old living lives that are more limited and isolated than they need to be. Even gerontologists sometimes expect the elderly to be less healthy and active than they actually are.

2. Contrary to the stereotype, most of the aged are happy, healthy, and active. Fortunately, ageism is weakening as gerontologists study more of the elderly who live in the community, and as the sheer number of the aged in the general population increases.

The Aging Process

3. The many apparent changes in the skin, hair, and body shape that began earlier in adulthood continue. In addition, most older people are somewhat shorter and weigh less than they did, and walk more stiffly. Such changes in appearance can affect the self-concept of the older person.

4. Vision and hearing are almost always impaired by late adulthood, to the point that nine out of ten of the elderly need glasses, and one out of three would benefit from a hearing aid. Most who need glasses get them, but, unfortunately, most who need a hearing aid do not get one, fearing that it would make them appear old.

5. The age-related declines of the major body systems and organ reserve eventually reach a point—different for everyone—at which some of the routines of daily life need adjusting. For example, although exercise is just as important during late adulthood as earlier, its pace needs to be slower.

6. By late adulthood, the brain has become smaller and works more slowly, the result of cell death and overall circulatory slowdown. However, the implications of these changes are controversial. Because of research on den-

drite development and on learning in old age, many developmentalists think that, barring disease, the brain continues to function well in late adulthood.

7. Sleep patterns in late adulthood are quite different from those in early adulthood, with several naps sometimes needed to augment an interrupted night's sleep.

Aging and Disease

8. The aging process is not synonymous with the disease process. We should not assume that illness is an expected, and thus an accepted, companion during the later years. Unfortunately, many of the elderly attempt to overlook problems that need medical attention because they believe their symptoms are just part of growing old.

9. It is true, however, that aging makes people more susceptible to most chronic and critical diseases, and makes recovery slower. At some point, usually in late-late adulthood, multiple problems may occur, and death will result.

10. The decline in the immune system also contributes to the elderly's increasing vulnerability to disease. As the thymus shrinks and production of both B- and T-cells decreases, the body becomes less able to fight against diseases.

Causes of the Aging Process

11. There are many theories that address the environmental and genetic causes of aging. One theory is that as we use our bodies we wear them out, just as a machine wears out with extended use. This wear-and-tear theory does not, however, explain much of what research finds—that activity promotes longer life and healthier aging.

12. Cellular theories of aging seem more plausible. Perhaps the DNA duplication and repair processes are affected by radiation and other factors, leading to an accumulation of errors when new cells are made. These errors may eventually produce an error catastrophe, in which the body suddenly reaches the point that it is much more vulnerable to disease and death. It is also possible that cross-linkage between protein molecules, reducing tissue elasticity, and an increase in free radicals, bombarding and damaging normal cells, are instrumental in the aging process.

13. The maximum life span may well be fixed by a kind of genetic clock that switches the aging process on at some point. The theory that genes may be responsible for aging is buttressed by evidence that several conditions that are accompanied by premature aging, such as Down syndrome and progeria, are caused by genetic abnormalities.

14. Further evidence for programmed senescence is found in the Hayflick limit. Even in ideal conditions, cells in the laboratory stop reproducing themselves after a certain number of divisions. This number decreases as the age of the cell donor increases.

Life Extension

15. Theoretically, once the cause of the aging process is understood, it can be altered so that the maximum human life span will increase. Whether or not this is a good idea depends on what quality of life very old people could be expected to have.

16. Wherever they live, those who live to be 100 typically continue to live happy and active lives. In three regions of the world, parts of the Soviet Union, Pakistan, and Peru, large numbers of people seem to live to be very old. Moderate diet, high altitude, hard work, and traditional respect for the aged characterize all three places.

KEY TERMS

ageism *(526)*	free radicals *(537)*
young-old *(526)*	error catastrophe *(538)*
old-old *(526)*	cross-linkage *(538)*
gerontology *(527)*	programmed
demography *(528)*	senescence *(539)*
demographic	maximum life span *(539)*
pyramid *(528)*	average life
senescence *(529)*	expectancy *(539)*
neurotransmitters *(532)*	genetic clock *(539)*
B-cells *(535)*	progeria *(539)*
T-cells *(535)*	Hayflick limit *(541)*
wear-and-tear	
theory *(536)*	

KEY QUESTIONS

1. To what extent are the elderly inactive, lonely, and frail?

2. What are the advantages of exercise in old age?

3. What changes occur in the brain in old age and how do they affect thinking?

4. What is the relationship between aging and disease?

5. How is the immune system affected by aging?

6. Why is the wear-and-tear theory not very helpful in explaining the aging process?

7. In what ways do cellular theories of aging seem plausible?

8. What does Leonard Hayflick's research on cells from individuals of various ages tend to show?

9. What are some of the characteristics of people who live to a very old age?

CHAPTER 24

Late Adulthood:
Cognitive Development

*My brain is much more easily fatigued now and I
have forgotten lots of things but I think on the
whole that it is much clearer than it was. The
simile is that, if the level of a sea of forgetfulness
rises over a landscape, the peaks of the
mountains that remain above the surface stand
in a clearer relationship than when you could see
all the intervening landscape.*

Konrad Lorenz
quoted in Singleton, 1983

As you will remember from Chapter 21, there is substantial controversy regarding
the trajectory of IQ throughout the years of early and middle adulthood. Some
researchers say that it falls, others that it rises, and others that it varies so much
that no general trend can be discerned. The trend for late adulthood, however, is
clear: average IQ falls with every decade.

What are the reasons for this general decrement? Even more important, how do
the declines that show up on IQ tests or in laboratory measurements affect older
people in daily life? First let us look specifically at what changes researchers have
found to occur.

Intellectual Decline in the
Processing of Information

In Schaie's Seattle Longitudinal Study, described in Chapter 21, the average scores
on some abilities fell only a bit in middle age, but beginning in the 60s, there was a
distinct downward shift. On all five of the Primary Mental Abilities—verbal comprehension, spatial visualization, reasoning, mathematical ability, and word fluency—adults aged 64 and over averaged lower than they had seven and fourteen
years before (Schaie and Herzog, 1983). Similar results have been reported by researchers at Duke University, who now have thirty years' worth of longitudinal
data on adults, some of whom are among the oldest-old (including two centenarians): the overall trend is for IQ scores to decline with each decade after age
70 (Manton et al., 1986).

Comparison of the rates of decline of various cognitive abilities shows that fluid
intelligence, including memory and abstract reasoning, falls faster than crystallized intelligence. This finding suggests that declines in cognitive functioning are
related to changes in the basic components of information-processing—**input** (the
transmission of information into the brain), **storage** (the placing of information into
memory), **program** (the organization and interpretation of information).

Input

There are a number of differences in the way information comes into the brain of an old person and that of a younger one (Verrillo and Verrillo, 1985). On the simplest level, the old person's receptors of information, especially the eyes and ears, are less adept at picking up sensory stimuli. The vision and hearing losses that become increasingly common with age mean that older people cannot even begin to process some information—like the details of a dimly lit room or a soft conversation spoken against a noisy background—because the stimuli in question go undetected.

Figure 24.1 *Young adults tend to enjoy lively, noisy parties, while older adults prefer quieter gatherings. One reason is that, in addition to coping with diminished sensory abilities, many of the elderly have difficulty processing several perceptions at once, so socializing at a large get-together with multiple conversations, a buzz of activity, and many new faces may be a strain.*

In addition, perceptual processes slow down with age. This can be demonstrated experimentally by showing a person two pictures on a screen, one a fraction of a second after the other. If the second picture follows too closely after the first, the person will "see" only one image. On such experiments, older adults need significantly more time between pictures than younger adults do in order to perceive both images (Kline and Szafran, 1975). Other laboratory research shows similar results, suggesting that, in old age, the speed at which the brain registers new information declines, so if information is presented too rapidly, a substantial portion is lost (Fozard, 1980; Poon, 1985). Thus an older adult who is searching for a friend in a busy airport, or who is trying to understand instructions from a fast-talking instructor, will be likely to miss some of the input that a younger person would receive.

The slowing of the perceptual processes may be related to another factor that impedes input in the elderly—a decrease in selective attention. Especially when older people must pay attention to several things at once, they are more likely than younger people to be distracted by irrelevant stimuli (Hoyer and Plude, 1980; Welford, 1985). Overall, then, input becomes slower and less efficient with age.

Storage

Once information is perceived, it must be stored in the memory so that it can be used. Research has shown that, in certain ways, information storage is less efficient in old age. Not all aspects of memory decline equally; nevertheless, some theorists think that memory deficits may be *the* crucial element of declines in cognition (Salthouse, 1985).

Sensory Register As you recall from Chapter 12, sensory register (also called sensory store) is the momentary afterimage, visual or auditory, that occurs in the brain after perception. For example, if you close your eyes, the image of the page you are now reading will remain in your mind for a fraction of a second; if you hear someone say something and then realize the person was speaking to you, you can usually remember what the person said even though you weren't listening at the time.

Research on whether or not the sensory register declines with age is inconclusive: it may well be that it takes longer with age for information to register in the senses, but, once it registers, it may take longer to fade. However, even if sensory register does fade more quickly with aging, the loss is minimal—certainly not enough to account for the overall decline in information-processing (Craik, 1977; Poon, 1985).

Short-term Memory Short-term memory (sometimes called primary memory), which is memory that is stored for less than a minute, shows the greatest loss with age (Poon, 1985). Laboratory research indicates that older adults are not quite as good as younger adults at remembering a short series of numbers they have just heard, and they have greater difficulty remembering the numbers long enough to repeat them backward. The same trend is seen for more meaningful material, such as the content of a paragraph or an essay (Rice and Meyer, 1986; Zelinkski et al., 1984). Older adults have more trouble than younger adults remembering the significant details of a passage they have just read or heard, and even more trouble remembering it some time later. The most important reason for this is that keeping something in short-term memory long enough for the transfer to long-term memory to occur is a particular problem for the elderly, at least on laboratory-test tasks such as these.

It is surprising to see that even a few seconds is long enough for these age-related deficits in memory to become apparent. For example, in one study (Light and Capps, 1986), younger and older adults (averaging ages 24 and 71) were asked to listen to short sentences and designate the probable antecedent for ambiguous pronouns. In each case, the context provided clear clues to the likely answer. For example, in *"Henry spoke at a meeting while John drove to the beach. He brought along a surfboard,"* it is likely that "He" refers to John, whereas in *"Henry spoke at a meeting while John drove to the beach. He lectured on the administration,"* it is likely that "He" refers to Henry. On tasks such as these, older and younger adults scored equally high.

However, when a sentence or two separated the pronoun and its antecedent—as in *"Henry spoke at a meeting while John drove to the beach. It was a nice day, and there was the sound of activity in the streets. He brought along a surfboard"*—older adults were significantly less adept at figuring out who "He" referred to, because the intervening sentence made them forget who did what in the first sentence. As the authors of this study point out, one general conclusion that can be drawn is that "under conditions of high memory load . . . , memory problems may masquerade as comprehension problems." If memory overload or interference (here caused by the intervening sentences) is not a factor, then older adults understand the meanings just as well as younger adults.

A critical element that causes decline in memory is the processing time required to get information into temporary storage. For example, one study tested subjects of various ages on their ability to recall forty-eight slides (half of them pictorial, half of them single sentences) presented at the rate of one every 8 seconds. College-age subjects performed the best, elderly subjects, the worst. However, when the older adults were given 15 seconds per slide, their performance increased markedly. This suggests that older adults may need more time to actively process various items to be remembered, but that, given that time, their thinking processes can work quite well (Pezdek and Miceli, 1982).

Other research also shows that the time it takes older adults to get information into storage is a critical variable in explaining short-term memory difficulties. In one variation of the chess-game experiment described in Chapter 18 (Charness, 1981), chess players of varying ages but equal ability were asked to watch a chess game for a few seconds and then write down the positions of all the pieces. When the subjects wrote down the positions immediately after turning away from the game, younger adults did significantly better than older adults. However, when the subjects were made to wait 15 seconds before writing down the positions, the rate of forgetting was no greater for the older adults than for the younger ones. Since the memories of the older adults did not fade faster than those of the younger adults after the 15-second interval, the author of the study concluded that difficulty getting the positions of the chess pieces into short-term memory, rather than trouble keeping them there, was the main reason for the older adults' poorer showing on the first test.

Figure 24.2 *Judging from the superior positioning of the white pieces, the younger player will soon lose. This should come as no surprise: the caution, experience, and especially the strategy typical of an older player are rewarded in a game like chess, while impatient risk-taking, typical of younger players, leads to defeat. To even the odds, this pair could set a time limit of a move every minute. Then the slower cognitive processes, including memory, of the older man would become a substantial handicap.*

Long-term Memory   Compared with short-term memory, long-term memory, especially for information stored years ago, is quite good for older adults. Of course, definitive research comparing this type of long-term memory in older and younger adults is difficult (Erber, 1982; Poon, 1985), because verification of personal recollections is almost impossible, especially on a large, representative sample. One approach has been to compare people's memories of public events. But this approach may give an advantage to adults of one age group or another, depending on the particular question. For example, asking for the date of Franklin Delano Roosevelt's death might favor people who were politically aware in 1945, whereas asking for the names of the Beatles may favor those who were teenagers in the 1960s. Since the interests and experiences of older and younger cohorts vary a great deal, it is hard to find any item that should be equally accessible from [the] memory of the average 70-year-old and the average 20-year-old.

Nonetheless, when attempts are made to compare long-term memory a[cross] groups, the results have been quite favorable to the aged (Botwinick and St[one,] 1980). One study compared how well Americans of various ages remembered [the] Spanish they had studied in high school or college, and had used very little since. As would be expected, those young adults who had studied Spanish within the past three years remembered it best, but thereafter the amount of forgetting was very gradual—with the elderly who studied Spanish fifty years earlier remembering about 80 percent of what young adults who had studied it a mere five years earlier remembered. The crucial variable in the degree of language retention was not how long ago a person had studied Spanish but how well the person had learned it in the first place: those who got As in Spanish fifty years earlier outscored those who got Cs a mere twelve months before (Bahrick, 1984).

Program

The research on memory points to a third crucial aspect of information-processing—the individual's program, the way one characteristically organizes and connects new information with old. For example, what do you do when you want to learn the name of the new head of a foreign country in which the names are unfamiliar and thus hard to remember? Do you connect the sounds or the letters of the name with something else you know well? Do you rehearse the name, saying it out loud to yourself several times, perhaps repeating the name of the country as well in order to link the two mentally? These are among the many ways to encode new information. When you are unable to spontaneously retrieve information from memory, do you begin a systematic "search" for it, or do you just wait, hoping it will spring to mind?

By adulthood, everyone has many characteristic processes, or programs, for assimilating and accessing knowledge. Such mental structures and strategies may be quite different in older and younger people, and this may account for some of the differences in cognition that we find. At least in laboratory research, the memory strategies used by older people to encode and retrieve information are less efficient than those used by younger people. For example, older adults are less likely to rehearse new information, or to chunk items together, or to try to link a new piece of information with something familiar (Bruce et al., 1982; Rice and Meyer, 1986).

Figure 24.3 *Over the years, interests, areas of expertise, and personal style contribute to a person's characteristic ways of organizing and understanding information— ways that might be useless and even incomprehensible to others.*

"A penny for your thoughts."

Drawing by Chas. Addams; © The New Yorker Magazine, Inc.

Problem-Solving   A similar difference in the use of strategies is found when adults are given abstract problems to solve. Older adults are not as good at abstract problem-solving as younger adults are, primarily because their strategies are less efficient, and they tend to follow their first hunches rather than systematically reasoning out a problem (Arenberg, 1982; Reese and Rodeheaver, 1985).

One standard test of problem-solving ability is the game of Twenty Questions. As you probably know, the object of this game is to guess the person, place, or thing the other player has in mind, using no more than twenty questions that can be answered "yes" or "no." The best strategy is to begin with broad, general questions and then ask increasingly specific ones, progressively narrowing the categories to which the object might belong. Asking if the object is animate or inanimate, human or nonhuman, living or dead, male or female, and so forth, is clearly a better problem-solving method than starting off with specific guesses such as "Is it this chair?" "Is it Uncle Glen?" "Is it Canada?" Laboratory tests on Twenty Questions reveal distinct age-related differences in the logical strategies used to discover the secret object (Denney and Palmer, 1981). Children, for instance, tend to guess impulsively and redundantly; teenagers and young adults are much more systematic in their approach. However, as Denney and Palmer found, in concurrence with many other studies of problem-solving, skill level falls as adulthood progresses. Older adults are less adept at logically narrowing the possibilities than are middle-aged adults, who themselves are less skilled than younger adults. This finding is one of many that have been taken to prove that problem-solving ability declines with age, beginning in early adulthood.

Similar results were found in a more specific test of deductive reasoning and problem-solving. Young, middle-aged, and older adults were asked to figure out which foods at an imaginary buffet banquet made some of the diners ill, ostensibly because someone had deliberately poisoned some of the food (Hartley, 1981). The subjects were given the menus of all the guests, each of whom had selected one of four appetizers, one of four main courses, and one of four desserts, thus choosing one of sixty-four possible meals. The menus were so constructed that the only way to solve the mystery was to figure out which of the food items had not been tampered with.

To ensure that the experiment was a test of reasoning rather than of some other variable, the author of the study took care to control for possible declines in memory and speed of processing: all the information remained available while the person was trying to solve the problem, and there was no time limit. Younger people tended to be much better at this task than older people: of the younger adults, 85 percent solved the problem correctly; of the middle-aged adults, 70 percent did; of the older adults, only 54 percent picked the correct combination of poisoned portions. Those of the elderly who failed tended not to weigh all the evidence as carefully as they should have, leaping to a wrong conclusion or saying that there was no correct answer.

The researcher concluded his report: "Clear differences in deductive inferences during problem-solving are consistent with the claim that reasoning ability declines with age." However, like almost all research on problem-solving, this study was cross-sectional. Acknowledging this limitation, the author added, "More precisely, ability declines as a function of some factor indexed by chronological age."

Reasons for Age-Related Changes

Of course, the declines in cognitive functioning that occur with age may in many ways be tied to the aging process itself—a result of the cellular changes or programmed senescence described in the previous chapter. However, there are many

Figure 24.4 *Practice, or the lack of it, is one reason cognitive abilities remain strong in some of the elderly and fade in others. Dr. Eileen Gersh, shown here in her 70s, continued to teach college biology as she had done much of her life, and published her first book,* The Biology of Women, *after her 70th birthday.*

other factors that are relevant to cognitive performance and indexed by chronological age, including motivation, practice, and specific learning experiences. Two of the most important are health and education.

Health Virtually every study has found that good cognitive functioning and good health are positively correlated (Labouvie-Vief, 1985); middle-aged and older adults in good physical health have a clear intellectual advantage over their less well contemporaries. Longitudinal research also shows that intellectual quickness and acuteness fall when health fades, or, more accurately, that they follow the same trajectory. In fact, some research suggests that a drop in cognitive functioning precedes a drop in physical health (see A Closer Look, page 554).

Rather than speculating about whether good cognitive functioning depends on good physical health or vice versa, it is probably more valid to assume that each affects the other. Thus, keeping one's mind active generally helps maintain one's health, and keeping one's body active generally helps maintain one's mind. As the Duke researchers found from their longitudinal data spanning several decades, "excellent cognitive performance" was sustained even in the 80s "in persons who retained their physical and mental health" (Manton et al., 1986).

One category of health problems that seems to be particularly detrimental to cognitive functioning involves the circulatory system. People who have heart disease or untreated hypertension, for example, also tend to show reduced intellectual ability, probably because the same circulatory problems that caused those conditions also affect delivery of blood to the brain (Eisdorfer, 1977; Schaie and Willis, 1986). Although most research on the intellectual abilities of older adults screens out those who are known to have cardiovascular disease, it is quite possible that undetected slowdowns in the cardiovascular system may affect untold numbers of older research subjects. Indeed, as one study concludes, "because it has been estimated that approximately 50 percent of older adults have some sort of cardio-vascular problem, it may be that much of the general, age-related, memory decrement typically observed in memory research is due to the undetected presence of mild cardio-vascular problems in the older group" (Berrett and Watkins, 1986).

RESEARCH REPORT   Terminal Decline and Death

Several longitudinal studies of the elderly have tested subjects' cognitive abilities at frequent intervals and compared the scores of those who died in between testings with those of the subjects who were still alive. Such research has shown a telling pattern: people who are going to die fairly soon perform less well on various tests of intellectual functioning than do people who are not approaching death. This is true for elderly patients in nursing homes (Kleemeier, 1962; Lieberman, 1965) and for residents of senior-citizen housing (Botwinick et al., 1978), as well as for the population at large (Riegel and Riegel, 1972). Even when those who will die in the near future are matched on age, sex, education, social class, and health with those who will not soon die, their thinking has been shown to be less complex and reflect greater depression.

In fact, some studies show a *terminal decline,* that is, a noticeable dip in intellectual functioning that predicts death as well as, or even better than, traditional measures of physiological health do. For instance, in one study (Lieberman, 1965) twenty-five residents of a nursing home were tested repeatedly over a year on two tasks, copying eight simple designs (the Bender-Gestalt test) and drawing a person. Eight of the residents died within three months after completion of the series of tests, and the other seventeen were still alive a year later. Six of the eight who died performed notably less well on the last tests they took than on the first; and in some cases the changes were dramatic (the following drawings show one man's decline over the six months preceding his death).

By comparison, only three of the seventeen survivors dropped in their scores on the Bender-Gestalt, and only two showed less complexity in their drawings of a person.

The researcher notes that neither chronological age nor physical illness differentiated the two groups as well as the tests did. In fact, some of the survivors experienced life-threatening illnesses during the period of study, but their scores did not usually show the same decline as the scores of those who were soon to die.

Similar trends were found in a study of 380 men and women who were representative of the general population (Riegel et al., 1967; Riegel and Riegel, 1972). All

the subjects were given a battery of cognitive and psychological tests. Five years and ten years later, as many subjects as possible were retested. Again, a terminal drop was noted, not only in tests of intelligence but also in measures of personality. Those who were to die were less tolerant and more rigid than those who were to continue living. In addition, of those who refused to be retested at the second testing, 43 percent had died by the time of the third testing. Of those who willingly were retested, only 25 percent had died by the third testing. People under age 65 who refused to be retested were especially likely to die, compared with their more willing contemporaries. This suggested to the authors of this study that psychological as well as cognitive factors are predictive of death. In fact, they go so far as to say that "at lower age levels nonsurvivors can be described as a subgroup which differs sociopsychologically from the rest in almost the same sense as victims of cancer or heart attacks may differ from healthy persons" (Riegel et al., 1967).

Similar conclusions can be drawn from another study that examined the relationship between cognitive and psychological functioning and death (Botwinick et al., 1978). In this case the subjects were 380 residents of two housing projects for senior citizens. They were given a battery of tests twice, a year apart, and then the scores of the 83 people who died within the next five years were compared with those of the people who survived. No terminal decline per se was observed, probably because two tests taken a year apart were not sufficient to spot it. But when the deceased were matched with survivors of the same age, sex, prior health, and so forth, there were significant differences between the two groups on fourteen measures. In fact, 70 percent of those who died had scores below a certain point, and 63 percent of the living were above that point. Six of the fourteen predictive measures were quite simple tests of psychomotor ability, such as the Bender-Gestalt, and five of the measures were directly psychological, assessing such characteristics as depression and neuroticism.

The implication of these studies is that cognitive functioning and emotional state may foretell a person's death as well as, or better than, measurable physical condition does. If an elderly person declines in intellectual sharpness and zest for life, steps should be taken to improve the level of functioning, just as steps are taken to correct blood pressure if it becomes too high. Riegel believes that this may apply at every age, and suggests that better understanding of the relationship between physical, cognitive, and psychological health, and more cognitive stimulation at every point in the life span, would add years to all our lives (Riegel and Riegel, 1972).

Another proven detriment to cognitive functioning is impaired hearing. Oddly enough, virtually no studies of intellectual performance in the aged screen out those who are somewhat hard of hearing (Obler and Albert, 1985). Other perceptual disabilities, such as poor vision, may also affect cognition, although that is not known.

Education On many cognitive measures, years of education are a more powerful influence than age. In general, 70-year-olds with advanced college degrees will outscore much younger high-school dropouts, and sometimes even middle-aged high-school graduates. This obviously has important implications for cross-sectional studies, since, for example, less than 10 percent of older adults in the United States have completed four years of college, while more than 25 percent of those under age 40 have done so (U.S. Bureau of the Census, 1986). If cross-sectional research does not take years of education into account, older adults will undoubtedly perform less well than younger adults, for reasons that may have nothing to do with age per se.

Even if years of education are held constant, many studies compare older adults who attended college decades ago with younger adults who are currently in college. Obviously, since the current students are using learning strategies and memory techniques on a regular basis, they are likely to be more practiced in these intellectual skills than older adult are. Indeed, when researchers compare the learning strategies of students curr ntly in college with those of nonstudents of the same age, past education, and ability, they find that the strategies of current students are much more efficient, both for understanding and remembering new material (Ratner et al., 1987; Zivian and Darjes, 1983).

Figure 24.5 *The academic performance of elderly students suggests that once they are back in the classroom, they soon remaster learning strategies they may not have had to use for forty or fifty years.*

These two factors, health and education, are certainly part of the explanation for the declines in cognitive ability as measured by the various psychological tests and laboratory procedures we have examined. However, even when great care is taken to compare healthy older adults with younger adults of equal education, some of the declines remain. The critical question is, however, how important these declines really are.

Cognitive Functioning in the Real World

When older people are asked about their memory, most readily acknowledge that it is not as keen as it used to be, corroborating to some extent the laboratory research that finds marked deficits in short-term memory with age. However, very few of the elderly consider memory loss even a minor handicap in daily life (Sunderland et al., 1986). What's more, when older adults are asked about their problem-solving ability, they say that they are better at solving "everyday problems" than they used to be (Williams et al., 1983). How do we explain this?

Memory

In the typical laboratory experiment on memory, subjects are given items to remember within a specific time, and then are tested on the accuracy and speed of their retrieval processes. In order to control for the unfair advantage that would result if a subject were very familiar with, or unusually interested in, the material to be remembered, memory tests usually consist of items that are fairly meaningless, perhaps a string of unrelated words or numbers. This control, in combination with time limits, also has the effect of making it difficult for subjects to repeat and review the test material. This is particularly deleterious to older adults, for they are more likely than their younger counterparts to benefit from repeated practice. Indeed, practice in various cognitive tasks can bring the performance of older people beyond that of younger, less practiced people (Beres and Baron, 1981; Bruce and Herman, 1986).

The typical memory experiment may also put the elderly at a disadvantage with respect to spontaneous **priming.** Priming refers to the fact that bringing one item forth from memory makes it quicker and easier to bring forth other, related items. We all use priming in daily life: if you misplace your wallet, for instance, you may mentally review the activities of your day, hoping to trigger a memory of the place you put your wallet last; if you forget someone's name, you may try to recall other things about that person, or go through the letters of the alphabet, until the name comes to you. Most memory experiments, however, are constructed to exclude much opportunity for spontaneous priming, thereby depriving older adults of a cognitive tool that is very useful to them in daily life (Howard et al., 1986), and one that shows no age-related deficits (Rabinowitz, 1986).

Finally, motivation may be a critical variable in tests of memory. Young college students might have an advantage in a laboratory, because, still being in what Schaie described as the acquiring stage of cognitive development (see Chapter 18), they are accustomed to learning material that is not immediately relevant to their lives. Older adults may be much more likely to question the purpose of such learning, and therefore may try less hard to commit test items to memory. In daily life, however, older people, especially if they are aware of possible memory problems, are more motivated to work at remembering by using memory aids (Perlmutter, 1986).

In many cases, the cumulative effect of practice, priming, and motivation offsets the pure memory loss that laboratory research has so often found. One study designed to mimic the memory demands of daily life illustrates this point well. Older and younger adults (all living busy lives) were asked to call a telephone answering service every day for two weeks, at a designated time they had picked as conven-

Figure 24.6 *Representative Claude Pepper, age 88, has remained one of the most powerful members of Congress partly because his area of expertise involves issues concerning the elderly. His commitment and knowledgeability in this area have given him not only a powerful constituency, but also strong motivation in the performance of his role as spokesman for both the old and the young. As Pepper has said of himself: "I am as full of ideas about what to do in the future as a dog has fleas."*

ient. One reason this activity was chosen as the basic test procedure is that remembering appointments is something everyone must do in daily life, and is also something many older people say they have difficulty with. The researchers were surprised to find that only 20 percent of the younger adults remembered to make every call, and that 90 percent of the elderly did.

Why the dramatic difference? Younger adults, it seems, were likely to put excessive trust in their memories ("I have an internal alarm that always goes off at the right time") and therefore were less likely to use mnemonic devices. Older adults, with a heightened awareness of the vagaries of memory, used reminders, such as a note on the telephone or a shoe near the door, and thus almost always called on time.

The experimenters then attempted to increase the rate of forgetting in the older adults by asking for only one call per week, at a time designated by the researcher, and by making the subjects promise to avoid using any visible reminders of the appointment. This time the results were, indeed, much different: about half the elderly and an equal proportion of the young missed calling at the appointed time. And it seems likely that the number of old people who forgot the appointment might have been greater, since some of the older subjects continued to use some memory-priming measures of sorts, carrying the phone number in a visible place in their wallet, for instance, despite instructions to keep it out of sight. One of the researchers concludes:

> With more effort, we are sure we can bring old people's memory to its knees . . . , but that hardly seems to be the point of this research. The main lesson of this venture into the dangerous real world is that old people have learned from experience what we have so consistently shown in the laboratory—that their memory is getting somewhat poorer—and they have structured their environment to compensate. [Moscovitch, 1982]

Learning in Late Adulthood

The evidence that older adults are much more cognitively skilled than the results from laboratory research would indicate raises an important question: Can older adults be taught to remedy some of the deficits that once seemed inevitable?

Experimental attempts to teach older people skills that will improve their memories have had mixed results (Erber, 1982; Labouvie-Vief, 1985). In some studies, short-term memory abilities seem to improve; in others, they do not; in still others, they improve temporarily and then decline again.

However, attempts to teach older adults more general cognitive skills have had much more success. Research projects with two quite different sample populations illustrate this point. One population was made up of people who, overall, were much healthier and better educated than the population as a whole—participants in Schaie's Seattle Longitudinal Study. In 1984, Schaie retested 229 individuals between the ages of 64 and 94 on the same five Primary Mental Abilities they had been tested on seven and fourteen years before. On two of the five abilities, spatial understanding and reasoning ability, scores of almost half (47 percent) of the group had remained relatively stable while the other half fell significantly on one or both of the abilities.

After the 1984 testing, all the participants were given five sessions of individual tutoring in strategies that are useful in solving problems involving these abilities (Schaie and Willis, 1986). Everyone was then retested. Average scores improved on both abilities, for those who had previously remained stable as well as for those who had declined. In fact, 40 percent of those who had previously shown a decline improved so much that they were at least at the level they had scored fourteen years before. In other words, the seemingly inexorable decline in basic mental abilities had been reversed through training—training that lasted only five hours!

The other sample population involves a group who are among the least healthy and least educated of the elderly—residents of nursing homes. To understand the sig-

nificance of this study, you need to know that many nursing homes encourage passive, dependent, and predictable behavior, and discourage behavior that is active, independent, or innovative. For example, elderly residents who do not manage their own personal care or hygiene—who, say, just sit staring at their food when it is placed before them—are likely to receive help and attention from the staff; those who manage for themselves are likely to be ignored. Similarly, those who stick to the nursing home's schedules and routines are much more likely to be praised than those who, against the rules, attempt to get a midnight snack, or want to go shopping on a day not designated as a shopping day, or try to keep a pet in their room. When elderly patients ask for an explanation of their being given some medicine or therapy, or, worse, refuse to cooperate with some aspect of their medical treatment, they are likely to be labeled as mentally impaired and disruptive, and to be treated accordingly. (Similar behavior in younger persons might be looked on as a sign of alertness.) One recent review sums up the picture in many nursing homes with frightening clarity:

> . . . the individual . . . gives up control over the most mundane daily activities, when to sleep, wake, visit, perform toileting activities, bathe, and shop. The patient is exposed to infantilization and numbing bureaucratic and health routines that are of obscure purpose due to the invariably poor communication and misinformation given to placate the patient. Information is withheld or distorted under the assumption that it will not be understood or well-tolerated by the patient. [White and Janson, 1986]

Research has shown that when the opposite approach is used, and nursing-home patients are encouraged to manage on their own as much as possible, many in fact can learn to take more control of their activities, developing their own schedules and social lives as well as becoming more responsible for their daily care (Piper and Langer, 1986).

Other research shows that nursing-home residents can learn to improve their perspective-taking skills. One commonly cited characteristic of nursing-home residents is their egocentrism: on Piagetian tasks, they often show themselves to be limited to their own perspective; in daily

Problem-Solving   Research on reasoning ability also shows that problem-solving in daily life is far less impaired with age than problem-solving in the scientist's laboratory (Cavenaugh et al., 1985; Reese and Rodeheaver, 1985). For example, after giving adults of various ages the test of Twenty Questions, and finding, as expected, that the youngest adults did best, Denney and Palmer (1981) gave these same adults problems from real life—such as what to do if you are caught in a blizzard, or if your refrigerator feels warm, or if your 8-year-old daughter is not home from school

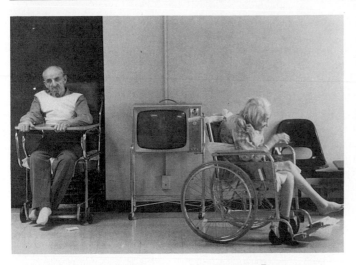

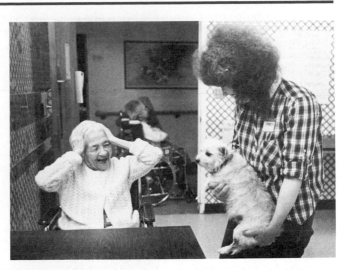

When nursing-home residents are isolated and dependent (note the physical restraints that prevent the people on the left from trying to stand), their cognitive abilities are likely to *deteriorate. By contrast, residents can find zest in life when they have the opportunity to socialize and to maintain their interests or develop new ones. In some nursing* *homes, "visiting pet" programs provide the elderly with an undemanding companionship that can lighten spirits and improve cognitive and social functioning. These lessons* *learned from the minority of the elderly who are in nursing homes apply also to the majority of the elderly in the community at large.*

life, they typically seem greatly absorbed in their own problems and needs, with little awareness of those of their fellow residents. The goal of one research project was to reduce this egocentrism (Zaks and Labouvie-Vief, 1980). The researchers tested the ability of nursing-home residents to understand other points of view and then divided the residents into three groups that were equal in terms of age, health, and performance on the tests. One group was given special training in social understanding. The members participated in discussions and role-playing, centering on problems that might occur in the home (such as what could be done if one roommate liked to watch television late at night and the other liked to sleep). A second group discussed such problems but did not role-play. The third group, the control group, had no special training at all.

On retesting after the training period, the residents who had had the most active social-learning training were markedly less egocentric than the other two groups, and they improved in their ability to communicate with each other. As the authors of the study conclude, a substantial part of the egocentrism of older institutionalized adults may be caused by their lack of social interaction, rather than by their cognitive inability to see other points of view.

Together, these two research projects—one with the well elderly and one with the frail—lead to a clear conclusion. Overall, when cognitive deficits appear in the elderly, trying to find a way to remedy them is a better approach than sympathetically accepting the deficits as inevitable.

when you expect her. On these problems, the highest scores were attained by adults in their 40s, with only a very gradual decline in the scores after that. Adults in their 60s scored only slightly below the adults in their 20s.

Other studies using real-life contexts have found similar results. For example, in a study comparing young, middle-aged, and elderly adults (average ages 19, 46, and 68) on answers to complex real-life problems (such as how to solve a hostage dilemma), the oldest were the most dialectical and least mechanistic in their solu-

Figure 24.7 *The idea that reasoning and moral judgment may remain strong even as the body weakens is illustrated in many real-life cases. Popes, heads of state, and Supreme Court Justices (shown here) often deliver their most memorable and powerful pronouncements when they are over age 70.*

tions (Kramer and Woodruff, 1986). Another study compared levels of moral thought in two groups of adults, one between the ages of 30 and 49, and the other between the ages of 63 and 85. When years of education were taken into account, the two groups showed no age-related differences in moral thinking overall. However, the elderly were better at dilemmas that involved an older person, such as what should be done with an elderly man who lived alone on a fixed income and who occasionally shoplifted items of food (Chap, 1985–1986). This difference suggests, once again, that whether a person finds cognitive tests interesting or personally relevant affects his or her performance on them.

Of course, even these examples are artificial. What about the real-life problems that people actually face? One researcher asked 405 people how they coped with their problems (McCrae, 1982). Contrary to the popular stereotype of the elderly as rigid, stubborn thinkers, the older people were as likely as younger adults to think rationally about their problems, to seek help, and to express their feelings to someone else. At the same time, their thinking differed from that of younger people in several constructive ways: they were less likely to use escapist fantasies, to react with hostility, or to blame themselves.

When all the evidence is considered, then, it seems that there is no doubt that information-processing of all kinds slows down in late adulthood (Hale et al., 1987; Herzog et al., 1986), and that short-term memory and abstract problem-solving ability probably decline in other ways as well. However, an older adult's ability to cope with the cognitive demands of daily life may not be significantly impaired by these changes. While declines are observable for most people in their 60s, and for virtually everyone by age 80, these declines are limited in scope and severity, are more apparent in the laboratory than in daily life, and are affected a great deal by factors other than age—health, education, and specific learning experiences among them.

In short, the picture of cognitive ability in old age appears to be quite good—with one notable, heartrending exception: dementia.

Dementia

In ordinary conversation, pathological loss of intellectual ability in elderly people is often referred to as "senility." However, this term inaccurately emphasizes the factor of age. A better and more precise term for pathological loss of intellectual functioning is **dementia**—literally, demented or severely impaired thinking. Although dementia is more likely to occur among the aged, it can occur before old age. At the same time, it is by no means an inevitable occurrence of old age, even among the very old (see Table 24.1).

TABLE 24.1 **Estimated Prevalence of Dementia**

Age	Prevalence
65–69	2%
70–74	3%
75–79	6%
80–84	12%
85–89	22%
90 +	41%

Based on studies in Japan, Australia, New Zealand, Britain, Sweden, and Denmark (Preston, 1986).

Dementia can be caused by a variety of diseases and circumstances, but the general symptoms are similar: severe memory loss, rambling conversation and language lapses, confusion about place and time, and changes in personality. Traditionally, when dementia occurred before age 60, it was called *presenile dementia;* when it occurred after age 60, it was referred to as *senile dementia* or *senile psychosis.* This age-based distinction was quite arbitrary, for the symptoms, causes, and treatments are the same no matter what the person's age.

Stages of Dementia

Dementia is usually progressive, with identifiable stages (Reisberg et al., 1981). It begins with a general forgetfulness, particularly of names and places. A common problem is putting something away and then forgetting where it is.

In this early stage, most people recognize that they have a memory problem and try to cope with it, writing down names, addresses, appointments, shopping lists, and so forth, much more than they once did. This first stage is often indistinguishable from the "benign forgetfulness" that many older people experience, according to a leading researcher (Reisberg, 1981). Many people reach this first stage and remain somewhat forgetful for the rest of their lives, but never get any worse (Roth et al., 1985).

In the second stage, more general confusion occurs, and there are noticeable deficits in concentration and short-term memory. People at this stage are often aimless and repetitious in conversation, and frequently mix up words, using "tunnel" when they mean "bridge," for instance. They are likely to read a newspaper article and forget it completely the next moment, or to put down their keys or glasses and within seconds have no idea where they could be. Personality changes also occur, with the individual becoming somewhat withdrawn and out of touch— except for occasional overreactions, such as a sudden outburst of temper or tears.

The accumulation of symptoms in the second stage makes it quite clear to others that something is seriously amiss, but most of the people who are in this stage of dementia deny that anything is wrong, to themselves and to others. When asked a question that tests their memory, say, "Who is president now?" they are likely to answer evasively and defensively, with something like "I don't follow politics any more" (Roth et al., 1985). One woman, a lawyer, insisted that it was obvious she had no serious memory loss: "Otherwise I couldn't work as an attorney. I couldn't go to court. I couldn't prepare a case." However, as her husband and law partner explained the actual situation:

> She doesn't really have clients anymore. I let her come to the office, but she hasn't tried a case in years. . . . She keeps misplacing and losing documents. But I'd rather have her at the office. It makes her feel better. She's always worked, she wouldn't know what to do at home. [quoted in Reisberg, 1981]

The third stage begins when memory loss becomes truly dangerous and people are no longer able to take care of their basic needs. They may take to eating a single food, like bread, exclusively, or they may forget to eat at all. Often they fail to dress properly, going out barefoot in winter or walking about the neighborhood half-naked. They are likely to turn away from a lighted stove or a hot iron and completely forget about it for the rest of the day. They might go out on some errand and then lose track not only of the errand but also of the way back home.

Eventually, demented persons need full-time care. They not only cannot take care of themselves; they also do not respond normally to others, sometimes becoming irrationally angry or paranoid. At the end, they no longer can put even a few words together to communicate, and they fail to recognize their closest relatives.

While these general symptoms and stages are the same for all forms of dementia, there actually are several distinct types, each with different risk factors and treatment. We will discuss each in turn.

Alzheimer's Disease

Approximately 70 percent of all people who suffer from dementia are afflicted with **Alzheimer's disease,** a disorder characterized by the proliferation of certain abnormalities in the cerebral cortex, called plaques and tangles, that destroy normal brain functioning (Roth et al., 1985). Until recently, the term "Alzheimer's disease" was reserved for symptoms of dementia in people under age 60, while the same symptoms in people over age 60 were termed "senility." However, new techniques for analyzing brain tissue on autopsy show that no matter what the age of the victim, the brain damage takes the same form, and that the amount of plaques and tangles correlates, not with the victim's age, but with the degree of intellectual impairment before death.

There are some age-related characteristics, however. When Alzheimer's disease appears in middle age, it usually progresses more quickly, reaching the last phase within three to five years, while in late adulthood it can take ten years or more to run its course. Alzheimer's disease in middle age is relatively rare. By late adulthood it affects one in every twenty adults (mostly over age 75), altogether affecting an estimated 3 million people in the United States. This number is expected to rise dramatically in the next decades, when the baby boomers reach late adulthood (Holden, 1987).

While Alzheimer's disease does not usually appear until late adulthood, the disease probably originates decades earlier. In fact, one form is inherited, caused directly by a dominant gene on chromosome 21. Other forms may be multifactorial, caused by the combination of a genetic vulnerability and other life circumstances—although no one knows what these circumstances are (St George-Hyslop et al.,

Figure 24.8 *"What day is it? Where am I?"—These are two questions that are frequently asked by Alzheimer's victims. In the early stages of this disorder, memory for dates, places, and names fades much more quickly than memory for basic skills, such as reading, so visual aids like the one shown here are likely to help when memory falters.*

1987). It is believed that one likely causative agent might be a "slow" virus, perhaps one contracted in childhood but kept at bay by the immune system until old age; another might be the cumulative effect of toxins derived from one's diet or the environment.

Despite intense research efforts to control Alzheimer's disease with special diets or drugs, at the moment nothing seems to prevent the disorder or slow its progress. The victim of Alzheimer's disease is powerless: inexorably, day by day, month by month, memory fades, until the individual cannot remember how to do something as simple as using a fork; cannot recognize loved ones; cannot even control body functions.

The victims of Alzheimer's disease include not only the patient but the patient's family, who typically are the main source of care until the last stages of the disease. It is particularly hard for family members to be understanding and patient with a person who seems in good health but is unable to behave or think normally, and who sometimes becomes angry at efforts to help (Deimling and Bass, 1986). Almost always, the time comes when the family must seek help in care-giving, but paying for home-care aides or for nursing-home care is usually problematic: few insurance plans cover the long-term care of those with Alzheimer's disease.

The burden that Alzheimer's disease places on the family makes it difficult not to blame the victim. One man writes movingly about his feelings of helplessness as his father shouted at his mother to "think harder, think harder" when she forgot where she had placed something. She, sufficiently early in the disease to be distressed at his shouting and at her loss of memory, wandered about the house close to tears searching randomly for whatever she had lost (Sayre, 1979). Understandably, for family members, feelings of sadness and anger are often intermingled with feelings of guilt.

Younger family members suffer in another way as well. Knowing that Alzheimer's disease is, at least sometimes, inherited, inevitably makes them wonder if they, too, will be victims some day. Although genetic mapping may soon make it possible for some of them to find out for sure if they will develop the disease, probably many of them would rather not know unless a cure can be found.

Multi-infarct Dementia

The second major type of dementia is **multi-infarct dementia (MID)**, which by itself is responsible for about 15 percent of all dementia, and in combination with Alzheimer's disease, accounts for another 25 percent. MID occurs because an *infarct*, or temporary obstruction of the blood vessels, prevents a sufficient supply of blood from reaching an area of the brain. This causes destruction of brain tissue, in what is commonly called a stroke, or ministroke, often so small the person is unaware that it has occurred.

The underlying cause of such obstructions is general arteriosclerosis (hardening of the arteries). People who have problems with their circulatory systems, including people with heart disease, hypertension, numbness or tingling in their extremities, and people with diabetes are at risk for arteriosclerosis and MID. Therefore, measures to improve circulation, such as exercise, or to control hypertension and diabetes, such as diet and drugs, help to prevent MID and to slow or halt the progression of the disease if it occurs.

The progression of MID is quite different from that of Alzheimer's disease (see Figure 24.9). Typically, the person with MID shows a sudden drop in intellectual functioning following an infarct. Then, as other parts of the brain take over some of the functions of the damaged area, the person becomes better.

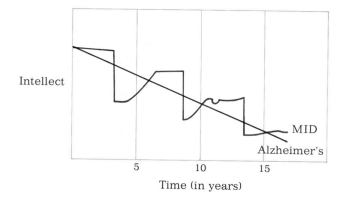

Figure 24.9 *As shown on this chart, cognitive decline is apparent in both Alzheimer's disease and multi-infarct dementia. However, the pattern of decline for each disease is different. Victims of Alzheimer's disease show steady, gradual decline while those who suffer from MID get suddenly much worse, improve somewhat, and then experience another serious loss.*

However, as the name of the disease denotes, multiple infarcts typically occur, making it harder and harder for the remaining parts of the brain to compensate. If heart disease, major stroke, or other illnesses do not kill the MID victim, and ministrokes continue to occur, the person's behavior eventually becomes indistinguishable from that of the person suffering from Alzheimer's disease. On autopsy, however, it is clear that parts of the brain have become destroyed while other parts seem normal; the many plaques and tangles characteristic of Alzheimer's disease are not present.

Other Organic Causes of Dementia

Several other diseases also can cause dementia. The best known is **Parkinson's disease,** which often produces dementia as well as the distinctive rigidity and/or tremor of the muscles. Parkinson's disease is more likely to strike the aged than the young, but it is not exclusively an old person's disease: an estimated 8 percent of the victims are under age 40 when the disease is first diagnosed (Lewin, 1987). Parkinson's disease is related to the degeneration of neurons in an area of the brain that produces dopamine, a neurotransmitter essential to normal brain functioning. The underlying reason for this degeneration is not known. Among the factors implicated as contributors to Parkinson's disease are genetic vulnerability and certain viruses. An interesting finding is that dementia in Parkinson's is not usually evident until the destruction of brain cells has reached a certain threshold. It is likely that forgetfulness and confusion are caused when, and only when, the normal ability of the aging brain to compensate for neuron loss is overloaded (Roth et al., 1985). Other diseases that can result in dementia are Huntington's disease, Pick's disease, and AIDS.

Another cause of dementia is Down syndrome. Individuals with Down syndrome who survive to adulthood often develop the same patterns of memory losses and personality changes that occur in victims of Alzheimer's disease (Kolata, 1985). Since Down syndrome usually involves an extra chromosome at pair 21, the same location where the genetic defect for Alzheimer's disease is located, the same underlying problem probably causes dementia in both cases (Barnes, 1987a).

Brain tumors can also cause dementia, as can head injuries that result in an excess of fluid pressing on the brain. In these cases, surgery can often remedy the problem and restore normal cognitive functioning.

Nonorganic Causes of Dementia

It is not uncommon for the elderly to be thought to be suffering from one form or another of brain disease when, in fact, their "symptomatic" behavior is caused by some other factor such as medication, alcohol, or depression.

Medication and Confusion   Drug-related changes in intellectual functioning are common in the elderly, for several reasons. First, the appropriate dose of most prescription drugs is usually determined by tests on younger adults, yet the correct dose for a 30-year-old may be an overdose for the elderly, whose physiological ability to get rid of excess drugs is impaired. Not surprisingly, adverse reactions to various drugs are at least twice as common in adults over age 60 as they are in younger adults (Vestal and Dawson, 1985).

Second, many of the elderly, both at home and in institutions, take several prescription drugs every day, and a number of the drugs commonly taken (such as most of those to reduce high blood pressure, or combat Parkinson's disease, or mitigate pain) can, by themselves, slow down the mental processes (La Rue et al., 1985). The intermixing of drugs can also have a deleterious effect. One survey found that 22 percent of the elderly took six or more drugs, often using drugs that interact with each other, exacerbating or negating their effects (Krupka and Vener, 1979).

In addition, malnutrition (see Chapter 23), especially deficiencies of B-vitamins and folic acid, can cause confusion and loss of memory, and can further exaggerate the effects of various drugs. Finally, if an older adult is already forgetful, he or she may take several pills on one day and none the next, compounding any drug-induced mental disorder.

Alcohol   One drug known to have serious effects on memory and other aspects of functioning is taken by an unknown number of the elderly—alcohol. While the rate of alcoholism is lower among elderly men than among younger men, the same cannot necessarily be said for women. It seems that a significant number of women, especially widows who live alone, become alcoholics in their later years (Nathan, 1983). Alcoholism among the elderly is more likely to take the form of steady, measured drinking rather than heavy drinking at a single sitting (La Rue et al., 1985). This type of "maintenance" drinking may be difficult for others to notice, since they might assume that some unsteady movement or slightly slurred speech or lapse of memory is attributable to age rather than to alcohol (Craik and Byrd, 1982).

Figure 24.10 *Many of the elderly who seem to be demented are suffering from curable maladies that have symptoms similar to those of organic dementia. Among these diseases are malnutrition, depression, and alcoholism.*

Unfortunately, at the same blood alcohol levels, older adults show greater impairment of reaction time, memory, and decision-making than younger adults. When these functions are already affected by age, the further declines produced by alcohol may reach the point where serious errors of judgment are likely to occur. As at every age, alcohol is a major contributor to accidental and suicidal death in old age.

Psychological Illness   In general, psychological illnesses such as schizophrenia and personality disorders are less common in the elderly than in younger adults (La Rue et al., 1985). Nonetheless, about 10 percent of the elderly who are diagnosed as demented are actually experiencing psychological, rather than physiological, illness.

In some cases the person is unusually anxious, which, as anyone who has taken a final exam under pressure knows, can make even a young, bright, healthy person forget important information. For many elderly people, the anxiety that occurs on arrival at a hospital or nursing home is sufficient to cause substantial disorientation and loss of memory. If the anxious new arrival is tested immediately, a misdiagnosis of organic brain damage is possible.

When depression occurs in late adulthood, the symptoms sometimes resemble those of dementia. Depressed people are usually slow-moving, often do not seem to care about such basics as eating and dressing, ignore or become distracted during conversation, and, in late adulthood, may talk about losing their mental ability. In fact, talk about fading intellect is one clue that depression, not dementia, is the problem. People who are suffering from organic brain disease usually are ashamed of their loss, and often use denial as a coping mechanism, while those suffering from depression tend to exaggerate any loss of mental functions (Reisberg, 1981).

Depression in late adulthood, as at younger ages, is one of the most treatable mental illnesses (La Rue et al., 1985). Psychotherapy and chemotherapy usually effect notable improvement in a few weeks. However, even more than at younger ages, most depressed elderly people are not treated because no one recognizes their depression as a curable disease. Instead, many care-givers consider depression a natural consequence of aging, or they confuse the symptoms with those of brain disease. If the depressed person has recently lost a loved one, the symptoms of depression may mistakenly be attributed to bereavement. It is normal, of course, for the elderly who are in mourning to be sad and to have difficulty eating and sleeping, but the symptoms of bereavement do not normally include strong feelings of guilt and self-deprecation, or last longer than a few months (Breckenridge et al., 1986).

Sadly, one consequence of untreated depression among the elderly is that the suicide rate is higher for those over age 60 than for any other age group, and the rate has increased over the past decade (Sainsbury, 1986). In the United States, the suicide rate for white males at age 85 is three times that of white males at age 20 (Manton et al., 1987). Comparable statistics in other countries show an even wider disparity: in Japan, the ratio of suicide for those over age 75 is five times that of those between ages 15 and 24; in France, the rate is six times higher (Padmore and Meada, 1985).

Figure 24.11 *One cause of depression in the elderly is social isolation—when too many moments in the day are like this one. Over the past five years, depression has become one of the easiest psychological disorders to treat, thanks to advances in psychotherapy and chemotherapy. Unfortunately, because feelings of hopelessness characterize depression, victims are unlikely to believe that help is possible.*

In most cases, the precipitating event for the suicide is a social loss, retirement or widowhood being the most common. A related cause is illness, particularly cancer or illnesses that affect the brain (Whitlock, 1986). About half of the elderly men, and a third of the elderly women, who commit suicide are physically ill. Even in cases of illness, however, severe depression, generally arising from fears about the possible consequences of the disease rather from the disease itself, is a factor in the suicide. On the whole, old people's fears about their illness are worse than the likely outcome. They do not realize that chronic pain, disability, and depression can usually be treated with drugs and other therapy (Whitlock, 1986). Depression is not an inevitable reaction to either the losses of age or the problems of illness, and thus it should not be accepted as such.

New Cognitive Development in Later Life

So far in this chapter we have mainly considered possible declines in the intellectual functioning of the elderly. What about positive changes? Can the elderly develop new interests, new patterns of thought, a "criterion shift" (Cavenaugh et al., 1985), a deeper wisdom? Most of the major theorists on human development believe that they can. For example, Erik Erikson finds that the older generation are more interested in the arts, in children, and in the whole of human experience. They are the "social witnesses" to life, and thus are more aware of the interdependence between one generation and another (Erikson et al., 1986). Abraham Maslow maintains that older adults are much more likely than younger people to reach self-actualization, which, as you remember from Chapter 2, includes heightened aesthetic, creative, philosophical, and spiritual sensitivity (Maslow, 1970). Let us look, then, at these areas of life during late adulthood.

Aesthetic Sense

Many people seem to experience nature and the arts in a deeper, more appreciative way, as they get older. As two leading gerontologists explain, "healthy late life is frequently a time for greater enjoyment of all the senses—colors, sights, sounds, smells, touch—and less involvement in the transient drives of achievement, possession, and power" (Butler and Lewis, 1982).

For many older people, this heightened appreciation leads to active expression. They may begin gardening, bird-watching, pottery, painting, or playing a musical instrument—and not simply because they have nothing better to do. Among those who painted in earnest after age 60 were Winston Churchill, Dwight Eisenhower, and Henry Fonda—all of whom had many other activities (such as heading a nation or winning an Oscar) with which to fill up their days. The importance that creativity can have for some in old age is wonderfully expressed by a 79-year-old man, unfamous, little educated, yet joyful at his workbench:

Figure 24.12 *While an enhanced appreciation of aesthetic beauty is typical of the elderly, the particular expression varies depending on the individual's talents, interests, and personality. An extrovert, for example, would be more likely to join some sort of group activity, such as singing in a chorus, while the introvert might choose a solitary activity, such as photographing tiny plant growth on a rock.*

This is the happiest time of my life. . . . I wish there was twenty-four hours in a day. Wuk hours, awake hours. Yew can keep y' sleep; plenty of time for that later on . . . That's what I want all this here time for now—to make things. I draw and I paint too . . . I don't copy anything. I make what I remember. I tarn wood. I paint the fields. As I say, I've niver bin so happy in my whole life and I only hope I last out. [quoted in Blythe, 1979]

For this man, and for many people, the impulse to create did not suddenly arise in late adulthood; it was present, although infrequently expressed, in earlier years. What does seem to occur in late adulthood is a deepening need to express and develop that impulse, perhaps because, as the years left to live become fewer, those people who were "bearers of a secret dream" decide to defer that dream no longer (McLeish, 1976). This was the case with Laura Ingalls Wilder, who did not begin to write her first book, *Little House in the Big Woods,* until she was 64. By the time she was 75, she had produced seven more novels in the world-famous "Little House" series (Kerber, 1980).

One of the most remarkable examples of late creative development is found in Anna Moses, a farm wife and mother of ten. For most of her life, she expressed her artistic sensitivity by stitching quilts and doing embroidery during the long farm winters, when little outside work could be done. At age 75, arthritis made needlework impossible, so she took to "dabbling in oil" instead. Four years later, three of her oil paintings, displayed in a local drugstore, caught the eye of a New York City art dealer who happened to be passing by. He bought them, drove to her house to buy fifteen more, and began to exhibit them in the city. One year later, at age 80, "Grandma Moses" had her first one-woman show in New York, receiving international recognition for her unique "primitive" style. She continued to paint, "incredibly gaining in assurance and artistic discretion" into her 90s (Yglesias, 1980).

Figure 24.13 *Instead of working on expensive canvas, Grandma Moses used ordinary wood, which she prepared with an undercoat of white house paint. Because she had arthritis, she had to rest her arm on a table rather than use an easel. These two working methods became virtues, contributing to her unusual colors and perspective.*

Philosophical Turns of Mind

Many older people become more reflective and philosophical than they once were. In most cases, this reflectivity is personally centered, as the individual attempts to put his or her life in perspective, assessing accomplishments and failures in terms of what the person perceives to be the overall scheme of life. Neugarten (1973) refers to this turn of mind as **interiority,** a heightening of the older person's propensity for introspection. Erikson points to a similar pattern in his depiction of the final stage of psychosocial development, integrity versus despair, in which the individual attempts to see his or her life as a meaningful whole.

Another formulation of this attempt to put one's life into perspective is called the **life review** (Butler, 1963), in which an older person recalls and recounts various aspects of his or her life, remembering the highs and the lows, and comparing the past with the present. In general, the life-review process connects one's own life with the future as one tells one's story to younger generations; at the same time, links with past generations are renewed as one remembers what one's parents, grandparents, and even great-grandparents did and thought. One's relationship to humanity, to nature, to the whole of life also becomes a topic for reflection. For some, the life review becomes a psychological survey that leads to resolution of past conflicts and reintegration of the entire life course.

Often the person reflects on the relationship between his or her personal history and the social history of the times, a reflection of great interest to certain historians, who now recognize the life-review process as oral history, an important primary source of historical data (Davis, 1985). The process can take written form as well: many of the famous and not so famous write their autobiographies in their later years.

Sometimes the life review takes the simple form of nostalgia, reminiscence, or story-telling, which may be quite helpful to the older person, although not always easy to listen to. According to Butler and Lewis,

> one of the greatest difficulties for younger persons (including mental health personnel) is to listen thoughtfully to the reminiscences of older people. We have been taught that this nostalgia represents living in the past and a preoccupation with self and that it is generally boring, meaningless, and time-consuming. Yet as a natural healing process it represents one of the underlying human capacities on which all psychotherapy depends. The life review as a necessary and healthy process should be recognized in daily life as well as used in the mental health care of older people. [Butler and Lewis, 1982]

In some cases, the reflectivity of old age may lead to, or intensify, attempts to put broader historical, social, and cultural contexts of life into perspective. It is interesting to note that when Wayne Dennis (1966) studied the production of professionals in sixteen different fields, he found that in two of them—history and philosophy—production peaked in the 60s and 70s. Certainly one of the most famous examples is Will and Ariel Durant's *The Story of Civilization*—a monumental, ten-volume history of civilization written mostly in late adulthood—followed by *The Lessons of History* and *Interpretations of Life,* published when the Durants were in their 80s. The "philosophical bent" of old age is also reflected in the growing number of the elderly showing up in the classroom to take courses in ethnic roots, history, and philosophy, sometimes in greater numbers than the young. One elderly New York University student, noting that three-quarters of his class in Existentialism and Modern Jewish Philosophy were over 65, remarked: "I'm afraid young people aren't interested in philosophy. They're interested in making a living."

Spiritual Values

Closely related to the idea that the old are more philosophical than the young is the idea that the old are more spiritual. In many cultures over many centuries, the very old have been the spiritual leaders (Clayton and Birren, 1980). They are the "elders" of the church. More popes, for instance, have been elected in their 80s than in any other decade of life.

Obviously, it is hard to measure spirituality, just as it is hard to measure wisdom. We do have data, however, on the importance of religion in people's lives. As can be seen in the chart below, those 65 and older are likely to value religion more highly than people under age 65 do. These data are cross-sectional, and thus may reflect cohort differences. However, a review of all the research on the importance of religion during old age, including longitudinal studies, confirms a general increase in religious faith, prayer, and spirituality in later years (Achenbaum, 1985; Moberg, 1965).

Measures of the Significance of Religion in Adulthood

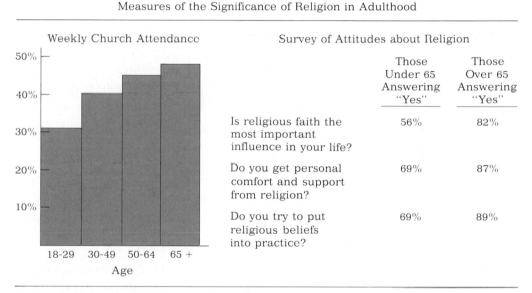

	Those Under 65 Answering "Yes"	Those Over 65 Answering "Yes"
Is religious faith the most important influence in your life?	56%	82%
Do you get personal comfort and support from religion?	69%	87%
Do you try to put religious beliefs into practice?	69%	89%

Source: Schick, 1986; U.S. Bureau of the Census, 1986.

There is some indication that the substance of religion, not the formal trappings, is of more interest to the elderly (Achenbaum, 1985). For example, the numbers agreeing that religion is the most important influence in their lives show a greater age contrast than the numbers attending church. Perhaps by late adulthood, one's own religious beliefs are sufficiently strong that following particular rituals or rules of a particular faith becomes less critical than following the spiritual dictates of that faith.

Summing Up

On balance, then, it seems fair to conclude that the mental processes in late adulthood can be adaptive and creative, not necessarily better than thinking at younger ages, but more appropriate to the final period of life (Labouvie-Vief, 1985). An illustrative and exemplary case in point is the following poem, written by Henry Wadsworth Longfellow at age 80.

But why, you ask me, should this tale be told
Of men grown old, or who are growing old?
Ah, Nothing is too late
Till the tired heart shall cease to palpitate;
Cato learned Greek at eighty; Sophocles
Wrote his grand *Oedipus,* and Simonides
Bore off the prize of verse from his compeers,
When each had numbered more than four score years,
And Theophrastus, at four score and ten,
Had just begun his *Characters of Men.*
Chaucer, at Woodstock with the nightingales,
At sixty wrote the *Canterbury Tales;*
Goethe at Weimar, toiling to the last,

Completed *Faust* when eighty years were past.
These are indeed exceptions, but they show
How far the gulf-stream of our youth may flow
Into the arctic regions of our lives
When little else than life itself survives.
Shall we then sit us idly down and say
The night hath come; it is no longer day?
The night had not yet come; we are not quite
Cut off from labor by the failing light,
Some work remains for us to do and dare;
Even the oldest tree some fruit may bear;
And as the evening twilight fades away
The sky is filled with stars, invisible by day.

SUMMARY

Intellectual Decline in the Processing of Information

1. Thinking processes, as measured by standardized tests, become slower and less sharp once a person reaches late adulthood. One reason is that less information reaches the brain because vision and hearing become less sensitive.

2. The sensory register declines relatively little in late adulthood. However, short-term memory shows notable declines, especially when memory for meaningless or uninteresting material is examined. One reason for this loss is that processing takes longer with age, and this makes it more difficult for older adults to get information into their memories.

3. Long-term memory, by contrast, appears to decline very little with age. It seems that, once information is securely placed in the memory bank, it tends to stay there, no matter how many years elapse.

4. On intelligence tests and laboratory studies of memory, older people use fewer and less efficient strategies to help them reason and remember. In tests of memory, for example, they are less likely to use mnemonic devices. When trying to solve abstract problems, they are likely to be distracted by irrelevant information, or to follow their hunches rather than use logical techniques.

5. One reason older adults, on average, do not perform as well as younger adults is that more of the older group have health problems, especially undetected cardiovascular difficulties, that slow down thinking. In addition, members of the older cohorts completed less formal education.

Cognitive Functioning in the Real World

6. In daily life, most of the elderly are not handicapped by memory difficulties, for several reasons. One is that they can use practice and priming, two memory techniques that are often excluded from the laboratory but are very helpful in daily life. Another is that most of them, once they recognize deficits in memory, develop strategies that aid their memory.

7. The decline in problem-solving ability, evidenced in the laboratory, is also much less apparent in daily life.

8. Older adults are able to learn new learning strategies, as research on the elderly in the community, as well as those in nursing homes, confirms.

Dementia

9. Dementia, whether it occurs in late adulthood or earlier, is characterized by memory loss—at first minor lapses, then more serious forgetfulness, and finally such extreme losses that recognition of closest family members fades.

10. The most common cause of dementia is Alzheimer's disease, an incurable ailment that becomes more prevalent with age. While some cases of Alzheimer's disease are genetic, for the most part the cause is unknown.

11. Multi-infarct dementia is caused by a series of ministrokes that occur when impairment of blood circulation destroys part of the brain tissue. Measures to improve circulation and to control hypertension can prevent or slow the course of this form of dementia.

12. In addition to Alzheimer's disease and multi-infarct dementia, other organic causes of dementia are brain tumors, Parkinson's disease, and Down syndrome. Non-organic problems that are frequently misdiagnosed as dementia include drug misuse, alcohol abuse, and psychological illness.

New Cognitive Development in Later Life

13. Many people become more responsive to nature, more interested in creative endeavors, and more philosophical as they grow older. The life review is a personal reflection that many older people undertake, remembering earlier

experiences and putting their entire lives into perspective. For the individual, this is a valuable psychological experience; others who listen to the life review can gain a new view of family and social history.

14. Religious concerns and spiritual awareness may also increase in old age.

KEY TERMS

input *(547)*

storage *(547)*

program *(547)*

priming *(556)*

dementia *(561)*

Alzheimer's disease *(562)*

multi-infarct dementia (MID) *(564)*

Parkinson's disease *(564)*

interiority *(570)*

life review *(570)*

KEY QUESTIONS

1. In terms of the information-processing approach, how are input, storage, and program affected by age?

2. What are some of the reasons for age-related declines in cognition?

3. What are some of the factors that help older adults show better memory skills in everyday life than on laboratory tests of memory?

4. In what ways do the problem-solving skills of older and younger adults differ?

5. What is the general pattern of cognitive decline in all types of dementia?

6. What are some of the other factors, besides brain disorders, that can cause older individuals to exhibit symptoms of dementia?

7. What are some of the positive cognitive developments that are likely to find expression in older adulthood?

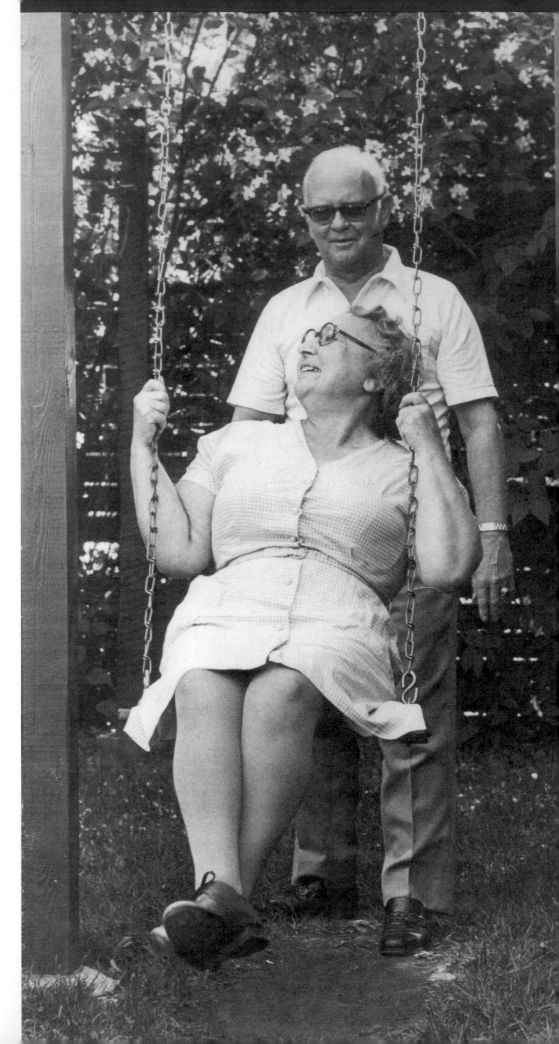

CHAPTER 25

Late Adulthood: Psychosocial Development

When grace is joined with wrinkles, it is adorable.
There is an unspeakable dawn in happy old age.

Victor Hugo
Les Misérables

It seems a reasonable hypothesis that psychosocial development would be dramatically affected by old age. Such things as retirement, maturing children and grandchildren, increased leisure time, decreased income, failing health, and the imminence of death all create new pressures that the elderly must face sooner or later. How do the elderly typically react to such pressures? Do they become conservative, withdrawn, and passive, allowing the world to pass them by, or do they begin to explore new vistas as they gain freedom from the responsibilities of career and parenthood? Or do they adjust to these changes in much the same way as they adjusted to the many changes earlier in life?

Theories of Psychosocial Aging

All the above questions have been the focus of theories of late-life development. Each theory points in a different direction, yet each seems plausible.

Disengagement versus Activity

The first, and most controversial, theory regarding development in late adulthood, **disengagement theory** (Cumming and Henry, 1961), maintains that, in old age, the individual and society mutually withdraw from each other. This uncoupling occurs in four steps:

1. Beginning in late middle age, a person's social sphere becomes increasingly narrow. Traditional roles, such as worker and parent, become less available or less important, and one's social circle shrinks because friends die or move away.

2. People anticipate, adjust to, and participate in this narrowing of the social sphere, relinquishing many of the roles they have played and accepting the gradual closing of their social circle.

3. As people become less role-centered, their style of interaction changes from an active to a passive one.

4. Because of this more passive style of interaction, older people are less likely to be chosen for new roles, and therefore are likely to disengage even more.

In this view, then, the elderly's participation in the disengagement process is voluntary. By old-old age, it is said, people prefer to be quite withdrawn from most social interaction, avoiding the noisy bustle and insistent demands of the younger person's world.

Disengagement theory has provoked a storm of protest and controversy among gerontologists, many of whom object strenuously to the implication that the elderly disengage and withdraw, or that when they do, they do so willingly (Lemon et al., 1972; Rosow, 1976; Sill, 1980). Acknowledging that disengagement does occur for some elderly people in some areas of life, a number of researchers claim that disengagement in one area can lead to reengagement in other areas (Maas and Kuypers, 1974; Sears, 1977; Sears and Barbee, 1977). Disengagement from work roles, for instance, may promote more involvement in social and community activity (Parnes and Less, 1985); the death of a spouse may reactivate involvement in friendship networks (Adams, 1987).

In addition, these researchers generally find that the more activities older people engage in, and the more roles they play, the greater their life satisfaction. They maintain that **activity theory** is more applicable to late adult development than disengagement theory. According to activity theory, if disengagement does occur, it should not be considered a part of normal aging but, rather, should be looked on as a sign that something is wrong, either in the person or in the society. Even when someone seems to be disengaging voluntarily, his or her withdrawal into a smaller social world may merely be an attempt to cope with an ageist society. As one critic explains, our society "systematically undermines the position of the elderly and

Figure 25.1 *Although some of the elderly obviously disengage, many others become active in new ways. This woman is telling a Black-history class about her personal experiences as a Black person in America, a way to interweave memories and thoughts about the past with concern for the future.*

deprives them of major institutional functions'' (Rosow, 1976), and then labels as ''natural'' the fact that the elderly are less engaged than they were. Consequently, whatever disengagement we observe is the result of our culture's attempt to strip the elderly of their social roles, powers, and status, rather than the cause of it.

The most unfortunate aspect of disengagement theory, say its critics, is that it was seized on as fact by ''important policy makers—usually politicians, nursing-home administrators, and the like.'' Because some of these people had ''unconscious gerontophobic attitudes . . . , they interpreted [disengagement] according to their personal motivation and limitations—often colored by financial considerations'' (Cath, 1975). As a result, mandatory retirement, reduced income, socially sterile nursing homes, and the like were justified on the grounds that the elderly, after all, want to withdraw.

Continuity versus Discontinuity

Another controversy involves, once again, the question of continuity versus discontinuity. As one group of researchers point out, many life events alter cognitive and behavioral patterns (Brim and Ryff, 1980; Danish et al., 1980; Eisdorfer, 1985). Several of the most transforming events, such as retirement, death of a spouse, and failing health, are likely to occur in late adulthood. Discontinuity seems therefore inevitable in late adulthood.

On the other side, some theorists point out that continuity, particularly of friends, interests, and personality, is even more likely in late adulthood than earlier because the patterns are well established. These theorists maintain that how a person reacts to the life events of old age is likely to be consistent with their responses to the various life events of their earlier years. As we have seen before, the underlying continuity of personality tends to produce a continuity of coping and adaptation patterns.

Figure 25.2 *Continuity theory would suggest that fifty years ago, this man probably spent a fair amount of time on the porch pushing forty. Discontinuity theory reminds us that time brings changes: while our friend was probably always a baseball fan, he could not always have supported the Mets. The Dodgers, who originally played in Brooklyn, were probably his youthful choice.*

"I'll be on the porch, pushing ninety."

Drawing by Price; © 1985 The New Yorker Magazine, Inc.

Diversity

Not surprisingly, then, as longitudinal research tracing the daily lives and personality characteristics of the elderly accumulates, what seems clear is that each of the foregoing theories applies in certain cases, but that none is sufficient to stand alone as a predictor or explanation of behavior patterns in old age. The reason is that, as researchers look longitudinally at aging people, they find that patterns of psycho-social development are at least as varied as in earlier periods. Disengagement is apparent in some of the elderly; increased activity is evident in others. Continuity and discontinuity can both be readily seen. In the words of Maas and Kuypers (1974), who studied the lives and personality patterns of middle-class adults at about age 30 and again at about age 70:

> The range and array of ways of life . . . seem to us re-markable. Stereotypes of old people as though they were homogeneous on any dimension other than chronological age obviously have no support from the findings of this study . . . the potential for variety in life style in old age—when economic circumstances are not constricting—seems almost boundless.

These researchers found that widowhood and retirement, two of the experiences thought to be most disruptive for adults, varied tremendously in their effect: after a period of adjustment, some individuals seemed liberated by these experiences; others turned bitterly inward; and still others seemed essentially unchanged.

Indeed, several leading gerontologists believe that variability and diversity in-crease with age, so that by late adulthood the multiplicity of personalities and pat-terns is greater than at any other age (McKenzie, 1980; Rowe and Kahn, 1987; Schaie, 1981). Consider the diversity one would find in a cross-section of American 65-year-olds. Most would be married, but a fourth would not be, the men being equally divided among the never-married, the single, and the widowed, and the women being mostly widows. Most would have children and grandchildren, and at least 10 percent would have great-grandchildren, but a fourth would have no living offspring. Most would be retired, but some would be still employed. A minority would be ill and feeble, with 2 percent destined to die within a year, but the average 65-year-old would have seventeen more years to live, and a minority would live to be 100. Most would be financially comfortable (see A Closer Look, next page), but some would be among the poorest of the poor.

In fact, all that a group of 65-year-olds would have in common is age, and, given the diversity of physical health and psychosocial experiences, age as an index to developmental patterns becomes increasingly less relevant as life goes on. Every careful study of old age has ended up having to use many contrasting labels to describe the elderly and their patterns of daily life—"hobbyist," "family-centered," "rocking chair men," "armored," "merry widow," "reorganizers," and so forth.

When we consider this diversity, we realize why there are many theories of psy-chosocial aging, each with some logic and research to support it. In any cross-section of elderly people, some are quite disengaged and some quite active; some seem much changed from their younger selves, and some hardly changed at all. As in middle adulthood, specific personality traits are one reason for this: the neurotic would be more likely to disengage, and the extrovert would be more likely to be active; those open to new experiences would most likely show greater discontinu-ity than the others. Equally important would be the specific events and conditions of a person's life. Achievement and affiliation continue to be central, so retirement or death of a spouse would necessitate some discontinuity. Health becomes criti-cal: notable disability is bound to limit activity and require some disengagement. Let us now look at the specific adjustments and changes that the elderly often make.

A CLOSER LOOK   Income and Age

More than twenty-five years ago, Michael Harrington's *The Other America* drew a chilling picture of the extent of poverty among adults who were then over age 65:

Fifty percent of the elderly exist below minimum standards of decency, and this is a figure much higher than that for any other age group. . . . We have given them bare survival, but not the means of living honorable and satisfying lives. [Harrington, 1962]

Harrington was not far wrong. In actuality, one out of every three elderly Americans was then living below the poverty line, and a substantial additional number were "near poor," living on the very edges of poverty. Since then, however, economic and political changes—including more extensive Social Security coverage with rates indexed to inflation, and a potpourri of social and medical programs—have helped raise the living standards of many of the elderly (Chen, 1985; Harris, 1986). By 1985, the percentage of individuals aged 65 and older with incomes below the poverty level was 12 percent; when the value of noncash benefits is included, the percentage falls to 8 percent (U.S. Bureau of the Census, 1986).

Overall, most of the elderly have sufficient income to meet their needs (Zopf, 1987). This is especially apparent when income, assets, and benefits are compared with expenses, debits, and needs. In fact, on one measure, net worth—which is the total value of property and investments minus mortgages and other debts—families headed by someone between the ages of 65 and 74 are wealthier than families headed by someone of any other age.

For many of the elderly, home ownership engenders pride, independence, and security. For others, however, it can also create financial burdens and physical strain—something the elderly may be reluctant to acknowledge.

Average Family Net Worth

Age of Family Head	Mean	Median
Under 25	$4,218	$5
25-34	$20,391	$3,654
35-44	$51,893	$28,721
45-54	$81,350	$43,797
55-64	$119,714	$55,181
65-74	$125,284	$50,181
75+	$72,985	$35,939

This is not to say that the elderly typically have all the income they would like, or even the income they had earlier. In many cases, net worth includes the value of the family home, which is likely to have increased considerably over the past decades, but is an asset that most of the elderly could ill-afford to convert to cash. On measures other than net worth, the elderly do not fare as well. After retirement, income typically goes down by at least a third.

Of all persons with some annual income, the median cash income of a man over age 65 is about $10,000, and of a woman, $5,600, compared with median income at age 40 of $23,000 for men, and $9,000 for women. As a result, while most healthy older adults have enough income to cover their basic needs, many find, to their disappointment and anger, that their preretirement plans to travel, entertain, or simply enjoy life need to be drastically curtailed.

We should also keep in mind that while the expenses of the typical older adult are much less than the expenses of the typical younger adult raising a family, a minority of older American adults have substantial medical bills, less than half of which are paid by the government. (Most other developed countries provide much more subsidy for medical care.) Finally, the minority of elderly who are poor also tend to have the most needs, and the fewest family resources. The poor elderly are much more likely to be among the old-old, or living alone, or from minority racial groups, or childless, or widows, or all of the above. (The specific problems of the poor elderly are considered later in this chapter.)

Employment and Retirement

Historically, the labor of early adulthood continued into late adulthood: adults worked until their health made work impossible. The reasons were primarily economic. First, most older adults had to go on working to support themselves, and second, there was a demand for their labor.

Now, as then, most older adults are quite capable of productive work. A review of age-related performance on the job finds that, while some reduction of efficiency occurs with age in tasks where physical labor, speed, and/or concentration are critical, for the most part older workers perform as well as younger workers. Further, those older adults who are still employed are more satisfied with their jobs, and therefore less likely to quit or miss work unless they are truly sick (Davies and Sparrow, 1985).

In industrialized countries today, however, relatively few older citizens are in the paid labor force. In fact, although mandatory retirement laws are lifting, the average age at retirement is younger than it was even ten years ago. In most of Western Europe and North America in the 1980s, fully one-third of the men between the ages of 55 and 65 are no longer in the labor force, and more than half the women in this age group are no longer, or never were, employed.

Figure 25.3 *Employment in late adulthood is an option that is rarely chosen but one that is usually enjoyed when it is taken. Among those who prefer to work are the three shown here: a security guard, a master book binder, and a blacksmith. The first job is the most typical: the majority of employed elderly work part-time, at a job related to their previous career (many retired policemen become involved in security), in which independence and conscientious work habits are an asset.*

Reasons to Retire

Almost as many people retire because they want to as because they have to. Currently, about 45 percent of all retirees could have continued working but decided against it, while 40 percent found they were required to give up their work because they had reached the legal or "understood" retirement age for their job or because their job was eliminated. Only 15 percent had to retire because of failing health (Zopf, 1986).

In large measure, the high rate of voluntary retirement is due to an improvement in the economic prospects of retirement. For instance, in 1950, only one out of every three older Americans received any pension income, including Social Security; by 1984 more than nine out of ten had Social Security and more than half had private pensions as well (Crystal, 1982; U.S. Bureau of the Census, 1986). Private pensions often can start at age 55 or even earlier, and thus are important in making early retirement desirable, especially for those blue-collar and service workers whose work is not particularly satisfying (Hardy, 1985). Such workers tend to retire as soon as they think they can afford to maintain a reasonable lifestyle.

Economic factors can also precipitate retirement before a worker really wants it. When a financial squeeze causes an industry to shrink or a factory to close, some older workers are offered financial incentives for early retirement. Others are simply laid off, unable to find another job, and then, in discouragement, retire. Since 1970, the employment rate of American men between the ages of 55 and 64 has fallen by about 20 percent, primarily because the job market has shrunk, not because early retirement has become more attractive (Robinson et al., 1985).

Adjustment to Retirement

For most people, retirement is, in the words of one gerontologist, "a major economic, social, and psychological event," and this is true whether it is welcomed or feared. For most older adults, including the housewives of employed husbands, retirement means that "the old center of life is gone and another must be carved out if days are not to be shapeless" (Glick, 1980).

For some, the transitions brought on by retirement are extremely difficult. Many initially share the feelings expressed by a recently retired policeman:

> What the hell do I do now? I thought I would feel differently now that I'm retired. I still want to work but they won't let me . . . I always thought that I wanted to rest and fish at this age. I never dreamed I would be in such good physical health. I thought I would be all dilapidated . . .
> I'm raring to go, but there is no place to go. Rest is O.K. for a couple of months. In fact, I've looked forward to this all my life, lying around, doing what I want to do. But now I know this is crazy. A person's got to be wanted and needed. A person's got to do something with people. This retirement stuff is a myth, a rip-off as the kids say.
> [quoted in Kornhaber and Woodward, 1981]

Problems like this are especially likely to occur when retirement is entered into involuntarily and abruptly. A person generally works full time until a certain age, and then wakes up one morning and has to begin an entirely new pattern. Consequently, the first year of retirement tends to be difficult, because it takes time to develop a new set of routines and interests. Juanita Kreps (1979) put this problem into a wider perspective when she suggested that it would be better if work and leisure activities were distributed more equitably throughout the life span:

> By concentrating free time on the old and the young, we have lengthened two role periods, which have many of the same characteristics: minimum incomes, excessive amounts of free time, lack of status. Meanwhile those in their middle years are harnessed by heavy work commitments. Work satisfaction during this stage of life is diminished by the length of the work year, just as the joy of free time declines when it becomes endless.

Rather than being subjected to sudden retirement, many gerontologists suggest, workers should have the option of gradually retiring, perhaps working part-time or taking longer vacations (Lazarus and Lauer, 1985). In fact, many of the elderly arrange to do this on their own; after leaving one job, they find another with shorter hours. Most of the 10 percent of those older than 65 who are in the labor force work part-time.

Fortunately, when a newly retired person feels discarded, disrupted, and disoriented, such feelings are usually temporary. Most retired persons are able "to carve out a new center" for their lives (Mutran and Reitzes, 1981). In fact, most older Americans quickly find more than enough to do with their new-found freedom, and are limited only if failing health or greatly reduced income accompany retirement (Zopf, 1986). Sometimes retirement leads not only to fullness but to unexpected richness as well, as shown in the journal of one newspaperman who had vowed he would never retire, and then was forced to do so. In spite of the many problems retirement created for him and his wife, he concluded at the end of the first six months of his new life:

> I don't believe I have ever been happier. I don't believe Carol and I have ever found more pleasure in each other's presence. If this kind of life doesn't work [because of financial problems]—and the next six months may demonstrate that it can't, it should. I know now that at least it is worth fighting for, worth putting everything else into. I know it is an opportunity, not a fate; not a deprivation or a diminishment but an enrichment; a beginning no less engaging because it must be the beginning of an end. The most frequent thought I have had in the first half year is that I should have begun doing this a long time ago. [Olmstead, 1975]

New Achievement Patterns

As you learned in earlier chapters, a sense of achievement is a basic psychosocial need in adulthood. This is no less true at 60 or 80 than it is in younger years. As the 80-year-old writer and critic Malcolm Cowley (1978) explains:

> Man or woman, artist or not, every old person needs a work project to keep himself more alive. The project can be one of long standing that asks to be rounded off, or it can be something completely new. . . . It should be big enough to demand their best efforts, yet not so big as to dishearten them and lead them to subside into apathy.

However, once the two main sources of achievement, employment and child-rearing, are completed, what kinds of projects do the elderly undertake? Usually they involve activities of the person's own choosing (Parnes and Less, 1985) (see Table 25.1). As we saw in Chapter 24, some of the elderly become involved in aesthetic pursuits, such as painting or playing a musical instrument. For others, the practical arts become a center of daily activity. Gardening, carpentry, sewing, knitting, and cooking are favorite pastimes of many of the elderly. While the young

TABLE 25.1 **Leisure Activities (Hours per Year)**

	Workers	Retirees
Exercise	167	304
Reading	437	518
Hobbies	193	285
Visiting	218	283
Home maintenance	115	191
Volunteer work	97	178

Figure 25.4 *For the elderly who are among the young-old, a wide variety of achievement patterns is available. Shown here are a recently elected Texas legislator evaluating a proposed law, gardeners tending their crop, and Elderhostel students searching for tropical fish in the waters around Bermuda.*

often look on such activities as busywork or puttering, we should note that these same tasks are the life work of many people who live in labor-intensive societies.

Sometimes the practical interests of the elderly can lead to new careers, such as that of a handyman, gardener, caterer, or quilt maker, for example. More often, they simply come in handy at home, and at the homes of nearby friends and family members.

Learning At any given time, about one out of every twenty adults aged 60 and older is enrolled in classes of some sort. The demand for courses at the college level has become so great among the elderly that over 1,000 universities in the United States have made provisions for senior citizens to take courses for credit or audit them (Berger, 1987). Some colleges have installed programs generally known as Elderhostel, in which retired adults live on campus and take classes during college vacation periods.

While every imaginable kind of study is probably pursued by some of the elderly, they are more likely than younger adults to study literature, art, history, or religion (Cross and Florio, 1978). They study for the joy of learning, or for some specific goal, rather than for the more general advancement sought by younger adults (Willis, 1985). Informal learning is common as well. Most older adults spend more time reading, writing, and thinking than they did since their formal education ended some fifty years earlier (Parness and Less, 1985).

Children and Grandchildren   For many of the elderly, the younger generations are a source of pride, comfort, and concern. Most older adults view their relationships with their grown children as loving and close, and continue to take an interest in their children's development. In their hearts, parents do not stop raising their children simply because the children are full-grown, married, or parents themselves. As one 82-year-old woman succinctly put it: "No matter how old a mother is, she watches her middle-aged children for signs of improvement" (Scott-Maxwell, 1968).

About 90 percent of the elderly who have grown children give them gifts, advice, or practical help, and 90 percent receive such aid (Shanas, 1980). Financial support, perhaps surprisingly, is more likely to go from the older generation to the younger, while care-giving in times of sickness is more likely to go from the younger generation to the older. (Some of the problems of such care-giving are discussed later in this chapter.)

While support is usually forthcoming when needed, family closeness does not necessarily mean family consensus. In fact, a "generation gap" between the elderly and their children and grandchildren is common (Hagestad, 1985). When the generations differ in their views on such areas as politics, sex, child-rearing, and religion, family members, in the name of "good relations," may confine their conversation to "demilitarized zones," involving topics that are unlikely to provoke anger or hurt feelings. Frequently this results in superficial relations, with both generations hesitating to offer constructive criticism or advice, even when it might be needed. The generations may also be divided in terms of their daily concerns and patterns. As Arling (1976) points out, even when they are deeply concerned about each other, members of different generations may have difficulty "truly sharing their experiences or empathizing with each other." Thus the price of intergenerational harmony may be intergenerational distance.

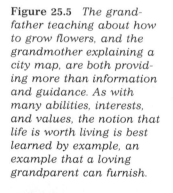

Figure 25.5 *The grandfather teaching about how to grow flowers, and the grandmother explaining a city map, are both providing more than information and guidance. As with many abilities, interests, and values, the notion that life is worth living is best learned by example, an example that a loving grandparent can furnish.*

A CLOSER LOOK   **Grandparents Then and Now**

Today's pattern of detachment between grandparents and grandchildren is quite different from the one that was typical in the first half of the century when many grandparents were central to the lives of their children's children. What caused this change? Researchers have suggested several reasons (Bengston and Robertson, 1985; Kornhaber and Woodward, 1981; Rodeheaver and Thomas, 1986; Troll, 1980).

1. Increased geographical mobility of offspring has meant that many grandparents live quite far from their children and grandchildren. This makes visiting infrequent and the relationship less intimate.

2. Because grandparents of today are more financially independent than earlier generations of grandparents were, they are more likely to have homes, friends, and interests apart from those of their children and grandchildren.

3. Grandmothers are more often employed than they were earlier, so they have less time, energy, or need to make their grandchildren central to their lives.

4. Because of the rising divorce rate, some grandparents rarely see their grandchildren, who are living with their former daughter- or son-in-law.

5. Finally, the relationship between parents and children has become more egalitarian in recent decades, with the result that the deference once automatically accorded to the older generation has diminished: since their advice regarding child-rearing is likely to go unheeded, the elders take a background role in order to keep family harmony.

Most grandparents are comfortable with their reduced role: they like to play with their grandchildren, give them presents, and baby-sit occasionally—and they are quite happy to leave parenting up to the parents. In fact, those who feel responsible for advising their grandchildren (which grandfathers do more than grandmothers, and young grandparents more than older ones) tend to be less satisfied with the grandparenting role than those who feel their role is simply to enjoy their grandchildren (Thomas, 1986).

These generalities should not obscure the diversity of grandparenting style. As we have seen, some grandparents are so involved with their grandchildren as to be surrogate parents; others are very distant, and rarely heard from; and most are only modestly involved. Styles of grandparenting also vary by sex, age, and ethnic group. In general, grandmothers are more active and more comfortable in the grandparenting role than grandfathers, and grandparents in their 60s are able to be more active than younger or older ones. In the United States, black Americans, Asian-Americans, Italian-Americans, and Hispanic-Americans are likely to be more involved in the lives of their children and grandchildren than members of other groups are (McCready, 1985).

Gone is the stern matriarch or patriarch of yesteryear. Now grandparents are more often fun-loving, indulgent playmates.

Differences within these broad ethnic groupings are apparent as well. For example, Italian-American grandmothers tend to be much more satisfied and involved with grandparenting than Italian-American grandfathers, who tend to be more distant. Among Hispanics, Cuban-Americans are least, and Mexican-Americans most, likely to be involved with the daily lives of their descendants (Bengston, 1986). The timing of grandparenthood can be crucial to satisfaction with the role, and this is particularly true for black grandmothers, who are traditionally very involved in their grandchildren's lives. Black grandmothers who are under age 40 or so often feel pressured to provide care for a new grandchild they were not eager for, while those over age 60 tend to feel that they are fulfilling a cherished role.

It is important to note that specific styles of grandparenting vary greatly. This fact is highlighted by the case of an 18-month-old child who had grandparents from very different ethnic backgrounds, one pair Latin, the other Nordic:

Her Latin grandparents tickled, frolicked with, and cajoled her. Her Nordic grandparents (who love her no less) let her "be." Her Latin mother thought her in-laws were "cold and hard" while her Nordic father thought his in-laws were "driving her crazy." The youngster was perfectly content with both sets of grandparents. [Kornhaber, 1986]

Developmentalists are convinced that the particular style of grandparenting is probably not relevant, so long as it is loving, and comfortable for those involved.

Most grandparents today also watch their grandchildren's development with interest. The mere fact that they have grandchildren and great-grandchildren is an affirmation of their own lives, giving older adults a sense of biological renewal, self-fulfillment, and even immortality. However, most of the elderly do not focus much of their energy or time on their grandchildren. As one review explains, "grandparent-grandchild interchanges are usually peripheral to the lives of both" (Troll, 1980). According to a study of 300 grandparents and 300 grandchildren of many ethnic and economic backgrounds, only about 5 percent were closely involved with each other. Another 5 percent had no contact at all. The remaining 90 percent visited occasionally but did not develop intimate relationships (Kornhaber and Woodward, 1981).

While many grandparents have moments of regret that they are not closer to their grandchildren, few would give up their independence and freedom in order to become more involved in their lives. In general, the older a grandparent is, the less likely he or she is to see the grandparent role as central (Cherlin and Furstenberg, 1986). The majority of grandparents seem in accord with the one who confessed of his own grandchildren, "Glad to see them come, and glad to see them go."

Although most older adults today are reasonably satisfied with their less active grandparenting role, whether it be relatively detached or involved, about a third feel a certain amount of discontent. Some are not pleased with the way their grandchildren are being raised, but feel unable to do much about it. Others feel disappointed that their relationship with their grandchildren is not more affectionate. This sense of distance is especially likely to occur if their own memories of their grandparents are positive and warm. As one woman said:

> I would have loved for my grandchildren to grow up the way I grew up. A close family, lots of people around all the time. I loved my grandparents. I would go over to their house all the time. It was wonderful. I would so love to have that feeling of closeness to the children. But there's a wall, a barrier. . . . I've lost my grandchildren and they've lost me. [quoted in Kornhaber and Woodward, 1981]

Some might regard such a view as old-fashioned and sentimental, asserting that the contemporary roles grandparents and grandchildren play in each other's lives is better than the past roles: with less involvement comes less conflict, especially between parents and grandparents. Given the vast differences between the childhood experiences of the oldest generation and the youngest, there may be some merit to this position. At the same time, however, a close relationship with a grandparent can often be a buffer against family stresses (Cherlin and Furstenberg, 1986; Kornhaber and Woodward, 1981). Grandparents can provide warmth, love, confidence, and a positive image of family continuity to their grandchildren. When grandparents accept this role, their grandchildren feel proud and fortunate. As one young girl in a study of grandparents' practices and attitudes expressed it:

> Grandma has got a bad leg, so she can't walk around without her cane. I can sit on her lap, though, and she tells me stories about when she was young and I can cuddle up with her . . . She is so cozy. She can't walk too well, but she can talk. And she is the best back rubber in the world. [quoted in Kornhaber and Woodward, 1981]

When the bond is close between grandparents and grandchildren, both generations are likely to benefit. As one grandmother in the same study explains:

I guess my grandchildren are a big part of my life right now. They come over very often, spend the night . . . In some way it is my job to spoil the kids a bit when they are young, listen to them when they are older and do things for them that their parents do not have time to do. It's different from being a parent. Special for me and for the kids . . . Let's get it straight—I need them.

At least in the opinion of many researchers, the more remote grandparenting that is increasingly common in today's families provides more independence for each generation but at considerable cost. The sense of continuity and awareness of human interdependence that every generation needs may be lost.

Service in the Community   Another important avenue for continued achievement is volunteer work, which is a part of life for 32 percent of older adults (Volunteer, 1984). Perhaps because of their perspective on life, or because of their patience and experience, they are particularly likely to work with the very young, the very old, or the sick.

Much of this volunteer work is informal. In stable neighborhoods that have a sizable number of elderly, the more capable run errands, fix meals, repair broken appliances, and generally make it possible for the disabled elderly to continue to live at home. For example, in one four-block residential area, routine sharing and socializing allowed three widows in their late 80s, one of them blind, one obviously failing cognitively, and one physically disabled, to remain in the community. Their neighbors shopped for them, cooked for them, and simply checked in on them often enough to make them feel part of the neighborhood, while still allowing them some privacy and independence (Rosel, 1986).

While such informal volunteer help is the most frequent, some of the elderly also become involved with more structured volunteering, often through churches or

Figure 25.6 *Obviously the boy at left, whose brain damage requires him to wear a protective helmet, appreciates the comfort his Foster Grandmother provides. Just as clearly, the boy on the right is pleased to have a Foster Grandfather to play basketball with. No one would dispute the benefits Foster Grandparents can provide, but not everyone realizes the benefits they receive: volunteering tends to make an elderly person healthier and happier.*

public schools. Several volunteer organizations now appeal directly to the retired senior citizen. One such organization is SCORE (Service Corps of Retired Executives), whose volunteers provide advice to owners of small businesses. Another is Foster-Grandparents, which is designed to help an older person "adopt" a child who needs the help a grandparent could provide. Volunteering would probably be more common if other agencies and the elderly themselves realized how useful they could be. The response of one older woman when she and her husband were asked to work with stroke victims is not unusual:

> Of what possible use could we be, a seventyish couple who had none of the seemingly necessary skills, no knowledge of speech or physical therapy, no special social skills either?

Several months after this couple overcame their self-doubt and began the volunteer work, the woman wrote:

> The real question was [not] "What can we do?" [but] "What can we be?" Can we be warm, caring unstroke-damaged human beings, to meet . . . with stroke-damaged ones, exchanging concerns, playing word games, encouraging them to feel at ease, to talk and tease and laugh with us and with each other . . . I wish everyone who feels so useless and lonely could have such a completely satisfying experience as Tom and I are having. [quoted in Vickery, 1978]

Figure 25.7 *While the politically active older citizen is more likely to vote, write letters, and make phone calls than to take to the streets, the Gray Panthers are an important exception. They participate in public demonstrations, organize economic boycotts, and even engage in civil disobedience. Perhaps their biggest success is in public relations: they jostle our ageist stereotypes about the elderly as passive, docile, and conservative.*

Many older adults become more involved in other community activities. Some become more active church members, not as volunteers, but simply as participants in worship, study, and fellowship. Others become more active citizens. In the United States, those over age 65 are more likely to be registered and to vote than members of any other age group. In fact, their participation rates are about twice that of the youngest voting group, those between 18 and 20 years old (U.S. Bureau of the Census, 1986). The major organization affecting the elderly, the American Association of Retired Persons (members must be over 50, but need not be retired), had a membership of 24 million in 1987. The political influence of this organization is one reason Social Security is usually immune to budget-cutting that affects most other domestic government programs.

Affiliation: The Social Convoy

The phrase **social convoy** highlights the truism that we travel our life course in the company of others (Antonucci, 1985). At various points, other people join and leave our convoy, but just like members of a wagon train headed West, we could never make the journey successfully by ourselves. Furthermore, the bonds formed as we journey together help us in good times and bad. It is more pleasant to share triumphs with those who know how important the victory is; it may be critical for our survival in times of defeat and sorrow to have familiar confederates whom we have helped in the past. For older adults particularly, the social network's continuity over time is an important affirmation of who they are and what they have been. Friends who "knew them when" are particularly valuable, as are family members (especially a spouse) who share a lifetime of experiences.

Friendship

By late adulthood, many members of the social network have been part of a person's convoy for decades. This helps explain a surprising finding: older people's satisfaction with life bears relatively little relationship to the quantity or quality of their contact with the younger members of their own family, but shows substantial correlation to the quantity and quality of their contact with friends (Antonucci, 1985; Essex and Nam, 1987; Larson, 1978; Lee and Ellithorpe, 1982).

Particularly important is the quality of friendship. Having at least one close friend in whom to confide acts as a buffer against the loss of status and roles that comes with such common experiences of the aged as retirement and widowhood.

In trying to understand why friends are so important, many researchers have hit upon the fact that older people are very concerned that they not be a burden on anyone, that they help others just as others help them. This reciprocity is an important keystone of friendship, particularly in late adulthood (Roberto and Scott, 1986). Especially if the friendship is of long duration, both members can remember times when each of them was indebted to the other. This sustains not only the friendship but also pride in oneself.

Given the importance of friendship, it is comforting to know that most elderly people have at least one close friend. In fact, older adults are less likely than younger ones to want more friends (Antonucci, 1985). As at all ages, women have larger social circles, and more intimate relationships with their friends, than men do.

Figure 25.8 *Sitting in the park on a sunny day, with good friends, watching the world go by—a favorite activity for the young and old. Looking at the posture and expressions of these four, it is not hard to imagine them sixty years earlier, as young teenagers, hanging out on the same bench.*

Neighborhood

The importance of the ongoing social convoy also helps explain a notable characteristic of most older adults: they like to stay put. Most of them have a well-established niche in their neighborhood, with a network of friends and acquaintances that they are reluctant to move away from. Often the web that holds this network together has been carefully constructed over a period of years, built on everything from friendly conversation and mutual assistance in daily errands to major support and crucial understanding in time of trouble (Rosel, 1986; Wentowski, 1981). When older Americans do move, they more often move to another part of town than to a new and different community (Warnes et al., 1985). Only one out of every six of those who move crosses state lines (Lawton, 1985).

The need to be in a neighborhood of like-minded people is particularly important for the current cohort of older Americans, many of whom are immigrants or first-generation Americans. Having access to friends, churches, temples, cultural sources, and social clubs of one's own ethnic background is often the difference between social interaction and lonely isolation (Cantor, 1979).

Age-Segregated Communities   Most elderly people are not immediately attracted to the idea of living in **age-segregated communities** restricted to senior citizens. The main reason many people finally decide to move into such facilities is financial: for a variety of reasons, the cost of a housing unit designed for the elderly is much less than comparable housing designed for mixed age groups.

Once people have actually settled in this type of housing, they usually find that it has a number of other advantages to its credit (Lawton, 1980). For one thing, such communities generally offer a combination of independence and security, usually with medical help nearby, all of which are very important to the well-being of older people. Even more important, age-segregated housing provides a natural setting for making friends, since all the residents are about the same age and often share similar problems. Jennie Ross (1977), a young social scientist who lived in a new housing project for the elderly, found an openness and comradery among residents that initially surprised her. She was particularly struck by their readiness to discuss the troubles of old age with humor and sympathy. Ross also noted a sense of community and mutual concern that pervaded the residence. For example, when the first resident died, the young director, thinking that the death would upset the other residents, attempted to have the body whisked away before anyone knew what had happened. However, rumors of a death had circulated, and many residents gathered in the lobby to pay their last respects, managing to assemble only moments before the hearse arrived to remove the corpse by a side door. Subsequently, the residents petitioned the director that henceforth all deaths should be announced, and that those who wanted to attend the funeral should be provided transportation.

Similar findings were reported by Arlie Hochschild (1973), who studied a small apartment-house residence for senior citizens in California. One indication of the mutual concern that existed in this community was the fact that the residents informally but conscientiously checked on each other's health every day—at the same time respecting established signs that indicated when one did not want to be disturbed. An assured balance between the opportunities for social interaction and for privacy is, in fact, a crucial element in housing for the elderly (Fennell, 1985).

Figure 25.9 *Two retirement villages, one in Maryland and one in Iowa, both illustrate the same point. In such communities, social relations are easy to establish and maintain. For instance, the opportunity for a friendly chat or a group activity is usually just a short walk away.*

Given their many advantages, it is surprising, indeed, that so few of the elderly consider moving into retirement communities or senior-citizen housing projects. This reluctance may well be the consequence of the ageism of our culture, which derides these communities as "geriatric ghettos" or "fogey farms" (Ross, 1977), while ignoring the security, affordability, and friendship they can provide.

Marriage

Most of the current cohort of older adults have, or had, a spouse. In fact, the current cohort of older adults, born in the beginning of the twentieth century, may well be the most and longest married ever. Their marriage rate was higher than earlier and later cohorts, and fewer of their marriages ended before old age, either through death, as it did in many earlier marriages, or through divorce, as it did in many later ones (Norton and Moorman, 1987).

This cohort, then, would seem the best suited to inform us about marriage of long duration. In general, the reports are encouraging. Older husbands and wives tend to be happier with each other, and with their marriages, than they have been since they were newlyweds (Anderson et al., 1983; Maas and Kuypers, 1974). Older marriages are particularly less likely to be characterized by serious disagreements: apparently, the conflicts that many couples experienced in younger years have either disappeared, diminished, or no longer seem worth disputing.

Figure 25.10 *Browning's famous lines "Grow old along with me/The best is yet to be" seem prophetic of many marriages that last forty years or more.*

However, while age of the spouses and length of marriage are both positively correlated with satisfaction, these factors alone are not very powerful predictors of whether or not a particular marriage will be satisfying. Indeed, not every study finds this positive correlation (Ade-Ridder and Brubaker, 1983). Much more important to marital satisfaction is that the spouses be committed to each other as individuals—and therefore prepared to let their relationship evolve—rather than simply being rigidly committed to marriage as an institution (Swensen and Trahaug, 1985). It seems that a satisfying marriage, even in late adulthood, requires continual adjustment as life circumstances change. Let us look at two common, and possibly upsetting, changes, retirement and poor health.

The Effect of Retirement on Marriage   We have already noted that retirement can create stress in the retiree. Often retirement provokes stress in the marriage relationship as well, since both partners must decide anew how to spend their time and money together, agreeing on such things as daily routines, vacations, change of residence, and new budget patterns. Even the best of partnerships experience some turbulence in the resolving of these questions. If the marriage is a troubled one that held together over the years because work allowed the spouses to avoid each other, then retirement sometimes precipitates divorce (Weinstein, 1979).

Problems commonly arise for many couples approaching retirement when both partners are wage-earners. Usually, husbands are older and have been working longer than wives, and consequently are ready to retire first. Typically, the husband's retirement precipitates the wife's (Davies and Sparrow, 1985). If the wife enjoys her job, she may resent having to leave it; and if she does not leave, her husband may resent her continued employment.

Women who have always worked in the home rather than outside it may also have problems with their husband's retirement. Once he no longer works, she loses some privacy and independence (Keating and Cole, 1980). For example, many older women are used to being an autocrat in "their" kitchen, or to having regular daily patterns—mid-morning coffee with a neighbor, afternoon shopping at the local stores—that might conflict with their husband's demands and wishes. One woman complained bitterly that she was made to watch "like a dumbbell" while her husband showed her how to wash a glass, hold a paring knife, peel a hard-boiled egg, and perform dozens of other little chores that she had done perfectly well for forty years (Sorenson, 1977).

A more common pattern is that the husband helps out a good deal around the house, doing chores that the wife once did. Couples typically evolve a clear division of labor, so that the husband always does the cooking or dish-washing, or, more often, the tasks that are considered more appropriate to a man, such as vacuuming and shopping (Atchley and Miller, 1983; Dobson, 1983). With time, both partners usually adjust to new patterns, establishing routines that allow them to share more activities than they once did, yet maintaining enough independence to keep them both happy.

Poor Health   The other problem that eventually occurs in most long-term marriages is the serious disability of one or both spouses. As we will see later in this chapter, the enormous burden of caring for a seriously ill person over a long period of time cannot be minimized. However, if the demands on the healthier spouse are not too great, most older couples adjust fairly well. Indeed, one study of seventy-six older couples in which one spouse or the other was ill found that although sickness did change the marriage relationship somewhat, it did not affect overall marital satisfaction. As the researcher learned from her interviews:

> Successful marriage in old age is neither a romantic relationship nor one laden with positive or negative emotions. Instead, there is a muted quality . . . where the fact of the marriage's mere survival connotes success. The many years of shared experiences, of hardships as well as successes, are usually viewed as a source of cohesion. With the illness of one spouse, when additional demands are placed on the marriage, the interdependence that had developed over the years appears to provide the means to meet these needs, usually without reservations. . . . some conflict was observed in these interviews but . . . disagreements are handled with a good-humored joking, sarcastic remarks, or teasing, rather than overt arguments. [Johnson, 1985]

In sickness or in health, then, the overall finding is that older people are usually better off married than not. Married older adults tend to be healthier, wealthier, happier, and less lonely than those who never married, or who are divorced or widowed (Haring-Hidore et al., 1985; Peplau and Goldston, 1984; Smith and Zick, 1986; Ward, 1979).

The Single Older Adult

While marriage is generally preferable, almost half (47 percent) of all adults over age 65 are not presently married. Contrary to the custom during earlier decades in the United States, most of this group live alone. Indeed, the number of elderly women living alone has more than tripled since 1960: in 1985, 40 percent of all women over age 65 lived alone (U.S. Bureau of the Census, 1986). Less than 10 percent of the unmarried elderly live with younger family members. (The same trends are apparent in virtually ever industrialized country except Japan, where most of the elderly still live with their adult children and only 6 percent live alone [Lawton, 1982; Palmore and Maeda, 1985].) Does this mean that most of the unmarried aged are sad and lonely most of the time? Not at all, as we will now see.

The Never-Married About one in twenty older males, and one in eighteen older females, have never married. Since they have spent a lifetime without a spouse, they usually have developed long-standing, alternative social patterns and activity preferences that continue to keep them active and happy in late adulthood as long as their health is reasonably good. The women, particularly, are often much involved with their relatives, caring for an aged parent, living with a sibling, or actively helping nieces and nephews (Allen and Pickett, 1987). In fact, one study of loneliness found that never-married women compared favorably with married ones (Essex and Nam, 1987).

Loneliness in Elderly Women

HOW OFTEN LONELY?	Never	Hardly Ever	Some-times	Often
Married women	38%	33%	23%	6%
Never married	29%	44%	24%	3%
Formerly married	15%	33%	36%	16%

Of course, not every single person is socially inclined. There are some "lifetime loners," usually men who chose long ago to avoid close ties. In urban areas, many live in single-room occupancy (SRO) hotels, where, as one observer explains, "isolation and loneliness is a price they are prepared to pay to maintain their independence" (Stephens, 1976). Yet even here, an informal network to get odd jobs and to survive with little income evolves. For instance, one SRO resident obtained merchandise suitable for sidewalk sale and then hired several of his coresidents to help him sell it. Another discovered that an iron turned upside down between two Gideon Bibles made a good hot plate, an innovation he quickly passed on to his coresidents.

The Divorced Older Adult Good adjustment also seems to be typical of those elders who are divorced, as about 3 percent of the men and 5 percent of the women are (Maas and Kuypers, 1974; Perlman and Peplau 1984). Especially if the divorce occurred long ago, by late adulthood, most divorced persons have worked out satisfactory patterns of interaction and lifestyle. However, two problems are more frequent for divorced older adults (Hennon, 1983). First, they tend to have smaller incomes than their married or single contemporaries who have had comparable education. Second, for an unfortunate minority, family interaction is problematic, because children and other relatives sometimes continue to blame the divorced person for the family breakup.

In Chapter 22, we learned that divorce after a long marriage is more difficult than divorce earlier. This is probably true in late adulthood as well, so a divorce after fifty years of marriage would be particularly hard. Fortunately, this is rare: only one in a hundred divorces involves a spouse aged 65 or over, and a disproportionate number of these occur in marriages that are relatively recent (Uhlenberg and Myers, 1981).

Widowhood   Half of all married older adults will, obviously, experience the death of a spouse, one of the most serious life stresses a person can undergo. Most surviving spouses will then spend several years, even decades, alone. This is particularly likely for widows: because the average adult woman lives six years longer than the average man, and the average husband is three years older than his wife, a wife is much more likely to be the surviving spouse. In fact, there are five times as many widows as widowers over the age of 65.

For both widows and widowers, the first months alone are generally hardest, for obvious reasons. The death of a mate usually means not only the loss of a close friend and lover but also a lower income, less status, a broken social circle, and disrupted daily routines. It is not surprising, then, that widows and widowers are more likely to be physically ill in the months following the death of their spouse than their married contemporaries are. Widowers particularly are also at a markedly increased risk of death, either by suicide or by natural causes (Stroebe and Stroebe, 1983; Svanborg, 1985).

Sex Differences in Adjustment.   In general, living without a spouse is somewhat easier for widows than for widowers, especially in the first few months (Stroebe and Stroebe, 1983). One reason is that, most likely, elderly women expect to outlive their husbands and, to some degree, anticipate and make arrangements for some of the adjustments that widowhood will require. In addition, the recently widowed usually have friends and neighbors who themselves are widows and who are ready to provide sympathy and support.

In most communities, widows can also get help from formal widow-support groups.* One longitudinal study found that widows who participated in a widow-to-widow support program were, two years after their husband's death, significantly less anxious, less distressed, and less impelled to hide their true emotions than were the control-group widows who had not been in the program and instead relied solely on friends and family for help (Vachon et al., 1980).

Another sex difference also favors widows. Many of the men who now are elderly grew up with restrictive notions of masculine behavior, and tended to depend on their wives to perform the basic tasks of daily living (such as cooking and cleaning) and to be their main source of emotional support and social interaction. When their wives die, they often find it hard to reveal their feelings of weakness and sorrow to another person, to ask for help, or even to invite someone over to chat.

---

*Addresses for self-help groups of various kinds, including those for widows, can be obtained from the Self-Help Clearinghouse, City University of New York, 33 West 42nd Street, New York, NY, 10036.

While women have an easier time coping with the emotions of losing a spouse, they have a harder time with the financial consequences. They tend to have smaller pensions (they are eligible for only half of their husband's Social Security benefits), and less knowledge about savings and investments, than their husbands did. Widowhood often precipitates poverty (Smith and Zick, 1986).

Widows must also come to grips with a hard truth: they are probably never again going to find a man to be their close friend and lover. Men, however, once they have gotten over their initial depression, discover they are a much-sought-after item—the unmarried, mature male. There are many women who would happily fix their meals, clean their houses, and marry them, if possible. Statistics on remarriage bear this out: widowers are far more likely to remarry (often to women considerably younger than themselves) than widows are.

Remarriage

Figure 25.11 *Remarriage between widow and widower in late adulthood is likely to benefit them both, financially, socially, and sexually. The only likely problem is grown children, who sometimes complain about their parent's choice of partner in words quite similar to those used by parents to complain about their children's mates.*

Not surprisingly, remarriage in late adulthood tends to be happier than remarriage in earlier adulthood (Campbell, 1981). One reason is that both partners are usually widowed rather than divorced. Their former experience of a lasting love relationship helps them build another such relationship. In addition, since any children from former marriages are usually full-grown, neither spouse needs to take on an active, and often troubling, stepparent role.

This does not mean, of course, that remarriage in late adulthood is problem-free. Some widows and widowers idealize their first spouse and fault the second by comparison. Sometimes, children, grandchildren, or siblings-in-law also make comparisons that can hurt the "interloper." In some instances, inheritance is a bone of contention if grown children think money that should have gone to them is being spent on an elderly stepparent.

Yet even when such problems exist, marriage in late adulthood is usually much better for the couple than are the alternatives (Kohn and Kohn, 1978). In most respects, it is less expensive and less lonely than living alone. For many older people, it also provides sexual affirmation. As one 82-year-old widower, recently married, explained: "Everyone tells me about companionship. It gets me mad. Sure, I want companionship, but I also got married for sex. I always had an active sex life and still do" (quoted in Starr and Weiner, 1981).

The Frail Elderly

So far in this chapter we have emphasized the majority of the elderly—those who are active, financially stable, supported by friendship and family ties. These are the capable ones, the young-old. They are quite different from the group we focus on now—the physically infirm, the very ill, or the psychologically impaired, who are called the **frail elderly.**

Although the frail elderly are a minority, they are a growing minority, for three reasons. First, as average life expectancy continues to increase, the fastest-growing age group in the industrialized nations are those over age 85, among whom only a small minority are completely free from chronic illness or disability.

The second reason is that our medical institutions are generally geared toward dramatic life-saving intervention for acute illness: organ transplants and possible cancer cures are what capture the medical profession's attention. However, most of the problems of the elderly are chronic, needing long-term ameliorative care.

Relatively little research money or time goes to such problems, so the ill elderly who cannot be quickly cured, and who do not quickly die, are destined for years of frailty. Obviously, the long-term solution is better understanding, treatment, and prevention of the conditions that lead to frailty, so that, as life expectancy increases, the months or years of frailty do not continue to increase as well.

The third reason the number of frail elderly is increasing is financial. In ageist societies that tend to see poverty in old age as a private problem rather than a public responsibility, inadequate health care and the circumstances of their daily life often make those who are old and poor become frail as well. Let us look at some of the links between poverty and frailty.

Poverty and Old Age

The statistics on wealth and the elderly presented on page 579 are reassuring. However, it is important to keep in mind that, even today in the United States, one out of every eight older persons is poor. For this minority, life may be very hard indeed, for poverty exacerbates the problems of old age while putting many benefits out of reach. Furthermore, looking at the characteristics of those who are most likely to be poor suggests that many of them might have to cope with compounding problems as well. The elderly poor are more likely to be female (mostly widowed) than male, and much more likely to be black or Hispanic than white. Black elderly people have one chance in three of being poor; Hispanics, one chance in five (U.S. Bureau of the Census, 1986).

Like the younger poor, the impoverished elderly are frequently blamed for their plight. They are criticized for squandering their money, for not having planned ahead. This "myth of improvidence" (Butler, 1975) ignores the facts. Many of the elderly poor have been at the edge of poverty all their lives: they never had a chance to save for their old age. And many of those who were able to save have had their financial position severely compromised by the high inflation of the 1970s (whatever savings one had in 1970, for instance, would have decreased in buying power by more than 50 percent by 1980) and by the rising cost of medical care, which has tripled over the past fifteen years.

Figure 25.12 *The impoverished elderly are far from being a typical cross-section of the American population. A disproportionate number are females from nonwhite groups: they tend to live in large cities and to be without close relatives nearby.*

It is sometimes claimed that there would be no poverty in late adulthood if adult children met their obligations to their parents. Once again, this ignores the facts. Because the current cohort of elderly were of childbearing age during the Great Depression, many had relatively few children. Indeed, an estimated one out of every four older adults has no living children at all, and the rate is even higher among blacks.

People often overestimate the effectiveness of the many special services offered to the elderly. Some even think that with such programs as subsidized housing, senior-citizen centers, reduced fares on public transportation, visiting-nurse services, Meals on Wheels, and various commercial discounts, all older Americans have their basic needs met. In fact, many of these resources are less likely to reach the elderly who are poor, alone, and in ill health than to reach their more self-sufficient contemporaries (Zopf, 1986). As one gerontologist explains, most housing for the elderly

> is filled with the relatively less needy seniors. Most services . . . (senior centers, meal sites) are captured by the less needy who often actively discriminate against the needy, the culturally different, the eccentric, and the mentally ill. The senior in need is thus faced with a series of "no vacancy" signs and overt and covert discouragement to participate. The tragedy is compounded by the realization that often the places are taken by people who could afford to compete in the real world market for meals, housing, and nursing service. [Knight, 1982]

Inadequate housing is a particular problem. Most of the poor elderly live either in rural areas or inner cities. In the cities, they usually live in substandard housing in crime-ridden neighborhoods, a fact of life that necessitates a type of disengagement. For instance, one study of the black elderly in Washington, D.C., found that virtually all of them were significantly affected by their fear of crime. In fact, two-thirds were reluctant to leave their homes even in the daytime (McAdoo, 1979). Interestingly, the victimization rate in Washington, and elsewhere, is less for the elderly than for younger adults. However, fear of crime is greater (Lawton, 1985), perhaps because of the physical vulnerability that almost all the elderly feel.

The rural poor have less reason to fear crime, but their housing is generally far worse than even that of the urban poor (Lawton, 1980). Inadequate plumbing, leaky roofs, and antiquated heating and lighting are difficult for anyone to cope with, but they are especially so for the aged. In addition, since few rural areas have adequate public transportation, getting around becomes a major problem, as both walking and driving grow increasingly difficult with age.

Poverty and Health   A major consequence of poverty in old age is inadequate health care. In the United States, many people mistakenly believe that **Medicare,** the federal health insurance plan for the elderly, covers all the health expenses of senior citizens. In fact, it covers only about half of them. Those individuals who cannot pay the remainder must apply for **Medicaid,** the health-care plan for the poor. In order to qualify for Medicaid, the elderly person must have gone through all his or her assets (such as a home that has provided low-cost shelter and security for decades) to pay for medical care. It should be noted that, on the whole, the quality of care received through Medicaid is much less than that privately paid for, largely because the low fee scale and slow payment discourages many good clinics, doctors, and nursing homes from accepting Medicaid patients. In addition, neither

Medicaid nor Medicare pays for transportation to and from the doctor, for special diets, for all prescriptions, or for many other health-related expenses. Not surprisingly, then, the poor elderly are in worse health than their wealthier contemporaries. Poverty and frailty go hand in hand.

Caring for the Frail Elderly

When thinking about the care of the frail elderly, it must be kept in mind that their main goal is not necessarily to keep on living as long as possible but to keep on living as independently as possible. Quality of life is crucial. Because of their wish to be independent, most do not want to enter nursing homes, but they do not want to "burden" their children either. When a sample of older Americans were asked "Would you like to live with one of your children if something happens that you cannot live alone?" 96 percent of the whites, 90 percent of blacks, and 50 percent of Mexican-Americans answered no (Bengston, 1986).

In fact, for many of the very old and unwell, maintaining a measure of independence despite their advancing age and infirmity is a source of great pride (Rosel, 1986). Thus the first steps to be taken to aid an increasingly frail person should be measures that allow continued independence: rearranging home routines to fit the frail person's abilities, finding a part-time housekeeper or visiting nurse, installing communication for summoning emergency help, and so on.

Care by Relatives   What should be done when the elderly person can no longer manage self-care, when even the basic **activities of daily life (ADLs),** such as eating and bathing, are impossible? Most of the frail elderly have been traditionally, and still are, cared for, day and night, by their closest family members. If they are married, their spouse usually provides the care; if they are widowed, their daughters or daughters-in-law usually care for them. Note that, since many of these frail elderly are aged 85 or older, their care-giver is often elderly too—a wife who is 80 herself, for instance, or a 65-year-old daughter.

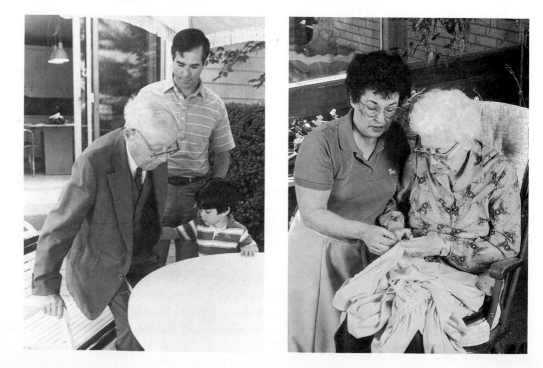

Figure 25.13 *Although the popular press sometimes bemoans the indifference of the younger generations, the facts are otherwise: most of the frail and widowed elderly, like these two, are cared for by their children at home. Only when aged people are very ill, or have no descendants able to help, do they enter a nursing home.*

The demands of providing care for a frail elderly relative should not be underestimated. In many cases, the care-giver must forgo all other activities, because the physical work and psychological stress are overwhelming. One daughter reports on the strain she and her elderly father experienced when her mother developed Alzheimer's disease:

> I worked the entire time through four pregnancies . . . returning to work within six weeks of delivery. It was a piece of cake compared to trying to cope with a combative, frustrated adult who cannot dress, bathe, feed herself; who wanders constantly. A person faced with this situation . . . having to work a full day, raise a family, and take care of an "impaired" relative would be susceptible to suicide, "parent-abuse" . . . possibly murder.

> My father tried very hard to take care of her, but a man 84 years old cannot go without sleep, and cannot force her to take care of her personal cleanliness. Up until two years ago, she was taking care of the finances and household. Her signature was beautiful . . . now it's just a wavy line. An 84-year-old man does not learn to cook and balance the budget very easily, and he becomes bitter. He did not want to put her in the nursing homes he visited in, and so he reluctantly sold his house and moved to a city he didn't like so that his children could help with her care. It has been a nightmare . . . she obviously belonged in a secondary-care facility because no one can give her 24-hr. care and still maintain their sanity and families. But the real victim is Dad . . . his meager income eaten away by the nursing home . . . separated from his wife of 50 years . . . stripped of his house, car, acquaintances . . . dignity. He is the real victim.

It is eloquent testimony to the power of human love and selflessness that most family care-givers continue to provide care for months and years, with little respite or reward. Unfortunately, those who might provide relief for the primary care-giver, including other relatives and public agencies, sometimes assume that it is not really necessary. Yet the amount of stress on the primary care-giver is more strongly correlated with the degree of assistance from others than with any other variable, even the variable that might seem most important—how impaired the aged person is (Zarit et al., 1980).

Elder Abuse   The stress on care-givers to the frail elderly helps explain an unfortunate problem that has recently captured public attention: **elder abuse.**

There is no doubt that a number of the frail and confused elderly are mistreated by their care-givers. However, the extent of the problem is unknown, and, indeed, even the definition is unclear. Obviously, physical mistreatment is abuse, but such abuse is rare in the case of the elderly. What about locking the door so a confused elderly person cannot wander from his or her bedroom at night? What about forcing an older person to eat, or to take medication? What about yelling in anger, or neglecting to talk with, respect, or transport an elderly person as much as he or she would like? One study of adult children who cared for their parents found that while only 1 percent had ever hit or shaken a parent, 14 percent had forced medication, and 43 percent had yelled (Steinmetz and Amsden, 1983). Since these data are based on the self-reports of the care-givers, it is likely that the actual percentages are higher.

In trying to mitigate elder abuse, it is important to understand the stress the care-giver is under. Typically, one person is the chief care-giver twenty-four hours a day, and often the elderly person not only requires a great deal of help but also makes the situation worse than it might be—interfering with privacy, making special demands, evoking sympathy and guilt.

In fact, in the study cited above, care-givers reported that their parents did not respect their privacy (40 percent), refused to eat (16 percent), and hit or shook them (19 percent). Few adult children have any preparation for helping a frail, dependent, and demanding parent who behaves in an "abusive" manner.

Clearly, help for the care-givers and the frail is needed to protect both members of the dyad. Unfortunately, in many cases, the solution to an untenable situation at home seems to be a difficult one for all concerned—institutional care.

Nursing Homes

Many older Americans and their relatives feel that nursing homes should be avoided at all costs, usually because they believe that all nursing homes are horrible.

Some nursing homes are indeed horrible. The worst tend to be those profit-making ventures where most patients are subsidized entirely by Medicaid. The only way for these institutions to make a profit is to cut down on expenses. Consequently, they are staffed by overworked, poorly trained aides who provide minimal, often dehumanizing care. Typical are the circumstances described by a participant-observer who worked in such a home:

> The rush imposed by a heavy work load leads aides to treat all the patients in the same fashion: It becomes legitimate to stuff food down their throats because the goal has become serving the meal, not nourishing the patient, or to lift them in and out of bed as if they were inanimate dummies because the goal is making the bed, not making the patient comfortable. [Fontana, 1977]

However, some profit-making nursing homes are excellent. Consider the case of Laura Hunter (Hunter and Memhard, 1981), for example. She had reached the age of 80 "bedridden, arthritic, and crotchety . . . relying on drugs, and incapacitated by fears of impending change, illness, and death, [she] clung to the radio for company and turned away even her closest family and friends." A period of hospitalization convinced her children that something had to be done. After much soul-searching, they decided to place her in a nursing home, one of the best in the country. Residents there had their own rooms and were able to carry on private lives, having friends in to visit, for instance, or going to sleep when they wanted to. The staff encouraged the residents to stay active, and the residents had their own council, which helped determine residence policies.

Initially, Laura was despondent and uncommunicative, but gradually she came out of her shell and joined the community life around her. Among other things, she made friends, joined a book-review club and an exercise class, won election to the residence council, worked as a reporter for the residence newspaper, and developed a romance. She also kept a journal of all those activities. Reading it, one gets the impression of a spunky, good-humored lady with a love of life. She needed good nursing-home care, surrounded by people who could become her friends, to help her express that love.

Fortunately, most nursing homes are better than the one in the first example; unfortunately, few are as good as the one in the second. One problem with many homes is that they concentrate almost exclusively on the physical maintenance of their residents and give insufficient attention to the residents' psychological needs, such as social interaction and a feeling of social control.

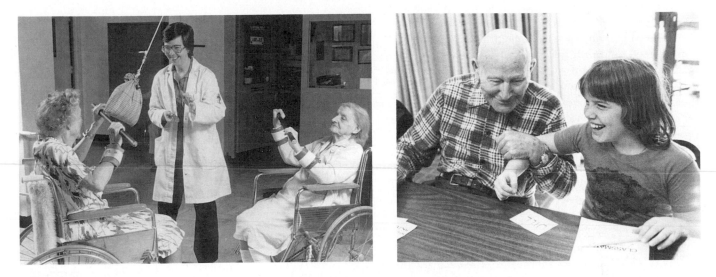

Figure 25.14 *Good nursing homes encourage residents to participate in regular physical exercise, and the best provide physical therapy for those who need it. Good nursing homes also encourage contact with the outside community, sometimes including visits from schoolchildren to allow friendship between the generations. This type of social contact is especially important, since a disproportionate number of nursing-home residents have no living relatives.*

As we saw in Chapter 24 (pages 558–559), too often the staff pay more attention to those patients who docilely wait for their needs to be met, thereby reinforcing the dependence of many patients who could learn to fill some of these needs themselves. Correspondingly, the more independent patients suffer from inattention and, even worse, they learn that the best way to cope with life is by becoming passive, relinquishing control. An immediate consequence is that they become less active and lose self-esteem; in the longer term, they become less healthy and die earlier than they would have (Baltes and Reisenzein, 1986).

One simple research project illustrates this pattern. Forty-seven residents on one floor of a nursing home were asked to help in their own care as much as possible, and were given plants to care for. Forty-five residents on another floor were told that the staff was responsible for their care, and even for the care of their plants. Three weeks later, 93 percent of the first group showed improvement on measures of activity level, alertness, sociability, and self-reliance. In marked contrast, 71 percent of the second group showed declines on the same measures (Langer and Rodin, 1976). The initial experimental treatment apparently set in motion a series of positive reactions that were self-perpetuating. A follow-up study found that even two years later, the group that had been encouraged to be more independent was healthier than the conventional-care group (Rodin and Langer, 1977).

Increasingly, professionals are becoming involved in developing good nursing-home care, where the goal is to help each patient gain as much independence, control, and self respect as possible (Burger et al., 1986). Thus it is possible to find good care, if one knows what to look for, and can afford it. Quality and suitability of nursing-home care can make the difference between a long and happy late adulthood and a short, sad one (see A Closer Look, page 602).

What to Look for in a Nursing Home

The first question that needs to be asked in the search for a good nursing home is, What kind of care is needed? Some homes provide "skilled nursing care," such as that required by people with acute illnesses, as well as rehabilitative services for individuals recovering from surgery, strokes, or other serious medical problems. A second category of nursing home provides "basic nursing care" to residents who might have some health problems but mainly need help with ADLs (activities of daily living). Often an institution provides both skilled and basic care, in different wings, so that residents need not leave as their condition improves or worsens. Finally, about a quarter of the homes in the United States provide some meals, supervision, social activities, and precautionary care to the elderly who are able to manage their ADLs on their own. Strictly speaking, these are not nursing homes (the British call them sheltered housing), but they are usually included in discussions of them (Lawton, 1980).

Staff

The crucial element in quality care is the number and quality of staff. Since staff salaries are the easiest expense to cut, particular attention needs to be given to this variable in choosing a home (Burger et al., 1986). Nurses, social workers, occupational and physical therapists, and dietitians should be licensed and trained to work in gerontology. Aides should be well paid and enjoy their work. A low turnover rate is generally a sign that this is the case. At the typical American nursing home, almost half (47 percent) of the staff have been there less than a year (Lemke and Moos, 1986). Ideally, less than 20 percent should be that new to the job.

The ratio of aides to residents should be about one to eight in the daytime (Wallace, 1982), although this figure will vary depending on the kind of care required and on the number of other staff members available to deliver it.

The Facility

Safety, rather than superficial appearance, is the first consideration. Fire exits should be clearly marked; smoke alarms should be in every room, railings and bright lights, in every hall; call buttons should be in easy reach of every bed. Most, but not all, nursing homes are licensed and inspected by state and federal agencies, and safety reports are available. Cleanliness, obviously, is also an essential factor: the home should look and smell clean, preferably without conveying the antiseptic, unhomey atmosphere of a hospital.

Another important question is whether the physical facility encourages independence and individuality. Do the residents arrange and decorate their own rooms? Are

there visual aids to help those with poor vision or failing memory to go from one part of the building to another? Clearly numbered rooms and directional logos pointing the way to the dining room, the lobby, the nursing station, and so on, are good signs in more than one sense.

Activities

Outings of various kinds (shopping, attending church, sightseeing) should be scheduled and encouraged, as should a variety of active and passive recreational activities. Flexible scheduling that allows each resident to decide when, as well as whether, to engage in any given activity is important. Day rooms and dining areas should be conducive to socializing and observing other people, both favorite pastimes for many elderly people. (Generally, television-watching is antithetical to socializing, so beware a day room that appears to be organized around the TV set.) At the same time, residents should be able to watch television, or to get off by themselves, when they want to.

Links to the Outside World

Frequent visiting is not only good for the morale of all concerned; it also tends to make the staff more attentive, thereby improving the quality of care. It is therefore helpful if the home is conveniently located near old friends and family members, and has extensive visiting hours, accessible phones, and policies that make it easy for residents to leave for a visit at the home of family or friends.

Costs

Unfortunately, one of the key factors in selecting care for the elderly is cost. In the United States at least, less than half of the cost of nursing-home care is covered by private or public insurance. Medicare, being designed primarily for acute hospital care, pays only 2 percent of national nursing-home expenses. (Only when all of a person's assets have been depleted does Medicaid pick up the bill, often at such reduced rates that the person must move to an inferior home.) The actual cost of nursing homes varies greatly, with most charging between $20,000 and $50,000 a year. Nonprofit homes generally provide better care than for-profit ones, but they are not necessarily less expensive, and, in the United States, they are harder to find. Eighty percent of all American nursing-home beds are in profit-making ventures, most of them part of national chains (Burger et al., 1986).

Unfortunately, the emotions stirred up by the thought of nursing-home care often make it hard for rational planning to occur. Yet one conclusion is obvious: since nursing homes vary enormously in facilities, staffing, and policies, as well as in cost, taking the time to discuss and investigate this form of care may, in the long run, help the development of all concerned.

Prepared with the help of Beverly Reed, M.S.W.

Integrity, Identity, and Continuity

Figure 25.15 *Eugene Lang (left), a self-made millionaire, was invited to address the graduating class of the public grade school he had attended in Harlem. During the ceremony, he reflected on the outlook for the graduates, many of whom had grown up in poverty, and suddenly made an astonishing offer: he would pay for their college education if they graduated from high school. Virtually all, in fact, went on to college. In his gesture of extraordinary generosity, and integrity, Lang affirmed his personal history and made it meaningful for the next generation.*

Erikson (1963) believes that the crisis of late adulthood is *integrity versus despair.* He maintains that some of the elderly develop a sense of pride and contentment with their past and present lives, while others experience despair, "feeling that the time is now short, too short for the attempt to start another life and to try out alternate roads to integrity."

Recent research, based on Erikson's theory, finds that the process of achieving integrity is much like the process of achieving identity in adolescence (Walaskay, 1983–1984). The person must adjust to the changes of the body and to new demands from life while maintaining a sense of self. For the elderly, this is best done by affirming their individuality as well as recognizing their ties with others—not only with family and friends but also with the human race as a whole.

As we have seen in this chapter, there is no one path, no one lifestyle, no one cultural route to integrity: "A wise Indian, a true gentleman, and a mature peasant share the final stage of integrity" (Erikson, 1963). Yet, whatever life pattern a person exemplifies, those who attain integrity must believe that it is best for them. As Erikson expresses it:

> Although aware of the relativity of all the various life styles which have given meaning to human striving, the possessor of integrity is ready to defend the dignity of his own life style against all physical and economic threats. For he knows that an individual life is the accidental coincidence of but one life cycle with but one segment of history; and that for him all human integrity stands or falls with the one style of integrity of which he partakes.

In other words, instead of comparing themselves with others, those who achieve integrity become self-actualized, able to judge their lives by their own standards, and find it good. This is a goal we all could seek at whatever stage of development we are now in: to finally reach integrity, affirming our own unique life as worthwhile.

SUMMARY

Theories of Psychosocial Aging

1. Several theories have been offered to explain how the elderly react to their changing experiences with age. According to disengagement theory, late adulthood is a time when mutual withdrawal occurs between the elderly and society, as work and family roles become less available or less important.

2. Activity theory holds that the less older people disengage, the happier they are. According to this theory, most of the elderly replace one form of activity with another; if an elderly person becomes less active, it is because society or ill health requires it, not because the older person chooses to withdraw.

3. As they do regarding other stages of life, researchers disagree about how much continuity and discontinuity there is between late adult development and patterns ear-

lier in life. It is clear that some events, such as retirement or death of a spouse, necessitate some discontinuity, while enduring personality characteristics and social networks promote continuity.

4. Overall, the elderly are more diverse, in lifestyle, income, and personality than adults at earlier periods of life. Thus each of the major theories is true for some people some of the time, but not for all the people all of the time.

Employment and Retirement

5. More and more people are retiring, at earlier ages than ever before, partly because the financial incentives to do so have increased, and partly because the jobs available have decreased. After a period of transition, most retirees enjoy retirement.

6. Many retired people do volunteer work, attend classes, pursue hobbies, join organizations, and extend

their social activities. Which of these activities a particular person chooses depends largely on lifelong interests that the person always wanted to devote time and energy to.

7. The activities of children and grandchildren also continue to be a source of pride and concern. While each generation is interested in the activities of the other, and is willing to help out if necessary, relationships between the generations are not as intimate as they once were. One important reason is that today's grandparents are more likely to have homes, social lives, and interests of their own.

Affiliation: The Social Convoy

8. Friendship continues to be important in late adulthood, as a source of happiness and as a buffer against trouble. If the elderly have lived in the same neighborhood for many years, their friends are likely to be neighbors. If the elderly move to a senior-citizen community, they are also likely to find social support there.

9. Marriage also provides social support. While many of the elderly must adjust to changes in their spouse's work patterns or health, for the most part they still value each other's company.

10. The single older adult, and the divorced older adult, usually have long-standing friendships and interests that keep them active in late adulthood. Health problems, however, might be particularly discouraging.

11. About 80 percent of elderly married women, and 20 percent of elderly married men, must adjust to being widowed at some point in late adulthood, usually before age 75. The ease of this adjustment is affected by the gender of the surviving spouse: men are more likely to experience health problems as widowers, but they are also much more likely to remarry. Women are more likely to have financial difficulties, and to be comforted by friends who are also widows.

12. Remarriage in late adulthood is usually a happy event, helping both spouses feel younger, less lonely, happier, and sexier than they did before.

The Frail Elderly

13. A minority of the elderly are frail, too feeble or ill to care for themselves. The frail are likely to be poor, over age 80, female, and widowed.

14. When older persons are unable to care for themselves, they are usually cared for by a close relative—typically their spouse, daughter, or daughter-in-law. Despite the personal sacrifices this entails, most relatives consider such care part of being a close family member.

15. For a minority, however, caring for a dependent and needy older person leads to frustration, anger, or abuse. The scope of the problem is not yet known, but we do know that greater community services are needed for the frail elderly living at home.

16. When family members are unable to continue providing care for a frail elderly person, a nursing home is the usual alternative. For the minority of the elderly who must be placed in a nursing home, the quality of their final years of life can vary enormously, depending on the quality of the home.

Integrity, Identity, and Continuity

17. In Erikson's final stage of life, the older person attempts to accept and appreciate the essential worth of his or her own particular life, seeing the continuity of that life in relationship to the lives of all humankind.

KEY TERMS

disengagement theory *(575)*	Medicare *(597)*
activity theory *(576)*	Medicaid *(597)*
social convoy *(588)*	activities of daily life (ADLs) *(598)*
age-segregated communities *(590)*	elder abuse *(599)*
frail elderly *(595)*	

KEY QUESTIONS

1. Which theory of late adulthood do you think is most accurate and why?

2. Which theory of late adulthood do you think is least accurate and why?

3. What are some of the factors that lead to a satisfying adjustment to retirement?

4. What factors are associated with dissatisfaction with retirement?

5. How and why has the experience of being a grandparent changed in recent times?

6. When you are elderly, what do you hope the schedule of your typical day will include?

7. What specific needs of the elderly do friends fill?

8. What are the changes that occur in marriages of long duration in late adulthood?

9. What are some of the important factors in providing good nursing-home care for the elderly?

Part VIII

The Developing Person So Far: Late Adulthood, Age 60 Onward

Physical Development

Senescence

Changes in appearance and decline of the sense organs and major body systems continue at a gradual but steady pace. Because of declines in organ reserve, the immune system, and overall muscle strength, older adults are at greater risk of chronic and acute diseases, heart disease, and cancer. However, risk is also related to long-standing health habits.

The Aging Brain

The brain becomes physically smaller and slower in its functioning. Slowing of brain processes may be due in part to reduced production of neurotransmitters and to reduced blood flow. However, compensatory mechanisms in brain functioning may help to maintain the quality of thinking in older adults, despite these quantitative losses.

Cognitive Development

Decline in Information-Processing Capacity

Experimental testing of older adults reveals deficits in their ability to receive information, store it in memory, and organize and interpret it. However, in the tasks of real life, most older adults develop ways to compensate for memory loss and slower thinking.

Dementia

Dementia, with its progressive impairment of cognitive functioning, is not inevitable in old age but it does become more common, especially in the very old. Symptoms of dementia may be caused by Alzheimer's disease, problems in the circulatory system, other diseases, depression, or drugs.

New Cognitive Development

Many older individuals develop or intensify their aesthetic, philosophical, and spiritual interests and values in later life.

Psychosocial Development

Activities

Retirement is a major economic, social, and psychological event that can be either stressful or benign for both the worker and his or her spouse, depending on an individual's circumstances, including finances, health, outside interests, and new opportunities for achievement.

Affiliation

Older adults' satisfaction with life depends in large part on continuing contact with friends. Despite some usually temporary problems that may be triggered by retirement, marital satisfaction generally continues to improve. The death of a spouse can cause extreme stress for both sexes, but men generally have more difficulty than women in coping with the loss of their partner.

CHAPTER 26

Death and Dying

*Death's stamp gives value to the coin
of life; making it possible to buy with
life what is truly precious.*

Rabindranath Tagore
Stray Birds

One goal of the study of human development, as outlined in Chapter 1, is to help each person realize his or her full potential. According to many developmental theorists, including Freud, Erikson, and Maslow, achieving an understanding of death and dying is essential to the complete realization of self. The ability to accept death and to work through one's grief over the death of others is crucial if life is to be lived to the fullest.

A Historical View

Death can have many meanings: it can be seen as "a biological event, a rite of passage, an inevitability, a natural occurrence, a punishment, extinction, the enforcement of God's will, the absurd, separation, reunion, . . . a reasonable cause for anger, depression, denial, repression, guilt, frustration, relief . . . " (Kalish, 1985). The specific meanings that actually are attached to death, and the reactions that death prompts, vary, of course, from individual to individual and case to case. They also vary culturally and historically.

Throughout most of history, death, like birth, was an accepted, familiar event that usually occurred at home (Ariès, 1981). Family members of all ages had intimate contact with death that resulted from childbirth, from disease and infection, from accidents, and from the consequences of old age. In general, they were the ones who tended to the dying person and then to the corpse: in most cases they built the coffin, dug the grave, and buried the body themselves.

In twentieth-century North America and Western Europe, however, death came to be withdrawn from everyday life. More and more, people died alone in hospitals rather than at home among family. The disposition of the deceased passed into the hands of professionals, who sanitized and euphemized death in an effort to dis-

Figure 26.1 *Hindu funerals typically include placing the corpse on a specially built funeral pyre, which is set aflame and then pushed into the river Ganges. This ceremony is a public everyday event, a marked contrast to cremation in North America and Western Europe.*

guise its reality. They embalmed and made up the corpse to give it a normal and "healthy" look; they coined terms like "slumber room" that gave no hint of death; they supervised the burial and formalized the grieving.

This denial of death likewise came to permeate the medical profession. Doctors and nurses routinely resisted telling terminal patients the truth about their condition, and in fact, avoided the dying as much as possible. Before the end of biological life, the dying, in effect, experienced a "social death" of limited contact, muted voices, and averted eyes (Kastenbaum, 1986). The focus of the medical staff came to be "on the technical, not the human, aspects of dying" (Sinacore, 1981). Even the definition of biological death was called into question, since a person could be kept "alive" on a life-support system even though the brain had ceased to function.

Research on Death and Dying

Recently there has been a shift away from the denial of death to a more accepting view. A major factor leading to this change in attitude about death was the pioneering work of Elizabeth Kübler-Ross, a physician who was asked in 1965 by four seminary students for help in doing research on people close to death. When Kübler-Ross approached her professional colleagues for permission to interview the dying, they responded with anger and shock, and even denied that any of their patients were terminally ill. "It suddenly seemed that there were no dying patients in this huge hospital" (Kübler-Ross, 1969). Finally, the first interview was obtained, and thereafter Kübler-Ross found many other dying people who were eager to talk about their feelings, and many others who were ready to listen to the truth about their condition.

One of the first things Kübler-Ross learned from her interviews was how important informing the dying of their condition can be. She discovered that doctors sometimes told the immediate family of a member's terminal illness and then explicitly instructed them to keep "the facts from the patient in order to avoid an emotional outburst." In many cases the patients eventually guessed their fate but were unable to talk about their feelings because family and staff continued to pretend that all would be well. The result was increased isolation and sorrow for both the patients and their families. In other instances, patient and family were told of the probability of death in such an abrupt and insistent manner that all hope was destroyed. And sometimes the truth was hidden from everyone until the last moment, allowing the dying no time to put their affairs in order or to share final expressions of love with their family.

Fortunately, Kübler-Ross found some doctors "who quite successfully present the patient with the awareness of a serious illness without taking away all hope" (Kübler-Ross, 1969). To ensure that this approach would be more typical in the future, she instituted seminars designed to help health professionals interact with the terminally ill with sensitivity, honesty, and understanding. Her seminars became increasingly popular, for they apparently met a need that had previously been ignored. **Thanatology,** as the study of death is called, became a respected field of research. By the end of the 1970s, an estimated 20,000 death-education programs were underway all over North America, sensitizing schoolchildren, college students, medical personnel, and adults of all ages and walks of life to the issues surrounding death (Durlak, 1978–1979). Subsequent treatment of the topic of death and dying by the popular media has brought the subject to national attention.

A Life-Span Perspective

Thinking about Death

Research has shown that in addition to being influenced by cultural and social factors, how individuals conceptualize the realities of death is affected by the stage of development they are in (Lonetto, 1980).

Childhood Conceptions of Death

For example, children younger than 6 rarely think that death is universal, inevitable, and final. Instead, they often believe that only people who want to die, or who are cowardly, evil, or careless, actually succumb, and that dead things can be revived, perhaps by giving them hot food, or keeping them warm. Between the ages of 5 and 7, these ideas fade, and children become more aware of the reality of death.

That attitudes toward death are closely tied to age was first reported in 1948, when Maria Nagy studied attitudes toward death among Hungarian children. She found three distinct stages. Most children between ages 3 and 5 denied the permanence of death. Those between 6 and 9 believed that death exists but tended to cope with this idea by thinking of death as a person who carries away only certain people, especially people who are bad. Children older than 9 recognized death as an inevitable, biological phenomenon.

Similar shifts were reported by Gerald Koocher (1973), who first tested seventy-five North American children, aged 6 to 15, to determine the cognitive stage they were in, and then queried them about death. When asked "What makes things die?" preoperational children gave egocentric or illogical answers, such as "When God reads your name in a book" or "Going swimming alone." By contrast, concrete operational children were likely to think of specific violent causes for death, such

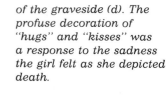

(c)   (d)

(b)

(a)

Figure 26.2 *As these drawings reveal, younger children's views of death are much less realistic or threatening than those of older children. Compare the 5-year-old's portrait of* death as a "kind of clown" (a) with the frightening mask (b) by a 12-year-old. Similarly, a 6-year-old's view of her grandmother's funeral (c) is much cheerier than a 9-year-old's version *of the graveside (d). The profuse decoration of "hugs" and "kisses" was a response to the sadness the girl felt as she depicted death.*

as "Guns," "Bows and arrows," "Eating rat poison," "Getting beat up." Finally, children at the formal operational stage of thinking realized that old age, illness, and a "worn-out body" are the most likely causes of death.

Koocher did not find any evidence of the personification of death, as Nagy had. Instead he found that children in middle childhood became interested in the specific details about death as a way of mastering their fears. For instance, if Grandpa dies, a school-age child might want to know what the exact cause of death was and if the death could have been prevented; the child might also ask about each detail of the funeral arrangements.

Families sometimes find it hard to respond honestly to a child's inquisitiveness about death. Thanatologists agree that parents who provide a good model for the child are those who are truthful in discussing death with the child—avoiding euphemism ("Grandpa passed away") or, worse, falsities ("He went away on a long trip")—and who are willing to express their own emotions and fears. With death, as with other topics, honest discussion aids development.

Conceptions of Death in Adulthood

Although adolescents and young adults realize that death is final, they seldom think of that finality as being a possibility in their own immediate lives. As discussed in chapters 15 and 17, young people are often risk-takers, because they tend to believe that harm won't come to them personally. Though they know that they will die someday, they don't believe it could be today. Their anxiety about death is extremely low—unhealthily so, since, as we saw earlier, young people are more likely to die from an accident than from any other cause.

Once adulthood is well underway, risk-taking declines, and anxiety about death depends in part on the particular circumstances of one's life. For example, one study of death anxiety in adults found that unmarried men were more anxious about death than married men were, and that married men with children were less anxious about death than married men without children. Exactly the opposite was true for women, with mothers having the highest death anxiety.

In many studies, the elderly are less anxious about death than younger adults are (Gesser et al., 1987–1988; Reker et al., 1987). In fact, on the whole, the elderly seem to view death with equanimity. For example, one study of 171 people aged 65 and older found that 35 percent of them thought about death rarely, 49 percent occasionally, and 16 percent frequently. Most of these adults categorized themselves as "death-accepters" (67 percent), and only 2 percent described their attitude toward death as fear (Wass et al., 1978–1979).

Acceptance of death in general, as well as of their own death, whenever it might occur, is characteristic of the elderly who have reached the integrity stage (Veroff and Veroff, 1980). The elderly who have a strong religious faith find it easier to accept death (Kalish, 1981).

Facing One's Own Death

Thus far we have been discussing thinking about death in general, and as a hypothetical possibility. We will now turn to what has been learned about the thoughts of people who know they are terminally ill.

Kübler-Ross's Stages of Dying

Kübler-Ross's research on death and dying (1969, 1975) led her to propose that the dying go through five emotional stages in confronting their impending death. The first is *denial*, in which they refuse to believe that their condition is terminal. Typically, they convince themselves that their laboratory tests were inaccurate, or that the disease will have an unexpected remission. The second stage is *anger*—at everyone else for not caring, or for caring too much, or simply for being alive and well. The third stage is *bargaining*, in which a person tries to negotiate away the death, promising God or fate to, say, pray more or to live a better life. When bargaining appears to have failed, *depression* sets in, causing the dying person to mourn his or her own impending death and to be unwilling to make any plans or to take any interest in medical treatment. Finally, *acceptance* can occur. Death is understood as the last stage of this life and, perhaps, the beginning of the next—a transition, not a trauma. Kübler-Ross (1969) writes:

> Acceptance should not be mistaken for a happy stage. It is almost void of feelings. It is as if the pain had gone, the struggle is over, and there comes a time for "the final rest before the long journey" as one patient phrased it. This is also the time during which the family usually needs more help, understanding, and support than the patient . . .

Evaluations of Kübler-Ross's Stages

Kübler-Ross's findings have been investigated by many other researchers, few of whom have found the same five stages occurring in sequence. More typically, denial, anger, and depression appear and reappear during the dying process (Kastenbaum, 1986), depending largely on the specific context of the death. For example, denial often occurs when the illness is one of the forms of cancer that has periods of remission. Anger often predominates when the dying feel that others are responsible for their condition, or are not sympathetic to it, as many victims of AIDS feel. Some studies show that depression increases as death nears, but it should be noted that what often appears to be depression related to dying may, in many cases, be a side effect of pain killers or other drugs.

The age of the dying person also affects the way he or she feels, according to an experienced clinician who works with the dying (Pattison, 1977). Young children, not understanding the concept of death, are usually upset by the thought of dying because it suggests the idea of being separated from those they love. A dying child therefore needs constant companionship and reassurance. The developing cognitive competencies of the school-age child often lead the very ill young person to become absorbed with learning the facts about his or her illness and treatment and about the "mechanics" of dying. Adolescents tend to think not about the distant future but about the quality of present life. Thus, to the dying or seriously ill adolescent, the effect of their condition on their appearance and social relationships may be of primary importance. For the young adult, coping with dying often produces great rage and depression at the idea that, just as life is about to begin in earnest, it must end. For the middle-aged adult, death is an interruption of important obligations and responsibilities, so most middle-aged people who know they are dying need to make sure that others will take over those obligations. An older adult's feelings about dying depend a great deal on the particular situation. If one's spouse has already died, and if the terminal illness brings pain and infirmity, acceptance of death is relatively easy.

Finally, it is critically important that the stage concept not be misused, as when a doctor or nurse trivializes a dying person's anger by offering comfort such as "There, there, you're just experiencing the anger stage" (Kalish, 1981).

Nevertheless, Kübler-Ross deserves great credit for helping many professionals, as well as lay people, realize that the dying need honesty, respect, and the opportunity to openly express their emotions. Her work has brought to prominence the idea of a good death, one the dying and their families may come closest to accepting.

Dying a Good Death

What is "a good death"? Most people would say that it is one that occurs swiftly and allows the individual to die with dignity surrounded by loved ones. The elderly are particularly agreed about this; they fear lengthy and lonely dying much more than they fear death itself (Rogers, 1980; Ross, 1977; Scott-Maxwell, 1968).

In earlier centuries, when little could be done medically to postpone death, it was more likely that people would die a good death (Ariès, 1981). Now, however, because modern medicine can frequently sustain a life beyond its "time," many people are deprived of such a death. Those who die in hospitals are likely to die in pain or semiconsciousness, attached to an assortment of machines, tubes, and intravenous drips (Smyser, 1982). In addition, visitors are sometimes prohibited entirely, or may be allowed "to visit" only for a few minutes on the condition that they do not talk.

And when the dying are maintained on life-support systems, medical personnel may have difficulty predicting when death will occur, or even determining that it has occurred,[*] with the result that loved ones who want to be present in the last moments are often unable to be. Today, most people die among strangers or alone (Ariès, 1981).

---

[*] Although medical and legal definitions of death differ from state to state, one definition that is gaining increasing acceptance is the cessation of all functions of the entire brain.

Figure 26.3 *Life-support equipment, like that pictured here, can often work miracles in sustaining life. However, if this extension of life merely prolongs a person's dying and suffering, it may amount to what one nurse calls "needless torture."*

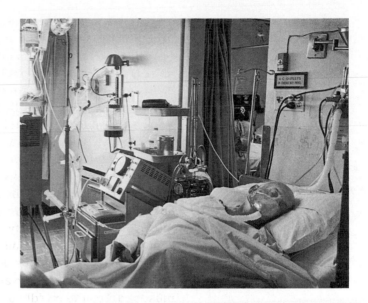

This situation is often compounded by the attitudes of some doctors who view death as an enemy to be fended off at all costs. Their blind zeal is reflected in this man's bitter account of his mother-in-law's final months:

> She had wasted down to less than half her normal weight. One evening my wife got a call that she was failing and drove to the hospital. "I think I have saved your mother," a young doctor beamed to my wife. He had applied heart and lung resuscitation. My wife loved her mother very much, but her immediate response to the young man was "Why?" [Smyser, 1982]

Fortunately, there is a growing awareness in the medical profession and among the general public of the undesirability of using extraordinary measures that prolong the life of terminal patients and deprive them of the opportunity of a good death. Increasingly popular is the **living will,** a legal document drawn up by a person who is not yet near death, detailing how much medical care they want to receive if they become ill (Humphrey and Wickett, 1986). As one 74-year-old physician expressed it in his own behalf:

> I should like to make a plea that death in the old should be accepted as something always inevitable and sometimes as positively desirable. Once I reach the stage of pre-death, all I ask is that I go on to the end with as much dignity and as little pain as possible—length of further survival becomes immaterial. [Burnet, 1974]

Doctors, nurses, and other professionals who take this point of view would like to see **passive euthanasia** become common practice. Unlike *active euthanasia* (sometimes called mercy-killing), in which someone must perform an act that kills the individual (administering a lethal injection, for example), passive euthanasia involves inaction—such as not using a respirator or withholding special drugs—that allows someone to die as his or her condition dictates.

While almost all professionals who care for the terminally ill, as well as most lay people, agree with the concept of passive euthanasia, the issue is not a simple one. One problem is that the quality of life is impossible to measure, so one person's opinion about how aggressively a particular death should be postponed will differ from another's.

While the living will is one factor in making this decision, it is not considered legally binding: the hospital and the state reserve the right to decide when to allow someone to die. These constraints on personal decision-making were dramatized by the case of Elizabeth Bouvia, a divorced California woman in her 20s who had incurable multiple sclerosis in a form so severe that she was unable even to feed herself. Her disease was considered chronic, not fatal, so rather than face many years of what she considered intolerable disability, she requested that the hospital in which she was staying withhold food, allowing her to starve herself to death. The hospital objected, and the state intervened, causing her to be force-fed on the grounds that, since her death was not imminent, her self-starvation would not be easing the dying process but altering the life course.

A related problem is that prediction about the course of an illness is inexact. Often some of those involved with a dying person still hope for a cure, or at least for a long remission, while others are convinced that death is near and should be made as easy and painless as possible. Even doctors disagree—perhaps not always on the basis of the evidence alone. One study found that the doctors who are more likely to deny the imminence of death are those who are most anxious about their own death. In this research, doctors who had revealed a personal fear of dying managed to extend the lives of their terminal patients by an average of five days beyond those of patients whose doctors were more accepting of death (Schulz and Aderman, 1978–1979).

Understandably, then, passive euthanasia often becomes a painful controversy among family members and hospital staff, rather than a simple, humane decision. One person might accuse another of not caring enough, and of giving up too soon, while another might blame the first for not allowing the patient to die peacefully. Ideally, the patient should decide when and how to die, but many dying patients are unable to make rational decisions because they are in a state of semiconsciousness due to their condition and/or medication. Furthermore, unless the concept of a good death is clearly established in the minds of many medical personnel, its application is far from universal. As one critic of standard hospital practices points out, "all too many terminal patients" continue to be "allowed or made to endure avoidable, useless (sometimes constant) pain, nausea and vomiting, constipation, personal indignity, hospitalization, surgery, chemotherapy, injections, and harnessing to tubes and other disabling instruments" (paraphrased in Smyser, 1982). (This particular critic is a nurse who works in a **hospice,** a facility where terminally ill patients are helped to die in dignity and without pain. However, as the Closer Look on the next page points out, a hospice death is available only to a minority.)

Much needs to be done to assure that when death does come, it will be a good death, not only for the sake of the dying person but for the sake of those who live after them. This idea is reflected in the thoughts of one college student following the death of a grandmother:

> We had enabled my grandmother to die at home. This brought all the family members home from all over the U.S. . . . She accepted death—which in turn made us accept death. When she died we all had the feeling of emptiness/loneliness because she would no longer be with us. But on the other hand we were very happy because she no longer had to suffer physical pain. We are a very close family and being with her when she died tied the knots even tighter. [Lagrand, 1981]

A CLOSER LOOK   **The Hospice**

One effort to help ensure that more of the terminally ill die a good death began in London during the 1950s, when a dedicated woman named Cecily Saunders opened the first hospice, a place where the terminally ill could come to die in peace.

Conceived in response to the dehumanization of the typical hospital death, hospices provide the dying with skilled medical care—which includes pain-killing medication but shuns artificial life-support systems—and a setting where their dignity as human beings is respected.

In many cases, the person's home can become the hospice. Skilled doctors and nurses visit often, to give comfort as well as medication and therapy and to instruct family members in how to provide daily care. Expert help and guidance are available, by phone or in person, twenty-four hours a day.

If the hospice is a separate facility, the atmosphere is more like that of a friendly dormitory than that of a hospital. The patients usually wear their own clothes, move about the hospice as they choose, and socialize with each other and with the staff. The kitchen is always open for individually prepared food as well as for conversation around the coffee pot.

Visitors are encouraged day and night. One close friend or family member, called a *lay primary care-giver,* is present much of the time and is responsible for some of the routine care. This arrangement makes the dying person feel less alone and helps the care-giver to be involved, rather than excluded as he or she would be in most hospital settings. The hospice staff direct their care to both the patient and the patient's family. When death comes, the staff continue to tend to the family's psychological and other needs.

Obviously, the hospice has much to recommend it. However, the hospice concept is not accepted uncritically (Aiken and Marx, 1982; Klagsbrun, 1982). First of all, to be accepted by a hospice, patients must be diagnosed as terminally ill; that is, they must have no reasonable chance of recovery. Such a diagnosis can be made for only a minority of the dying. Second, they and their family must accept this diagnosis, agreeing that longer life and cure are virtually impossible, and that a good death is the only remaining choice. Understandably, even for some who are extremely ill, hope is so crucial that they would rather have one last operation, with all odds against success, than wait for death in a hospice.

Another reservation about the hospice concept is that some patients who choose to enter a hospice might be giving in to death prematurely, perhaps losing years of life that vigilant hospital care might have given them. In addi-

A dying man has been granted his wish—to spend his last days at home. This is possible because a hospice worker—whose occu-pation was nonexistent twenty years ago—provides medical and emotional support to him and his family.

tion, all manner of legal questions arise when a life that could have been prolonged is, by passive euthanasia or by extensive use of addictive pain-killers, shortened.

There is also the potential danger that if the hospice concept spreads too quickly, such rapid growth will outstrip the number of available well-trained and well-adjusted doctors, nurses, psychologists, social workers, volunteers, and clergy essential to a well-run hospice. In addition, the stress of working intimately with dying patients and their families can cause burn-out in both professionals and volunteers, resulting in a yet smaller pool of those who are needed for the smooth functioning of hospice teamwork (Grey-Toft and Anderson, 1987–1988; Patchener and Finn, 1987–1988; Vachon, 1987).

All these criticisms have some validity, and the people who are most directly involved in the hospice movement are well aware of them (More, 1987). However, the problem that most concerns them is a more practical one—namely, that a hospice death is available to only a limited number of even those people whose condition qualifies them for acceptance. For example, many hospices admit only patients who have a primary care-giver available on a twenty-four-hour basis. For this reason, a disproportionate number of hospice patients are married and relatively young—only 17 percent are over age 75 (More, 1987).

Finally, many parts of the world as yet have no hospices available. This is changing rapidly as the hospice concept becomes more widely accepted. At the moment, however, the good death that the hospice can provide is a privilege available to only a relative few.

Bereavement

In keeping with the denial of death in the modern Western world, *mourning,* as all the ways of expressing grief at the death of a loved one are called, seemed by the mid-twentieth century to have gone out of fashion (Ariès, 1981). In earlier times, the bereaved were encouraged to express their emotions openly and fully. After the funeral itself, mourners followed age-old rituals to help release and control their grief. For example, they wore black, pulled down their window shades, and refused social invitations for a defined mourning period. During this time, friends and family members were expected to visit, bringing food and drink, to help the bereaved talk about their emotions without being overcome by them. When the mourning period was over, people were helped to pick up the pieces of their lives again, neither forgetting nor dwelling on their loss.

However, in recent times, much of this has changed for many people. The bereft are now often encouraged to "bear up"; friends and relatives often do less consoling than advising—to keep busy, to remarry, to look on the bright side (Caine, 1974); the large funeral has generally given way to a small memorial service; the deceased is less likely to be buried in a commemorative family plot and more likely to be cremated without ceremony. If current trends continue, according to one observer, we may eventually reach the point where death becomes little more than a minor annoyance, to be handled as efficiently and unemotionally as possible (Kastenbaum, 1979).

What are the results of these trends? Certainly they do not abolish the grief; they merely stifle its expression. Many psychologists and psychiatrists have warned that if grief cannot be expressed openly, its indirect manifestations may cripple a person's life (Osterweis et al., 1984). Gorer (1973), for example, describes instances of **mummification,** in which the bereaved leave intact the belongings of the dead. Widowers polish their late wives' knickknacks and clean their clothes; widows burn their husbands' pipe tobacco and air out their suits. Many widowers and widows sanctify their late spouses to the point where they would not consider remarriage (Kalish, 1985). Death of the mourner, either from suicide or a refusal to eat or care for oneself, is common among the elderly. It may be that, in many such cases, the pain of loss was compounded by the feeling that, since no one else seemed to care for the deceased, no one cared for the mourner either.

The Stages of Mourning

To be able to be sensitive to a bereaved person's needs, it is useful to know the emotional progression that seems typical of the mourning process. Four distinct phases have been described (Kalish, 1985; Kastenbaum, 1986).

First there is *shock,* during which some people seem very calm and rational and others seem dazed and distant. The second phase is an intense *longing* to be with the deceased: memories, thoughts, dreams, and even hallucinations of the dead person flood the mind. The third major phase is *depression and despair,* often characterized by irrational anger and confused thinking. This is the period when friends and family members need to be especially understanding, since the mourner's behavior may at times be bewildering. One widow, for instance, bought a large house that she had no need of and could ill afford, in a new neighborhood where she had no friends (Caine, 1974). Another woman divorced her husband because she felt that he was not sufficiently sympathetic when her mother died. Four years later she deeply regretted her decision (Wallerstein and Kelly, 1980). Finally, the death is put into perspective, and *recovery,* the last stage of mourning, occurs.

How long it takes to achieve recovery depends in part on how well the particular culture provides for the needs of the mourner. According to Helena Lopata (1979), contemporary North America does not meet these needs very well. Among the essentials she finds lacking are meaningful rituals that keep the bereaved from feeling isolated; an opportunity to grieve openly; memorial customs that ease the transition back to normal life; emotional support and companionship in the months and years (not just days) after the death; and practical help to gradually overcome the many problems that result from the death of an important person in one's life.

The stages of grief, like the stages of dying, do not follow a schedule. While a degree of numbness almost always occurs at first, longing, sorrow, and acceptance seem to come in waves. So a bereaved person might seem to accept the death and then might suddenly be overwhelmed with sadness when the realization surfaces again that an attachment that has helped to sustain life has been broken (Jacobs et al., 1987–1988). Such responses are particularly likely to occur on holidays, or birthdays, or the anniversary of the death and are accompanied by a new period of mourning. These **anniversary reactions** should be expected and accepted: the mourner who lights a candle or visits a grave on an anniversary is better off psychologically than the mourner who is surprised, and depressed or troubled, by a new surge of sadness.

Figure 26.4 *Visiting the grave comforts many of the bereaved. If such visits are part of a tradition, occurring on Christmas, Memorial Day, or, as shown in the first photo, on the anniversary of the death, mourners have an accepted opportunity to express their grief and are less likely to experience pathological grief.*

The crucial problem for many mourners in our society is that outsiders sometimes do not understand and sympathize with their grief. For example, it might seem that an elderly person should accept the death of a very elderly mother or father, or that a young parent should not be unduly distressed over the death of an unborn baby. However, every death—especially a death that changes the generational line—is potentially grievous. The older person who now becomes part of the oldest generation of the family, or the parent whose hopes for the next generation are unfulfilled, can be, temporarily at least, devastated by their loss.

While overall the elderly are accepting of death, they are particularly vulnerable to **bereavement overload,** as a second or third death starts the mourning process up again before the earlier ones have been completed. When this happens, other people need to be particularly sympathetic.

**Expected and
Unexpected Death**

Death is somewhat easier to cope with if it is expected. Sometimes the diagnosis of terminal illness allows a period of **anticipatory grief,** when both the dying person and the mourners can cry together and can share their affection for each other (Rando, 1986). Having time for anticipation does not necessarily ease the pain of loss, since attachment is often strengthened during the period of anticipatory grief. However, the emotions expressed at this time can make the leave-taking less conflicted than when such exchanges never occur. When individuals achieve the cognitive maturity to realize that death is final and universal, this awareness of mortality often leads them to anticipate eventual grief and make a special effort to express affection and appreciation of their elderly relatives, even when there is no evidence of impending death.

The sudden death of someone who is not "supposed to" die is the most difficult to bear. The clearest example is the death of a child, especially one who has lived long enough to have a distinct personality and position in the family. If the death is a violent and sudden one, as most young people's deaths are, the loss is particularly devastating (Shanfield et al., 1986–1987). Parents and siblings are often racked by powerful and personal emotions of guilt, denial, and anger, as well as sorrow. One protective impulse is to blame someone—perhaps oneself for not having been more careful or more loving, perhaps a spouse, and perhaps even the dead child. For instance, the mother of a fatally injured eight-year-old could not bring herself to enter his hospital room. Eventually

she confessed to a sympathetic nurse that she desperately wanted to see the dying youngster, but could not stand by his bedside without the overwhelming urge to "beat the hell out of him." She had told her son repeatedly to keep away from the farm machinery. He had disobeyed her orders and been fatally injured in the moving mechanism. How could he do this to himself or to her? [Easson, 1977]

Figure 26.5 *Family celebrations like this man's 80th birthday party are often particularly cherished during a loved relative's late adulthood; everyone appreciates the good fortune of being able to be together one more time.*

Blame and guilt, however, can destroy a family, just when family members need each other most. The Compassionate Friends, a self-help group for parents of dead children, finds that many married couples are driven apart by their separate reactions to the death (Klass, 1986–1987). One parent might need to talk about the death at the time the other cannot bear to hear the child's name. Brothers and sisters of a dead child suffer, too, partly because parents are so involved in their own grief that the surviving children are deprived of attention, and partly because a child's grief may follow a different course from that of adults. Denial and regression are common at first, with sorrow and acceptance coming much later than for adults (Osterweis et al., 1984; Rosen, 1986). Each family member should make an effort to understand and accept the many possible individual forms and paces of mourning that may be exhibited by others. Children in particular need to know that all questions and feelings are acceptable.

Recovery

Many parents and siblings never fully recover from the death of a child—even if that child was fully grown (Lehman, 1987; Rosen, 1986). Epidemiological studies indicate a higher divorce rate among parents of a dead child, and also a higher rate of illness and death among surviving parents and siblings. In fact, everyone who loses someone close to them is at risk for psychological and physical illness. According to longitudinal research, a child whose parent dies, or a husband whose wife dies, is particularly vulnerable (Osterweis et al., 1984).

What can others do, then, to help the bereaved person? The first step is simply to be aware that powerful and complicated emotions are likely: a friend should listen, sympathize, and not ignore the pain. The second step is to understand that bereavement is often a lengthy process, demanding sympathy, honesty, and social support for months or even years. As time passes, the bereaved person should become involved in other activities, but should not be expected to forget the person to whom they were attached: sorrow and memory usually continue. Interestingly, sadness does not necessarily correlate with depression (Jacobs et al., 1987–1988). In fact, several studies have found that those who take steps to remember a dead person—to save mementos or photographs, to talk about experiences shared, to visit the gravesite—cope better with death than those who do not (Kastenbaum, 1986; Murphy, 1986–1987). If the emotions of grief have been given expression, the time will come when the person "feels a sense of weary relief in having worked through the bitter emotions of grief, and is ready to approach . . . the new situation more calmly" (Bowlby, 1974).

After working through the emotions of grief, the bereaved may develop a deeper appreciation of human relationships. As one young widow wrote to her friends, "Live out your love for one another now . . . Don't be afraid to touch and share deeply and openly all the tragic and joyful dimensions of life" (quoted in Cassem, 1975). Robert Kastenbaum (1979) puts the same idea into a broader context when he expresses the hope that, in the future, "thoughts and feelings about the dead (will be) accepted as a vital thread of continuity that symbolizes our existence as members of the human race rather than as solitary individuals." Death, when accepted, and grief, when allowed expression, give meaning to birth, growth, development, and all human relationships.

SUMMARY

A Historical View

1. Contrary to the custom of earlier centuries, Western civilization in the twentieth century has attempted to deny, as much as possible, the reality of death. Instead of occurring naturally at home, death came to be an often unnatural and dehumanizing process occurring in a hospital.

Research on Death and Dying

2. In recent years, there has been a trend toward an increased awareness, and a more accepting view, of death. As Kübler-Ross revealed, dying people, and those who love them, benefit from honesty and openness.

3. How an individual views death is influenced by the developmental stage he or she is in. Before the age of 6, children do not understand that death is universal, inevitable, and final. In adulthood, one's view of death is affected by the circumstances of one's life, including age, sex, and marital status. In general, the aged are quite accepting of death, but very fearful of prolonged and dependent dying.

Facing One's Own Death

4. Kübler-Ross's depiction of the stages of dying, although not applicable to everyone, has helped health professionals and family members to better understand and cope with the experiences of the dying.

5. Most people want to die "a good death," a swift, dignified death among loved ones. The living will, passive euthanasia, and the hospice movement allow an individual to make personal choices concerning the manner of his or her own death. Advanced medical technology and legal restrictions, however, often interfere with an individual's decision about these matters.

Bereavement

6. The rituals of funerals and mourning serve to help the bereaved express and eventually overcome their grief. In modern times, however, grief is often stifled and unexpressed. Mourners experience several phases of grief and require the understanding and emotional support of friends and family during these periods.

7. If death is expected, anticipatory grief can help dull the pain of loss. Unexpected death often can be the most painful, leading to blame and guilt when family members do not understand the individual forms that mourning may take in others. It often requires a long time, sometimes a year or more, before those who lose a loved one recover from their grief.

KEY TERMS

thanatology *(609)*

living will *(613)*

passive euthanasia *(613)*

hospice *(614)*

mummification *(616)*

anniversary
  reactions *(617)*

bereavement
  overload *(617)*

anticipatory grief *(618)*

KEY QUESTIONS

1. What are the differences between people's experience of death in earlier times and today?

2. What are some of the factors that affect how people think about their impending death?

3. What are some of the ways that have been used to help the dying achieve "a good death"?

4. What ethical issues surround passive euthanasia?

5. What benefits does hospice care offer the dying?

6. How is a chapter on death relevant to a book on life-span development?

Reading Resources for Further Study

When your interests or assignments require further investigation of a topic discussed in this text, you can begin your research by noting the references cited in the text discussion and then looking them up in the bibliography (page B-1) to obtain the name of the journal article or book in which material on the topic will be found. Many of the sources cited in this text will be found in your college or university library or in a large public library. Some of the books and articles will be easy to understand; others, because they are written for professionals and for the research community, will be difficult for most undergraduates.

The sources of information below, categorized as either "beginning" or "advanced," can help you to gather information on any topic. However, keep in mind that no matter which sources you choose, one of the most important factors in coming to valuable conclusions is your own ability to think critically—to use your knowledge, understanding, and experience to evaluate the material you read, and to note and try to account for differences among the findings and opinions of researchers.

BEGINNING READING RESOURCES

Journals

The following respected and widely available journals are well written and easily accessible to nonprofessional readers.

Psychology Today, published by the American Psychological Association. Covers a wide range of topics and includes many articles about development.

American Journal of Orthopsychiatry, published by the American Society of Orthopsychiatry. Intended for those in the helping professions, including psychologists, social workers, doctors, and nurses.

Generations, published by the American Society on Aging. Designed for those who work with the aged, and also for the interested layperson.

Guides to Journal and Magazine Articles

The following indexes list articles by topic:

The Reader's Guide to Periodic Literature, the **Newspaper Index,** and the **Magazine Index.**

It is important to keep in mind that not all the articles you find will report information accurately or reflect reliable research. Informed by your reading of this text, including the discussion of the scientific method in Chapter 1, you should always try to distinguish between those discussions that are supported by substantial evidence and sound scientific principles and those that are not. Be sure to note any questions raised by the articles you read or any inconsistencis in the

ADVANCED READING RESOURCES

The following journals and books are more difficult because they are written primarily for researchers. Advanced discussions of theory, methodology, and recent research may be included.

Journals

There are literally hundreds of scholarly journals relevant to the study of human development. The following are among the most highly respected and readily available sources of information and research:

Developmental Psychology

Human Development

Child Development

Journal of Marriage and the Family

Journal of Gerontology

When you find an article that seems particularly relevant to your area of interest, read it carefully, identify and write down the most important points that were made, and make note of significant details. You may then want to select several of the references cited and read them as well.

Guides to Journal and Magazine Articles

To gather the most current information on any topic, ask the librarian to direct you to the current abstracts in child development, education, psychology, sociology, and so on. Abstracts list recent scholarly articles by topic; a brief description of the content of each article is included.

(continued next page)

(continued next page)

BEGINNING READING RESOURCES

information you find. Formulating your own questions and attempting to resolve them is one of the best ways to further your thinking.

Guides to Current Books

Current books on topics that interest you are another important resource. For reviews of books relevant to human development, any of these three sources would be helpful:

Contemporary Psychology, a journal published by the American Psychological Association.

Readings, a publication of the American Society of Orthopsychiatry.

Child Development Abstracts, published by the Society for Research in Child Development. The book reviews are located at the end of each volume.

ADVANCED READING RESOURCES

Annual Reviews

There are several series of books published annually that provide in-depth articles on the latest research, theories, and controversies. Among them are:

Annual Review of Psychology (Palo Alto, CA: Annual Reviews).

Annual Review of Sociology (Palo Alto, CA: Annual Reviews).

Lifespan Developmental Psychology (San Diego, CA: Academic Press).

Advances in Child Development and Behavior (San Diego, CA: Academic Press).

Handbooks

Several excellent handbooks provide overviews of theory and research on topics of significance. Because of their extensiveness and depth, many, though not all, of their articles may be daunting to the undergraduate. However, looking up a topic in a handbook provides information about the kind of research that has been done, names important researchers in the field, and can serve as a guide to other sources of information.

Handbook of Aging and the Social Sciences. Edited by Robert H. Binstock and Ethel Shanas. New York: Van Nostrand Rhinehold, 1985.

Handbook of the Biology of Aging. Edited by Caleb E. Finch and Edward L. Schneider. New York: Van Nostrand Rhinehold, 1985.

Handbook of Child Psychology. Volumes I–IV. Edited by Paul H. Mussen et al. New York: Wiley, 1983.

Handbook of Developmental Psychology. Edited by Benjamin B. Wolman. Englewood Cliffs, NJ: Prentice-Hall, 1982.

Handbook of Intelligence. Edited by Benjamin B. Wolman. New York: Wiley, 1985.

Handbook of the Psychology of Aging. Edited by James Birren and K. Warner Schaie. New York: Van Nostrand Rhinehold, 1985.

Glossary

This glossary provides brief definitions of the most important terms used in this book. To understand the terms more fully in context, consult the Index and read about the terms in the pages on which they first appear.

accommodation The process of shifting or enlarging usual modes of thinking, or schemas, in order to encompass new information. For example, many Americans would have to expand their concept of food in order to be able to consider eating octopus, even though it is a delicacy in some cultures. *(50)*

achieving stage The first of Schaie's adult stages, in which young adults move away from the indiscriminate acquisition of knowledge and become more discerning about gaining the knowledge they need to establish themselves in the world. *(413)*

activity theory A theory of psychosocial development in late adulthood that was developed to counteract the theory of disengagement. According to activity theory, those of the elderly who maintain active social lives are happier and better adjusted than those who do not. *(576)*

adaptation Piaget's term for the cognitive processes through which a person adjusts to new ideas or experiences. Adaptation takes two forms, assimilation and accommodation. *(49)*

adolescence The period between childhood and adulthood. Adolescence usually begins at puberty and ends at the age at which the culture first assigns a person adult responsibilities. *(321)*

adolescent egocentrism A characteristic of adolescent thought that leads the young person to think that he or she is unlike other people in many ways. (See also *personal fable* and *imaginary audience*.) *(344)*

ageism A term coined in 1965 by Robert Butler to refer to prejudice against the aged. Like racism and sexism, ageism works to prevent elderly people from being as happy and productive as they could be. *(526)*

agency According to David Bakan, the motivation of young adults to be active agents in their lives, pushing for their own success and personal accomplishments. Coworkers are sometimes seen as obstacles to advancement, and assignments are taken to show off one's talents. (See also *communion*.) *(511)*

age of viability The age at which a fetus can survive outside the mother's uterus if optimal care is available (usually between twenty and twenty-six weeks after conception). *(80)*

age-segregated communities Housing designed exclusively for older adults. Age-segregated communities offer independence, security, medical care, and a setting for interaction among older people. *(590)*

alienation In vocational psychology, a feeling of distance from, and lack of interest in, one's work. Employees who have dull, repetitive jobs, where there is little opportunity to experience a sense of pride and completion in a job well done, are thought to be especially likely to experience alienation. *(513)*

Alzheimer's disease The most common form of dementia, characterized by gradual deterioration of memory and personality, caused by plaques and tangles in the brain. The underlying cause is unknown, as is the cure. Alzheimer's disease is not part of the normal aging process. *(562)*

amniocentesis A prenatal diagnostic procedure in which a sample of amniotic fluid is withdrawn by syringe and tested to determine if the fetus is suffering from problems such as Down syndrome or Tay-Sachs disease. *(70)*

anal stage Sigmund Freud's term for the second stage of psychosexual development (occurring during toddlerhood), in which the anus becomes the main source of bodily pleasure, and defecation and toilet training are therefore important activities. *(33)*

androgyny The tendency to incorporate both male and female qualities (as conventionally defined) into the personality. For example, an androgynous person might be both independent and nurturing. *(242)*

animism The belief that inanimate objects are alive, and therefore have emotions or intentions; an idea held by many young children. *(203)*

anniversary reactions Following the death of a loved one, overwhelming emotions of grief and sorrow that occur on holidays, birthdays, or the anniversary of the death and that initiate a new period of mourning. *(617)*

anorexia nervosa A rare disorder in which a person refuses to eat and may consequently starve. Most victims are adolescent girls. *(325)*

anoxia A temporary lack of fetal oxygen during the birth process; if prolonged, it can cause brain damage. *(92)*

anticipatory grief A period of mourning occurring before a death. Sometimes, the diagnosis of a terminal illness allows a period of anticipatory grief during which the dying person and the mourners can cry together and express affection for one another. *(618)*

Apgar scale A quick assessment of a newborn's heart and respiratory rate, muscle tone, color, and reflexes. This simple method is used to determine whether a newborn needs immediate medical care following birth. *(81)*

assimilation Piaget's term for the inclusion of new information into already existing categories, or schemas. For example, a person may eat a new food and be unable to name it, but may nevertheless be able to guess by its taste, smell, or texture that it is, say, a fruit. In this way, a new object is placed in the preexisting category "fruit." *(50)*

associative play A form of social play in which two or more children play together, but are involved in their own separate activities. They do not cooperate with each other. *(223)*

attachment An affectional bond between a person and other people, animals, or objects that endures over time and produces a desire for consistent contact and feelings of distress during separation. *(160)*

attention deficit disorder A state of excessive activity, usually accompanied by an inability to concentrate and impulsive behavior. (Also called *hyperactivity*.) *(284)*

authoritarian parenting A style of child-rearing in which the parents' word is law, and misconduct is punished—no excuses accepted. *(228)*

authoritative parenting A style of child-rearing in which the parents set limits and provide guidance, and at the same time are willing to listen to the child's ideas and make compromises. *(228)*

autism A serious psychological disturbance that first becomes apparent during early childhood, when a child does not initiate normal social contact. Most autistic children prefer to play with objects or by themselves, unlike normal children who enjoy company. *(236)*

autonomy versus shame and doubt Erikson's term for the toddler's struggle between the drive for self-control, and feelings of shame and doubt about oneself and one's abilities. This is the second of Erikson's eight stages of development. *(151)*

autosomes The twenty-two pairs of chromosomes that are identical in human males and females; they direct the development and functioning of much of the body, but do not determine sex. *(56)*

average life expectancy The number of years the average newborn of a particular species is likely to live. In humans, this age has tended to increase over time; in the United States in 1985, the average life expectancy at birth was 71 years for men and 78 for women. *(539)*

Babinski reflex A normal, neonatal reflex that causes the child's toes to fan upward when the sole of the foot is stroked. *(111)*

baby talk A term for the special form of language typically used by adults to speak with infants. Adults' baby talk is high-pitched, with many low-to-high intonations; it is simple in vocabulary and employs many questions and repetitions. (Also called *Motherese*.) *(139)*

bereavement overload An emotional state that may result when a person experiences the death of several loved ones over a relatively short period of time and is unable to reach acceptance of the first death before having to mourn the second. *(617)*

B-cells Cells created in the bone marrow and designed for isolating and destroying foreign substances in the immune system. B-cells create antibodies to disease. *(535)*

behaviorism A major theory of psychology that holds that most human behavior is learned, or conditioned. Behaviorists have formulated laws of behavior that are applicable to animals and to people of all ages. Behaviorism is also called *learning theory*. *(38)*

Black English A form of English with its own rules of grammar, at variance with some rules of standard English. It is called Black English because it is a dialect spoken by many black Americans. However, many black Americans do not speak Black English, and some white Americans do. *(277)*

blind experimenters Researchers who collect data without knowing what results to expect or without knowing which individuals have been subjected to special experimental conditions. Being unaware of the expected outcome of research, blind experimenters can be objective and reliable in reporting their findings. *(19)*

body image A person's concept of his or her physical appearance. *(330)*

brain waves Electrical activity in the brain as recorded on an electroencephalogram (EEG). *(108)*

Brazelton Neonatal Behavior Assessment Scale A rating of a newborn's responsiveness to people and the strength of his or her reflexes. *(112)*

breathing reflex A normal reflex that ensures that newborns (as well as older children and adults) maintain an adequate supply of oxygen by inhaling and exhaling air. *(111)*

bulimia A dangerous eating disorder characterized by compulsive binge-eating followed by compulsive purging by inducing vomiting or abusing laxatives. This eating pattern can lead to damage to the gastrointestinal system and electrolyte imbalances severe enough to cause heart attacks. *(406)*

burn-out In vocational psychology, a feeling of emotional and physical exhaustion that can come when a person works with great dedication and high ideals on a difficult task and then becomes disillusioned. *(512)*

career establishment The second stage of a person's work history, usually occurring around age 30, when a worker demonstrates increasing competence and commitment, and consequently, is given more responsibility and respect at work. *(451)*

career exploration The first stage of a person's work history, during which a person tries to match skills, talents, and needs to a specific occupation and job. *(451)*

carrier An individual who possesses a particular gene or chromosomal configuration as part of his or her genotype (total genetic make-up). A carrier can pass on a recessive gene to his or her children, but unless the child inherits the same gene from both parents, the child will not develop the characteristic. *(60)*

case study The research method in which the scientist reports and analyzes the life history, attitudes, behavior, and emotions of a single individual in much more depth than is usually done with a large group of people. *(25)*

centration The focusing of attention on one aspect of a situation or object to the exclusion of other aspects. Young children, for example, have difficulty realizing that a mother is also a daughter, because they concentrate, or center, on one role to the exclusion of others. *(200)*

cephalo-caudal development Growth proceeding from the head downward (literally, from head to tail). Human growth, from the embryonic period throughout childhood, follows this pattern. *(78)*

Cesarean section A surgical procedure in which an obstetrician cuts open the abdomen and uterus to deliver the baby. This technique is used if the fetus is unable to travel safely through the birth canal. *(92)*

child abuse A term used to describe the many forms of maltreatment inflicted on children by parents and other care-givers; includes physical injury, emotional maltreatment, sexual abuse, and neglect of the child's basic needs. *(168)*

childhood schizophrenia An emotional disturbance that can develop during early or middle childhood. Its symptoms include an unusual difficulty with social play, conversation, and emotional expression. *(237)*

chromosome Molecules in every cell that carry the genetic material transmitted from parents to offspring, determining their inherited characteristics. *(55)*

chunking A memory technique that consists of grouping items to be memorized into categories. *(269)*

classical conditioning The process by which an animal or person learns to associate a neutral stimulus (for example, a bell) with a meaningful one (for example, food). After training, the subject will respond in the same way to the neutral stimulus as to the meaningful one. *(39)*

classification The concept that objects can be sorted into categories or classes, as in sorting foods according to whether they are fruits, vegetables, or dairy products. According to Piaget, this concept is mastered during the period of concrete operational thought. (See also *class inclusion*.) *(263)*

class inclusion The idea that a particular object or person may belong to more than one class. For example, a father can also be someone's brother. *(263)*

climacteric Any period of life accompanied by relatively sudden physiological changes, and often by psychological ones as well. The months preceding and following menopause are the best-known climacteric of the human life span. *(478)*

closed system A cognitive system in which a finite number of known variables results in a specific outcome. In a closed system, the relationships among variables can be analyzed until a single, correct answer is found, making formal thought adequate for solving problems in such a system. *(414)*

code-switching A pragmatic communication skill that comes into play when a person switches appropriately from one form of speech to another. *(276)*

cognition The mental processes by which the individual obtains knowledge or becomes aware of the environment, for example, perception, memory, imagination, and use of language. *(121)*

cognitive domain That domain of human development that includes all the mental processes through which the individual obtains knowledge or becomes aware of the environment. *(4)*

cognitive theory The theory that states that the way people understand and think about their experiences is an important determinant of their behavior and personality. *(47)*

cohabitation Literally, "cohabitation" means living together. It is used primarily to refer to unrelated adults of the opposite sex who share the same house or apartment, presumably sharing the same bed as well. *(438)*

cohort A group of people who, because they were born at the same time, experience the same historical and social conditions. Differences between cohorts often complicate research that attempts to focus on differences caused by development alone, a complication known as "the cohort effect." *(27)*

collective monologue A "conversation" between two or more children in which the participants are talking in turn, as though in conversation, but are not actually responding to the content of each other's speech. *(210)*

communion According to David Bakan, the concern of middle-aged workers for the welfare of coworkers, their company as a whole, and the company within the larger community. (See also *agency*.) *(511)*

concrete operational thought In Piaget's theory, that stage of cognitive development in which a person understands specific logical ideas and can apply them to concrete problems but has difficulty with abstract, hypothetical thought and logic. This period usually begins at about age 7 and tends to continue to adolescence (though it sometimes ends later, or not at all). *(261)*

conditioning In learning theory, the process of learning that occurs either through the association of two stimuli or through the use of positive or negative reinforcement. *(39)*

congenital Present at birth, either as a result of genetic or prenatal influences or specific complications of the birth process. Not all congenital characteristics are apparent at birth. For example, diabetes is inherited and therefore congenital, but it does not appear until later in life. *(66)*

conservation   In Piagetian theory, the idea that certain properties of a given quantity of matter (volume, weight, etc.) remain constant despite changes in shape, length, or position. Preschool children usually cannot understand the concept of conservation. *(201)*

conservation of liquids   The idea that the volume of a liquid remains constant, even if the containers into which it is poured are quite different in size and shape. A child who has entered the stage of concrete operational thought will understand that when 6 ounces of juice in a short, squat glass is poured into a tall, narrow glass, the total amount of juice remains the same. *(48)*

conservation of matter   The idea that the volume or weight of an object remains the same even if the form is changed. For example, when two balls of clay have the same volume, rolling one ball into a long rope will not increase the amount of clay. *(201)*

conservation of number   The idea that the number of a set does not change even if the objects in it are repositioned. For example, if two sets have the same number of coins, spacing out one set so that it is distributed over a larger area will not increase the number of coins in that set. *(202)*

continuity   A term used to label development that is seen as gradual and steady. *(13)*

control group   In research, a group of subjects who are similar to the experimental group on all relevant dimensions (e.g., sex, age, educational background) but who do not experience special experimental conditions or procedures. *(19)*

cooperative play   Play in which two or more children cooperate while playing, for example, by taking turns or following rules. This form of play is difficult for many preschoolers. *(223)*

corpus callosum   A network of nerves connecting the two halves (the left side and the right side) of the brain. *(182)*

correlation   A statistical term that indicates that two variables are somehow related. Whenever one variable changes in the same direction as another (for example, both decrease), the correlation is *positive*. Whenever one variable increases as another decreases, the correlation is *negative*. *(20)*

crawling   A way in which babies move by getting onto their hands and knees and coordinating movement of their arms and legs to achieve locomotion. *(114)*

critical period   Any period during which a person is especially susceptible to certain harmful or, in some instances, beneficial influences. During prenatal development, for example, the critical period is usually said to occur during the first eight weeks, when the basic organs and body structures are forming and are therefore particularly vulnerable. *(83)*

cross-linkage   The biochemical process in which one molecule becomes attached, or linked, to another. Cross-linkage may be an underlying cause of the aging process. *(538)*

cross-sectional research   Research involving the comparison of groups of people who are different in age but similar in other important ways (e.g., sex, socioeconomic status, level of education). Differences among the groups—as, for instance, between a group of 12-year-olds and a group of 15-year-olds—are presumably the result of development, rather than some other factor. *(26)*

crystallized intelligence   Cattell's term for those types of intellectual ability that reflect accumulated learning. Vocabulary and general information are examples. Some developmental psychologists think crystallized intelligence increases with age, while fluid intelligence declines. *(489)*

decenter   To simultaneously focus on more than one aspect of a situation or consider more than one point of view. If given only a side view of an object, for example, older children are usually able to imagine how the object might look if observed from above. *(262)*

deep structure   Noam Chomsky's term for the underlying rules of grammar and inherent meaning in each language. According to Chomsky, a child's ability to understand this structure is innate (understanding, for example, that different sentence structures indicate either a statement or a question). *(137)*

defense mechanisms   Behavioral or thought patterns that distort one's feelings or perceptions in order to avoid unbearable inner conflicts. In psychoanalytic theory, the ego is thought to institute these defenses, involuntarily and unknowingly, when a real or imagined threat is perceived. *(35)*

deferred imitation   The ability to re-create an action, or mimic a person, one has witnessed some time in the past. According to Piaget, infants are usually first able to do this between 18 and 24 months of age. *(128)*

dementia   Irreversible loss of intellectual functioning caused by organic brain damage or disease. Dementia becomes more common with age, but even in the very old, dementia is abnormal and pathological. Sometimes dementia is misdiagnosed, since reversible conditions such as depression and drug overdose can cause the symptoms of dementia. *(561)*

demographic pyramid   The shape that results when populations are graphed by numbers of individuals in each age group. In the past, the largest population group was the youngest and the smallest group was the oldest, giving the graph the shape of a pyramid. Currently, however, the pyramid is becoming more square, with equal numbers of older and younger persons in the population. *(528)*

demography   The study of populations and social statistics associated with these populations. *(528)*

developmental psychology   The branch of psychology that scientifically studies the changes in behavior, personality, social relationships, thought processes, and body and motor skills that occur as the individual grows older. *(3)*

dialectical thought   Thought that is characterized by understanding the pros and cons, advantages and disadvantages, and possibilities and limitations inherent in every idea and course of action. Unlike Piaget's stages of cognitive development, dialectical thinking is produced less by maturation than by extended education and life experiences. *(414)*

discontinuity A term used to label development that is seen to occur in stages, or which is characterized by abrupt or uneven changes. Many developmental psychologists emphasize the discontinuity of development. (Also called the stage view of development.) *(13)*

disengagement theory A theory of psychosocial development in late adulthood that holds that elderly people voluntarily withdraw from involvement in society and that society responds by withdrawing as well. The shrinking social world caused by disengagement is said to be mutual and gradual, so that by old old age most people enjoy a much smaller social circle than they once did. (See *activity theory*.) *(575)*

disequilibrium Piaget's term for the state of conflict that results from difficulties with integrating new information into existing schemas. *(48)*

displacement A defense mechanism in which a feeling toward one object is shifted to another, less threatening, one (as when a person becomes angry at a friend rather than at his or her boss, who is the original object of the anger). *(35)*

dizygotic twins Simultaneously born offspring who develop from two separate zygotes, each the product of a different sperm and ovum. These twins are no more similar genetically than any other two children born to the same parents. *(57)*

dominant gene A gene that exerts its full phenotypic effect in the offspring regardless of whether it is paired with another dominant gene or with a recessive gene. *(60)*

Down syndrome A genetic abnormality caused by an extra chromosome in the twenty-first chromosome pair. Individuals with this syndrome have round faces, short limbs, and are underdeveloped physically and intellectually. (Also called *trisomy-21*.) *(65)*

dramatic play Mutual fantasy play that occurs when two or more children choose roles and cooperate in acting them out. Dramatic play is an important social development of the preschool years. *(224)*

drug abuse The use of a drug to the extent that it impairs one's physical, cognitive, or social well-being. *(400)*

dyscalcula A specific learning disability involving unusual difficulty in arithmetic. *(282)*

dysgraphia A specific learning disability involving unusual difficulty in handwriting. *(282)*

dyslexia A specific learning disability involving unusual difficulty in reading. *(281)*

echolalia The word-for-word repetition of what another person has just said—a speech characteristic of many autistic children. *(237)*

eclampsia A serious disease that can occur during the last weeks of pregnancy when a mother may have difficulty in ridding her system of fetal wastes. If not promptly treated, eclampsia can cause fetal brain damage and even death to the child and mother. *(85)*

eclectic perspective A view incorporating what seems to be the best, or most useful, from various theories, rather than working from a single perspective. *(52)*

ecological approach A way of looking at human development that emphasizes the impact of society, culture, physical setting, and other people on the development of each individual. (Also called the *systems approach*.) *(5)*

ego As conceptualized by Freud, the rational, reality-oriented part of the personality. *(34)*

egocentrism Thought processes that are governed solely by one's own point of view. In the egocentrism of early childhood, many children believe that other people think exactly as they themselves do. *(202)*

ejaculation The release of seminal fluid from the penis. *(326)*

elaborated code This is a form of speech used by children in school and in other formal situations and is characterized by extensive vocabulary, complex syntax, lengthy sentences, and conformity to other middle-class norms for language. *(276)*

elective mutism A psychological disorder of childhood, in which a child who is able to speak refuses to speak in nearly all social situations. *(237)*

Electra complex The female version of the Oedipus complex. According to psychoanalytic theory, at about age 4, girls have sexual feelings for their father and accompanying hostility toward their mother. *(239)*

elder abuse The physical and emotional mistreatment of older adults, most often by the care-giver. *(599)*

embryo The human organism from about two to eight weeks after conception, when basic body structures and organs are forming. *(77)*

encapsulation The concentration of intellectual strengths into one small area. Expertise, through encapsulation, gathers intelligence into specific areas, with the result that there may be a decline in other areas. Thus, encapsulation may result in individuals becoming experts in some areas and remaining novices in others. *(494)*

endometriosis A condition in which fragments of the uterine lining become implanted and grow on the surface of the ovaries or the Fallopian tubes, blocking the reproductive tract and leaving many women with fertility problems. *(399)*

environment The external forces, including physical surroundings, social institutions, or other individuals, that impinge on human development. *(62)*

equilibrium Piaget's term for the state of mental balance achieved through the assimilation and accommodation of conflicting experiences and perceptions. *(48)*

error catastrophe A key idea in a theory of aging that holds that, while the body can isolate and repair a certain number of errors in cell duplication, at some point, accumulating errors can no longer be controlled and fatally impair the body's ability to function. *(538)*

erythroblastosis A condition that occurs when antibodies produced by the mother's blood damage the fetal blood supply. This disease can now be prevented. (Also called *RH disease*.) *(67)*

estrogen A hormone produced primarily by the ovaries that regulates sexual development in puberty. Although boys' adrenal glands produce some estrogen, it is chiefly a female hormone. *(321)*

estrogen replacement therapy (ERT) A medical treatment for the symptoms of menopause in which women are given artificial estrogen to compensate for the lower production of estrogen by the body. ERT was very common a decade ago, but has become controversial because it has been linked to cancer of the uterus. *(478)*

executive stage In Schaie's stages of cognitive development, the individual's concern for, and involvement with, some aspects of the larger society, such as one's company or town government. *(413)*

experiment The research method in which the scientist brings people into a controlled setting, and then manipulates a variable and observes the results. For instance, children might be brought into a well-equipped playroom, and then told they can play with one, and only one, of the many toys available. *(22)*

experimental group In research, a group of subjects who experience special experimental conditions or procedures. *(19)*

expertise The acquisition of knowledge in a specific area. As individuals grow older, they concentrate their learning in certain areas which are of the most importance to them, and encapsulation results. *(494)*

exosystem In the ecological approach to studying development, this term refers to the neighborhood and community structures, for example, local government agencies and newspapers, that affect the functioning of the smaller systems. *(6)*

extended family The network formed by several generations of a family in close and frequent contact. *(505)*

extroversion A personality dimension characterized by a tendency to be outgoing, active, and assertive. *(515)*

fear of strangers An infant's distress when confronted with a new person, especially an adult who looks unusual or who acts in an unusual way. This emotion is first noticeable at about 6 months, and is full-blown at a year. *(146)*

fetal alcohol syndrome (FAS) A congenital condition characterized by a small head, abnormal eyes, malproportioned face, and retardation in physical and mental growth, that sometimes appears in children whose mothers used alcohol during pregnancy. *(86)*

fetal monitor A sensing device commonly used during labor that measures and records the fetal heartbeat and the strength and frequency of the mother's contractions. The fetal monitor helps determine whether medical intervention, such as a Cesarean section, is necessary to protect the health of the mother and/or the infant. *(92)*

fetoscopy A procedure that uses a narrow tube inserted into the uterus to view the fetus and the placenta directly. *(70)*

fine motor skills Skills involving small body movements, especially with the hands and fingers. Drawing, writing, and tying a shoelace demand fine motor skills. *(187)*

5-to-7 shift The change in cognitive development between ages 5 to 7 that allows the elementary-school child to think, learn, and remember in a more mature way than had been previously possible. *(262)*

fluid intelligence Cattell's term for those types of basic intelligence that make learning of all sorts quick and thorough. Underlying abilities such as short-term memory, abstract thought, and speed of thinking are all usually considered part of fluid intelligence. (See *crystallized intelligence*.) *(489)*

foreclosure Erikson's term for premature identity formation, in which the young person does not explore all the identities that are available. *(364)*

formal operational thought Piaget's term for the last period of cognitive development, characterized by hypothetical, logical, and abstract thought. This stage is not reached until adolescence, if at all. *(339)*

frail elderly Older people who are physically infirm, very ill, or psychologically impaired. *(595)*

free radicals Atoms or molecules that have an unpaired electron. Free radicals are believed to damage cells, affect organs, accelerate diseases, and decrease the ability of DNA to maintain and repair the body. *(537)*

full term A baby born at about 38 weeks after conception (often calculated as 40 weeks after the beginning of the woman's last menstrual period, since ovulation normally occurs 2 weeks after a period begins). *(88)*

game of thinking Flavell's term for the adolescent's ability to suspend knowledge of reality and think creatively about hypothetical possibilities. *(341)*

gamete A human reproductive cell. Female gametes are called ova, or eggs; male gametes are called spermatozoa, or sperm. *(56)*

gene The basic unit of heredity carried by the chromosomes. Genes direct the growth and development of every organism. *(55)*

generation stake The different ways in which each generation views the family, which depends in part on the developmental stages of the members of a particular generation. *(374)*

generativity versus stagnation Erikson's seventh stage of development, in which adults seek to be productive through vocation, avocation, or child-rearing. Without such productive work, adults stop developing and growing. *(432)*

genetic clock According to one theory of aging, a regulatory mechanism in the DNA of cells regulates the aging process. *(539)*

genetic counseling A program of consultation and testing through which couples learn about their genetic inheritance in order to make informed decisions about childbearing. *(67)*

genital stage Freud's term for the last stage of psychosexual development, in which the primary source of sexual satisfaction is an erotic relationship with another adult. *(33)*

genotype A person's entire genetic heritage, including those characteristics carried by the recessive genes but not expressed in the phenotype. *(60)*

germinal period The first two weeks after conception, during which rapid cell division occurs. (Also called the *period of the ovum*.) *(77)*

gerontology The study of the elderly and of old age. This is one of the fastest-growing special fields in the social sciences. *(527)*

glaucoma An eye disease that begins without apparent symptoms and often causes eventual blindness. Early detection and treatment can prevent vision impairment from glaucoma. *(466)*

gradual desensitization A technique, often used by behavior therapists, to reduce a person's fear of something by gradually exposing that person to the feared object. *(236)*

grammar Structures, techniques, and rules that languages use to communicate meaning, including word order, tense, and voice. *(207)*

grasping reflex A normal, neonatal reflex that causes newborns to grip tightly when something touches their palms. *(111)*

gross motor skills Those physical skills that use large body movements. Running, jumping, and climbing involve gross motor skills. *(187)*

growth spurt The relatively rapid physical growth that occurs during puberty. *(322)*

habituation A process whereby a particular stimulus becomes so familiar that physiological responses initially associated with it are no longer present. For instance, a newborn might initially stare wide-eyed at a mobile, but gradually look at it less often as habituation occurs. *(109)*

Hayflick limit The number of times a human cell is capable of dividing into two new cells. Leonard Hayflick determined that the limit for most human cells is approximately fifty divisions, suggesting that the life span is limited by our genetic program, which does not allow cells to reproduce themselves indefinitely. *(541)*

heterogamy Marriage between individuals who are dissimiliar in age, socioeconomic background, religion, or ethnic group. (See *homogamy*.) *(439)*

holistic development A view of human development as unified and whole; this perspective emphasizes the interaction among the various physical, cognitive, and psychosocial aspects of growth. *(5)*

holophrase A single word that is intended to express a complete thought. Young children (usually about 1 year of age) use this early form of communication. *(139)*

HOME A method which measures how well the home environment of a child fosters learning. HOME looks at maternal responsiveness and involvement with the child, the child's freedom of movement, the play environment, the play materials, and the variety of activities in the child's day. *(163)*

homeostasis The adjustment of the body's systems to keep physiological functions in a state of equilibrium. As the body ages, it takes longer for these homeostatic adjustments to occur, making it harder for older bodies to adapt to stresses, such as pregnancy or missing a night's sleep. *(394)*

homogamy Marriage between individuals who are similiar in age, socioeconomic background, religion, and ethnic group. *(439)*

hospice Originally a religious way station where poor travelers could obtain lodging, the hospice is now a place where the dying can die painlessly and with dignity, surrounded by friends and family. *(614)*

humanist theory A theory in psychology that emphasizes the potential of each human being to be fully functional, self-actualizing, and truly healthy. *(44)*

hyperactivity See *attention deficit disorder*.

id As conceptualized by Freud, that part of the personality containing primitive, unconscious sexual and aggressive impulses. *(34)*

identification A defense mechanism through which a person feels like, or adopts the perspective of, someone else. Children identify with their parents for many reasons, one of them, according to psychoanalytic theory, is to cope with the powerful emotions of the Oedipus (or Electra) complex. *(239)*

identity As a Piagetian term, the principle of logic which states that a given quantity of matter remains the same if nothing is added to or subtracted from it, no matter what changes occur in its shape or appearance. Before they enter the concrete operational period, children do not recognize this principle. *(202)*

identity achievement Erikson's term for a person's achievement of a sense of who he or she is as a unique individual. The main task of adolescence, according to Erikson, is the establishment of the young person's identity, including sexual, moral, political, and vocational identity. *(364)*

identity diffusion Erikson's term for the experience of a young person who is uncertain what path to take toward identity formation, and therefore becomes apathetic and disoriented. *(364)*

imaginary audience A term referring to the constant scrutiny that many adolescents typically imagine themselves to be under—from nearly everyone. *(344)*

implantation After conception, the burrowing of the organism into the lining of the uterus where it can be nourished and protected during growth. *(78)*

industry versus inferiority The fourth of Erikson's eight "crises," in which the school-age child busily masters many skills or develops a sense of incompetence. *(291)*

infantile sexuality   The idea, held by psychoanalytic theorists, that sexual pleasures and fantasies occur in childhood. *(33)*

infertility   The inability to conceive a child after one year of trying. *(397)*

information processing   A model of human learning that uses the functioning of the computer as an analogy for the functioning of the human mind. *(267)*

initiative versus guilt   The third of Erikson's eight "crises" of psychosocial development. During this stage, the preschool child begins, or initiates, new activities—and feels guilt when efforts result in failure or criticism. *(220)*

input   In the information-processing theory of intelligence, the information that is transmitted to the brain via the five senses. *(547)*

insecure attachment   A parent-child bond marked by the child's overdependence on, or lack of interest in, the parents. Insecurely attached children are not readily comforted by their parents and are less likely to explore their environments than are children who are securely attached. *(161)*

integrity versus despair   The last of Erikson's eight stages of development. During late adulthood, according to Erikson, people either feel that their lives have had meaning, and look back on their past experiences with a sense of integrity and wholeness, or they despair at their past and therefore dread the future. *(603)*

interindividual variation   Differences between individuals that are the result of the uniqueness of each person's genetic make-up and particular environment. *(493)*

interiority   According to Neugarten, the increased introspection and reflection that many people develop as they grow older. *(570)*

interview   The research method in which the scientist asks people specific questions designed to discover their opinions or experiences pertaining to a particular topic. Attitudes about sex, religion, or politics are often assessed through an interview. *(24)*

intimacy versus isolation   The sixth of Erikson's eight stages of development. Adults seek to find someone with whom to share their lives, in an enduring and self-sacrificing commitment. Without such commitment, they become loners, isolated from their fellow humans. *(431)*

invincibility fable   The feeling of many adolescents that they are invulnerable to laws of mortality and probability. As a result of egocentric thought, adolescents often take life-threatening risks with the belief that they will never be hurt. *(346)*

in vitro fertilization   A technique in which ova (egg cells) are surgically removed from a woman and fertilized with sperm. After the original fertilized cells (the zygotes) have divided several times, they are implanted in a woman's uterus (usually, but not necessarily, the woman who provided the ova). This method of reproduction is used to bypass problems that cause infertility, such as blocked Fallopian tubes. *(399)*

IQ   A number, or score, on an intelligence test that is designed to indicate the aptitude of a particular person for learning, especially learning in school. The average IQ is 100. *(284)*

kinkeepers   The people who celebrate family achievements, gather the family together, and keep in touch with family members who no longer live nearby. *(506)*

kwashiorkor   A disease resulting from protein-calorie deficiency in children. The symptoms include thinning hair, paleness, and bloating of the stomach, face, and legs. *(116)*

Lamaze method   A technique of childbirth that involves breathing and relaxation exercises during labor. *(95)*

language acquisition device (LAD)   Noam Chomsky's term for an infant's inborn ability to acquire language according to a relatively stable sequence and timetable. *(137)*

latency   Freud's term for the period between the phallic stage and the genital stage. During latency, which lasts from about age 7 to age 11, the child's sexual drives are relatively quiet. *(33)*

learned helplessness   The assumption that one is unable to do anything to improve one's performance or situation. For example, children who continually fail in school sometimes respond with learned helplessness. *(297)*

learning theory   A major theory of psychology which maintains that most human behaviors are learned, or conditioned, and which formulates laws of behavior that are applicable to animals and to people of all ages. Learning theory is also called *behaviorism*. *(38)*

life events   Notable occurrences in life that trigger new patterns of thinking and cognitive development. The birth of a child, for example, tends to propel parents into more "adult" thinking, and, as a result, new cognitive growth. *(425)*

life review   The examination of one's own past life that many elderly people engage in. According to Butler, the life review is therapeutic, for it helps the older person to come to grips with aging and death. *(570)*

living will   A legal document drawn up by a person not yet near death, detailing how much medical care should be administered if the person is near death. Living wills are designed to prevent doctors from taking extraordinary measures, beyond the patient's wishes, to prolong the life of the dying. *(613)*

longitudinal research   A study of the same people over a long period of time. Longitudinal research is designed to measure both changes and continuity in behavior and personality over time. *(26)*

long-term memory   A memory storage system in which information can remain for days, months, or years. *(268)*

low-birth-weight infant   A newborn who weighs less than 5½ pounds (2,500 grams) at birth. (See also *preterm infant, small-for-dates infant.*) *(88)*

macrosystem In the ecological or systems approach to development, the larger patterns of culture, politics, economy, and society that affect the individual. *(6)*

mainstreaming The practice of assigning handicapped children to regular classrooms, rather than segregating them in special classes. *(257)*

marasmus A disease that afflicts infants suffering from severe malnutrition. Growth stops, body tissues waste away, and eventually death occurs. *(116)*

mastery play Any form of play that leads to a mastering of new skills. During the play years, mastery play tends to develop physical skills (such as skipping or using scissors). Later, mastery play includes intellectual activities such as play with words and ideas. *(190)*

maximum life span The oldest age to which members of a species have been known to live. For humans, that age is 115 years. *(539)*

memory capacity A measure of how much information the brain can hold, and how well it can be processed and stored. (See also *sensory register, short-term memory,* and *long-term memory.*) *(267)*

menarche A female's first menstrual period. This is taken as a sign, or even as *the* sign, of puberty. *(326)*

menopause The time in middle age, usually around age 50, when a woman's menstrual periods cease completely. Physical symptoms, such as vasomotor instability, typically accompany menopause; psychological symptoms, such as depression, typically do not. *(478)*

mental combinations The mental playing-out of various courses of action before actually exercising one of them. According to Piaget, this ability usually becomes apparent between 18 and 24 months of age. *(128)*

mentor In career development, a mentor is a more experienced coworker or supervisor who provides advice, instruction, and support. In many professions, the mentor system operates informally, with great impact. Finding a good mentor may make the difference between success or failure in a new career. *(451)*

mesosystem In the ecological or systems approach to development, the interlocking relationships, such as the parent-teacher communications, or the employment practices impinging on family life, that link one microsystem to another. *(5)*

metacognition The general principles and techniques of thinking and organizing information. *(270)*

metamemory The ability to use and explain techniques that aid memory. *(268)*

microsystem In the ecologic or systems approach to development, the immediate systems, such as the family and the classroom, that affect an individual's daily life. *(5)*

middle age The years from age 40 to age 60. *(499)*

midlife The point, at about age 40, when the adult has about as many years of life ahead as have already past. Midlife ushers in middle age, which lasts until about age 60. *(499)*

midlife crisis A time in middle age when adults question their goals and achievements, perhaps becoming depressed or engaging in atypical behavior. For instance, the seemingly happy husband might suddenly demand a divorce, or the seemingly contented mother might want to leave her children and travel to a distant place. Not everyone has a midlife crisis: it seems more typical of middle-class American men than of other people. *(500)*

mnemonic device A memory-aiding device. (See also *rehearsal* and *chunking.*) *(269)*

modeling The patterning of one's behavior after that of someone else. New responses can be learned, and old ones modified, through modeling. *(41)*

monologue Speech delivered out loud but with no intent to communicate to others. This is one of the types of egocentric speech typical of preschool children. (See also *collective monologue.*) *(210)*

monozygotic twins Two offspring who began development as a single zygote (formed from one sperm and one ovum) that subsequently divided into two zygotes. They have the same genetic make-up, are of the same sex, and look alike. *(57)*

moratorium Erikson's term for the informal pause in identity formation that allows young people to explore alternatives without making final choices. For many young people, college or military service provides such a moratorium. *(365)*

Moro reflex A normal neonatal reflex in response to a sudden, intense noise or movement. In this reflex, newborns fling their arms outward, then bring them together; they may also cry with eyes wide open. *(112)*

mosaicism A condition in which a person's cells, including reproductive cells, are a patchwork, or mosaic, of different patterns, some normal in their numbers of chromosomes, some not. A parent with this condition has a higher than average probability of having a child with a chromosomal defect. *(65)*

motility The ability of sperm to swim quickly and far enough to reach an ovum for fertilization. Deficiencies in sperm motility diminish male fertility. *(398)*

motor skills Those abilities which involve body movement and physical coordination, such as walking and reaching. *(112)*

multidimensionality A description of cognitive abilities emphasizing that intelligence is composed of several distinct elements, all of which must be studied independently. *(492)*

multidirectionality A way of describing variations in specific cognitive abilities over time. *(493)*

multifactorial characteristics Those abilities or qualities that are determined by the interaction among several genetic and environmental influences. Characteristics such as intelligence, personality, and talent, are multifactorial. *(62)*

multi-infarct dementia (MID) The form of dementia characterized by sporadic, and progressive, memory loss. The cause is repeated instances of insufficient blood reaching the brain. Each infarct destroys some brain tissue. The underlying cause is an impaired circulatory system. *(564)*

mummification A reaction to the death of a loved one in which an individual acts as if the dead person were still alive, by preserving the deceased's belongings, setting the deceased's dinner place, and the like. *(616)*

myelin A fatty substance that insulates the neurons. As the nervous system matures, myelin coating helps to conduct neural impulses more quickly and efficiently, enabling growing children greater control over motor and perceptual functions. *(105)*

naturalistic observation The research method in which the scientist tests a hypothesis by observing people in their usual surroundings (home, school, work place). Specific methods of data collection and special training for the observers are generally used to make this method more objective than our usual daily observations of each other. *(17)*

nature Those innate factors that affect development. *(11)*

nature-nurture controversy The debate within developmental psychology over the relative importance of genetically inherited capacities ("nature") and environmental influences ("nurture") in determining an individual's various traits and characteristics. *(10)*

negative identity Erikson's term for a chosen identity that is the opposite of the identity preferred by one's parents or society. *(364)*

negative reinforcer The removal of an unpleasant stimulus in response to a particular behavior, such removal serving to increase the likelihood that the behavior will occur again. *(40)*

neglect A form of child abuse in which parents or caregivers fail to provide adequate or proper nutritional, supervisional, or physical care of a child. (See also *child abuse.*) (169)

neonate A newborn baby. Infants are neonates from the moment of birth to the end of the first month of life. *(81)*

neural tube The fold of cells that appears in the embryo about two weeks after conception and later develops into the head and spine. *(78)*

neurons Nerve cells. *(105)*

neuroticism Designates a personality dimension characterized by feelings of anxiety, worry, hostility, and depression. *(515)*

neurotransmitters Chemicals in the brain that communicate nerve impulses from one nerve cell to another. *(532)*

norms Statistical averages based on the results of research derived from a large, representative sample of a given population. Norms are not to be taken as implying "the best." For instance, the norm for an infant's first step is 12 months of age, but the infant who doesn't walk until 14 months is not necessarily less smart or less healthy than the infant who walks at 12 months. *(114)*

nuclear family One married couple and their dependent children. *(505)*

nurture Environmental influences that affect development. *(11)*

obesity Overweight to a degree that is deemed unhealthy, given the person's height and age. *(251)*

object permanence The understanding that objects and people continue to exist even when they cannot be seen. This concept develops gradually between 6 and 18 months of age. *(124)*

Oedipus complex In psychoanalytic theory, both the sexual desire that boys in the phallic stage have for their mother and the associated feelings of hostility they have toward their father. This complex is named after Oedipus, a character in ancient Greek legend who unwittingly killed his father and married his mother. *(238)*

old-old Older adults who suffer from severe physical, mental, or social deficits and thus require supportive services such as nursing homes and hospital stays. *(526)*

onlooker play "Play" that consists of one child's watching another's active play. *(223)*

openness Designates a personality dimension characterized by receptiveness to new experiences, new ideas, and changing one's life as well as one's mind. *(515)*

open system A cognitive system characterized by ambiguity and a large number of overlapping variables. Problems in an open system are best solved using postformal thought. (See *closed system.)* *(414)*

operant conditioning A learning process, conceptualized by B.F. Skinner, through which a person or animal is more likely to perform or refrain from performing a certain behavior because of past reinforcement or punishment. (Also called *instrumental conditioning.*) *(39)*

oral stage Freud's term for the first stage of psychosexual development, when the infant gains both nourishment and pleasure through sucking and biting. *(33)*

organization Piaget's term for the process of synthesizing and analyzing perceptions and thoughts. At every stage of cognitive development, according to Piaget, people actively organize their existing ideas and adapt to new experiences. *(49)*

organ reserve The extra capacity of the heart, lungs, and other organs that makes it possible for the body to withstand moments of intense stress, or years of poor health practices, without collapse. With age, organ reserve is gradually depleted, but the rate of depletion depends on how good the individual's health habits are. *(394)*

osteoporosis A loss of calcium within the bone that makes the bone more porous and fragile. It occurs somewhat in everyone with aging, but serious osteoporosis is more common in elderly women than men. Osteoporosis is the main reason the elderly suffer broken hip bones much more often than the young. *(479)*

overextension The overuse of a given word to describe several objects that share a particular characteristic. For example, toddlers often use "doggie" to label all four-legged animals. *(138)*

overregularization The tendency of young children to apply grammatical rules and forms without recognizing exceptions and irregularities. Overregularization might, for example, lead a child to use the suffix "ed" to form the past tense of all verbs and say "bringed" instead of "brought." *(208)*

parental monitoring Parental checking into their child's activities and whereabouts. The worse the social milieu, the more likely it is that monitoring will be strict to keep the child from possibly dangerous situations. *(376)*

parent-infant bond The strong feelings of attachment between parents and infants. *(97)*

Parkinson's disease A chronic progressive nervous disease that is characterized by muscle tremors and rigidity, caused by a reduction of dopamine production in the brain. *(564)*

passive euthanasia The practice whereby a person is allowed to die by withholding some procedure or drug that would have allowed life to continue a bit longer. Passive euthanasia is practiced in many hospitals and hospices, when the extension of life seems only to prolong misery. *(613)*

pelvic inflammatory disease (PID) A common result of recurring pelvic infections in women. Pelvic inflammatory disease often leads to blocked Fallopian tubes, which, in turn, can lead to sterility. *(398)*

penis envy The psychoanalytic idea that, beginning at about age 4, girls realize that boys have a penis and become jealous because they do not. *(239)*

perception The processing or interpretation of sensations in order to make them comprehensible. *(108)*

period of acquisition Schaie's characterization of cognitive development in childhood and adolescence. During this period, information is absorbed and problem-solving techniques are learned. *(412)*

period of the embryo From approximately the second to the eighth week after conception, during which time the rudimentary forms of all anatomical structures develop. *(77)*

period of the fetus From two months after conception until birth. In a full-term pregnancy, this period lasts seven months. *(77)*

permissive parenting A style of child-rearing in which parents allow their children to do virtually anything they want to do. Permissive parents rarely punish, guide, or control their children. *(228)*

personal fable The idea, held by many adolescents, that one is special—destined for great accomplishments and immune to normal troubles. *(347)*

personality An individual's usual way of reacting to people and experiences. *(153)*

phallic stage The third stage of psychosexual development, according to Freud, in which the penis, or phallus, is the focus of psychological concern as well as of physiological pleasure. *(33)*

phenotype An individual's physical characteristics, which are the result of the interaction of the genes and the environment. (See also *genotype*.) *(60)*

phenylketonuria (PKU) A genetic disease, now easily detected, in which the individual is unable to properly metabolize protein. If left untreated, mental retardation and hyperactivity result. *(67)*

phobia An irrational fear that interferes with a person's normal functioning. Many phobias have specific names, such as claustrophobia (fear of enclosed places), aquaphobia (fear of water), and agoraphobia (fear of open spaces). *(236)*

physical domain That domain of development that includes changes which are primarily biological. For instance, increases in height and weight, improvements in motor skills, and the development of sense organs are usually considered aspects of physical development. *(4)*

placenta An organ made up of blood vessels leading to both the mother's and the fetus's bloodstream and having membranes to prevent mixture of the two bloodstreams. These membranes serve as screens through which oxygen and nourishment pass to the fetus and wastes pass from the fetus to the mother to be excreted through her system. *(79)*

plasticity In developmental psychology, a term used to indicate that a particular characteristic is shaped by many environmental influences. Many human characteristics, once thought to be firmly fixed before adulthood, are now known to have much more plasticity than was once believed. *(494)*

play face A facial expression, such as a smile or laugh, that accompanies playful activity. The play face helps distinguish rough-and-tumble play from real hostility. *(191)*

polygenic inheritance The interaction of many genes to produce a particular characteristic. For example, skin color, body shape, and memory are all polygenic. *(61)*

positive reinforcer A reward, or something pleasant, that is given in response to a particular behavior and which increases the likelihood that that behavior will occur again. *(40)*

postformal thought A type of adult thinking that is suited to solving real-world problems. Postformal thought is less abstract and absolute than formal thought, more adaptive to life's inconsistencies, and more dialectical—capable of combining contradictory elements into a comprehensive whole. (See *dialectical thought*.) *(414)*

pragmatics In the study of language, a term for the practical aspect of communication, for example, the skill a person shows in adjusting vocabulary and grammar to fit the social context. *(209)*

preoperational thought Piaget's term for the second period of cognitive development. Children in this stage of thought, which usually occurs between the ages of 2 and 7, are unable to grasp logical concepts such as conservation, reversibility, or classification. *(200)*

preterm An infant born more than three weeks before the due date or, according to the World Health Organization, who weighs less than 5½ pounds (2500 grams) at birth. Also called premature baby. *(89)*

primary sex characteristics Those aspects of the body directly related to sexual reproduction, such as the ovaries and the testicles. *(326)*

priming The technique of calling up related items from memory with the goal of recalling a specific piece of information. For example, when a wallet is lost, it is often helpful to review the day's events, in hope of triggering the memory of where the wallet was last placed. *(556)*

progeria A rare genetic disease which causes young children to age prematurely. *(539)*

program In the information-processing theory of intelligence, the organization and interpretation of information. *(547)*

programmed senescence The belief that the human body is genetically programmed to age. Many researchers believe that the timetable for aging, like the timetable for biological maturation, is contained within the DNA. *(539)*

prosocial behaviors Any behavior that benefits other people. Cooperation, helping, sharing, and generosity are all prosocial behaviors. *(301)*

provocation ecologies Classroom environments that provoke or exacerbate hyperactive behavior in some children. *(288)*

proximo-distal development Growth proceeding from the center (spine) toward the extremities (literally, from near to far). Human growth, from the embryonic period through childhood, follows this pattern. *(78)*

psychoanalytic theory A theory of psychology, originated by Sigmund Freud, that stresses the influence of unconscious motivation and drives on all human behavior. *(32)*

psychosexual stages The idea, held by psychoanalytic theorists, that development occurs in a series of stages (oral, anal, phallic, and genital), each of which is characterized by the focusing of sexual interest and gratification on one part of the body. *(33)*

psychosocial domain The domain of human development involving emotions, personality characteristics, and relationships with other people. *(4)*

psychosocial theory A theory emphasizing social and cultural effects on the individual. *(36)*

puberty The period of early adolescence characterized by rapid physical growth and the attainment of the physiological capability of sexual reproduction. Puberty usually begins at about age 10 or 11 for girls, and 11 or 12 for boys, although there is much variation caused by genes and nutrition. *(321)*

punishment An unpleasant event, such as a slap, which when administered in response to a particular behavior, makes it less likely that the behavior will be repeated. *(40)*

rarefaction ecologies Classroom environments that ameliorate or diminish hyperactive behavior in some children. *(288)*

reaction time The time it takes a person to react to a stimulus. *(254)*

recessive gene A gene that affects the expression of a particular phenotypic characteristic only when it is paired with another of the same type of recessive gene, not with a dominant gene. *(60)*

reciprocity The logical principle that a change in one dimension of an object results in a change in another dimension. For example, a ball of clay, rolled out, will be both longer and thinner. According to Piaget, children begin to understand these relationships during the period of concrete operational thought. *(263)*

reflex An automatic response, such as an eye blink, involving one part of the body. *(111)*

regression A defense mechanism in which an individual under stress will temporarily revert to a more immature form of behavior (such as bed-wetting by a 12-year-old) *(35)*

rehearsal A memory technique involving repetition of the material to be memorized. *(269)*

reinforcement In operant conditioning, the process whereby a particular behavior is strengthened, making it more likely that the behavior will be repeated. *(40)*

reinforcer Anything (for example, food, money, a smile) that increases the likelihood that a given response will occur again. For example, giving a child a warm hug for being polite will increase the chances that that behavior will be repeated. (See also *positive reinforcer, negative reinforcer.)* *(40)*

reintegrative stage Schaie's last stage of cognitive development, during which the individual develops wisdom and focuses on the larger purposes of life and existence. *(413)*

replicate To repeat or duplicate. Scientists describe their experiments in detail sufficient to allow others to replicate their test procedures. *(17)*

representation The ability, usually first evident between 18 and 24 months of age, to remember (through the creation of a mental image) an object, event, or person that has been seen or experienced at an earlier time. *(128)*

representative sample A group of subjects in a research project who have the relevant characteristics (e.g. sex, race, socioeconomic level) of the general population or of a particular segment of the population to which the experimental results are most applicable. *(18)*

repression A defense mechanism in which anxiety-provoking thoughts and fantasies are excluded from consciousness. *(35)*

resource room A classroom equipped with special learning materials designed to teach children who have learning difficulties. *(257)*

response A behavior (either instinctual or learned) following a specific cue. (See *stimulus.)* *(39)*

responsible stage   In Schaie's stages of adult cognitive development, a period during which the adult cares for the well-being of not only the self, but of others, too, such as a family and friends. *(413)*

restricted code   A form of speech characterized by limited vocabulary and syntax. Meaning is communicated by gestures, intonation, and shared understandings. *(276)*

reversibility   The idea, described by Piaget, that something that has been changed can be returned to its original state simply be reversing the process of change. For example, a ball of clay that has been rolled out into a long, thin rope can be rerolled into a ball. Preschoolers cannot regularly apply the rule of reversibility. *(199)*

Rh disease   See *erythroblastosis*.

rite of passage   An anthropological term for a ritual that marks the transition from one stage of life to another. The initiation ceremonies at puberty are examples of a rite of passage; weddings and funerals are others. *(367)*

rooting reflex   A normal neonatal reflex that helps babies find a nipple by causing them to turn their heads toward the stimulus and start to suck whenever something brushes against their cheek. *(111)*

rough-and-tumble play   Wrestling, chasing, and hitting that occurs purely in fun, with no intent to harm. *(191)*

rubella (German measles)   A virus which, if contracted during pregnancy, can cause the fetus to develop serious handicaps, among them blindness and deafness. *(85)*

sample size   In research, a group of individuals that is studied so that conclusions may be drawn about a larger group or segment of the population. (Also called *sample population*.) *(18)*

sandwich generation   The generation "in between," having both adolescent children and elderly parents. Many middle-aged people feel pressured by the needs and demands of their children, on the one side, and of their elderly (and perhaps ailing or widowed) parents on the other. *(506)*

schema   Piaget's term for a general way of thinking about, or interacting with, ideas and objects in the environment. *(48)*

scientific method   A procedural model used to formulate questions, collect data, test hypotheses, and draw conclusions. Use of the scientific method helps researchers overcome biases, test assumptions, and in short, be "scientific." *(16)*

scientific reasoning   The ability to understand and use the principles of science. *(340)*

secondary sex characteristics   Sexual features other than the actual sex organs, such as a man's beard or a woman's breasts, that distinguish male from female. *(327)*

secular trend   The tendency of each recent new generation to grow taller and to experience puberty earlier than their parents. *(330)*

secure attachment   A healthy parent-child bond in which the child feels comfort when the parent is present, experiences moderate distress at the parent's absence, and quickly reestablishes contact when the parent returns. *(160)*

selective attention   The ability to focus attention on particular stimuli and ignore distractions. *(269)*

self-actualization   The ultimate goal of human development, according to humanist theories. A self-actualized person has fulfilled his or her potential to the maximum, making *actual*, or real, the unique and creative self that each person can become. *(44)*

self-awareness   A person's sense of himself or herself as a separate person, with particular characteristics. The development of this sense of self begins at birth, but only between 1 and 2 years of age do children begin to truly differentiate themselves from others. *(147)*

self-theory   The older child's or adult's complex theory about the self, based on the evidence of experience, the opinion of others, and untested assumptions about self. *(296)*

senescence   The state of physical decline, in which the body gradually becomes less strong and efficient with age. *(529)*

sensation   The process by which the senses detect stimuli within the environment. *(108)*

sensorimotor intelligence   Piaget's term for the first stage of cognitive development (from birth to about 2 years old). Children in this stage primarily use the senses and motor skills (i.e., grasping, sucking, etc.) to explore and manipulate the environment. *(122)*

sensorimotor play   Play that captures the pleasures of using the senses, including the primary senses (touching, tasting, hearing, etc.) and the sense of motion and balance. For example, children who mash their food, or whirl their bodies around for the pure fun of it are engaging in sensorimotor play. *(190)*

sensory register   A memory system that functions for only a fraction of a second during sensory processing, retaining a fleeting impression of the stimulus that has just impinged on a particular sense organ (e.g., the eyes). If a person looks at an object, for example, and then closes his or her eyes, the visual image of the object is briefly maintained. *(268)*

separation anxiety   A child's fear of being left or abandoned by the mother or other care-giver. This emotion emerges at about 8 or 9 months, peaks at about 14 months, then gradually subsides. *(146)*

separation-individuation   A term used by Margaret Mahler to describe the period during which the child gradually develops a sense of self, apart from the mother. This period extends from about 5 months to 3 years, and is marked by the child's increasingly secure attempts to achieve psychological separation from the mother. *(152)*

sequential research   Research designed to take into account cohort effects. First, a cross-sectional study is done. Then, months or years later, the same persons are tested again, as in a longitudinal study. At the same time, new people from each age group are added and tested to control for changes in the original group. *(28)*

seriation   The concept that items can be arranged in a logical series, as by sorting a group of sticks from longest to shortest, or arranging a group of crayons from lightest to darkest. This concept is mastered during the period of operational thought, according to Piaget. *(264)*

sexually transmitted diseases (STDs)   Diseases spread by sexual contact. Such diseases include syphilis, gonorrhea, herpes simplex, and AIDS. *(353)*

short-term memory   The memory system in which information is kept for a brief time, no longer than a minute. *(268)*

small-for-date   An infant who is born weighing less than the average baby born after the same number of weeks of gestation. *(89)*

social clock   Neugarten's term for the idea that the stages of life in adulthood are set by social standards rather than by biological maturation. For instance, "middle age" begins when the culture believes it does, rather than at a particular age in all societies. *(434)*

social cognition   A person's understanding of the dynamics of human interaction. *(294)*

social convoy   Each person's family members, friends, and acquaintances who move through life with him or her are called the social convoy. While peripheral members of the convoy can drop out or join without much disruption, the arrival or departure of central members, as in the birth of an offspring or the loss of a spouse, can have great impact on the individual. People whose social convoy is large and close are better able to endure the various physical and psychological assaults that life brings. *(588)*

social learning theory   The theory that learning occurs through imitation of, and identification with, other people. *(41)*

social reinforcers   Rewards that come from other people in response to particular behavior and increase the likelihood that that behavior will occur again. For instance, if people smile and babble in response to the infant's first babbling, the infant is likely to babble again. *(40)*

social smile   An infant's smile in response to seeing another person. In full-term infants, this kind of smile first appears at about 6 weeks after birth. *(146)*

society of children   The culture of games, sayings, and traditions passed down from one generation of children to the next. *(299)*

socioeconomic status (SES)   A measure that takes into account family income, parents' education, and father's occupation and employment. Socioeconomic status has been found to influence physical health, academic performance, and many other aspects of an individual's life. *(304)*

solitary play   A form of play in which the child plays alone, seemingly unaware of other children playing nearby. *(223)*

sonogram   A method of determining the size and position of the fetus by means of sound waves. *(70)*

specific learning disability   Any of a number of particular difficulties in mastering basic academic skills, without apparent deficit in intelligence or impairment of sensory functions. *(281)*

stepping reflex   A normal neonatal reflex that causes newborns to move their legs as if to walk when their feet touch a flat surface. *(111)*

stimulus   An external condition or event that elicits a bodily response or prompts a particular action. For example, the sight or aroma of an appetizing meal is a stimulus to which the response is usually salivation. *(39)*

storage   In the information-processing theory of intelligence, the placing of information into memory. Storage consists of sensory register (the momentary afterimage after perception), short-term memory, and long-term memory. *(547)*

stress   The adverse physical and emotional reaction to demands put on the individual by unsettling conditions or experiences, also known as stressors. Adverse reactions to stress have been linked to increased incidence of disease and health problems. *(471)*

stressors   Unsettling conditions or experiences that cause stress. *(471)*

sucking reflex   A normal neonatal reflex that causes newborns to suck anything that touches their lips. *(111)*

superego   Freud's term for that part of the personality that contains the conscience, including the internalization of moral standards set by one's parents. *(34)*

surface structure   Noam Chomsky's term for the particular vocabulary and rules of grammar that differ from one language to another. The surface structure is distinct from the deep structure of language, which includes the general rules that are shared by most languages. *(137)*

swimming reflex   A normal neonatal reflex that causes the newborn to make swimming motions when held aloft horizontally. *(111)*

symbiosis   A term used by Margaret Mahler to describe the period of infancy, from about 2 to 5 months, during which the infant is so entirely dependent on the mother that he or she feels literally a part of her. *(152)*

symbol   Sounds, written words, drawings, actions, or objects that stand for or signify something else. For example, a flag symbolizes the spoken word "dog," which, in turn, symbolizes a particular kind of animal. *(198)*

synchrony   Carefully coordinated interaction between infant and parent (or any other two people) in which each is exquisitely, often unknowingly, attuned to the other's verbal and nonverbal cues. *(158)*

systems approach   An approach to human development that emphasizes the effects of the environment and that conceptualizes the environment as consisting of several levels of systems, ranging from small systems, such as the family, to large systems, for example, the cultural and political systems. (Also called the *ecological approach*.) *(6)*

T-cells Cells created in the thymus that produce substances that attack infected cells in the body. *(535)*

temperament The characteristic way a person responds to things and people; for example, his or her usual quality of mood, activity level, intensity of reaction, and attention span. (153)

teratogens External agents, such as viruses, drugs, chemicals, and radiation, which can cross the barrier of the placenta and harm the embryo or fetus. *(82)*

teratology The scientific study of birth defects caused by genetic or prenatal problems, or by birth complications. *(82)*

test of significance A statistical test, using a specific mathematical formula, applied to research in order to determine the probability of the results of an experiment occurring by chance. *(19)*

testosterone Hormones that are produced primarily by the testes and regulate sexual development in puberty. Although girls' adrenal glands produce some testosterone, it is chiefly a male hormone. *(322)*

thanatology A field of research that studies death. *(609)*

theory A systematic statement of hypotheses and general principles that provides a framework for future research and interpretation. *(31)*

toddler A child, usually between the ages 1 and 2, who has just begun to master the art of walking. *(114)*

toxoplasmosis A mild disease caused by a parasite often found in uncooked meat and in cat feces. If a pregnant woman contracts this disease, her fetus may suffer eye or brain damage. *(85)*

trimester One of the three-month periods in the nine months of pregnancy. *(80)*

trust versus mistrust Erik Erikson's term for the infant's basic experience of the world as either good and comfortable or as threatening and uncomfortable. Early care-giving experiences usually mold the child's viewpoint. This is the first of Erikson's eight stages of development. *(150)*

vasomotor instability The temporary inefficiency of the homeostatic mechanisms of the vascular system, which usually constrict or dilate the blood vessels to maintain body temperature. Vasomotor instability causes moments of feeling suddenly hot or cold, a typical experience during the climacteric at menopause. *(478)*

X-linked genes Genes that are carried on the X chromosome exclusively. X-linked genes account for the fact that certain recessive genetic diseases or conditions are more likely to occur in males, who have only one X chromosome, than in females, who have two. *(60)*

wear-and-tear theory A theory of aging that states that the human body wears out merely by being lived-in and by being exposed to environmental stressors. The wear-and-tear theory sees the human body as a machine that wears out over time. *(536)*

young-old Healthy, vigorous, financially secure older adults who are well-integrated into the lives of their families and their communities. *(526)*

zygote The one-celled organism formed from the union of a sperm and an ovum. *(55)*

Bibliography

Abraham, Kitty G., and Christopherson, Victor A. (1984). Perceived competence among rural middle school children: Parental antecedents and relation to locus of control. *Journal of Early Adolescence, 4,* 343–351.

Abraham, Sidney, Lowenstein, Frank W., and Johnson, Clifford, L. (1974). *First health and nutrition survey, United States, 1971–1972: Dietary intake and biochemical findings.* Washington, DC: National Center for Health Statistics.

Abravanel, Eugene, and Sigafoos, Ann D. (1984). Exploring the presence of imitation during early infancy. *Child Development, 55,* 381–392.

Achenbach, Thomas M. (1982). *Developmental psychopathology* (2nd ed.). New York: Wiley.

Achenbach, Thomas M., and Edelbrock, Craig S. (1981). Behavioral problems and competencies reported by parents of normal and disturbed children aged four through sixteen. *Monographs of the Society for Research in Child Development, 46* (Serial No. 188).

Achenberg, W. Andrew. (1985). Religion in the lives of the elderly. In Gari Lesnoff-Caravalglia (Ed.), *Values, ethics and aging.* New York: Human Sciences Press.

Ackerman, B.P. (1978). Children's comprehension of presupposed information: Logical and pragmatic inferences to speaker belief. *Journal of Experimental Child Psychology, 26,* 92–114.

Adams, Cynthia, and Labouvie-Vief, Gisela. (1986). *Modes of knowing and language processing.* Symposium on developmental dimensions of adult adaptation: Perspectives on mind, self, and emotion. Presented at 1986 meeting of the Gerontological Association of America, Chicago, November 20.

Adams, G.R., and Jones, R.M. (1981). Imaginary audience behavior: A validation study. *Journal of Early Adolescence, 1,* 1–10.

Adams, Rebecca, G. (1987). Patterns of network change: A longitudinal study of friendship of elderly women. *Journal of Gerontology, 27,* 222–227.

Ade-Ridder, Linda, and Brubaker, Timothy II. (1983). The quality of long-term marriages. In Timothy H. Brubaker (Ed.), *Family relationships in later life.* Beverly Hills, CA: Sage.

Adelson, Joseph. (1979). Adolescence and the generalization gap. *Psychology Today. 12*(9), 33–37.

Ahrons, Constance, and Wallisch, L. (1986). The relationship between former spouses. In Daniel Perlman and Steve Duck (Eds.), *Intimate relationships: Development, dynamics, and deterioration.* Beverly Hills, CA: Sage.

Aiken, Linda H., and Marx, Martita M. (1982). Hospices: Perspectives on the public policy debate. *American Psychologist, 37,* 1271–1279.

Ainsworth, Mary D. Salter. (1973). The development of infant-mother attachment. In Bettye M. Caldwell and Henry N. Ricciuti (Eds.), *Review of child development research.* Vol. III. Chicago: University of Chicago Press.

Ainsworth, Mary D. Salter, and Bell, Silvia M. (1970). Attachment, exploration, and separation: Illustrated by the behavior of one-year-olds in a strange situation. *Child Development, 41,* 49–67.

Ainsworth, Mary D.S., Blehar, M., Waters, Everett, and Wall, S. (1978). *Patterns of attachment.* Hillsdale, NJ: Erlbaum.

Aldis, Owen. (1975). *Play fighting.* New York: Academic Press.

Aldous, Joan. (1978). *Family careers: Developmental change in families.* New York: Wiley.

Alexander, James F., and Malouf, Roberta E. (1983). Problems in personality and social development. In Paul H. Mussen (Ed.), *Handbook of child psychology: Vol. 4. Socialization, personality and social development.* New York: Wiley.

Ali, Z., and Lowry, M. (1981). Early maternal-child contact: Effects on later behavior. *Developmental Medicine and Child Neurology, 23,* 337–345.

Allen, Cynthia Clapp. (March, 1987). *Career plans and life patterns of college-educated women: A twenty-year follow-up study.* Paper presented at the annual conference of the Association for Women in Psychology, Denver, Colorado.

Allen, J.R., Barsotti, D.A., and Carsten, L.A. (1980). Residual effect of polychlorinated biphenyls on adult nonhuman primates and their offspring. *Journal of Toxicology and Environmental Health, 6,* 55–66.

Allison, Clara (1985). Developmental direction of action programs: Repetitive action to correction loops. In Jane E. Clark and James H. Humphrey (Eds.), *Motor development: Current selected research.* Princeton, NJ: Princeton Book Company.

Alpert, Judith L., and Richardson, Mary Sue. (1980). Parenting. In Leonard W. Poon (Ed.), *Aging in the 1980's: Psychological issues.* Washington, DC: American Psychological Association.

Alwin, Duane F. (1976). Socioeconomic background, colleges, and post-college achievement. In William H. Sewell, Robert M. Hauser, and David L. Featherman (Eds.), *Schooling and achievement in American society.* New York: Academic Press.

Alwin, Duane F., Converse, Philip E., and Martin, Steven S. (1985). Living arrangements and social integration. *Journal of Marriage and the Family, 47,* 319–334.

Amabile, Teresa M. (1983). *The social psychology of creativity.* New York: Springer-Verlag.

Amato, Paul R. (1987). Family processes in one-parent, stepparent, and intact families: The child's point of view. *Journal of Marriage and the Family, 49,* 327–337.

Ambert, Anne-Marie. (1982). Differences in children's behavior toward custodial mothers and custodial fathers. *Journal of Marriage and the Family, 44,* 73–86.

Ambert, Anne-Marie. (1986). Being a stepparent: Live-in and visiting stepchildren. *Journal of Marriage and the Family, 48,* 795–804.

American Council on Education. (1984). *Fact book on higher education.* New York: Macmillan.

Anastasi, Anne. (1982). *Psychological testing* (5th ed.). New York: Macmillan.

Anderson, Stephen A., Russell, Candyce S., and **Schumm, Walter R.** (1983). Perceived marital quality and family life-cycle categories: A further analysis. *Journal of Marriage and the Family, 45,* 127–139.

Andres, Reubin. (1981). Influence of obesity on longevity in the aged. In D. Danon, N. W. Schock, and M. Marios (Eds.), *Aging: A challenge to science and society.* Vol. I. London: Oxford University Press.

Anglin, Jeremy M. (1977.) *Word, object, and conceptual development.* New York: Norton.

Anolik, S. A. (1981). Imaginary audience behavior and perceptions of parents among delinquent and nondelinquent adolescents. *Journal of Youth and Adolescence, 10,* 443–454.

Anolik, Steven A. (1983). Family influence upon delinquency: Biosocial and psychosocial perspectives. *Adolescence, 18,* 489–498.

Antonovsky, Aaron. (1986). Intergenerational networks and transmitting the sense of coherence. In Nancy Datan, Anita L. Greene, and Hayne W. Reese (Eds.), *Life-span developmental psychology: Intergenerational relations.* Hillsdale, NJ: Erlbaum.

Antonucci, Toni C. (1985). Personal characteristics, social support, and social behavior. In Robert H. Binstock and Ethel Shanas (Eds.), *Handbook of aging and the social sciences.* New York: Van Nostrand.

Apgar, Virginia. (1953). A proposal for a new method of evaluation in the newborn infant. *Current Research in Anesthesia and Analgesia, 32,* 260.

Apgar, Virginia, and **Beck, Joan.** (1973). *Is my baby all right?* New York: Trident Press.

Applebaum, Mark I., and **McCall, Robert B.** (1983). Design and analysis in developmental psychology. In Paul H. Mussen (Ed.), *Handbook of child psychology: Vol 1. History, theory, and methods.* New York: Wiley.

Archer, Cynthia. (1984). Children's attitudes toward sex-role divisions in adult occupational roles. *Sex Roles, 10,* 1–10.

Arenberg, David. (1982). Changes in age with problem-solving. In Fergus I. M. Craik and Sandra Trehub (Eds.), *Aging and cognitive processes.* New York: Plenum.

Arend, Richard, Gove, Frederick, L., and **Sroufe, L. Alan.** (1979). Continuity of individual adaptation from infancy to kindergarten: A predictive study of ego-resiliency and curiosity in preschoolers. *Child Development, 50,* 950–959.

Ariès, Philippe. (1962). *Centuries of childhood: A social history of family life.* Robert Baldick (trans). New York. Knopf.

Ariès, Philippe. (1981). *The hour of our death.* New York: Knopf.

Arlin, Greg. (1976). The elderly widow and her family, neighbors, and friends. *Journal of Marriage and the Family, 38,* 757–768.

Arlin, P.K. (1984). Adolescent and adult thought: A structural interpretation. In M. L. Commons, F. A. Richards, and C. Armon (Eds.), *Beyond formal operations: Late adolescent and adult cognitive development.* New York: Praeger.

Armstrong, Louise. (1978). *Kiss daddy goodnight.* New York: Hawthorn Books.

Asher, Steven R., and **Renshaw, Peter D.** (1981). Children without friends: Social knowledge and social skill training. In Steven R. Asher and John M. Gottman (Eds.). *The development of children's friendships.* Cambridge, Eng.: Cambridge University Press.

Åstrand, Per-Olof. (1976). The child in sport and physical activity: Physiology. In J. G. Albinson and G. M. Andrew (Eds.), *Child in sport and physical activity.* Baltimore, MD: University Park Press, 19–33.

Atchley, Robert C., and **Miller, Sheila J.** (1983). Types of elderly couples. In Timothy H. Brubaker (Ed.), *Family relationships in later life.* Beverly Hills, CA: Sage.

Auster, C. J., and **Auster, D.** (1981). Factors influencing women's choice of non-traditional careers: The role of family, peers and counselors. *The Vocational Guidance Quarterly, 29,* 253–263.

Ausubel, David P., Montemayor, Raymond R., and **Svajian, Pergrouhi (Najarian).** (1977). *Theory and problems of adolescent development* (2nd ed.). New York: Grune and Stratton.

Axia, Giovanna, and **Baroni, Rosa.** (1985). Linguistic politeness at different age levels. *Child Development, 56,* 918–927.

Bachrach, Christine, A. (1984). Contraceptive practice among American women. 1973–1982. *Family Planning Perspectives, 16,* 253–259.

Bahrick, H.P. (1984). Semantic memory content in permastore: Fifty years of memory for Spanish learned in school. *Journal of Experimental Psychology: General, 113,* 1–35.

Bailey, D. A. (1977). The growing child and the need for physical activity. In Russell C. Smart and Mollie S. Smart (Eds.), *Readings in child development and relationships* (2nd ed.). New York: Macmillan.

Bailey, Lynn B. (1986). Nutritional anemias: Iron, folacin, vitamin B 12. In Linda H. Chen (Ed.), *Nutritional aspects of aging:* Vol. 2. Boca Raton, FL: CRC Press.

Bakan, David. (1966). *The duality of human existence.* Chicago: Rand-McNally.

Bakeman, R., and **Brown, J.V.** (1980). Early interaction. Consequences for social and mental development at three years. *Child Development, 51,* 437–447.

Baker, J.L., and **Gottlieb, Jay.** (1980). Attitudes of teachers toward mainstreaming. In Jay Gottlieb (Ed.), *Educating mentally retarded persons in the mainstream.* Baltimore: University Park Press.

Baker, S. Josephine. (1923). Healthy children. Boston: Little, Brown.

Ball, Jean A. (1987). *Reactions to motherhood.* New York: Cambridge University Press.

Baltes, Margret M., and **Reisenzein, Rainer.** (1986). The social world in long-term care institutions: Psychosocial control toward dependency? In Margret M. Baltes and Paul B. Baltes (Eds.), *The psychology of control and aging.* Hillsdale, NJ: Erlbaum.

Baltes, Paul B., and **Schaie, K. Warner.** (1974). Aging and IQ: The myth of the twilight years. *Psychology Today, 7*(10), 35–40.

Bandura, Albert. (1977). *Social learning theory.* Englewood Cliffs, NJ: Prentice-Hall.

Bandura, Albert. (1981). Self-referent thought: A developmental analysis of self-efficacy. In John H. Flavell and Lee Ross (Eds.), *Social cognitive development: Frontiers and possible futures.* Cambridge, Eng.: Cambridge University Press.

Bandura, Albert. (1986). *Social foundations of thought and action: A social cognitive theory.* Englewood Cliffs, NJ: Prentice-Hall.

Bank, Stephen P., and Kahn, Michael D. (1982). *The sibling bond.* New York: Basic Books.

Bard, B., and Sachs, J. (1977). *Language acquisition patterns in two normal children of deaf parents.* Paper presented at the meeting of the Boston University Conference on Language Acquisition.

Bardin, C. Wayne. (1986). The Pituitary-Testicular Axis. In Samuel S. C. Yen and Robert B. Jaffe (Eds.), *Reproductive endocrinology: Physiology, pathophysiology, and clinical management* (2nd ed.). Philadelphia: Saunders.

Barnard, Kathyrn, E., and Bee, Helen L. (1983). The impact of temporally patterned stimulation on the development of preterm infants. *Child Development, 54,* 1156–1167.

Barnes, Deborah M. (1987a). Defect in Alzheimer's is on chromosome 21. *Science, 235,* 846–847.

Barnes, Deborah M. (1987b). Brain damage by AIDS under active study. *Science, 235,* 1574–1577.

Barnes, Grace M. and Welte, John W. (1986). Adolescent alcohol abuse: Subgroup differences and relationship to other problem behaviors. *Journal of Adolescent Research, 1,* 79–94.

Barnett, Mark A. (1986). Sex bias in the helping behavior presented in children's picture books. *Journal of Genetic Psychology, 147,* 343–351.

Barnett, Rosaline C., and Baruch, Grace K. (1987). Determinants of fathers' participation in family work. *Journal of Marriage and the Family, 49,* 29–40.

Barrera, Maria E., and Maurer, Daphne. (1981a). The perception of facial expressions by the three-month-old. *Child Development, 52,* 203–206.

Barrera, Maria E., and Maurer, Daphne. (1981b). Discrimination of strangers by the three-month-old. *Child Development, 52,* 558–563.

Barrett, Martyn D. (1986). Early semantic representations and early world-usage. In Stan A. Kuczaj and Martyn D. Barrett (Eds.), *The development of word meaning: Progress in cognitive developmental research.* New York: Springer-Verlag.

Barrett, Terry, R., and Watkins, Sandy R. (1986). World familiarity and cardiovascular health as determinants of age-related declines. *Journal of Gerontology, 41,* 222–224.

Bartel, Nettie R. (1971). Locus of control and achievement in middle- and lower-class children. *Child Development, 42,* 1099–1107.

Baruch, Grace. (1984). The psychological well-being of women in the middle years. In Grace K. Baruch and Jeanne Brooks-Gunn (Eds.), *Women in midlife.* New York: Plenum.

Baruch, G., Barnett, R., and Rivers, C. (1983). *Lifeprints: New patterns of work and love for today's women.* New York: McGraw-Hill.

Baskin, Yvonne. (1984). *The gene doctors.* New York: Morrow.

Basseches, Michael. (1984). *Dialectical thinking and adult development.* Norwood, NJ: Ablex.

Baumrind, Diana. (1967). Child-care practices anteceding three patterns of preschool behavior. *Genetic Psychology Monographs, 75,* 43–88.

Baumrind, Diana. (1971). Current pattens of parental authority. *Developmental Psychology, 4* (Monograph I), 1–103.

Baumrind, Diana. (1982). Are androgynous individuals more effective persons and parents? *Child Development, 53,* 44–75.

Bayer, Leona M., and Snyder, Margaret M. (1971). Illness experience of a group of normal children. In Mary Cover Jones, Nancy Bayley, Jean Walker Macfarlane, and Marjorie Pyles Honzik (Eds.), *The course of human development.* Waltham, Mass.: Xerox College Publishing, 91–103.

Bayley, Nancy. (1935). The development of motor abilities during the first three years. *Monograph of the Society for Research in Child Development, 1.*

Bayley, Nancy. (1955). On the growth of intelligence. *American Psychologist, 10,* 805–818.

Bayley, Nancy. (1966). Learning in adulthood: The role of intelligence. In Herbert J. Klausmeier and Chester W. Harris (Eds.), *Analysis of concept learning.* New York: Academic Press.

Bayley, Nancy, and Oden, Melita. (1955). The maintenance of intellectual ability in gifted adults. *Journal of Gerontology, 10,* Section B (1), 91–107.

Beal, Carole R., and Flavell, John H. (1982). Effect of increasing the salience of message ambiguities on kindergartners' evaluations of communication success and message adequacy. *Developmental Psychology, 18,* 43–48.

Beal, Carole R., and Flavell, John H. (1983). Young speakers' evaluation of their listeners' comprehension in a referential communication task. *Child Development, 54,* 148–153.

Beckwith, Leila, and Parmelee, Arthur H. (1986). EEG patterns of preterm infants, home environment, and later I.Q. *Child Development, 57,* 777–789.

Behrman, Richard E., and Vaughn, Victor C. III. (1983). *Pediatrics.* Philadelphia: Saunders.

Bell, Richard Q., and Harper, Lawrence, V. (1977). *Child effects on adults.* Hillsdale, NJ: Erlbaum.

Bell, Richard Q., and Waldrop, M.F. (1982). Temperament and minor physical anomalies. In R. Porter and G.M. Collins (Eds.), *Temperamental differences in infants and young children.* London: Pittman.

Beller, E. Kuno. (1979). Early intervention programs. In Joy D. Osofsky (Ed.), *Handbook of infant development.* New York: Wiley.

Belsky, Jay. (1984). The determinants of parenting: A process model. *Child Development, 55,* 83–96.

Belsky, Jay, Steinberg, Laurence D., and Walker, Ann. (1982). The ecology of day care. In Michael Lamb (Ed.), *Nontraditional families: Parenting and child development.* Hillsdale, NJ: Erlbaum.

Belsky, Jay, Gilstrap, Bonnie, and Rovine, Michael. (1984b). The Pennsylvania Infant and Family Development Project I: Stability and change in mother-infant and father-infant interaction in a family setting at one, three, and nine months. *Child Development, 55,* 692–705.

Belsky, Jay, Robins, Elliot, and Gamble, Wendy. (1984c). The determinants of parental competence: Toward a contextual theory. In Michael Lewis (Ed.), *Beyond the dyad.* New York: Plenum.

Belsky, Jay, Taylor, Dawn G., and Rovine, Michael. (1984a). The Pennsylvania Infant and Family Development Project II: The development of reciprocal interaction in the mother-infant dyad. *Child Development, 55,* 706–717.

Bem, Sandra L. (1974). The measurement of psychological androgyny. *Journal of Consulting and Clinical Psychology, 42,* 155–162.

Bem, Sandra L. (1985). Androgyny and gender schema theory: A conceptual and empirical integration. In T. B. Sondegegger (Ed.), *Nebraska symposium on motivation 1984: Psychology and gender.* (Vol. 32). Lincoln: University of Nebraska.

Benbow, C.P., and Stanley, J. C. (1980). Sex differences in mathematical ability. Fact or artifact. *Science, 210,* 1262.

Bender, Lauretta. (1973). The life course of children with schizophrenia. *American Journal of Psychiatry, 130,* 783–786.

Benet, Sula. (1974). *Abkhasians: The long-lived people of the Caucasus.* New York: Holt, Rinehart and Winston.

Bengston, Vern L. (1975). Generation and family effects in value socialization. *American Sociological Review, 40,* 358–371.

Bengston, Vern L. (1985). Diversity and symbolism in grandparental roles. In Vern L. Bengston and Joan F. Robertson (Eds.), *Grandparenthood*. Beverly Hills, CA: Sage.

Bengston, Vern L., Reedy, Margaret N., and Gordon, Chad. (1985). Aging and self-conceptions: Personality processes and social contexts. In James E. Birren and K. Warner Schaie (Eds.), *Handbook of the psychology of aging* (2nd ed.). New York: Van Nostrand.

Bengston, Vern L., and Robertson, Joan F. (Eds.). (1985). *Grandparenthood*. Beverly Hills, CA: Sage.

Bengston, Vern L., and Troll, Lillian E. (1978). Youth and their parents: Feedback and intergeneration. In Richard M. Lerner and Gerald B. Spanier (Eds.), *Child influences on marital and family interaction: A life-span perspective*. New York: Academic Press.

Bennett, Neville. (1976). *Teaching styles and pupil progress*. Cambridge, MA: Harvard University Press.

Benson-von der Ohe, E. (1984). First and second marriages: The first three years of married life. In Sarnoff A. Mednick, Michele Harway, and Karen M. Finello (Eds.), *Handbook of longitudinal research: Vol 2. Teenage and adult cohorts*. New York: Praeger.

Beres, Cathryn A., and Baron, Alan. (1981). Improved digit symbol substitution by older women as a result of extended practice. *Journal of Gerontology, 36,* 591–597.

Bergler, Edmund. (1985). *The revolt of the middle-aged man*. New York: International Universities Press.

Bergman, Abraham B. (1986). *The discovery of sudden infant death syndrome: Lessons in the practice of political medicine*. New York: Praeger.

Berman, Sandra, and Richardson, Virginia. (1986-1987). Social change in the salience of death among adults in America: A projective assessment. *Omega, 17,* 195–207.

Berndt, Thomas J. (1981). Relations between social cognition, nonsocial cognition, and social behavior. In John H. Flavell and Lee Ross (Eds.), *Social cognitive development: Frontiers and possible futures*. Cambridge, Eng.: Cambridge University Press.

Berndt, Thomas J. (1982). The features and effects of friendship in early adolescence. *Child Development, 53,* 1447–1460.

Bernstein, Basil. (1971, 1973). *Class, codes, and control* (Vols. 1, 2). London: Routledge and Kegan Paul.

Berry, Ruth E., and Williams, Flora L. (1987). Assessing the relationship between quality of life and marital and income satisfaction: A path analytic approach. *Journal of Marriage and the Family, 49,* 107–116.

Bierman, Karen Lynn, and Furman, Wyndol. (1984). The effects of social skills training and peer involvement on the social adjustment of preadolescents. *Child Development, 55,* 151–162.

Bigelow, Ann E. (1983). Development of the use of sound in the search behavior of infants. *Developmental Psychology, 19,* 317–321.

Bigelow, B.J. (1977). Children's friendship expectations: A cognitive developmental study. *Child Development, 48,* 246–253.

Bigelow, B.J., and La Gaipa, J.J. (1975). Children's written descriptions of friendship: A multidimensional analysis. *Developmental Psychology, 11,* 857–858.

Biggs, John B., and Collis, K. F. (1982). *Evaluating the quality of learning: The SOLO taxonomy (Structure of the Observed Learning Outcome)*. New York: Academic Press.

Bijou, Sidney W., and Baer, Donald M. (1978). *Behavioral analysis of child development*. Englewood Cliffs, NJ: Prentice Hall.

Billinghan, Robert E., and Sack, Alan R. (1986). Courtship and violence the interactive status of the relationship. *Journal of Adolescent Research, 1,* 315–326.

Binet, Alfred. (1909). *Les idées modernes sur les enfants*. Paris: Schleicher.

Bing, Elizabeth D. (1983). Dear Elizabeth Bing: We've had our baby. New York: Pocket Books.

Birren, James E., Woods, Anita M., and Williams, M. Virtrue. (1980). Behavioral slowing with age: Causes, organization, and consequences. In Leonard W. Poon (Ed.), *Aging in the 80's: Psychological issues*. Washington, D.C.: American Psychological Association.

Black, Rebecca, and Mayer, Joseph. (1980). Parents with special problems: Alcoholism and opiate addiction. In C. Henry Kempe and Ray E. Helfer (Eds.), *The battered child* (3rd ed.). Chicago: The University of Chicago Press.

Blake, J. (1981). The only child in America: Prejudice versus performance. *Population and Development Review, 1,* 25–44.

Blasi, Augusto. (1980). Bridging moral cognition and moral action: A critical review of the literature. *Psychological Bulletin, 88,* 593–637.

Blewitt, Pamela. (1982). Word meaning acquisition in young children. A review of theory and research. In H.W. Reese and L.P. Lipsitt (Eds.), *Advances in child development and behavior*. New York: Academic Press.

Block, Jack. (1981). Some enduring and consequential structures of personality. In A.I. Rabin (Ed.), *Further explorations in personality*. New York: Wiley.

Block, Jack H., and Block, Jeanne. (1980). The role of ego control and ego-resiliency in the organization of behavior. In W. Andrew Collins (Ed.), *Development of cognitive affect and social relations: Minnesota symposia on child psychology* (Vol. 13). Hillsdale, NJ: Erlbaum.

Block, Jeanne. (1971). *Lives through time*. Berkeley, CA: Bancroft Books.

Bloom, Bernard, L., and Caldwell, Robert A. (1981). Sex differences in adjustment during the process of marital separation. *Journal of Marriage and the Family, 43,* 693–701.

Bloom, Lois. (1975). Language development. In Frances Degen Horowitz (Ed.), *Review of child development research* (Vol. IV). Chicago: University of Chicago Press, 245–303.

Bloom, Lois, Merkin, Susan, and Wootten, Janet. (1982). Wh-Questions: Linguistic factors that contribute to the sequence of acquisition. *Child Development, 53,* 1084–1092.

Blythe, Ronald. (1979). *The view in winter: Reflections on old age*. New York: Penguin.

Boffey, Philip M. (1983, August 2). "Safe" form of radiation arouses new worry. *The New York Times*, p. C1.

Bogin, Barry, and MacVean, Robert B. (1983). The relationship of socioeconomic status and sex to body size, skeletal maturation, and cognitive status of Guatemala City schoolchildren. *Child Development, 54,* 115–128.

Bombardieri, Merle. (1981). *The baby decision*. New York: Rawson, Wade.

Bond, Guy L., Tinker, Miles A., Wasson, Barbara B. (1979). *Reading difficulties: Their diagnosis and correction* (4th ed.). Englewood Cliffs, NJ: Prentice Hall.

Boocock, Sarane Spence. (1978). Getting started in adult life. In John C. Flanagan, (Ed.), *Perspectives on improving education: Project TALENTS young adults look back*. New York: Praeger.

Boring, E.G. (1950). *A history of experimental psychology* (2nd ed.). New York: Appleton-Century-Crofts.

Borstelmann, L.J. (1983). Children before psychology: Ideas about children from antiquity to the late 1800's. In Paul H. Mussen (Ed.), *Handbook of child psychology* (4th ed.). (Vol. 1) *History, theory, and methods*. New York: Wiley.

Bortz, W.M. (1980). Effects of exercise on aging—effect of aging on exercise. *Journal of the American Gerontological Society, 28,* 49–51.

Botwinick, Jack. (1977). Intellectual abilities. In James E. Birren and

K. Warner Schaie (Eds.), *Handbook of the psychology of aging*. New York: Van Nostrand Reinhold.

Botwinick, Jack, and Storandt, Martha. (1980). Recall and recognition of old information in relation to age and sex. *Journal of Gerontology, 35*, 70–76.

Botwinick, Jack, West, Robin, and Storandt, Martha. (1978). Predicting death from behavioral test performance. *Journal of Gerontology, 33*, 755–762.

Bouchard, Thomas. (1981, August). *The Minnesota study of twins reared apart: Description and preliminary findings*. Paper presented at the annual meeting of the American Psychological Association.

Bower, Bruce. (1987a). The character of cancer. *Science News, 131*, 120–121.

Bower, Bruce. (1987b). IQ's generation gap. *Science News, 132*, 108–109.

Bower, T.G.R. (1977). *A primer of infant development*. San Francisco: Freeman.

Bower, T.G.R. (1979). *Human Development*. San Francisco: Freeman.

Bower, T.G.R., and Wishart, J.G. (1979). Towards a unitary theory of development. In E.B. Thomas (Ed.), *Origins of the infant's social responsiveness*. Hillsdale, NJ: Erlbaum.

Bowerman, Melissa. (1982). Reorganizational processes in lexical and syntactic development. In Eric Wanner and Lila R. Gleitman (Eds.), *Language acquisition: The state of the art*. Cambridge, England: Cambridge University Press.

Bowlby, John. (1974). Psychiatric implications in bereavement. In A.A. Kutscher (Ed.), *Death and bereavement*. Chicago: Charles C. Thomas.

Boxer, Andrew M., Gershenson, Harold P., and Offer, Daniel. (1984). Historical time and social change in adolescent experience. *New Directions for Mental Health Services, 22*, 83–95.

Bracha, H. Stephan. (1986). On concordance for tuberculosis and schizophrenia. *American Journal of Psychiatry, 143*, 1634.

Brackbill, Yvonne, McManus, Karen, and Woodward, Lynn. (1985). *Medication in maternity: Infant exposure and maternal information*. Ann Arbor: University of Michigan Press.

Bradbard, Marilyn, R., and Endsley, Richard C. (1986). Sources of variance in young working mothers' satisfaction with child care. In Sally Kilmer (Ed.), *Advances in early education and day care*. Vol. 4. Greenwich, CT: JAI.

Bradley, Robert H., and Caldwell, Bettye M. (1980). The relation of home environment, cognitive competence, and IQ among males and females. *Child Development, 51*, 1140–1148.

Bradley, Robert H., and Caldwell, Bettye M. (1984). The HOME inventory and family demographics. *Developmental Psychology, 20*, 315–320.

Bradley, Robert H., and Rock, Stephen L. (1985). The HOME inventory: Its relation to school failure and development of an elementary-age version. In William K. Frankenberg, Robert N. Emde, and Joseph W. Sullivan (Eds.), *Early identification of children at risk: An international prospective*. New York: Plenum.

Bradshaw, John L. (1983). *Human cerebral asymmetry*. Englewood Cliffs, NJ: Prentice-Hall.

Braine, Martin D.S., and Rumain, Barbara. (1983). Logical reasoning. In Paul H. Mussen (Ed.), *Handbook of child psychology: Vol. 3. Cognitive development*. New York: Wiley.

Brainerd, Charles J. (1983). Working memory systems and cognitive development. In Charles J. Brainerd (Ed.), *Recent advances in cognitive-developmental theory: Progress in cognitive development research*. New York: Springer-Verlag.

Brandes, Stanley. (1985). *Forty: The age and the symbol*. Knoxville: University of Tennesee.

Brannon, Robert B. (1976). "No 'Sissy Stuff'": The stigma of anything vaguely feminine. In Deborah S. David and Robert B. Brannon (Eds.), *The Forty-nine percent majority*. Reading, MA: Addison Wesley.

Bray, Douglas, W., and Howard, Ann. (1980). Career success and life satisfactions of middle-aged managers. In Lynne A. Bond and James C. Rosen (Eds.), *Competence and coping during adulthood*. Hanover, NH: University Press of New England.

Bray, D.W., and Howard, A. (1983). The AT & T longitudinal studies of managers. In K. Warner Schaie (Ed.), *Longitudinal studies of adult psychological development*. New York: Guilford.

Brazelton, T. Berry, Yogman, Michael, Als, Heidelise, and Tronick, Edward. (1979). The infant as a focus for family reciprocity. In Michael Lewis and Leonard A. Rosenblum (Eds.), *The child and his family*. New York: Plenum.

Breckenridge, James N., Dallagher, Dolores, Thompson, Larry W., and Peterson, James. (1986). Characteristic depressive symptoms of bereaved elders. *Journal of Gerontology, 41*, 163–168.

Brent, Robert L. (1986). The complexities of solving the problem of human malformations. In John L. Sever and Robert L. Brent (Eds.), *Teratogen update: Environmentally induced birth defect risks*. New York: Liss.

Bretherton, Inge, and Waters, Everett. (1985). Growing points of attachment theory and research. *Monographs of the Society for Research in Child Development, 50* (1, 2, Serial No. 209).

Breznitz, Zvia, and Sherman, Tracy. (1987). Speech patterning of natural discourse of well and depressed mothers and their young children. *Developmental Psychology, 58*, 395–400.

Brim, Orville G. (1976). Life span development of the theory of oneself: Implications for child development. In Hayne W. Reese (Ed.), *Advances in child development and behavior* (Vol. 2). New York: Academic Press.

Brim, Orville G., Jr., and Ryff, Carol D. On the properties of life events. In Paul B. Baltes and Orville G. Brim, Jr. (Eds.), *Life-span development and behavior*. Vol. III. New York: Academic Press, 1980.

Broman, S.H., Nichols, P.L., and Kennedy, W.A. (1975). *Preschool I.Q.: Prenatal and early development correlates*. New York: Erlbaum.

Bronfenbrenner, Urie. (1974). *A report on the longitudinal evaluations of preschool programs. Vol. II.: Is early intervention effective?* U.S. Department of Health, Education, and Welfare. DHEW Publication No. OHD 74–24. Washington, DC: Office of Child Development, 1974.

Bronfenbrenner, Urie. (1975). Is early intervention effective? In M. Guttenberg and E.L. Struening (Eds.), *Handbook of evaluation research* (Vol. 2). Beverly Hills, CA: Sage.

Bronfenbrenner, Urie. (1977). Toward an experimental ecology of human development. *American Psychologist, 32*, 513–531.

Bronfenbrenner, Urie. (1979). *The ecology of human development: Experiments by nature and design*. Cambridge, MA: Harvard University Press.

Bronfenbrenner, Urie. (1986). Ecology of the family as a context for human development research perspectives. *Developmental Psychology, 22*, 723–42.

Bronfenbrenner, Urie, and Crouter, Ann C. (1983). The evolution of environmental models in developmental research. In Paul H. Mussen (Ed.), *Handbook of child psychology: Vol. 1. History, theory, and methods*. New York: Wiley.

Bronson, Wanda C. (1985). Growth in the organization of behavior over the second year of life. *Developmental Psychology, 21*, 108–117.

Brooks, George A., and Fahey, Thomas, D. (1984). *Exercise physiology: Human bioenergetics and its application*. New York: Wiley.

Brooks-Gunn, Jeanne, and **Kirsch, Barbara.** (1984). Life events and the boundaries of midlife for women. In Grace K. Baruch and Jeanne Brooks-Gunn (Eds.), *Women in midlife.* New York: Plenum.

Brooks-Gunn, Jeanne, Warren, Michelle P., Samelson, Marion, and **Fox, Richard.** (1986). Physical similarity of and disclosure of menarchal status to friends: Effects of grade and pubertal status. *Journal of Early Adolescence, 6,* 3–14.

Brown, Ann L., and **De Loache, Judy S.** (1978). Skills, plans, and self-regulation. In Robert S. Siegler (Ed.), *Children's thinking: What devel ops.* Hillsdale, NJ: Erlbaum.

Brown, Ann L., Bransford, John D., Ferrara, Roberta, and **Campione, Joseph.** (1983). Learning, Remembering, and Understanding. In Paul H. Mussen (Ed.), *Handbook of child psychology: Vol. 3. Cognitive development.* New York: Wiley.

Brown, B. Bradford, Lohr, Mary Jane, and **McClenahan, Eben L.** (1986). Early adolescents' perception of peer pressure. *Journal of Early Adolescence, 6,* 139–154.

Brown, B.J., and **Lloyd, H.A.** (1975). A controlled study of children not speaking at school. *Journal of the Association of Workers for Maladjusted Children, 3,* 49–63.

Brown, Bertram, S. (1983). The impact of political and economic changes upon mental health. *American Journal of Orthopsychiatry, 53,* 583–592.

Brown, Judith K. (1975). Adolescent initiation rites: Recent interpretations. In Robert E. Grinder (Ed.), *Studies in adolescence: A book of readings in adolescent development* (3rd ed.). New York: Macmillan.

Brown, W.T., Jenkins, E.C., Gross, A.C., Chan, C.B., Krawczun, M.S., Duncan, C.J., Sklower, S.L., and **Fisch, G.S.** (1987). Further evidence for genetic heterogenity in the Fragile X syndrome. *Human Genetics, 75,* 311–321.

Brownell, Kelly D., Marlatt, G. Alan, Lishtenstein, Edward, and **Wilson, G. Terrence.** (1986). Understanding and preventing relapse. *American Psychologist, 41,* 765–782.

Bruce, Patricia R., Coyne, Andrew C., and **Botwinick, Jack.** (1982). Adult age differences in metamemory. *Journal of Gerontology, 37,* 354–357.

Bruce, Patricia, R., and **Herman, James P.** (1986). Adult age differences in spatial memory. *Journal of Gerontology, 41,* 774–777.

Bruch, Hilde. (1978). *The golden cage: The enigma of anorexia nervosa.* Cambridge, MA: Harvard University Press.

Bruner, Jerome Seymour. (1964). The course of cognitive growth. *American Psychologist, 19,* 1–15.

Bruner, Jerome Seymour. (1973). *Beyond the information given: Studies in the psychology of knowing.* Jeremy M. Anglin (Ed.) New York: Norton.

Bryant, Peter E. (1985). The distinction between knowing when to do a sum and knowing how to do it. *Educational Psychology, 5,* 207–215.

Buck-Morss, Susan. (1987). Piaget, Adorno, and dialectical operations. In John M. Broughton (Ed.), *Critical theories of psychological development.* New York: Plenum.

Buhrmester, Duane, and **Furman, Wyndol.** (1987). The development of companionship and intimacy. *Child Development, 58,* 1101–1113.

Burelson, Brant R. (1982). The development of comforting communication skills in childhood and adolescence. *Child Development,* 1578–1588.

Burger, Sarah G., Miller, Brenoa H.S., and **Mauney, Brenda Fay.** (1986). *A guide to management and supervision in nursing homes.* Springfield, IL: Thomas.

Burnet, Frank M. (1974). *The biology of aging.* Aukland, New Zealand: Aukland University Press.

Buskirk, Ellsworth R. (1985). Health maintenance and longevity: Exer-

cise. In Caleb E. Finch and Edward L. Schneider (Eds.), *Handbook of the biology of aging* (2nd ed.). New York: Van Nostrand.

Buss, Ray R., Yussen, Steven R., Mathews, Samuel R. II, Miller, Gloria E., and **Rembold, Karen L.** (1983). Development of children's use of a story scheme to retrieve information. *Developmental Psychology, 19,* 22–28.

Busse, Ewald, W. (1985). Normal aging: The Duke longitudinal studies. In M. Bergener, Marco Ermini, and H.B. Stahelin (Eds.), *Thresholds in aging.* London: Academic Press.

Butler, John A., Starfield, Barbara, and **Stemark, Suzanne.** (1984). Child health policy. In Harold W. Stevenson and Alberta E. Siegel (Eds.), *Child development research and social policy.* Chicago: University of Chicago Press.

Butler, N.R., and **Golding, Jean.** (1986). *From birth to five: A study of the health and behaviour of Britain's 5-year-olds.* Oxford: Pergamon.

Butler, Robert N. (1963). The life review: An interpretation of reminiscence in the aged. *Psychiatry, 26,* 65–76.

Butler, Robert N. (1975). *Why survive: Growing old in America.* New York: Harper & Row.

Butler, Robert N., and **Lewis, Myrna I.** (1982). *Aging and mental health: Positive psychosocial and biomedical approaches* (3rd ed.), St. Louis: Mosby.

Cadoret, R.J., and **Gath, A.** (1978). Inheritance of alcoholism in adoptees. *British Journal of Psychiatry, 132,* 252–258.

Cain, Barbara S. (1982, December 12). Plight of the grey divorcee. *New York Times Magazine,* p. 89–90, 92, 95.

Caine, Lynn. (1974). *Widow.* New York: Morrow.

Cairns, Robert B. (1983). The emergence of developmental psychology. In Paul H. Mussen (Ed.), *Handbook of child psychology: Vol. 1. History, theory, and methods.* New York: Wiley.

Campbell, Angus. (1981). *The sense of well-being in America: Recent patterns and trends.* New York: McGraw Hill.

Campos, Joseph J., Barrett, Karen C., Lamb, Michael L., Goldsmith, H. Hill, and **Stenberg, Craig.** (1983). Socioemotional development. In Paul H. Mussen (Ed.), *Handbook of child psychology: Vol. 2. Infancy and developmental psychobiology.* New York: Wiley.

Cantor, M.H. (1979). Life space and social support. In T.O. Byerts, S.C. Howell, and L.A. Pastalan (Eds.), *The environmental context of aging.* New York: Garland STPM Press.

Cantwell, Hendrika, B. (1980). Child neglect. In C. Henry Kempe and Ray E. Helfer (Eds.), *The battered child* (3rd ed.). Chicago: University of Chicago Press.

Caplow, Theodore. (1982). *Middletown families: Fifty years of change and continuity.* Minneapolis: University of Minnesota Press.

Cárdenas, Jose A. (1977). Response I. in Noel Epstein (Ed.), *Language, ethnicity and the schools.* Washington DC: Institute for Educational Leadership.

Carew, Jean V. (1980). Experience and the development of intelligence in young children at home and in day care. *Monographs of the Society for Research in Child Development, 45* (Serial No. 187).

Carey, Susan. (1978). The child as word learner. In Morris Halle, J. Bresman, and G.A. Miller (Eds.), *Linguistic theory and psychological reality.* Cambridge, MA: MIT Press.

Carey, William B., and **McDevitt, Sean C.** (1978). Stability and change in individual temperament diagnoses from infancy to early childhood. *Journal of the American Academy of Child Psychiatry, 17,* 331–337.

Cargen Leonard. (1981). Singles: An examination of two stereotypes. *Family Relations, 30,* 377–385.

Carlberg, C. and Kavale, Kenneth. (1980). The efficacy of special versus regular class placement for exceptional children: A meta-analysis. *Journal of Special Education, 14,* 296–309.

Carlson, Bonnie E. (1984). The father's contribution to child care: Effects on children's perceptions of parental roles. *American Journal of Orthopsychiatry, 54,* 123–136.

Carlsson, S.G., Fagerberg, H., Horneman, G. Hwang, P., Larson, K., Rodholm, M., Schaller, J., Daniellson, B., and Gundewal, C. (1979). Effects of various amounts of contact between mother and child on the mother's nursing behavior: A follow-up study. *Infant Behavior and Development, 2,* 209–214.

Caron, Albert J., and Caron, Rose F. (1981). Processing of relational information as an index of infant risk. In S.L. Friedman and M. Sigman (Eds.), *Preterm birth and psychological development.* New York: Academic Press.

Caron, Albert J., and Caron, Rose F. (1982). Cognitive development in infancy. In Tiffany M. Field, Aletha Huston, Herbert C. Quay, Lillian Troll, and Gordon E. Finley (Eds.), *Review of Human Development.* New York: Wiley.

Caron, Albert J., Caron, Rose F., Caldwell, Roberta C., and Weiss, Sandra J. (1973). Infant perception of the structural properties of the face. *Developmental Psychology, 9,* 385–399.

Case, Robbie. (1985). *Intellectual development: Birth to adulthood.* Orlando: Academic Press.

Casem, Ned H. (1975). Bereavement as indispensible to growth. In Bernard Schoenberg et al. (Eds.), *Bereavement: Its psychosocial aspects.* New York: Columbia University Press.

Caspi, Avshalom, Elder, Glen H., and Bem, Daryl B. (1987). Moving against the world: Life-course patterns of explosive children. *Developmental Psychology, 23,* 308–313.

Cassill, Kay. (1982). *Twins reared apart.* New York: Atheneum.

Cath, Stanley H. (1975). The orchestration of disengagement. *International Journal of Aging and Human Development, 6,* 199–213.

Cavenaugh, John L., Kramer, Deirdre A., Sinnott, Jan C., Camp, Cameron J., and Markley, Robert P. (1985). On missing links and such: interfaces between cognitive research and everyday problem-solving. *Human Development, 28,* 146–168.

Cazden, Courtney B. (1976). The neglected situation in child language research and education. In Arlene Skolnik (Ed.), *Rethinking childhood.* Boston: Little, Brown (originally published in *Language and poverty,* Frederick Williams (Ed.), Madison: University of Wisconsin Press, 1970).

Celluci, Tony. (1984). The presentation of alcohol abuse: Conceptual and methodological issues. In Peter M. Miller and Ted D. Nirenberg (Eds.), *Prevention of alcohol abuse.* New York: Plenum.

Centers for Disease Control. (November, 1986). *Homicide surveillance: High-risk racial and ethnic groups—Blacks and Hispanics, 1970–1983.* Atlanta: Centers for Disease Control.

Centers for Disease Control. (1987). Postservice mortality among Vietnam veterans. *JAMA, 257,* 790–795.

Cervantes, Lucius F. (1965). *The dropout: Causes and cures.* Ann Arbor: University of Michigan Press.

Chap, Janet Blum. (1985–1986). Moral judgment in middle and late adulthood: The effects of age-appropriate moral dilemmas and spontaneous role-taking. *International Journal of Aging and Human Development, 22,* 161–172.

Chapman, Michael. (1979). Listening to reason: Children's attentiveness and parental discipline. *Merrill-Palmer Quarterly, 25,* 251–263.

Charlesworth, William R. (February 27, 1984). Personal correspondence.

Charness, Neil. (1981). Visual short-term memory and aging in chess players. *Journal of Gerontology, 36,* 615–619.

Charness, N. (1986). Expertise in chess, music, and physics: A cognitive perspective. In L.K. Obler and D.A. Fein (Eds.), *The neuropsychology of talent and special abilities.* New York: Guilford Press.

Chen, Linda H. (1986). Biomedical influences on nutrition of the elderly. In Linda H. Chen (Ed.), *Nutritional aspects of aging:* Vol. 1. Boca Raton, FL: CRC Press.

Chen, Yung-Ping. (1985). Economic status of the aging. In Robert H. Binstock and Ethel Shanas (Eds.), *Handbook of aging and the social sciences.* New York: Van Nostrand Reinhold.

Cherlin, Andrew. (1981). *Marriage, divorce, and remarriage.* Cambridge, MA: Harvard University Press.

Cherlin, Andrew, and Furstenberg, Frank F. Jr. (1986). *The new American grandparent: A place in the family, a life apart.* New York: Basic Books.

Cherniss, Cary. (1980). *Professional burnout in human services organizations.* New York: Praeger.

Cherry, Louise, and Lewis, Michael. (1976). Mothers and two-year-olds: A study of sex-differentiated aspects of verbal interaction. *Developmental Psychology, 12,* 278–282.

Chess, Stella, Korn, Sam J., and Fernandez, Paulina B. (1971). *Psychiatric disorders of children with congenital rubella.* New York: Brunner/Mazel.

Chess, Stella, and Thomas, Alexander. (1982). Infant bonding: Mystique and reality: *American Journal of Orthopsychiatry, 52,* 213–222.

Chess, Stella, and Thomas, Alexander. (1986). The New York longitudinal study. In Robert Plomin and Judith Dunn (Eds.), *The study of temperament: Changes, continuities, and challenges.* Hillsdale, NJ: Erlbaum.

Chi, Michelene T.H. (1985). Developmental perspectives on content specific knowledge and memory performance. In F. Weinart and Marion Perlmutter (Eds.), *Memory development: Universal changes and individual differences.* Hillsdale, NJ: Erlbaum.

Chickering, Arthur W. (1981). Conclusion. In Arthur W. Chickering (Ed.), *The modern American college: Responding to the new realities of diverse students and a changing society.* San Francisco: Jossey-Bass.

Chilman, Catherine S. (1983). *Adolescent sexuality in a changing American society* (2nd ed.). New York: Wiley.

Chira, Susan. (1984, February 11). Town experiment cuts TV. *The New York Times.*

Chiriboga, David A. (1982). Adaptation to marital separation in later and earlier life. *Journal of Gerontology, 37,* 109–114.

Chomsky, Carol. (1969). *The acquisition of syntax in children from five to ten.* Cambridge, MA: MIT Press.

Chomsky, Noam. (1968). *Language and mind.* New York: Harcourt, Brace, World.

Chomsky, Noam. (1980). *Rules and representations.* New York: Columbia University Press.

Chukovsky, Kornei Ivanovich. (1968). *From two to five.* Berkeley: University of California Press.

Cicourel, Aaron V., Jennings, Kenneth H., Jennings, Sybillyn H.M., Leiter, Kenneth C.W., Mackay, Robert, Mehan, Hugh, and Roth, David R. (1974). *Language use and school performance.* New York: Academic Press.

Clark, Eve V. (1982). The young word maker: A case study of innovation in the child's lexicon. In Eric Wanner and Lila R. Gleitman (Eds.), *Language acquisition: The state of the art.* Cambridge, England: Cambridge University Press.

Clark, Jane E., and Phillips, Sally J. (1985). A developmental sequence of the standing long jump. In Jane E. Clark and James H. Humphrey (Eds.), *Motor development: Current selected research.* Princeton, NJ: Princeton Book Company.

Clark, Samuel D. Jr., Zabin, Laurie S., and Hardy, Janet B. (1984). Sex, contraception and parenthood: Experiences and attitudes among urban black young men. *Family Planning Perspectives, 16,* 77–82.

Clarke-Stewart, K. Alison. (1978). And daddy makes three: The father's impact on mother and young child. *Child Development, 49,* 466–478.

Clarke-Stewart, K. Alison. (1984). Day Care: A new context for research and development. In Marion Perlmutter (Ed.), *Parent-child interactions and parent-child relations in child development: The Minnesota symposia on child psychology,* Vol. 17, Hillsdale, NJ: Erlbaum.

Clarke-Stewart, K. Alison, and Gruber, Christian P. (1984). Day care forms and features. In Ricardo C. Ainslie (Ed.), *The child and the day care setting.* New York: Praeger.

Clarkson, Marsha G., and Berg, W. Keith. (1983). Cardiac orienting and vowel discrimination in newborns: Crucial stimulus parameters. *Child Development, 54,* 162–171.

Clarkson, Marsha G., Clifton, Rachel K., and Morrongiello, Barbara A. (1985). The effects of sound duration on newborns' head orientation. *Journal of Experimental Child Psychology, 39,* 20–36.

Clayton, Richard R., and Voss, Harwin, L. (1977). Shacking up: Cohabitation in the 1970's. *Journal of Marriage and Family, 39,* 273–284.

Clayton, Vivian P., and Birren, James E. (1980). The development of wisdom across the lifespan: A reexamination of an ancient topic. In Paul B. Baltes and Orville G. Brim (Eds.), *Life-span development and behavior.* Vol. III. New York: Academic Press.

Cleek, Margaret Buminski, and Pearson, T. Allan. (1985). Perceived causes of divorce: An analysis of interrelationships. *Journal of Marriage and the Family, 47,* 179–191.

Clifford, Edward. (1971). Body satisfaction in adolescence. *Perceptual and Motor Skills, 33,* 119–125.

Clingempeel, W.G., and Reppucci, N.D. (1982). Joint custody after divorce: Major issues and goals for research. *Psychological Bulletin, 91,* 102–127.

Coates, Thomas J., Petersen, Anne C., and Perry, Cheryl. (1982). Crossing the barriers. In Thomas J. Coates, Anne C. Petersen, and Cheryl Perry (Eds.). *Promoting adolescent health: A dialogue on research and practice.* New York: Academic Press.

Cohen, Leslie B., DeLoache, Judy S., and Strauss, Mark S. (1979). Infant visual perception. In Joy D. Osofsky (Ed.), *Handbook of infant development.* New York: Wiley.

Cohn, Anne H. (1983). The prevention of child abuse: What do we know about what works. In Jerome E. Levitt (Ed.), *Child abuse and neglect: Research and innovation.* The Hague: Martinus Nijhoff.

Cohn, Jeffrey F., and Tronick, Edward Z. (1983). Three-month-old infants' reaction to simulated maternal depression. *Child Development, 54,* 185–193.

Cohn, Jeffrey F., and Tronick, Edward Z. (1987). Mother-infant face to face interaction: The sequence of dyadic states at 3, 6, and 9 months. *Developmental Psychology, 23,* 68–77.

Cohn, Jeffrey P. (1987). The molecular biology of aging. *Bioscience, 37,* 99–102.

Coie, John D., and Krehbiel, Gina. (1984). Effects of academic tutoring on the social status of low-achieving, socially rejected children. *Child Development, 55,* 1465–1478.

Colby, Anne, Kohlberg, Lawrence, Gibbs, John, and Lieberman, Marcus. (1983). A longitudinal study of moral development. *Monographs of the Society for Research in Child Development, 48* (1–2, Serial No. 200).

Coleman, John C. (1980). Friendship and the peer group in adolescence. In Joseph Adelson (Ed.), *Handbook of adolescent psychology.* New York: Wiley.

Coles, Robert. (1970). *Erik H. Erikson: The growth of his work.* Boston: Little, Brown.

Collins, J.K., and LaGanza, S. (1982). Self-recognition of the face: A study of adolescent narcissism. *Journal of Youth and Adolescence, 11,* 317–328.

Comfort, Alexander. (1979). *The biology of senescence,* 3rd ed. New York: Elsevier.

Condron, John C., and Bode, Jerry G. (1982). Rashomon, working wives, and family division of labor: Middletown, 1980. *Journal of Marriage and the Family, 44,* 421–439.

Connors, C.K. (1980). *Food additives and hyperactive children.* New York: Plenum.

Cook, Thomas D., Appleton, Hilary, Connor, Ross F., Shaffer, Ann, Tamkin, Gary, and Weber, Stephen J. (1975). *"Sesame Street" revisited.* New York: Russell Sage.

Coopersmith, Stanley. (1967). *The antecedents of self-esteem.* San Francisco: Freeman.

Corballis, Michael C. (1983). *Human Laterality.* New York: Academic Press.

Cordes, Colleen. (1985). Fields cooperate to study surveys. *APA Monitor, 16 (6),* 32.

Cordes, Colleen. (1986). Drug use and pregnancy link sought. The *APA Monitor, 17,* 28–29.

Cornell, Edward H. (1980). Distributed study facilitates infants' delayed recognition memory. *Memory and Cognition, 8,* 539–542.

Corrigan, Roberta L. (1978). Language development as related to Stage 6 object permanence development. *Journal of Child Language, 5,* 173–189.

Corrigan, Roberta L. (1983). The development of representational skills. In Kurt W. Fischer (Ed.), *Levels and transitions of children's development.* New Directions for Child Development, No. 21. San Francisco: Jossey-Bass.

Costello, E.J., Edelbrock, C.S., and Costello, A.J. (1984). Validity of the NIMH Diagnostic Interview Schedule for Children: A comparison between pediatric and psychiatric referrals. *Journal of Abnormal Child Psychology, 7,* 136–143.

Cotman, Carl W., and Holets, Vicky R. (1985). Structural changes at synapses with age: Plasticity and regeneration. In Caleb E. Finch and Edward L. Schneider (Eds.), *Handbook of the biology of aging* (2nd ed.). New York: Van Nostrand.

Cottle, Thomas J. (1979). Adolescent voices. *Psychology Today, 12(9),* 43–44.

Cowan, Philip A. (1978). *Piaget, with feeling: Cognitive, social, and emotional dimensions.* New York: Holt, Rinehart and Winston.

Cowley, Malcolm. The view from 80. *Life,* December 1978.

Cowley, Malcolm. *The view from 80.* New York: Viking, 1980.

Cox, Maureen V. (1986). *The child's point of view: The development of cognition and language.* New York: St. Martin's Press.

Craik, F.I.M. (1977). Age differences in human memory. In James E. Birren and K. Warner Schaie (Eds.), *Handbook of the psychology of aging.* New York: Van Nostrand Reinhold.

Craik, Fergus I.M., and Byrd, Mark. (1982). Aging and cognitive deficits: The role of attentional resources. In Fergus I. M. Craik and Sandra Trehub (Eds.), *Aging and Cognitive Processes.* New York: Plenum.

Cregler, Louis L., and Mark, Herbert. (1986). Cardiovascular dangers of cocaine abuse. *American Journal of Cardiology, 57,* 1185–1186.

Cross, W., and Florio, C. *You are never too old to learn.* New York: Academy for Educational Development, 1978.

Crouter, Ann C., Perry-Jenkins, Maureen, Huston, Ted L., and McHale, Susan M. (1987). Processes underlying father involvement in dual-earner and single-earner families. *Developmental Psychology, 23,* 431–440.

Crnic, Keith, A., Ragozin, Arlene S., Greenberg, Mark T., Robinson, Nancy M., and Basham, Robert B.(1983). Social interaction and developmental competence of preterm and full term infants during the first year of life. *Child Development, 54,* 1199–1210.

Crum, Julie F., and Eckert, Helen M. (1985). Play patterns of primary school children. In Jane E. Clark and James H. Humphrey (Eds.), *Motor development: Current selected research.* Princeton, NJ: Princeton Book Company.

Crystal, Stephen. (1982). *America's old age crisis: Public policy and the two worlds of aging.* New York: Basic Books.

Culliton, Barbara. (1987). Osteoporosis reexamined: Complexity of bone biology is a challenge. *Science, 235,* 833–834.

Cumming, Elaine, and Henry, William H. (1961). *Growing old: The process of disengagement.* New York: Basic Books.

Cunningham, Walter R., and Owens, William A. Jr. (1983) The Iowa State study of the adult development of intellectual abilities. In K. Warner Schaie (Ed.), *Longitudinal studies of adult psychological development.* New York: Guilford.

Curtiss, Susan R. (1977). *Genie: A linguistic study of a modern-day "wild child."* New York: Academic Press.

Cutler, Richard G. (1984). Antioxidants and longevity. In Donald Armstrong, R.S. Sohol, Richard G. Cutler, and Trevor F. Slater (Eds.), *Free radicals in molecular biology, aging, and disease.* New York: Raven Press.

Damon, William. (1984). Self understanding and moral development from childhood to adolescence. In William M. Kurtines and Jacob L. Gewirtz (Eds.), *Morality, moral behavior, and moral development.* New York: Wiley.

Daniels, Denise, and Plomin, Robert. (1985). Origins of individual differences in infant shyness. *Developmental Psychology, 21,* 118–121.

Danish, Steven, J., Smyer, Michael A., and Nowak, Carol A. (1980). Developmental intervention: Enhancing life-event processes. In Paul B. Baltes and Orville G. Brim (Eds.), *Life-span development and behavior.* Vol. III. New York: Academic Press.

Danforth, David N. (Ed.) (1977). *Obstetrics and gynecology,* 3rd ed. New York: Harper & Row.

Dasen, P.R. (Ed.). (1977). *Piagetian psychology: Cross-cultural contributions.* New York: Gardner.

Datan, Nancy. (1986). Oedipal conflict, Platonic love: Centrifugal forces in intergenerational relations. In Nancy Datan, Anita L. Greene, and Hayne W. Reese (Eds.), *Life-span developmental psychology: Intergenerational relations.* Hillsdale, NJ: Erlbaum.

Davidson, Lucy. (1986). Is teenage suicide contagious? *Atlanta Constitution,* March 7, B1, B7–B9.

Davies, D. Roy, and Sparrow, Paul R. (1985). Age and work behavior. In Neil Charness (Ed.), *Aging and human performance.* New York: Wiley.

Davis, Elizabeth. (1983). *A Guide to midwifery: Heart and hands.* New York: Bantam.

Davis, G. Cullom. (1985). Accounts of lives and times. In Gari Lesnoff-Caravalglia (Ed.), *Values, ethics, and aging.* New York: Human Sciences Press.

Davis, Janet M., and Rovee-Collier, Carolyn K. (1983). Alleviated forgetting of a learned contingency in 8-week-old infants. *Developmental Psychology, 19,* 353–365.

Davis, Kingsley. (1985). The future of marriage. In Kingsley Davis (Ed.), *Contemporary marriage: Comparative perspectives on a changing institution.* New York: Russell Sage.

Dawis, Rene V. (1984). Job satisfaction: Worker aspirations, attitudes and behavior. In Norman C. Gysbers (Ed.), *Designing careers.* San Francisco: Jossey-Bass.

Dawood, M. Yusoff. (1985). Overall approach to the management of dysmenorrhea. In M. Yusoff Dawood, John L. McGuire, and Laurence M. Demers (Eds.), *Premenstrual syndrome and dysmenorrhea.* Baltimore MD: Urban and Schwartzenberg.

de Beauvoir, Simone. (1964). *Force of circumstance.* Translated by Richard Howard. New York: G.P. Putnam's Sons.

DeCasper, Anthony J., and Fifer, William P. (1980). Of human bonding: Newborns prefer their mothers' voices. *Science, 208,* 1174–1175.

de Hirsch, Katrina, Jansky, Jeanette Jefferson, and Langford, William S. (1966). *Predicting reading failure: A preliminary study of reading, writing, and spelling disabilities in preschool children.* New York: Harper and Row.

Deimling, Gary T., and Bass, David M. (1986). Symptoms of mental impairment among elderly adults and their effects on family caregivers. *Journal of Gerontology, 41,* 778–784.

DeLorey, Catherine. (1984). Health care and midlife women. In Grace Baruch and Jeanne Brooks-Gunn (Eds.), *Women in midlife.* New York: Plenum.

DeMaris, Alfred, and Leslie, Gerald R. (1984). Cohabitation with the future spouse: Its influence upon marital satisfaction and communication. *Journal of Marriage and the Family, 46,* 77–84.

deMause, Lloyd. (1975). The evolution of childhood. In Lloyd deMause (Ed.), *The history of childhood.* New York: Harper and Row.

Dement, William, Richardson, Gary, Prinz, Patricia, Carskadon, Mary, and Kripke, Daniel. (1985). Changes in sleep and wakefulness with age. In Caleb E. Finch and Edward L. Schneider (Eds.), *Handbook of the biology of aging* (2nd ed.). New York: Van Nostrand.

Demopoulus, Harry B. (1982). Oxygen free radicals in the central nervous system. In A.P. Autor (Ed.), *Pathology of oxygen.* New York: Academic Press.

Denny, B. Michael, and Thomas, Susan. (1986). The relationship of proportional reasoning ability to self-concept: A cognitive developmental approach. *Journal of Early Adolescence, 6,* 45–54.

Denny, Nancy Wadsworth, and Palmer, Ann M. (1981). Adult age differences on traditional and practical problem-solving measures. *Journal of Gerontology, 36,* 323–328.

Denny, Nancy Wadsworth, and Thisson, David M. (1983–1984). Determinants of cognitive ability in the elderly. *International Journal of Aging and Human Development, 16,* 29–41.

Denny, Nancy Wadsworth, Zeytinoglu, Sezen, and Selzer, S. Claire. (1977). Conservation training in four-year-old children. *Journal of Experimental Child Psychology, 24,* 129–146.

Dennis, Wayne. (1966). Creative productivity between the ages of 20 and 80 years. *Journal of Gerontology, 21,* 1–8.

Dennis, Wayne. (1973). *Children of the crèche.* New York: Appleton-Century-Crofts.

Deno, E.N. (Ed.). (1973). *Instructional alternatives for exceptional children.* Arlington, VA: Council for Exceptional Children.

Deutsch, Helene. (1944–1945). *The psychology of women: A psychoanalytic interpretation.* (Vol. 2). New York: Grune and Stratton.

de Villiers, J.G. (1980). The process of rule learning in children: A new look. In K.E. Nelson (Ed.), *Children's language* (Vol. 2). New York: Gardner Press.

de Villiers, Jill G., and de Villiers, Peter A. (1978). *Language acquisition.* Cambridge, MA: Harvard University Press.

de Villiers, Peter A., and de Villiers, Jill G. (1979). *Early language.* Cambridge, MA: Harvard University Press.

Diamond, Jared. (November, 1986). I want a girl just like the girl. . . . *Discover, 7 (11)*, 65–68.

Diaz, Rafael M. (1985). Bilingual cognitive development: Addressing three gaps in current research. *Child Development, 56*, 1376–1388.

Dick-Read, Grantly. (1972). *Childbirth without fear: The original approach to natural childbirth.* Rev. ed. Helen Wessel and Harlan F. Ellis (Eds.). New York: Harper & Row.

Dietrich, Richard A., and Spuhler, Karen. (1984). Genetics of alcoholism and alcohol actions. In Reginald G. Smart, Howard D. Cappell, Frederick B. Glaser, Yedy Israel, Harold Kalent, Robert E. Popham, Wolfgang Schmidt, and Edward M. Sellers (Eds.), *Research advances in alcohol and drug problems:* Vol. 8. New York: Plenum.

Dietz, William H., Jr., and Gortmaker, Steven L. (1985). Do we fatten our children at the television set? Obesity and television viewing in children and adolescents. *Pediatrics, 75*, 807–812.

DiPietro, Janet Ann. (1981). Rough and tumble play. A function of a gender. *Developmental Psychology, 17*, 50–58.

Dixon, Roger A., Kramer, Dierdre A., and Baltes, Paul B. (1985). Intelligence: A life-span developmental perspective. In Benjamin B. Wolman (Ed.), *Handbook of intelligence: Theories, measurements, and applications.* New York: Wiley.

Dobson, Cynthia. (1983). Sex-role and marital role expectations. In Timothy H. Brubaker (Ed.), *Family relationships in later life.* Beverly Hills, CA: Sage.

Dockrell, J., Campbell, R., and Neilson I. (1980). Conservation accidents revisited. *International Journal of Behavioral Development, 3*, 423–439.

Dodge, Kenneth A., Murphy, Roberta R., and Buchsbaum, Kathy. (1984). The assessment of intention-cue detection skills in children: Implications for developmental psychopathology. *Child Development, 55*, 163–173.

Dodge, Kenneth A., and Somberg, Daniel R. (1987). Hostile attributional biases among aggressive boys are exacerbated under conditions of threats to self. *Child Development, 58*, 213–224.

Doman, Glenn. (1980) *Teach your baby to read.* London: Chatto Bodley Jonathan.

Donaldson, Margaret. (1978). *Children's minds.* London: Fontana.

Donaldson, Margaret, Grieve, Robert, and Pratt, Chris. (1983). General introduction. In Margaret Donaldson, Robert Grieve, and Chris Pratt (Eds.), *Early childhood development and education: Readings in psychology.* New York: Guilford.

Donovan, Bonnie. (1977). *The Cesarean birth experience: A practical, comprehensive, and reassuring guide for parents and professionals.* Boston: Beacon Press.

Dornbush, S.M., Carlsmith, J.M., Gross, R.T., Martin, J.A., Jennings, D., Rosenberg, A., and Duke, P. (1981). Sexual development, age and dating: A comparison of biological and social influences upon one set of behaviors. *Child Development, 52*, 179–185.

Dryfoos, Joy G. (1982). Contraceptive use, pregnancy intentions, and pregnancy outcomes among U.S. women. *Family Planning Perspectives, 14 (2)*, 81–94.

Dryfoos, Joy G. (1984). A new strategy for preventing unintended teenage childbearing. *Family Planning Perspectives, 16*, 193–195.

Dubey, Dennis R., O'Leary, Susan G., and Kaufman, Kenneth F. (1983). Training parents of hyperactive children in child management: A comparative outcome study. *Journal of Abnormal Child Psychology, 11*, 229–246.

Dubow, Eric F., Huesmann, L. Rowell, and Eron, Leonard D. (1987). Childhood correlates of adult ego development. *Childhood Development, 58*, 859–869.

Dunham, Richard M., Kidwell, Jeannie S., and Wilson, Stephen M. (1986). Rites of passage at adolescence: A ritual process paradigm. *Journal of Adolescence Research, 1*, 139–154.

Dunn, Judy, and Munn, Penny. (1985). Becoming a family member: Family conflict and the development of social understanding in the second year. *Child Development, 56*, 480–492.

Dunn, Lloyd M. (Ed.). (1973). *Exceptional children in the schools: Special education in transition* (2nd ed.). New York: Holt, Rinehart and Winston.

Dunphy, Dexter C. (1963). The social structure of urban adolescent peer groups. *Sociometry, 26*, 230–246.

Du Randt, Rosa. (1985). Ball-catching proficiency among 4-, 6-, and 8-year-old girls. In Jane E. Clark and James H. Humphrey (Eds.), *Motor development: Current selected research.* Princeton, NJ: Princeton Book Company.

Duvall, Evelyn M. (1971). *Family development.* Philadelphia: Lippincott.

Dweck, Carol S., and Elliott, Elaine S. (1983). Achievement motivation. In Paul H. Mussen (Ed.), *Handbook of child psychology: Vol. 4. Socialization and personality development.* New York: Wiley.

Eakins, Pamela A. (1986). *The American way of birth.* Philadelphia: Temple.

Easson, William M. (1977). Accidents and trauma. In E. Mansell Pattison (Ed.), *The experience of dying.* Englewood Cliffs, NJ: Prentice-Hall.

East, Whitfield, B., and Hensley, Larry D. (1985). The effects of selected sociocultural factors upon the overhand-throwing performance of prepubescent children. In Jane E. Clark and James H. Humphrey (Eds.), *Motor development: Current selected research.* Princeton, NJ: Princeton Book Company.

Eaton, Warren O. (1983). *Motor activity from fetus to adult.* Paper.

Edgerton, Robert B. (1979). *Mental Retardation.* Cambridge, MA: Harvard University Press.

Edwards, D.D. (1987). Alcohol-breast cancer link. *Science News, 131*, 292.

Eagan, Jane (1976). Object-play in cats. In Jerome S. Bruner, Alison Jolly, and Kathy Sylva (Eds.), *Play.* New York: Basic Books.

Egeland, Brian, and Farber, Ellen A. (1984). Infant-mother attachment: Factors related to its development and changes over time. *Child Development, 55*, 753–771.

Egeland, B., and Sroufe, L. Alan. (1981). Attachment and early maltreatment. *Child Development, 52*, 44–52.

Eichorn, Dorothy H. (1979). Physical development: Current foci of research. In Joy D. Osofsky (Ed.), *Handbook of infant development.* New York: Wiley.

Eichorn, Dorothy H., Hunt, Jane V., and Honzik, Marjorie P. (1981). Experience, personality, and IQ: Adolescence to middle age. In Dorothy H. Eichorn, John A. Clausen, Marjorie P. Honzik, and Paul H. Mussen (Eds.), *Present and past in middle life.* New York: Academic Press.

Eidison, Bernice T., and Forsythe, Alan B. (1983). Life change events in alternate family styles. In Edward J. Callahan and Kathleen A. McClusky (Eds.), *Life-span developmental psychology: Non-normative life events.* New York: Academic Press.

Eiger, Marvin S. (1987). The feeding of infants and children. In Robert A. Hoekelman, Saul Blatman, Stanford B. Friedman, Nicholas M. Nelson, and Henry M. Seidel (Eds.), *Primary pediatric care.* St. Louis: Mosby.

Eimas, Peter D., Sigueland, Einar, R., Juscyzk, Peter, and Vigorito, James. (1971). Speech perception in infants. *Science, 171*, 303–306.

Eisdorfer, Carl. (1977). Intelligence and cognition in the aged. In Ewald W. Busse and Eric Pfeiffer (Eds.), *Behavior and adaptation in late life.* 2nd ed. Boston: Little, Brown.

Eisdorfer, Carl. (1985). The conceptualization of stress and a model for further study. In Michael R. Zales (Ed.), *Stress in health and disease.* New York: Brunner/Mazel.

Eisenberg, Nancy. (1982). *The development of prosocial behavior.* New York: Academic Press.

Eisenberg, Nancy, Lunch, Teresa, Shell, Rita, and **Roth, Karlsson.** (1985). Children's justifications for their adult and peer-direction compliant (prosocial and nonprosocial) behaviors. *Developmental Psychology, 21,* 325–331.

Eisenberg-Berg, Nancy, Boothby, Rita, and **Matson, Tom.** (1979). Correlates of preschool girls' feminine and masculine toy preferences. *Developmental Psychology, 48,* 1411–1416.

Eisenson, Jon. (1986). *Language and speech disorders in children.* New York: Pergamon.

Ekholm, Mats. (1984). Readiness to help others and tolerance: Attitude development during the school years and a ten-year comparison. *Scandinavian Journal of Educational Research, 28,* 71–86.

Elardo, Richard, Bradley, Robert, and **Caldwell, Bettye M.** (1975). The relation of infants' home environments to mental test performance from six to thirty-six months. A longitudinal analysis. *Child Development, 46,* 71–76.

Elardo, Richard, Bradley, Robert, and **Caldwell, Bettye M.** (1977). A longitudinal study of the relation of infants' home environments to language development at age three. *Child Development, 48,* 595–603.

Elder, Glen H., Jr. (1962). Structural variations in the childrearing relationship. *Sociometry, 25,* 241–262.

Elder, Glen H. (1986). Military times and turning points in men's lives. *Developmental Psychology, 22,* 233–245.

Elevenstar, D. (1980, January 8). Happy couple a tribute to old-fashioned virtues. *The Los Angeles Times,* p. 2.

Elkind, David. (1974). *Children and adolescents: Interpretive essays on Jean Piaget.* New York: Oxford University Press.

Elkind, David. (1978). *The child's reality: Three developmental themes.* Hillsdale, NJ: Erlbaum.

Elkind, David. (1979). *The child and society.* New York: Oxford University Press.

Elkind, David. (1981). *The hurried child: Growing up too fast too soon.* Reading, MA: Addison-Wesley.

Elkind, David, and **Bowen, R.** (1979). Imaginary audience behavior in children and adolescents. *Developmental Psychology, 15,* 38–44.

Elsayed, Mohamed, Ismail, A.H., and **Young, R. John.** (1980). Intellectual differences of adult men related to age and physical fitness before and after an exercise program. *Journal of Gerontology, 35,* 383–387.

Elstein, A.S., Shulman, L.S., and **Sprafka, S.A.** (1978). *Medical problem-solving: An analysis of clinical expertise.* Cambridge, MA: Harvard University Press.

Emde, Robert N., and **Harmon, R.J.** (1972). Endogenous and exogenous smiling systems in early infancy. *Journal of the American Academy of Child Psychiatry, 11,* 77–100.

Emery, Alan E.H. (1983). *Elements of medical genetics* (6th ed.) Edinburgh: Churchill Livingstone.

Emery, Donald G. (1975). *Teach your preschooler to read.* New York: Simon and Schuster.

Emery, Robert E., Hetherington, E. Mavis, and **Dilalla, Lisabeth F.** (1984). Divorce, children, and social policy. In Harold W. Stevenson and Alberta E. Siegel (Eds.), *Child development research and social policy.* Chicago: University of Chicago Press.

Endo, R., Sue, S., and **N.N. Wagner** (Eds.) (1980). *Asian-Americans: Social and psychological perspectives.* Palo Alto, CA: Science and Behavior Books.

Engstrom, Paul F. (1986). Cancer control objectives for the year 2000. In Lee E. Mortenseon, Paul F. Engstrom, and Paul N. Anderson (Eds.), *Advances in cancer control.* New York: Alan R. Liss.

Entwisle, Doris R., and **Baker, David P.** (1983). Gender and young children's expectations for performance in arithmetic. *Developmental Psychology, 19,* 200–209.

Erber, Joan T. (1982). Memory and age. In Tiffany M. Field, Aletha Huston, Herbert C. Quay, Lillian Troll, and Gordon E. Finley (Eds.), *Review of human development.* New York: Wiley.

Erikson, Erik H. (1963). *Childhood and society.* (2nd ed.). New York: Norton.

Erikson, Erik H. (1964). A memorandum on identity and Negro youth. *Journal of Social Issues, 20,* 29–42.

Erikson, Erik H. (1968). *Identity, youth, and crisis.* New York: Norton.

Erikson, Erik H. (1975). *Life history and the historical moment.* New York: Norton.

Erikson, Erik H., and **Erikson, J.M.** (1981). On generativity and identity. *Harvard Educational Review, 51,* 249–269.

Erikson, Erik H., Erikson, Joan M., and **Kivnick, Helen Q.** (1986). *Vital involvement in old age.* New York: Norton.

Eriksson, Bengt O. (1976). The child in sport and physical activity: Medical aspects. In J.G. Albinson and G.M. Andrew (Eds.), *Child in sport and physical activity.* Baltimore: University Park Press, 43–66.

Essex, Marilyn J., and **Nam, Sunghee.** (1987). Marital status and loneliness among older women. *Journal of Marriage and the Family, 49,* 93–106.

Estes, R., and **Wilensky, R.** (1978). Life cycle squeeze and the morale curve. *Social Problems, 25,* 278–292.

Etaugh, Claire. (1974). Effects of maternal employment on children: A review of recent research. *Merrill-Palmer Quarterly, 20,* 71–98.

Evans, Richard I., and **Raines, Bettye E.** (1982). Control and prevention of smoking in adolescents. In Thomas J. Coates, Anne C. Petersen, and Cheryl Perry (Eds.), *Promoting adolescent health: A dialogue on research and practice.* New York: Academic Press.

Eveleth, Phillis B., and **Tanner, James M.** (1976). *Worldwide variation in human growth.* Cambridge, England: Cambridge University Press.

Fabricius, William V., and **Wellman, Henry M.** (1983). Children's understanding of retrieval cue utilization. *Developmental Psychology, 19,* 15–21.

Fagot, Beverly L. (1978). The influence of sex of child on parental reactions to toddler children. *Child Development, 49,* 459–465.

Falbo, Toni. (1984). Only children: A review. In Toni Falbo (Ed.), *The single-child family.* New York: Guilford.

Fantz, Robert. (1961). The origin of form perception. *Scientific American, 204* (5), 66–72.

Farnham-Diggory, Sylvia. (1978). *Learning disabilities.* Cambridge, MA: Harvard University Press.

Farrell, Michael P., and **Rosenberg, Stanley D.** (1981). *Men at midlife.* Boston: Auburn House.

Fasick, Frank A. (1984). Parents, peers, youth culture and autonomy in adolescence. *Adolescence, 19,* 143–157.

Feather, Norman T. (1980). Values in adolescence. In Joseph Adelson (Ed.), *Handbook of adolescent psychology.* New York: Wiley.

Featherstone, Helen. (1980). *A difference in the family.* New York: Basic Books.

Feiring, Candice, and Lewis, Michael. (1984). Changing characteristics of the U.S. family: Implications for family networks, relationships, and child development. In Michael Lewis (Ed.), *Beyond the dyad.* New York: Plenum.

Feldman, Ben H., and Rosenkrantz, Arthur L. (1977). Drug use by college students and their parents. *Addictive Diseases: An International Journal, 3,* 235–241.

Feldman, D.H. (1984). A follow-up of subjects scoring above 180 I.Q. in Terman's "Genetic Studies of Genius." *Exceptional Children, 50,* 518–523.

Feldman, Harold. (1981). A comparison of intentional parents and intentionally childless couples. *Journal of Marriage and the Family, 43,* 593–600.

Feldman, S. Shirley, Biringen, Zeynap C., and Nash, Sharon Churnin. (1981). Fluctuations of sex-related self-attributions as a function of stage of family life cycle. *Developmental Psychology, 17,* 24–35.

Feldstein, Jerome H., and Feldstein, Sandra. (1982). Sex differences on televised toy commercials. *Sex Roles, 8,* 581–587.

Felton, Gary, and Segelman, Florrie. (1978). Lamaze childbirth training and change in belief about personal control. *Birth and the Family Journal, 5,* 141–150.

Fennell, G. (1985). Sheltered housing: Some unanswered questions. In Alan Butler (Ed.), *Aging: Recent advances and creative responses.* London: Croom Helm.

Ferguson, Charles A. (1977). Baby talk as a simplified register. In Catherine E. Snow and Charles A. Ferguson (Eds.), *Talking to children: Language input and requisition.* Cambridge, England: Cambridge University Press.

Fernald, Anne. (1985). Four-month-old infants prefer to listen to motherese. *Infant Behavior and Development, 8,* 181–195.

Ferree, Myra M. (1976). Working class jobs: Housework and paid work as sources of satisfaction. *Social Problems, 23,* 431–441.

Field, Dorothy. (1981). Can preschool children really learn to conserve? *Child Development, 52,* 326–334.

Field, Tiffany M. (1980). Interactions of high risk infants: Quantitative and qualitative differences. In D.B. Sawin, R.C. Hawkins, L.P. Walker, and J.H. Penticuff (Eds.), *Exceptional infant: Vol. 4. Psychosocial risks in infant environmental transactions.* New York: Brunner/Mazel.

Field, Tiffany M. (1982). Individual differences in the expressivity of neonates and young infants. In R. Feldman (Ed.), *Development of nonverbal behavior in children.* New York: Springer-Verlag.

Field, Tiffany M., Dempsey, J.R., and Shuman, H.H. (1981). Developmental follow-up of preterm and post term infants. In S.L. Friedman and Marian Sigman (Eds.), *Preterm birth and psychological handicap.* New York: Academic Press.

Field, Tiffany M., Woodson, R., Greenberg, R., and Chen, D. (1982). Discrimination and imitation of facial expression by neonates. *Science, 218,* 179–181.

Field, Tiffany, Gewirtz, Jacob L., Cohen, Debra, Garcia, Robert, Greenberg, Reena, and Kerry, Collins. (1984). Leavetakings and reunions of infants, toddlers, preschoolers, and their parents. *Child Development, 55,* 628.

Field, Tiffany M., Schanberg, Saul M., Scafidi, Frank, Bauer, Charles R., Vega-Lahr, Nitza, Garcia, Robert, Nystrom, Jerome, and Kuhn, Cynthia M. (1985). Effects of tactile/kinesthetic stimulation on preterm neonates. *Pediatrics,* in press.

Finkelhor, David. (1979a). *Sexually victimized children.* New York: Free Press.

Finkelhor, David. (1979b). What's wrong with sex between adults and children? Ethics and the problems of sexual abuse. *American Journal of Orthopsychiatry, 49,* 692–697.

Finkelhor, David. (1984) *Child sexual abuse: New theory and practice.* New York: Free Press.

Fischer, Judith L., Sollie, Donna L., and Morrow, K. Brent. (1986). Social networks in male and female adolescents. *Journal of Adolescent Research, 1,* 1–14.

Fischer, Kurt W. (1980). A theory of cognitive development: The control of hierarchies of skill. *Psychological Review, 87,* 477–531.

Fischman, Joshua. (1987). Type A on trial. *Psychology Today, 21 (2),* 42–50.

Fisher, Helen E. (October, 1987). The four-year itch. *Natural History, 96 (10),* 22–33.

Fitzgerald, Hiram E., and Brackbill, Yvonne. (1976). Classical conditioning in infancy: Development and constraints. *Psychological Bulletin, 83,* 353–376.

Fitzpatrick, Mary Ann. (1984). A typological approach to marital interaction: Recent theory and research. In Leonard Berkowitz (Ed.), *Advances in experimental social psychology,* Vol 18. New York: Academic Press.

Flavell, John H. (1963). *The developmental psychology of Jean Piaget.* Princeton, NJ: Van Nostrand.

Flavell, John H. (1970). Cognitive changes in adulthood. In L.R. Goulet and Paul B. Baltes (Eds.), *Life-span developmental psychology: Research and theory.* New York: Academic Press.

Flavell, John H. (1975). *The development of role-taking and communication skills in children.* Huntington, NY: Krieger (originally published by Wiley, 1968).

Flavell, John H. (1982). Structures, stages, and sequences in cognitive development. In W. Andrew Collins (Ed.), *The concept of development: The Minnesota symposia on child psychology* (Vol. 15). Hillsdale, NJ: Erlbaum.

Flavell, John H., (1985). *Cognitive development* (2nd ed.). Englewood Cliffs, NJ: Prentice-Hall.

Flavell, John H., and Ross, Lee (Eds.). (1981). *Social cognitive development: Frontiers and possible futures.* Cambridge, Eng.: Cambridge University Press.

Flavell, John H., Speer, James Ramsey, Green, Frances L., August, Diane L. (1981). The development of comprehension monitoring and knowledge about communication. *Monograph of the Society for Research in Child Development.* Serial no. 192, 46 (5).

Fleming, Jacqueline. (1984). *Blacks in college.* San Francisco: Jossey-Bass.

Flint, Marsha. (1982). Male and female menopause: A cultural put-on. In A.M. Voda, M. Dinnerstein, and S.R. O'Donnell (Eds.), *Changing perspectives on menopause.* Austin: University of Texas.

Flynn, T.M., and Flynn, L.A. (1978). Evaluation of the predictive ability of five screening measures administered during kindergarten. *Journal of Experimental Education, 46,* 65–69.

Foege, William H. (1985). Preface. National Research Council (Ed.). *Injury in America: A continuing public health problem.* Washington DC: National Academy Press.

Fonda, Jane. (1984). *Women coming of age.* New York: Simon and Schuster.

Fontanta, Vincent J. (1977). *The maltreated child: The maltreatment syndrome in children* (2nd ed). Springfield, IL: Thomas.

Fowler, James W. (1981). *Stages of faith: The psychology of human development and the quest for meaning.* New York: Harper and Row.

Fowler, James W. (1986). Faith and the structuring of meaning. In Craig Dykstra and Sharon Parks (Eds.), *Faith development and Fowler.* Birmingham, AL: Religious Education Press.

Fox, Greer Litton. (1982). *The childbearing decision: Fertility attitudes and behavior.* Beverly Hills, CA: Sage.

Fox, Margery, Gibbs, Margaret, and Auerbach, Doris. (1985). Age and gender dimensions of friendship. *Psychology of Women Quarterly, 9,* 489–502.

Fozard, James L. (1980). The time for remembering. In Leonard W. Poon (Ed.), *Aging in the 1980's: Psychological issues.* Washington, D.C.: American Psychological Association.

Frances, Carol. (1980). *College enrollment trends: Testing the conventional wisdom against the facts.* Washington, D.C.: American Council on Education.

Frankenburg, W.K., Frandal, A., Sciarillo, W., and Burgess, D. (1981). The newly abbreviated and revised Denver Developmental Screening Test. *Journal of Pediatrics, 99,* 995–999.

French, Doran C. (1984). Children's knowledge of the social functions of younger, older, and same-age peers. *Child Development, 55,* 1429–1433.

French, Lucia A. (1986). Acquiring and using words to express logical relationships. In Stan A. Kuczaj and Martyn D. Barrett (Eds.), *The development of word meaning: Progress in cognitive developmental research.* New York: Springer-Verlag.

Freud, Anna. (1968). Adolescence. In A.E. Winder and D.L. Angus (Eds.), *Adolescence: Contemporary studies.* New York: American Books.

Freud, Sigmund. (1935). *A general introduction to psychoanalysis.* Joan Riviare (Trans.). New York: Modern Library.

Freud, Sigmund. (1938). *The basic writings of Sigmund Freud.* A.A. Brill (Ed. and Trans.). New York: Modern Library.

Freud, Sigmund. (1963). *Three case histories.* New York: Collier. (Originally published 1918.)

Freud, Sigmund. (1960). *A general introduction to psychoanalysis,* Joan Riviare (Trans.). New York: Washington Square Press. (Original work published 1935.)

Freud, Sigmund. (1964). *An outline of psychoanalysis: Vol. 23, The standard edition of the complete psychological works of Sigmund Freud.* J. Strachey (Ed. and Trans.), London: Hogarth Press. (Original work published 1940.)

Freud, Sigmund. (1965). *New introductory lectures on psychoanalysis.* James Strachey (Ed. and Trans.). New York: Norton, 1965. (Original work published 1933.)

Friedman, Meyer, and Rosenman, R.H. (1974). *Type A behavior and your heart.* New York: Knopf.

Friedrich-Cofer, Lynette, and Huston, Aletha C. (1986). Television violence and aggression: The debate continues. *Psychological Bulletin, 100,* 364–371.

Fries, James F., and Crapo, Lawrence M. (1981). *Vitality and aging.* San Francisco: Freeman.

Frisch, Rose E. (1983). Fatness, puberty, and fertility: The effects of nutrition and physical training on menarche and ovulation. In Jeanne Brooks-Gunn and Anne C. Petersen (Eds.), *Girls at puberty: Biological and psychosocial aspects.* New York: Plenum.

Fritz, Janet, and Wetherbee, Sally. (1982). Preschoolers' beliefs regarding the obese individual. *Canadian Home Economics Journal, 33,* 193–196.

Fuhrmann, Walter, and Vogel, Friedrich. (1983). *Genetic counseling* (3rd. ed.). New York: Springer-Verlag.

Furman, Wyndol. (1987). Acquaintanceship in middle childhood. *Developmental Psychology, 23,* 563–570.

Furstenberg, Frank F., Jr. (1976). *Unplanned parenthood: The social consequences of teenage childbearing.* New York: Free Press.

Furstenberg, Frank F., Jr., and Nord, Christine Winquist. (1985). Parenting apart: Patterns of childbearing after marital disruption. *Journal of Marriage and the Family, 47,* 893–912.

Furstenberg, Frank F., Jr., Spanier, G.B., and Rothschild, N. (1982). Patterns in parenting in the transition from divorce to remarriage. In P.W. Berman and E.R. Ramey (Eds.), *Women in a developmental perspective.* (NIH publication no. 82–2298) Washington, D.C.

Gaddis, Alan, and Brooks-Gunn, Jeanne. (1985). The male experience of pubertal change. *Journal of Youth and Adolescence. 14,* 61.

Gadverry, Sharon. (1980). Effects of restricting first graders' TV viewing on leisure time use, IQ change, and cognitive style. *Journal of Applied Developmental Psychology, 1,* 45–57.

Gaensbauer, Theodore J. (1980). Anaclitic depression in a three-and-a-half month old child. *American Journal of Psychiatry. 137,* 841–842.

Gaff, Jerry G., and Gaff, Sally Shake. (1981). Student-faculty relationships. In Arthur W. Chickering (Ed.), *The modern American college: Responding to the new realities of diverse students and a changing society.* San Francisco: Jossey-Bass.

Galinsky, Ellen. (1981). *Between generations: The six stages of parenthood.* New York: Berkley.

Gallatin, Judith. (1980). Political thinking in adolescence. In Joseph Adelson (Ed.), *Handbook of adolescent psychology.* New York: Wiley.

Gardner, Howard. (1980). *Artful scribbles: The significance of children's drawings.* New York: Basic Books.

Gardner, Howard. (1982). *Art, mind and brain: A cognitive approach to creativity.* New York: Basic Books.

Gardner, Howard. (1983). *Frames of mind: The theory of multiple intelligences.* New York: Basic Books.

Gardner, Lytt I. (1972). Deprivation dwarfism. *Scientific American, 227* (1), 76–82.

Gardner, William I. (1977). *Learning and behavior characteristics of exceptional children and youth: A humanistic behavioral approach.* Boston: Allyn & Bacon.

Garmezy, Norman. (1976). Vulnerable and invulnerable children: Theory, research, and intervention. Abstracted in the Journal Supplement Abstract Service, *Catalog of Selected Documents in Psychology, 6* (4), 96.

Garmezy, Norman, Masten, Ann, Nordstrom, Lynn, and Ferrarese, Michael. (1979). The nature of competence in normal and deviant children. In Martha Whalen Kent and Jon E. Rolf (Eds.), *Primary prevention of psychopathology: Vol. III. Social competence in children.* Hanover, NH: University Press of New England.

Garner, D.M. Garfinkel, P.E., Schwartz, D., and Thompson, M. (1980). Cultural expectations of thinness in women. *Psychological Reports, 47,* 483–491.

Garson, Barbara. (1979). Luddites in Lordstown. In Rosabeth Moss Kanter and Barry A. Stein (Eds.), *Life in organizations.* New York: Basic Books.

Garvey, Catherine. (1976). Some properties of social play. In Jerome S. Bruner, Alison Jolly, and Kathy Sylva (Eds.), *Play.* New York: Basic Books.

Garvey, Catherine. (1977). *Play.* Cambridge, MA: Harvard University Press.

Garvey, Catherine. (1984). *Children's talk.* Cambridge, MA: Harvard University Press.

Gasser, R.D., and Taylor, C.M. (1976). Role adjustment of single parent fathers with dependent children. *Family Coordinator, 25,* 397–401.

Geiger, Roger L. (1986). *Private sectors in higher education: Structure, function, and change in eight countries.* Ann Arbor: University of Michigan Press.

Gelles, Richard J. (1975). Violence and pregnancy: A note on the extent of the problem and needed services. *Family Coordinator, 24,* 81–86.

Gelles, Richard J. (1978). Violence toward children in the United States. *American Journal of Orthopsychiatry, 48,* 580–592.

Gelman, Rochel, and **Baillargeon, Renee.** (1983). A review of some Piagetian concepts. In Paul H. Mussen (Ed.), *Handbook of child psychology: Vol. 3. Cognitive development.* New York: Wiley.

Gelman, Rochel, and **Gallistel, C.R.** (1978). *The child's understanding of number.* Cambridge, MA: Harvard University Press.

Gelman, Rochel and **Spelke, Elizabeth.** (1981). The development of thoughts about animate and inanimate objects: Implications for research on social cognition. In John H. Flavell and Lee Ross (Eds.), *Social cognitive development: Frontiers and possible futures.* Cambridge, England: Cambridge University Press.

Gelman, Rochel, Maccoby, Eleanor, and **LeVine, Robert.** (1982). Complexity in development and developmental studies. In W. Andrew Collins (Ed.), *The concept of development: Minnesota symposia on child psychology* (Vol. 15). Hillsdale, NJ: Erlbaum.

Genesse, F. (1983). Bilingual education of majority language children: The immersion experiments in review. *Applied Linguistics, 4,* 1–46.

Genishi, Celie, and **Dyson, Anne Haas.** (1984). *Language assessment in the early years.* Norwood, NJ: Ablex.

George, Victor, and **Wilding, Paul.** (1972). *Motherless families.* London: Routledge and Kegan Paul.

Gesell, Arnold. (1926). *The mental growth of the pre-school child: A psychological outline of normal development from birth to the sixth year including a system of developmental diagnosis.* New York: Macmillan.

Gesell, Arnold, Ames, Louise Bates, and **Ilg, Frances L.** (1977). *The child from five to ten.* (Rev. ed.). New York: Harper & Row.

Gesell, Arnold, and **Ilg, Frances L.** (1946). *The child from five to ten.* New York: Harper.

Gibson, Eleanor J. (1982). The concept of affordances in development: The renascence of functionalism. In W. Andrew Collins (Ed.), *The concept of development: The Minnesota symposia on child psychology* (Vol. 15). Hillsdale, NJ: Erlbaum.

Gibson, Eleanor J., and **Levin, Harry.** (1975). *The psychology of reading.* Cambridge, MA: MIT Press.

Gibson, Eleanor J., and **Walk, Richard D.** (1960). The "visual cliff." *Scientific American, 202* (4), 64–72.

Gibson, James J. (1979). *The ecological approach to visual perception.* Boston: Houghton-Mifflin.

Giles-Sims, Jean, and **Finkelhor, David.** (1984). Child abuse in stepfamilies. *Family Relations, 33,* 407–413.

Gilligan, Carol. (1981). Moral development. In Arthur W. Chickering (Ed.), *The modern American college: Responding to the new realities of diverse students and a changing society.* San Francisco: Jossey-Bass.

Gilligan, Carol. (1982). *In a different voice: Psychological theory and women's development.* Cambridge, MA: Harvard University Press.

Gilmartin, Brian G. (1985). Some family antecedents of severe shyness. *Family Relations, 34,* 429–438.

Ginsburg, Harvey J., and **Miller, Shirley M.** (1982). Sex differences in children's risk-taking behavior. *Child Development, 53,* 426–428.

Gittelman, R., Mannuzza, S. Shenker, R., and **Bonagura N.** (1985). Hyperactive boys almost grown up: Psychiatric status. *Archives of General Psychiatry, 42,* 937–947.

Giulian, Gary G., Gilbert, Enid F., and **Moss, Richard L.** (1987). Elevated fetal hemoglobin levels in Sudden Infant Death syndrome. *New England Journal of Medicine, 318,* 1122–1126.

Glass, Robert B. (1986). Infertility. In Samuel S.C. Yen and Robert B. Jaffe (Eds.), *Reproductive endocrinology: Physiology, pathophysiology, and clinical management* (2nd ed.). Philadelphia: Saunders.

Gleason, Jean Berko. (1967). Do children imitate? *Proceedings of the International Conference on Oral Education of the Deaf, 2,* 1441–1448.

Glenn, Norval D., and **Weaver, Charles N.** (1978). The marital happiness of remarried divorced persons. *Journal of Marriage and the Family, 40,* 269–282.

Glick, Paul C., and **Lin, Sung-Ling.** (1986b). Recent changes in divorce and remarriage. *Journal of Marriage and the Family, 48,* 737–748.

Glick, Ruth. (1980). Promoting competence and coping through retirement planning. In Lynne A. Bond and James C. Rosen (Eds.), *Competence and coping during adulthood.* Hanover, N.H.: University Press of New England.

Goertzel, Victor, and **Goertzel, Mildred George.** (1962). *Cradles of eminence.* Boston: Little, Brown.

Golding, Jean, Limerick, Sylvia, and **Macfarlane, Aidan.** (1985). *Sudden infant death: Patterns, puzzles, and problems.* Somerset, England: Open Books.

Golden, M., and **Birns, B.** (1976). Social class and infant intelligence. In Michael Lewis (Ed.), *Origins of intelligence.* New York: Plenum.

Goldman, Ronald, and **Goldman, Juliette.** (1982). *Children's sexual thinking.* London: Routledge and Kegan Paul.

Goldscheider, Frances Kobrin, and **Waite, Linda J.** (1987). Nest-leaving patterns and the transition to marriage for young men and women. *Journal of Marriage and the Family, 49,* 507–516.

Goldsmith, H. Hill. (1983). Genetic influences on personality from infancy to adulthood. *Child Development, 54,* 331–355.

Goldsmith, H. Hill, Buss, Arnold H., Plomin, Robert, Rothbart, Mary Klevjord, Thomas, Alexander, Chess, Stella, Hinde, Robert A., and **McCall, Robert B.** (1987). Roundtable: What is temperament? Four approaches. *Child Development, 58,* 505–529.

Goldsmith, H. Hill, and **Campos, Joseph J.** (1982). Toward a theory of infant temperament. In R.N. Emde and R.J. Harmon (Eds.), *The development of attachment and affiliative systems: Psychobiological aspects.* New York: Plenum.

Goldsmith, H. Hill, and **Gottesman, I.I.** (1981). Origins of variation in behavioral style: A longitudinal study of temperament in young twins. *Child Development, 52,* 91–103.

Goldstein, Harold. (1984). Changing structure of work: Occupational trends and implications. In Norman C. Gysbers (Ed.), *Designing careers.* San Francisco: Jossey-Bass.

Goode, W.J. (1956). *After divorce.* Glencoe, IL: Free Press.

Goodman, Ellen. (1979). *Turning points.* New York: Fawcett Crest.

Goodwin, Donald W. (1984). Biological predictors of problem drinking. In Peter M. Miller and Ted D. Nirenberg (Eds.), *Prevention of alcohol abuse.* New York: Plenum.

Goodwin, J. (1982). *Sexual abuse: Incest victims and their families.* Boston: John Wright.

Gordon, David, Southall, David P., Kelly, Dorothy H., Wilson, Adrian, Akselrod, Solange, Richards, Jean, Kennet, Barney, Kennet, Robert, Cohen, Richard J., and **Shannon, Daniel C.** (1986). Analysis of heart rate and respiratory patterns in sudden infant death victims and control infants. *Pediatric Research, 20,* 680–684.

Gorer, Geoffrey. (1973). Death, grief, and mourning in Britain. In James E. Anthony and Cyrille Koupernik (Eds.), *The child in his family: The impact of disease and death.* New York: Wiley.

Gottlieb, Benjamin H. (1975). The contribution of natural support systems to primary prevention among four social subgroups of adolescent males. *Adolescence, 10,* 207–220.

Gottlieb, Benjamin H. (1983). Social support as a focus for integrative research in psychology. *American Psychologist, 38*, 278–287.

Gottlieb, Gilbert, and Krasnegor, Norman. (1985). Measurement of audition and vision in the first year of postnatal life: A methodological overview. Cambridge, England: Cambridge University Press.

Gottlieb, Jay, and Leyser, Yona. (1981). Friendship between mentally retarded and nonretarded children. In Steven R. Asher and John M. Gotman (Eds.), *The development of children's friendships*. Cambridge, England: Cambridge University Press.

Gottman, John M. (1983). How children become friends. *Monographs of the Society for Research in Child Development, 48* (3, Serial No. 201).

Gottman, John M., and Parkhurst, J.T. (1980). A developmental theory of friendship and acquaintanceship processes. In W. Andrew Collins (Ed.), *Development of cognitive affect and social relations: Minnesota Symposia on Child Psychology* (Vol. 13). Hillsdale, NJ: Erlbaum.

Gould, Roger L. (1978). *Transformations: Growth and change in adult life*. New York: Simon and Schuster.

Grant, James P. (1982). *The state of the world's children: 1982–1983*. New York: UNICEF and Oxford University Press.

Grant, James. (1986). *The state of the world's children*. Oxford, England: Oxford University Press.

Gratch, Gerald. (1979). The development of thought and language in infancy. In Joy D. Osofsky (Ed.), *Handbook of infant development*. New York: Wiley.

Gray, S.W., and Wandersman, L.P. (1980). The methodology of home-based intervention studies: Problems and promising strategies. *Child Development, 51*, 993–1009.

Gray, William M., and Hudson, Lynne M. (1984). Formal operations and the imaginary audience. *Developmental Psychology, 20*, 619–627.

Green, A.L., and Boxer, Andres M. (1986). Daughters and sons as young adults. In Nancy Datan, Anita Greene, and Hayne W. Reese (Eds.), *Life-span developmental psychology: Intergenerational relations*. Hillsdale, NJ: Erlbaum.

Green, Lawrence W., and Horton, Denise. (1982). Adolescent health: Issues and challenges. In Thomas J. Coates, Anne C. Petersen, and Cheryl Perry (Eds.), *Promoting adolescent health: A dialogue on research and practice*. New York: Academic Press.

Greenspan, Stanley, and Greenspan, Nancy T. (1985). *First feelings*, New York: Viking.

Greer, Douglas, Potts, Richard, Wright, John C., and Huston, Aletha C. (1982). The effects of television commercial form and commercial placement on children's social behavior and attention. *Child Development, 53*, 611–619.

Greer, Germaine. (May, 1986). Letting go. *Vogue, 176*, 141–143.

Gresham, Frank M. (1982). Misguided mainstreaming: The case for social skills training with handicapped children. *Exceptional Children, 48*, 422–433.

Grinker, Joel A. (1981). Behavioral and metabolic factors in childhood obesity. In Michael Lewis and Leonard A. Rosenblum (Eds.), *The uncommon child*. New York: Plenum Press.

Grobstein, Clifford, Flower, Michael, and Mendeloff, John. (1983). External human fertilization: An evaluation of policy. *Science, 222*, 127–133.

Grossman, John H. (1986). Congenital syphilis. In John L. Sever and Robert L. Brent (Eds.), *Teratogen update: Environmentally induced birth defect risks*. New York: Liss.

Guilleminault, C., Boeddiker, Margaret Owen, and Schwab, Deborah. (1982). Detection of risk factors for "near miss SIDS" events in full-term infants. *Neuropediatrics, 13*, 29–35.

Gump, Paul V. (1975). Ecological psychology and children. In E. Mavis Hetherington (Ed.), *Review of child development research* (Vol. V). Chicago: University of Chicago Press.

Gupta, Chhanda, Yaffe, Sumner J., and Shapiro, Bernard H. (1982). Prenatal exposure to phenobarbital permanently decreases testosterone and causes reproductive dysfunction. *Science, 216*. 640–642.

Gurin, G., Veroff, J., and Feld, S. (1960). *Americans view their mental health*. New York: Basic Books.

Guthrie, D.M. (1980). *Neuroethology*. New York: Halsted Press.

Guthrie, Robert V. (1976). *Even the rat was white: A historical view of psychology*. New York: Harper and Row.

Gutmann, David. (1975). Parenthood: Key to the comparative psychology of the life cycle. In Nancy Datan and L. Ginsberg (Eds.), *Life span developmental psychology*, New York: Academic Press.

Gutmann, David. (1977). The cross-cultural perspective: Notes toward a comparative psychology of aging. In James E. Birren and K. Warner Schaie (Eds.), *Handbook of the psychology of aging*. New York: Van Nostrand Reinhold.

Gutmann, David L. (1985). The parental imperative revisited: Towards a developmental psychology of later life. *Contributions to Human Development, 14*, 31–60.

Haaf, Robert A., Smith, P. Hull, and Smitley, Suzanne. (1983). Infant response to facelike patterns under fixed-trial and infant-control procedures. *Child Development, 54*, 172–177.

Haan, Norma. (1981). Adolescents and young adults as producers of their development. In Richard M. Lerner and Nancy A. Busch-Rossnagel (Eds.), *Individuals as producers of their development: A life-span approach*. New York: Academic Press.

Haan, Norma. (1985a). Common personality dimensions or common organizations across the life span. In Joep M.A. Munnichs, Paul Mussen, Erhard Olbrich, and Peter G. Coleman (Eds.), *Life span and change in a gerontological perspective*. Orlando, FL: Academic Press.

Haan, Norma. (1985b). Processes of moral development: Cognitive or social disequilibrium. *Developmental Psychology, 21*, 996–1006.

Haas, Linda. (1981). Domestic role sharing in Sweden. *Journal of Marriage and the Family, 43*, 957–967.

Hagen, John W., Jongeward, Robert H., Jr., and Kail, Robert V., Jr. (1975). Cognitive perspectives on the development of memory. In Haynes W. Reese (Ed.), *Advances in child development and behavior*. Vol. X. New York: Academic Press.

Haggstrom, Gus W., Kanouse, David E., and Morrison, Peter A. (1986). Accounting for educational shortfalls of mothers. *Journal of Marriage and the Family, 48*, 175–186.

Haith, Marshall M., Bergman, Terry, and Moore, Michael J. (1977). Eye contact and face scanning in early infancy. *Science, 198*, 853–854.

Hale, Janice E. (1982). *Black children: Their roots, culture, and learning styles*. Provo, UT: Brigham Young University Press.

Hale, Sanora, Myerson, Joel, and Wagstaff, David. (1987). General slowing of non-verbal information processing: Evidence for a power law. *Journal of Gerontology, 42*, 131–136.

Hall, David A. (1976). *The aging of connective tissue*. London: Academic Press.

Hall, G. Stanley. (1883). The contents of children's minds. *Princeton Review, 2*, 249–272.

Hall, G. Stanley. (1904). *Adolescence: Its psychology and its relations to physiology, anthropology, sociology, sex, crime, religion and education*. New York: Appleton.

Halliday, M.A.K. (1979). One child's protolanguage. In Margaret Bullowa (Ed.), *Before speech: The beginning of interpersonal communication*. Cambridge, England: Cambridge University Press.

Halme, Jouko. (1985a). Endometriosis and infertility. In Mary G. Hammond and Luther M. Talbert (Eds.), *Infertility* (2nd ed.). Ordell, NJ: Medical Economics Books.

Halme, Jouko. (1985b). In vitro fertilization. In Mary G. Hammond and Luther M. Talbert (Eds.), *Infertility* (2nd ed.). Ordell, NJ: Medical Economics Books.

Halpern, Robert. (1984). Lack of effects for home-based early intervention? Some possible explanations. *American Journal of Orthopsychiatry, 54,* 33–42.

Hardy, Melissa. (1985). Occupational structure and retirement. In Zena Smith Blau (Ed.), *Current perspectives on aging and the life cycle: Work retirement and social policy.* Greenwich, CT: JAI.

Hareven, Tamara K. (1979). The last stage: Historical adulthood and old age. In David D. Van Tassel (Ed.), *Aging, death, and completion of being.* Philadelphia: University of Pennsylvania Press.

Harford, Thomas C. (1984). Situational factors in drinking: A developmental perspective on drinking contexts. In Peter M. Miller and Ted D. Nirenberg (Eds.), *Prevention of alcohol abuse.* New York: Plenum.

Haring-Hidore, Marilyn, Stock, William A., Okun, Morris A., and **Witter, Robert A.** (1985). Marital status and subjective well-being: A research synthesis. *Journal of Marriage and the Family, 47,* 947–953.

Harlow, Harry F., and **Mears, Clara.** (1979). *The human model: Primate perspectives.* New York: Wiley.

Harlow, Harry F., and **Suomi, Stephen J.** (1971). Social recovery by isolation reared monkeys. *Proceedings of the National Academy of Science, 68,* 1534–1538.

Harper, Lawrence V., and **Huie, Karen S.** (1985). The effects of prior group experience, age, and familiarity on the quality and organization of preschoolers' social relationships. *Child Development, 56,* 704–717.

Harper, Lawrence V., and **Sanders, Karen M.** (1975). Preschool children's use of space: Sex differences in outdoor play. *Developmental Psychology, 11,* 119.

Harman, Denham. (1984). Free radicals and the origination, evolution, and present status of the free radical theory of aging. In Donald Armstrong, R.S. Sohol, Richard G. Cutler, and Trevor P. Slater (Eds.), *Free radicals in molecular biology, aging, and disease.* New York: Raven Press.

Harman, S.M. (1978). Clinical aspects of the male reproductive system. In E.L. Schneider (Ed.), *Aging: Vol 4. The aging reproductive system.* New York: Raven Press.

Harpaz, Itzhak. (1985). Meaning of working: Profiles of various occupational groups. *Journal of Vocational Behavior, 26,* 25–40.

Harper, Rita G., and **Yoon, Jing Ja.** (1987). *Handbook of neonatology.* (2nd Ed.), Chicago: Year Book Medical Publishers.

Harrington, Michael. (1962). *The other America: Poverty in the United States.* New York: Macmillan.

Harris, I.D., and **Howard, K.I.** (1984). Parental criticism and the adolescent experience. *Journal of Youth and Adolescence, 13,* 113–121.

Harris, Raymond. (1986). *Clinical geriatric cardiology.* Phildelphia: Lippincott.

Harris, Richard J. (1986). Recent trends in the relative economic status of older adults. *Journal of Gerontology, 41,* 401–407.

Harrison, Algea, Serafica, Felicisima, and **McAdoo, Harriette.** (1984). Ethnic families of color. In Ross D. Parke (Ed.), *Review of Child Development Research* (Vol. 7). Chicago: University of Chicago Press.

Harrison, James B. (1984). Warning: The male sex role may be dangerous to your health. In Janice M. Swanson and Katherine A. Forrest (Eds.), *Men's reproductive health.* New York: Springer-Verlag.

Harter, Susan. (1982). Children's understanding of multiple emotions: A cognitive developmental approach. In W.F. Overton (Ed.), *The relationship between social and cognitive development.* Hillsdale, NJ: Erlbaum.

Harter, Susan. (1983). Developmental perspectives on the self-system. In Paul H. Mussen (Ed.), *Handbook of child psychology: Vol. 4. Socialization, personality and social development.* New York: Wiley.

Harter, Susan, and **Pike, Robin.** (1984). The pictorial scale of perceived competence and social acceptance for young children. *Child Development, 55,* 1969–1982.

Harter, Susan, and **Ward, C.** (1978). A factor-analysis of Coopersmith's self-esteem inventory. Unpublished manuscript, University of Denver. (Cited in Harter, 1983)

Hartley, Alan A. (1981). Adult age differences in deductive reasoning processes. *Journal of Gerontology, 36,* 700–706.

Hartshorne, Hugh, May, Mark A., and **Maller, J.B.** (1929). *Studies in service and self-control.* New York: Macmillan.

Hartsough, Carolyn S., and **Lambert, Nadine M.** (1985). Medical factors in hyperactive and normal children: Prenatal, developmental, and health history findings. *American Journal of Orthopsychiatry, 55,* 190–201.

Hartup, Willard W. (1983). Peer relations. In Paul H. Mussen (Ed.), *Handbook of child psychology: Vol. 4. Socialization, personality and social development.* New York: Wiley.

Hass, Aaron. (1979). *Teenage sexuality: A survey of teenage sexual behavior.* New York: Macmillan.

Hatfield, Elaine, and **Walster, Elaine.** (1978). *A new look at love.* Reading, MA: Addison-Wesley.

Hausman, Perrie B., and **Weksler, Marc E.** (1985). Changes in the immune response with age. In Caleb E. Finch and Edward L. Schneider (Eds.), *Handbook of the biology of aging* (2nd ed.). New York: Van Nostrand.

Hawkins, J., Pea, R.D., Glick, J., and **Scribner, S.** (1984). "Merds that laugh don't like mushrooms": Evidence for deductive reasoning in preschoolers. *Developmental Psychology, 20,* 584–594.

Hay, D.F., and **Ross, H.S.** (1982). The social nature of early conflict. *Child Development, 53,* 105–113.

Hayachi, Y., and **Endo, S.** (1982). All-night sleep polygraphic recordings of healthy aged persons: REM and slow wave sleep. *Sleep, 5,* 277–283.

Hayden-Thompson, Laura, Rubin, Kenneth H., and **Hymel, Shelley.** (1987). Sex preferences in sociometric choices. *Developmental Psychology, 23,* 558–562.

Hayflick, Leonard. (1979). Cell aging. In Arthur Cherkin (Ed.), *Physiology and cell biology of aging.* New York: Raven Press.

Hayflick, Leonard, and **Moorhead, Paul S.** (1961). The serial cultivation of human diploid cell strains. *Experimental Cell Research, 25,* 585.

Heald, Felix P. (1975). Juvenile obesity. In Myron Winick (Ed.), *Childhood obesity.* New York: Wiley.

Held, Richard. (1985). Binocular vision—behavioral and neuronal development. In Jacques Mehler and Robin Fox (Eds.), *Neonate cognition: Beyond the blooming confusion.* Hillsdale, NJ: Erlbaum.

Helson, Ravenna, and **Moane, Geraldine.** (1987). Personality change in women from college to midlife. *Journal of Personality and Social Psychology, 52,* 1176–186.

Henderson, Edmund. (1985). *Teaching spelling.* Boston: Houghton Mifflin.

Hennon, Charles B. (1983). Divorce and the elderly: A neglected area of research. In Timothy H. Brubaker (Ed.), *Family relationships in later life.* Beverly Hills, CA: Sage.

Herman, Judith. (1981). *Father-daughter incest.* Cambridge, MA: Harvard University Press.

Hermelin, Beate, and **O'Connor, N.** (1971). Functional asymmetry in the reading of Braille. *Neuropsychologia, 9,* 431–435.

Hershorn, Michael, and **Rosenbaum, Alan** (1985). Children of marital violence: A closer look at the unintended victims. *American Journal of Orthopsychiatry, 55,* 260–266.

Herzog, Christopher, Raskind, Cheryl L., and **Cannon, Constance J.** (1986). Age-related slowing in semantic information-processing speed: An individual difference analysis. *Journal of Gerontology, 41,* 500–512.

Hess, Robert D., and **Shipman, Virginia C.** (1965). Early experience and the socialization of cognitive modes in children. *Child Development, 36,* 869–886.

Hetherington, E. Mavis. (1972). Effects of parental absence on personality development of adolescent daughters. *Developmental Psychology, 7,* 313–326.

Hetherington, E. Mavis. (1987). *Presidential Address.* Society for Research in Child Development. Biennial Meeting, Baltimore, MD: April 25, 1987.

Hetherington, E. Mavis, and **Camara, Kathleen A.** (1984). Families in transition: The process of dissolution and reconstitution. In Ross D. Parke (Ed.), *Review of child development research* (Vol. 7). Chicago: University of Chicago Press.

Hetherington, E. Mavis and **McIntyre, C.W.** (1975). Developmental psychology. In M.R. Rosenzweig and L.W. Porter (Eds.), *Annual Review of Psychology,* Palo Alto, CA: Annual Reviews.

Hetherington, E. Mavis, Camara, Kathleen A., and **Featherman, D.L.** (1981). Achievement and intellectual functioning of children in one-parent households. In J. Spence (Ed.), *Assessing achievement.* New York: Freeman.

Hetherington, E. Mavis, Cox, Martha, and **Cox, Roger.** (1982). Effects of divorce on parents and children. In Michael E. Lamb (Ed.), *Nontraditional families: Parenting and child development.* Hillsdale, NJ: Erlbaum.

Higgins, Anne T., and **Turnure, James E.** (1984). Distractability and concentration of attention in children's development. *Child Development, 55,* 1799–1810.

Higgins, E. Tory. (1981). Role taking and social judgment: Alternative developmental perspectives and processes. In John H. Flavell and Lee Ross (Eds.), *Social cognitive development: Frontiers and possible futures.* Cambridge, England: Cambridge University Press.

Higham, Eileen. (1980). Variations in adolescent psychohormonal development. In Joseph Adelson (Ed.), *Handbook of adolescent psychology.* New York: Wiley.

Higher Education Research Institute. (1985). *American freshmen: National norms for Fall 1984.* Los Angeles, CA: UCLA Graduate School of Education.

Hillerich, Robert L. (1983). *The principal's guide to improving reading instruction.* Newton, MA: Allyn and Bacon.

Hirschi, Travis. (1969). *Causes of delinquency.* Berkeley, CA: University of California Press.

Hochschild, Arlie Russell. (1973). *The unexpected community: Portrait of an old age subculture.* Berkeley: University of California Press.

Hoffereth, Sandra L., and **Phillips, Deborah H.** (1987). Child care in the United States, 1970 to 1995. *Journal of Marriage and the Family, 49,* 559–571.

Hoffman, Lois Wladis. (1977). Changes in family roles, socialization, and sex differences. *American Psychologist, 32,* 644–657.

Hoffman, Lois Wladis. (1986). Work, family, and the child. In Michael S. Pallak and Robert O. Perloff (Eds.), *Psychology and work: Productivity, change, and employment.* Washington, DC: American Psychological Association.

Hoffman, Lois Wladis, and **Nye, F. Ivan.** (1974). *Working mothers.* San Francisco: Jossey-Bass.

Hoffman, Martin L. (1981). Perspectives on the difference between

understanding people and understanding things: The role of affect. In John H. Flavell and Lee Ross (Eds.), *Social cognitive development: Frontiers and possible futures.* Cambridge, Eng.: Cambridge University Press.

Hoffman, Martin L. (1984). Empathy, its limitations, and its role in a comprehensive moral theory. In William M. Kurtines and Jacob L. Gewirtz (Eds.), *Morality, moral behavior, and moral development.* New York: Wiley.

Holden, Constance. (1980). Identical twins reared apart. *Science, 207,* 1323–1328.

Holden, Constance. (1987). Alcoholism and the medical cost crunch. *Science, 235,* 1132–1133.

Holden, Constance. (1987). Female math anxiety on the wane. *Science, 236,* 660–661.

Holden, Constance. (1987). OTA cites financial disaster of Alzheimer's. *Science, 236,* 253.

Holden, George W. (1983). Avoiding conflict: Mothers as tacticians in the supermarket. *Child Development, 54,* 233–240.

Holdern, William A. (1980). *Sexual abuse of children.* Englewood, CO: American Humane Society.

Hollander, P. (1982). Legal context of educational testing. In National Research Council, Committee on Ability Testing, *Ability testing: Uses, consequences and controversies.* Washington, DC: National Academy Press.

Holmberg, M.C. (1980). The development of social interchange patterns from 12 to 42 months. *Child Development, 51,* 448–456.

Holstein, C. (1976). Development of moral judgment: A longitudinal study of males and females. *Child Development, 47,* 51–61.

Holzman, Mathilda. (1983). The language of children: Development in home and in school. Englewood Cliffs, NJ: Prentice-Hall.

Honzik, Marjorie P. (1986). The role of the family in the development of mental abilities: A 50-year study. In Nancy Datan, Anita L. Greene, and Hayne W. Reese (Eds.), *Life-span developmental psychology: Intergenerational relations.* Hillsdale, NJ: Erlbaum.

Hooker, J.G., Lucas, M., Richards, B.A., Shirley, I.M., Thompson, B.D., and **Ward, R.H.** (1984). Is maternal alpha-fetoprotein screening still of value in a low risk area for neural tube defects? *Prenatal Diagnosis, 4,* 29–33.

Hooper, Frank H., Hooper, Judith O., and **Colbert, Karen K.** (1984). *Personality and memory correlates of intellectual functioning: Young adulthood to old age.* Basel: Karger.

Hoorweg, Jan. (1976). *Protein-energy malnutrition and intellectual abilities: A study of teen-age Ugandan children.* The Hague: Mouton.

Horn, John L. (1976). Human abilities: A review of research and theory in the early 1970's. *Annual Review of Psychology, 27,* 437–485.

Horn, John L. (1982). The aging of human abilities. In Benjamin B. Wolman (Ed.), *Handbook of developmental psychology.* Englewood Cliffs, NJ: Prentice-Hall.

Horn, John L. (1985). Remodeling old models of intelligence. In Benjamin B. Wolman (Ed.), *Handbook of intelligence: Theories, measurements, and applications.* New York: Wiley.

Horn, John L., and **Donaldson, Gary.** (1976). On the myth of intellectual decline in adulthood. *American Psychologist, 31,* 701–719.

Horn, John L., and **Donaldson, Gary.** (1977). Faith is not enough: A response to the Baltes-Schaie claim that intelligence does not wane. *American Psychologist, 32,* 369–373.

Horner, Thomas M. (1985). The psychic life of the young infant: Review and critique of the psychoanalytic concepts of symbiosis and infantile omnipotence. *American Journal of Orthopsychiatry, 55,* 324–344.

Horney, Karen. (1967). *Feminine psychology.* Harold Kelman (Ed.), New York: Norton.

Horowitz, Frances Degen, and Paden, Lucile York. (1973). The effectiveness of environmental intervention programs. In Bettye M. Caldwell and Henry N. Ricciuti (Eds.)., *Review of child development research:* Vol. III. *Child development and social policy.* Chicago: University of Chicago Press, 331–402.

Hoving, K.L., Spencer, T., Robb, K., and Schulte, D. (1978). Developmental changes in visual information processing. In P.A. Ornstein (Ed.), *Memory development in children.* Hillsdale, NJ: Erlbaum.

Howard, A. (August, 1984). *Cool at the top: Personality characteristics of successful executives.* Paper presented at the annual meeting of the American Psychological Association, Montreal, Canada.

Howard, Darlene V., Heisy, Jane Gillette, and Shaw, Raymond J. (1986). Aging and the priming of newly learned associations. *Developmental Psychology, 22,* 78–85.

Hoyer, William J., and Plude, Dana J. (1980). Attentional and perceptual processes in the study of cognitive aging. In Leonard W. Poon (Ed.), *Aging in the 1980's: Psychological issues.* Washington, DC: American Psychological Association.

Hsu, Jeng M., and Smith, James C. (1984). B-Vitamins and ascorbic acid in the aging process. In J. Marc Ordy, Denham Harman, Roslyn B. Alfin-Slater (Eds.), *Nutrition in gerontology.* New York: Raven Press.

Hubert, Nancy C., Wachs, Theodore D., Peters-Martin, Patricia, and Gandour, Mary Jane. (1982). The study of early temperament: Measurement and conceptual issues. *Child Development, 53,* 571–600.

Hudson, Judith A. (1986). Memories are made of this: General event knowledge and development of autobiographic memory. In Katherine Nelson (Ed.), *Event knowledge: Structure and function in development.* Hillsdale, NJ: Erlbaum.

Hughes, A., and Trudgill, Peter. (1979). *English accents and dialects.* London: Edward Arnold.

Hughes, Martin, and Donaldson, Margaret. (1979). The use of hiding games for studying coordination of viewpoints. *Educational Review, 31,* 133–140.

Hughes, Martin, and Grieve, Robert. (1980). On asking children bizarre questions. *First Language, 1,* 149–160.

Humphrey, Derek, and Wickett, Ann. (1986). *The right to die.* New York: Harper and Row.

Humphreys, Anne P., and Smith, Peter K. (1987). Rough and tumble, friendship, and dominance in schoolchildren: Evidence for continuity and change with age. *Child Development, 58,* 201–212.

Hunter, Fumiyo, McCarthy, Mary E., MacTurk, Robert H., and Vietze, Peter M. (1987). Infants' social constructive interactions with mothers and fathers. *Developmental Psychology, 23,* 249–254.

Hunter, Fumiyo Tao. (1985). Adolescents' perception of discussion with parents and friends. *Developmental Psychology, 21,* 433–440.

Hunter, Laura Russell, with Membard, Polly Hunter. (1981). *The rest of my life.* Stamford, CT: Growing Pains Press.

Huston, Aletha C. (1983). Sex-typing. In Paul H. Mussen (Ed.), *Handbook of child psychology: Vol. 4. Socialization, personality and social development.* New York: Wiley.

Huston, Aletha C. (1985). The development of sex-typing: Themes from recent research. *Developmental Review, 5,* 1–17.

Huyck, Margaret Hellie. (1982). From gregariousness to intimacy: Marriage and friendship over the adult years. In Tiffany M. Field, Aletha Huston, Herbert C. Quay, Lillian Troll, and Gordon E. Finley (Eds.), *Review of human development.* New York: Wiley.

Ilg, Frances L., and Ames, Louise Bates. (1965). *School readiness: Behavior tests used at the Gesell Institute.* New York: Harper and Row.

Ilg, Frances L., Ames, Louise Bates, and Baker, Sidney M. (1981). *Child behavior* (Rev. Ed.). New York: Harper and Row.

Imbert, Michel. (1985). Physiological underpinnings of perceptual development. In Jacques Mehler and Robin Fox (Eds.), *Neonate cognition: Beyond the blooming confusion.* Hillsdale, NJ: Erlbaum.

Ingleby, David. (1987). Psychoanalysis and ideology. In John M. Broughton (Ed.), *Critical theories of psychological development.* New York: Plenum.

Inhelder, Bärbel, and Piaget, Jean. (1958). *The growth of logical thinking from childhood to adolescence.* New York: Basic Books.

Inhelder, Bärbel, and Piaget, Jean. (1970). *The early growth of logic in the child: Classification and seriation.* New York: Humanities Press. (Original work published 1964.)

Institute of Medicine. (1985). *Preventing low birthweight.* Washington, DC: National Academy Press.

Irwin, Kathleen L., Peterson, Herbert B., Hughes, Joyce M., and Gill, Sara W. (1986). Hysterectomy among women of reproductive age, United States, update for 1981–1982. *Morbidity and Mortality Weekly, 35,* 1SS–6SS.

Isaacs, N. (1974). *A brief introduction to Piaget.* New York: Schocken.

Izard, C.E. (1978). On the ontogenesis of emotions and emotion-cognition in infancy. In Michael Lewis and Leonard Rosenblum (Eds.), *The development of affect.* New York: Plenum.

Izard, C.E., Hembree, E.A., Dougherty, L.M., and Spizzire, C.C. (1983). Changes in facial expression of 2- to 19-month-old infants following acute pain. *Developmental Psychology, 19,* 418–426.

Jacklin, Carol Nagy, and Maccoby, Eleanor E. (1978). Social behavior at 33 months in same-sex and mixed-sex dyads. *Child Development, 49,* 557–569.

Jackson, B. (1984). *Fatherhood.* London: George Allen and Unwin.

Jackson, James S., and Gibson, Rose C. (1985). Work and retirement among the black elderly. In Zena Smith Blau (Ed.), *Current perspectives on aging and the life cycle: Work, retirement, and social policy.* Greenwich, CT: Jai.

Jackson, Sonia. (1987). Great Britain. In Michael E. Lamb (Ed.), *The father's role: Cross cultural perspectives.* Hillsdale, NJ: Erlbaum.

Jacobs, Blanche S., and Moss, Howard A. (1976). Birth order and sex of sibling as determinants of mother-infant interaction. *Child Development, 47,* 315–322.

Jacobs, Jerry. (1971). *Adolescent suicide.* New York: Wiley.

Jacobs, Selby C., Kosten, Thomas R., Kasl, Stanislav V., Ostfeld, Adrian M., and Berkman, Lisa. (1987–1988). Attachment theory and multiple expressions of grief. *Omega, 18,* 41–52.

Jacobson, Joseph L., Boersma, David C., Fields, Robert B., and Olson, Karen L., (1983). Paralinguistic features of adult speech to infants and small children. *Child Development, 54,* 436–442.

Jacobson, Joseph L., Jacobson, Sandra W., Fein, Greta G., Schwartz, Pamela M., and Dowler, Jeffrey K. (1984). Prenatal exposure to an environmental toxin: A test of multiple effects. *Developmental Psychology, 20,* 523–532.

Jahoda, Marie. (1981). Work, employment, and unemployment. *American Psychologist, 36,* 184–191.

James, William. (1950). *The principles of psychology.* (Vol. I) New York: Dover (Original work published 1890)

Janos, Paul M., and Robinson, Nancy M. (1985). Psychosocial development in intellectually gifted children. In Frances Degen Horowitz and Marion O'Brien (Eds.), *The gifted and talented: Developmental perspectives.* Washington, DC: American Psychological Association.

Jeavons, Sally. (1984). Small people on big estates. *Children's Environments Quarterly, 1,* 55–60.

Jelliffe, Derrick B., and **Jelliffe, E.F. Patrice.** (1977). Current concepts in nutrition: "Breast is best": Modern meanings. *New England Journal of Medicine, 297,* 912–915.

Jiles, Darrel. (1980). Problems in the assessment of sexual abuse referrals. In William A. Holdern (Ed.), *Sexual abuse of children.* Englewood, CO: American Humane Society.

Jirásek, Jan E. (1976). Principles of reproductive embryology. In J.L. Simpson (Ed.), *Disorders of sexual differentiation: Etiology and clinical delineation.* New York: Academic Press, 52–109.

John-Steiner, Vera. (1986). *Notebooks of the mind: Explorations of thinking.* Albuquerque: University of New Mexico.

Johnson, Craig, and **Conners, Mary E.** (1987). *The etiology and treatment of bulimia nervosa: A biopsychosocial perspective.* New York: Basic Books.

Johnson, Elmer J. (Ed.). (1983). *International handbooks of contemporary developments in criminology: Europe, Africa, The Middle East, and Asia.* Westport, CT: Greenwood Press.

Johnson, Russell R., Greenspan, Stephen, and **Brown, Gwyn M.** (1984). Children's ability to recognize and improve upon socially inept communications. *Journal of Genetic Psychology, 144,* 255–264.

Johnson, Timothy G., and **Goldfinger, Stephen E.** (1981). *The Harvard Medical School health letter book.* Cambridge, MA: Harvard University Press.

Johnston, Lloyd D., Bachman, Jerald G., and **O'Malley, Patrick M.** (1986). *Drug use among American high school students, college students, and other young adults: Nationwide trends through 1985.* Washington, DC: National Institute of Drug Abuse.

Johnston, Lloyd D., O'Malley, Patrick M., and **Bachman, Jerald G.** (1985). *Use of licit and illicit drugs by American high school students, 1975–1984.* National Institute of Drug Abuse, United States Department of Health and Human Services. Washington, DC.

Johnston, Robert B. (1976). Motor function: Normal development and cerebral palsy. In Robert B. Johnston and Phyllis R. Magrab (Eds.), *Developmental disorders: Assessment, treatment, education.* Baltimore, MD: University Park Press.

Jones, Celeste Pappas, and **Adamson, Lauren B.** (1987). Language use in mother-child and mother-child-sibling interactions. *Child Development, 58,* 356–366.

Jones, Harold E., and **Conrad, Herbert S.** (1983). The growth and decline of intelligence: A study of a homogeneous group between the ages of ten and sixty. *Genetic Psychology Monographs, 13,* 223–298.

Jones, Mary Cover. (1957). The later careers of boys who were early- or late-maturing. *Child Development, 28,* 113–128.

Jones, Mary Cover. (1965). Psychological correlates of somatic development. *Child Development, 36,* 899–911.

Jones, Mary Cover, and **Bayley, Nancy.** (1950). Physical maturing among boys as related to behavior. *Journal of Educational Psychology, 41,* 129–248.

Jones, N. Burton. (1976). Rough-and-tumble play among nursery school children. In Jerome S. Bruner, Alison Jolly, and Kathy Sylva (Eds.), *Play.* New York: Basic Books.

Joreskog, K.G. (1973). A general method for estimating a linear structural equation system. In A.S. Goldberger and O.D. Duncan (Eds.), *Structural equation models in the social sciences.* New York: Seminar Press.

Judd, Debra M., Siders, James A., Siders, Jane Z., and **Atkins, Kathleen R.** (1986) D. Sex-related differences on fine motor tasks at grade 1. *Perceptual and Motor Skills, 62,* 307–312.

Juel-Nielsen, Neils. (1980). *Individual and environment: Monozygotic twins reared apart.* New York: International Universities Press.

Justice, Blair, and **Justice, Rita.** (1979). *The broken taboo: Sex in the family.* New York: Human Sciences Press.

Kadushin, Alfred. (1970). *Adopting older children.* New York: Columbia University Press.

Kagan, Jerome. (1971). *Change and continuity in infancy.* New York: Wiley.

Kagan, Jerome. (1978). The baby's elastic mind. *Human Nature. 1.* 66–73.

Kagan, Jerome. (1979). Overview: Perspectives on human infancy. In Joy D. Osofsky (Ed.), *Handbook of infant development.* New York: Wiley.

Kagan, Jerome. (1983). Preface to the second edition. In Jerome Kagan and Howard Moss, *Birth to maturity* (2nd ed.). New Haven: Yale University Press.

Kagan, Jerome, Reznick, J. Steven, and **Snidman, Nancy.** (1986). Temperamental inhibition in early childhood. In Robert Plomin and Judith Dunn (Eds.), *The study of temperament: Changes, continuities, and challenges.* Hillsdale, NJ: Erlbaum.

Kalish, Richard A. (1981). *Death, grief, and caring relationships.* Monterey, CA: Brooks/Cole.

Kalish, Richard A. (1985). The social context of death and dying. In Robert H. Binstock and Ethel Shanas (Eds.), *Handbook of aging and the social sciences.* New York: Van Nostrand Reinhold.

Kandel, Denise. (1974). Inter- and intragenerational influences on adolescent marijuana use. *Journal of Social Issues. 30* (2), 107–135.

Karmel, Marjorie. (1959). *Thank you Dr. Lamaze.* Philadelphia: Lippincott.

Karp, David A. (1985–1986). Academics beyond midlife: Some observations on changing consciousness in the fifty to sixty year decade. *Aging and Human Development, 22,* 81–103.

Kart, Cary S., and **Metress, Seamus P.** (1984). *Nutrition, the aged, and society.* Englewood Cliffs, NJ: Prentice-Hall.

Kastenbaum, Robert. (1979). Exit and existence: Societies' unwritten script for old age and death. In David D. Van Tassel (Ed.), *Aging, death and the completion of being.* Philadelphia: University of Pennsylvania Press.

Kastenbaum, Robert J. (1986). *Death, society, and the human experience.* Columbus, OH: Merrill.

Katchadourian, Herant A. (1977). *The biology of adolescence.* San Francisco: Freeman.

Katchadourian, Herant. (1987). *Fifty: Midlife in perspective.* New York: Freeman.

Kaufman, Sharon R. (1986). *The ageless self.* Madison, WI: University of Wisconsin Press.

Kauser, Donald H. (1985). Episodic memory: Memorizing performance. In Neil Charness (Ed.), *Aging and human performance.* New York: Wiley.

Kaye, Kenneth. (1982). *The mental and social life of babies: How parents create persons.* Chicago: University of Chicago Press.

Keating, N., and **Cole, P.** (1980). What do I do with him 24 hours a day? Changes in the housewife role after retirement. *The Gerontologist, 20,* 84–89.

Keele, Reba. (1986). Mentoring or networking? Strong and weak ties in career development. In Linda L. Moore (Ed.), *Not as far as you think.* Lexington, MA: Heath, 1986.

Keller, W.D., Hildebrandt, K.A., and **Richards, M.** (1981, April). *Effects of extended father-infant contact during the newborn period.* Paper presented at the biennial meeting of the Society for Research in Child Development, Boston.

Kelley and Wallerstein. (1984). *The affects of parental divorce.* Paper presented at the 61st annual meeting of the American Orthopsychiatric Association, Toronto.

Kelly, Joan Berlin. (1982). Divorce: The adult perspective. In Benjamin B. Wolman (Ed.), *Handbook of developmental psychology.* Englewood Cliffs, NJ: Prentice-Hall.

Kelvin, Peter, and **Jarret, Joanna A.** (1985). *Unemployment: Its social and psychological effects.* Cambridge, England: Cambridge University Press.

Kempe, C. Henry, and **Helfeer, Ray E.** (1980). *The battered child syndrome* (3rd ed). Chicago: University of Chicago Press.

Kempe, Ruth S., and **Kempe, C. Henry.** (1978). *Child abuse.* Cambridge, MA: Harvard University Press.

Kempe, Ruth S., and **Kempe, C. Henry.** (1984). *The common secret: Sexual abuse of children and adolescents.* New York: Freeman.

Keniston, Kenneth, and **The Carnegie Council on Children.** (1977). *All our children: The American family under pressure.* New York: Harcourt, Brace, Jovanovich.

Kennel, John H., Jerauld, Richard, Wolfe, Harriet, Chesler, David, Kreger, Nancy C., McAlpine, Willie, Steffa, Meredith, and **Klaus, Marshall H.** (1974). Maternal behavior one year after early and extended post-partum contact. *Developmental Medicine and Child Neurology, 16,* 172–179.

Kerber, Linda K. (1980). Laura Ingalls Wilder. In Barbara Sicherman and Carol Hurd Green (Eds.), *Notable American women: The modern period.* Cambridge, MA: Belknap Press.

Kerr, Robert. (1985). Fitts' law and motor control in children. In Jane E. Clark and James H. Humphrey (Eds.), *Motor development: Current selected research.* Princeton, NJ: Princeton Book Company.

Kett, Joseph F. (1977). *Rites of passage: Adolescence in America, 1790 to the present.* New York: Basic Books.

King, P.M., Kitchner, K.S., Davison, M.L., Parker, C.A., and **Wood, P.K.** (1983). The justification of beliefs in young adults: A longitudinal study. *Human Development, 26,* 106–116.

Kinsbourne, Marcel, and **Hiscock, Merrill.** (1983). The normal and deviant development of functional lateralization of the brain. In Paul H. Mussen (Ed.), *Handbook of child psychology: Vol. 2. Infancy and developmental psychobiology.* New York: Wiley.

Kinsey, Alfred C., Pomeroy, Wardell B., and **Martin, Clyde E.** (1953). *Sexual behavior in the human male.* Philadelphia: Saunders.

Kirkwood, Thomas B.L. (1985). Comparative and evolutionary aspects of longevity. In Caleb E. Finch and Edward L. Schneider (Eds.), *Handbook of the biology of aging* (2nd ed.). New York: Van Nostrand.

Kitson, Gay C., and **Sussman, Marvin B.** (1982). Marital complaints, demographic characteristics, and symptoms of mental distress in divorce. *Journal of Marriage and the Family, 44,* 87–101.

Kitzinger, Sheila. (1983). *The complete book of pregnancy and childbirth.* New York: Knopf.

Klagsbrun, Samuel C. (1982). Ethics in hospice care. *American Psychologist, 37,* 1263–1265.

Klass, Dennis. (1986–1987). Marriage and divorce among bereaved parents in a self-help group. *Omega, 17,* 237–249.

Klaus, Marshall H., and **Kennell, John H.** (1976). *Maternal-infant bonding. The impact of early separation or loss on family development.* St. Louis: Mosby.

Klaus, Marshall H., and **Kennell, John H.** (1982). *Parent-infant bonding.* St. Louis: Mosby.

Kleemeier, Robert W. (1962). Intellectual changes in the senium. In *Proceedings, Social Statistic Section, American Statistical Association.* Washington, DC: American Statistical Association.

Klein, Melanie. (1957). *Envy and gratitude.* New York: Basic Books.

Klein, Robert P. (1985). Caregiving arrangements by employed women with children under 1 year of age. *Developmental Psychology, 21,* 403–406.

Kligman, Albert M., Grove, Gary L., and **Balin, Arthur K.** (1985). Aging of the human skin. In Caleb E. Finch and Edward L. Schneider (Eds.), *Handbook of the biology of aging* (2nd ed.). New York: Van Nostrand.

Kline, D.W., and **Szafran, J.** (1975). Age differences in backward monoptic visual noise masking. *Journal of Gerontology, 30,* 307–311.

Klug, William S., and **Cummings, Michael R.** (1983). *Concepts of genetics.* Columbus, OH: Merrill.

Knight, Bob G. (1982). There is no continuum of care. *Newsletter of Adult Development and Aging.* Division 20, American Psychological Association, *10,* 1, 7, 10–11.

Koch, Kenneth. (1970). *Wishes, lies, and dreams: Teaching children to write poetry.* New York: Chelsea House.

Koenig, Michael A., and **Zelnik, Melvin.** (1982). The risk of premarital first pregnancy among metropolitan-area teenagers: 1976 and 1979. *Family Planning Perspectives, 14,* 239–247.

Kohlberg, Lawrence. (1963). Development of children's orientation towards a moral order (Part I). Sequence in the development of moral thought. *Vita Humana, 6,* 11–36.

Kohlberg, Lawrence. (1966). A cognitive developmental analysis of children's sex-role concepts and attitudes. In Eleanor Maccoby (Ed.), *The development of sex differences.* Stanford, CA: Stanford University Press, 82–172.

Kohlberg, Lawrence. (1969). Stage and sequence. The cognitive developmental approach to socialization. In D.A. Goslin (Ed.), *Handbook of socialization theory and research.* Chicago: Rand McNally, 347–408.

Kohlberg, Lawrence. (1971). Stages of moral development as a basis for moral education. In C.M. Beck, B.S. Crittenden, and E.V. Sullivan (Eds.), *Moral education: Interdisciplinary approaches.* Toronto: University of Toronto Press.

Kohlberg, Lawrence. (1973). Continuities in childhood and adult moral development revisited. In Paul B. Baltes and K. Warner Schaie (Eds.), *Life-span developmental psychology: Personality and socialization.* New York: Academic Press.

Kohlberg, Lawrence. (1981). *The philosophy of moral development.* New York: Harper and Row.

Kohlberg, Lawrence, and **Elfenbein, Donald.** (1975). The development of moral judgments concerning capital punishment. *American Journal of Orthopsychiatry, 45,* 614–640.

Kohlberg, Lawrence, and **Ullian, Dorothy Z.** (1974). Stages in the development of psychosexual concepts and attitudes. In Richard C. Friedman, Ralph M. Richart, and Raymond L. VandeWiele (Eds.), *Sex differences in behavior: A conference.* New York: Wiley, 209–231.

Kohn, Robert R. (1977). Heart and cardiovascular system. In Caleb E. Finch and Leonard Hayflick (Eds.), *Handbook of the biology of aging.* New York: Van Nostrand Reinhold.

Kohn, Robert R. (1979). Biomedical aspects of aging. In David D. Van Tassel (Ed.), *Aging, death, and the completion of being.* Philadelphia: University of Pennsylvania Press.

Kohn, Willard K., and **Kohn, Jane Burgess.** (1978). *The widower.* Boston: Beacon.

Kolata, Gina Bari. (1981). Fetal alcohol advisory debated. *Science, 214,* 642–645.

Kolata, Gina. (1985). Down Syndrome—Alzheimer's linked. *Science, 230,* 1152–1153.

Kolata, Gina. (1986). Heart attacks at 9 a.m. *Science, 233,* 417–418.

Kolodny, Robert C. (1985). The clinical management of sexual problems in substance abusers. In Thomas E. Bratter and Gary G. Forrest (Eds.), *Alcoholism and substance abuse: Strategies for clinical intervention.* New York: The Free Press.

Koo, Helen P., Suchindran, C.M., and **Griffith, Janet D.** (1987). The completion of childbearing: Change and variation in timing. *Journal of Marriage and the Family, 49,* 281–294.

Kontos, Donna. (1978). A study of the effects of extended mother infant contact on maternal behavior at one and three months. *Birth and Family Journal, 5,* 133–140.

Kornhaber, Arthur, and Woodward, Kenneth L. (1981). *Grandparents/grandchildren: The vital connection.* Garden City, NJ: Anchor.

Kornhaber, R.C., and Schroeder, H.E. (1975). Importance of model similarity on extinction of avoidance behavior in children. *Journal of Consulting and Clinical Psychology, 43,* 601–607.

Korones, Sheldon B. (1986). Congenital rubella: An encapsulated review. In John L. Sever and Robert L. Brent (Eds.), *Teratogen update: Environmentally induced birth defect risks.* New York: Liss.

Koslowski, Barbara. (1980). Quantitative and qualitative changes in the development of seriation. *Merrill-Palmer Quarterly, 26,* 391–405.

Kotre, John. (1984). *Outliving the self: Generativity and the interpretation of lives.* Baltimore: Johns Hopkins University Press.

Kowles, Richard V. (1985). *Genetics, society, and decisions.* Columbus, OH: Merrill.

Kozel, Nicholas J., and Adams, Edgar H. (1986). Epidemiology of drug abuse: An overview. *Science, 234,* 970–974.

Kraft, David P. (1984). A prevention program for college students. In Peter M. Miller and Ted. D. Nirenberg (Eds.), *Prevention of alcohol abuse.* New York: Plenum.

Kramer, Deirdre A., and Woodruff, Diana S. (1986). Relativistic and dialectical thought in three adult age groups. *Human Development, 29,* 280–290.

Kramer, Judith A., Hill, Kennedy, T., and Cohen, Leslie B. (1975). Infants' development of object permanence: A refined methodology and new evidence for Piaget's hypothesized ordinality. *Child Development, 46,* 149–155.

Kratochwill, T.R. (1981) *Selective mutism: Implications for research and treatment.* Hillsdale, NJ: Erlbaum.

Krim, Kathy E. (1985). *Mentoring at work: Developmental relationships in organizational life.* Glenview, IL: Scott Foresman.

Kropp, Joseph P., and Haynes, O. Maurice. (1987). Abusive and nonabusive mothers' ability to identify general and specific emotion signals of infants. *Child Development, 58,* 187–190.

Krupka, Lawrence, and Vener, Arthur M. (1979). Hazards of drug use among the elderly. *The Gerontologist, 19,* 90–95.

Kübler-Ross, Elisabeth. (1969). *On death and dying.* New York: Macmillan.

Kübler-Ross, Elisabeth. (1975). *Death: The final stage of growth.* Englewood Cliffs, NJ: Prentice-Hall.

Kuczaj, Stan A. (1986). Thoughts on the intentional basis of early object word extension: Evidence from comprehension and production. In Stan A. Kuczaj and Martyn D. Barrett (Eds.), *The development of word meaning: Progress in cognitive developmental research.* New York: Springer-Verlag.

Kuczaj, Stan A., and Lederberg, A.R. (1977). Height, age and function: Differing influences on children's comprehension of "older" and "younger." *Journal of Child Language, 4,* 395–416.

Kuhlen, Raymond G. (1968). Developmental changes in motivation during the adult years. In Bernice L. Neugarten (Ed.), *Middle age and aging: A reader in social gerontology.* Chicago: University of Chicago Press.

Kuhn, Deanna. (1976). Short-term longitudinal evidence for the sequentiality of Kohlberg's early stages of moral judgment. *Developmental Psychology, 12,* 162–166.

Kuhn, Deanna. (1978). Mechanisms of cognitive and social development: One psychology or two? *Human Development, 25,* 233–249.

Kuhn, Deanna. (Ed.). (1979). *Intellectual development beyond childhood. New Directions in Child Development: Vol. 5.* San Francisco: Jossey-Bass.

Kuhn, Deanna, Nash, Sharon Churnin, and Brucken, Laura. (1978). Sex role concepts of two- and three-year-olds. *Child Development, 49,* 445–451.

Kurdek, Lawrence A., and Schmitt, J. Patrick. (1986). Early development of relationship quality in heterosexual married, heterosexual cohabiting, gay, and lesbian couples. *Developmental Psychology, 22,* 305–309.

Kurokawa, Minako. (1969). Acculturation and childhood accidents among Chinese and Japanese Americans. *Genetic Psychology Monographs, 79,* 89–159.

Labouvie-Vief, Gisela. (1985). Intelligence and cognition. In James E. Birren and K. Warner Schaie (Eds.), *Handbook of the psychology of aging* (2nd ed.). New York: Van Nostrand Reinhold.

Labouvie-Vief, Gisela. (1986). *Mind and self in life-span development.* Symposium on developmental dimensions of adult adaptation: Perspectives on mind, self, and emotion. Presented at 1986 meeting of the Gerontological Association of America, Chicago, November 20.

Labov, William. (1972). *Language in the inner city: Studies in the black English vernacular.* Philadelphia: University of Pennsylvania Press.

La Freniere, Peter, Strayer, F.F., and Gauthier, Roger. (1984). The emergence of same-sex affiliative preferences among preschool peers: A developmental/ethological perspective. *Child Development, 55,* 1958–1965.

Lagrand, Louis E. (1981). Loss reactions of college students: A descriptive analysis. *Death Education, 5,* 235–248.

Lamb, David R. (1984). *Physiology of exercise: Response and adaptation* (2nd ed.). New York: Macmillan.

Lamb, Michael E. (1981). The development of father-infant relationships. In Michael E. Lamb (Ed.), *The role of the father in child development* (Rev. Ed.). New York: Wiley.

Lamb, Michael E. (1982). Maternal employment and child development: A review. In Michael E. Lamb (Ed.), *Nontraditional families: Parenting and child development.* Hillsdale, NJ: Erlbaum.

Lamb, Michael E. (1986). The emergent American father. In Michael E. Lamb (Ed.), *The father's role: Applied perspectives.* New York: Wiley.

Lamb, Michael E. (1987). *The father's role: Cross-cultural perspectives.* New York: Wiley.

Lamb, Michael, and Elster, Arthur B. (1985). Adolescent mother-father-infant relationships. *Developmental Psychology, 21,* 768–773.

Lamb, Michael E., Frodi, Majt, Hwang, Carl-Philip, and Frodi, Ann M. (1983). Effects of paternal involvement on infant preferences for mothers and fathers. *Child Development, 54,* 450–458.

Lamb, Michael, Gaensbauer, T.J., Malkin, C.M., and Shults, L.A. (1985). The effects of child maltreatment on the security of infant-adult attachment. *Infant Behavior and Development, 8,* 34–45.

Lamb, Michael E., and Hwang, C.P. (1982). Maternal attachment and mother-neonate bonding: A critical review. In Michael E. Lamb and Ann L. Brown (Eds.), *Advances in developmental psychology* (Vol. 2). Hillsdale, NJ: Erlbaum.

Lamb, Michael E., and Sutton-Smith, Brian (Eds.). (1982). *Sibling relationships: Their nature and significance across the life-span.* Hillsdale, NJ: Erlbaum.

Lambert, Nadine M., and Hartsough, Carolyn S. (1984). Contribution of predispositional factors to the diagnosis of hyperactivity. *American Journal of Orthopsychiatry, 54,* 97–109.

Lambert, Nadine M., Sandoval, Jonathan, and Sassone, Dana. (1978). Prevalence of hyperactivity in elementary school children as a function of social system definers. *American Journal of Orthopsychiatry, 48*, 446–463.

Langer, E.J., and Rodin, J. (1976). The effects of choice and enhanced personal responsibility for the aged: A field experiment in an institutional setting. *Journal of Personality and Social Psychology, 34*, 191–198.

Langlois, J.H., and Downs, A.C. (1980). Mothers, fathers, and peers as socialization agents of sex-typed play behaviors in young children. *Child Development, 51*, 1237–1247.

Langone, John. (1981). Too weary to go on. *Discover, 2* (11), 72–77.

Laosa, Luis. (1980). Maternal teaching strategies in Chicano and Anglo-American families: The influence of culture and education on maternal behavior. *Child Development, 51*, 759–765.

Laosa, Luis. (1984). Social policies toward children of diverse ethnic racial and language groups in the United States. In Harold W. Stevenson and Alberta E. Siegel (Eds.), *Child development research and social policy*. Chicago: University of Chicago Press.

Larsen, John W. (1986). Congenital toxoplasmosis. In John L. Sever and Robert L. Brent (Eds.), *Teratogen update: Environmentally induced birth defect risks*. New York: Liss.

Larsen, Spencer A., and Homer, Daryl R. (1978). Relation of breast versus bottle feeding to hospitalization for gastroenteritis in a middle-class U.S. population. *The Journal of Pediatrics, 92*, 417–418.

La Rue, Asenath, Dessonville, Connie, and Jarvik, Lissy F. (1985). Aging and mental disorders. In James E. Birren and K. Warner Schaie (Eds.), *Handbook of the psychology of aging*. New York: Van Nostrand Reinhold.

Lawton, M. Powell. (1980). *Environment and aging*. Monterey, CA: Brooks/Cole.

Lawton, M. Powell. (1982). Environments and living arrangements. In Robert H. Binstock, W.S. Chow, and J.H. Shultz (Eds.), *International perspectives on aging*. New York: United Nations Fund for Population Activities.

Lawton, M. Powell. (1985). Housing and living environments of older people. In Robert H. Binstock and Ethel Shanas (Eds.), *Handbook of aging and the social sciences*. New York: Van Nostrand Reinhold.

Lazarus, Maurice, and Lauer, Harvey. (1985). Working past retirement: Practical and motivational issues. In Robert N. Butler and Herbert P. Gleason (Eds.), *Productive aging: Enhancing vitality in later life*. New York: Springer.

Lazarus, Richard S., and DeLongis, Anita. (1983). Psychological stress and coping in aging. *American Psychologist, 38*, 245–254.

Leadbeater, B. (1986). The resolution of relativism in adult thinking: Subjective, objective, or conceptual. *Human Development, 29*, 291–300.

Leaf, Alexander. (1975). *Youth in old age*. New York: McGraw-Hill.

Leaper, D.J., Gill, P.W., Staniland, J.R., Horrocks, J.C., and deDombal, F.T. (1973). Clinical diagnostic process: An analysis. *British Medical Journal, 3*, 569–574.

Leavitt, Jerome E. (1983). Preface. In Jerome E. Leavitt (Ed.), *Child abuse and neglect: Research and innovation*. The Hague: Martinus Nijhoff.

Lee, Gary R., and Ellithorpe, Eugene. (1982). Intergenerational exchange and subjective well-being among the elderly. *Journal of Marriage and the Family, 44*, 217–224.

Lee, James Michael. (1980). Christian religious education and moral development. In Brenda Munsey (Ed.), *Moral development, moral education and Kohlberg: Basic issues in philosophy, psychology, religion and education*. Birmingham, AL: Religious Education Press.

Lee, Lee C. (1976). *Personality development in childhood*. Monterey, CA: Brooks/Cole.

Lehman, Darrin R. (1987). Long-term effects of losing a spouse or child in a motor vehicle accident. *Journal of Personality and Social Psychology, 52*, 218–231.

Leifer, A.D., Leiderman, P.H., Barnett, C.R., and Williams, J.A. (1972). Effects of mother-infant separation on maternal attachment behavior. *Child Development, 43*, 1203–1218.

Leifer, Myra. (1980). *Psychological effects of motherhood: A study of first pregnancy*. New York: Praeger.

Lemon, B.W., Bengtson, V.L., and Peterson, J.A. (1972). An exploration of the activity theory of aging: Activity types and life satisfaction among in-movers to a retirement community. *Journal of Gerontology, 27*, 511–523.

Lenneberg, Eric H. (1967). *Biological foundations of language*. New York: Wiley.

Lepper, M.R., and Greene, D. (Eds.). (1978). *The hidden costs of reward*. Hillsdale, NJ: Erlbaum.

Lerner, H.E. (1978). Adaptive and pathogenic aspects of sex-role stereotypes: Implications for parenting and psychotherapy. *American Journal of Psychiatry, 135*, 48–52.

Lerner, Richard M., Karson, M., Meisels, M., and Knapp, J.R. (1975). Actual and perceived attitudes of late adolescents and their parents: The phenomenon of the generation gap. *Journal of Genetic Psychology, 126*, 195–207.

Lerner, Richard M., Schroeder, C., Rewitzer, M., and Weinstock, A. (1972). Attitude of high-school students and their parents toward contemporary issues. *Psychology Reports, 31*, 255–258.

Lerner, Richard M., and Spanier, G.B. (1978). *Child influences on marital and family interaction*. New York: Academic Press.

Lesser, Gerald S. (1984). A world of difference. *Action for Children's Television Magazine, 13*, 8.

Lester, Barry M., Als, Heidelise, and Brazelton, T. Berry. (1982). Regional obstetric anesthesia and newborn behavior: A reanalysis toward synergistic effects. *Child Development, 53*, 687–692.

Lester, Barry M., Hoffman, Joel, and Brazelton, T. Berry. (1985). The rhythmic structure of mother-infant interaction in term and preterm infants. *Child Development, 56*, 15–27.

Lester, Barry M., and Zeskind, Philip Sanford. (1982). A biobehavioral perspective on crying in early infancy. In H. Fitzgerald, Barry Lester, and Michael W. Yogman (Eds.), *Theory and research in behavioral pediatrics* (Vol. 1). New York: Plenum.

Lever, Janet. (1976). Sex differences in the games children play. *Social Problems, 23*, 478–487.

Levine, Laura E. (1983). Mine: Self-definition in 2-year-old boys. *Developmental Psychology, 19*, 544–549.

Levinson, Daniel J. (1978). *The season's of a man's life*. New York: Knopf.

Levinson, Daniel J. (1986). A conception of adult development. *American Psychologist, 41*, 3–13.

Lewin, Roger. (1987). More clues to the cause of Parkinson's disease. *Science, 237*, 978.

Lewis, Michael, and Brooks, Jeanne. (1978). Self-knowledge and emotional development. In Michael Lewis and Leonard A. Rosenblum (Eds.), *The development of affect*. New York: Plenum, 205–226.

Lewis, Michael, and Michalson, Linda. (1983). *Children's emotions and moods*. New York: Plenum.

Lewis, Richard. (1977). *Miracles: Poems by children of the English-speaking world*. New York: Bantam.

Lewis, Richard. (1978). *Journeys*. New York: Bantam.

Liben, Lynn S. (1982). The developmental study of children's memory. In Tiffany M. Field, Aletha Huston, Herbert C. Quay, Lillian Troll, and

Gordon E. Finley (Eds.), *Review of human development*. New York: Wiley.

Lidz, Theodore. (1976). *The person: His and her development throughout the life cycle.* (Rev. Ed.). New York: Basic Books.

Lieberman, Morton A. (1965). Psychological correlates of impending death: Some preliminary observations. *Journal of Gerontology, 20,* 181–190.

Liebert, Robert M. (1984). What develops in moral development. In William M. Kurtines and Jacob L. Gewirtz (Eds.), *Morality, moral behavior, and moral development.* New York: Wiley.

Liebert, Robert M., Neale, John M., and **Davidson, Emily S.** (1973). *The early window: Effects of television on children and youth.* New York: Pergamon Press.

Light, Leah L., and **Capps, Janet L.** (1986). Comprehension of pronouns in young and older adults. *Developmental Psychology, 22,* 580–585.

Light, P.H., Buckingham, N., and **Robbins, A.H.** (1979). The conservation task as an interactional setting. *British Journal of Educational Psychology, 49,* 304–310.

Lindblad-Goldberg, Marion, and **Dukes, Joyce Lynn.** (1985). Social support in black, low-income, single-parent families: Normative and dysfunctional patterns. *American Journal of Orthopsychiatry, 55,* 42–58.

Linde, Eleanor Vander, Morrongiello, Barbara A., and **Rovee-Collier, Carolyn.** (1985). Determinants of retention in 8-week-old infants. *Developmental Psychology, 21,* 601–613.

Lindquist, G.T. (1982). Preschool screening as a means of predicting later reading achievement. *Journal of Learning Disabilities, 15,* 331–332.

Lipschultz, Larry I., and **Howards, Stuart S.** (Eds.). (1983). *Infertility in the male.* New York: Churchill Livingstone.

Lipsitt, Lewis P. (1982). Infant learning. In Tiffany M. Field, Aletha Huston, Herbert C. Quay, Lillian Troll, and Gordon E. Finley (Eds.), *Review of human development.* New York: Wiley.

Lisle, Laurie. (1980). *Portrait of an artist: A biography of Georgia O'Keeffe.* New York: Seaview.

Livesley, W.J., and **Bromley, D.B.** (1973). *Person perception in childhood and adolescence.* London: Wiley.

Livson, Florence B. (1976). Patterns of personality development in middle-aged women: A longitudinal study. *International Journal of Aging and Human Development, 7,* 107–115.

Livson, Norman, and **Peskin, Harvey.** (1980). Perspectives on adolescence from longitudinal research. In Joseph Adelson (Ed.), *Handbook of adolescent psychology.* New York: Wiley.

Locksley, A., and **Colten, M.E.** (1979). Psychological androgyny: A case of mistaken identity. *Journal of Personality and Social Psychology, 37,* 1017–1031.

Loehlin, John C., Willerman, Lee, and **Horn, Joseph M.** (1982). Personality resemblances between unwed mothers and their adopted-away offspring. *Journal of Personality and Social Psychology, 42,* 1089–1099.

Lonetto, Richard. (1980). *Children's conceptions of death.* New York: Springer.

Lopata, Helena Znaniecka. (1979). *Women as widows: Support systems.* New York: Elsevier North-Holland.

Lopata, Helena Z., and **Barnewolt, Debra.** (1984). The middle years: Changes and variations in social role commitments. In Grace Baruch and Jeanne Brooks-Gunn (Eds.), *Women in midlife.* New York: Plenum.

Lovell, Kenneth, and **Shayer, Michael.** (1978). The impact of the work of Piaget on science curriculum development. In Jeanette McCarthy Gallagher and J.A. Easley Jr. (Eds.), *Knowledge and development: Vol. 2. Piaget and education.* New York: Plenum Press.

Lowe, Marianne. (1975). Trends in the development of representational play in infants from one to three years—An observational study. *Journal of Child Psychology, 16,* 33–48.

Lowrey, George H. (1978). *Growth and development of children* (7th ed.). Chicago: Year Book Medical Publishers.

Lueptow, Lloyd B. (1984). *Adolescent sex role and social change.* New York: Columbia University Press.

Luria, Zella, and **Meade, Robert G.** (1984). Sexuality and the middle-aged woman. In Grace Baruch and Jeanne Brooks-Gunn (Eds.), *Women in midlife.* New York: Plenum.

Lynch, Margaret A., and **Roberts, Jacquie.** (1983). A follow-up study of abused children and their families. In Jerome E. Leavitt (Ed.), *Child abuse and neglect: Research and innovation.* The Hague: Martinus Nijhoff.

Lytton, Hugh, Conway, Dorice, and **Sauvé, Reginald.** (1977). The impact of twinship on parent-child interaction. *Journal of Personality and Social Psychology, 35,* 97–107.

Maas, Henry S., and **Kuypers, Joseph A.** (1974). *From thirty to seventy: A forty-year longitudinal study of adult life styles and personality.* San Francisco: Jossey-Bass.

McAdoo, Harriette. (1979, May). Black kinship. *Psychology Today.*

McAdoo, John Lewis. (1979). Well being and fear of crime among the black elderly. In Donald E. Gelfand and Alfred J. Kutzik (Eds.), *Ethnicity and aging.* New York: Springer.

McAuliffe, Kathleen, and **McAuliffe, Sharon.** (1983, November 6). Keeping up with the genetic revolution. *The New York Times Magazine,* pp. 40–44, 92–97.

McCabe, Maryann. (1985). Dynamics of child sexual abuse. In Maryann McCabe, Ronald E. Cohen, and Victor Weiss (Eds.), *Child sexual abuse.* New York: Goldner Press.

McCall, Robert B. (1981). Nature-nurture and the two realms of development: A proposed integration with respect to mental development. *Child Development, 52,* 1–12.

McCartney, Kathleen. (1984). Effect of quality of day care environment on children's language development. *Developmental Psychology, 20,* 244–260.

McCauley, Elizabeth, Kay, Thomas, Ito, Joanne, and **Treder, Robert.** (1987). The Turner Syndrome: Cognitive deficits, affective discrimination, and behavior problems. *Child Development, 58,* 464–473.

McClearn, Gerald, and **Foch, Terry T.** (1985). Behavioral genetics. In James E. Birren and K. Warner Schaie (Eds.), *Handbook of the psychology of aging.* New York: Van Nostrand Reinhold.

McClelland, Kent A. (1982). Adolescent subculture in the schools. In Tiffany Field, Aletha Huston, Herbert C. Quay, Lillian Troll, and Gordon E. Finley (Eds.), *Review of human development.* New York: Wiley.

Maccoby, Eleanor Emmons. (1980). *Social development: Psychological growth and the parent-child relationship.* New York: Harcourt Brace Jovanovich.

Maccoby, Eleanor Emmons. (1984). Socialization and developmental change. *Child Development, 55,* 317–328.

Maccoby, Eleanor Emmons, and **Hagen, John W.** (1965). Effect of distraction upon central versus incidental recall: Developmental trends. *Journal of Experimental Child Psychology, 2,* 280–289.

Maccoby, Eleanor Emmons, and **Jacklin, Carol Nagy.** (1974). *The psychology of sex differences.* Stanford: Stanford University Press.

Maccoby, Eleanor Emmons, and **Martin, John A.** (1983). Socialization in the context of the family: Parent-child interaction. In Paul H. Mussen (Ed.), *Handbook of child psychology: Vol. 4. Socialization, personality and social development.* New York: Wiley.

McCrae, Robert R. (1982). Age differences in the use of coping mechanisms. *Journal of Gerontology, 37,* 454–460.

McCrae, Robert R., and Costa, Paul T. (1984). *Emerging lives, enduring dispositions: Personality in adulthood.* Boston: Little, Brown.

McCrae, Robert R., and Costa, Paul T. (1987). Validation of the five factor model of personality across instruments and observors. *Journal of Personality and Social Psychology, 52,* 81–90.

Macfarlane, Aidan. (1977). *The psychology of childbirth.* Cambridge, MA: Harvard University Press.

McGurk, Harry, and Lewis, Michael. (1974). Space perception in early infancy: Perception within a common auditory-visual space? *Science, 186,* 649–650.

McKenzie, S.C. (1980). *Aging and old age.* Glenview, Ill.: Scott, Foresman.

McKusick, Victor A. (1986). *Mendelian inheritance in man* (7th ed.). Baltimore: Johns Hopkins University Press.

Mackworth, N.H., and Bruner, Jerome S. (1970). How adults and children search and recognize pictures. *Human Development, 13,* 149–177.

McLaughlin, Barry. (1984). *Second language acquisition in childhood: Vol. 1. Preschool children* (2nd ed.). Hillsdale, NJ: Erlbaum.

McLaughlin, Barry. (1985). *Second language acquisition in childhood: Vol. 2. School-age children* (2nd Ed.). Hillsdale, NJ: Erlbaum.

McLeish, John A.B. (1976). *The Ulyssean adult.* Toronto: McGraw-Hill, Ryerson Limited.

McNeill, David. (1970). Language development in children. In Paul Mussen (Ed.), *Handbook of child psychology* (3rd ed.). New York: Wiley.

McPherson, B.D., and Kozlik, C.A. (1979). Canadian leisure patterns by age: Disengagement, continuity or ageism. In V.M. Marshall (Ed.), *Aging in Canada: Social perspectives.* Pickering, Ontario: Fitzhenry and Whiteside.

Maehr, Martin L. (1974). Culture and achievement motivation. *American Psychologist, 29,* 887–896.

Magnusson, D., Duner A., and Zetterblom, G. (1975). *Adjustment.* New York: Wiley.

Mahler, Margaret. (1968). *On human symbiosis and the vicissitudes of individuation.* New York: International Universities Press.

Mahler, Margaret S., Pine, Fred, and Bergman, Anni. (1975). *The psychological birth of the human infant: Symbiosis and individuation.* New York: Basic Books.

Maisto, Stephen A., and Caddy, Glenn R. (1981). Self-control and addictive behavior: Present status and prospects. *The International Journal of the Addictions, 16,* 104–133.

Main, Mary, and George, Carol. (1985). Responses of abused and disadvantaged toddlers to distress in agemates: A study in the day care setting. *Developmental Psychology, 21,* 407–412.

Makinson, Carolyn. (1985). The health consequences of teenage fertility. *Family Planning Perspectives, 17,* 132–139.

Malcolm, Andrew H. (1985, October 20). New generation of poor youths emerges in U.S. *The New York Times,* p. 1.

Malson, M. (1983). The social support systems of black families. *Marriage and Family Review, 5,* 37–57.

Manton, Kenneth G., Siegler, Ilene C., and Woodbury, Max A. (1986). Patterns of intellectual development in later life. *Journal of Gerontology, 41,* 486–499.

Maqsud, M. (1983). Relationship of locus of control to self-esteem, academic performance, and prediction of performance among Nigerian secondary school pupils. *British Journal of Educational Psychology, 53,* 215–221.

Marcia, James E. (1966). Development and validation of ego identity status. *Journal of Personality and Social Psychology, 3,* 551–558.

Marcia, James E. (1980). Identity in adolescence. In Joseph Adelson (Ed.), *Handbook of adolescent psychology.* New York: Wiley.

Margolin, Leslie, and White, Lynn. (1987). The continuing role of physical attractiveness in marriage. *Journal of Marriage and the Family, 49,* 21–27.

Margolis, Lewis H., and Runyan, Carol W. (1983). Accidental policy: An analysis of the problem of unintended injuries of childhood. *American Journal of Orthopsychiatry, 53,* 629–644.

Markell, Richard A., and Asher, Steven R. (1984). Children's interactions in dyads: Interpersonal influence and sociometric status. *Child Development, 55,* 1412–1424.

Markstrom, Carol A., and Mullis, Ronald L. (1986). Ethnic differences in imaginary audience. *Journal of Adolescence Research, 1,* 289–301.

Martin, Barclay. (1975). Parent-child relations. In Frances Degen Horowitz (Ed.), *Review of child development research* (Vol. IV.) Chicago: University of Chicago Press.

Martin, Clyde E. (1977). Sexual activity in the aging male. In John Money and Herman Musaph (Eds.), *Handbook of sexology.* Great Britain: Elsevier/North-Holland Biomedical Press.

Martin, Harold P. (1980). The consequences of being abused and neglected: How the child fares. In C. Henry Kempe and Ray E. Helfer (Eds.), *The battered child* (3rd ed.). Chicago: University of Chicago Press.

Martin, John A. (1981). A longitudinal study of the consequences of early mother-infant interaction: A microanalytic approach. *Monographs of the Society for Research in Child Development, 46* (3, Serial No. 190).

Martinez, Marco Antonio. (1986). Family socialization among Mexican-Americans. *Human Development, 29,* 264–279.

Maruyama, Fudeko. (1986). Dietary characteristics of the elderly. In Linda H. Chen (Ed.), *Nutritional aspects of aging:* Vol. 2. Boca Raton, FL: CRC Press.

Marx, Jean L. (1987). Oxygen-free radicals linked to many diseases. *Science, 235,* 529–531.

Maslow, Abraham, H. (1968). *Toward a psychology of being* (2nd ed.). Princeton, NJ: Van Nostrand.

Maslow, Abraham, H. (1970). *Motivation and personality* (2nd ed.). New York: Harper & Row.

Masnick, George S., and Bane, Mary Jo. (1980). *The nation's families: 1960–1980.* Boston: Auburn House.

Mass, Henry S., and Kuypers, Joseph A. (1974). *From thirty to seventy: A forty-year longitudinal study of adult life styles and personality.* San Francisco: Jossey-Bass.

Masters, William H., and Johnson, Virginia E. (1966). *Human sexual response.* Boston: Little, Brown.

Masters, William H., and Johnson, Virginia E. (1970). *Human sexual inadequacy.* Boston: Little, Brown.

Matthews, Ralph, and Matthews, Anne Martin. (1986). Infertility and involuntary childlessness: The transition to nonparenthood. *Journal of Marriage and the Family, 48,* 641–649.

Mayer, Jean. (1968). *Overweight: Causes, costs, and control.* Englewood Cliffs, NJ: Prentice-Hall.

Mazess, R.B., and Forman, S.H. (1979). Longevity and age exaggeration in Vilcabamba, Ecuador. *Journal of Gerontology, 34,* 94–98.

Mazor, Miriam D., and Simons, Harriet F. (Eds.). (1984). *Infertility: Medical, emotional and social considerations.* New York: Human Sciences Press.

Medvedev, Zhores. (1974). Caucasus and Atlay longevity: A biological or social problem? *The Gerontologist, 14,* 381–387.

Meilman, Phillip W. (1979). Cross-sectional age changes in ego identity status during adolescence. *Developmental Psychology, 15,* 230–231.

Meisel, C. Julius (Ed.). (1986). *Mainstreaming handicapped children: Outcomes, controversies and new directions.* Hillsdale, NJ: Erlbaum.

Mellendick, George J.D. (1983). Nutritional issues in adolescence. In Adele D. Hofmann (Ed.), *Adolescent medicine.* Reading, MA: Addison-Wesley.

Meltzoff, Andrew N., and **Moore, M. Keith.** (1983). Newborn infants imitate adult facial gestures. *Child Development, 54,* 702–709.

Menaghan, Elizabeth G., and **Lieberman, Morton A.** (1986). Changes in depression following divorce: A panel study. *Journal of Marriage and the Family, 48,* 319–328.

Menken, Jane, Trussell, James, and **Larsen, Ulla.** (1986). Age and infertility. *Science, 233,* 1389–1394.

Meredith, Howard V. (1978). Research between 1960 and 1970 on the standing height of young children in different parts of the world. In Hayne W. Reese and Lewis P. Lipsitt (Eds.), *Advances in child development and behavior* (Vol. 12, pp. 2–59). New York: Academic Press.

Mervis, Carolyn B., and **Mervis, Cynthia A.** (1982). Leopards are kittycats: Object labeling by mothers for their thirteen-month-olds. *Child Development, 53,* 267–273.

Metropolitan Life. (1983). Table published by Metropolitan Life Insurance Company, New York.

Michels, Robert. (1985). The role of the psychiatrist in health promotion. In Michael R. Zales (Ed.), *Stress in health and disease.* New York: Brunner/Mazel.

Miller, Brent C., McCoy, J. Kelly, and **Olson, Terrance D.** (1986). Dating age and stage as correlates of adolescent sexual attitudes and behavior. *Journal of Adolescent Research, 1,* 361–371.

Miller, George L. (1985, June). Cocaine. In U.S. Public Health Services, *Patterns and trends in drug abuse: A national and international perspective.* Department of Health and Human Services, Washington, DC.

Miller, Louise B., and **Bizzell, Rondeall P.** (1983). Long-term effects of four preschool programs: Sixth, seventh, and eighth grades. *Child Development, 54,* 727–741.

Miller, Louise B., and **Dyer, Jean L.** (1975). Four preschool programs: Their dimensions and effects. *Monographs of the Society for Research in Child Development, 40* (5&6).

Miller, Patricia H. (1983). *Theories of developmental psychology.* San Francisco: Freeman.

Miller, Peter M. (1979). Behavioral strategies for reducing drinking among young adults. In Howard T. Blane and M.E. Chafets (Eds.), *Youth, alcohol and social policy.* New York: Plenum.

Miller, William R., and **Hester, Reid K.** (1986). Inpatient alcoholism treatment: Who benefits? *American Psychologist, 41,* 794–805.

Milne, Conrad, Seefeldt, Vern, and **Reuschlein, Philip.** (1976). Relationship between grade, sex, race, and motor performance in young children. *Research Quarterly, 47,* 726–730.

Minuchin, Patricia P. (1977). *The middle years of childhood.* Monterey, CA: Brooks/Cole.

Minuchin, Patricia, and **Shapiro, Edna K.** (1983). The school as a context for social development. In Paul H. Mussen (Ed.), *Handbook of child psychology: Vol. 4. Socialization, personality and social development.* New York: Wiley.

Minuchin, Salvador, Rosman, Bernice L., and **Baker, Lester.** (1978). *Psychosomatic families: Anorexia nervosa in context.* Cambridge, MA: Harvard University Press.

Mischel, Walter. (1970). Sex typing and socialization. In Paul H. Mussen (Ed.), *Carmichael's manual of child development* (Vol. II). New York: Wiley.

Mischel, Walter. (1977). On the future of personality measurement. *American Psychologist, 32,* 246–254.

Mischel, Walter. (1979). On the interface of cognition and personality: Beyond the person-situation debate. *American Psychologist, 34,* 740–754.

Mischel, Walter, and **Peake, P.K.** (1983). Analyzing the construction of consistency in personality. In M.M. Page (Ed.), *Personality: Current theory and research.* Lincoln: University of Nebraska Press.

Mitchell, G., and **Shively, C.** (1984). Naturalistic and experimental studies of nonhuman primate and other animal families. In Ross D. Parke (Ed.), *Review of child development research: Vol. 7. The family.* Chicago: University of Chicago Press.

Mitchell, Sandra K., Bee, Helen L., Hammond, Mary A., and **Barnard, Kathryn, E.** (1985). Prediction of school and behavior problems in children followed from birth to age eight. In William K. Frankenberg, Robert N. Emde, and Joseph W. Sullivan (Eds.), *Early identification of children at risk: An international perspective.* New York: Plenum.

Moberg, David O. (1965). Religiousity in old age. *The Gerontologist, 5,* 78–87.

Moen, Phylis, and **Dempster-McClain, Donna I.** (1987). Employed parents: Role strain, work time, and preferences for working less. *Journal of Marriage and the Family, 49,* 579–590.

Montemeyer, Raymond. (1983). Parents and adolescents in conflict: All families some of the time and some families most of the time. *Journal of Early Adolescence, 3,* 83–103.

Montemeyer, Raymond. (1986). Family variation in parent-adolescent storm and stress. *Journal of Adolescent Research, 1,* 15–31.

Moore, Keith L. (1982). *The developing human: Clinically oriented embryology* (3rd ed.). Philadelphia: Saunders.

Moore, Raymond S., and **Moore, Dorothy N.** (1985). *Better late than early.* New York: Reader's Digest Press.

Moos, Rudolf H. (1986). Work as a human context. In Michael S. Pallak and Robert O. Perloff (Eds.), *Psychology and work: Productivity, change, and employment.* Washington, DC: American Psychological Association.

More, Vin. (1987). *Hospice care systems.* New York: Springer.

Mortimer, J.T., and **Kumka D.** (1982). A further examination of the occupational linkage hypothesis. *Sociological Quarterly, 23,* 241–256.

Morton, Teru. (1987). Childhood aggression in the context of family interaction. In David H. Crowell, Ian M. Evans, and Clifford R. O'Donnell (Eds.), *Childhood aggression and violence: Sources of influence, prevention, and control.* New York: Plenum.

Moscovitch, Morris. (1982). Neuropsychology of perception and memory in the elderly. In Fergus I. M. Craik and Sandra Trehub (Eds.), *Aging and cognitive processes.* New York: Plenum.

Moss, H.A., and **Susman, E.J.** (1980). Longitudinal study of personality development. In Orville G. Brim, Jr., and Jerome Kagan (Eds.), *Consistency and change in human development.* Cambridge, MA: Harvard University Press.

Mukherjee, Anil B., and **Hodgen, Gary D.** (1982). Maternal ethanol exposure induces transient impairment of umbilical circulation and fetal hypoxia in monkeys. *Science, 218,* 700–702.

Mulvihill, John J. (1986). Fetal alcohol syndrome. In John L. Sever and Robert L. Brent (Eds.), *Teratogen update: Environmentally induced birth defect risks.* New York: Liss.

Munoz, Rodrigo A., and **Amado, Henry.** (1986). Anorexia nervosa: An affective disorder. In Felix E. F. Larocca (Ed.), *Eating disorders.* San Francisco: Jossey-Bass.

Murphy, John M., and **Gilligan, Carol.** (1980). Moral development in late adolescence and adulthood: A critique and reconstruction of Kohlberg's theory. *Human Development, 23,* 77–104.

Murphy, Lois Barclay, and **Moriarty, Alice E.** (1976). *Vulnerability, coping, and growth: From infancy to adolescence.* New Haven: Yale University Press.

Murphy, Patricia Ann. (1986–1987). Parental death in childhood and loneliness in young adults. *Omega, 17,* 219–228.

Murray, Ann D., Dolby, Robyn M., Nation, Roger L., and **Thomas, David B.** (1981). Effects of epidural anesthesia on newborns and their mothers. *Child Development, 52,* 71–82.

Murstein, Bernard I. (1982). Marital choice. In Benjamin B. Wolman (Ed.), *Handbook of developmental psychology.* Englewood Cliffs, NJ: Prentice-Hall.

Mussen, Paul, Eichorn, Dorothy, Honzik, M.P., Bieber, S.L., and **Meredith, W.M.** (1980). Continuity and change in women's characteristics over four decades. *International Journal of Behavioral Development, 3,* 333–334.

Mussen, Paul Henry, and **Eisenberg-Berg, Nancy.** (1977). *Roots of caring, sharing, and helping: The development of prosocial behavior in children.* San Francisco: Freeman.

Mussen, Paul Henry, and **Jones, Mary Cover.** (1957). Self-conceptions, motivations, and interpersonal attitudes of late- and early-maturing boys. *Child Development, 28,* 243–256.

Mutran, E., and **Reitzes, D.C.** (1981). Retirement, identity and well-being. Realignment of role relationships. *Journal of Gerontology, 36,* 733–740.

Myers, Barbara J. (1984). Mother-infant bonding: The status of this critical period hypothesis. *Developmental Review, 4,* 240–274.

Myers, H.F. (1982). Research on the Afro-American family: A Critical review. In B. Bass, G. Wyatt, and G. Powell (Eds.), *The Afro-American family: Assessment, treatment and research issues.* New York: Grune and Stratton.

Naisbitt, John. (1982). *Megatrends: Ten new directions for transforming our lives.* New York: Warner Books.

Nathan, Peter E. (1983). Failures in prevention: Why we can't prevent the devastating effect of alcoholism and drug abuse. *American Psychologist, 38,* 459–467.

National Center for Health Statistics. (1976). NCHS Growth Charts, *Vital Statistics, 253* (Supp.), Washington, DC: U.S. Department of Health, Education, and Welfare.

National Center for Health Statistics. (1984a). *Vital statistics of the United States, 1979: Vol. 2. Mortality (Part A).* United States Department of Health and Human Services, Hyattsville, MD.

National Center for Health Statistics. (1984b). Advance report of final mortality statistics, 1982. *Monthly Vital Statistics, 33,* (9).

National Center for Health Statistics. (1985). *Health: United States 1985.* Washington, DC: United States Department of Health and Human Services.

National Council on Aging. (1975). *The myth and reality of aging in America.* Washington, DC: National Council on Aging, Inc.

National Institute of Child Health. (1979). *Antenatal diagnosis.* United States Department of Health, Education, and Welfare, NIH publication No. 80-1973.

National Research Council. (1980). *Recommended dietary allowances.* (9th rev. ed.). Washington, DC: National Academy of Sciences.

National Research Council. (1986). *Environmental tobacco smoke: Measuring exposures and assessing health effects.* Washington, DC: National Academy Press.

Neimark, Edith D. (1975). Intellectual development during adolescence. In Frances Degen Horowitz (Ed.), *Review of research in child development* (Vol. IV). Chicago: University of Chicago Press.

Neimark, Edith D. (1982). Cognitive development in adulthood: Using what you've got. In Tiffany M. Field, Aletha Huston, Herbert C. Quay, Lillian Troll, and Gordon E. Finley (Eds.), *Review of human development.* New York: Wiley.

Nelson, Katherine. (1973). Structure and strategy in learning to talk. *Monographs of the Society for Research in Child Development, 38* (1 & 2, Serial No. 149).

Nelson, Katherine. (1986). Preface. In Katherine Nelson (Ed.), *Event knowledge: Structure and function in development.* Hillsdale, NJ: Erlbaum.

Nelson-Le Gall, Sharon A., and **Gumerman, Ruth A.** (1984). Children's perceptions of helpers and helper motivation. *Journal of Applied Developmental Psychology, 5,* 1–12.

Neugarten, Bernice L., and **Neugarten, Dail A.** (1986). Changing meanings of age in the aging society. In Alan Pifer and Lynda Bronte (Eds.), *Our aging society: Paradox and promise.* New York: Norton.

Neugarten, Bernice L., Wood, Vivian, Kraines, Ruth J., and **Loomis, Barbara.** (1968). Women's attitudes toward the menopause. In Bernice L. Neugarten (Ed.), *Middle age and aging: A reader in social psychology.* Chicago: University of Chicago Press.

New York Times. (September 19, 1980). *Mistaken identity leads to surprising discovery,* p. 17.

New York Times. (December 10, 1986). *More women postponing marriage.* I, 22, col. 4.

Newberger, Carolyn Moore, Newberger, Eli H., and **Harper, Gordon P.** (1976). The social ecology of malnutrition in childhood. In John D. Lloyd-Still (Ed.), *Malnutrition and intellectual development* (pp. 160–186). Littleton, MA: Publishing Sciences Group.

Newberry, Phillis, Weissman, Myrna, and **Myers, Jerome K.** (1979). Working wives and housewives: Do they differ in mental status and social adjustment? *American Journal of Orthopsychiatry, 49,* 282–291.

Newbrun, Ernest. (1982). Sugar and dental caries: A review of human studies. *Science, 217,* 418–423.

Newton, Robert A. (1984). The medical work up: Male problems. In Miriam D. Mazor and Harriet F. Simons (Eds.), *Infertility: Medical, emotional, and social considerations.* New York: Human Sciences Press.

Nisbett, R.E., and **Ross, Lee.** (1980). *Human inference: Strategies and shortcomings of social judgment.* Englewood Cliffs, NJ: Prentice-Hall.

Nix, Gary W. (Ed.). (1976). *Mainstream education for hearing impaired children and youth.* New York: Grune & Stratton.

Nock, Steven L. (1981). Family life transitions: Longitudinal effects on family members. *Journal of Marriage and the Family, 43,* 703–714.

Norris, Joan E., and **Rubin, Kenneth H.** (1984). Peer interaction and communication: A life-span perspective. In Paul B. Baltes and Orville G. Brim, Jr. (Eds.), *Life-span development and behavior:* Vol. 6. Orlando, FL: Academic Press.

Norton, Arthur J., and **Moorman, Jeanne E.** (1987). Current trends in marriage and divorce among American women. *Journal of Marriage and the Family, 49,* 3–14.

Notman, Malkah. (1980). Adult life cycles: Changing roles and changing hormones. In Jacquelynne E. Parsons (Ed.), *The psychobiology of sex differences and sex roles.* New York: McGraw-Hill.

Nottelmann, Edith D., and **Welsh, C. Jean.** (1986). The long and the short of physical stature in early adolescence. *Journal of Early Adolescence, 6,* 15–27.

Novak, M.A., and **Harlow, Harry F.** (1975). Social recovery of monkeys isolated for the first year of life: 1. Rehabilitation and therapy. *Developmental Psychology, 11,* 453–465.

Nurco, David N., Wegner, Norma, Stephenson, Philip, Makofsky, Abraham, and **Shaffer, John W.** (1983). *Ex-addicts self-help groups: Potentials and pitfalls.* New York: Praeger.

Nydegger, Corinne N. (1986). Asymmetrical kin and the problematic son-in-law. In Nancy Datan, Anita L. Greene, and Hayne W. Reese (Eds.), *Life-span developmental psychology: Intergenerational relations*. Hillsdale, NJ: Erlbaum.

Oakley, Ann. (1980). *Women confined: Toward a sociology of childbirth*. Oxford: Martin Robertson.

Oberman, Albert. (1980). The role of exercise in preventing coronary heart disease. In Elliot Rapaport (Ed.), *Current controversies in cardiovascular disease*. Philadelphia: Saunders.

Obler, Loraine K., and **Albert, Martin L.** (1985). Language skills across adulthood. In James E. Birren and K. Warner Schaie (Eds.), *Handbook of the psychology of aging*. New York: Van Nostrand Reinhold.

O'Brien, Marion, Huston, Aletha C., and **Risley, Todd R.** (1983). Sex-typed play of toddlers in a day care center. *Journal of Applied Developmental Psychology, 4,* 1–9.

O'Brien, Thomas E., and **McManus, Carol E.** (1978). Drugs and the fetus. A consumer's guide by generic and brand name. *Birth and the Family Journal, 5,* 58–86.

Obrist, W.D. (1980). Cerebral blood flow and EEG changes associated with aging and dementia. In E.W. Busse and D.G. Blazer (Eds.), *Handbook of geriatric psychiatry*. New York: Van Nostrand Reinhold.

Ockene, J.K. (1984). Toward a smoke-free society. *American Journal of Public Health, 74,* 1198–2000.

O'Donnell, Lydia N. (1985). *The unheralded majority: Contemporary women as mothers*. Lexington, MA: Heath.

Offer, Daniel, and **Offer, Judith.** (1975). *From teenage to young manhood*. New York: Basic Books.

Olmstead, Alan H. (1975). *Threshold: The first days of retirement*. New York: Harper and Row.

Olsho, Lynne Werner, Harkins, Stephen W., and **Lenhardt, Martin L.** (1985). Aging and the auditory system. In James E. Birren and K. Warner Schaie (Eds.), *Handbook of the psychology of aging*. New York: Van Nostrand Reinhold.

Olson, David H., and **McCubbin, Hamilton.** (1983). *Families: What makes them work*. Beverly Hills, CA: Sage.

O'Malley, J.M. (1982). Instructional services for limited English proficiency students. *NABE Journal, 7,* 21–36.

Opie, Iona, and **Opie, Peter.** (1959). *The lore and language of schoolchildren*. Oxford: The Clarendon Press.

Orford, Jim. (1985). *Excessive appetites: A psychological view of addictions*. New York: Wiley.

Ornitz, Edward M., and **Ritvo, Edward R.** (1976). The syndrome of autism: A critical review. *American Journal of Psychiatry, 133,* 609–621.

Osherson, Daniel N., and **Markman, Ellen.** (1974–1975). Language and the ability to evaluate contradictions and tautologies. *Cognition, 3,* 213–226.

Osterweis, Marian, Soloman, Fredric, and **Green, Morris.** (Eds.). (1984). *Bereavement: Reactions, consequences, and care*. Washington DC: National Academy Press.

Ostrea, Enrique M., Jr., and **Chavez, Cleofe, J.** (1979). Perinatal problems (excluding neonatal withdrawal) in maternal drug addiction: A study of 830 cases. *The Journal of Pediatrics, 94,* 292–295.

Ostrow, Andrew C. (1983). Age role stereotyping: Implications for physical activity participation. In Graham D. Rowles and Russell J. Ohta (Eds.), *Aging and milieu*. New York: Academic Press.

Ouchi, William G. (1981). *Theory Z*. Reading, MA: Addison-Wesley.

Padmore, Erdman B., and **Maeda, Daisaku.** (1985). *The honorable elders revisited*. Durham, NC: Duke University Press.

Paige, Karen Eriksen. (1983). A bargaining theory of menarcheal responses in preindustrial cultures. In Jeanne Brooks-Gunn and Anne C. Petersen (Eds.), *Girls at puberty: Biological and psychosocial perspectives*. New York: Plenum.

Palkovitz, Rob. (1985). Fathers' birth attendence, early contact, and extended contact with their newborns: A critical review. *Child Development, 50,* 392–406.

Palmer, Frances H. (1978). The effects of early childhood intervention. In Bernard Brown (Ed.), *Found: Long-term gains from early intervention*. Boulder, CO: Westview Press.

Parke, Ross D. (1977). Punishment in children: Effects, side effects, and alternative strategies. In Harry L. Hom, Jr. and Paul A. Robinson (Eds.), *Psychological processes in early education* (pp. 71–97). New York: Academic Press.

Parke, Ross D., and **Slaby, Ronald G.** (1983). The development of aggression. In Paul H. Mussen (Ed.), *Handbook of child psychology: Vol. 4. Socialization and personality development*. New York: Wiley.

Parke, Ross D., and **Tinsley, Barbara R.** (1981). The father's role in infancy: Determinants of involvement in caregiving and play. In Michael E. Lamb (Ed.), *The role of the father in child development* (2nd ed.). New York: Wiley.

Parlee, Mary Brown. (1984). Reproductive health, including menopause. In Grace Baruch and Jeanne Brooks-Gunn (Eds.), *Women in midlife*. New York: Plenum.

Parloff, Morris B., London, Perry, and **Wolfe, Barry.** (1986). Individual psychotherapy and behavioral change. *Annual Review of Psychology, 37,* 321–349.

Parmelee, Arthur H., Jr., and **Sigman, Marian D.** (1983). Perinatal brain development and behavior. In Paul H. Mussen (Ed.), *Handbook of child psychology: Vol. 2. Infancy and developmental psychobiology*. New York: Wiley.

Parnes, Herbert S., and **Less, Lawrence.** (1985). Variation in selection forms of leisure activity among elderly males. In Zena Smith Blau (Ed.), *Current perspectives on aging and the life cycle: Work retirement and social policy*. Greenwich, CT: JAI.

Parry, Ruth S., Broder, Elsa A., Schmitt, Elizabeth A.G., Saunders, Elisabeth B., and **Hood, Eric** (Eds.). (1986). *Custody disputes: Evaluation and intervention*. Lexington, MA: Lexington Books.

Parten, Mildred B. (1932). Social participation among preschool children. *Journal of Abnormal and Social Psychology, 27,* 243–269.

Patterson, David. (1987). The causes of Down's syndrome. *Scientific American, 257 (2),* 52–60.

Patterson, Gerald R. (1980). Mothers: The unacknowledged victims. *Monographs of the Society for Research in Child Development, 45,* (5, Serial No. 186).

Patterson, Gerald R. (1982). *Coercive family processes*. Eugene, OR: Castalia Press.

Patterson, Gerald R., Littman, R.A., and **Bricker, W.** (1967). Assertive behavior in children: A step toward a theory of aggression. *Monographs of the Society for Research in Child Development, 32* (Serial No. 113).

Pattison, E. Mansell (Ed.). *The experience of dying*. Englewood Cliffs, NJ: Prentice-Hall, 1977.

Pearl, David. (1987). Familial, peer, and television influences on aggressive and violent behavior. In David H. Crowell, Ian M. Evans, and Clifford R. O'Donnell (Eds.), *Childhood aggression and violence: Sources of influence, prevention, and control*. New York: Plenum.

Pearlin, Leonard I. (1982). Discontinuities in the study of aging. In Tamara K. Hareven and Kathleen J. Adams (Eds.), *Aging and life course transitions: An interdisciplinary perspective*. New York: Guilford.

Pederson, Frank A. (1981). Father influences viewed in a family context. In Michael E. Lamb (Ed.), *The role of the father in child development*. 2nd ed. New York: Wiley.

Peel, E.A. (1971). *The nature of adolescent judgment*. New York: Wiley.

Pelton, Leroy H. (1978). Child abuse and neglect: The myth of classlessness. *American Journal of Orthopsychiatry, 48,* 608–617.

Peplau, Letitia, and **Gordon, S.L.** (1985). Women and men in love: Sex differences in close heterosexual relationships. In Virginia O'Leary, Rhosa K. Unger, and Barbara S. Wallston (Eds.), *Women, gender, and social psychology*. Hillsdale, NJ: Erlbaum.

Perlman, Daniel, and **Peplau, Letitia.** (1984). Loneliness research: A survey of empirical findings. In Letitia S. Peplau and E.D. Goldston (Eds.), *Preventing the harmful consequences of severe and persistent loneliness*. National Institute of Mental Health, Washington DC: U.S. Government Printing Office.

Perlmutter, Marion. (1986). A life-span view of memory. In Paul B. Baltes, David L. Featherman, and Richard M. Lerner (Eds.), *Life-span development and behavior*. Hillsdale, NJ: Erlbaum.

Perry, William G., Jr. (1981). Cognitive and ethical growth: The making of meaning. In Arthur W. Chickering (Ed.), *The modern American college: Responding to the new realities of diverse students and a changing society*. San Francisco: Jossey-Bass.

Pervin, Lawrence A. (1985). Personality: Current controversies, issues, and directions. *Annual Review of Psychology, 36,* 83–114.

Pesce, Rosario, and **Harding, Carol Gibb.** (1986). Imaginary audience behavior and its relationship to operational thought and social experience. *Journal of Early Adolescence, 6,* 83–94.

Peters-Martin, Patricia, and **Wachs, Theodore D.** (1984). A longitudinal study of temperament and its correlates in the first 12 months. *Infant Behavior and Development, 7,* 285–298.

Petersen, Anne. (1987). Those gangly years. *Psychology Today, 21 (9),* 28–35.

Petzel, Sue V., and **Riddle, Mary.** (1981). Adolescent suicide: Psychosocial and cognitive aspects. In Sherman Feinstein, John Looney, Allan Schwartzberg, and Arthur Sorosky (Eds.), *Adolescent psychiatry: Developmental and clinical studies*. Chicago and London: The University of Chicago Press (pp. 343–398).

Pezdek, Kathy, and **Miceli, Laura.** (1982). Life-span differences in memory integration as a function of processing time. *Developmental Psychology, 18,* 485–490.

Pfeiffer, Eric. (1977). Sexual behavior in old age. In Ewald W. Busse and Eric Pfeiffer (Eds.), *Behavior and adaptation in late life*. 2nd ed. Boston: Little, Brown.

Pfeiffer, Eric, and **Davis, Glenn C.** (1972). Determinants of sexual behavior. *Journal of American Geriatrics Society, 20,* 151–158.

Pfeiffer, Eric, Verwoerdt, Adriaan, and **Davis, Glenn C.** (1972). Sexual behavior in middle life. *American Journal of Psychiatry, 128,* 1262–1267.

Piaget, Jean. (1952a). *The origins of intelligence in children*. Margaret Cook (Trans.). New York: International University Press.

Piaget, Jean. (1952b). *The child's conception of number*. London: Routledge and Kegan Paul.

Piaget, Jean. (1959). *The language and thought of the child* (3rd ed.). Marjorie and Ruth Gabain (Trans.). London: Routledge and Kegan Paul.

Piaget, Jean. (1962). *Judgment and reasoning in the child*. Marjorie Warden (Trans.). London: Routledge and Paul. (Original work published 1928.)

Piaget, Jean. (1965). *The child's conception of number*. G. Gattegno and F.M. Ho (Trans.). New York: Norton.

Piaget, Jean. (1967). *Six psychological studies*. New York: Random House. (Originally published as *Six Etudes de Psychologie,* 1964).

Piaget, Jean. (1970a). *The child's conception of time,* A.J. Pomerans (Trans.). New York: Basic Books.

Piaget, Jean. (1970b). *The child's conception of movement and speed,* G.E.T. Holloway and M.J. Mackenzie (Trans.). New York: Basic Books.

Piaget, Jean. (1972). Intellectual evolution from adolescence to adulthood. *Human Development, 15,* 1–12.

Piaget, Jean. (1976). *The grasp of consciousness: Action and concept in the young child,* Susan Wedgwood (Trans.). Cambridge, MA: Harvard University Press.

Piaget, Jean, and **Inhelder, Bärbel.** (1963). *The child's conception of space,* F.J. Langdon and J.L. Lunzer (Trans.). London: Routledge and Paul.

Piaget, Jean, and **Inhelder, Bärbel.** (1969). *The psychology of the child,* Helen Weaver (Trans.). New York: Basic Books.

Piaget, Jean, and **Inhelder, Bärbel.** (1974). *The child's construction of quantities: Conservation and atomism*. London: Routledge and Kegan Paul.

Piaget, Jean, and **Szeminska, Aline.** (1965). *The child's conception of number,* Caleb Gattegno and F. Hodgson (Trans.). New York: Norton.

Pick, Anne D., Frankel, Daniel G., and **Hess, Valerie.** (1975). Children's attention: The development of selectivity. In E. Mavis Hetherington (Ed.), *Review of child development research* (Vol. V). Chicago: University of Chicago Press.

Pifer, Alan, and **Bronte, Lynda.** (1986a). Introduction: Squaring the pyramid. In Alan Pifer and Lynda Bronte (Eds.), *Our aging society: Paradox and promise*. New York: Norton.

Pifer, Alan, and **Bronte, Lynda.** (Eds.). (1986b). *Our aging society: Paradox and promise*. New York: Norton.

Pines, Ayala M., and **Aronson, Elliot.** (1981). *Burnout: From tedium to personal growth*. New York: Free Press.

Pines, Maya. (1978). Invisible playmates. *Psychology Today, 12 (4),* 38–42, 106.

Pines, Maya. (1984). In the shadow of Huntington's. *Science 84, 5, (4),* 32–39.

Piotrkowcski, Chaya S. (1979). *Work and the family system*. New York: Free Press.

Piper, Alison I., and **Langer, Ellen J.** (1986). Aging and mindful control. In Margret M. Baltes and Paul B. Baltes (Eds.), *The psychology of control and aging*. Hillsdale, NJ: Erlbaum.

Pissanos, Becky W., Moore, Jane B., and **Reeve, T. Gilmour.** (1983). Age, sex, and body composition as predictors of children's performance on basic motor abilities and health-related fitness items. *Perceptual and Motor Skills, 56,* 71–77.

Pitskhelauri, G.Z. (1982). *The long-living of Soviet Georgia*. (Trans. Gari Lesnoff-Caravaglia). New York: Human Sciences Press.

Placek, Paul J. (1986). Commentary: Cesarean rates still rising. *Statistical Bulletin, 67 (3),* 9.

Plude, Dana J., and **Hoyer, William J.** (1985). Attention and performance: Identifying and localizing age. In Neil Charness (Ed.), *Aging and human performance*. New York: Wiley.

Poffenberger, Thomas. (1981). Child rearing and social structure in rural India: Toward a cross-cultural definition of child abuse and neglect. In Jill E. Korbin (Ed.), *Child abuse and neglect: Cross-cultural perspectives*. Berkeley: University of California Press.

Polit, Denise. (1984). The only child in single-parent families. In Toni Falbo (Ed.), *The single-child family*. New York: Guilford.

Poon, Leonard W. (1985). Differences in human memory with aging: Nature, causes, and clinical implications. In James E. Birren and K. Warner Schaie (Eds.), *Handbook of the psychology of aging*. New York: Van Nostrand Reinhold.

Pope, H.G., Hudson, J.I., Yurgelun-Todd, D., and Hudson, M.S. (1984). Prevalence of anorexia nervosa and bulimia in three student populations. *International Journal of Eating Disorders, 3,* 45–51.

Porjesz, Bernice, and Begleitner, Henri. (1985). Human brain electrophysiology and alcoholism. In Ralph Tarter and David Thiel (Eds.), *Alcohol and the brain.* New York: Plenum Publishing Co.

Powers, Stephen, and Wagner, Michael J. (1984). Attributions for school achievement of middle school students. *Journal of Early Adolescence, 4,* 215–222.

Preston, George A.N. (1986). Dementia in elderly adults: Prevalence and institutionalization. *Journal of Gerontology, 41,* 261–267.

Public Health Services. (1985, June). *Patterns and trends in drug abuse: A national and international perspective.* U.S. Department of Health and Human Services, Washington, D.C.

Quinn, Joan L. (1987). Attitude of professionals toward the aged. In George L. Maddox (Ed.), *The encyclopedia of aging.* New York: Springer.

Rabinowitz, Jan C. (1986). Priming in episodic memory. *Journal of Gerontology, 41,* 204–213.

Radin, Norma. (1982). Primary caregiving and role-sharing fathers. In Michael E. Lamb (Ed.), *Nontraditional families: Parenting and child development.* Hillsdale, NJ: Erlbaum.

Radke-Yarrow, Marian, Zahn-Waxler, Carolyn, and Chapman, Michael. (1983). Children's prosocial dispositions and behavior. In Paul H. Mussen (Ed.), *Handbook of child psychology: Vol. 4. Socialization and personality development.* New York: Wiley.

Rahe, R.H. (1979). Life change events and mental illness: An overview. *Journal of Human Stress, 5,* 2–10.

Raloff, Janet. (1986). Even low levels in mom affect baby. *Science News, 130,* 164.

Ralston, Davis A., and Flanagan, Michael E. (1985). The effect of flextime on absenteeism and turnover for male and female employees. *Journal of Vocational Behavior, 26,* 206–217.

Ramey, C.T., and Haskins, Ron. (1981). The causes and treatment of school failure: Insights from the Carolina Abecedarian Project. In M.J. Begab, H. Garber, and H.C. Haywood (Eds.), *Psychosocial influences in retarded performance. Vol. 2. Strategies for improving competence.* Baltimore: University Park Press.

Ratner, Hilary Horn, Schell, David A., Crimmins, Anne, Mittelman, David, and Baldinelli, Laurie. (1987). Changes in adult prose recall: Aging or cognitive demands. *Developmental Psychology, 23,* 521–525.

Ravitch, Diane. (1983). *The troubled crusade: American education 1945–1980.* New York: Basic Books.

re:act. (1984). Viewing goes up. *Action for Children's Television Magazine, 13,* 4.

Read, M.H. and Graney, A.S. (1982). Food supplement use by the elderly. *Journal of American Dietetic Association, 80,* 250.

Rebelsky, Freda G., Starr, Raymond H., Jr., and Luria, Zella. (1967). Language development: The first four years. In Yvonne Brackbill (Ed.), *Infancy and early childhood.* New York: The Free Press.

Reese, Hayne W. and Rodeheaver, Dean. (1985). Problem solving and complex decision making. In James E. Birren and K. Warner Schaie (Eds.), *Handbook of the psychology of aging* (2nd ed.). New York: Van Nostrand Reinhold.

Reid, B.V. (1984). An anthropological reinterpretation of Kohlberg's stages of moral development. *Human Development, 27,* 56–74.

Reilly, R.R., Brown, B., Blood, M.R., and Maltesta, C.Z. (1981). The effects of realistic previews: A study and discussion of the literature. *Personnel Psychology, 34,* 832–834.

Reinke, Barbara J., Ellicott, Abbie M., Harris, Rochelle, L., and Hancock, Emily. (1985). Timing of psychosocial changes in women's lives. *Human Development, 28,* 259–280.

Reisberg, Barry. (1981). *Brain failure. An introduction to current concepts of senility.* New York: Free Press.

Reisberg, Barry, Ferris, Steven, de Leon, Mony J., and Crook, Thomas. (1985). Age-associated cognitive decline and Alzheimer's disease: Implications for assessment and treatment. In M. Bergener, Marco Ermini, and H.B. Stahelin (Eds.), *Thresholds in aging.* London: Academic Press.

Reker, Gary T., Peacock, Edward J., and Wong, Paul T.P. (1987). Meaning and purpose in life and well-being: A life-span perspective, *Journal of Gerontology, 42,* 44–49.

Resnick, Lauren B. (1983). Mathematics and science learning: A new conception. *Science, 220,* 477–478.

Rest, James R. (1983). Morality. In Paul H. Mussen (Ed.), *Handbook of child psychology: Vol. 3. Cognitive development.* New York: Wiley.

Rest, James R., and Thoma, Stephen J. (1985). Relation of moral judgment development to formal education. *Developmental Psychology, 21,* 709–714.

Rholes, William S., Blackwell, Janette, Jordan, Carol, and Walters, Connie. (1980). A developmental study of learned helplessness. *Developmental Psychology, 16,* 616–624.

Rhyne, Darla. (1981). Basis of marital satisfaction among men and women. *Journal of Marriage and the Family, 43,* 941–955.

Rice, G. Elizabeth, and Meyer, Bonnie J.F. (1986). Prose recall: Effects of aging, verbal ability, and reading behavior. *Journal of Gerontology, 41,* 469–480.

Rice, Mabel, L. (1982). Child language: What children know and how. In Tiffany Field, Aletha Huston, Herbert C. Quay, Lillian Troll, and Gordon E. Finley (Eds.), *Review of human development.* New York: Wiley.

Rice, Mabel L. (1984). Cognitive aspects of communicative development. In Richard L. Schiefelbusch and Joanne Pickar (Eds.), *The acquisition of communicative competence.* Baltimore, MD: University Park Press.

Riegel, Klaus F. (1975). Toward a dialectical theory of development. *Human Development, 18,* 50–64.

Riegel, Klaus F., and Riegel, Ruth M. (1972). Development, drop, and death. *Developmental Psychology, 6,* 306–319.

Riegel, Klaus F., Riegel, Ruth M., and Meyer, Günther. (1967). A study of the dropout rates in longitudinal research on aging and the prediction of death. *Journal of Personality and Social Psychology, 5,* 342–348.

Riese, Marilyn L. (1987). Temperament stability between the neonatal period and 24 months. *Developmental Psychology, 23,* 216–222.

Riesman, David. (1980). *On higher education.* San Francisco: Jossey-Bass.

Rikaus, Lora. (1986). Toxicological factors affecting nutritional status: Drugs. In Linda H. Chen (Ed.), *Nutritional aspects of aging:* Vol. 2. Boca Raton, FL: CRC Press.

Riordan, Jan. (1983). *A practical guide to breastfeeding.* St. Louis, MO: Mosby.

Ritchie, Jane, and Ritchie, James. (1981). Child rearing and child abuse: The Polynesian context. In Jill E. Korbin (Ed.), *Child abuse and neglect: Cross-cultural perspectives.* Berkeley: University of California Press.

Roberto, Karen A., and Scott, Jean Pearson. (1986). Equity considerations in the friendships of older adults. *Journal of Gerontology, 41,* 241–247.

Robins, L.N. (1974). The Vietnam drug user returns. *Special Action Office Monograph* (Series A, No. 2).

Robins, L.N., Helzer, J.E., Hesselbrock, M., and Wish, E. (1980). Vietnam veterans three years after Vietnam. In L. Brill and C. Winick (Eds.), *The yearbook of substance use and abuse.* Vol. II. New York: Human Sciences Press.

Robins, Lee N., Helzer, John E., Weissman, Myrna M., Orvaschel, Helen, Gruenberg, Ernest, Burke, Jack D., and Regier, Darrel A. (1984). Life-time prevalence of specific psychiatric disorders in three sites. *Archives of General Psychiatry, 41,* 949–958.

Robinson, E.J., and Robinson, W.P. (1981). Ways of reacting to communication failure in relation to the development of the child's understanding about verbal communication. *European Journal of Social Psychology, 11,* 189–208.

Robinson, Halbert B. (1981). The uncommonly bright child. In Michael Lewis and Leonard A. Rosenblum (Eds.), *The uncommon child.* New York: Plenum.

Robinson, Ira E., and Jedlicka, Davor. (1982). Change in sexual behavior of college students from 1965 to 1980: A research note. *Journal of Marriage and the Family, 44,* 237–240.

Robinson, Pauline K., Coberly, Sally, and Paul, Carolyn E. (1985). Work and retirement. In Robert R. Binstock and Ethel Shanas (Eds.), *Handbook of aging and the social sciences.* New York: Van Nostrand Reinhold.

Roche, Alex F. (1981). The adipocyte-number hypothesis. *Child Development, 52,* 31–43.

Rodeheaver, Dean, and Thomas, Jeanne L. (1986). Family and community networks in Appalachia. In Nancy Datan, Anita L. Greene, and Hayne W. Reese (Eds.), *Life-span developmental psychology: Intergenerational relations.* Hillsdale, NJ: Erlbaum.

Rodholm, M. (1981). Effects of father-infant post-partum contact on their interaction 3 months after birth. *Early Human Development, 5,* 79–86.

Rodin, Judith, Striegel-Moore, Ruth H., and Silberstein, Lisa. (1985). Women and weight: A normative discontent. In T.B. Sonderegger (Ed.), *Nebraska symposium on motivation: Vol. 32. Psychology and gender.* Lincoln: University of Nebraska Press.

Rogan, Walter J. (1986). PCB's and Cola-colored babies: Japan 1968 and Taiwan 1979. In John J. Sever and Robert L. Brent (Eds.), *Teratogen update: Environmentally induced birth defect risks.* New York: Liss.

Rogers, Carl R. (1961). *On becoming a person.* Boston: Houghton Mifflin.

Rogers, Carl R. (1980). *A way of being.* Boston: Houghton Mifflin.

Rogers, Gayle Thompson. (1985). Nonmarried women approaching retirement: Who are they and when do they retire. In Zena Smith Blau (Ed.), *Current perspectives on aging and the life cycle: Work retirement and social policy.* Greenwich, CT: JAI.

Rogers, Sinclair. (1976). The language of children and adolescents and the language of schooling. In Sinclair Rogers (Ed.), *They don't speak our language.* London: Edward Arnold.

Romaine, Suzanne. (1984). *The language of children and adolescents: The acquisition of communication competence.* Oxford: Blackwell.

Roopnarine, Jaipual L. (1984). Sex-typed socialization in mixed-age preschool classrooms. *Child Development, 55,* 1078–1084.

Roper, R., and Hinde, R.A. (1978). Social behavior in a play group: Consistency and complexity. *Child Development, 49,* 570–579.

Rose, Susan A. (1984). Developmental changes in hemispheres specialization for tactile processing in very young children: Evidence from cross-modal transfer. *Developmental Psychology, 20,* 568–574.

Rosel, Natalie. (1986). Growing old together: Neighborhood, communality among the elderly. In Thomas R. Cole and Sally A. Gadow (Eds.), *What does it mean to grow old?* Durham, NC: Duke University Press.

Rosen, Helen. (1986). *Unspoken grief: Coping with childhood sibling loss.* Lexington, MA: Lexington Books.

Rosenfeld, Albert. (1985). *Prolongevity II: An updated report on the scientific prospects for adding good years to life.* New York: Knopf.

Rosenman, Ray H., and Chesney, Margaret A. (1983). Type A behavior patterns and coronary heart disease. In Richard N. Podell and Michael M. Stewart (Eds.), *Primary prevention of coronary heart disease.* Menlo Park, CA: Addison Wesley.

Rosenman, Ray H., and Chesney, Margaret A. (1985). Type A behavior pattern: Its relationship to coronary heart disease and its modification by behavioral and pharmacological approaches. In Michael R. Zales (Ed.), *Stress in health and disease.* New York: Brunner/Mazel.

Rosow, Irving. (1976). Status and role change throughout the life span. In Robert Binstock and Ethel Shanas (Eds.), *Handbook of aging and the social sciences.* New York: Van Nostrand Reinhold.

Ross, Bruce M., and Kerst, Stephen M. (1978). Developmental memory theories: Baldwin and Piaget. In Hayne W. Reese and Lewis P. Lipsitt (Eds.), *Advances in child development and behavior* (Vol. XII). New York: Academic Press.

Ross, Dorothea M., and Ross, Sheila A. (1982). *Hyperactivity: Current issues, research, and theory* (2nd ed.). New York: Wiley.

Ross, H. Laurence. (1984). *Deterring the drinking driver* (rev. ed.). Lexington, MA: Lexington Books.

Ross, Jennie Keith. (1977). *Old people, new lives: Community creation in a retirement residence.* Chicago: University of Chicago Press.

Ross, Lee. (1981). The "intuitive scientist" formulation and its developmental implications. In John H. Flavell and Lee Ross (Eds.), *Social cognitive development: Frontiers and possible futures.* New York: Cambridge University Press.

Rossi, Alice S. (1980). Life-span theories in women's lives. *Signs, 6,* 4–32.

Rotberg, Iris C. (1981). Federal policy issues in elementary and secondary education. In Robert A. Miller (Ed.), *The federal role in education: New directions for the eighties.* Washington, DC: Institute for Educational Leadership.

Rotberg, Iris C. (1982). Some legal and research considerations in establishing federal policy in bilingual education. *Harvard Educational Review, 52,* 149–168.

Roth, Martin, Wischik, Claude M., Evans, Nicholas, and Mountjoy, Christopher. (1985). Convergence and cohesion of recent neurobiological findings in relation to Alzheimer's disease and their bearing on its aetiological basis. In M. Bergener, Marco Ermini, and H.B. Stahelin (Eds.), *Thresholds in aging.* London: Academic Press.

Roth, Susan, and Cohen, Lawrence J. (1986). Approach, avoidance, and coping with stress. *American Psychologist, 41,* 813–819.

Routh, Donald K., Schroeder, Carolyn S., and O'Tuama, Lorcan A. (1974). Development of activity level in children. *Developmental Psychology, 10,* 163–168.

Rowe, John W., and Kahn, Robert L. (1987). Human aging: Usual and successful. *Science, 237,* 143–149.

Rowe, John W., and Minaker, Kenneth L. (1985). Geriatric medicine. In Caleb E. Finch and Edward L. Schneider (Eds.), *Handbook of the biology of aging* (2nd ed.). New York: Van Nostrand.

Rubin, Kenneth H., Fein, Greata G., and Vandenberg, Brian. (1983). Play. In Paul H. Mussen (Ed.), *Handbook of child psychology: Vol. 4. Socialization, personality and social development.* New York: Wiley.

Rubin, Zick. (1980). *Children's friendships.* Cambridge, MA: Harvard University Press.

Ruff, Holly A. (1982). The development of object perception in infancy. In Tiffany M. Field, Aletha Huston, Herbert C. Quay, Lillian Troll, and Gordon E. Finley (Eds.), *Review of human development.* New York: Wiley.

Russell, Diana E.H. (1984). Sexual exploitation: Rape, child sexual abuse, and workplace harassment. *Sage Library of Social Research, 155*, Beverly Hills, CA: Sage Publications.

Rutter, D.R., and **Durkin, K.** (1987). Turn-taking in mother-infant interaction: An examination of vocalization and gaze. *Developmental Psychology, 23*, 54–61.

Rutter, Michael. (1975). *Helping troubled children.* London: Penguin.

Rutter, Michael. (1979). Protective factors in children's responses to stress and disadvantage. In Martha Whalen Kent and Jon E. Rolf (Eds.), *Primary prevention of psychopathology: Vol. III. Social competence in children.* Hanover, NH: University Press of New England.

Rutter, Michael. (1980). *Changing youth in a changing society: Patterns of development and disorder.* Cambridge, MA: Harvard University Press.

Rutter, Michael. (1981). *Maternal deprivation reassessed.* 2nd ed. Middlesex, England: Penguin.

Rutter, Michael. (1982). Epidemiological-longitudinal approaches to the study of development. In W. Andrew Collins (Ed.), *The concept of development: Minnesota symposia on child psychology* (Vol. 15). Hillsdale, NJ: Erlbaum.

Rutter, Michael. (1982). Socio-emotional consequences of day care for preschool children. In E.F. Zigler and E.W. Gordon (Eds.), *Day care: Scientific and social policy issues.* Boston: Auburn House.

Rutter, Michael. (1987). Psychosocial resilience and protective mechanisms. *American Journal of Orthopsychiatry, 57*, 316–331.

Rutter, Michael, and **Garmezy, Norman.** (1983). Developmental psychopathology. In Paul H. Mussen (Ed.), *Handbook of child psychology: Vol. 4. Socialization, personality and social development*, New York: Wiley.

Rutter, Michael, and **Giller, Henri.** (1984). *Juvenile delinquency: Trends and perspectives.* New York: Guilford.

Rutter, Michael, Maughan, Barbara, Mortimore, Peter, and **Ouston, Janet.** (1979). *Fifteen thousand hours: Secondary schools and their effects on children.* Cambridge, MA: Harvard University Press.

Rybash, John M., Hoyer, William J., and **Roodin, Paul A.** (1986). *Adult cognition and aging: Developmental changes in processing, knowing, and thinking.* New York: Pergamon.

Ryff, Carol D. (1984). Personality development from the inside: The subjective experience of change in adulthood and aging. In Paul B. Baltes and Orville G. Brim Jr. (Eds.), *Life-span development and behavior:* Vol. 6. Orlando, FL: Academic Press.

Sainsbury, Peter. (1986). The epidemiology of suicide. In Alec Roy (Ed.), *Suicide.* Baltimore: Williams and Wilkins.

St. George-Hyslop, Peter H., Tanzi, Rudolph E. Polinsky, Ronald J., and **others.** (1987). Absence of duplication of chromosome 21 genes in familiar and sporadic Alzheimer's disease. *Science, 238*, 664–666.

St. Leger, A.S., Cochrane, A.L., and **Moore, F.** (1979). Factors associated with cardiac mortality in developed countries with particular reference to the consumption of wine. *Lancet, 1*, 8124, 1017–1020.

Salapatek, Philip. (1977). Stimulus determinants of attention in infants. In Benjamin B. Wolman (Ed.), *International encyclopedia of psychiatry, psychology, psychoanalysis, and neurology* (Vol. X). New York: Aesculpaius.

Salthouse, Timothy A. (1985a). *A theory of cognitive aging.* Amsterdam: North-Holland.

Salthouse, Timothy A. (1985b). Speed of behavior and its implications for cognition. In James E. Birren and K. Warner Schaie (Eds.), *Handbook of the psychology of aging* (2nd ed.). New York: Van Nostrand Reinhold.

Salthouse, Timothy A. (1986). Effects of age and skill in typing. *Journal of Experimental Psychology: General, 113*, 345–371.

Sameroff, Arnold J. (1982). Development and the dialectic: The need for a systems approach. In W. Andrew Collins (Ed.), *The concept of development: Minnesota symposia on child psychology* (Vol. 15.). Hillsdale, NJ: Erlbaum.

Sameroff, Arnold J. (1983). Developmental systems: Contexts and evolution. In Paul H. Mussen (Ed.), *Handbook of child psychology: Vol. 1. History, theory, and methods.* New York: Wiley.

Sameroff, Arnold J., and **Seifer, Ronald.** (1983). Familial risk and child competence. *Child Development, 54*, 1254–1268.

Sanford, Nevitt. (1979). Freshman personality: A stage in human development. In Nevitt Sanford and Joseph Axelrod (Eds.), *College and character.* Berkeley, CA: Montaigne.

Santrock, John W., Warshak, Richard A., and **Elliott, Gary L.** (1982). Social development and parent-child interaction in father-custody and stepmother families. In Michael E. Lamb (Ed.), *Nontraditional families: Parenting and child development.* Hillsdale, NJ: Erlbaum.

Santrock, John W. (1972). Relation of type and onset of father absence to cognitive development. *Child Development, 43*, 455–469.

Satz, P., Taylor, A.G., Friel, L., and **Fletcher, J.** (1978). Some developmental and predictive precursor of reading disabilities: A six year follow-up. In A.L. Benton and D. Pearl (Eds.), *Dyslexia: An appraisal of current knowledge.* New York: Oxford University Press.

Saville-Troike, Muriel, McClure, Erica, and **Fritz, Mary.** (1984). Communicative tactics in children's second language acquisition. In Fred R. Eckman, Lawrence H. Bell, and Diane Nelson (Eds.), *Universals of second language acquisition.* Rowley, MA: Newbury House.

Savin-Williams, Ritch C., and **Demo, David H.** (1984). Developmental change and stability in adolescent self-concept. *Developmental Psychology, 20*, 1100–1110.

Sayre, Robert F. (1979). The parents' last lessons. In David D. Van Tassel (Ed.), *Aging, death, and the completion of being.* Philadelphia: University of Pennsylvania Press.

Scanlan, James V. (1975). *Self-reported health behavior and attitudes of youths 12–17 years.* U.S. Department of Public Health Service. Health Education and Welfare, Series 11, Number 147, Publication No. (HRA) 75–1629.

Scarr, Sandra. (1984). *Mother care/Other care.* New York: Basic Books.

Scarr, Sandra. (1985). Constructing psychology: Making facts and fables for our times. *American Psychologist, 40*, 499–512.

Scarr, Sandra, and **McCartney, Kathleen.** (1983). How people make their own environments. A theory of genotype/environmental effects. *Child Development, 54*, 424–435.

Scarr, Sandra, and **Weinberg, Richard A.** (1980). Calling all camps! The war is over. *American Sociological Review, 45*, 859–865.

Schachter, Rubin J., Pantel, Ernestine S., Glassman, George M., and **Zweibelson, Irving.** (1971). Acne Vulgaris and psychological impact on high school students. *New York State Journal of Medicine, 71*, 2886–2890.

Schachter, Stanley. Recidivism and self-cure of smoking and obesity. *American Psychologist, 1982, 37*, 436–444.

Schaffer, H. Rudolf. (1984). *The child's entry into a social world.* New York: Academic Press.

Schaie, K. Warner. (1958). Rigidity-flexibility and intelligence. A cross-sectional study of the adult life-span from 20 to 70. *Psychological Monographs, 72*, No. 462 (Whole no. 9).

Schaie, K. Warner. (1981). Psychological changes from midlife to early old age: Implications for the maintenance of mental health. *American Journal of Orthopsychiatry, 51*, 199–218.

Schaie, K. Warner. (1982). Toward a stage theory of adult cognitive development. In K. Warner Schaie and James Geiwitz (Eds.), *Readings in adult development and aging*. Boston: Little, Brown. (Originally published in *Journal of Aging and Human Development*, 1977–1978.)

Schaie, K. Warner. (1983). The Seattle longitudinal study. A twenty-one year investigation of psychometric intelligence. In K. Warner Schaie (Ed.), *Longitudinal studies of adult psychological development*. New York: Guilford.

Schaie, K. Warner, and Baltes, Paul B. (1977). Some faith helps to see the forest: A final comment on the Horn and Donaldson myth of the Baltes-Schaie position on adult intelligence. *American Psychologist*, 1977, 32, 1118–1120.

Schaie, K. Warner, and Herzog, Christopher. (1983). Fourteen-year cohort-sequential studies of adult intelligence. *Developmental Psychology*, 19, 531–543.

Schaie, K. Warner, and Herzog, Christopher. (1985). Measurement in the psychology of aging. In James E. Birren and K. Warner Schaie (Eds.), *Handbook of the psychology of aging*. New York: Van Nostrand Reinhold.

Schaie, K. Warner, and Labouvie-Vief, Gisela. (1974). Generational versus ontogenetic components of change in adult cognitive behavior: A fourteen-year cross-sequential study. *Developmental Psychology*, 10, 305–320.

Schaie, K. Warner, and Willis, Sherry L. (1986). Can decline in adult intellectual functioning be reversed? *Developmental Psychology*, 22, 223–232.

Schanberg, Saul M., Evonick, G., and Kuhn, Cynthia M. (1984). Tactile and nutritional aspects of maternal care: Specific regulators of neuroendocrine function and cellular development. *Proceedings of the Society for Experimental Biology and Medicine*, 175, 135.

Schanberg, Saul M., and Kuhn, Cynthia M. (1980). Maternal deprivation: An animal model of psychosocial dwarfism. In E. Usdin, T.L. Sourkes, and M.B.H. Youdim (Eds.), *Enzymes and neurotransmitters in mental disease*. New York: Wiley.

Scher, Jonathan, and Dix, Carol. (1983). *Will my baby be normal?: Everything you need to know about pregnancy*. New York: Dial Press.

Schick, Bela, and Rosenson, William. (1932). *Child care today*. New York: Greenberg.

Schick, Frank L. (Ed.). (1986). *Statistical handbook on aging Americans*. Phoenix, AZ: Oryx Press.

Schiefelbusch, Richard L. (1984). Assisting children to become communicatively competent. In Richard L. Schiefelbusch and Joanne Pickar (Eds.), *The acquisition of communicative competence*. Baltimore, MD: University Park Press.

Schiefelbusch, Richard L, and Pickar, Joanne (Eds.). (1984). *The acquisition of communicative competence*. Baltimore, MD: University Park Press.

Schieffelin, Bambi, B., and Eisnenberg, Ann R. (1984). Cultural variation in children's conversations. In Richard L. Schiefelbusch and Joanne Pickar (Eds.), *The acquisition of communicative competence*. Baltimore, MD: University Park Press.

Schlesinger, Benjamin. (1982). *Sexual abuse of children*. Toronto: University of Toronto Press.

Schneider, D.M. (1980). *American kinship: A cultural account*. Englewood Cliffs, NJ: Prentice-Hall.

Schneider, Edward L., and Reed, John D. (1985). Modulations of the aging process. In Caleb E. Finch and Edward L. Schneider (Eds.), *Handbook of the biology of aging* (2nd ed.). New York: Van Nostrand.

Schneidman, Edwin S. (1978). Suicide. In Gardner Lindzey, Calvin S. Hall, and Richard F. Thompson, *Psychology* (2nd ed.). New York: Worth.

Schoof-Tams, Karin, Schlaegel, Jürgen, and Walezak, Leonhard. (1976). Differentiation of sexual morality between 11 and 16 years. *Archives of Sexual Behavior*, 5, 353–370.

Schulz, Richard, and Aderman, David. (1978–1979). Physicians' death anxiety and patient outcomes. *Omega*, 9, 327–332.

Schumm, Walter R., and Bugaighis, Margaret A. (1986). Marital quality over the marital career. *Journal of Marriage and the Family*, 48, 165–168.

Schwartz, Marvin. (1978). *Physiological psychology* (2nd ed.). Englewood Cliffs, N.J.: Prentice Hall.

Schwartz, M., and Day, R.H. (1979). Visual shape perception in infancy. *Monographs of the Society for Research in Child Development*, 44 (7, Serial No. 182).

Schwertzbeck. (1983). In Jerome E. Leavitt (Ed.), *Child abuse and neglect: Research and innovation*. The Hague: Martinus Nijhoff.

Scott-Maxwell, Florida. (1968). *The measure of my days*. New York: Knopf.

Scribner, S., and Cole, M. (1981). *The consequences of literacy*. Cambridge, MA: Harvard University Press.

Sears, P.S., and Barbee, A.H. (1977). Career and life satisfaction among Terman's gifted women. In J.C. Stanley, W.C. George, and C.H. Solano (Eds.), *The gifted and the creative: A fifty-year perspective*. Baltimore, MD: Johns Hopkins University Press.

Sears, Robert R. (1977). Sources of life satisfaction of the Terman gifted men. *American Psychologist*, 32, 119–138.

Sears, Robert R., Rau, Lucy, and Alpert, Richard. (1965). *Identification and child rearing*. Stanford, CA: Stanford University Press.

Seashore, Marjorie J., Leifer, Aimee Dorr, Barnett, Clifford R., and Leiderman, P. Herbert. (1973). The effects of denial of early mother-infant interaction on maternal self-confidence. *Journal of Personality and Social Psychology*, 26, 369–378.

Segall, M.H. (1980). *Cross-cultural psychology: An introduction*. Monterey, CA: Brooks/Cole.

Segerberg, Osborn. (1982). *Living to be 100: 1,200 who did and how they did it*. New York: Scribners.

Segre, Diego, and Smith, Lester. (1981). *Immunological aspects of aging*. New York: Dekker.

Seitz, Victoria, Rosenbaum, Laurie K., and Apfel, Nancy H. (1985). Effects of family support intervention: A ten year follow-up. *Child Development*, 56, 376–391.

Serbin, L.A., Tronick, I.J., and Sternglanz, S. (1977). Shaping cooperative cross-sex play. *Child Development*, 48, 924–929.

Sever, John L., and Brent, Robert L. (Eds.). (1986). *Teratogen update: Environmentally induced birth defect risks*. New York: Liss.

Seyle, Hans. (1982). History and present status of the stress concept. In L. Goldberger and S. Breznitz (Eds.), *Handbook of stress: Theoretical and clinical aspects*. New York: Free Press.

Shanas, Ethel. (1980). Older people and their families. The new pioneers. *Journal of Marriage and the Family*, 42, 9–15.

Shantz, Carolyn Uhlinger. (1983). Social cognition. In Paul H. Mussen (Ed.), *Handbook of child psychology: Vol. 3. Cognitive development*. New York: Wiley.

Shea, John D.C. (1981). Changes in interpersonal distances and categories of play behavior in the early weeks of preschool. *Developmental Psychology*, 17, 417–425.

Sheehy, Gail. (1976). *Passages: Predictable crisis of adult life*. New York: Dutton.

Sheehy, Gail. (1981). *Pathfinders*. New York: Morrow.

Shell, Ellen Ruppel. (1982). The guinea pig town. *Science 82*, 3, 58–63.

Shephard, Roy J. (1976). Physiology–Comment. In J.G. Albinson and G.M. Andrew (Eds.), *Child in sport and physical activity* (pp. 35–40). Baltimore: University Park Press.

Sherman, Julia. (1982). Continuing in mathematics: A longitudinal study of the attitudes of high school girls. *Psychology of Women Quarterly, 7*, 132–140.

Sherrod, Kathryn B., O'Connor, Susan, Vietze, Peter M., and **Altermeier, William A.** (1984). Child health and maltreatment. *Child Development, 55*, 1174–1183.

Shields, James. (1962). *Monozygotic twins, brought up apart and brought up together. An investigation into the genetic and environmental causes of variation in personality.* London: Oxford University Press.

Shinn, Marybeth. (1978). Father absence and children's cognitive development. *Psychological Bulletin, 85*, 295–324.

Shipp, E.E. (1985, November 4). Teen-agers taking risks: When pregnancy is the result. *The New York Times*, p. A 16.

Shirley, Mary M. (1933). The first two years: A study of twenty-five babies. *Institute of Child Welfare Monograph No. 8.* Minneapolis: University of Minnesota Press.

Shock, Nathan W. (1985). Longitudinal studies of aging in humans. In Caleb E. Finch and Edward L. Schneider (Eds.), *Handbook of the biology of aging* (2nd ed.). New York: Van Nostrand.

Shock, Nathan W., Greulich, Richard C., Andres, Reuben, Arenberg, David, Costa, Paul T. Jr., Lakatta, Edward G., and **Tobin, Jordan D.** (1984). *Normal human aging: The Baltimore longitudinal study.* Washington, DC: U.S. Government Printing Office. NIH Publication, No. 84-2450.

Sigel, Irving E., Dreyer, Albert S., and **McGillicuddy-DeLisi, Ann V.** (1984). Psychological perspectives on the life course. In Ross D. Parke (Ed.), *Review of child development research: Vol. 7. The family.* Chicago: University of Chicago Press.

Sill., J.S. (1980). Disengagement reconsidered: Awareness of finitude. *Gerontologist, 20*, 457–462.

Silverman, W.A. (1980). *Retrolental fibroplasia: A modern parable.* New York: Grune and Stratton.

Simenauer, Jacqueline, and **Carroll, David.** (1982). *Singles: The new Americans.* New York: Simon and Schuster.

Simmons, Roberta G., Blyth, Dale A., and **McKinney, Karen L.** (1983). The social and psychological effects of puberty on white females. In Jeanne Brooks-Gunn and Anne C. Petersen (Eds.), *Girls at puberty: Biological and psychosocial aspects.* New York: Plenum.

Simmons, Roberta G., Rosenberg, Florence, and **Rosenberg, Morris.** (1973). Disturbance in the self-image at adolescence. *American Sociological Review, 38*, 553–568.

Sinclair, Caroline B. (1973). *Movement of the young child: Ages two to six.* Columbus, Ohio: Merrill.

Sinclair, David. (1978). *Human growth after birth* (3rd ed.). London: Oxford University Press.

Singer, Dorothy G., and **Singer, Jerome L.** (1977). *Partners in play: A step-by-step guide to imaginative play in children.* New York: Harper & Row.

Sinnott, Jan D. (1986). *Sex roles and aging: Theory and research from a systems perspective.* Basel, Switzerland: Karger.

Sizer, Theodore R. (1985). *Horace's compromise: The dilemma of the American high school.* Boston: Houghton Mifflin.

Skinner, B.F. (1953). *Science and human behavior.* New York: Macmillan.

Skinner, B.F. (1957). *Verbal behavior.* New York: Appleton-Century-Crofts.

Skinner, B.F. (1972). *Beyond freedom and dignity.* New York: Knopf.

Skinner, B.F. (1980). The experimental analysis of operant behavior: A history. In R.W. Riebes and K. Salzinger (Eds.), *Psychology: Theoretical-historical perspective.* New York: Academic Press.

Skinner, B.F. (1983). *A matter of consequences: Part 3 of an autobiography.* New York: Knopf.

Skunkard, A.J. (1972). New treatments for obesity. In G.A. Bray and J.E. Bethune (Eds.), *Treatment and management of obesity.* New York: Harper & Row.

Slade, Arietta. (1987). A longitudinal study of maternal involvement and symbolic play during the toddler period. *Child Development, 58*, 367–375.

Slaughter, Diana T. (1983). Early intervention and its effects on maternal and child development. *Monographs of the Society for Research in Child Development, 48* (Serial No. 202).

Sloane, Ethel. (1985). *Biology and women* (2nd ed.). New York: Wiley.

Slobin, Dan I. (1971). *Psycholinguistics.* Glenview, IL: Scott Foresman.

Slocum, John W., and **Cron, William L.** (1985). Job attitudes and performance during three career stages. *Journal of Vocational Behavior, 26*, 126–145.

Smith, Ken R., and **Zick, Cathleen D.** (1986). The incidence of poverty among the recently widowed: Mediating factors in the life course. *Journal of Marriage and the Family, 48*, 619–630.

Smith, M. Brewster. (1983). Hope and despair: Keys to the sociopsychodynamics of youth. *American Journal of Orthopsychiatry, 53*, 388–399.

Smith, Peter K. (1978). A longitudinal study of social participation in preschool children: Solitary and parallel play reexamined. *Developmental Psychology, 12*, 517–523.

Smith, Peter K. (1984). *Play in animals and humans.* Basil: Blackwood.

Smith, R. Jeffrey. (1978a). Agency drags its feet on warning to pregnant women. *Science, 199*, 748–749.

Smokler, C.S. (1982). Self-esteem in preadolescent and adolescent females. Unpublished doctoral dissertation, University of Michigan, 1975. Cited in Michael E. Lamb, Maternal employment and child development: A review. In Michael E. Lamb (Ed.), *Nontraditional families: Parenting and child development.* Hillsdale, NJ: Erlbaum.

Smyser, A.A. (1982). Hospices: Their humanistic and economic value. *American Psychologist, 37*, 1260–1262.

Snarey, John R., Reimber, Joseph, and **Kohlberg, Lawrence.** (1985). Development of social-moral reasoning among Kibbutz adolescents: A longitudinal cross-cultural study. *Developmental Psychology, 21*, 3–17.

Snow, Catherine E. (1984). Parent-child interaction and the development of communicative ability. In Richard L. Schiefelbusch and Joanne Pickar (Eds.), *The acquisition of communicative competence.* Baltimore: University Park Press.

Snow, Catherine E., and **Ferguson, Charles A.** (Eds.). (1977). *Talking to children.* Cambridge, England: Cambridge University Press.

Snyder, J., Dishion, T.J., and **Patterson, Gerald R.** (1986). Determinants and consequences of associating with deviant peers during preadolescence and adolescence. *Journal of Early Adolescence, 6*, 29–43.

Snyder, Lynn S. (1984). Communicative competence in children with delayed language development. In Richard L. Schiefelbusch and Joanne Pickar (Eds.), *The acquisition of communicative competence.* Baltimore: University Park Press.

Snyder, S. (1978). Dopamine and schizophrenia. In L. Wynne, R. Cromwell, and S. Matthysse (Eds.), *The nature of schizophrenia: New approaches to research and treatment.* New York: Wiley.

Solnit, Albert J. (1980). Preface to the third edition. In C. Henry

Kempe and Ray E. Helfer (Eds.), *The battered child* (3rd ed.). Chicago: The University of Chicago Press.

Song, Myung-Ja, Smetana, Judith G., and Kim, Sang Yoon. (1987). Korean children's conception of moral and conventional transgressions. *Developmental Psychology, 23,* 577–582.

Sonnenschein, Susan. (1984). How feedback from a listener affects children's referential communication skills. *Developmental Psychology, 20,* 287–292.

Sorenson, Marie. (1977). *Move over, mama.* Modesto, CA: Osmar Press.

Sosa, R., Kennell, J., Klaus, Marshall, and Urutia, J. (1976). The effects of early mother-infant contact on breast-feeding, infection, and growth. In *Breast-feeding and the mother. CIBA Foundation Symposium No. 45.* Amsterdam: Associated Scientific.

Southard, B. (1985). Interlimb movement control and coordination in children. In Jane E. Clark and James H. Humphrey (Eds.), *Motor development: Current selected research.* Princeton, NJ: Princeton Book Company.

Spearman, Charles. (1927). *The abilities of man.* New York: Macmillan.

Spelke, Elizabeth S., and Owsley, Cynthia. (1979). Intermodal exploration and knowledge in infancy. *Infant Behavior and Development, 2,* 13–27.

Spence, Janet T., and Helmreich, Robert L. (1978). *Masculinity and femininity: Their psychological dimensions, correlates, and antecedents.* Austin: University of Texas Press.

Spitz, René Arpad. (1945). Hospitalism: An inquiry into the genesis of psychiatric conditions in early childhood. *Psychoanalytic Study of the Child, 1,* 53–74.

Sprague, R.L., and Ullman, R.K. (1981). Psychoactive drugs and child management. In J.M. Kaufman and D.P. Hallahan (Eds.), *Handbook of special education.* New York: Prentice-Hall.

Squire, S. (1983). *The slender balance: Causes and cures for bulimia, anorexia, and the weight loss/weight gain seesaw.* New York: Putman.

Sroufe, L. Alan. (1978). Attachment and the roots of competence. *Human Nature, 1978, 1,* 50–57.

Sroufe, L. Alan. (1979). Socioemotional development. In Joy Osofsky (Ed.), *Handbook of infant development.* New York: Wiley.

Sroufe, L. Alan. (1985). Attachment classification from the perspective of infant-caregiver relationships and infant temperament. *Child Development, 56,* 1–14.

Sroufe, L. Alan, Fox, Nancy E., and Pancake, Van R. (1983). Attachment and dependency in developmental perspective. *Child Development, 54,* 1615–1627.

Sroufe, L. Alan, Jacobvitz, Deborah, Mengelsdorf, Sarah, DeAngelo, Edward, and Ward, Mary Jo. (1985). Generational boundary dissolution between mothers and their preschool children: A relationship systems approach. *Child Development, 56,* 317–325.

Sroufe, L. Alan, and Ward, Mary Jo. (1980). Seductive behavior of mothers of toddlers. Occurrence, correlates, and family origins. *Child Development, 51,* 1222–1229.

Sroule, L., and Fischer, A.K. (1980). The midtown Manhattan longitudinal study vs. the mental paradise lost doctrine. *Archives of General Psychiatry, 37,* 209–221.

Sroule, L., Langner, T.S., Michael, S.T., Opler, M.K., and Rennie, T. (1962). *Mental health in the metropolis: The midtown study.* New York: McGraw-Hill.

Staines, Graham L., Pottick, Kathleen J., and Fudge, Deborah A. (1986). Wives' employment and husbands' attitude toward work and life. *Journal of Applied Psychology, 71,* 118–128.

Stanton, M. Duncan. (1985). The family and drug abuse: Concepts and rationale. In Thomas E. Bratter and Gary G. Forrest (Eds.), *Alcoholism and substance abuse: Strategies for clinical intervention.* New York: The Free Press.

Stark, C.R., Orleans, M., Haverkamp, A.D., and Murphy, J. (1984). Short and long term risks after exposure to diagnostic ultrasound in utero. *Obstetrical Gynecology, 63,* 194–200.

Starr, Bernard D., and Weiner, Marcella Baker. (1981). *Sex and sexuality in the mature years.* New York: Stein and Day.

Steele, Brandt. (1980). Psychodynamic factors in child abuse. In C. Henry Kempe and Ray E. Helfer (Eds.), *The battered child* (3rd ed.). Chicago: The University of Chicago Press.

Stein, Aletha Huston, and Friedrich, Lynette Kohn. (1975). Impact of television on children and youth. In E. Mavis Hetherington (Ed.), *Review of child development research* (Vol. V). Chicago: University of Chicago Press.

Steinberg, Laurence D. (1977). *A longitudinal study of physical growth, intellectual development, and family interaction in early adolescence.* Unpublished doctoral dissertation. Cornell University.

Steinberg, Laurence. (1986). Stability (and instability) of Type A behavior from childhood to young adulthood. *Developmental Psychology, 22,* 393–402.

Steinberg, Laurence. (1987). Impact of puberty on family relations: Pubertal status and pubertal timing. *Developmental Psychology, 23,* 451–460.

Steinberg, Laurence. (1987a). Single parents, stepparents, and the susceptibility of adolescents to antisocial peer pressure. *Child Development, 58,* 269–275.

Steinmetz, Suzanne K., and Amsden, Devorah J. (1983). Dependent elders, family stress, and abuse. In Timothy H. Brubaker (Ed.), *Family relationships in later life.* Beverly Hills, CA: Sage.

Stenberg, Craig, and Campos, Joseph J. (1983). The development of the expression of anger in human infants. In Michael Lewis and Saarni, Carolyn (Eds.), *The socialization of affect.* New York: Plenum.

Stephens, Joyce. (1976). *Loners, losers, and lovers: Elderly tenants in a slum hotel.* Seattle: University of Washington Press.

Stephens, Richard C. (1987). *Mind-altering drugs: Use, abuse, and treatment.* Newbury Park, CA: Sage.

Stern, Daniel. (1977). *The first relationship: Mother and infant.* Cambridge, MA: Harvard University Press.

Stern, J.A., Oster, P.J., and Newport, K. (1980). Reaction time measures, hemispheric specialization, and age. In Leonard W. Poon (Ed.), *Aging in the 80's: Psychological issues.* Washington, DC: American Psychological Association.

Stern, Leonard. (1985). *The structures and strategies of human memory.* Homewood, IL: Dorsey Press.

Sternberg, Robert J. (1986). A triangular theory of love. *Psychological Review, 93,* 119–135.

Stevenson, Harold. (1983). How children learn—the quest for a theory. In Paul H. Mussen (Ed.), *Handbook of child psychology: Vol. 1. History, theory, and methods.* New York: Wiley.

Stewart, A., and Kneale, G.W. (1970). Radiation dose effects in relation to obstetric X-rays and childhood cancers. *Lancet, I,* 1495.

Stipek, Deborah J. (1984). Sex differences in children's attributions of success and failure on math and spelling tests. *Sex Roles, 11,* 969–981.

Stipek, Deborah J., and Hoffman, J. (1980). Development of children's performance-related judgments. *Child Development, 51,* 912–914.

Stipek, Deborah J., Roberts, Theresa A., and Sanborn, Mary E. (1984). Preschool-age children's performance expectations for themselves and another child as a function of the incentive value of success and the salience of past performance. *Child Development, 55,* 1983–1989.

Stoddart, Trish, and **Turiel, Elliot.** (1985). Children's concepts of cross-gender activities. *Child Development, 56,* 1241–1252.

Stone, L. Joseph, and **Church, Joseph.** (1973). *Childhood and adolescence: A psychology of the growing person* (3rd ed.). New York: Random House.

Strain, Phillip S., and **Shores, Richard E.** (1983). A reply to "misguided mainstreaming." *Exceptional Children, 50,* 271–273.

Strauss, Cyd, Smith, Karen, Frame, Cynthia, and **Forehand, Rex.** (1985). Personal and interpersonal characteristics associated with childhood obesity. *Journal of Pediatric Psychology, 10,* 337–344.

Strehler, Bernard L. (1977). *Time, cells, and aging.* New York: Academic Press.

Strehler, Bernard L. (1979). The future and aging research. In Arthur Cherkin (Ed.), *Physiology and cell biology of aging.* New York: Raven Press.

Streissguth, Ann Pytkowicz, Barr, Helen M., and **Martin, Donald C.** (1983). Maternal alcohol use and neonatal habituation assessed with the Brazelton scale. *Child Development, 54,* 1109–1118.

Streissguth, Ann Pytkowicz, Martin, Donald C., Barr, Helen M., Sandman, Beth MacGregor, Kirshner, Grace L., and **Darby, Betty L.** (1984). Intrauterine alcohol and nicotine exposure: Attention and reaction time in 4-year-old children. *Developmental Psychology, 20,* 533–541.

Striegel-Moore, Ruth H., Silberstein, Lisa, and **Rodin, Judith.** (1986). Understanding of risk factors for bulimia. *American Psychologist, 41,* 246–263.

Stroebe, Margaret S., and **Stroebe, Wolfgang.** (1983). Who suffers more? Sex differences in health risks of the widowed. *Psychology Bulletin, 93,* 279–301.

Stueve, Ann, and **O'Donnell, Lydia.** (1984). The daughter of aging parents. In Grace K. Baruch and Jeanne Brooks-Gunn (Eds.), *Women in midlife.* New York: Plenum.

Sullivan, Edmund V. (1977). A study of Kohlberg's structural theory of moral development: A critique of liberal social science ideology. *Human Development, 20,* 352–376.

Sullivan, Margaret Wolan. (1982). Reactivation: Priming forgotten memories in human infants. *Child Development, 53,* 516–523.

Sun, Marjorie. (1983). FDA draws criticism on prenatal test. *Science, 221,* 440–442.

Sunderland, Alan, Watts, Kathryn, Baddeley, Alan D., and **Harris, John E.** (1986). Subjective memory assessment and test performance in elderly adults. *Journal of Gerontology, 41,* 376–384.

Suomi, Stephen J., and **Harlow, Harry F.** (1976). Monkeys without play. In Jerome S. Bruner, Alison Jolly, and Kathy Sylva (Eds.), *Play.* New York: Basic Books.

Super, Charles M., and **Harkness, Sara.** (1982). The development of affect in infancy and early childhood. In Daniel A. Wagner and Harold W. Stevenson (Eds.), *Cultural perspectives on child development.* San Francisco: Freeman.

Super, Donald E. (1957). *The psychology of careers.* New York: Harper & Row.

Surgeon General of the United States. (1982). *The health consequences of smoking: Cancer.* Washington, DC: United States Department of Health and Human Services.

Sutcliffe, D. (1982). *British Black English.* Oxford: Blackwell.

Sutton-Smith, Brian. (1986). *Toys as culture.* New York: Gardner Press.

Swenson, Clifford H., Eskew, Ron W., and **Kohlhepp, Karen A.** (1981). Stages of the family life cycle, ego development, and the marriage relationship. *Journal of Marriage and the Family, 43,* 841–853.

Swensen, Clifford H., and **Tranaug, Geir.** (1985). Commitment and the long-term marriage relationship. *Journal of Marriage and the Family, 17,* 939–945.

Svanborg, Alvar. (1985). Biomedical and environmental influences on aging. In Robert N. Butler and Hebert P. Gleason (Eds.), *Productive aging: Enhancing vitality in later life.* New York: Springer.

Tanfer, Koray. (1987). Patterns of premarital cohabitation among never-married women in the United States. *Journal of Marriage and the Family, 49,* 483–497.

Tannenbaum, Abraham J. (1983). *Gifted children: Psychological and educational perspectives.* New York: Macmillan.

Tanner, James M. (1970). Physical growth. In Paul H. Mussen (Ed.), *Carmichael's manual of child psychology* (3rd ed.). Vol. I. New York: Wiley.

Tanner, James M. (1971). Sequence, tempo, and individual variation in the growth and development of boys and girls aged twelve to sixteen. *Daedalus, 100,* 907–930.

Tanner, James M. (1978). *Fetus into man: Physical growth from conception to maturity.* Cambridge, MA: Harvard University Press.

Tanzer, Deborah, and **Block, Jean Libman.** (1976). *Why natural childbirth? A psychologist's report on the benefits to mothers, fathers and babies.* New York: Schocken.

Taylor, P.M., Taylor, F.H., Campbell, S.B., Maloni, J., and **Dickey, D.** (1979, March). *Effects of extra contact on early maternal attitudes, perceptions, and behaviors.* Paper presented at the biennial meeting of the Society for Research in Child Development, San Francisco.

Taylor, Robert Joseph. (1986). Family support among black Americans. *Journal of Marriage and the Family, 48,* 67–78.

Telegdy, G.A. (1975). The effectiveness of four readiness tests as predictors of first grade academic achievement. *Psychology in the Schools, 12,* 4–11.

Terkel, Studs. (1974). *Working: People talk about what they do all day and how they feel about what they do.* New York: Pantheon.

Terman, Lewis M., and **Oden, Melita H.** (1959). *Genetic studies of genius Vol. 5: The gifted group at mid-life: Thirty-five years' follow-up of the superior child.* Stanford, CA: Stanford University Press.

Tesfaye, Andargatchew. (1976). Ethiopia. In Dea H. Chang (Ed.), *Criminology in a cross-cultural perspective.* Durham, NC: Carolina Academic Press.

Thomas, Alexander. (1981). Current trends in developmental theory. *American Journal of Orthopsychiatry, 51,* 580–609.

Thomas, Alexander, and **Chess, Stella.** (1977). *Temperament and development.* New York: Brunner/Mazel.

Thomas, Alexander, and **Chess, Stella.** (1980). *The dynamics of psychological development.* New York: Brunner/Mazel.

Thomas, Alexander, Chess, Stella, and **Birch, Herbert G.** (1963). *Behavioral individuality in early childhood.* New York: New York University Press.

Thomas, Alexander, Chess, Stella, and **Birch, Herbert G.** (1968). *Temperament and behavior disorders in children.* New York: New York University Press.

Thomas, Alexander, Chess, Stella, and **Mendez, O.** (1974). Cross-cultural study of behavior in children with special vulnerabilities to stress. In D. Ricks, Alexander Thomas, and M. Roff (Eds.), *Life history research in psychopathology* (Vol. 3). Minneapolis: University of Minnesota Press.

Thomas, Eugene L. (1982). Sexuality and aging: Essential vitamin or popcorn? *The Gerontologist, 22,* 240–243.

Thomas, Jeanne L. (1986). Age and sex differences in perceptions of grandparents. *Journal of Gerontology, 41,* 417–423.

Tomasello, Michael, and **Farrar, Michael Jeffrey.** (1986). Joint attention and early language. *Child Development, 57,* 1454–1463.

Thompson, M.E., Hartsock, G., and **Farson, C.** (1979). The importance of immediate postnatal contact: Its effect on breast-feeding. *Canadian Family Physician, 25,* 1374–1378.

Thompson, Michael. (1983). Organizing a human services network for primary and secondary prevention of the emotional and physical neglect and abuse of children. In Jerome E. Leavitt (Ed.), *Child abuse and neglect: Research and innovation.* The Hague: Martinus Nijhoff.

Thompson, Spencer K. (1975). Gender labels and early sex role development. *Child Development, 46,* 339–347.

Thornburg, Hershel D., and **Aras, Ziya.** (1986). Physical characteristics of developing adolescents. *Journal of Adolescent Research, 1,* 47–78.

Tice, Raymond R., and **Setlow, Richard B.** (1985). DNA repair and replication in aging organisms and cells. In Caleb E. Finch and Edward L. Schneider (Eds.), *Handbook of the biology of aging* (2nd ed.). New York: Van Nostrand.

Tognoli, J. (1980). Male friendship and intimacy across the life span. *Family Relations, 29,* 273–279.

Torrence, F. Paul. (1972). Characteristics of creatively gifted children and youth. In E. Philip Trapp and Philip Himelstein (Eds.), *The exceptional child.* New York: Appleton Century Crofts, pp. 273–291.

Tough, Allen. (1982). *Intentional changes: A fresh approach to helping people change.* Chicago: Follet.

Touliatos, John, and **Compton, Norma H.** (1983). *Approaches to child study.* Minneapolis, MN: Burgess.

Tower, Roni Beth, Singer, Dorothy G., Singer, Jerome L., and **Biggs, Ann.** (1979). Differential effects of television programming on preschoolers' cognition, imagination, and social play. *American Journal of Orthopsychiatry, 49,* 265–281.

Traupmann, Jane, and **Hatfield, Elaine.** (1981). Love and its effect on mental and physical health. In Robert W. Fogel, Elaine Hatfield, Sara B. Kiesler, and Ethel Shanas (Eds.), *Aging: Stability and change in the family.* New York: Academic Press.

Troll, Lillian. (1980). Grandparenting. In Leonard W. Poon (Ed.), *Aging in the 80's: Psychological issues.* Washington, DC: American Psychological Association.

Troll, Lillian E. (1986). Parents and children in later life. *Generations, 10,* (4), 23–25.

Troll, Lillian E., and **Perron, E.M.** (1981). Age changes in sex roles amid changing sex roles: The double shift. In Carl Eisdorfer (Ed.), *Annual Review of Gerontology and Geriatrics,* Vol. 2, pp. 118–143. New York: Springer.

Tronick, Edward Z., Als, H., Adamson, L., Wise, S., and **Brazelton, T.B.** (1978). The infant's response to entrapment between contradictory measures in face-to-face interaction. *Journal of the American Academy of Child Psychiatry, 17,* 1–13.

Turiel, Elliot. (1974). Conflict and transition in adolescent moral development. *Child Development, 45,* 14–29.

Turiel, Elliot. (1983). *The development of social knowledge: Morality and convention.* Cambridge, England: Cambridge University Press.

Turiel, Elliot, and **Smetana, Judith G.** (1984). Social knowledge and action: The coordination of domains. In William M. Kurtines and Jacob L. Gewirtz (Eds.), *Morality, moral behavior, and moral development.* New York: Wiley.

Udry, J. Richard. (1981). Marital alternatives and marital disruption. *Journal of Marriage and the Family, 43,* 889–897.

Uhlenberg, P., and **Myers, M.A.** (1981). Divorce and the elderly. *The Gerontologist, 21,* 276–282.

Ulrich, Beverly D., and **Ulrich, Dale A.** (1985). The role of balancing ability in performance of fundamental motor skills in 3-, 4-, and 5-year-old children. In Jane E. Clark and James H. Humphrey (Eds.), *Motor development: Current selected research.* Princeton, NJ: Princeton Book Company.

UNESCO. (1985). *Demographic Yearbook.* New York: United Nations.

Ungerer, Judy A., and **Sigman, Marian.** (1983). Developmental lags in preterm infants from one to three years. *Child Development, 54,* 1217–1228.

United States Bureau of the Census. (1976). *Historical statistics of the United States: Colonial times to 1970.* Washington, DC: United States Department of Commerce.

United States Bureau of the Census. (1986). *Statistical abstract of the United States, 1986.* Washington, DC: United States Department of Commerce.

United States Bureau of the Census. (1987). *Statistical abstract of the United States, 1987.* Washington, DC: United States Department of Commerce.

United States Department of Commerce. (July, 1987). Money, income, and poverty status of families and persons in the United States. Washington, DC: *Current population reports: Consumer income.* Series P-60, No. 157.

United States Department of Justice. (1984). *Crime in the United States.* Washington, DC: Federal Bureau of Investigation.

Uzgiris, Ina C., and **Hunt, J. McVicker.** (1975). *Assessment in infancy: Ordinal scales of psychological development.* Urbana: University of Illinois Press.

Vachon, M.L.S., Lyall, W.A.L., Rogers, J., Freedman-Letofsky, K., and **Freeman, S.J.A.** (1980). A controlled study of self-help intervention for widows. *American Journal of Psychiatry, 137,* 1380–1384.

Vaillant, George E. (1977). *Adaptation to life.* Boston: Little, Brown.

Vandenberg, Brian. (1978). Play and development from an ethological perspective. *American Psychologist, 33,* 724–738.

Vandenberg, Steven G. (1987). Sex differences in mental retardation and their implications for sex differences in ability. In June Machover Reinisch, Leonard A. Rosenblum, and Stephanie A. Sanders (Eds.), *Masculinity/femininity: Basic perspectives.* New York: Oxford University Press.

Vandenberg, Steven G., Singer, Sandra Manes, and **Pauls, David L.** (1986). *The heredity of behavior disorders in adults and children.* New York: Plenum.

Van Keep, Pieter A., and **Gregory, Ann.** (1977). Sexual relations in the ageing female. In John Money and Herman Musaph (Eds.), *Handbook of sexology.* Great Britain: Elsevier/North-Holland Biomedical Press.

Van Maanen, John, and **Schein, Edgar H.** (1977). Career development. In J. Richard Hackman and J. Lloyd Suttle (Eds.), *Improving life at work.* Santa Monica, CA: Goodyear.

Van Oeffelen, Michiel P, and **Vos, Peter G.** (1984). The young child's processing of dot patterns: A chronometric and eye movement analysis. *International Journal of Behavioral Development, 7,* 53–66.

Vaughan, Victor C. III. (1983). Developmental pediatrics. In Richard E. Behrman and Victor C. Vaughan, III (Eds.), *Pediatrics.* Philadelphia: Saunders.

Vaughn, Brian E. (1987). Maternal characteristics measured prenatally are predictive of ratings of temperamental "difficulty" on the

Carey Infant Temperament Questionnaire. *Developmental Psychology, 23*, 152–161.

Veevers, Jean E. (1980). *Childless by choice.* Toronto: Butterworth.

Vega, William A., Kolody, Bohdan, and **Valle, Juan Ramon.** (1986). The relationship of marital status, confidant support, and depression among Mexican immigrant women. *Journal of Marriage and the Family, 48*, 597–605.

Verbrugge, L.M. (1979). Marital status and health. *Journal of Marriage and the Family, 41*, 267–285.

Verbrugge, L.M. (1983). Multiple roles and physical health of women and men. *Journal of Health and Social Behavior, 24*, 16–30.

Verillo, Ronald T., and **Verillo, Violet.** (1985). Sensory and perceptual performance. In Neil Charness (Ed.), *Aging and human performance.* New York: Wiley.

Veroff, Joseph, and **Veroff, Joanne B.** (1980). *Social incentives: A life-span developmental approach.* New York: Academic Press.

Vestal, Robert E., and **Dawson, Gary W.** (1985). Pharmacology and aging. In Caleb E. Finch and Edward L. Schneider (Eds.), *Handbook of the biology of aging* (2nd ed.). New York: Van Nostrand.

Vickery, Florence E. (1978). *Old and growing.* Springfield, Ill.: Thomas.

Visher, J.S., and **Visher, E.B.** (1982). Stepfamilies and stepparenting. In R. Walch (Ed.), *Normal family processes.* New York: Guilford.

Volunteer. (1984). *Voluntary action leadership.* Arlington, VA: National Center for Citizen Involvement.

von Hofsten, Claes. (1983). Catching skills in infancy. *Journal of Experimental Psychology: Human Perception and Performance, 9*, 75–85.

Vosniadou, Stella. (1987). Children and metaphors. *Child Development, 58*, 870–885.

Vurpillot, Elaine. (1968). The development of scanning strategies, and their relation to visual differentiation. *Journal of Experimental Child Psychology, 6*, 632–650.

Waber, Deborah P. (1976). Sex differences in mental abilities: A function of maturation rate? *Science, 192*, 572–574.

Wachs, Theodore D. (1975). Relation of infant performance on Piaget scales between twelve and twenty-four months and their Stanford-Binet performance at thirty-one months. *Child Development, 46*, 929–935.

Wadsworth, M.E.J. (1979). *Roots of delinquency: Infancy, adolescence, and crime.* Oxford: Martin Robertson.

Walaskay, M., Whitbourne, S.K., and **Nehrke, M.F.** (1983–1984). Construction and validation of an ego-integrity status interview. *International Journal of Aging and Human Development, 18*, 61–72.

Waldrop, M.F., and **Halverson, C.F. Jr.** (1971). Minor physical anomalies and hyperactive behavior in young children. In J. Hellmuth (Ed.), *The exceptional infant* (Vol. 2). New York: Brunner/Mazel.

Walford, Roy L. (1983). *Maximum life span.* New York: Norton.

Walker, Elaine, and **Emory, Eugene.** (1983). Infants at risk for psychopathology: Offspring of schizophrenic parents. *Child Development, 54*, 1254–1285.

Walker, Lawrence J. (1982). The sequentiality of Kohlberg's stages of moral development. *Child Development, 53*, 1330–1336.

Walker, Lawrence J. (1984). Sex differences in the development of moral reasoning: A critical review. *Child Development, 55*, 677–691.

Walker, Lawrence J., de Vries, Brian, and **Trevethan, Shelley D.** (1987). Moral stages and moral orientations in real-life and hypothetical dilemmas. *Child Development, 58*, 842–858.

Wallace, James R., Cunningham, Thomas F., and **Del Monte, Vickie.** (1984). Change and stability in self-esteem between late childhood and early adolescence. *Journal of Early Adolescence, 4*, 253–257.

Wallace, Marsha T. (1982). Before you consider a nursing home. *National Women's Health News, 9.*

Wallach, Michael A. (1985). Creativity testing and giftedness. In Frances Degen Horowitz and Marion O'Brien (Eds.), *The gifted and talented: Developmental perspectives.* Washington, DC: American Psychological Association.

Wallerstein, Judith S. (1984). Children of divorce: Preliminary report of a ten year follow-up of young children. *American Journal of Orthopsychiatry, 54*, 444–458.

Wallerstein, Judith S., and **Kelly, Joan Berlin.** (1980). *Surviving the breakup: How children and parents cope with divorce.* New York: Basic Books.

Wallwork, Ernest. (1980). Morality, religion and Kohlberg's theory. In Brenda Munsey (Ed.), *Moral development, moral education and Kohlberg: Basic issues in philosophy, psychology, religion and education.* Birmingham, AL: Religious Education Press.

Walsh, R.N. (1981). *Towards an ecology of the brain.* New York: SP Medical and Scientific Books.

Ward, Russell A. (1979). The never-married in later life. *Journal of Gerontology, 34*, 861–869.

Waterman, Alan S. (1985). Identity in the context of adolescent psychology. In Alan S. Waterman (Ed.), *Identity in adolescence: Processes and contents. New directions in child development* (Vol. 30). San Francisco: Jossey-Bass.

Watson, John B. (1925). *Behaviorism.* New York: Norton.

Watson, John B. (1928). *Psychological care of the infant and child.* New York: Norton.

Watson, John B. (1967). *Behaviorism* (rev. ed.). Chicago: University of Chicago Press. (Original work published 1930.)

Watson, Malcolm W. (1981). The development of social roles: A sequence of social-cognitive development. In Kurt W. Fischer (Ed.), *Cognitive development. New directions for child development* (Vol. 12). San Francisco: Jossey-Bass.

Watson, Malcolm W. (1984). Development of social role understanding. *Developmental Review, 4*, 192–213.

Watson, Malcolm W., and **Amgott-Kwan, Terry.** (1983). Transitions in children's understanding of parental rules. *Developmental Psychology, 19*, 659–666.

Weber, Hans U., and **Miquel, Jamie.** (1986). Nutritional modulation of the aging process. Part II: Antioxidant supplementation and longevity. In Linda H. Chen (Ed.), *Nutritional aspects of aging: Vol. 1.* Boca Raton, FL: CRC Press.

Webster, Harold, Freedman, Mervin B., and **Heist, Paul.** (1979). Personality change in students. In Nevitt Sanford and Joseph Axelrod (Eds.), *College and character.* Berkeley, CA: Montaigne.

Weil, William B., Jr. (1975). Infantile obesity. In Myron Winick (Ed.), *Childhood obesity.* New York: Wiley.

Weinraub, M., and **Lewis, Michael.** (1977). The determinants of children's responses to separation. *Monographs of the Society for Research in Child Development, 42* (4, Serial no. 172.)

Weinstein, Grace W. (1979). *Life plans: Looking forward to retirement.* New York: Holt, Rinehart and Winston.

Weis, Lois. (1985). *Between two worlds: Black students in an urban community college.* Boston: Routledge and Kegan Paul.

Weiss, Gabrielle, and **Hechtman, Lily Trokenberg.** (1986). *Hyperactive children grow up: Empirical findings and theoretical considerations.* New York: Guilford.

Weiss, Joan O., Bernhardt, Barbara A., and Paul, Natalie W. (Eds.). (1984). *Genetic disorders and birth defects in families and society: Toward interdisciplinary understanding.* White Plains, NY: March of Dimes.

Weitzman, Lenore J. (1985). *The divorce revolution: The unexpected social and economic consequences for women and children in America.* New York: Free Press.

Welford, A.T. (1980). On the nature of higher-order skills. *Journal of Occupational Psychology, 53,* 107–110.

Welford, Alan T. (1985). Changes of performance with age: An overview. In Neil Charness (Ed.), *Aging and human performance.* New York: Wiley.

Wells, J.C. (1982). *Accents of English* (Vols. 1–3). New York: Cambridge University Press.

Wender, Paul H. (1987). *The hyperactive child, adolescent and adult: Attention deficit disorder through the lifespan.* New York: Oxford.

Wentowski, Gloria J. (1981). Reciprocity and the coping strategies of older people. Cultural dimensions of network building. *The Gerontologist, 21,* 600–609.

Werler, Martha M., Pober, Barbara R., and Holmes, Lewis B. (1986). Smoking and pregnancy. In John L. Sever and Robert Brent (Eds.), *Teratogen update: Environmentally induced birth defect risks.* New York: Liss.

Werner, Emmy E., Bierman, J.M., and French, F.E. (1971). *The children of Kauai: A longitudinal study from the prenatal period to age ten.* Honolulu: University of Hawaii Press.

Werner, Emmy E., and Smith, Ruth S. (1982). *Vulnerable but invincible: A study of resilient children.* New York: McGraw-Hill.

Werner, J.S., and Perlmutter, Marion. (1979). Development of visual memory in infants. In Hayne W. Reese and Lewis P. Lipsitt (Eds.), *Advances in child development and behavior* (Vol. 14). New York: Academic Press.

Werry, John S. (1977). The use of psychotropic drugs in children. *American Academy of Child Psychiatry, 16,* 446–468.

Westbrook, Mary T. (1978). The effects of the order of a birth on women's experience of childbearing. *Journal of Marriage and the Family, 40,* 165–172.

Whalen, C.K., Henker, B., Collins, B.E., Finck, D., and Dotemoto, S. (1979). A social ecology of hyperactive boys: Medication effects in systematically structured classroom environments. *Journal of Applied Behavioral Analysis, 12,* 65–81.

Whitbourne, Susan Krauss. (1985a). *The aging body.* New York: Springer-Verlag.

Whitbourne, Susan Krauss. (1985b) The psychological construction of the life span. In James E. Birren and K. Warner Schaie (Eds.), *Handbook of the psychology of aging* (2nd ed.). New York: Van Nostrand.

White, Burton L. (1975). *The first three years of life.* Englewood Cliffs, NJ: Prentice-Hall.

White, Charles B., and Janson, Philip. (1986). Helplessness in institutional settings: Adaptation or iatropic disease. In Margaret M. Baltes and Paul B. Baltes (Eds.), *The psychology of control and aging.* Hillsdale, NJ: Erlbaum.

White, Lynn K. and Booth, Alan. (1985). The quality and stability of remarriages: The role of stepchildren. *American Sociological Review, 50,* 689–698.

White, Robert W. (1959). Motivation reconsidered. The concept of competence. *Psychological Review, 66,* 297–333.

White, Robert W. (1979). Competence as an aspect of personal growth. In Martha Whalen Kent and Jon E. Rolf (Eds.), *Primary prevention of psychopathology:* Vol. III. *Social competence in children* (pp. 5–22). Hanover, NH: University Press of New England.

White, Sheldon H. (1965). Evidence for a hierarchical arrangement of learning processes. In Lewis P. Lipsitt and Charles C. Spiker (Eds.), *Advances in child development and behavior.* Vol. II. New York: Academic Press.

Whiteley, John. (1982). *Character development in college students.* Schenectady, NY: Character Education Press.

Whiten, A. (1977). Assessing the effects of perinatal events on the success of the mother-infant relationship. In H.R. Shaffer (Ed.), *Studies of mother-infant interaction.* New York: Academic Press.

Whiting, Beatrice B., and Whiting, John W.M. (1975). *Children of six cultures.* Cambridge, MA: Harvard University Press.

Whitlock, F.A. (1986). Suicide and physical illness. In Alec Roy (Ed.), *Suicide.* Baltimore, MD: Williams and Wilkins.

Widholm, Olaf. (1985). Epidemiology of premenstrual tension syndrome and primary dysmenorrhea. In M. Yusoff Dawood, John L. McGuire, and Laurence M. Demers (Eds.), *Premenstrual syndrome and dysmenorrhea.* Baltimore, MD: Urban and Schwartzenberg.

Wiggins, Jerry S., and Holzmuller, Ana. (1978). Psychological androgyny and interpersonal behavior. *Journal of Consulting and Clinical Psychology, 46,* 40–52.

Wilk, Carole. (1986). *Career women and childbearing.* New York: Van Nostrand Reinhold.

Willatts, Peter. (1984). Stages in the development of intentional search by young infants. *Developmental Psychology, 20,* 389–396.

Williams, R. Sanders, Logue, Everett E., Lewis, James L., Stead, Nancy W., Wallace, Andrew G., and Pizzo, Salvatore V. (1980). Physical conditioning augments the fibrinolytic response to venous occlusion in healthy adults. *New England Journal of Medicine, 302,* 987–991.

Williams, Sharon, Denney, Nancy Wadsworth, and Schadler, Margaret. (1983). Elderly adults' perception of their own cognitive development during the adult years. *International Journal of Aging and Human Development, 16,* 47–158.

Willis, Sherry L. (1985). Towards an educational psychology of the older adult learner: Intellectual and cognitive bases. In James E. Birren and K. Warner Schaie (Eds.), *Handbook of the psychology of aging* (2nd ed.). New York: Van Nostrand Reinhold.

Willis, Sherry L., and Baltes, Paul B. (1980). Intelligence in adulthood and aging: Contemporary issues. In Leonard W. Poon (Ed.), *Aging in the 80's: Psychological issues.* Washington, DC: American Psychological Association.

Wilson, Ronald S. (1983). The Louisville twin study: Developmental synchronies in behavior. *Child Development, 54,* 298–316.

Wilson, Ronald S., and Matheny, Adam P. (1986). Behavior-genetics research in infant temperament: The Louisville Twin Study. In Robert Plomin and Judith Dunn (Eds.), *The study of temperament: Changes, continuities, and challenges.* Hillsdale, NJ: Erlbaum.

Winer, G.A. (1980). Class inclusion reasoning in children. *Child Development, 51,* 309–328.

Winick, Myron (Ed.). (1975). *Childhood obesity.* New York: Wiley.

Winn, Marie. (1977). *The plug-in drug.* New York: Viking.

Wolff, Jurgen M., and Lipe, Dewey. (1978). *Help for the overweight child.* New York: Penguin.

Wolfram, W., and Christian, D. (1976). *Appalachian speech.* Center for Applied Linguistics: Washington, DC.

Wolock, Isabel, and Horowitz, Bernard. (1984). Child maltreatment as a social problem: The neglect of neglect. *American Journal of Orthopsychiatry, 54,* 530–543.

Wolpe, J. (1969). *The practice of behavior therapy.* Elmsford, NY: Pergamon.

Wong Fillmore, Lily. (1987). *Becoming bilingual: Social processes in second language learning.* Paper presented at Society for Research in Child Development, Baltimore, MD, April 25, 1987.

Woodruff, Diana. (1983). The role of memory in personality continuity: A 25-year follow-up. *Experimental Aging Research, 9,* 31–34.

Yakovlev, P.I., and Lecours, A.R. (1967). The myelogenetic cycles of regional development of the brain. In A. Minkowski (Ed.), *Regional development of the brain in early life: Symposium.* Oxford: Blackwell.

Yalisove, Daniel. (1978). The effect of riddle-structure on children's comprehension and appreciation of riddles. Doctoral dissertation. New York University. *Dissertation Abstracts International, 36,* 6.

Yarrow, Marian Radke, Campbell, John D., and Burton, Roger V. (1968). *Child-rearing: An inquiry into research and methods.* San Francisco: Jossey-Bass.

Yerkes, R.M. (1923). Testing and the human mind. *Atlantic Monthly, 131,* 358–370.

Yglesias, Helen. (1980). Moses, Anna Mary Roberson (Grandma). In Barbara Sicherman and Carol Hurd Green (Eds.), *Notable American women: The modern period.* Cambridge, MA: Belknap Press.

Yost, J. Kelley, and Mines, Robert A. (1985). Stress and alcoholism. In Thomas E. Bratter and Gary G. Forrest (Eds.), *Alcoholism and substance abuse: Strategies for clinical intervention.* New York: The Free Press.

Younger, Alastair J., Schwartzman, Alex E., and Ledingham, Jane E. (1985). Age-related changes in children's perceptions of aggression and withdrawal in their peers. *Developmental Psychology, 21,* 70–75.

Youniss, James. (1980). *Parents and peers in social development: A Sullivan-Piaget perspective.* Chicago: University of Chicago Press.

Zaks, Peggy M., and Labouvie-Vief, Gisela. (1980). Spatial perspective taking and referential communication skills in the elderly: A training study. *Journal of Gerontology, 35,* 217–224.

Zales, Michael R. (1985). *Stress in health and disease.* New York: Brunner/Mazel.

Zarit, Steven H., Eiler, John, and Hassinger, Marl. (1985). Clinical assessment. In James E. Birren and K. Warner Schaie (Eds.), *Handbook of the psychology of aging.* New York: Van Nostrand Reinhold.

Zegiob, Leslie E., and Forehand, Rex. (1975). Maternal interactive behavior as a function of race, socioeconomic status, and sex of the child. *Child Development, 46,* 564–568.

Zelazo, P.R. (1979). Infant reactivity to perceptual-cognitive events: Application for infant assessment. In Richard B. Kearsley and Irving E. Sigel (Eds.), *Infants at risk: Assessment of cognitive functioning.* Hillsdale, NJ: Erlbaum.

Zelinski, Elizabeth M., Light, Leah L., and Gilewski, Michael J. (1984). Adult age differences in memory for prose: The question of sensitivity to passage structure. *Developmental Psychology, 20,* 1181–1192.

Zeskind, Philip Sanford, and Lester, Barry M. (1978). Acoustic features and auditory perceptions of the cries of newborns with prenatal and perinatal complications. *Child Development, 49,* 580–589.

Zigler, Edward, Abelson, Willa D., and Seitz, Victoria. (1973). Motivational factors in the performance of economically disadvantaged children on the Peabody Picture Vocabulary Test. *Child Development, 44,* 294–303.

Zigler, Edward, Abelson, Willa D., Trickett, Penelope, K., and Seitz, Victoria. (1982). Is an intervention program necessary in order to improve economically disadvantaged children's I.Q. scores? *Child Development, 53,* 340–348.

Zigler, Edward, and Berman, Winnie. (1983). Discerning the future of early childhood intervention. *American Psychologist, 38,* 894–906.

Zigler, Edward, and Farber, Ellen A. (1985). Commonalities between the intellectual extremes: Giftedness and mental retardation. In Frances Degen Horowitz and Marion O'Brien (Eds.), *The gifted and talented: Developmental perspectives.* Washington, DC: American Psychological Association.

Zill, N. (1983). *Happy, healthy, and insecure.* New York: Doubleday.

Zimmerman, Barry J. (1983). Social learning theory: A contextualist account of cognitive functioning. In Charles J. Brainerd (Ed.), *Recent advances in cognitive developmental theory.* New York: Springer-Verlag.

Zimmerman, Burke K. (1986). Epilogue: Looking ahead. In Raymond A. Zilinskas and Burke K. Zimmerman (Eds.), *The gene splicing wars: Reflections on the recombinent DNA controversy.* New York: Macmillan.

Zivian, Marilyn T., and Darjes, Richard W. (1983). Free recall by in-school and out-of-school adults: Performance and metamemory. *Developmental Psychology, 19,* 513–520.

Zollar, Ann Creighton, and Williams, J. Sherwood. (1987). The contribution of marriage to the life satisfaction of black adults. *Journal of Marriage and the Family, 49,* 87–92.

Zopf, Paul E. (1986). *America's older population.* Houston: Cap and Gown Press.

Zucker, Robert A., and Gomberg, Edith S. Lisansky. (1986). Etiology of alcoholism reconsidered: The case for a biopsychosocial process. *American Psychologist, 41,* 783–793.

Acknowledgments

PART OPENERS

PART I *clockwise from upper left,* ©Erika Stone; ©Susan Lapides 1981/Design Conceptions; Dean Abramson/Stock, Boston; ©Joel Gordon; ©Alan Carey/The Image Works; ©Susan Lapides 1985/Design Conceptions; Bob Daemmrich/Stock, Boston; ©Elizabeth Crews
PART II *all,* ©Joel Gordon
PART III *all except top right,* ©Elizabeth Crews; *top right,* ©Elizabeth Crews/The Image Works
PART IV *all,* ©Pam Hasegawa/Taurus Photos
PART V *all,* ©Elizabeth Crews
PART VI *all,* ©Hazel Hankin
PART VII *all but bottom,* ©Brent Jones; *bottom,* ©Dwight Jones
PART VIII *all,* ©Elizabeth Crews

CHAPTER 1

Opener, ©Erika Stone   **1.2** ©Shirley Zeiberg   **1.3** ©Therese Frare/The Picture Cube   **1.4** ©Elizabeth Crews   **1.5** ©Susan Lapides 1986/Design Conceptions   **1.6** ©Hazel Hankin   **1.7** ©John Chiasson/Gamma-Liaison   **A Closer Look pp. 14-15,** *top left,* The Bettmann Archive; *top center,* ©James Carroll; *top right,* Gale Zucker/Stock, Boston; *bottom left,* Culver Pictures   **1.8** Charles Gatewood/Stock, Boston   **1.9** Based on data from the U.S. Bureau of the Census (1986)   **A Closer Look p. 21,** ©Ira Berger/Black Star   **1.12** ©Roe Di Bona   **1.14** *Discover Magazine* ©1987

CHAPTER 2

Opener, © Erika Stone   **2.1** *left,* Sybil Shelton/Peter Arnold, Inc.; *right,* ©Shirley Zeiberg   **2.2** Culver Pictures   **2.3** Fredrik D. Bodin/Stock, Boston   **2.5** UPI/Bettmann Newsphotos   **2.6** *left,* Paul S. Conklin; *right,* Alan Carey/The Image Works   **2.7** Sovfoto   **2.8** Christopher S. Johnson/Stock, Boston   **2.9** *left,* ©Mimi Cotter/Int'l. Stock Photo; *right,* ©Hazel Hankin   **2.10** *left,* ©Erika Stone; *right,* ©Elizabeth Crews/Stock, Boston   **2.11** The Bettmann Archive   **2.13** S. Nozizwe, Center for the Study of the Person   **2.14** Frank Siteman/Taurus Photos   **2.15** Yves de Braine/Black Star   **2.16** ©Hazel Hankin 1987   **2.17** David A. Krathwohl/Stock, Boston

CHAPTER 3

Opener, ©Petit Format/Photo Researchers   **3.1** Courtesy The March of Dimes   **3.2** *left,* ©Erika Stone; *right,* ©Michal Heron/Woodfin Camp & Associates   **A Closer Look pp. 58-59,** *left,* Courtesy Jack Solomon, P. C.; *right,* Courtesy Thomas J. Bouchard   **3.3** Drawing by Angie Lloyd   **3.4** ©Elizabeth Crews   **3.5** UPI/Bettmann Newsphotos   **3.6** ©Eric Kroll/Taurus Photos   **Research Report p. 70,** ©Bill Strode/Woodfin Camp & Associates   **3.7** Victor A. McKusick   **3.8** ©Robert McElroy/Woodfin Camp & Associates   **3.9** *photo,* ©Erika Stone

CHAPTER 4

Opener, from Lennart Nilsson, *A child is born.* Copyright ©1978 Delacorte Press, New York. Photograph courtesy Lennart Nilsson, Bonnier Fakta, Stockholm.   **4.1** ©Per Sundström/Gamma-Liaison   **4.2** (a) (b) (c) ©Claude Edelmann/Science Source, Photo Researchers; (d) ©Donald Yeager/Camera MD Studios   **4.3** Carolina Biological Supply Co.   **4.4** Milwaukee Sentinel/UPI, Bettmann Newsphotos   **4.5** Drawing by Margaret G. Garrison   **4.6** ©Joel Gordon 1984   **4.7** Adapted from Keith L. Moore (1982). *The developing human* (3rd ed.), Fig. 8.14, p. 152. Philadelphia: W. B. Saunders Company.   **4.8** Carnegie Institute of Washington, Department of Embryology, Davis Division   **4.9** ©Stephen Ferry/Gamma-Liaison   **4.10** From J. W. Hanson. Fetal alcohol syndrome experience with 41 patients. *Journal of the American Medical Association,* 235 (14). Courtesy James W. Hanson, M. D., Department of Pediatrics, College of Medicine, the University of Iowa.   **4.11** ©Robin Schwartz/Int'l. Stock Photo   **Research Report p. 90,** ©Jim Anderson/Woodfin Camp & Associates   **4.12** ©Elizabeth Crews   **4.13** ©Chuck Fishman/Woodfin Camp & Associates   **A Closer Look p. 93,** ©David York/Medichrome   **4.14** ©Mariette Pathy Allen, courtesy Maternity Center Association/Peter Arnold, Inc.   **A Closer Look p. 96,** ©Martin A. Levick

CHAPTER 5

Opener and p. 175, ©Hazel Hankin   **5.1** National Center for Health Statistics   **5.2** Adapted from W. J. Robbins and others. *Growth.* Copyright ©1929 by Yale University Press. Reprinted by permission.   **5.3** Drawing by Margaret G. Garrison   **A Closer Look pp. 106-107,** ©Erika Stone   **5.4** From Paul H. Mussen (ed.) (1983). *Handbook of child psychology* (4th ed.), Vol. II. Marshall M. Haith and Joseph J. Campos (volume eds.). *Infancy and developmental psychobiology.* New York: Wiley.   **5.5** ©Jason Lauré/Woodfin Camp & Associates   **5.6** Adapted from Robert L. Fantz (1961). *The origin of form perception.* Copyright ©1961 by Scientific American, Inc. All rights reserved.   **5.7** Drawing by Margaret G. Garrison   **5.8** ©Ken Karp   **5.9** ©Elizabeth Crews   **5.10** Alice Kandell/Rapho, Photo Researchers   **A Closer Look p. 116,** *left,* ©Erika Stone; *right,* ©Michal Heron/Woodfin Camp & Associates   **5.11** *left,* AP/Wide World Photos; *right,* Reuters/Bettmann Newsphotos

CHAPTER 6

Opener and p. 175, ©Elizabeth Crews   **6.1** ©Erika Stone   **6.2** ©Mimi Cotter/Int'l. Stock Photo   **6.3** ©Erika Stone   **A Closer Look pp. 126-127,** ©Hazel Hankin   **6.4** ©Elizabeth Crews   **6.5** Abigail Heyman/Archive Pictures   **6.6** Charles Gatewood/The Image Works   **6.7** J. Berndt/The Picture Cube   **6.8** Jason Lauré/Woodfin Camp & Associates   **Research Report p. 133,** from C. K. Rovee-

Collier (1980). Reactivation of infant memory. *Science*, 6 June 1980, *208*(4448), 1159-1161. Copyright ©1980 by the American Association for the Advancement of Science. **6.10** ©Erika Stone **6.11** ©Shirley Zeiberg **6.12** ©Erika Stone

CHAPTER 7
Opener and p. 175, Michael Hayman/Stock, Boston **7.1** ©Alan Carey/The Image Works **7.2** ©Ken Karp **7.3** ©Erika Stone **7.4** ©Erika Stone **7.5** ©Jean-Claude Lejeune/Stock, Boston **7.6** ©Elizabeth Crews **7.7** Frank Siteman/Taurus Photos **7.9** ©Linda Ferrer/Woodfin Camp & Associates **7.10** ©Elizabeth Crews/The Image Works **7.11** ©Marion Bernstein **7.12** *left,* ©Elizabeth Crews; *right,* ©Erika Stone **Research Report pp. 166-167,** *left,* Michael Hayman/Stock, Boston; *right,* ©Mimi Cotter/Int'l. Stock Photo **7.13** AP/Wide World Photos **7.14** ©Beatriz Schiller/Int'l. Stock Photo **7.15** ©Elizabeth Crews

CHAPTER 8
Opener and p. 245, Michael Weisbrot/Stock, Boston **8.1** National Center for Health Statistics **8.2** ©Shirley Zeiberg/Taurus Photos **8.3** David A. Krathwohl/Stock, Boston **A Closer Look pp. 184-185,** ©Elizabeth Crews **8.4** Jeff Albertson/Stock, Boston **8.5** *left,* ©Robin Schwartz/Int'l. Stock Photo; *right,* ©Joel Gordon **8.6** *both,* ©Elizabeth Crews **8.8** Bob Krist/Black Star **8.9** ©Elizabeth Crews **8.10** ©George Ancona/Int'l. Stock Photo **8.11** ©Susan Lapides 1984/Design Conceptions **Research Report p. 194,** Ginger Chih/Peter Arnold, Inc.

CHAPTER 9
Opener and p. 245, ©Shirley Zeiberg **9.1** From Celia Genishi and Anne Haas Dyson (1984). *Language assessment in the early years.* Copyright ©1984 Ablex Publishing Corporation. Reprinted by permission of the publisher. **9.2** *left,* ©Elizabeth Crews; *right,* Judy S. Gelles/Stock, Boston **Chart p. 291,** adapted from Howard Gardner (1978). *Developmental psychology: An introduction.* Copyright ©1978 by Little, Brown and Company. Reprinted by permission of the publisher. **9.4** ©Hazel Hankin **9.5** Adapted by permission from Howard Gardner, *Developmental psychology: An introduction* (Fig. 7.3). Copyright ©1978 by Little, Brown and Company, Inc. **9.6** Adapted from Martin Hughes and Margaret Donaldson (1979). The use of hiding games for studying coordination of viewpoints. *Educational Review, 31* 133-140. **A Closer Look pp. 210-211,** from Ruth Krauss (1952). *A hole is to dig,* illustrated by Maurice Sendak. Text ©1952 by Ruth Krauss. Illustrations ©1952 by Maurice Sendak. Reproduced by permission of Harper & Row Publishers, Inc. **9.8** ©James Carroll **9.9** ©Rae Russel/Int'l Stock Photo **9.10** ©Elizabeth Crews

CHAPTER 10
Opener and p. 245, ©Elizabeth Crews **10.1** Steve Takatsuno **10.2** David M. Grossman **Research Report pp. 224-225,** Harry F. Harlow, University of Wisconsin Primate Laboratory **10.3** (a) (b) (c) (e) ©Elizabeth Crews; (d) ©Alan Carey/The Image Works **Research Report p. 230,** ©Alan Carey/The Image Works **10.5** *left,* Ed Lettau/Photo Researchers; *right,* ©Erika Stone **10.6** *left,* J. Berndt/Stock, Boston; *right,* ©Shirley Zeiberg **10.7** Movie Still Archives **10.8** Joanne Leonard/Woodfin Camp & Associates **10.9** ©Susan Lapides 1985/Design Conceptions **10.10** *left,* Steve Takatsuno; *right,* ©Burk Uzzle/Woodfin Camp & Associates **10.11** *left,* ©Jean-Claude Lejeune; *right,* Jeffry W. Myers/Stock, Boston

CHAPTER 11
Opener and p. 317, Steve Takatsuno **11.1** *left,* ©Joel Gordon; *right,* ©Elizabeth Crews **11.2** ©Shirley Zeiberg **11.3** *left,* ©Mike Mazzaschi/Stock, Boston; *right,* ©Elizabeth Crews **11.4** ©Alan Carey/The Image Works **11.5** *left,* ©R. S. Uzzell III/Woodfin Camp & Associates; *right,* ©Spencer Grant/Taurus Photos **11.6** David S. Strickler/Monkmeyer Press Photo Service

CHAPTER 12
Opener and p. 317, ©Susan Lapides 1986/Design Conceptions **12.1** ©Joel Gordon 1984 **12.2** Ingeborg Lippmann/Peter Arnold,
Inc. **12.3** ©Marion Bernstein **12.4** *left,* Robert Kalman/The Image Works; *right,* ©Michal Heron/Woodfin Camp & Associates **12.5** ©Ken Karp **12.6** Michael Weisbrot/Stock, Boston **12.8** ©Susan Lapides 1986/Design Conceptions **12.9** ©Susan Lapides 1984/Design Conceptions **12.11** Charles Harbutt/Archive Pictures **12.12** ©Paul S. Conklin **12.13** ©Elizabeth Crews/Stock, Boston **12.14** ©Elizabeth Crews **Research Report p. 280,** ©Susan Lapides 1986/Design Conceptions **12.15** Will McIntyre/Photo Researchers **12.16** ©Beryl Goldberg **A Closer Look p. 285,** ©Susan Lapides 1985/Design Conceptions **12.17** Alice Kandell/Photo Researchers

CHAPTER 13
Opener and p. 317, ©Erika Stone **13.1** ©Joan Menschenfreund/Taurus Photos **13.2** Charles Gupton/Stock, Boston **13.4** *left,* George Ancona/Int'l. Stock Photo; *right,* ©Paul S. Conklin **13.5** (a) ©Scott Thode/Int'l. Stock Photo; (b) ©Erika Stone; (c) ©Charles Marden Fitch/Taurus Photos; (d) ©Martin Rogers/Stock, Boston; (e) ©David Austen/Stock, Boston **13.6** *left,* ©Hazel Hankin; *right,* ©Lenore Weber/Taurus Photos **13.7** ©Jean-Claude Lejeune/Stock, Boston **A Closer Look p. 303,** ©Elizabeth Crews **13.8** *left,* Michael Weisbrot/Int'l. Stock Photo; *right,* ©Hazel Hankin **13.9** ©Bryce Flynn/Stock, Boston **A Closer Look p. 309,** David M. Grossman **13.10** ©Joel Gordon **13.11** Michael Weisbrot/Int'l. Stock Photo **13.12** From Michael Rutter (1970). Protective factors in children's responses to stress and disadvantage. In Whalen and Rolf (eds.), *Primary prevention of psychopathology* (Vol. III): *Social competence in children.* Reprinted by permission of the University Press of New England. Copyright ©1970 by the Vermont Conference on the Primary Prevention of Psychopathology. **13.13** ©Alan Carey/The Image Works

CHAPTER 14
Opener and p. 387, Bill Ross/West Light, Woodfin Camp & Associates **14.1** ©Alan Carey/The Image Works **14.2** Donald Dietz/Stock, Boston **14.3** Wilson North/Int'l. Stock Photo **14.4** ©Elizabeth Crews **14.5** AP/Wide World Photos **14.6** Charles Gatewood/The Image Works **A Closer Look p. 331,** *left,* ©James Carroll; *right,* ©Erika Stone **14.7** ©Jean-Claude Lejeune **14.8** ©Elizabeth Crews **14.9** ©Elizabeth Crews

CHAPTER 15
Opener and p. 387, ©Michal Heron/Woodfin Camp & Associates **15.1** ©Elizabeth Crews **15.2** ©Susan Lapides 1987/Design Conceptions **A Closer Look p. 343,** ©Michal Heron/Woodfin Camp & Associates **15.3** Howard Dratch/The Image Works **15.4** ©Jean-Claude Lejeune **15.6** ©Susan Lapides 1981/Design Conceptions **15.7** ©Susan Meiselas/Magnum Photos **15.8** Stuart Boss/Shooting Star **15.9** ©Elizabeth Crews **15.10** Courtesy Planned Parenthood of Central Ohio **15.11** National Institute of Drug Abuse, U.S. Department of Health and Human Services **15.12** ©Joel Gordon 1983 **15.13** *left,* Steven Stone/The Picture Cube; *right,* ©Alan Carey/The Image Works **15.14** Courtesy SADD **15.15** Billy E. Barnes/Stock, Boston

CHAPTER 16
Opener and p. 387, ©Susan Lapides 1981/Design Conceptions **16.1** ©Gerard Fritz/Taurus Photos **16.2** *left,* Charles Gatewood/The Image Works; *right,* ©James Carroll **16.3** (a) The Bettmann Archive; (b) Elliott Erwitt/Magnum Photos; (c) Michael Abramson/Black Star; (d) ©Shirley Zeiberg/Taurus Photos **A Closer Look p. 368,** *left,* Fritz Goro, *Life Magazine* ©1955 Time, Inc.; *right,* Peter Southwick/Stock, Boston **16.4** *left,* ©Erika Stone; *right,* ©Lenore Weber/Taurus Photos **16.5** *left,* ©Shirley Zeiberg; *right,* Ellis Herwig/The Picture Cube **16.6** Steve Takatsuno/The Picture Cube **16.7** Peter Vandermark/Stock, Boston **16.8** ©Shirley Zeiberg **16.9** ©Pam Hasegawa/Taurus Photos **16.10** from Oralee Wachter (1983). *No more secrets for me,* illustrated by Jane Aaron. Text copyright ©1983 by Oralee Wachter. Illustrations copyright ©1983 by Jane Aaron. Boston: Little, Brown and Company. Reproduced by permission of the publisher. **16.11** ©Shirley Zeiberg/Taurus Photos **16.12** National Center for Health Statistics **16.13** O. Banger/Int'l. Stock Photo

CHAPTER 17
Opener and p. 459, ©Alan Carey/The Image Works **17.1** Adapted from D. B. Bromley (1974). *The psychology of human ageing* (2nd ed.). Copyright ©D. B. Bromley, 1966, 1974. Reprinted by permission of Penguin Books, Ltd. **17.2** ©John Coletti/Stock, Boston **17.3** ©Peter Russell Clemens/Int'l. Stock Photo **A Closer Look p. 396,** *top,* ©Andy Hayt/Sports Illustrated ©Time, Inc.; *bottom,* ©Richard Mackson/Sports Illustrated ©Time, Inc. **17.4** ©Carol Palmer/The Picture Cube **17.5** Adapted from H. Begleiter and others (1987). Auditory recovery function and P3 in boys at high risk for alcoholism. *Alcohol, 4,* 317. **17.6** ©Alan Carey/The Image Works **17.7** Based on data from the U. S. Bureau of the Census (1986). **17.8** Courtesy National Council on Alcoholism **17.9** Janice Fullman/The Picture Cube **17.10** Thelma Shumsky/The Image Works **17.11** Peter Arnold, Inc.

CHAPTER 18
Opener and p. 459, Peter Vandermark/Stock, Boston **18.1** ©Joel Gordon 1986 **18.2** *left,* ©Erika Stone/Peter Arnold, Inc.; *right,* ©Susan Lapides 1986/Design Conceptions **18.3** ©Hazel Hankin **A Closer Look p. 419,** *left,* Margaret Bourke-White, *Life Magazine* ©1946 Time, Inc.; *right,* Neal Boenzi/NYT Pictures **18.4** ©Margot Granitsas/The Image Works **18.5** Based on data from UNESCO (1985). **18.6** ©Alan Carey/The Image Works **18.7** Ellis Herwig/The Picture Cube **18.8** *left,* ©Alan Carey/The Image Works; *right,* Midland Reporters/Gamma-Liaison

CHAPTER 19
Opener and p. 459, ©Susan Lapides 1985/Design Conceptions **19.1** UPI/Bettmann Newsphotos **19.2** Owen Franken/Stock, Boston **19.3** *left,* ©Joe McNally/Wheeler Pictures; *right,* ©Alan Carey/The Image Works **A Closer Look p. 437,** *left,* ©Nancy Bates/The Picture Cube; *right,* ©Karen Kasmauski/Wheeler Pictures **19.4** Mike Rizza/Stock, Boston **A Closer Look p. 441,** Michael Malyszko/Stock, Boston **19.6** ©Therese Frare/The Picture Cube **A Closer Look p.443,** *left,* ©Edward Lettau/Peter Arnold, Inc.; *right,* ©Hazel Hankin **19.7** ©David M. Grossman/Photo Researchers **19.8** ©Elizabeth Crews **19.9** *left,* ©Spencer Grant/The Picture Cube; *right,* ©Mark Antman/The Image Works **19.10** ©Joel Gordon 1982 **19.11** *left,* Billy E. Barnes/Stock, Boston; *right,* ©Paul S. Conklin **19.12** ©Spencer Grant/Taurus Photos **A Closer Look p. 456,** ©Richard Howard/Black Star

CHAPTER 20
Opener and p. 521, Frank Siteman/The Picture Cube **20.1** ©Joel Gordon 1983 **Research Report p. 469,** ©John Griffin/The Image Works **20.4** ©John Chiasson/Gamma-Liaison **20.5** ©Eric Kroll/Taurus Photos **20.6** *left,* ©Liane Enkelis/Stock, Boston; *right,* Neal Boenzi/NYT Pictures **20.7** AP/Wide World Pictures **A Closer Look p. 481,** chart from Bernice L. Neugarten and others (1963). Women's attitudes toward the menopause. *Vita Humana, 6,* 140-151. Reprinted with permission of *Vita Humana* (now *Human Development*), published by S. Karger AG, Basel. **20.8** *left,* ©Erika Stone; *right,* ©Frank Siteman/Taurus Photos

CHAPTER 21
Opener and p. 521, ©Alan Carey/The Image Works **21.1** Brown Brothers **21.3** ©Susan Lapides 1982/Design Conceptions **21.4** *left,* Laimute E. Druskis/Taurus Photos; *right,* ©Karen R. Preuss/Taurus Photos **21.5** *left,* Lionel Delevingne/Stock, Boston; *right,* ©Joel Gordon 1981 **21.6** ©J. Y. Rabeuf/The Image Works

CHAPTER 22
Opener and p. 521, Bob Daemmrich/Stock, Boston **22.1** *left,* Movie Still Archives; *right,* Reuters/Bettmann Newsphotos **22.4** ©Phil Huber/Black Star **22.6** *left,* Cary Wolinsky/Stock, Boston; *right,* ©Phil Huber/Black Star **22.7** ©Bob Krist/Black Star **22.8** Adapted from Helen E. Fisher (1987). The four-year itch. *Natural History,* October 1987, p. 24. **22.9** ©Bob Krist/Black Star **22.10** *left,* ©Tom Lankes/TexaStock; *right,* ©Susan Lapides 1986/Design Conceptions **22.11** Adapted from A. Howard and D. W. Bray (1980). Career motivation in mid-life managers. Paper presented at the annual meeting of the American Psychological Association, Montreal, Canada. Reprinted by permission. **22.12** ©Alan Carey/The Image Works **22.13** ©Jim Pickerell/Black Star **22.14** ©Jodi Cobb/Woodfin Camp & Associates **22.15** ©Eric Kroll/Taurus Photos **22.16** Jack Spratt/The Image Works **22.17** ©Michal Heron/Woodfin Camp & Associates

CHAPTER 23
Opener and p. 605, ©Robert Kalman/The Image Works **23.1** ©Elizabeth Crews **23.2** ©Lynn Johnson/Black Star **A Closer Look p. 528,** chart based on data from the U. S. Bureau of the Census (1984). **23.3** ©Elizabeth Crews **23.4** *left,* ©John Coletti/Stock, Boston; *center,* ©Paul S. Conklin; *right,* George Bellerose/Stock, Boston **23.5** ©David Strick/Black Star **23.7** ©Ira Berger **23.8** *left,* ©Claus Meyer/Black Star; *right,* ©Kathryn Dudek/Photo News, Int'l. Stock Photo **A Closer Look p. 540,** ©Sal di Marco/Black Star **23.9** *all,* ©John Launois/Black Star

CHAPTER 24
Opener and p. 605, Dean Abramson/Stock, Boston **24.1** Laimute E. Druskis/Taurus Photos **24.2** ©Timothy Eagan/Woodfin Camp & Associates **24.4** Abigail Heyman/Archive Pictures **Research Report p. 554,** drawings from Morton A. Lieberman (1965). Psychological correlates of impending death: Some preliminary observations. *Journal of Gerontology, 20* (April). Reprinted by permission of the *Journal of Gerontology.* **24.5** ©Timothy Eagan/Woodfin Camp & Associates **24.6** ©Alan Carey/The Image Works **Research Report p. 559,** *left,* Rick Smolan/Stock, Boston; *right,* ©Elizabeth Crews **24.7** Jim Harrison/Stock, Boston **24.8** ©Ira Wyman/Sygma **24.9** From Barry Reisberg (1981). *Brain failure.* New York: The Free Press. Copyright ©1981 by Barry Reisberg. Reprinted by permission of Macmillan Publishing Co. **24.10** Ann Marie Rousseau/The Image Works **24.11** ©Ira Berger/Woodfin Camp & Associates **24.12** *left,* Paul Fortin/Stock, Boston; *right,* George Bellerose/Stock, Boston **24.13** AP/Wide World Photos **24.14** Based on data from Schick (1986) and the U.S. Bureau of the Census (1986)

CHAPTER 25
Opener and p. 605, ©Joel Gordon **25.1** ©Elizabeth Crews **A Closer Look p. 579,** ©Judi Benvenuti/Taurus Photos **25.3** *left,* ©Joel Gordon; *center,* ©Steve Hansen/Stock, Boston; *right,* Edward L. Miller/Stock, Boston **25.4** *top left,* Bob Daemmrich/Stock, Boston; *top right,* ©Paul S. Conklin/Monkmeyer Press Photo Service; *bottom,* ©Howard Chapnick/Black Star **25.5** *left,* ©Glyn Cloyd/Taurus Photos; *right,* ©Pam Hasegawa/Taurus Photos **A Closer Look p. 585,** Janice Fullman/The Picture Cube **25.6** *left,* Frances M. Cox/Stock, Boston; *right,* ©Elizabeth Crews **25.7** ©Paul S. Conklin **25.8** Owen Franken/Stock, Boston **25.9** *left,* Stacy Pick/Stock, Boston; *right,* ©Joe McNally/Wheeler Pictures **25.10** Ginger Chih/Peter Arnold, Inc. **25.11** ©Joel Gordon 1986 **25.12** ©Spencer Grant/Taurus Photos **25.13** *left,* ©Susan Lapides 1986/Design Conceptions; *right,* ©Phil Huber/Black Star **25.14** *left,* ©Joel Gordon 1982; *right,* ©Elizabeth Crews **25.15** Chester Higgins, Jr./NYT Pictures

CHAPTER 26
Opener, Robert Eckert/Stock, Boston **26.1** ©Frank Grant/Int'l. Stock Photo **26.2** Richard Lonetto (1980). *Children's conceptions of death* (Figs. 2-12, 3-7, 4-3, and 4-12). New York: Springer Publishing Co. ©1980. Reproduced by permission. **26.3** George W. Gardner/Stock, Boston **A Closer Look p. 615,** ©Linda Bartlett **26.4** *left,* Michael Weisbrot/Int'l. Stock Photo, *right,* ©Julian Calder/Woodfin Camp & Associates **26.5** ©Elizabeth Crews

Name Index

Subject Index

See Name Index for page references for authors cited in text; see Glossary for definitions of key terms.